Flow Injection Analysis of Food Additives

Food Analysis & Properties

Series Editor
Leo M.L. Nollet
University College Ghent, Belgium

Flow Injection Analysis of Food Additives   **(2015)**
Edited by Claudia Ruiz-Capillas and Leo M.L. Nollet

Food Analysis & Properties Series

Flow Injection Analysis of Food Additives

EDITED BY **CLAUDIA RUIZ-CAPILLAS** • **LEO M. L. NOLLET**

CRC Press
Taylor & Francis Group
Boca Raton London New York

CRC Press is an imprint of the
Taylor & Francis Group, an **informa** business

CRC Press
Taylor & Francis Group
6000 Broken Sound Parkway NW, Suite 300
Boca Raton, FL 33487-2742

First issued in paperback 2019

ISBN-13: 978-1-4822-1819-0 (hbk)
ISBN-13: 978-0-367-37716-8 (pbk)

Library of Congress Cataloging-in-Publication Data

Flow injection analysis of food additives / Claudia Ruiz-Capillas and Leo M.L. Nollet, editors.
    pages cm
    "A CRC title."
    Includes bibliographical references and index.
    ISBN 978-1-4822-1819-0 (alk. paper)
    1. Food additives. 2. Food--Analysis. I. Ruiz-Capillas, Claudia, editor. II. Nollet, Leo M. L., 1948- editor.

TX553.A3F47 2016
664'.06--dc23
                                                    2015019707

Visit the Taylor & Francis Web site at
http://www.taylorandfrancis.com

and the CRC Press Web site at
http://www.crcpress.com

Contents

SECTION IX   ANTIMICROBIAL EFFECTS

SECTION X   ACIDITY

Series Preface

There will always be a need for analyzing methods of food compounds and properties. Current trends in analyzing methods are automation, increasing the speed of analyses, and miniaturization. The unit of detection has evolved over the years from micrograms to picograms.

A classical pathway of analysis is sampling, sample preparation, cleanup, derivatization, separation, and detection. At every step, researchers are working and developing new methodologies. A large number of papers are published every year on all facets of analysis. So there is need for books gathering information on one kind of analysis technique or on analysis methods of a specific group of food components.

The scope of the CRC series on Food Analysis and Properties aims at bringing out a range of books edited by distinguished scientists and researchers who have significant experience in scientific pursuits and critical analysis. This series is designed to provide state-of-the-art coverage on topics such as

1. Recent analysis techniques of a range of food components
2. Developments and evolutions in analysis techniques related to food
3. Recent trends in analysis techniques of specific food components and/or a group of related food components
4. The understanding of physical, chemical, and functional properties of foods

The book *Flow Injection Analysis of Food Additives* is the first volume of this series. I am happy to be a series editor of such books for the following reasons:

- I am able to pass on my experience in editing high-quality books related to food.
- I get to know colleagues from all over the world more personally.
- I continue to learn about interesting developments in food analysis.

A lot of work is involved in the preparation of a book. I have been assisted and supported by a number of people, all of whom I would like to thank. I would especially like to thank the team at CRC Press, with a special word of thanks to Steve Zollo, senior editor.

Many, many thanks to all the editors and authors of this volume and future volumes. I appreciate very much all their effort, time, and willingness to do a great job.

I dedicate this series to

- My wife, for her patience with me (and all the time I spent on my computer).
- All patients suffering from prostate cancer. Knowing what this means, I am hoping they will have some relief.

Leo M.L. Nollet

Foreword

Flow injection analysis is more than analytical technique. It is a platform for the majority of analytical methods.

G. D. Christian and A. Townshend

It is not by accident that the majority of the contributors to this outstanding monograph come from Spain. And it is not a coincidence that from 23 monographs published worldwide to date, the only ones that address the real-life applications of FIA, namely, *Flow Injection Analysis of Pharmaceuticals* (J. M. Clatayud), *Flow Injection Analysis of Marine Samples* (M. C. Yebra-Biurrun), and now *Flow Injection Analysis of Food Additives*, originate from Spain, where, 30 years ago, the pioneering work of M. Valcarcel and Luque de Castro laid the foundation for the acceptance of FIA as an important tool for chemical analysis.

While the discovery of a new analytical technique is exciting and self-rewarding, the real test of any novel technique is its acceptance by those who use it in real-life situations. And this is why the contribution of the Spanish research community is so important—it documents the applicability of the FIA technique in diverse fields of research and technology.

This monograph is truly unprecedented in its scope, both in terms of content and the breadth of the collective expertise. Its editors, Claudia Ruiz-Capillas and Leo M.L. Nollet, assembled a team of 80 authors from 14 countries to compose a work in 36 chapters of which the first six deal with FIA techniques, while the body of the work is focused on the ways in which a wide variety of samples and analytes can be assayed. The methods listed include preservatives, antioxidants, sweeteners, colorants, flavor enhancers, and other species. And this is where the lasting value of this exceptional book is to be found—in the detailed description of the amazing variety of underlying chemistries on which the individual assay protocols are based. As flow-based methodology and instrumentation evolve, their older versions are becoming obsolete; however, the chemistry of the underlying chemical reactions will remain the cornerstone on which the selective and sensitive assay protocols are based.

It must have been a Herculean task to orchestrate this multitude of talents spread across the globe and communicate in multiple languages. The editors are to be congratulated for bringing all these contributions to a successful conclusion and creating this work. I am sure that the result will be appreciated by many, not only in the current generation, but also in future generations of analytical chemists.

Jaromir (Jarda) Ruzicka
Department of Oceanography
University of Hawaii, Honolulu

Preface

During the past few years, the study of additives has become of great interest in different areas of research (nutrition, health, food technology, agricultural, etc.) worldwide. In the food industry, the use of additives is widespread, including a wide range of preservatives, antioxidants, sweeteners, colorants, flavor enhancers, emulsifiers, etc. Additives are used every day in the elaboration, processing, and conservation of food products for their different functions. The useful functions of these additives are important for maintaining food and beverage quality and characteristics that consumers demand, keeping food safe from farm to fork. However, some of these additives, such as nitrite and sulfite, present potential implications for human health. Therefore, the use of additives in foods and beverages is regulated by national and international authorities. The European Union and the U.S. Food and Drug Administration (FDA) have established a list of additives or a Food Additive Status List, respectively, according to their function. These agencies regulate additives in the manufacture of food products, assigning them an E number. This regulation tries to fix the levels of use so that these compounds fulfill a useful purpose (are effective) and at the same time are safe for the consumer. On the other hand, the European Food Safety Authority has established an acceptable daily intake (ADI) of different additives, fixing the amount that can be ingested without adverse effects on human health. In this sense, the levels of additives used should also be rigorously controlled and/or tested to maintain the characteristic of the foods to which they are added and to ensure consumer safety.

There are a number of methodologies for determining the different kinds of additives in foods and beverages, among them ion chromatography, gas chromatography, high-performance liquid chromatography, capillary electrophoresis, and enzymatic reactions. Most of these methods are expensive or less precise, or require complex instrumentation to work successfully, and this limits their use to the laboratory for routine analysis. However, among these existing methodologies, flow injection analysis (FIA) has many advantages, with great potential as an analytical methodology but also with some limitations and challenges. The FIA system was first described by Ruzicka and Hansen in 1975 and offers interesting advantages such as versatility, flexible accuracy, and suitability for a fast routine analysis of a large number of samples with high precision. Affordably priced and easy to use, it requires minimum intervention by the operator and lower consumption of samples and reactants. This is especially important when working with toxic or expensive reactants. It can be installed online and is compatible with a large number of detectors. This methodology responds to the growing demand for the mechanization and automation of analysis made by many laboratories faced with the challenge of the increasing number of routine analyses required. These advantages make FIA an important alternative to conventional methods for routine laboratory analysis, official institutions, and industry applications.

This book was proposed based on the importance of the determination of additives and the possible application of FIA for their determination. The objective of this book is to provide an overview/review of the possible applications of FIA as a useful tool for analyzing additives in foods and beverages. It is intended to be a compilation of useful FIA techniques as important alternatives to conventional methods in routine laboratory analysis.

No previous book has addressed both FIA and food additives. FIA is not a new topic, and there are some books on FIA, but these generally focus mainly on the theoretical basis and principles of FIA and on the design of equipment, instrumentation, manifold, setting mechanism, and so on. On the other hand, there are many books on food additives, but both these topics have been approached separately. However, the advantage of this book is the combination (link) of these two topics. This book aims to combine the analytical technique description itself and the direct application in the determination of food additives. It addresses the more important additives used in foods and beverages, and additives used every day in the food industry all over the world. This book provides the first review of measurements of additives and other substances by FIA in relation to the use of additives in food. It is intended to be an important manual for research, industry, and official administration laboratories.

This book consists of ten sections. The first section provides an introduction to the topic; reviews the origin of FIA, including recent developments and future trends. It discusses technical aspects of the study of SIA, miniaturization, and automation: nanotechnology; multicommutated FIA, and the combination of FIA and other analytical systems. The next six sections discuss the determination of additives. These sections are divided based on the official classification of additives according to function by the EU. Section II covers the study of preservatives with chapters on nitrates, nitrites, sulfites, sorbic acid and sorbates, benzoic acid, benzoates, parabens, acetic acid, acetates and lactic acid. Sections III and IV deal with the study of antioxidants; Section III with synthetic antioxidants (with chapters on BHT, BHA, TBHQ, propyl, octyl, dodecyl, gallates, phosphoric acid and phosphates, and lactates); and Section IV with natural antioxidants (with chapters on tartaric acid, ascorbic acid (vitamin C), vitamin E, phenolic compounds, flavonoids, and glutathione). Section V discusses sweeteners (with chapters on aspartame, sorbitol, acesulfame-K, cyclamate, and saccharin). Section VI discusses colorants and dyes (with chapters on riboflavin and quinoline and sunset yellow). Section VII discusses flavor enhancers (with chapters on monosodium glutamate, IMP, and GMP). Finally, in Sections VIII, IX, and X, the reader finds a review of the determination of antioxidant capacity (FRAP, ABTS, etc.), antimicrobial effects (TVBN, TMA, biogenic amines, etc.), and acidity by FIA, respectively. These determinations are not proper determinations of additives (with E numbers or without). However, these determinations are in relation to the use of additives in food (e.g., antioxidant or antimicrobial). These parameters or substances are very important in research and in the industry; some of them are used as quality control indices in food.

All chapters are organized in the same way: The first part begins with a small introduction in relation to the importance of the additive or compound in question in foods and beverages and discusses the legislation affecting their use and control. The second part focuses on the determination of the compound or additive by FIA in different foods or beverages, with sample preparation and extraction from a food/beverage on the one hand and FIA methods used for the separation and detection on the other.

We hope that the information in this book will be novel and will serve as a useful guide and reference source for research scientists in the food industry and other scientific areas, academia, government, and official laboratories, as well as graduate and postgraduate students from various disciplines (veterinary medicine, chemistry, pharmacy, biology, food technology, etc.); regulatory agencies that use very affordable and easy-to-use technologies such as FIA; and those who are actively engaged in the analysis of foods and beverages to ensure both their safety and quality. Also, the book is meant for individuals interested in learning more about the determination of additives by FIA. In addition, we hope this book can help in the development of some of these methodologies.

The editors would like to thank all the contributors, who are well known in the world of FIA. They have collaborated to communicate their knowledge and experience to serve the readers of this book. We greatly appreciate their cooperation. We would also like to thank CRC Press, Taylor & Francis Group, for their encouragement and help in editing.

Claudia Ruiz-Capillas would like to thank Dr. Leo M.L. Nollet for granting her the opportunity to be his coeditor on this book and to share his extensive knowledge on this subject. Claudia would also like to thank Professors Jaromir Ruzicka and Elo Harald Hansen, who were the first authors of a publication on FIA. Professor Ruzicka encouraged Claudia all the time and agreed to write a foreword for this book; this was an honor. Thanks to F. Jiménez-Colmenero and A. M. Herrero for their advice and suggestions.

Finally, Claudia thanks her family, especially her mother, for her tolerance and time.

Claudia Ruiz-Capillas
Leo M.L. Nollet

Editors

Claudia Ruiz-Capillas earned her BSc degree from the Faculty of Veterinary Sciences of the Complutense University of Madrid (UCM), Spain and her MSc degree in food science (Faculty of Veterinary Sciences of the UCM). She has been working for several years on the production of biogenic amines and free amino acids in fish and meat products stored under different manufacturing and storage conditions. Dr. Ruiz-Capillas received her PhD degree in veterinary science from the UCM in 1997. She was a postdoc scholar (1997–1999) at the International Fisheries Institute in Hull University in the United Kingdom. The focus of her research was on flow injection analysis (FIA), and was done as part of the FAIR CT 96.3253 "QUALPOISS 2 project" (title: The Evaluation of a Simple, Cheap, Rapid Method of Non-Protein Nitrogen Determination in Fish Products through the Processing/Merchandising Chain). Presently, she is a research scientist at the Department of Products at the Meat and Meat Products Science and Technology Laboratory at the Institute of Food Science Technology and Nutrition (ICTAN-CSIC) of the Spanish Science Research Council (CSIC). Dr. Ruiz-Capillas's current research is focused on developing healthy meat products and improving their quality (reducing additives, biogenic amines, etc.). She has participated in various national and international research projects, taking the lead on some of them, and has more than 90 SCI scientific contributions in food science and technology to her credit. She has coauthored various chapters in published books, and is a guest editor of the *Special Issue Journal of Chemistry*. Dr. Ruiz-Capillas has supervised PhD theses and given lectures at conferences, seminars, courses, and master classes.

Leo M.L. Nollet earned his MS (1973) and PhD (1978) degrees in biology from the Katholieke Universiteit Leuven, Belgium.

Dr. Nollet is the editor and associate editor of a number of books. He edited the first, second, and third editions of *Food Analysis by HPLC* and *Handbook of Food Analysis* (Marcel Dekker, New York—now CRC Press, Taylor & Francis Group). The last edition of *Handbook of Food Analysis* is a two-volume book. He also edited *Handbook of Water Analysis* (first, second, and third editions) and *Chromatographic Analysis of the Environment*, 3rd edition (CRC Press).

He has coedited two books with F. Toldrá: *Advanced Technologies for Meat Processing* (CRC Press, 2006) and *Advances in Food Diagnostics* (Blackwell Publishing—now Wiley, 2007). He has coedited with M. Poschl the book *Radionuclide Concentrations in Foods and the Environment* (CRC Press, 2006).

He has coedited several books with Y. H. Hui and other colleagues: *Handbook of Food Product Manufacturing* (Wiley, 2007), *Handbook of Food Science, Technology and Engineering* (CRC Press, 2005), *Food Biochemistry and Food Processing* (first and second editions; Blackwell Publishing–Wiley, 2006, 2012), and *Handbook of Fruit and Vegetable Flavors* (Wiley, 2010).

He edited *Handbook of Meat, Poultry and Seafood Quality* (first and second editions; Blackwell Publishing–Wiley, 2007, 2012).

From 2008 to 2011, he published with F. Toldrá five volumes in animal product–related books: *Handbook of Muscle Foods Analysis, Handbook of Processed Meats and Poultry Analysis, Handbook of Seafood and Seafood Products Analysis, Handbook of Dairy Foods Analysis*, and *Handbook of Analysis of Edible Animal By-Products*. Also with F. Toldrá, he coedited two volumes in 2011: *Safety Analysis of Foods of Animal Origin* and *Sensory Analysis of Foods of Animal Origin* (both from CRC Press). In 2012, they were coauthors of *Handbook of Analysis of Active Compounds in Functional Foods*, CRC Press.

He coedited with Hamir Rathore *Handbook of Pesticides: Methods of Pesticides Residues Analysis* (2009), *Pesticides: Evaluation of Environmental Pollution* (2012), and *Biopesticides Handbook* (2015), CRC Press.

His other completed book projects are *Food Allergens: Analysis, Instrumentation, and Methods* with A. van Hengel (CRC Press, 2011) and *Analysis of Endocrine Compounds in Food* (Wiley–Blackwell, 2011).

His recent projects are *Proteomics in Foods* with F. Toldrá (Springer, 2013) and *Transformation Products of Emerging Contaminants in the Environment: Analysis, Processes, Occurrence, Effects and Risks* with D. Lambropoulou (Wiley, 2014).

Contributors

Carolina C. Acebal
Department of Chemistry
National University of the South
Buenos Aires, Argentina

Eduardo Santos Almeida
Institute of Chemistry
Federal University of Uberlandia
Minas Gerais, Brazil

Elisa Isabel Vereda Alonso
Department of Analytical Chemistry
University of Málaga
Málaga, Spain

Célia M.G. Amorim
Department of Chemical Sciences
University of Porto
Porto, Portugal

Alberto N. Araújo
Department of Chemical Sciences
University of Porto
Porto, Portugal

Mihaela Badea-Doni
Biotechnology Department
National Institute for Research and
    Development in Chemistry and
    Petrochemistry—ICECHIM
Bucharest, Romania

Luísa Barreiros
UCIBIO, REQUIMTE
Department of Chemical Sciences
University of Porto
Porto, Portugal

L.R. Braga
Institute of Chemistry
University of Brasilia-UnB
Brasilia, Brazil

Juan Carlos Bravo
Department of Analytical Sciences
National University of Distance
    Education (UNED)
Madrid, Spain

Susana Campuzano
Department of Analytical Chemistry
Complutense University of Madrid
Madrid, Spain

Dayene do Carmo Carvalho
Institute of Chemistry
Federal University of Uberlandia
Minas Gerais, Brazil

Jessica Avivar Cerezo
I+D Department
Sciware Systems, S.L.
Bunyola, Spain

Luciana M. Coelho
Department of Chemistry
Federal University of Goiás
Goiás, Brazil

Nívia M.M. Coelho
Institute of Chemistry
Federal University of Uberlandia
Minas Gerais, Brazil

Felipe Conzuelo
Department of Analytical Chemistry
Complutense University of Madrid
Madrid, Spain

Andrei Florin Danet
Department of Analytical Chemistry
University of Bucharest
Bucharest, Romania

Vitor de Cinque Almeida
Centro de Ciências Exatas
Departamento de Química
Universidade Estadual de Maringá
Paraná, Brazil

Hilda Ledo de Medina
Laboratorio de Química Ambiental
Universidad del Zulia
Estado Zulia, Venezuela

Sérgio Antônio Lemos de Moraes
Institute of Chemistry
Federal University of Uberlandia
Minas Gerais, Brazil

Alberto de Oliveira
Institute of Chemistry
Federal University of Uberlandia
Minas Gerais, Brazil

Helen Cristine de Rezende
Institute of Chemistry
Federal University of Uberlandia
Minas Gerais, Brazil

Amparo García de Torres
Department of Analytical Chemistry
University of Málaga
Málaga, Spain

A.C.B. Dias
Institute of Chemistry
Group of Automation
Chemometrics and Environmental
    Chemistry (AQQUA Group)
University of Brasilia-UnB
Brasilia, Brazil

Jesús Senén Durand
Department of Analytical Sciences
National University of Distance Education
    (UNED)
Madrid, Spain

Anastasios Economou
Department of Chemistry
National and Kapoditrian University of
    Athens
Athens, Greece

Pilar Fernández
Department of Analytical Sciences
National University of Distance Education
    (UNED)
Madrid, Spain

Alejandrina Gallego
Department of Analytical Sciences
National University of Distance Education
    (UNED)
Madrid, Spain

Rosa Mª Garcinuño
Department of Analytical Sciences
National University of Distance Education
    (UNED)
Madrid, Spain

Juan Godoy-Navajas
Analytical Chemistry Department
Institute of Fine Chemistry and
    Nanochemistry
University of Córdoba
Córdoba, Spain

Agustina Gómez-Hens
Analytical Chemistry Department
Institute of Fine Chemistry and
    Nanochemistry
University of Córdoba
Córdoba, Spain

David González-Gómez
Department of Analytical Sciences
National University of Distance Education
    (UNED)
Madrid, Spain

Hideki Hakamata
Department of Analytical Chemistry
School of Pharmacy
Tokyo University of Pharmacy and Life
    Sciences
Tokyo, Japan

Ana M. Herrero
Department of Products
Institute of Food Science, Technology and
    Nutrition (ICTAN-CSIC)
Madrid, Spain

Francisco Jiménez-Colmenero
Department of Products
Institute of Food Science, Technology and
    Nutrition (ICTAN-CSIC)
Madrid, Spain

Yon Ju-Nam
College of Engineering
Chemical Engineering
Swansea University
Swansea, United Kingdom

Isao Karube
School of Bioscience and Biotechnology
Tokyo University of Technology
Tokyo, Japan

Akira Kotani
Department of Analytical Chemistry
School of Pharmacy
Tokyo University of Pharmacy and Life
    Sciences
Tokyo, Japan

Fumiyo Kusu
Department of Analytical Chemistry
School of Pharmacy
Tokyo University of Pharmacy and Life
    Sciences
Tokyo, Japan

Adriana G. Lista
Department of Chemistry
National University of the South
Buenos Aires, Argentina

Luís M. Magalhães
UCIBIO, REQUIMTE
Department of Chemical Sciences
University of Porto
Porto, Portugal

Sara S. Marques
UCIBIO, REQUIMTE
Department of Chemical Sciences
University of Porto
Porto, Portugal

Víctor Cerdà Martín
Laboratory of Environmental and
    Analytical Chemistry (LQA2)
University of the Balearic Islands
Palma de Mallorca, Spain

Alessandro Campos Martins
Centro de Ciências Exatas
Departamento de Química
Universidade Estadual de Maringá
Paraná, Brazil

Kiyoshi Matsumoto
Department of Applied Microbial
    Technology
Sojo University
Kumamoto, Japan

M. Conceição B.S.M. Montenegro
Department of Chemical Sciences
University of Porto
Porto, Portugal

Basil K. Munjanja
Department of Applied Chemistry
National University of Science and
    Technology
Bulawayo, Zimbabwe

Rodrigo Alejandro Abarza Munoz
Institute of Chemistry
Federal University of Uberlandia
Minas Gerais, Brazil

Abdul Nabi
Department of Chemistry
University of Balochistan
Balochistan, Pakistan

Emine Nakilcioglu
Food Engineering Department
Ege University
Izmir, Turkey

Leo M.L. Nollet
Retired University College Ghent
Gent, Belgium

Jesus J. Ojeda
Experimental Techniques Centre
Institute of Materials and Manufacturing
Brunel University
Uxbridge, United Kingdom

Gracy Kelly Faria Oliveira
Institute of Chemistry
Federal University of Uberlandia
Minas Gerais, Brazil

Semih Otles
Food Engineering Department
Ege University
Izmir, Turkey

José Manuel Cano Pavón
Department of Analytical Chemistry
University of Málaga
Málaga, Spain

María Pedrero
Department of Analytical Chemistry
Complutense University of Madrid
Madrid, Spain

José M. Pingarrón
Department of Analytical Chemistry
Complutense University of Madrid
Madrid, Spain

Paula C.A.G. Pinto
Department of Chemical Sciences
University of Porto
Porto, Portugal

Inês Ramos
UCIBIO, REQUIMTE
Department of Chemical Sciences
University of Porto
Porto, Portugal

Salette Reis
UCIBIO, REQUIMTE
Department of Chemical Sciences
University of Porto
Porto, Portugal

A. Julio Reviejo
Department of Analytical Chemistry
Complutense University of Madrid
Madrid, Spain

Eduardo Mathias Richter
Institute of Chemistry
Federal University of Uberlandia
Minas Gerais, Brazil

Jose Manuel Estela Ripoll
Laboratory of Environmental and
    Analytical Chemistry (LQA2)
University of the Balearic Islands
Palma de Mallorca, Spain

Claudia Ruiz-Capillas
Department of Products
Institute of Food Science,
    Technology and Nutrition
    (ICTAN-CSIC)
Madrid, Spain

M. Lúcia M.F.S. Saraiva
Department of Chemical Sciences
University of Porto
Porto, Portugal

Marcela A. Segundo
UCIBIO, REQUIMTE
Department of Chemical Sciences
University of Porto
Porto, Portugal

L.K. Shpigun
Kurnakov Institute of General and
    Inorganic Chemistry
Russian Academy of Sciences
Moscow, Russia

Raquel Maria Ferreira Sousa
Institute of Chemistry
Federal University of Uberlandia
Minas Gerais, Brazil

Kiyoko Takamura
Department of Analytical Chemistry
School of Pharmacy
Tokyo University of Pharmacy and Life
    Sciences
Tokyo, Japan

Thiago Faria Tormin
Institute of Chemistry
Federal University of Uberlandia
Minas Gerais, Brazil

Ildikó V. Tóth
UCIBIO, REQUIMTE
Department of Chemical Sciences
University of Porto
Porto, Portugal

Marek Trojanowicz
Department of Chemistry
University of Warsaw
and
Laboratory of Nuclear Analytical Methods
Institute of Nuclear Chemistry and
    Technology
Warsaw, Poland

Mohammad Yaqoob
Department of Chemistry
University of Balochistan
Balochistan, Pakistan

M. Carmen Yebra-Biurrun
Department of Analytical Chemistry
University of Santiago de Compostela
Santiago de Compostela, Spain

Wataru Yoshida
School of Bioscience and Biotechnology
Tokyo University of Technology
Tokyo, Japan

Flow Injection Analysis
Theory and Trends

CHAPTER 1

Flow Injection Analysis
Origin and Development

Marek Trojanowicz

CONTENTS

Technological progress is a constant feature these days of any area of human activity. Chemical analysis, which is indispensable for quality control in all branches of industry, production of food and pharmaceuticals, medical diagnostics, and natural and technical sciences, is no exception. Among the numerous trends in development of analytical chemistry and chemical analysis, of especially great importance for routine applications are mechanization and automation of analytical measurements, and miniaturization of analytical instrumentation. In mechanized instruments one or several stages of analytical procedures are carried out by appropriate mechanical devices, whereas in automated systems at least some part of the whole procedure is carried out by a measuring system based on feedback loop computerized control, involving advanced software and artificial intelligence. The main attributes of mechanized and automated analytical instruments, compared to manual systems, are increased efficiency of analyses and improved precision of determinations.

1.1 ORIGIN

A particularly successful concept of mechanization of analytical procedures is carrying out part or whole of an analytical procedure in flow conditions, the first benefit being enhancement of the efficiency of analytical determinations. Measurements of various physicochemical parameters (e.g., conductivity, pH) in flowing streams were already employed in process monitoring in the first decades of the twentieth century, but the first approach to designing a laboratory flow analytical system was reported in the mid 1950s for clinical diagnostics (Skeggs, 1956, 1957). The development of appropriate flow-through modules and photometric flow-cells not only allowed continuous monitoring of the UV/Vis absorbance in a flowing solution, but also allowed the performance of several operations in a flowing stream, including dilution, heating, or dialysis. In order to limit convectional and diffusional dispersion of the analyte in a flow stream, the ingenious division of flowing liquid into small segments with air bubbles was developed. Figure 1.1 shows schematically the principle of its functioning and main parts of such a system. Within the following decade, flow measurements with segmented streams, also described as *continuous flow analysis* (CFA), became a rapidly growing field of instrumentation, first for medical analysis and then for many other areas such as environmental and agricultural analysis, food quality control, and also process analysis in various branches of industry (Furmann, 1976). This led to design and production of powerful clinical analyzers, allowing determinations of tens of different analytes in the same sample (Schwartz et al., 1974; Snyder et al., 1976). Figure 1.2a shows an example of the most advanced CFA instrument for clinical diagnostics, the SMAC (*sequential multiple analyzer with computer*) from Technicon, USA. Since the 1970s, however, such systems in medical laboratories were gradually replaced by different types of discrete and centrifugal analyzers (Fyffe et al., 1988). In recent decades continuous flow analyzers have found increasing applications in food control and environmental analysis (Hollar and Neele, 2008). An example of such a commercial analyzer available on the market is shown in Figure 1.2b. A perceptible decrease of interest in laboratory flow analysis around the early 1970s, especially for medical applications, can be attributed to certain of its disadvantages, which were especially evident when compared with discrete analyzers based on complex mechanical designs but having demonstrably lower sample volume requirements and shorter times of determination.

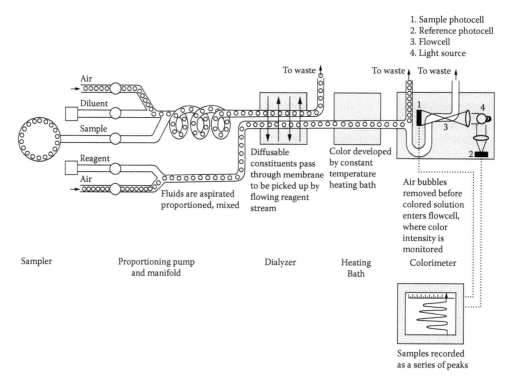

FIGURE 1.1   Schematic diagram of a continuous flow system with segmented stream and spectrophotometric detection.

(a)

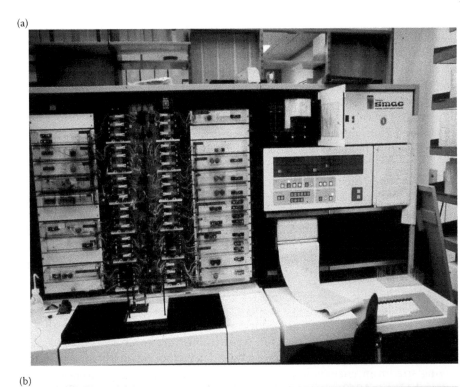

(b)

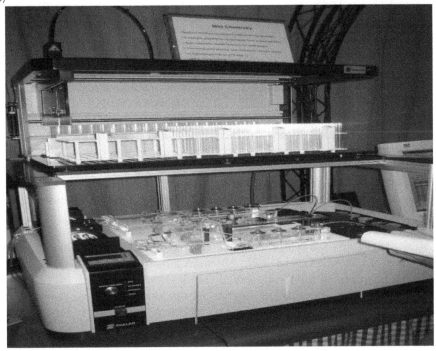

FIGURE 1.2 Examples of commercial continuous flow analyzers: (a) clinical analyzer model SMAC (Technicon, USA), (b) automated wet chemistry analyzer model San⁺⁺. (Adapted from Skalar, Netherlands, www.skalar.com/analyzers/automated-wet-chemistry-analyzers/.)

More or less in parallel to the invention and development of the CFA concept, flowing conditions were also employed in laboratory scale syntheses (e.g., Albeck and Rav-acha, 1970), which is currently described as flow chemistry (Hartman et al., 2011; Wegner et al., 2011). In more recent years, numerous microsystems have been and are being developed for this purpose (Illg et al., 2010).

An extraordinary impetus for further interest of the analytical community in laboratory flow measurements was given by series of research papers published in 1975–1977 by Ruzicka and Hansen from the Technical University of Denmark, and their coworkers, on the injection mode for carrying out laboratory flow measurements (Ruzicka and Hansen, 1975, 1976, 1978; Ruzicka and Stewart, 1975; Stewart et al., 1976; Ruzicka et al., 1976, 1977a,b; Stewart and Ruzicka, 1976; Hansen et al., 1977). This concept of measurement, described as *flow injection analysis* (FIA), is an ingenious invention that developed from earlier achievements and drawbacks of laboratory segmented flow measurements. Compared to CFA measurements, the main improvements were the elimination of the segmentation of the flowing stream that carries the sample to the detector, and the injection of a limited sample volume, which does not generate a steady-state plateau of signal from the detector.

Practically the same approach was reported in the same period by Stewart et al. (1974, 1976). In fact, the possibility of analytical flow measurements with segmentation of flowing streams, but with the introduction such a small sample volume that steady-state signal is not reached in the detector, was demonstrated even earlier (Reid and Wise, 1968), as well as application of laboratory flow determinations without segmentation of the flowing stream in enzymatic determination of glucose with photometric detection (Blaedel and Hicks, 1962). Injection of small volumes of sample into a flowing stream of the carrier solution, which delivers sample to detector, was also reported earlier with voltammetric detection (Pungor et al., 1970).

There is no doubt, however, that the most powerful impact on the further rapid development of injection methods of laboratory flow analysis was given by the series of papers published by Ruzicka and Hansen, and their coworkers, mentioned above. Their additional crucial message of great practical importance to analytical community, based on reported achievements, was that such systems can be assembled in virtually any analytical research laboratory using commercially available components such as pumps for delivering solutions, or different detectors, which can be simply converted to the flow-through mode of signal measurement. In following few years this resulted quickly in wide interest in the proposed methodology of analytical measurements, such that there was even comment on "the epidemiology of research on flow-injection analysis" (Braun and Lyon, 1984).

Although segmentation of a flowing stream as a means of limiting analyte dispersion is occasionally employed in typical flow injection systems (e.g., Pasquini and de Oliveira, 1985; Tian et al., 1990; Sanchez et al., 2010), in a hybrid of a segmented flow analysis system with FIA, termed mono-segmented continuous flow analysis (MCFA) (Pasquini and de Oliveira, 1985), the sample is introduced sandwiched between two air bubbles.

Other types of flow analyzers have also been in parallel development in recent decades. Besides laboratory flow analyzers, which are the main subject of this chapter, there are also different types of chemical analyzers for process analysis. Usually, in such instruments the volume of analyzed sample is not a significant factor, and measurements are carried out with continuous sample aspiration and during the flow of sample through the detector that is used. The variety of commercial offerings of such instrumentation and the variety of construction types are vast (Liptak, 2003). Scientific journals also contain reports of numerous examples of types of flow analyzers that are outside on the mainstream of flow injection analyzers. Such might be, for example, a flow–batch analyzer for the chemiluminescence determination of catecholamines in pharmaceutical preparations (Grunhut et al., 2011), or a flow analyzer employing optoelectronic detector for continuous monitoring of nitrate in waters (Cogan et al., 2013).

1.2 DEVELOPMENT OF BASIC FLOW INJECTION MEASURING SYSTEMS

The development over four decades of methods and instrumentation for conducting laboratory analytical determinations in flow conditions with injection of small samples volumes is the subject of about 20,000 original research papers published in scientific journals. The history and review of those achievements is found in a large number of research papers, and first of all several monographs and numerous multiple-authored books published since the early 1980s (Ruzicka and Hansen, 1981; Ueno and Kina, 1983; Karlberg and Pacey, 1989; Burguera, 1989; Fang, 1993, 1995; Valcarcel and Luque de Castro, 1994; Martinez Calatayud, 1996; Trojanowicz, 2000, 2008a; Kolev and McKelvie, 2008; Yebra-Biurrun, 2009; Zagatto et al., 2011).

The general features of conducting analytical determinations in flow injection systems can be summarized as follows:

1. Determination is carried out by injecting into the measuring system a small but precisely defined volume of sample for analysis. In an appropriate design of measuring system, the injection of sample can be preceded by a stage of mechanized sample treatment using much larger initial sample volume—for example, in order to preconcentrate the analyte or for sample cleanup in case of heavily loaded matrices of natural samples. Usually injected sample volumes are in the range 20–200 µL, but they can be much smaller in miniaturized flow injection systems.

2. The measurement of analytical signal takes place during the flow of sample segment through a flow detector of appropriate design. This allows a larger sampling rate than in similar determinations carried out manually. In many cases this enhances the transport of analyte to the sensing element of the detector, which may result in improvement of sensitivity expressed by the limit of detection (LOD). In some special cases this can also lead to obtaining favorable kinetic effects for modulation of selectivity of detection or can be utilized to design a multicomponent measuring systems. Also, in order to obtain some particular advantage, in such systems the signal measurement can be carried out during the stop of sample flow through the detector, but this is not a typical situation for flow analysis.

3. Different instrumental modules of appropriate design may allow various sample processing operations (e.g., extraction, sorption, dialysis, distillation, etc.) to be conducted online in flow conditions. This is favorable to mechanization of such operations, with first of all an improvement of their precision that makes an important contribution to the precision of whole determination.

Manifolds of the most typical measuring systems used in FIA are shown schematically in Figure 1.3. Both in the simplest system with a single line to transport the carrier

(a)

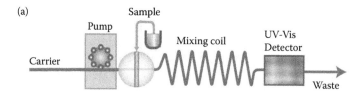

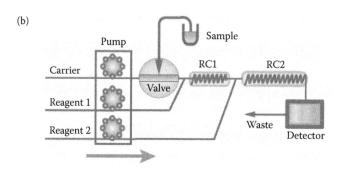

FIGURE 1.3 Schematic manifolds of typical flow injection analysis (FIA) systems: (a) single line manifold, (b) multiline configuration; RC1, RC2—reaction coils. (Adapted from Ruzicka, J., 2009, *Tutorial on Flow Injection Analysis* (4th ed.). *From Beaker to Microfluidics*, www.flowinjection.com.)

solution and also in multiline systems, different sample processing modules can be placed between the injection point (valve) and the flow-through detector. The basic instrumental elements of typical FIA systems are a pumping device, an injection valve, and a detector with appropriate continuous recording of measured signal. Such systems have to be considered mechanized measuring systems, regardless of the degree of computerization of the functions of particular components. Quite commonly in this case is misuse of the term "automated," which as was mentioned above should be used for measuring systems that are regulated by feedback loop, so that the apparatus is self-monitoring or/and self-adjusting (Kingston and Kingston, 1994). In order to limit the consumption of reagent employed in conducting the determination, and with continuous aspiration of analyzed samples, reverse flow injection analysis (rFIA) systems have been designed (Mansour and Danielson, 2012). Small volumes of reagent solutions are injected into the stream of aspirated sample. Then, if both the volume of injected sample and the volume of reagent solution are critical, a concept of merging zones can be employed in design of FIA systems (Bergamin et al., 1978; Schneider and Horning, 1993).

The device most commonly used for the pumping of solutions in FIA systems is the single-line or multiline peristaltic pump, although in constructing many systems a difficulty is encountered with pulsation of the stream caused by the rollers of the pump, and this requires special attention (see, e.g., López García et al., 1995). This problem can be eliminated by the use of syringe pumps. A reliable fluid propulsion system for FIA instrumentation can be also based on generation of electroosmotic flow in a high electric filed. This was shown to be applicable for single-line and double-line systems (Dasgupta and Liu, 1994). Additionally, in spectrophotometric determination of chloride, online preconcentration based on the electrostacking effect was employed. In such a capillary FIA system, injected sample volumes are fractions of microliter. The design of an appropriate dual-membrane module in such capillary FIA systems allows also electrodialytic introduction of ionizable reagents, which was demonstrated for trace spectrophotometric determination of zinc with introduction of 4-(2-pyridylazo)resorcinol (Mishra and Dasgupta, 2010).

Pumping of solutions in FIA systems can be also carried out by the use of gaspressurized reservoirs, which is simple, is inexpensive, and is not disturbed by pulsation (Ruzicka et al., 1982).

With the use of multichannel peristaltic pumps, so-called multicommutated FIA systems are also constructed for various applications (Reis et al., 1994). Manifolds for such systems are based on the use of three-way solenoid valves, which under computer control allow steering of the flow of solutions in the system and realize quite complex hydrodynamic operations in FIA systems. As examples may be mentioned the development of a complex system exploiting multizone trapping for spectrophotometric determination of boron in plants (Tumang et al., 1998), and a biparametric system for simultaneous determination of activity of acid phosphatase and alkaline phosphatase in human serum with optoelectronic spectrophotometric detection (Figure 1.4) (Tymecki et al., 2013). Such systems can be also used, for example, for producing concentration profiles in flow systems, which was demonstrated for complexometric titration with potentiometric detection (Wang et al., 1998).

Dispersion of analyte in an FIA system, which significantly affects the analytical parameters of determination, depends on numerous physicochemical and instrumental factors. The value of the LOD, as well as type of the detector used, depends on degree of dilution of the analyte in the flowing stream, which is associated with the flow of a sample segment from the injection valve to the detector. The degree of dispersion can be modulated according to the methodological needs of a particular determination; the

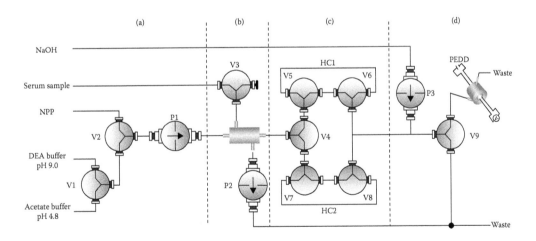

FIGURE 1.4 Manifold of a multicommutated flow injection system (MCFA) developed for simultaneous determination of acid phosphatase and alkaline phosphatase activity consisting of substrate/buffer delivery module (a), sample delivery cell module (b), reaction module (c), and detection module (d). V—valves, P—pumps, HC—holding coils, PEDD—paired emitter detector diode, NPP—*p*-nitrophenyl phosphate, DEA—diethanolamine. (Adapted from Tymecki, L., K. Strzelak, and R. Koncki. 2013. *Anal. Chim. Acta* 797:57–63.)

numerical value of the dispersion coefficient D is defined as the ratio of analyte concentration in the injected sample segment to the maximum concentration of analyte in the sample segment that reaches the flow-through detector (Ruzicka and Hansen, 1978). The relationship between the D value and several instrumental parameters of the FIA system has been given as follows (Ramsing et al., 1981):

$$D = 2\pi^{3/2}R^2(Lv\delta T)^{1/2}V_s^{-1}$$

where R is tubing diameter, L is length of tubing between injection point and detector, v is average flow rate of solution, V_s is injected sample volume, T is mean residence time of analyte in the system (measured to the point of maximum value of signal), and δ is dispersion number, given by the expression $\delta = (1/8)[8(\sigma^2 + 1)^{1/2} - 1]$, where σ is variance associated with recorded FIA peak broadening.

FIA systems with low value $D \leq 2$ are constructed mainly for such determinations, where the chemical form of analyte should not be changed as result of the dilution in the system. This may concern, for example, cases of speciation determinations. FIA systems with especially larger values $D \gg 10$ are designed, for example, for determinations of major components of analyzed samples occurring in large concentrations, or for carrying out titrations in flow systems. In the great majority of FIA systems, the D value is contained within the range $2 < D < 10$. Measurements are almost always carried out in conditions of laminar flow, which depends on the tubing diameter and flow rate employed. Basic sources of dispersion of analyte in a flowing stream are forced convection—in which fluid motion is generated by an external source, for example, a pump—and molecular diffusion. The contribution of convection is reflected in the asymmetry of the recorded FIA peak. In description of the dispersion of analyte in FIA system it can be also helpful to use the so-called reduced time $\tau = TD_m/R^2$, where D_m is the diffusion coefficient of analyte, and R is the tubing radius (Stewart, 1983). Values of the dimensionless quantity τ typical for FIA systems are within the range 0.1–0.6, which corresponds to comparable contributions to peak broadening by convection and diffusion. The complete description of analyte concentration at any point of the tubing as a function of time is described by the differential equation of Taylor (Horvai et al., 1987). An alternative description of the dispersion in FIA system can be given in terms of the width of the recorded peak at baseline (Vanderslice et al., 1981). Some theoretical attempts have also been reported of description of dispersion in rFIA systems (Mansour and Danielson, 2012).

The theoretical approaches mentioned above for describing dispersion in FIA systems related to straight tubing, as employed in the measuring system, and without any packing. Ways to limit the dispersion can include the use of coiled or knotted tubing in the FIA system, and also using tubing packed with granules of solid material (e.g., glass) of various diameters (Van der Linden, 1982). The dispersion of analyte in an FIA system will also be affected by various instrumental characteristics, such as imperfections of connections in the system, differences in diameters of various parts, or such intrinsic factors as the dead-volume of the flow-through detector (Reijn et al., 1980).

Besides the effects of the physical processes of forced convection and diffusion, the magnitude of the recorded signal can also be affected by various kinetic effects. These include both the rate of detector response and the rate of chemical reactions occurring during the determination, for product measured in the detector. In several cases good agreement was reported in the theoretical description of such effect compared to experimental data (Wada et al., 1986; Hungerford and Christian, 1987). Difference in reaction kinetics for different analytes can also find several interesting applications. They can be used, for example, in designing FIA systems for multicomponent determinations. For instance, FIA systems with spectrophotometric detection employing differences in kinetics were reported for simultaneous determination of calcium and magnesium, based on different rates of dissociation of cryptand complexes (Kagenov and Jensen, 1983), for Fe(II) and Fe(III) determinations based on the different kinetic–catalytic behavior in the redox reaction between leucomalachite green and peroxodisulfate (Muller et al., 1990), and also for simultaneous determination of copper and zinc based on the different kinetics of ligand-exchange between metal complexes with Zincon and EDTA (Shpigun et al., 2007). Simultaneous determinations with chemiluminescence detection exploiting kinetic effects were reported, for example, for determination of Co(II) and Cu(II) based on different kinetic characteristics in the luminal–H_2O_2 system (Li et al., 2006), and alkaloids codeine and noscapine based on kinetic differences in the reduction rate of Ru(phen)$_3^{2+}$–Ce(IV) system (Rezaei et al., 2009).

Carrying out analytical determinations in FIA systems has several advantages compared to similar determinations performed manually, although one can find some inconveniences as well. There is unquestionable improvement of efficiency and precision from mechanization of the whole measuring system, especially if entire procedure involves some two-phase processes of sample treatment conducted online. Fundamentally, however, carrying out analytical procedure in FIA system is associated with existing dispersion in the flow system, and this can be a source of worsening of the LOD in FIA system. This was demonstrated in one case in the response of potentiometric detectors in FIA systems compared to the manual procedure (Trojanowicz and Matuszewski, 1982). In another case, the application of a flow-through voltammetric detector might allow a significant mass-transfer enhancement of more than 2 orders of magnitude (Macpherson and Unwin, 1998), and this can directly improve the LOD of the analyte in amperometric detection. An essential difference has to be recognized in the methodology of the evaluation of LOD in FIA systems and in manual measurements. In manual procedures it is commonly related to the precision of blank measurements, whereas in FIA measurements—similarly to, for example, chromatographic ones—it is often related to the magnitude of noise in the baseline, which prevents a direct comparison.

A basic challenge in all analytical flow measurements on real samples is the selectivity of the determination of a given analyte or given analytes. This problem obviously does not concern typical column chromatographic systems, where also detection in flow conditions is employed, during the flow of analyte through the flow-through detector, but it is preceded by optimized online separation of sample constituents. In a typical flow system, and also in the FIA system, the selectivity of determination is obtained in different way. If one analyte is available, a specific detector can be used responding only to particular species, or such instrumental parameters of detector functioning can be exploited as provide specific response to the analyte. This can be, for example, the use of appropriate ion-selective electrodes (ISEs), selective biosensors, or molecular or atomic spectroscopy with spectral parameters providing response that is not affected by other components of the sample. This is a rather idealized situation. In the particular instance of potentiometric detection with membrane electrodes, conducting measurements in FIA conditions with recorded fast transient signal one can obtain a much better apparent selectivity than in steady-state manual determination (e.g., Trojanowicz and Matuszewski, 1983; Ferreira and Lima, 1994). This results from the difference between the rate of response of ISEs to the main ion and interfering ions, but this kinetic discrimination concerns only those interfering ions that exhibit a Eisenman–Nikolsky selectivity coefficient greater than 1. The kinetic discrimination effect has also been reported in FIA spectrophotometry of bromide (Emaus and Henning, 1993).

An often satisfactory method of creating the selectivity of response to particular analyte is derivatization of the analyte to obtain a product that can be selectively detected by the flow-through detector employed. This primarily concerns molecular spectroscopy in absorptive and luminescence modes. Incorporation into the flow system of appropriate an sample processing step is an especially effective way of creating the required selectivity of response. It may allow the extraction of the analyte from the sample matrix, or online cleanup of the sample. The method has been discussed and optimized in thousands of original research papers, dedicated books (Fang, 1993), and numerous chapters in the books mentioned above (Ruzicka and Hansen, 1981; Ueno and Kina, 1983; Burguera, 1989; Karlberg and Pacey, 1989; Fang, 1993, 1995; Valcarcel and Luque de Castro, 1994; Martinez Calatayud, 1996; Trojanowicz, 2000, 2008a; Kolev and McKelvie, 2008; Yebra-Biurrun, 2009; Zagatto et al., 2011).

In addition to optimization of hydrodynamic conditions of the determination in FIA system and eventual online sample processing, a factor of fundamental importance is appropriate selection and optimization of chemical conditions. Without exaggeration one can say that practically all of the modern instrumental methods of chemical analysis for measurements in liquids have been employed in FIA measurements, from simple photometry to mass spectrometry (Benkhedda et al., 2002) or nuclear magnetic resonance (NMR) (Keifer, 2003, 2007; Baker et al., 2012). Undoubtedly the most commonly employed detection method is molecular spectrophotometry in the UV/Vis range, because of the wide availability of instrumentation and the vast literature and numerous applications in determination in all fields of modern chemical analysis. Both absorptive and luminescence techniques are used. Besides optimization of chemical conditions of determinations, much attention is also devoted to instrumental improvement of FIA measuring systems; examples are the increasing interest in detectors based on miniaturized optoelectronic elements (Trojanowicz et al., 1991; O'Toole et al., 2005; Pokrzywnicka et al., 2012; Strzelak and Koncki, 2013), or the increasing use in such systems of long-pathlength capillary cells to enhance sensitivity (Pascoa et al., 2012). Increasing interest can be also noted in direct solid-phase optical measurements in flow systems, which allows minimization of reagent consumption as well as waste generation (Rocha et al., 2011). An example of an advanced spectral detector utilizing a rapid solid-phase fluoro-immunoassay, is seen in a transducer chip developed for simultaneous multianalyte determination of pharmaceuticals and pesticides in waters based on the use of a commercial module for construction of the manifold (Rodriguez-Mozaz et al., 2004) (Figure 1.5). The most common use of spectrophotometric detection is for the determination of inorganic analytes based on color reactions with different chromophores of fluorophores (see, e.g., Zagatto et al., 2011), and also chapters in other books mentioned above (Ruzicka and Hansen, 1981; Ueno and Kina, 1983; Burguera, 1989; Karlberg and Pacey, 1989; Fang, 1993; Valcarcel and Luque de Castro, 1994; Fang, 1995; Martinez Calatayud, 1996; Trojanowicz, 2000; Kolev and McKelvie, 2008; Trojanowicz, 2008a; Yebra-Biurrun, 2009), and also for determination of various groups of organic compounds such as the determination of pharmaceuticals in FIA systems (Tzanavaras and Themelis, 2007; Trojanowicz, 2012). Such systems can also be effective tools for the improved environmental analysis of organic compounds expressed as total indices (Maya et al., 2010). Numerous applications are also found of FIA methods of molecular luminescence spectroscopy (Fletcher et al., 2001). Fewer applications involve infrared spectroscopy, mostly

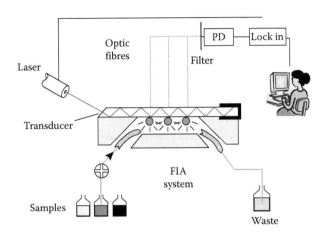

FIGURE 1.5 Schematic diagram of flow injection analysis (FIA) system with immuno-assay detection RIANA (abbreviation from "River ANAlyser") based on total internal reflection fluorescence with immobilized fluorescently labeled antibodies. The source of the excitation light is a He–Ne laser, and the collected fluorescent light is filtered and detected by photodiodes (PD). (Adapted from Rodriguez-Mozaz, S. et al. 2004. *Biosens. Bioelectron.* 19:633–640.)

because of the strong absorption of most of the common solvents, especially water, and also because of relatively poorer sensitivity compared to absorptive UV/Vis methods or fluorescence (Gallignani and Brunetto, 2004). There has been increasing application in recent years of thermal lens spectrometric detection, based on measurement of heat generated from nonradiative relaxation process, following absorption of optical radiation (Franko, 2008).

There is substantial interest in FIA methodology in analytical atomic spectroscopy (Tyson, 1991). Absorptive and emission modes with flame atomization were examined in FIA systems very early due to rather simple hyphenation of FIA systems with continuous aspiration of sample to nebulizers in spectrometric instruments (Zagatto et al., 1979). Significant improvements in precision and sampling rate with mechanized online sample processing resulted in a wide application of those methodologies in numerous branches of chemical analysis, including environmental, agricultural and food analysis (Burguera, 1989; Trojanowicz and Olbrych-Sleszyńska, 1992; Fang, 1995). Numerous systems have also been also developed for determination using electrothermal atomization (Burguera and Burguera, 2001), or hydride generation (e.g., Anthemidis and Kalogiuri, 2013).

Another group of detection technologies widely used in FIA systems are electrochemical methods, and application of potentiometric flow-through detectors with ISEs has appeared in pioneering works on FIA (Stewart et al., 1976). Broad development of FIA systems with voltammetric and amperometric detection methods was due to the existence of a good theoretical description of electroanalytical measurements in flowing liquids (Stulik and Pacakova, 1987). In the case of potentiometric detection it was due to intense development of membrane ISEs, which seemed to offer a satisfactory selectivity of response in their application with different configurations of flow-through detection cells (e.g., Davey et al., 1993). In both cases such applications are favored by wide availability of commercial instrumentation. As already mentioned, the selectivity of potentiometric determinations may sometimes be enhanced by performing determinations in FIA conditions (Trojanowicz and Matuszewski, 1983; Ferreira and Lima, 1994), or in amperometric detections by the use of, for example, chemically modified electrodes (Wang et al., 1991). Electrochemical detection methods can be satisfactorily used both in conventional laboratory FIA systems and in microfluidics (Trojanowicz et al., 2003; Trojanowicz, 2009; Felix and Angnes, 2010).

For environmental applications and monitoring of radionuclides in the nuclear energy industry, an increasing number of FIA methods are being developed with radiochemical detection (Fajardo et al., 2010; Kolacinska and Trojanowicz, 2014). Besides all the attributes of FIA systems discussed above, such as improvement of sampling rate and precision of determinations and also online sample processing, the mechanization of determination of radionuclides improves the safety of personnel performing those analyses. FIA determination with all of the detection methods discussed above have also been adopted in specific immunochemical determinations (Puchades and Maqueira, 1996; Fintschenko and Wilson, 1998).

Mass spectroscopy (MS) is among instrumental methods of chemical analysis with the greatest potential in terms of structure identification and quantitative determinations. Development of different methods for ionization of analytes and outstanding progress in construction of mass analyzers allows the application of MS to practically all types of analytes in liquid samples. Adding to such instrumentation a FIA system as another accessory, which can be hyphenated similarly to HPLC (high-performance liquid chromatography) systems, allows the mechanization of sampling, improves the precision of determinations, and above all enables online sample processing for samples of heavily loaded matrices or those requiring analyte preconcentration. Pioneering application of hyphenation of FIA with MS concerned its use in inductively coupled plasma–mass spectrometry (ICP-MS) systems for chemical speciation by reversed-phase liquid chromatography, where it was employed to estimate the dispersion in a sample delivery system (Thompson and Houk, 1986). In determination of organic compounds, hyphenation of electron impact ionization with tandem MS was employed for monitoring of products of fermentation (Hayward et al., 1990). FIA-ICP MS systems with different mass analyzers were used in online isotope dilution analysis in a rFIA system (Beauchemin and Specht, 1997), and with online preconcentration (Benkhedda et al., 2002). In determination of nonpolar analytes in a flow injection analysis–quadrupole-time-of-flight mass spectrometry (FIA-QTOF MS) configuration, atmospheric pressure photoionization has been reported (Gomez-Ariza et al., 2006) but MS with electrospray ionization is used much more often for this purpose. This was reported, for example, for a high-throughput chemical residue analysis with fast extraction and dilution (Nanita, 2011). It was shown that dilution of extracts may allow direct FIA-MS multiresidue analysis, without chromatography, which increases throughput in the sample preparation and instrumental analysis.

For many years FIA methods have found wide application in speciation analysis of elements in different matrices. In contrast to chromatography, it usually requires development of rather complex procedures and construction of an instrumental setup in order to obtain a suitable combination of the separation and conversion steps. A short time for carrying out those operations in the FIA manifold is advantageous in terms of avoiding the shift of chemical equilibria during the speciation measurement (Campanella et al., 1996; Cerda and Estela, 2005). Relatively simple if achievable, it seems to be a parallel application of several selective detectors allowing simultaneous determination of particular chemical forms of element. A more difficult construction challenge is development of such system with a single detector. In such case it is necessary to employ smart online sample processing: for example, the preconcentration of all forms and fractional elution or selective sorption; application online of appropriate redox processes; or kinetic discrimination. It is also possible to use differently constructed systems, such as multivalve systems with simultaneous injection of various reagents or with branched manifolds. Two such classic examples are shown in Figure 1.6. First is a FIA system with dual injection valve and spectrophotometric detection for the speciation of chromium in terms of oxidation state and acid–base equilibria (Ruz et al., 1986) (Figure 1.6a), and the second one is a branched FIA system with biamperometric detection for simultaneous determination of nitrate and nitrite (Trojanowicz et al., 1992) (Figure 1.6b). An important element of speciation analysis of elements bound in different complexes in natural samples is usually also the determination of the noncomplexed fraction of element, which can be carried out in FIA systems with potentiometric detection (e.g., Trojanowicz et al., 1998). Among numerous original research papers on different types of samples analyzed, different analytes, and different detection methods, one can also find review papers concerning, for example, FIA systems for speciation of inorganic nitrogen in waters (Rocha and Reis, 2000), or flow injection speciation of aluminum (Pyrzynska et al., 2000).

An improvement in construction of FIA systems for routine use that is important due to the limitation of volume of consumed solutions is the design of closed flow-through systems with circulating or (eventually) regenerated reagent. This concept of FIA system design has been initiated by the development of a measuring system for enzymatic determination of glucose with amperometric detection that was based on biocatalytic destruction of the hydrogen peroxide produced in the presence of catalase (Wolff and Mottola, 1978). Examples of such FIA systems developed with spectrophotometric detection include determination of Fe(III) employing series reactions and kinetic effects (Eswara Dutt et al., 1977); the determination of chloride based on a precipitation reaction and the dissolution of the precipitate (Zenki and Iwadou, 2002); and also the determination of chemical oxygen demand based on online regeneration of consumed permanganate (Zenki et al., 2006).

Another technical and functional improvement is utilization of programmed transport of micrometer (10–200 μm) beads in FIA systems. This was initially employed, for example, in immunoassay in CFA systems; in FIA measurements it was used for the formation of a renewable surface by a bed of particles that can be used for direct spectroscopic measurements in a special design of jet ring cell (Ruzicka et al., 1993). Determinations carried out in FIA systems with movable microbeads require the formation of a stable suspension, its transfer within a measuring system, and temporary immobilization in order to detect the measured signal (Hartwell et al., 2004). The beads can be applied in sample pretreatment such as preconcentration, isolation and separation, and also for accommodation of chemical reactions. A particular feature of those operations is their potential for the simplification and mechanization of bed exchange, when sorbent is degraded during multiple measurements or when the washing of retained components is very difficult.

The wide interest of the analytical community in this measurement methodology is reflected in increasing commercial offers of such instruments from several manufacturers. Some examples of commercial instrumentation currently available are shown in Figure 1.7.

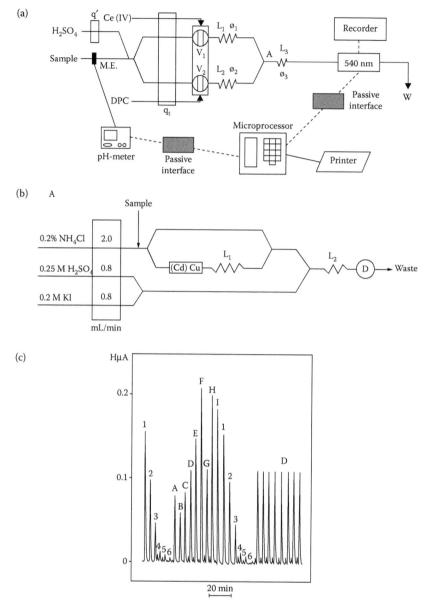

FIGURE 1.6 Example of manifolds developed for flow injection speciation analysis. (a) Manifold of the reversed FIA system for speciation of chromium with pH measurements and spectrophotometric detection. q—peristaltic pumps, V—injection valves, L—reaction coil, M.E.—pH glass electrode. (Adapted from Ruz, J. et al. 1986. *J. Autom. Chem.* 8:70–74.) (b) Manifold of branched FIA system for simultaneous biamperometric determination of nitrate and nitrite. (Cd)Cu—reductive column with copperized cadmium, L—mixing coils. (Adapted from Trojanowicz, M., W. Matuszewski, and B. Szostek. 1992. *Anal. Chim. Acta* 261:391–398.) (c) Example signal recordings obtained in the FIA system shown in (b): 1–6—standard solutions, A–I—natural water samples. Concentration in standard solutions: 1—0.075; 2—0.050; 3—0.025 mM nitrate; and 4—7.5; 5—5.0; 6—2.5 μM nitrite.

(a)
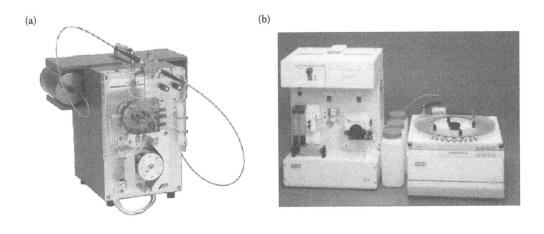

(b)

(c)
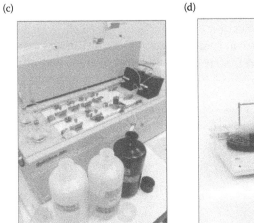

(d)
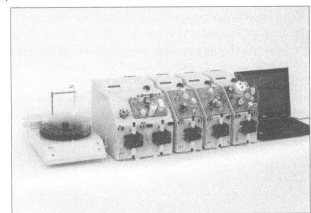

FIGURE 1.7 Examples of commercially available flow injection analysis instruments: (a) FIAlab2500 analyzer from FIAlab Inc., USA, www.flowinjection.com, (b) FIA2000 multichannel analyzer from Burkard, United Kingdom, www.burkardscientific.co.uk, (c) QuikChem 8500 Series Flow Injection Analyzer from Lachat, USA, www.lachatinstruments.com, and (d) flow injection analysis system from MLE company, Dresden, Germany, www.mle-dresden.com.

1.3 OTHER DESIGNS OF FLOW INJECTION SYSTEMS

The basic methodology of FIA measurements with the injection of small sample volumes into flowing stream of carrier or reagent solution has one significant drawback, which is quite a large consumption of continuously pumped solutions during the measurement. This is most often cited as a reason for further modifications of instrument manifolds for FIA.

In a relative simple methodology it was shown that, in appropriate design of measuring cell, a very small sample volume can be injected using an automatic pipette directly onto the sensing surface of the detector. This concept of measurement, also described as batch injection analysis (BIA) was described for the first time for amperometric detection, where sample was injected directly onto the surface of the working electrode (Wang and Taha, 1991a). An example of the construction of the detection cell for such measurements is shown in Figure 1.8 (Brett et al., 1997). Injected sample solution flows over the surface of the electrode, and during this a transient detector signal is recorded as a peak; hence this methodology can be incorporated into flow analysis although it is conducted without continuous pumping or gravity-based flow of liquid. It was suggested to name this method tube-less FIA (Trojanowicz et al., 2005). The basic limitation of analytic potential of such measurements is the lack of ability to carry out online sample processing in such systems, although it was shown that by if a small sorbent bed is placed in the pipette tip a solid-phase extraction can be carried out for the preconcentration of analyte or sample cleanup (Trojanowicz et al., 2005). It has also been shown that similar measurements can be carried out also with potentiometric (Wang and Taha, 1991b), spectrophotometric (Wang and Angnes, 1993), or thermometric (Thavarungkul et al., 1999) detections, but this methodology has not found wider application.

One of the milestones in the development of laboratory methods of flow analysis was without doubt invention of sequential injection analysis (SIA) (Ruzicka and Marshall, 1990). A schematic diagram showing main parts of the instrumentation and the principle of measurement is shown in Figure 1.9a; the general idea of this concept is replacement of continuous pumping of solutions by introduction into a single line a sequence of small segments of sample and reagent(s). Basic instrumental elements of the measuring system are a single-channel syringe pump allowing reversal of pumping direction, and also a multiposition switching valve and the detector. In the first stage of measurement, with different positions of the switching valve, a sequence of solution segments is introduced into a holding coil; then, after reversing the flow, segments are transported to the detector. Determined by dispersion in the system, overlapping of segments leads to the course

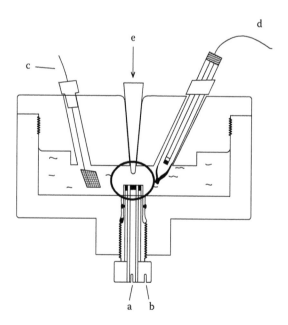

FIGURE 1.8 Scheme of a Perspex cell used for amperometric detection in batch injection analysis: (a) working disk electrode, (b) ring electrode, (c) auxiliary electrode, (d) saturated calomel reference electrode, and (e) micropipette tip. (Adapted from Brett, C. M. A., A. M. O. Brett, and L. C. Mitoseriu. 1997. *Electroanalysis* 7:225–234.)

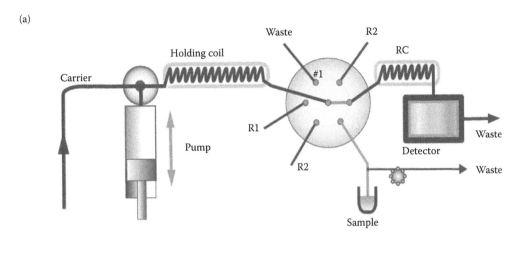

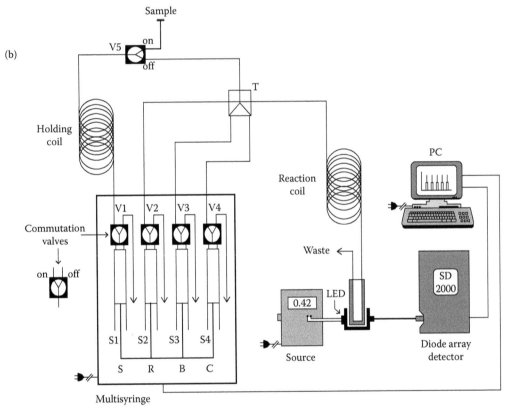

FIGURE 1.9 (a) Schematic manifold of sequential injection analysis (SIA) system. R—reagents, RC—reaction coil. (Adapted from Ruzicka, J., 2009, *Tutorial on Flow Injection Analysis* (4th ed.). *From Beaker to Microfluidics*, www.flowinjection.com.) (b) Schematic diagram of multisyringe flow injection analysis (MSFIA) system with commutated manifold developed for determination of uranium. V—commutation valves, T—confluence junction, LED—light emitting diode. (Adapted from Albertus, F. et al. 1999. *Analyst* 124:1373–1381.)

of the chemical reaction between analyte and reagent, if they are needed for the detection. The flow system is simplified and there is a significant decrease of the solution consumption, although much greater attention must be given to the computerized control of the mechanized system. Soon after its conception, various configurations of such systems for different applications, and with different detectors were published. Valuable progress was made in these systems with the introduction of moveable bead suspensions (Ruzicka et al., 1993; Pollema and Ruzicka, 1994; Egorov and Ruzicka, 1995). Such a methodology allows, for example, renewable surface immunoassay with fluorimetric detection (Pollema et al., 1994), or the use of such a suspension in a renewable spectrophotometric sensor for analytes retained on beads (Egorov and Ruzicka, 1995). Further improvement in such systems with multiposition switching valves can be derived from incorporation into a rotary part of the valve some detecting elements or channels playing the role of minicolumns for temporary retention of a bead suspension (Ruzicka, 2000). This additional modification of SIA systems, described as *lab-on-valve* (LOV) is finding increasing numbers of applications (e.g., Hansen and Miro, 2008; Yu et al., 2011; Anthemidis et al., 2014).

The concept of multisyringe flow injection analysis systems (MSFIA) developed by Cerda and coworkers (Albertus et al., 1999) has found wide acceptance as a way of improving the design of multiline SIA systems. The main element of such systems is the multisyringe burette, which is a multiple-channel piston pump, driven by a single motor. A two-way commutation valve is connected to the head of each syringe, allowing coupling to the manifold lines. Various flow rates are obtained by using syringes of different volumes. An example of a manifold for such a system developed for determination of uranium in water samples with UV/Vis detection is shown in Figure 1.9b (Guzman Mar et al., 2009). In this example, an alternative approach is shown wherein MSFIA system was hyphenated not to a typical SIA system but to a commutation system, providing precise control of the merging and reaction timing of the reagents and sample. It allows low consumption of reagent and increase of the sampling rate. Several other examples of MSFIA systems will be shown below.

The advantages of SIA methodology indicated above and the commercial availability of such instrumentation generated wide interest in it and led to the development of numerous different applications in various fields of chemical analysis. They are most commonly used with spectrophotometric detection but numerous electrochemical detections were also employed (Barnett et al., 1999; Perez-Olmos et al., 2005). Sequential injection analysis methods with various detections were developed for environmental analysis (Cerda et al., 2001; Miro and Hansen, 2006; Mesquite and Rangel, 2009), and also for speciation analysis (Van Staden and Stefan, 2004). These are considered a useful tool for clinical and biochemical analysis (Economou et al., 2007), and even as an alternative approach to process analytical chemistry (Barnett et al., 1999).

One can find numerous additional improvements and particular applications in works published about SIA methodologies in recent years. A reagent-free SIA system was reported for monitoring sugar content, color, and dissolved CO_2 in soft drinks (Teerasong et al., 2010). Light reflection at the liquid interface (Schlieren effect) of sucrose and water was utilized for sucrose determination. Measuring systems were designed in many studies, in which SIA setups were constructed in hyphenation with FIA systems, with multicommutated ones, and also with the above-mentioned MSFIA systems (Pinto et al., 2011). The main purpose of such hyphenated constructions is mechanization of additional operations of online sample processing prior to the introduction of sample into the SIA system, such as extraction, preconcentration, gas diffusion, or online sample digestion. By the implementation of the multicommutation concept in SIA system, it was possible to introduce samples in the milliliter volume range (Araujo et al., 1999), or with a rotating rod renewable microcolumn to perform solid-phase DNA hybridization studies (Brucker-Lea et al., 2000). In another example, such an approach was employed in an SIA/FIA setup for trace determination of lead with ion-exchange preconcentration and elimination of interferences (Mesquite et al., 2004). The hyphenation of SIA with an MSFIA system is shown schematically in Figure 1.10, which illustrates the system developed for online monitoring of orthophosphate in soils and sediments (Buanuam et al., 2007). Hyphenation allowed the use of the selection valve to select the appropriate extractant and accommodated several reagents and a soil microcolumn.

A very smart concept of modification of SIA was incorporation into a measuring system of a chromatographic column, which can be used without the need for a high-pressure system (Chocholous et al., 2007). Especially favorable for this are modern monolithic columns, which provide high efficiency of chromatographic separation without requiring high-pressure conditions. This enables chromatographic measurements without expensive HPLC apparatus, although drawbacks of such an approach include poorer precision of syringe pumps, lower sensitivity of spectrophotometric detectors, and much less advanced construction of SIA systems and software compared to modern HPLC systems (Chocholous et al., 2011). On the other hand, the advantage of chromatographic measurements carried out in SIA systems, is the possibility of easy exchange of eluents, both in single-syringe system and in MSFIA system with multiposition switching valve (Hartwell et al., 2013).

Another implementation of SIA systems in high-performance separation methods is instrumental hyphenation with HPLC or capillary electrophoresis instruments. This is

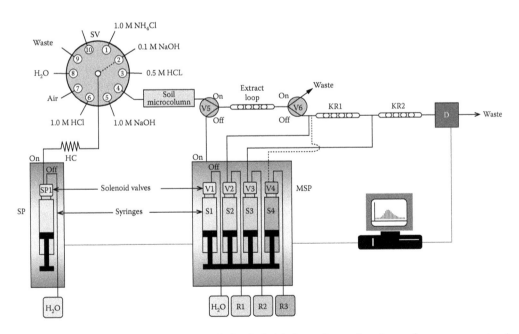

FIGURE 1.10 Schematic diagram of the hybrid flow-through microcolumn extraction/multisyringe flow injection analysis (MSFIA) system for fractionation and online determination of orthophosphate. SP—syringe pump, MSP—multisyringe pump, SV—selection valve, V—solenoid valves, KR—knotted reactors, D—detector. (Adapted from Buanuam, J. et al. 2007. *Talanta* 71:1710–1719.)

illustrated by schemes of such systems in Figure 1.11. In the first, the HPLC instrument was hooked up to a sequential injection analysis–lab-on-valve/multisyringe flow injection analysis system (SIA–LOV/MSFIA) (Quintana et al., 2006) (Figure 1.11a), and was optimized for the determination of trace residues of selected pharmaceuticals in wastes. In the SIA part of the setup, the solid-phase extraction was carried out with renewable commercial hydrophilic–lipophilic balance (HLB) resin beads type Oasis HLB from Waters, USA. The SIA system hyphenated with capillary electrophoresis (Figure 1.11b) was developed for determination of adenosine and adenosine monophosphate (Kulka et al., 2006). Such hyphenation allows for mechanized and reproducible liquid handling, without manual sample preparation or capillary conditioning or cleaning.

At least two other concepts for devices for flow injection measurements in laboratory scale developed in recent years should be mentioned. Although they are not yet widely employed, they exhibit some interesting features. The so-called *stepwise injection analysis* system is similar to a conventional SIA system in its main instrumental component (Bulatov et al., 2012), but instead of the holding coil and reaction coil (see Figure 1.9a) a cylindrical reaction vessel with a funnel-shaped inlet at the bottom is used (Figure 1.12a). Additionally, one of inlets of the switching valve is connected to a gas container used for intensive mixing of sample and reagents in the reaction vessel. A steady-state signal is recorded instead of transient analytical signal, and this results in simpler optimization of measuring procedure. Several online sample processing steps were also developed for such systems, but the sampling-rate that can be obtained is smaller than in conventional SIA measurements.

Another design of instrument for flow injection measurements similar to SIA mode was termed *cross injection analysis* (Nacapricha et al., 2013). The hydrodynamic module designed for this concept is shown schematically in Figure 1.12b. It is made from a rectangular acrylic block with crossing cylindrical channels drilled out perpendicularly. The main channel serves as the analytical flow path and is filled with the carrier solution, whereas four perpendicular channels are made for introducing segments of reagents' solutions and sample. Such system was employed, for example, for spectrophotometric determination of Fe(II), and the speciation of Fe(II) and Fe(III) (Nacapricha et al., 2013).

1.4 CONCLUSIONS: CURRENT TRENDS AND PERSPECTIVES

More than half a century of development of various concepts of laboratory flow analysis earned it solid place among instrumental methods of modern chemical analysis. Those methods find appropriate applications when, for various reasons (cost, efficiency, qualification of personnel), chromatographic methods cannot be employed and when there is a need for mechanization of previously manual determinations of selected analytes. The continuous evolution of the concepts and instrumentation of flow analysis is clear and is oriented toward simplification and scaling down of instrumentation, improvement of efficiency, and minimization of the necessary sample volume and the consumption of reagent(s). The same factors are also reasons for special interest in research and development laboratories in flow injection modes of flow analysis. On the other hand, CFA measurements still predominate among routine applications of flow analysis methods; after decades of wide application in clinical laboratories in 1960s and 1970s, they are now mostly employed in environmental analysis and food monitoring and control.

Continuous progress in the development of online methods of sample processing must also be acknowledged. This is certainly one of the most important aspects of

(a)

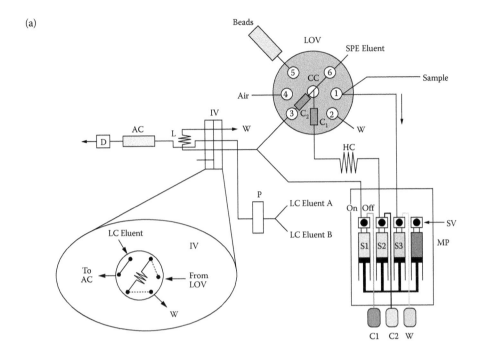

(b)

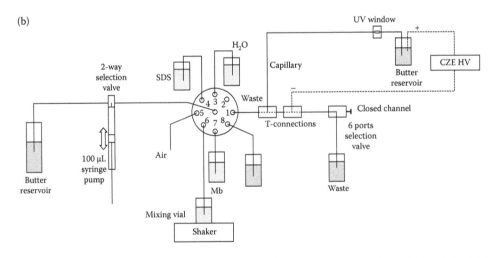

FIGURE 1.11 Systems hyphenating sequential injection analysis (SIA) with high-performance separation methods: (a) schematic diagram of multisyringe flow injection analysis–lab-on-valve (MSFIA–LOV) system with HPLC. LOV—lab-on-valve, MP—multisyringe pump, S—syringe pump, SV—solenoid valve, HC—holding coli, CC—communication channel, C_1 and C_2—cavities for beads, P—HPLC pump, AC—analytical column, IV—injection valve, L—loop, D—detector, W—waste, C1, C2—carriers, SPE—solid-phase extraction. (Adapted from Quintana, J. B. et al. 2006. *Anal. Chem.* 78:2832–2840.) (b) Manifold of the SIA-capillary electrophoresis system. 1—capillary, 2—waste, 3—water, 4—sodium dodecyl sulfate (SDS) solution, 5—air, 6—mixing vial, 7—myoglobin solution, 8—HCl solution, CZE HV—high-voltage supplier for capillary zone electrophoresis. (Adapted from Kulka, S., G. Quintas, and B. Lendl. 2006. *Analyst* 131:739–744.)

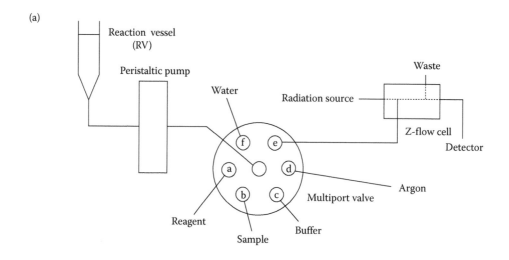

(a)

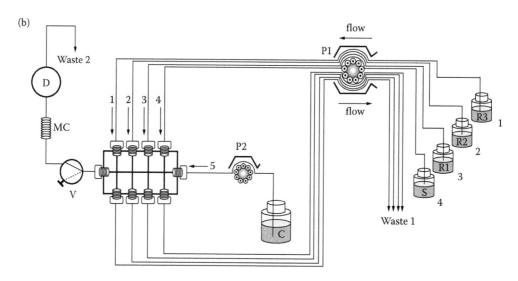

(b)

FIGURE 1.12   (a) Schematic manifold of the stepwise injection analysis. (Adapted from Bulatov, A. V. et al. 2012. *Talanta* 96:62–67.) (b) Cross injection analysis manifold. P—peristaltic pump, MC—mixing column, V—switching vale, C—carrier solution, S—sample, R—reagents. (Adapted from Nacapricha, D. et al. 2013. *Talanta* 110:89–95.)

research in the field of flow analysis because as well as typical applications in flow methods, the devices can be used as initial steps in other more complex instrumentation, such as atomic spectrometers or even mass spectrometers. Accordingly, the design of appropriate flow-through modules that can be employed as accessories for such instruments is an important trend in the development of flow analysis. One field of application widely investigated in research laboratories has been mentioned already: the hyphenation of typical flow analysis systems with instruments for high-performance separations, such as liquid and gas chromatography and also capillary electrophoresis. In the 1970s the first attempts at construction of such systems were reported for CFA systems, while in more recent years such systems have been reported mostly for different variants of the flow injection method (e.g., Almeida et al., 2011; Miro et al., 2011).

A particular opportunity for successful competition with other analytical instrumentation on the market is miniaturization of instruments for flow analysis in their different instrumental modes. Progress here is made in two alternative ways. In one case the components of conventional systems can be scaled down to produce portable equipment—for example, for field use. They can, for instance, be fully integrated instruments for environmental applications (Higuchi et al., 1999), or even devices with remote control such as submersible systems for analytical measurements in seawater (Gardolinski et al., 2002) is the alternative route involves transfer of the whole concept of flow analytical determinations down to the microfluidics scale, where all operations are carried out in capillary channels of diameters below 0.1 mm, with different sample injection methods, and also with miniaturized detectors. This is accompanied by online sample processing operations, which can be different from those in conventional flow analysis instrumentation. The first step in this evolution was the design of microconduits for FIA measurements, and these are often designed to incorporate detection elements (e.g., Ruzicka and Hansen, 1984). Progress in micromechanics and materials science produced a significant decrease of dimensions of such devices over the last two decades to allow authentic microfluidics for a wide potential of applications, for example, in environmental and clinical analysis (Gardeniers and Van Den Berg, 2004). Theoretically, they can even be programmable and reconfigurable systems (Renaudot et al., 2013). They constitute a special kind of analytical instrumentation that can compete with devices currently available on the market, for instance, for individualized personal care devices. These may concern personal monitoring of various human physiological parameters, or the detection and quantitation of environmental pollutants or the presence of harmful species in a workplace. For these purposes miniaturized chemical sensors and biosensors are most commonly used; however, often they are not sufficiently selective and sensitive, and hence development of miniaturized devices for sample pretreatment may result in a significant improvement.

These topics have been strongly connected in recent years with rapid increase nanotechnology applications in chemical analysis (Trojanowicz, 2006; Valcarcel et al., 2008), which can be also observed in flow analysis (Trojanowicz, 2008b). These trends feature in three main aspects: improvement of existing methods or introduction of novel methods of detection in flow conditions; improvement of online sample processing methods; and further miniaturization of measuring equipment. The last factor involves the construction of nanofluidics (Piruska et al., 2010). At the present stage of development, this mostly means chemical manipulation in nanopores, including, for example, nanopore-mediated separations, formation of three-dimensional networks, or fast diffusive mixing. A special version of miniaturization of flow systems to microfluidcs dimensions is the technique of droplet-based microfluidics (Song et al., 2006). This is based on manipulation of single drops of samples and reagents in microfluidics, which provides a rapid mixing of reagents, control of the timing of reactions over a wide time scale, and also control of interfacial processes.

REFERENCES

Albeck, M. and Ch. Rav-acha. 1970. An apparatus for investigation of heterogeneous reactions in a flow system. *Experientia* 26:1043–1045.

Albertus, F., B. Horstkotte, A. Cladera et al. 1999. A robust multisyringe system for process flow analysis. Part I. On-line dilution and single point titration of protolytes. *Analyst* 124:1373–1381.

Almeida, M. I. G. S., J. M. Estela, and V. Cerda. 2011. Multisyringe flow injection potentialities for hyphenation with different types of separation techniques. *Anal. Lett.* 44:360–373.

Anthemidis, A. N., E. I. Daftsis, and N. O. Kalogiouri. 2014. A sequential lab-on-valve (SIA-LOV) platform for hydride generation atomic absorption spectrometry (HG-AAS): On-line determination of inorganic arsenic. *Anal. Methods* 6:2745–2750.

Anthemidis, A. N. and N. P. Kalogiuri. 2013. Advances in on-line hydride generation atomic spectrometric determination of arsenic. *Anal. Lett.* 46:1672–1704.

Araujo, A. N., R. C. R. C. Costa, and J. F. C. Lima. 1999. Application of sequential injection analysis to the assay of lead retention characteristics by poly(vinylpyrrolidone): Trace analysis of lead in waters. *Anal. Sci.* 15:991–995.

Baker, J. M., J. L. Ward, and M. H. Beale. 2012. Combined NMR and flow injection ESI-MS for *Brassicaceae* metabolomics. *Methods Mol. Biol.* 860:177–191.

Barnett, N. W., C. E. Lenehan, and S. W. Lewis. 1999. Sequential injection analysis: An alternative approach to process analytical chemistry. *Trends Anal. Chem.* 18:346–353.

Beauchemin, D. and A. A. Specht. 1997. On-line isotope dilution analysis with ICPMS using reverse flow injection. *Anal. Chem.* 69:3183–3187.

Benkhedda, K., H. G. Infante, and F. C. Adams. 2002. Inductively coupled plasma mass spectrometry for trace analysis using flow injection on-line preconcentration and time-of-flight mass analyzer. *Trends Anal. Chem.* 21:332–340.

Bergamin, H., E. A. G. Zagatto, F. J. Krug et al. 1978. Merging zones in flow injection analysis: Part 1. Double proportional injector and reagent consumption. *Anal. Chim. Acta* 101:17–23.

Blaedel, W. J. and G. P. Hicks. 1962. Continuous analysis by measurement of the rate of enzyme catalyzed reactions. *Anal. Chem.* 34:388–394.

Braun, T. and W. S. Lyon. 1984. The epidemiology of research on flow-injection analysis: An unconventional approach. *Fresenius Z. Anal. Chem.* 319:74–77.

Brett, C. M. A., A. M. O. Brett, and L. C. Mitoseriu. 1997. Amperometric batch injection analysis: Theoretical aspects of current transients and comparison with wall-jet electrodes in continuous flow. *Electroanalysis* 7:225–234.

Brucker-Lea, C. J., M. S. Stottlemyre, D. A. Holman et al. 2000. Rotating rod renewable microcolumns for automated, solid-phase DNA hybridization studies. *Anal. Chem.* 72:4233–4141.

Buanuam, J., M. Miro, E. H. Hansen et al. 2007. A multisyringe flow-through sequential extraction system for on-line monitoring of orthophosphate in soils and sediments. *Talanta* 71:1710–1719.

Bulatov, A. V., A. V. Petrova, A. B. Vishnikin et al. 2012. Stepwise injection spectrophotometric determination of epinephrine. *Talanta* 96:62–67.

Burguera, J. L., ed. 1989. *Flow Injection Atomic Spectroscopy.* New York: Marcel Dekker.

Burguera, J. L. and M. Burguera. 2001. Flow injection-electrothermal atomic absorption spectrometry configurations: Recent developments and trends. *Spectrochim. Acta B* 56:1801–1829.

Campanella, L., K. Pyrzynska, and M. Trojanowicz. 1996. Chemical speciation by flow-injection analysis. A review. *Talanta* 43:825–838.

Cerda, V., A. Cerda, A. Cladera et al. 2001. Monitoring of environmental parameters by sequential injection analysis. *Trends Anal. Chem.* 20:407–418.

Cerda, V. and J. M. Estela. 2005. Automatic pre-concentration and treatment for the analysis of environmental samples using non-chromatographic flow techniques. *Int. J. Environ. Anal. Chem.* 85:231–253.

Chocholous, P., L. Kosarova, D. Satinsky et al. 2011. Enhance capabilities of separation in sequential injection chromatography—Fused-core particle column and its comparison with narrow-bore monolithic column. *Talanta* 85:1129–1134.

Chocholous, P., P. Solich, and D. Satinsky. 2007. An overview of sequential injection chromatography. *Anal. Chim. Acta* 600:129–135.

Cogan, D., J. Cleary, T. Phelan et al. 2013. Integrated flow analysis platform for the direct detection of nitrate in water using a simplified chromotropic acid method. *Anal. Methods* 5:4798–4804.

Dasgupta, P. K. and S. Liu. 1994. Electroosmosis: A reliable fluid propulsion system for flow injection analysis. *Anal. Chem.* 66:1792–1798.

Davey, D. E., D. E. Mulcahy, and G. R. O'Connel. 1993. Comparison of detector cell configurations in flow-injection potentiometry. *Electroanalysis* 5:581–588.

Economou, A., P. D. Tzanavaras, and D. G. Themelis. 2007. Sequential-injection analysis: A useful tool for clinical and biochemical analysis. *Curr. Pharm. Anal.* 3:249–261.

Egorov, O. and J. Ruzicka. 1995. Flow injection renewable fibre optic sensor system. Principle and validation on spectrophotometry of chromium(VI). *Analyst* 120:1959–1969.

Emaus, W. J. M. and H. J. Henning. 1993. Determination of bromide in sodium chloride matrices by flow-injection analysis using blank peak elimination by kinetic discrimination. *Anal. Chim. Acta* 272:245–250.

Eswara Dutt, V. V. S., D. Scheeler, and H. A. Mottola. 1977. Repetitive determinations of iron(III) in closed flow-through systems by series reactions. *Anal. Chim. Acta* 94:289–296.

Fajardo, Y., J. Avivar, L. Ferrer, E. Gomez, M. Callas, and V. Cerda. 2010. Automation of radiochemical analysis by applying flow techniques to environmental samples. *Trends Anal. Chem.* 29:1399–1410.

Fang, Z. 1993. *Flow Injection Separation and Preconcentration*, Weinheim: VCH.

Fang, Z. 1995. *Flow Injection Atomic Spectroscopy*, New York: Wiley.

Felix, F. S. and L. Angnes. 2010. Fast and accurate analysis of drugs using amperometry associated with flow injection analysis. *J. Pharm. Sci.* 99:4784–4804.

Ferreira, I. M. P. L. V., and J. L. F. C. Lima. 1994. Tubular potentiometric detector for flow-injection based on homogeneous crystalline membranes sensitive to copper, cadmium and lead. *Analyst* 119:209–212.

Fintschenko, Y. and G. S. Wilson. 1998. Flow injection immunoassays: A review. *Mikrochim. Acta* 129:7–18.

Fletcher, P., K. N. Andrew, A. C. Calocerinos, S. Forbes, and P. J. Worsfold. 2001. Analytical applications of flow injection with chemiluminescence detection—A review. *Luminescence* 16:1–23.

Franko, M. 2008. Thermal lens spectrometric detection in flow injection analysis and separation techniques. *Appl. Spectrosc. Rev.* 43:358–388.

Furmann, W. B. 1976. *Continuous Flow Analysis. Theory and Practice*. New York: Marcel Dekker.

Fyffe, J., P. Daga, and M. J. Roddis. 1988. A review of large biochemistry analysers. *J. Autom. Chem.* 10:43–61.

Gallignani, M. and M. R. Brunetto. 2004. Infrared detection in flow analysis—Developments and trends (review). *Talanta* 64:1137–1146.

Gardeniers, H. and A. Van Den Berg. 2004. Micro- and nanofluidic devices for environmental and biomedical applications. *Int. J. Environ. Anal. Chem.* 84:809–819.

Gardolinski, P. C. F. C., A. R. J. David, and P. J. Worsfold. 2002. Miniature flow injection analyzer for laboratory, shipboard and *in situ* monitoring of nitrate in estuarine and coastal waters. *Talanta* 58:1015–1027.

Gomez-Ariza, J. L., A. Arias-Borrego, and T. Garcia-Barrera. 2006. Use of flow injection atmospheric pressure photoionization quadrupole time-of-flight mass spectrometry for fast olive fingerprinting. *Rapid Commun. Mass Spectrom.* 20:1181–1186.

Grunhut, M., V. L. Martins, M. E. Centurion et al. 2011. Flow-batch analyzer for the chemiluminescence determination of catecholamines in pharmaceutical preparations. *Anal. Lett.*, 44:67–81.

Guzman Mar, J. L., L. L. Martinez, P. L. Lopez de Alba et al. 2009. Multisyringe flow injection spectrophotometric determination of uranium in water samples. *J. Radioanal. Nucl. Chem.* 281:433–439.

Hansen, E. H. and M. Miro. 2008. Interfacing microfluidic handling with spectroscopic detection for real-life applications via the lab-on-valve platform: A review. *Appl. Spectrosc. Rev.* 43:335–357.

Hansen, E. H., J. Ruzicka, and B. Rietz. 1977. Flow injection analysis. 8. Determination of glucose in blood-serum with glucose dehydrogenase. *Anal. Chim. Acta* 89:241–254.

Hartman, R. L., J. P. McMullen, and K. F. Jensen. 2011. Deciding whether to go with the flow: Evaluating the merits of flow reactors for synthesis. *Angew. Chem. Int. Ed.* 50:7502–7519.

Hartwell, S. K., G. D. Christian, and K. Grudpan. 2004. Bead injection with a simple flow-injection system: An economical alternative for trace analysis. *Trends Anal. Chem.* 23:619–628.

Hartwell, S. K., A. Kehling, S. Lapanantnoppakhun et al. 2013. Flow injection/sequential injection chromatography: A review of recent developments in low pressure with high performance chemical separation. *Anal. Lett.* 46:1640–1671.

Hayward, M. J., T. Kotiaho, A. K. Lister, R. G. Cooks, G. D. Austin, R. Narayan, and G. T. Tsao, 1990. On-line monitoring of bioreactions of *Bacillus polymyxa* and *Klabsiella oxytoca* be membrane introduction tandem mass spectrometry with flow injection analysis sampling. *Anal. Chem.* 62:1798–1804.

Higuchi. K., A. Inoue, H. Tamanouchi et al. 1999. On-site analysis for nitrogen oxides using a newly developed portable flow injection analyzer. *Bunseki Kagaku* 48:477–482.

Hollar, K. and B. Neele. 2008. Industrial and environmental applications of continuous flow analysis. In *Advances in Flow Analysis*, ed. M. Trojanowicz, pp. 639–661. Weinheim: Wiley-VCH Verlag.

Horvai, G., E. Pungor, and H. A. Mottola. 1987. Theoretical backgrounds of flow analysis. *CRC Crit. Rev. Anal. Chem.* 17:231–264.

Hungerford, J. M. and G. D. Christian. 1987. Chemical kinetics with reagent dispersion in single-line flow-injection systems. *Anal. Chim. Acta* 200:1–19.

Illg, T., P. Lob, and V. Hessel. 2010. Flow chemistry using milli- and microstructured reactors—From conventional to novel process windows. *Bioorg. Med. Chem.* 18:3707–3719.

Kagenow, H. and A. Jensen. 1983. Kinetic determination of magnesium and calcium by stopped-flow injection analysis. *Anal. Chim. Acta* 145:125–133.

Karlberg, B. and G. C. Pacey. 1989. *Flow Injection Analysis. A Practical Guide.* Amsterdam: Elsevier.

Keifer, P. A. 2003. Flow injection analysis NMR (FIA-NMR): A novel flow NMR technique that complements LC-NMR and direct injection NMR (DI-NMR). *Magn. Reson. Chem.* 41:509–510.

Keifer, P. A. 2007. Flow techniques in NMR spectroscopy. *Annu. Rep. NMR Spectrosc.* 62:1–47.

Kingston, H. M. and M. L. Kingston. 1994. Nomenclature in laboratory robotics and automation. *Pure Appl. Chem.* 66:609–630.

Kołacinska, K. and M. Trojanowicz. 2014. Application of flow analysis in determination of selected radionuclides. *Talanta* 125:131–145.

Kolev, S. D. and I. D. McKelvie, eds. 2008. *Advances in Flow Injection Analysis and Related Technique.* Amsterdam: Elsevier.

Kulka, S., G. Quintas, and B. Lendl. 2006. Automated sample preparation and analysis using a sequential-injection-capillary electrophoresis (SI-CE) interface. *Analyst* 131:739–744.

López García , I., P. Vinas, N. Campillo et al. 1995. Use of submicroliter-volme samples for extending the dynamic range of flow-injection flame atomic absorption spectrometry. *Anal. Chim. Acta* 308:85–95.

Li, B., D. Wang, J. Lv et al. 2006. Flow-injection chemiluminescence simultaneous determination of cobalt(II) and copper(II) using partial least squares calibration. *Talanta* 69:160–165.

Liptak, B. G. 2003. *Instrument Engineers' Handbook* (4th ed.), Vol. 1: *Process Measurement and Analysis.* Boca Raton, FL: CRC Press.

Macpherson, J. V. and P. R. Unwin. 1998. Radial flow microring electrode: Development and characterization. *Anal. Chem.* 70:2914–1921.

Mansour, F. R. and N. D. Danielson. 2012. Reverse flow-injection analysis. *Trends Anal. Chem.* 40:1–14.

Martinez Calatayud, J. 1996. *Flow Injection Analysis of Pharmaceuticals. Automation in the Laboratory.* New York: Taylor & Francis.

Maya, F., J. M. Estela, and V. Cerda. 2010. Flow analysis techniques as effective tools for the improved environmental analysis of organic compounds expressed as total indices. *Talanta* 81:1–8.

Mesquite, R. B. R., S. M. V. Fernandes, and A. O. S. S. Rangel. 2004. A flow system for the spectrophotometric determination of lead in different types of waters using ion-exchange for pre-concentration and elimination of interferences. *Talanta* 62:395–401.

Mesquite, R. B. R. and A. O. S. S. Rangel. 2009. A review on sequential injection methods for water analysis. *Anal. Chim. Acta* 648:7–22.

Miro, M. and E. H. Hansen. 2006. Solid reactors in sequential injection analysis: Recent trends in the environmental field. *Trends Anal. Chem.* 25:267–281.

Miro, M., H. M. Oliveira, and M. A. Segundo. 2011. Analytical potential of mesofluidic lab-on-valve as a front end to column-separation systems. *Trends Anal. Chem.* 30:153–161.

Mishra, S. K. and P. K. Dasgupta. 2010. Electrodialytic reagent introduction in flow systems. *Anal. Chem.* 82:3981–3984.

Muller, H., V. Muller, and E. H. Hansen. 1990. Simultaneous differential rate determination of iron(II) and iron(III) by flow-injection analysis. *Anal. Chim. Acta* 230:113–123.

Nacapricha, D., P. Sastranurak, T. Mantim et al. 2013. Cross injection analysis: Concept and operation for simultaneous injection of sample and reagents in flow analysis. *Talanta* 110:89–95.

Nanita, S. C. 2011. High-throughput chemical residue analysis by fast extraction and dilution flow injection mass spectrometry. *Analyst* 136:285–287.

O'Toole, M., K. T. Lau, and D. Diamond. 2005. Photometric detection in flow analysis systems using integrated PEDDs. *Talanta* 66:1340–1344.

Pascoa, R. N. M. J., I. V. Toth, and A. O. S. S. Rangel. 2012. Review on recent applications of the liquid waveguide capillary cell in flow based analysis techniques to enhance the sensitivity of spectroscopic detection methods. *Anal. Chim. Acta* 739:1–13.

Pasquini, C. and W. A. de Oliveira. 1985. Monosegmented system for continuous flow analysis. Spectrophotometric determination of chromium(VI), ammonia, and phosphorus. *Anal. Chem.* 57:2575–2579.

Perez-Olmos, R., J. C. Soto, N. Zarate et al. 2005. Sequential injection analysis using electrochemical detection: A review. *Anal. Chim. Acta* 554:1–16.

Pinto, P. C. A. G., M. Lucia, M. F. S. Saraiva et al. 2011. Sequential injection analysis hyphenated with other flow techniques. *Anal. Lett.* 44:374–397.

Piruska, A., M. Gong, J. V. Sweedler et al. 2010. Nanofluidics in chemical analysis. *Chem. Soc. Rev.* 39:1060–1072.

Pokrzywnicka, M., M. Fiedoruk, and R. Koncki. 2012. Compact optoelectronic flow-through device for fluorimetric determination of calcium ions. *Talanta* 93:106–110.

Pollema, Cy. H., and J. Ruzicka. 1994. Flow injection renewable surface immunoassay: A new approach to immunoanalysis with fluorescence detection. *Anal. Chem.* 66:1825–1831.

Puchades, R. and A. Maqueira. 1996. Recent developments in flow injection immunoanalysis. *Crit. Rev. Anal. Chem.* 26:195–218.

Pungor, E., Zs. Feher, and G. Nagy. 1970. Application of silicone rubber-based graphite electrodes for continuous flow measurements. *Anal. Chim. Acta* 52:47–54.

Pyrzynska, K., S. Gucer, and E. Bulska. 2000. Flow-injection speciation of aluminum. *Water. Resour.* 34:359–365.

Quintana, J. B., M. Miro, J. M. Estela et al. 2006. Automated on-line renewable solid-phase extraction-liquid chromatography exploiting multisyringe flow injection-bead injection lab-on-valve. *Anal. Chem.* 78:2832–2840.

Ramsing, A. U., J. Ruzicka, and E. H. Hansen. 1981. The principles and theory of high-speed titrations by flow injection analysis. *Anal. Chim. Acta* 129:1–17.

Reid, R. H. P. and L. Wise. 1968. A study of analyses done under non-steady state conditions. In *Advances in Automated Analysis. Technicon Symposia 1967*, pp. 159–165. White Plains: Mediad Inc.

Reijn, J. M., W. E. van der Linden, and H. Poppe. 1980. Some theoretical aspects of flow injection analysis. *Anal. Chim. Acta* 114:105–118.

Reis, B. F., M. F. Gine, E. A. G. Zagatto et al. 1994. Multicommutation in flow analysis. Party I. Binary sampling: Concepts, instrumentation and spectrophotometric determination of iron in plant digests. *Anal. Chim. Acta* 293:129–138.

Renaudot, R., V. Agache, Y. Fouillet et al. 2013. A programmable and reconfigurable micro-fluidic chip. *Lab Chip* 13:4517–4524.

Rezaei, B., T. Khayamian, and A. Mokhtari. 2009. Simultaneous determination of codeine and noscapine by flow-injection chemiluminescence method using N-PLS regression. *J. Pharm. Biomed. Anal.* 49:234–239.

Rocha, F. R. P., I. M. Raimundo Jr., and L. S. G. Taixeira. 2011. Direct solid-phase optical measurements in flow systems: A review. *Anal. Lett.* 44:528–559.

Rocha, F. R. P. and B. F. Reis. 2000. A flow system exploiting multicommutation for speciation of inorganic nitrogen in waters. *Anal. Chim. Acta* 409:227–235.

Rodriguez-Mozaz, S., S. Reder, M. Lopez de Alda, G. Gauglitz, and D. Barcelo. 2004. Simultaneous multi-analyte determination of estrone, isoproturon and atrazine in natural waters by the River ANAlyser (RIANA), and optical immunosensor. *Biosens. Bioelectron.* 19:633–640.

Ruz, J., A. Torres, A. Rios, M. D. Luque de Castro, and M. Valcarcel. 1986. Automation of a flow-injection system for multispeciation. *J. Autom. Chem.* 8:70–74.

Ruzicka, J. 2000. Lab-on-valve: Universal microflow analyzer based on sequential and bead injection. *Analyst* 125:1053–1060.

Ruzicka, J. 2009. *Tutorial on Flow Injection Analysis* (4th ed.). *From Beaker to Microfluidics*, www.flowinjection.com.

Ruzicka, J. and E. H. Hansen. 1975. Flow injection analyses. 1. New concept of fast continuous-flow analysis. *Anal. Chim. Acta* 78:145–157.

Ruzicka, J. and E. H. Hansen. 1976. Flow injection analysis. 6. Determination of phosphate and chloride in blood-serum by dialysis and sample dilution. *Anal. Chim. Acta* 86:353–363.

Ruzicka, J. and E. H. Hansen. 1978. Flow injection analysis. 10. Theory, techniques and trends. *Anal. Chim. Acta* 99:37–76.

Ruzicka, J. and E. H. Hansen. 1981. *Flow Injection Analysis*. New York: Wiley.

Ruzicka, J. and E. H. Hansen. 1984. Integrated microfluidics in flow injection analysis. *Anal. Chim. Acta* 161:1–15.

Ruzicka, J. and G. D. Marshall. 1990. Sequential injection: A new concept for chemical sensors, process analysis and laboratory assays. *Anal. Chim. Acta* 237:329–343.

Ruzicka, J., E. H. Hansen, and H. Mosbaek. 1977a. Flow injection analysis. 9. New approach to continuous-flow titrations. *Anal. Chim. Acta* 92:235–249.

Ruzicka, J., E. H. Hansen, and A. U. Ramsing. 1982. Flow injection analyzer for students, teaching and research: Spectrophotometric methods. *Anal. Chim. Acta* 134:55–71.

Ruzicka, J., E. H. Hansen, and E. A. G. Zagatto. 1977b. Flow injection analysis. 7. Use of ion-selective electrodes for rapid analysis of soil extracts and blood-serum—Determination of potassium, sodium and nitrate. *Anal. Chim. Acta* 88:1–16.

Ruzicka, J., Cy. H. Pollema, and K. M. Scudder. 1993. Jet ring cell: A tool for flow injection spectroscopy and microscopy on a renewable solid support. *Anal. Chem.* 65:3566–3570.

Ruzicka, J. and J. W. B. Stewart. 1975. Flow injection analysis. 2. Ultrafast determination of phosphorus in plant material by continuous-flow spectrophotometry. *Anal. Chim. Acta* 79:79–91.

Ruzicka, J., J. W. B. Stewart, and E. A. G. Zagatto. 1976. Flow injection analysis. 4. Stream sample splitting and its application to continuous spectrophotometric determination of chloride in brackish waters. *Anal. Chim. Acta* 81:387–396.

Sanchez, R., J. L. Todoli, C. P. Lienemann, and J. M. Mermet. 2010. Air-segmented, 5 µL flow injection associated with a 200°C heated chamber to minimize plasma loading limitations and difference of behaviour between alkanes, aromatic compounds and petroleum products in inductively coupled plasma atomic emission spectrometry. *J. Anal. At. Spectrom.* 25:1888–1894.

Schneider, J. A. and J. F. Horning. 1993. Spectrophotometric determination of lead in tap water with 5,10,15,20-tetra (4-N-sulfoethylpyridinium)porphyrin using merging zones flow injection. *Analyst* 118:933–936.

Schwartz, M. K., V. G. Bethune, M. Fleisher, G. Pennachia, C. J. Menendaz-Botet, and D. Lehman. 1974. Chemical and clinical evaluation of the continuous-flow analyzer "SMAC". *Clin. Chem.* 20:1062–1070.

Shpigun, L. K., Ya. V. Shushenachev, and P. M. Kamilova. 2007. Simultaneous spectrophotometric determination of copper(II) and zinc(II) based on their kinetic separation in flow-injection systems. *J. Anal. Chem.* 62:623–631.

Skeggs Jr., L. T. 1956. An automatic method for colorimetric analysis. *Clin. Chem.* 2:241–241.

Skeggs Jr., L. T. 1957. An automatic method for colorimetric analysis. *Am. J. Clin. Pathol.* 28:311–322.

Snyder, L., J. Levine, R. Stoy, and A. Conetta. 1976. Automated chemical analysis: Update on continuous-flow approach. *Anal. Chem.* 48:942A–956A.

Song, H., D. L. Chen, and R. F. Ismagilov. 2006. Reactions in droplets in microfluidic channels. *Angew. Chem. Int. Ed.* 45:7336–7356.

Stewart, J. W. B. and J. Ruzicka. 1976. Flow injection analysis. 5. Simultaneous determination of nitrogen and phosphorus in acid digests of plant material with a single spectrophotometer. *Anal. Chim. Acta* 82:137–144.

Stewart, K. K. 1983. Flow injection analysis. New tool for old analysis. New approach to analytical measurements. *Anal. Chem.* 55:931A–937A.

Stewart, K. K., G. R. Beeher, and P. E. Hare. 1974. Automated high-speed analyses of discrete samples—Use of nonsegmented, continuous-flow systems. *Fed. Proc.* 33:1439–1439.

Stewart, K. K., G. R. Beeher, and P. E. Hare. 1976. Rapid analysis of discrete samples: The use of nonsegmented, continuous flow. *Anal. Biochem.* 70:167–173.

Strzelak, K. and R. Koncki. 2013. Nephelometry and turbidimetry with paired emitter detector diodes and their application for determination of total urinary protein. *Anal. Chim. Acta* 788:68–73.

Stulik, K. and V. Pacakova. 1987. *Electroanalytical Measurements in Flowing Liquids.* Chichester: Ellis Horwood, 290pp.

Teerasong, S., S. Chan-Eam, K. Sereenonchai et al. 2010. A reagent-free SIA module for monitoring of sugar, color and dissolved CO_2 content in soft drinks. *Anal. Chim. Acta* 668:47–53.

Thavarungkul, P., P. Suppapitnarm, P. Kanatharana et al. 1999. Batch injection analysis for the determination of sucrose in sugar cane juice using immobilized invertase and thermometric detection. *Biosens. Bioelectron.* 14: 19–25.

Thompson, J. J. and R. S. Houk. 1986. Inductively coupled plasma mass spectrometric detection for multielement flow injection analysis and elemental speciation by reversed-phase liquid chromatography. *Anal. Chem.* 1986:2541–2548.

Tian, L. C., X. P. Sun, Y. Y. Xu, and Z. L. Zhi. 1990. Segmental flow-injection analysis: Device and applications. *Anal. Chim. Acta* 238:183–190.

Trojanowicz, M. 2000. *Flow Injection Analysis. Instrumentation and Applications.* Singapore: World Scientific.

Trojanowicz, M. 2006. Analytical applications of carbon nanotubes: A review. *Trends Anal. Chem.* 25:480–489.

Trojanowicz, M., ed. 2008a. *Advances in Flow Analysis.* Weinheim: Wiley-VCH.

Trojanowicz, M. 2008b. Nanostructures in flow analysis. *J. Flow Injection Anal.* 25:5–13.

Trojanowicz, M. 2009. Recent developments in electrochemical flow detection—A review. Part I. Flow analysis and capillary electrophoresis. *Anal. Chim. Acta* 653:36–58.

Trojanowicz, M. 2012. Flow-injection analysis as a tool for determination of pharmaceutical residues in aqueous environment. *Talanta* 96:3–10.

Trojanowicz, M., P. W. Alexander, and D. B. Hibbert. 1998. Flow-injection potentiometric determination of free cadmium ions with a cadmium ion-selective electrode. *Anal. Chim. Acta* 370:267–278.

Trojanowicz, M., P. Koźmiński, H. Dias et al. 2005. Batch-injection stripping voltammetry (tubeless flow-injection analysis) of tracer metals with on-line sample pretreatment. *Talanta* 68:394–400.

Trojanowicz, M. and W. Matuszewski. 1982. Limitation of lineał response in flow-injection systems with ion-selective electrodes. *Anal. Chim. Acta* 138:71–79.

Trojanowicz, M. and W. Matuszewski. 1983. Potentiometric flow-injection determination of chloride. *Anal. Chim. Acta* 151:77–84.

Trojanowicz, M., W. Matuszewski, and B. Szostek. 1992. Simultaneous determination of nitrite and nitrate in water using flow-injection biamperometry. *Anal. Chim. Acta* 261:391–398.

Trojanowicz, M. and E. Olbrych-Sleszyńska. 1992. Flow-injection sample processing in atomic absorption spectrometry. *Chem. Anal. (Warsaw)* 37:111–138.

Trojanowicz, M., M. Szewczyńska, and M. Wcisło. 2003. Electroanalytical flow measurements—Recent advances. *Electroanalysis* 15:347–355.

Trojanowicz, M., J. Szpunar-Lobńska, and Z. Michalski. 1991. Multicomponent analysis with a computerized flow injection system using LED photometric detection. *Mikrochim. Acta* 1:159–169.

Tumang, C. A., C. De Luca, R. N. Fernandes et al. 1998. Multicommutation in flow analysis exploiting a multizone trapping approach: Spectrophotometric determination of boron in plants. *Anal. Chim. Acta* 374:53–59.

Tymecki, L., K. Strzelak, and R. Koncki. 2013. Biparametric multicommutated flow analysis system for determination of human serum phosphoesterase activity. *Anal. Chim. Acta* 797:57–63.

Tyson, J. F. 1991. Flow-injection atomic spectroscopy. *Spectrochim. Acta Rev.* 14:169–233.

Tzanavaras, P. D. and D. G. Themelis. 2007. Review of recent applications of flow injection spectrophotometry to pharmaceutical analysis. *Anal. Chim. Acta* 588:1–9.

Ueno, K. and K. Kina. 1983. *Introduction to Flow Injection Analysis. Experiments and Applications.* Tokyo: Kodansha Scientific.

Valcarcel, M. and M. D. Luque de Castro. 1994. *Flow-Through (Bio)Chemical Sensors.* Amsterdam: Elsevier.

Valcarcel, M., B. M. Simonet, and S. Cardenas. 2008. Analytical nanoscience and nanotechnology today and tomorrow. *Anal. Bioanal. Chem.* 391:1881–1887.

Van der Linden, W. E. 1982. Flow injection analysis; the manipulation of dispersion. *Trends Anal. Chem.* 1:188–191.

Vanderslice, J. T., K. K. Stewart, A. G. Rosenfeld et al. 1981. Laminar dispersion in flow-injection analysis. *Talanta* 28:11–18.

Van Staden, J. F. and R. I. Stefan. 2004, Chemical speciation by sequential injection analysis: An overview. *Talanta* 64:1109–1113.

Wada, H., S. Hiraoka, A. Yuchi et al. 1986. Sample dispersion with chemical reaction in a flow-injection system. *Anal. Chim. Acta* 179:181–188.

Wang, E., H. Ji, and W. Hou. 1991. The use of chemically modified electrodes for liquid chromatography and flow-injection analysis. *Electroanalysis* 3:1–11.

Wang, J. and L. Angnes. 1993. Batch injection spectroscopy. *Anal. Lett.* 26:2329–2339.

Wang, J. and Z. Taha. 1991a. Batch injection analysis. *Anal. Chem.* 63:1053–1056.

Wang, J. and Z. Taha. 1991b. Batch injection with potentiometric detection. *Anal. Chim. Acta* 252:215–221.

Wang, X. D., T. J. Cardwell, R. W. Cattrall et al. 1998. Time-division multiplex technique for producing concentration profiles in flow analysis. *Anal. Chim. Acta* 368:105–111.

Wegner, J., S. Ceylan, and A. Kirschning. 2011. Ten key issues in modern flow chemistry. *Chem. Commun.* 47:4583–4592.

Wolff, Ch. M. and H. A. Mottola. 1978. Enzymic substrate determination in closed flow-through systems by sample injection and amperometric monitoring of dissolved oxygen levels. *Anal. Chem.* 50:94–99.

Yebra-Biurrun, M. C. 2009. *Flow Injection Analysis of Marine Samples*. Hauppauge: Nova Science Publishers.

Yu, Y. L., Y. Jiang, M. L. Chen, and J. H. Wang. 2011. Lab-on-valve in the miniaturization of analytical systems and sample processing for metal analysis. *Trends Anal. Chem.* 30:1649–1660.

Zagatto, E. A. G., F. J. Krug, H. F. Bergamin, S. S. Jorgensen, and B. F. Reis. 1979. Merging zones in flow injection analysis; Part 2. Determination of calcium, magnesium and potassium in plant material by continuous flow injection atomic absorption and flame emission spectrometry. *Anal. Chim. Acta* 104:279–284.

Zagatto, E. A. G., C. C. Oliveira, A. Townshend, and P. Worsfold. 2011. *Flow Analysis with Spectrophotometric and Luminometric Detection*. Amsterdam: Elsevier.

Zenki, M., S. Fujiwara, and T. Yokoyama. 2006. Repetitve determination of chemical oxygen demand by cyclic flow injection analysis using on-line regeneration of consumed permanganate. *Anal. Sci.* 22:77–85.

Zenki, M. and Y. Iwadou. 2002. Repetitive determination of chloride using the circulation of the reagent solution in closed flow-through system. *Talanta* 58:1055–1061.

CHAPTER 2

Sequential Injection Analysis

Anastasios Economou

CONTENTS

2.1 INTRODUCTION: A SNAPSHOT OF THE EVOLUTION OF FLOW-BASED ANALYSIS

Flow-based procedures were originally proposed as alternative approaches to batch analysis with the following aims: (a) to enhance the automation of chemical assays; (b) to shorten the analysis time; (c) to improve the accuracy and precision; (d) to reduce the cost (through unattended operation and lower reagent consumption) (Ruzicka, 2014). The temporal evolution of the different versions of flow analysis is illustrated in Figure 2.1.

The flow systems of the first generation were based on the concept of segmented flow analysis (SFA) (invented in the early 1950s) and were successfully employed as workhorse

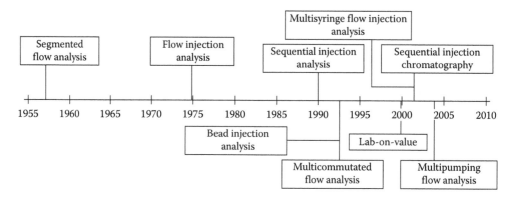

FIGURE 2.1   Historical evolution of flow analysis techniques (the techniques in bold are the topic of this chapter).

analyzers, mainly in clinical laboratories. In these systems, samples and reagents were continuously aspirated and segmented by air bubbles in order to minimize axial sample dispersion and mutual interference between adjacent samples (Skeggs, 2000). However, SFA manifolds were complicated, the consumption of reagents and sample was significant, the sample throughput was low, and carry-over effects were frequent.

In the second generation of flow analysis, termed flow injection analysis (FIA) introduced in 1975 (Ruzicka and Hansen, 1975), the sample was injected into a flowing carrier stream that merged downstream with one or more reagent streams to produce a measurable product. The introduction of this technique came to revolutionize the concept of automation in chemical analysis because FIA provided a way to automate a wide range of wet chemical assays, it addressed the drawbacks of SFA, and it provided wider scope for sample manipulation operations. FIA is described in more detail in Chapters 1 and 7.

Sequential injection analysis (SIA) (Ruzicka and Marshall, 1990), multicommutated flow analysis (MCFA) (Rocha et al., 2002), multisyringe flow injection analysis (MSFIA) (Miró et al., 2002) as well as various multipumping approaches (Lima et al., 2004) can be considered as an evolution of the FIA concept. These variants are more effective than FIA in minimizing reagent consumption, exploiting intermittent reagent addition, and providing wider scope for more sophisticated sample manipulation and pretreatment (Melchert et al., 2012).

The development of bead injection analysis (BIA), a variant of SIA, allowed the manipulation of surface-modified beads for chemical assays involving chemical processes in the heterogeneous phase (Ruzicka et al., 1993). The lab-on-valve (LOV) concept provided a further impetus toward miniaturization in flow analysis, minimization of reagent consumption, and scope for more sophisticated sample handling (Ruzicka, 2000). Finally, the introduction of sequential injection chromatography (SIC) made possible the direct hyphenation of flow manifolds with separation columns, enabling multiparametric assays, which have always been a challenge in flow-based analysis (Šatínský et al., 2003).

2.2 FLOW INJECTION ANALYSIS

The term FIA was coined by Ruzicka and Hansen (1975). The introduction of this technique revolutionized the concept of automation in chemical analysis by allowing instrumental measurements to be carried out in the absence of either physical equilibrium (i.e., without complete homogenization of sample and reagents) or chemical equilibrium (i.e., without completing chemical reactions). FIA became popular since it made possible to automate routine procedures in a way that was simple, rapid, economical, and precise (Ruzicka and Hansen, 1988; Hansen, 2008; Trojanowicz, 2008; Ruzicka, 2014). A very basic FIA manifold for the photometric detection of an analyte after reaction with a suitable reagent is illustrated in Figure 2.2. The FIA manifold consists of a (peristaltic) pump that continuously delivers a carrier/reagent stream, an injection valve used for introduction of a zone of sample into the flowing reagent, a reaction coil that serves as a mixing unit, a (photometric) detector, and tubing sufficient for connecting all the different components of the system. As the sample is injected and transported downstream by the flowing carrier/reagent, the sample zone starts to disperse; simultaneously a chemical reaction starts between the analyte in the sample and the reagent; and, finally, the product of the reaction is detected. FIA is based on a combination of: (a) reproducible sample introduction; (b) controlled sample dispersion, and; (c) absence of steady-state conditions (in terms of either physical equilibrium or chemical equilibrium).

Dispersion is measured by the conceptually simple and practically useful dispersion coefficient, D, ($D = C_o/C$, where C_o is the concentration of the analyte in the sample as the point of introduction (i.e., before the dispersion process begins) and C is the concentration of the analyte at the detection point) (McKelvie, 2008). Although this definition of dispersion does not take into consideration chemical reactions, the kinetic implications of chemical reactions were included in theoretical models developed in the 1980s (Kolev, 2008). Critical issues related to the description of mass transfer in FIA have been reviewed (Kolev, 2008; Inon and Tudino, 2008; Ruzicka, 2014).

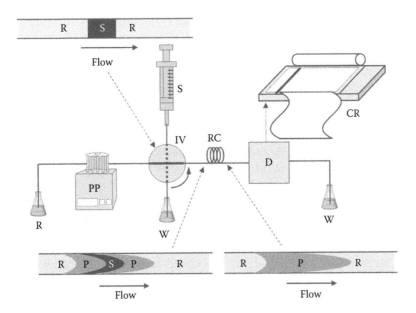

FIGURE 2.2 A single-channel FIA manifold based on a reaction between the analyte in the sample, S, and a reagent, R, used as a carrier to form the product, P (PP, peristaltic pump; IV, injection valve; RC, reaction coil; D, detector; W, waste; CR, chart recorder).

In practice, multichannel manifolds are normally employed in which the sample is injected into an inert flowing carrier and mixing with suitable reagent streams occurs downstream at confluence points. For more complex operations, additional components (such as mixing devices, gas-diffusion/dialysis units, packed columns) can be incorporated in the FIA manifold (McKelvie, 2008; Ruzicka, 2014).

Despite its numerous advantageous features, FIA has some disadvantages:

a. *Low versatility and high complexity of the manifold*. Usually a new manifold must be developed for each application while the optimization of the geometrical parameters requires physical reconfiguration of the manifold components. Multichannel FIA manifolds are also rather complex.

b. *High reagent consumption*. Reagents are pumped continuously irrespective of whether they are being used or not.

c. *Relatively long reaction coils*. A requirement for increased reaction time implies the use of longer reaction coils, which results in increased dispersion and lower sampling rates.

d. *Constant flow rate*. The requirement of constant flow conditions throughout each run precludes the possibility to exploit reaction rate differences for enhanced selectivity.

e. *Flow rate instability*. With peristaltic pumps it is difficult to maintain a constant flow rate for extended periods of time (due to differences in temperature, solution viscosity, and elasticity of pump tubing).

2.3 PRINCIPLES OF SEQUENTIAL INJECTION ANALYSIS

SIA was proposed by Ruzicka and Marshall in 1990 (Ruzicka and Marshall, 1990). As with FIA, this is a nonsegmented continuous flow technique whose mode of functioning is based on the concept of programmable flow or zone fluidics (Marshall et al., 2003).

A basic SIA manifold for the photometric detection of an analyte after reaction with a suitable reagent consists of a bidirectional pump, a holding coil, a multiposition selection valve, a reaction coil, a detector, appropriate tubing for connecting all the different components of the system, and a computer (Figure 2.3). The holding coil is connected to the central position of the selection valve, which, in turn, can be connected to any of the peripheral positions. By commutation of the selection valve, sample and reagent zones are aspirated into the holding coil. By inverting the direction of the flow, the stacked zones are delivered to the detector. During the aspiration step the sample and reagent zones start to overlap due to axial dispersion. Upon flow reversal, turbulent flow caused by acceleration promotes both radial and axial dispersion and more efficient mixing of the zones, resulting in the formation of a measurable product. The whole operation is under full computer control, allowing complete and precise synchronization of the steps required.

The introduction of SIA marked a new era in the development of flow analysis by addressing some important limitations associated with FIA. The advantages of SIA include:

a. *High versatility and simplicity of the manifold*. A SIA manifold is normally single channel and can be used for many applications involving different chemistries. No reconfiguration is required for method optimization (since any variation in the geometrical experimental conditions is achieved through the software).

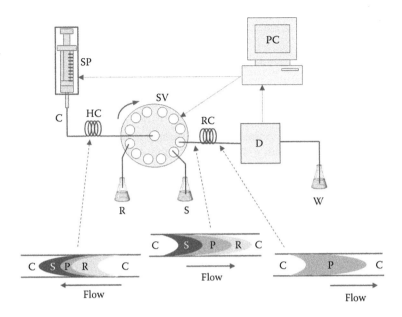

FIGURE 2.3   A simple SIA manifold based on a reaction between the analyte in the sample, S, and a reagent, R, to form the product P (SP, syringe pump; SV, selection valve; HC, holding coil; RC, reaction coil; D, detector; W, waste; PC, computer; C, inert carrier).

b. *Low consumption of reagents.* SIA offers economy in the use of reagents and reduction in waste generation because a permanent flow of carrier solution is not necessary and because the reagents and samples are used in a more efficient manner, given that only those volumes strictly necessary are actually used.

c. *Efficient mixing.* Oscillating the injected zones of sample and reagents back (aspiration) and forth (delivery) provides efficient mixing without the need for long reaction coils.

d. *Scope for kinetic studies.* SIA can be applied successfully in kinetic studies of chemical reactions, since the factors that affect the precision of the volumes and reaction times are rigorously controlled.

e. *Flow rate stability.* Syringe pumps normally used in SIA provide pulseless flow with minimal variation of the flow rate with time.

f. *Scope for sophisticated sample manipulation and processing.* The bidirectional nature of SIA makes it ideal for different sample pretreatment operations and for hyphenation to other analytical systems.

On the other hand, SIA also exhibits some disadvantages compared to FIA:

a. *Lower sampling rate.* The need for initial aspiration of the sample and reagents before delivery to the detector limits the sample throughput with respect to FIA.

b. *Need for computer control.* Since SIA is based on synchronized and accurate manipulation of zones of solutions, the use of dedicated control software is essential.

c. *Mixing.* When multireagent chemistries are utilized (involving the introduction of several reagent or buffer zones), mixing in SIA can be less efficient than in multichannel FIA manifolds.

In SIA the sample and reagent zones are sequentially stacked and, as they move through the manifold, the zones penetrate each other and become interdispersed. As a consequence, controlled partial dispersion and reagent sequencing are essential to understanding the operation of SIA (Ruzicka, 2014). With reference to the random walk model, it can be postulated that by moving the stack of reagents back and forth efficient mixing is achieved without actually travelling any net distance, an effect that is actually exploited for performing solution chemistry in SIA. Comprehensive studies have been carried out on the factors affecting dispersion, including sample volume, reagent volumes, tubing diameter, reactor geometry, pump speed, aspiration order, flow rate, and reactor and holding coil geometries (Lenehan et al., 2002; Ruzicka, 2014).

2.4 COMPONENTS OF A SIA MANIFOLD

2.4.1 Pumps

Requirements for the solution delivery system are stable and pulseless flow rate; bidirectional flow; potential for computer control of flow rate, start/stop, and direction of flow; and low power requirements. Flow rates typically range between 1 and 200 μL/s. while manifold back pressure is typically below 100 kPa. Initially, a sinusoidal pump was tested (Ruzicka and Marshall, 1990) but nowadays peristaltic, and more commonly, syringe pumps predominate (McKelvie, 2008). The main advantage of the former is higher sample throughput in contrast to syringe pumps, which have a limited reservoir capacity and need to be refilled periodically. However, peristaltic pumps produce pulsed flow and suffer from long-term flow rate instability. A positive displacement piston array pump designed specifically for use in FIA/SIA, the MilliGAT™ pump, has been marketed (Lenehan et al., 2002). This fully programmable pump has the advantages of bidirectional pulseless flow at a wide range of flow rates. Electroosmotic flow and gravity flow have also been reported in SIA.

2.4.2 Selection Valve

General requirements for the commutation device are scope for full computer control; low power requirements; and back pressure resistance of at least up to 700 kPa.

Commonly, in a SIA instrument the commutation device is a multiposition selection valve (with 6–14 peripheral positions, any one of which can be internally connected to the central position) operating in synchronization with the pump. The peripheral positions of the multiposition valve are coupled to various sample and reagent reservoirs as well as other flow devices via Teflon tubing. Common autosamplers can also be used as commutating units. A major conceptual and instrumental breakthrough associated with the commutation device is the introduction of the LOV configuration (described in detail in Section 2.7).

2.4.3 Detectors

Detectors for SIA simply require that the dimensions of the flow-through cell be such that they allow propulsion of the solution with low-pressure pumps (Lenehan et al., 2002;

McKelvie, 2008). Detection methods have included UV/Vis spectroscopy (the largest number of applications for its robustness, versatility, simplicity, and low cost), luminescence and chemiluminescence (CL) (which offer low detection limits and high sensitivity, being therefore especially favored for biological, biochemical, and trace analysis), atomic absorption and emission spectroscopy (which benefit enormously from automated sample pretreatment, used for matrix removal and analyte accumulation), electrochemistry (pH, fluoride ion selective electrodes, stripping voltammetry and conductivity), turbidimetry, vibrational spectroscopy (Fourier transform infrared spectroscopy [FTIR] and Raman) and mass spectrometry.

2.4.4 Other Units

There is a vast array of devices that can be employed in SIA fluidic manifolds such as reaction coils, phase separators, solid reactors, bead reactors, gas diffusion/permeation units, stirred mixers, solvent extraction units, renewable columns or additional pumps, valves and detectors (Marshall et al., 2003; Economou, 2005; McKelvie, 2008; Ruzicka, 2014). More complex units that can be hyphenated to basic SIA manifolds include complete FIA, MSFIA, MCFA, and multipumping systems (Pinto et al., 2011) and separation instrumentation (Section 2.11).

2.4.5 Software Control

While early FIA analyzers were operated manually and data acquisition and display was typically accomplished with chart recorders, the effective implementation of SIA requires microprocessor control of the timing and hardware synchronization steps. Use of a computer has the added benefit of digital data acquisition and processing of the results. Many dedicated software programs have been developed "in-house" in Turbo C++, Visual Basic, Basic, and LabVIEW (Lenehan et al., 2002). Nowadays, SIA analyzers are commercially available accompanied by dedicated control and acquisition software (Lenehan et al., 2002; Ruzicka, 2014).

2.5 APPLICATION AREAS OF SIA

SIA can be applied to environmental analysis for single species detection or for multiparametric determinations (Mesquita and Rangel, 2009); pharmaceutical analysis (Mervartová et al., 2007); food analysis (Pérez-Olmos et al., 2005); clinical analysis and life sciences (Chen and Wang, 2007; Ruiz-Medina and Llorent-Martínez, 2010); and industrial and process analysis (Barnett et al., 1999). Many illustrative applications of SIA are described in Ruzicka (2014).

2.6 BEAD INJECTION ANALYSIS

With the aid of an appropriate flow cell, small amounts of bead suspensions can be manipulated by SIA to form renewable microcolumns that act as disposable reaction surfaces (Hartwell et al., 2004; Miró et al., 2008; Ruzicka, 2014). BIA offers two main

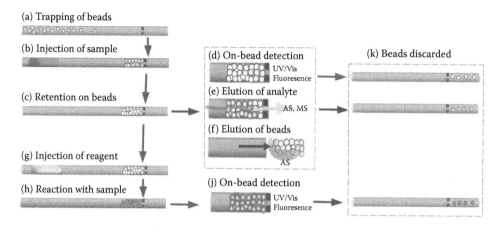

FIGURE 2.4 Different strategies for performing BIA. (AS, atomic spectroscopy; MS, mass spectrometry; UV/Vis, ultraviolet–visible spectroscopy.) (Modified with permission from J. Ruzicka, *Flow Injection Analysis*, 2014, http://www.flowinjectiontutorial.com.)

advantages: (a) the automatic renewal of the beads without any replacement of devices or physical reconfiguration, and (b) the accurate metering of bead quantity.

There are many strategies for performing BIA (Figure 2.4). Initially, a renewable column is formed by delivering and trapping the beads in a specially designed flow cell or column (Figure 2.4a). The sample is injected (Figure 2.4b) and the analyte is retained on the beads (Figure 2.4c). Then, four main detection approaches can be utilized. The first involves on-bead detection of the analyte provided that the flow cell is located close to the detector (Figure 2.4d). In the second approach, the analyte is released by injecting a suitable eluent and detected downstream (commonly by atomic spectroscopy or mass spectrometry) (Figure 2.4e). In the third scheme, the beads carrying the analyte are released and transported into the detector (which typically uses atomic spectroscopy) (Figure 2.4f). The fourth option is based on injection of a suitable reagent (Figure 2.4g) that reacts with the retained analyte on the beads (Figure 2.4h) and the product is monitored on-bead (Figure 2.4j). Finally, the used beads are discarded (Figure 2.4k). Of course, there are many more variations of the basic principle demonstrating the great versatility of the BIA approach (Ruzicka, 2014).

There are many examples using these approaches for SPE of trace metals using atomic spectrometry detection (Miró and Hansen, 2013; Ruzicka, 2014), column separations of radionuclides and affinity chromatography (Ruzicka, 2014), the determination of organic molecules in food or clinical samples (Chen and Wang, 2007; Miró et al., 2011; Ruzicka, 2014) and for DNA assays, bioligand interaction assays and cellular studies (Ruzicka, 2014). Over the last few years, BIA has been combined with the LOV platform (Chen and Wang, 2007; Miró and Hansen, 2012; Ruzicka, 2014).

2.7 LAB-ON-VALVE

The LOV system, introduced in 2000 (Ruzicka, 2000, 2014) and considered as the next generation of flow analysis, consists of a multichannel monolithic structure microconduit mounted atop a multiposition selection valve and designed to integrate different components, including detectors, mixing points, flow channels, reaction chambers, and column reactors. A typical LOV design incorporating two microcolumns is illustrated in Figure 2.5. LOV allows miniaturization of the flow channels: sample and reagent volumes are downscaled to the 10–20 µL range, categorizing SIA–LOV manifolds as mesofluidic systems (from their capability to manipulate fluids in the regime between "micro" and "macro") (Miró and Hansen, 2007, 2012). The micromachined unit is advantageously used as a "front end" to execute appropriate sample pretreatment such as dilution, pre-concentration, and derivatization reactions before introducing the processed sample into the detection device (Miró and Hansen, 2012; Ruzicka, 2014). The microfabricated channel system is amenable to coupling with conventional detectors (UV/Vis, electrochemical, fluorescence); to hyphenation with a plethora of modern detection instruments, such as different variants of atomic spectrometry and mass spectrometry (Miró and Hansen, 2013; Ruzicka, 2014); and most importantly to column separation systems, such as capillary electrophoresis (CE) or high-performance liquid chromatography (HPLC) (Miró et al., 2011; Ruzicka, 2014). A valuable asset of the microflow structure is the microfluidic handling of solid suspensions and beads in the BIA mode for exploitation of heterogeneous physical processes and chemical reactions. The LOV approach facilitates the in-valve manipulation of sorbent materials carrying suitable surface moieties in order to generate packed column reactors for micro-scale BIA (including ion-exchange, chelation, or hydrophobic interactions) in a permanent or a renewable flow fashion, depending on the particular chemical assay. Microcolumns are generated *in situ* by aspirating beads with particular surface characteristics. Following sample loading and cleanup protocols,

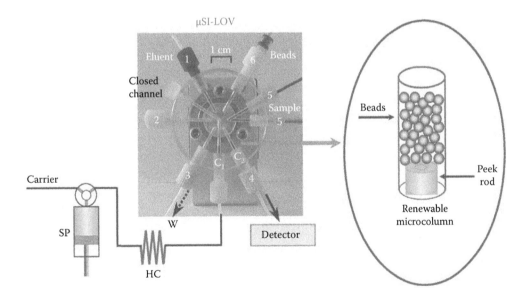

FIGURE 2.5 Illustration of a SIA-LOV microflow network as assembled for in-valve sorptive preconcentration using renewable sorbent materials prior to further detection via peripheral analytical instruments. SP, syringe pump; HC, holding coil. The inset at the right shows how the sorptive beads are retained within the column positions. (Reprinted with permission from Miró, M., and E. H. Hansen. 2007. Miniaturization of environmental chemical assays in flowing systems: The lab-on-a-valve approach vis-a-vis lab-on-a-chip microfluidic devices. *Anal. Chim. Acta* 600:46–57.)

appropriate eluents can be aspirated, and the eluate propelled to either the flow-through cell or an external detection device. The LOV system can be configured to perform bioanalytical assays serving as a platform for real-time monitoring of chemical events at solid surfaces, the exploration of cellular activities via immobilized living cells, performing enzyme-linked immunosorbent assays and bio-affinity chromatography as well as DNA assays and purification (Chen and Wang, 2007; Luque de Castro et al., 2008; Ruzicka, 2014).

2.8 SIA AS A SAMPLE PRETREATMENT APPROACH

In SIA, after aspiration of the sample zone into the holding coil via the sample line, the sample can be manipulated in different ways within the manifold by taking advantage of the stopped-flow, bidirectional nature of fluid handling in SIA. Therefore, a noteworthy feature of SIA is its inherent versatility to implement unit operations at will with no need for manifold reconfiguration (Marshall et al., 2003; Economou, 2005; McKelvie, 2008; Ruzicka, 2014).

Different sample pretreatment operations include dilution, membrane-extraction (gas diffusion, dialysis), liquid-phase extraction techniques (liquid/liquid extraction, liquid-phase microextraction, single-drop microextraction) and solid reactors and packed columns aiming to facilitate online chemical derivatization, chromatographic separation of target species, removal of interfering matrix compounds, enzymatic assays, or determination of trace levels of analyte via sorptive preconcentration procedures (Marshall et al., 2003; Economou, 2005; Miró and Hansen, 2006; Theodoridis et al., 2007; McKelvie, 2008; Ruzicka, 2014). In this context, BIA and the LOV configurations are particularly useful. Acid–base titrations can also be automated using simple SIA manifolds and potentiometric (van Staden et al., 2002) or photometric (Kozak et al., 2011) detection. Typically, a zone of the sample to be titrated is sandwiched between two zones of titrant by aspiration. In the case of photometric detection, an additional zone of a suitable pH-sensitive colored indicator is aspirated. The stacked zones are delivered to the detector and the width of the peaks is monitored and related to the pH of the solution.

2.9 HYPHENATION TO OTHER FLOW TECHNIQUES

Hyphenation of SIA to other flow techniques (FIA, MSFIA, MCFA, multipumping) has been reviewed recently (Pinto et al., 2011). The hyphenation of SIA with these flow techniques allows the automation of sample pretreatment procedures such as gas-diffusion, preconcentration, and dialysis. Combining SIA with FIA and MCFA enables the coupling of flow networks to particular detection techniques such as atomic and mass spectrometry or HPLC. In these cases, injection and solenoid valves, respectively, function as an interface between the detector and the SIA system. Hyphenation of MSFIA and SIA increases the versatility and flexibility of the resulting flow systems since it allows sample insertion on a time basis and so the analysis conditions can be changed without system reconfiguration. At the same time, the carry-over risk associated with the utilization of the syringes for sample introduction is reduced. Coupling of the multipumping and SIA concepts has demonstrated that it can improve the mixing of the stacked zones in SIA, while the use of solenoid micropumps obviates the need for the periodic replacement of the tubing in peristaltic pumps or the interruption of the analytical cycle for filling or emptying the syringe in syringe pumps.

2.10 HYPHENATION TO ATOMIC SPECTROSCOPY

Sample handling and introduction has been widely recognized as the weak point of atomic spectrometric techniques involving transport of aerosols into an atomizer. All the different variants of atomic spectroscopy (flame atomic absorption spectroscopy [FAAS], electrothermal atomic absorption spectrometry [ETAAS], inductively coupled plasma atomic emission spectrometry [ICP-AES], inductively coupled plasma mass spectrometry [ICP-MS], atomic fluorescence spectrometry [AFS], and hydride generation atomic emission spectrometry [HGAES]) can benefit from hyphenation to SIA. Flow analysis techniques can solve many of the problems associated with sample manipulation: (a) the sampling frequency can be increased; (b) the consumption of sample and reagents is lower; (c) the detector is exposed to the sample for only a short period of time, reducing the chance of clogging; and (d) sophisticated online sample processing procedures (including liquid–liquid extraction [LLE], liquid extraction/back-extraction, solid-phase extraction [SPE], single-drop microextraction [SDME] preconcentration in knotted reactors, micelle-mediated separation and preconcentration, and precipitation/co-precipitation) can be automated. The LOV systems and BIA are especially helpful in performing these operations (Chen et al., 2007; Wang et al., 2007; Miró and Hansen, 2013; Ruzicka, 2014).

2.11 HYPHENATION TO SEPARATION TECHNIQUES

In its simplest form, SIA, especially when coupled to other flow techniques, can be regarded as an advanced platform for reproducible, fast sample introduction technique to column separation systems (HPLC), affinity chromatography, and pressure-assisted CE. The possible interface modes for coupling a SIA-LOV assembly to column-separation systems are illustrated in Figure 2.6.

SIA is not suitable for direct coupling to packed chromatographic columns due to its inability to handle the high pressures developed by these columns, so that the interface between the low-pressure SIA manifold and the high-pressure HPLC apparatus is usually the injection valve of the latter. A more complex operation in the homogeneous phase involves online pre-column derivatization (Burakham and Grudpan, 2009). Heterogeneous chemistry is also possible before separation by GC and HPLC. This field is dominated by online SPE on solid or renewable bead reactors (Miró et al., 2011; Ruzicka, 2014).

The advantages of hyphenation of CE to SIA over stand-alone CE instruments are that (a) the separation electrolyte and the sample are continuously replenished in the interface; (b) the samples that are injected in the CE system are introduced into the separation capillary without the need to interrupt the separation voltage or to physically move the separation capillary; (c) repeated injections of samples can be made independently during the progress of the separation process inside the CE capillary; and (d) sample treatment (such as filtration, dilution, membrane-based separations, derivatization, preconcentration, etc., can be effectively carried out in the SIA part of the manifold during the separation of the previous sample in the CE instrument (Kubáň and Karlberg, 2009).

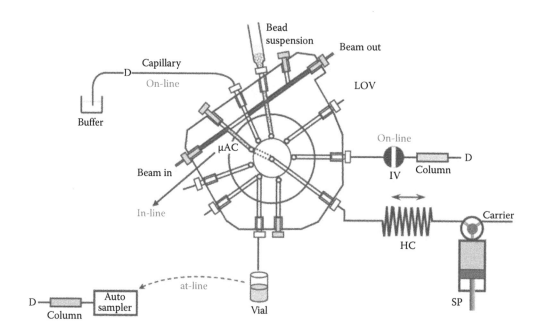

FIGURE 2.6 Sequential injection lab-on-valve (SIA-LOV) assembly as coupled to column-separation systems using various interface modes. SP, syringe pump; IV, injection valve; HC, holding coil; D, detector; μAC, microaffinity chromatograph. (Reprinted with permission from Miró, M., H. M. Oliveira, and M. A. Segundo. 2011. *TrAC* 30:153–164.)

In particular, SIA offers some additional advantages when coupled to CE when compared to other flow modes, such as the increased scope for sample pretreatment, lower sample volumes, and enhanced potential for hydrodynamic injection. A suitable interface is the most critical operational unit in achieving an efficient coupling of SIA to CE. The requirements for such an interface are as follows: (a) the grounding electrode should preferably be inserted in the interface, to prevent failure of the electronic parts of the flow system; (b) the capillary and the electrode should be positioned as close together as possible; (c) the waste/outflow channel in the interface should be relatively wide to avoid (or minimize) hydrodynamic flow into the CE separation capillary. Different interfaces, sample pretreatment procedures, SIA-microchip CE approaches, and practical applications of SIA-CE have been reviewed (Kubáň and Karlberg, 2009; Ruzicka, 2014).

2.12 SEQUENTIAL INJECTION CHROMATOGRAPHY

SIC was proposed in 2003 as simple, rapid, and inexpensive separation approach (Šatínský et al., 2003; Chocholouš et al., 2007; Solich, 2008; Ruzicka, 2014). Due to the low-pressure operation of SIA, direct coupling to conventional packed separation columns is impossible. SIC is actually a development of SIA with a view to achieving direct coupling to a separation column. The development of SIC was the result of the introduction of monolithic stationary phases. In a typical SIC manifold, a low-pressure monolithic separation column is introduced between the multiport selection valve and the detector (Figure 2.7). Monolithic columns consist of a single piece of high-purity

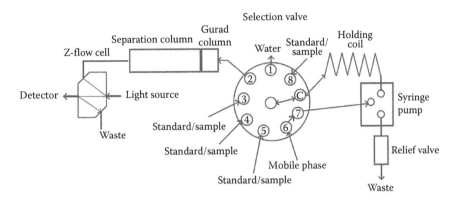

FIGURE 2.7   Schematic diagram of a SIC instrument. (Reprinted with permission from Idris, A. M. 2014. *Crit. Rev. Anal. Chem.* 44:220–232.)

polymeric silica gel rod with a bimodal pore structure (macropores with average size of 2 μm dramatically reduce the column backpressure and allow the use of higher flow rates, and mesopores with average size 13 nm increase the active surface enabling high performance chromatographic separations) (Chocholouš et al., 2007; Fernández et al., 2012). Monolithic rods also exhibit very high mechanical stability and long operative lifetime. However, their most critical favorable property for incorporation in SIC systems is the low backpressure they develop, which enables separations under low- or medium-pressure conditions. Further possibilities (e.g., ion chromatography and online derivatization) can also be implemented using SIC (Ruzicka, 2014). The applications of SIC extend from pharmaceutical analysis to biological, food, and environmental analysis (Fernández et al., 2012; Hartwell et al., 2013; Idris, 2014; Ruzicka, 2014).

2.13 APPLICATION OF SIA TO THE ANALYSIS OF FOOD ADDITIVES, ACIDITY, AND THE ANTIOXIDANT CAPACITY OF FOODS

2.13.1 Determination of Food Additives

Nitrite, in some cases in combination with nitrate, has long been used as an additive in different types of foods (curing of meat products to prevent food poisoning from *Clostridium botulinum* and to enhance the flavor, color, and texture of products). Once ingested by animals or humans, nitrate can be reduced to nitrite. Nitrite can react with dietary components to form nitrosamines, which have been shown to be potent carcinogens. Nitrite in the bloodstream converts oxyhemoglobin to methemoglobin, thereby interfering with oxygen transport in the blood (Zárate et al., 2009).

Most spectrophotometric methods consist of colorimetric detection of nitrite at 520–540 nm, based on the classical reaction with *N*-(1-naphthyl)ethylenediamine and sulfanilamide (Griess reaction). Nitrate is determined colorimetrically in the same way after reduction to nitrite by means of a cadmium column (ISO Standard 14673-1:2004/IDF Standard 189-1:2004). Different SIA systems based on the Griess reaction have been developed for the determination of nitrate and nitrite in infant formulas and milk powder, in dairy samples, in cured meat and infant formulas and milk powder (Oliveira et al., 2004, 2007; Reis Lima et al., 2006; Pistón et al., 2011).

A potentiometric SIA method was developed to quantify the nitrite contained in cured meat products. The assessment of nitrite concentration was accomplished by using the Gran's plot approach using a nitrite-selective electrode. A silver sulfate solution was used to remove the salt present in the samples since chloride ion is one of the most important interferents (Zárate et al., 2009).

Sulfites are added to wine and many other food products in order to preserve their freshness and shelf life. Sulfite is an antimicrobial agent that helps to halt the growth of undesirable yeasts and bacteria and acts as an antioxidant protecting wine against browning. In winemaking, this usually takes the form of sulfur dioxide. Since the 1980s, the use of sulfites has come under increased scrutiny due to potential health concerns. United States Food and Drug Administration (FDA) regulations require food and wine producers to indicate "contains sulfites" on the label of any product that has at least 10 mg/L. These regulations, which came into effect in 1986, were instituted because sulfite-sensitive individuals can experience allergic reactions (Tzanavaras et al., 2009).

A hybrid SIA–FIA spectrophotometric method for the determination of total sulfite in white and red wines has been reported (Tzanavaras et al., 2009). The assay was based on the reaction of sulfite with o-phthalaldehyde (OPA) and ammonium chloride. Upon online alkalization with NaOH, a blue product was formed having an absorption maximum at 630 nm. Sulfite was separated from the wine matrix through an online gas-diffusion process incorporated in the SI manifold, followed by reaction with OPA in the presence of ammonia. The reaction mixture merged online with a continuously flowing of NaOH prior to detection. The SIA–FIA manifold is illustrated in Figure 2.8.

A gas diffusion SIA system with amperometric detection using a boron-doped diamond electrode was developed for the determination of sulfite in wine (Chinvongamorn

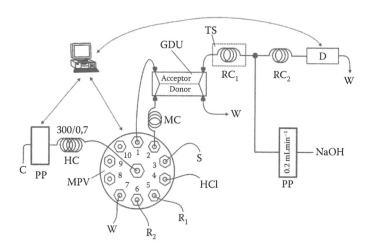

FIGURE 2.8 Schematic depiction of the hybrid SIA-FI setup for the determination of sulfite: C, water as carrier; PP, peristaltic pump; HC, holding coil; MPV, multiposition valve; S, sample; HCl, 1.0 mol/L HCl solution; R1, OPA ($c = 1 \times 10^{-2}$ mol/L); R2, ammonium chloride ($c = 2 \times 10^{-2}$ mol/L)/phosphate buffer (pH 8.5, 7.5×10^{-2} mol/L); W, waste; MC, mixing coil (30 cm/0.5 mm i.d.); GDU, gas-diffusion unit; TS, thermostat; RC_1 and RC_2, reaction coils (60 cm/0.5 mm i.d.), D, spectrophotometric detector ($\lambda_{max} = 630$ nm). (Reprinted with permission from Tzanavaras, P. D., E. Thiakouli, and D. G. Themelis. 2009. *Talanta* 77:1614–1619.)

et al., 2008). The gas diffusion unit was used to prevent interference from sample matrix. The sample was mixed with an acid solution to generate gaseous sulfur dioxide prior to its passage through the donor channel of the gas diffusion unit. The sulfur dioxide diffused through the polytetrafluoroethylene hydrophobic membrane into a carrier solution of 0.1 M phosphate buffer (pH 8)/0.1% sodium dodecyl sulfate in the acceptor channel and turned to sulfite. Then the sulfite was carried to the electrochemical flow cell and detected directly by amperometry using a boron-doped diamond electrode at 0.95 V (vs. Ag/AgCl).

Benzoic acid and sorbic acid are widely used as acceptable food preservatives. Salicylic acid is not a food preservative and is not allowed to be added to food; it is, however, illegally used in foods and beverage due to its superior effectiveness in controlling mold and inhibiting yeast growth. A SIC method for the determination of benzoic acid, sorbic acid, and salicylic acid has been investigated (Jangbai et al., 2012). Separation was performed on a monolithic C-18, (5 × 4.6 mm) column with 1% acetonitrile:ammonium acetate buffer pH 4.5 as a mobile phase at a flow rate of 1.20 mL/min at ambient temperature and with UV detection at 235 nm. Separation of the three compounds was achieved within less than 3 min. The developed procedure was demonstrated to be an effective alternative fast and simple method for the analysis of food, fruit juice, syrup, and soft drink samples.

Some sweeteners (aspartame, cyclamate, saccharin, and acesulfame K) were determined by CE-SIA with contactless conductivity detection (Stojkovic et al., 2013). The analyses were carried out in an aqueous running buffer consisting of 150 mM 2-(cyclohexylamino)ethanesulfonic acid and 400 mM tris(hydroxymethyl)aminomethane at pH 9.1 in order to render all analytes in the fully deprotonated anionic form. The four compounds were determined successfully in food samples; the experimental set-up and typical analysis results are illustrated in Figure 2.9. Another SIA system combined with solenoid valves was used to automate an enzymatic method for the determination of aspartame in commercial sweetener tablets. The method involves the enzymatic conversion of aspartame to hydrogen peroxide by the chymotrypsin–alcohol oxidase system, followed by the use of 2,2-azinobis(3-ethylbenzthiazoline-6-sulfonic acid) (ABTS) as electron donor for peroxidase. Chymotrypsin and alcohol oxidase enzymes were immobilized on activated porous silica beads (Peña et al., 2004).

In terms of food safety, L-glutamate is one of the most important amino acids because Chinese restaurant syndrome, Parkinson disease, and Alzheimer disease raise the argument of glutamate safety. Moreover, there is a growing body of literature of *in vivo* L-glutamate measurement because it is an important neurotransmitter in the mammalian central nervous system. A microfluidic BIA system has been developed for specific analysis of L-glutamate in food based on substrate recycling fluorescence detection. L-Glutamate dehydrogenase and D-phenylglycine aminotransferase, were covalently immobilized on polystyrene beads. The immobilized enzymes recycle L-glutamate by oxidation to 2-oxoglutarate followed by the transfer of an amino group from D-4-hydroxyphenylglycine to 2-oxoglutarate. The reaction was accompanied by reduction of nicotinamide adenine dinucleotide NAD^+ to NADH, which was monitored by fluorescence (Laiwattanapaisal et al., 2009).

A SIA system with attenuated total reflectance Fourier transform infrared spectroscopy (ATR-FTIR) detection has been studied for sugar and organic acid analysis of Belgian tomato samples (Vermeir et al., 2009). Prediction models for D-glucose, D-fructose, citric acid, and L-glutamic acid concentrations in tomato samples were successful.

Ascorbic acid (vitamin C) appears naturally in most fresh fruits and fruit juices but is often added during the manufacture of juices or soft drinks as a preservative or antioxidant. A simple, fast, and direct SIA method for ascorbic acid determination in juices was

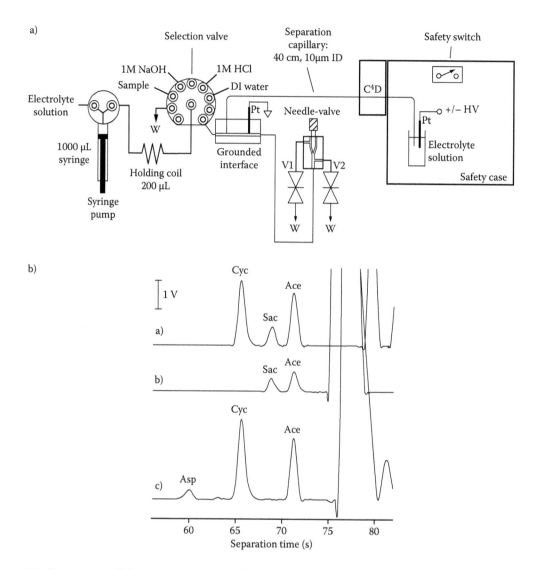

FIGURE 2.9 (a) Schematic drawing of the SIA-CE manifold with contactless conductivity detection employed for automated injection and superimposed hydrodynamic pumping. C⁴D, capacitatively coupled contactless conductivity detector; HV, high-voltage power supply. V1 and V2 are solenoid valves. (b) Electropherograms with hydrodynamic pumping for three different soft-drink samples after degassing and dilution employing the conditions for fast separations. Background electrolyte: 150 mM CHES and 400 mM Tris at pH 9.1. Capillary: 10 μm i.d., 40 cm and 32.5 cm total and effective lengths, respectively. Separation voltage: 25 kV. (Reprinted with permission from Stojkovic, M., T. D. Mai, and P. C. Hauser. 2013. *Anal. Chim. Acta* 787:254–259.)

developed based on the reaction between Dawson-type molybdophosphate and ascorbic acid with spectrometric detection at 814 nm (Vishnikin et al., 2011). Another SIA system was developed for the determination of ascorbic acid in juices using reduction of the guanidinium salt of 11-molybdobismuthophosphate. Absorbance of the reduced form of the compound was monitored at 720 nm (Vishnikin et al., 2010).

One of the most important phenols in olive oil is hydroxytyrosol (3,4-dihydroxy-phenylethanol). Several studies have reported its benefits for human health, such as antioxidant and antimicrobial properties and beneficial effects on the cardiovascular system. As a result, research has been carried out in the last few years in order to use this compound in enriched foods (vegetable oils, milk, or infant formulas). A SIA optosensor was developed for the fluorimetric determination of hydroxytyrosol in several foods. The use of a solid support in the flow cell increases the sensitivity and the selectivity of the system (Llorent-Martínez et al., 2013).

After the ban on the use of sulfites by the U.S. FDA in 1987, thiols (cysteine, glutathione, N-acetylcysteine, etc.) have emerged as a viable alternative for preventing browning of products caused by oxidase enzymes. The mechanism of their antibrowning properties is based on the reaction with quinones that are formed during the first stages of the enzyme-caused browning to produce colorless addition products or on the reduction of o-quinones to o-diphenols. SIA coupled to HPLC was applied for the determination of thiols (cysteine, glutathione, N-acetylcysteine, etc.) in fresh fruit samples with ethyl propiolate as a precolumn derivatization reagent and UV detection at 285 nm (Karakosta et al., 2011).

2.13.2 Determination of Acidity

Acetic acid is one of the most important aliphatic carboxylic acids, being the principal acid constituent of vinegars (5% w/v) and volatile acidity in wines. The food industry uses vinegar to preserve and season food at the same time. Determination of acidity by SIA is invariably based on titration with a base solution while the titration reaction is monitored by UV/Vis spectroscopy or potentiometry.

An online potentiometric SIA titration process analyzer for the determination of acetic acid in vinegar has been proposed (van Staden et al., 2002). Titration was achieved by aspirating the sample between two zones of base and flow reversal through a reaction coil to a potentiometric sensor. The method had a sampling frequency of 28 samples per hour. SIA titration systems with spectrophotometric detection have been developed for the assay of acetic acid content in vinegar (Lenghor et al., 2002), of acidity in fruit juices (Jackmunee et al., 2006), and of acetic, citric, and phosphoric acids in vinegars and various soft drinks (Kozak et al., 2011). The systems were based on acid–base titration of the acids with sodium hydroxide and phenolphthalein or thymolphthalein (in the case of phosphoric acid determination) as indicators. The decrease in the color intensity of the indicator was proportional to the acid content.

A new concept for microtitration using a LOV system with sequential injection of monosegmented flow has been proposed (Jakmunee et al., 2005). The performance of the system was demonstrated by the assay of acidity in fruit juices based on acid–base neutralization. Standard/sample solutions containing citric acid, indicator, and sodium hydroxide were sandwiched between air segments and then aspirated; upon flow reversal and removal of the air segments, the change in the absorbance of the indicator color was monitored.

2.13.3 Determination of Antioxidant Capacity

Antioxidant compounds contained in foods have received special attention in recent years. Antioxidants play a major role in scavenging the reactive oxygen species (ROS)

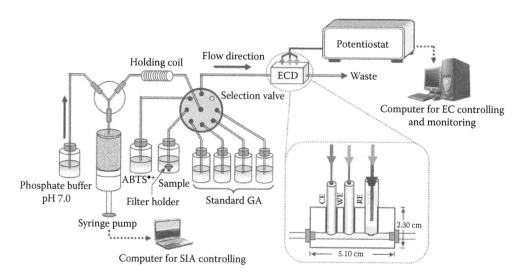

FIGURE 2.10 Schematic diagram of the SIA system for the evaluation of total antioxidant capacity with an in-house flow-through electrochemical detector (ECD) (1.9 cm width × 5.1 cm length × 2.3 cm height). CE, counter electrode; WE, working electrode; RE, reference electrode. (Reprinted with permission from Chan-Eam, S. et al. 2011. *Talanta* 84:1350–1354.)

that are implicated in several health issues. The ingestion of foods with high antioxidant activity can be important in the prevention of the oxidative stress due to ROS and, consequently, in the prevention of health disorders. A sequential injection system based on the ABTS methodology was developed to measure the total antioxidant activity of several beverages and foods (Reis Lima et al., 2005). A SIA method with CL detection was developed and validated for the rapid assay of the total antioxidant capacity in wines (Fassoula et al., 2011). The method exploited the Co(II)-catalyzed CL reaction of luminol with hydrogen peroxide in alkaline medium. Zones of sample, hydrogen peroxide, catalyst (Co(II) solution), and alkaline luminol were sequentially aspirated into the holding coil of the SIA manifold. Then, the flow was reversed and the stacked zones were directed to the CL detector. As the zones overlapped, antioxidants in the samples scavenged a portion of hydrogen peroxide and the decrease in the CL intensity was monitored and related to the antioxidant capacity. The sampling frequency was 60 samples/h. Finally, a new SIA method for the evaluation of the total antioxidant capacity in commercial instant ginger infusion beverages with amperometric detection has been reported (Chan-Eam et al., 2011). The method is based on monitoring the decrease of the cathodic current signal of the ABTS$^{\bullet+}$ radical after reaction with antioxidants. A schematic diagram of the SIA is illustrated in Figure 2.10.

2.14 CONCLUSIONS

In the more than two decades since its invention, SIA has matured into an exciting and versatile research tool. The programmable bidirectional flow on which SIA is based allows more sophisticated and flexible sample manipulation than FIA. SIA can be used for simple solution-phase chemistry involving chemical reactions enabling the minimization of reagent and sample consumption. In addition, SIA can be used for a plethora of online operations such as dilution, membrane-extraction, liquid-phase extraction and SPE. More significantly, SIA is ideal for hyphenation to other flow techniques, atomic spectrometry, and separation methods as a method of sample introduction and/or sample pretreatment. The ability to manipulate solid suspensions of beads (BIA) with judiciously selected surface properties facilitates the determination of both inorganic and organic species. The LOV concept enables further miniaturization of manifolds and detectors acting as a "front end" to spectroscopic detectors and separation instrumentation. Finally, SIC with monolithic materials has allowed direct coupling of SIA to separation columns, thus addressing the main limitation of traditional SIA in terms of multiparametric assays.

REFERENCES

Barnett, N. W., C. E. Lenehan, and S. W. Lewis. 1999. Sequential injection analysis: An alternative approach to process analytical chemistry. *TrAC* 5:346–353.

Burakham, R. and K. Grudpan. 2009. Flow injection and sequential injection on-line pre-column derivatization for liquid chromatography. *J. Chromatogr. Sci.* 47:631–635.

Chan-Eam, S., S. Teerasong, K. Damwand, D. Nacapricha, and R. Chaisuksant. 2011. Sequential injection analysis with electrochemical detection as a tool for economic and rapid evaluation of total antioxidant capacity. *Talanta* 84:1350–1354.

Chen, M. L, A. M. Zou, Y. L. Yu, and R. H. He. 2007. Hyphenation of flow injection/sequential injection with chemical hydride/vapor generation atomic fluorescence spectrometry. *Talanta* 73:599–605.

Chen, X. W. and J. H. Wang. 2007. The miniaturization of bioanalytical assays and sample pretreatments by exploiting meso-fluidic lab-on-valve configurations: A review. *Anal. Chim. Acta* 602:173–180.

Chinvongamorn, C., K. Pinwattana, N. Praphairaksit, T. Imato, and O. Chailapakul. 2008. Amperometric determination of sulfite by gas diffusion-sequential injection with boron-doped diamond electrode. *Sensors* 8:1846–1857.

Chocholouš, P., P. Solich, and D. Šatínský. 2007. An overview of sequential injection chromatography. *Anal. Chim. Acta* 600:129–135.

Economou, A. 2005. Sequential-injection analysis (SIA): A useful tool for on-line sample handling and pre-treatment. *TrAC* 24:416–425.

Fassoula, E., A. Economou, and A. Calokerinos. 2011. Development and validation of a sequential-injection method with chemiluminescence detection for the high throughput assay of the total antioxidant capacity of wines. *Talanta* 84:1350–1354.

Fernández, M., R. Forteza, and V. Cerdá. 2012. Monolithic columns in flow analysis: A review of SIC and MSC techniques. *Instrum. Sci. Technol.* 40:90–99.

Hansen, E. H. 2008. Flow injection, its origin and progress. In *Advances in Flow Injection Analysis and Related Techniques*, eds. S. D. Kolev, and I. D. McKelvie, pp. 3–21. Amsterdam: Elsevier.

Hartwell, S. K., G. D. Christian, and K. Grudpan. 2004. Bead injection with a simple flow-injection system: An economical alternative for trace analysis. *TrAC* 23:619–623.

Hartwell, S. K., A. Kehling, S. Lapanantnoppakhun, and K. Grudpan. 2013. Flow injection/sequential injection chromatography: A review of recent developments in low pressure with high performance chemical separation (review). *Anal. Lett.* 46:1640–1671.

Idris, A. M. 2014. The second five years of sequential injection chromatography: Significant developments in the technology and methodologies (review). *Crit. Rev. Anal. Chem.* 44:220–232.

Inon, F. A. and Tudino, M. B. 2008. Theoretical aspects of flow analysis. In *Advances in Flow Analysis*, ed. M. Trojanowitz, pp. 3–42. Weinheim: Wiley-VCH.

ISO Standard 14673-1:2004/IDF Standard 189-1:2004. *Milk and Milk Products—Determination of Nitrate and Nitrite Contents. Method Using Cadmium Reduction and Spectrometry.* Geneva, Switzerland: International Organization for Standardization, Brussels, Belgium: International Dairy Federation, 2004.

Jackmunee, J., T. Rujiralai, and K. Grudpan. 2006. Sequential injection titration with spectrophotometric detection for the assay of acidity in fruit juices. *Anal. Sci.* 22:157–160.

Jakmunee, J., L. Pathimapornlert, S. K., Hartwell, and K. Grudpan. (2005). Novel approach for mono-segmented flow micro-titration with sequential injection using a lab-on-valve system: A model study for the assay of acidity in fruit juices. *Analyst* 130:299–303.

Jangbai, W., W. Wongwilai, K. Grudpan, and S. Lapanantnoppakhun. 2012. Sequential injection chromatography as alternative procedure for the determination of some food preservatives. *Food Anal. Methods* 5:631–636.

Karakosta, T. D., P. D. Tzanavaras, and D. G. Themelis. 2011. Automated pre-column derivatization of thiolic fruit-antibrowning agents by sequential injection coupled to high-performance liquid chromatography using a monolithic stationary phase and an in-loop stopped-flow approach. *J. Sep. Sci.* 34:2240–2246.

Kolev, S. D. 2008. Theoretical basis of flow injection analysis. In *Advances in Flow Injection Analysis and Related Techniques*, eds. S. D. Kolev, and I. D. McKelvie, pp. 47–79. Amsterdam: Elsevier.

Kozak, J., M. Wójtowicz, N. Gawenda, and P. Kościelniak. 2011 An automatic system for acidity determination based on sequential injection titration and the mono-segmented flow approach. *Talanta* 84:1379–1383.

Kubáň, P. and B. Karlberg. 2009. Flow/sequential injection sample treatment coupled to capillary electrophoresis. A review. *Anal. Chim. Acta* 648:129–145.

Laiwattanapaisal, W., J. Yakovleva, M. Bengtsson et al. 2009. On-chip microfluidic systems for determination of L-glutamate based on enzymatic recycling of substrate. *Biomicrofluidics* 3:014104.

Lenehan, C. E., N. W. Barnett, and S. W. Lewis. 2002. Sequential injection analysis. *Analyst* 127:997–1020.

Lenghor, N., J. Jakmunee, M. Vilen, R. Sara, G. D. Christian, and K. Grudpan. 2002. Sequential injection redox or acid–base titration for determination of ascorbic acid or acetic acid. *Talanta* 58:1139–1144.

Lima, J. L. F. C., J. L. M. Santos, A. C. B. Dias, M. F. T. Ribeiro, and E. A. G. Zagatto. 2004. Multi-pumping flow systems: An automation tool. *Talanta* 64:1091–1098.

Llorent-Martínez, E. J., J. Jiménez-López, , M. L. Fernández-de Córdova, P. Ortega-Barrales, and A. Ruiz-Medina. 2013. Quantitation of hydroxytyrosol in food products using a sequential injection analysis fluorescence optosensor. *J. Food Compos. Anal.* 32:99–104.

Luque de Castro, M. D., J. Ruiz-Jiménez, and J. A. Pérez-Serradilla. 2008. Lab-on-valve: A useful tool in biochemical analysis. *TrAC* 27118–126.

Marshall, G., D. Wolcott, and D. Olson. 2003. Zone fluidics in flow analysis: Potentialities and applications. *Anal. Chim. Acta* 499:29–40.

McKelvie, I. D. 2008. Principles of flow injection analysis. In *Advances in Flow Injection Analysis and Related Techniques*, eds. S. D. Kolev, and I. D. McKelvie, pp. 81–109. Amsterdam: Elsevier.

Melchert, W. R., B. F. Reis, and F. R. P. Rocha. 2012. Green chemistry and the evolution of flow analysis. A review. *Anal. Chim. Acta* 714:8–19.

Mervartová, K., M. Polášek, and J. M. Calatayud. 2007. Recent applications of flow-injection and sequential-injection analysis techniques to chemiluminescence determination of pharmaceuticals. *J. Pharm. Biomed. Anal.* 45:367–381.

Mesquita, R. B. R. and A. O. S. S. Rangel. 2009. A review on sequential injection methods for water analysis. *Anal. Chim. Acta* 648:7–22.

Miró, M., V. Cerdà, and J. M. Estela. 2002. Multisyringe flow injection analysis: Characterization and applications. *TrAC* 21:199–210.

Miró, M. and E. H. Hansen. 2006. Solid reactors in sequential injection analysis: Recent trends in the environmental field. *TrAC* 25:267–281.

Miró, M. and E. H. Hansen. 2007. Miniaturization of environmental chemical assays in flowing systems: The lab-on-a-valve approach vis-a-vis lab-on-a-chip microfluidic devices. *Anal. Chim. Acta* 600:46–57.

Miró, M. and E. H. Hansen. 2012. Recent advances and future prospects of mesofluidic lab-on-a-valve platforms in analytical sciences—A critical review. *Anal. Chim. Acta* 750:3–15.

Miró, M. and E. H. Hansen. 2013. On-line sample processing involving microextraction techniques as a front-end to atomic spectrometric detection for trace metal assays: A review. *Anal. Chim. Acta* 782:1–11.

Miró, M., S. K. Hartwell, J. Jakmunee, K. Grudpan, and E. H. Hansen. 2008. Recent developments in automatic solid-phase extraction with renewable surfaces exploiting flow-based approaches. *TrAC* 27:749–761.

Miró, M., H. M. Oliveira, and M. A. Segundo. 2011. Analytical potential of mesofluidic lab-on-a-valve as a front end to column-separation systems. *TrAC* 30:153–164.

Oliveira, S. M., T. I. M. S. Lopes, and A. O. S. S. Rangel. 2004. Spectrophotometric determination of nitrite and nitrate in cured meat by sequential injection analysis. *J. Food Sci.* 69:C690–C695.

Oliveira, S. M., T. I. M. S. Lopes, and. A. O. S. S. Rangel. 2007. Sequential injection determination of Nitrate in vegetables by spectrophotometry with inline cadmium reduction. *Commun. Soil Sci. Plant Anal.* 38:533–544.

Peña, R. M., J. L. F. C. Lima, and M. L. M. F. S. Saraiva. 2004. Sequential injection analysis-based flow system for the enzymatic determination of aspartame. *Anal. Chim. Acta* 514:37–43.

Pérez-Olmos, R., J. C. Soto, N. Zárate, A. N. Araújo, J. L. F. C. Lima, and M. L. M. F. S. Saraiva. 2005. Application of sequential injection analysis (SIA) to food analysis. *Food Chem.* 90:471–490.

Pinto, P. C. A. G., M. Lúcia, M. F. S. Saraiva, and J. L. F. C. Lima. 2011. Sequential injection analysis hyphenated with other flow techniques: A review. *Anal. Lett.* 44:374–397.

Pistón, M., A. Mollo, and M. Knochen. 2011. A simple automated method for the determination of Nitrate and Nitrite in infant formula and milk powder using sequential injection analysis. *J. Autom. Methods Manage. Chem.*, Article ID 148183.

Reis Lima, M. J., S. M. V. Fernandes, and A. O. S. S. Rangel. 2006. Determination of nitrate and nitrite in dairy samples by sequential injection using an in-line cadmium-reducing column. *Int. Dairy J.* 16:1442–1447.

Reis Lima, M. J., I. V. Tóth, and A. O. S. S. Rangel. 2005. A new approach for the sequential injection spectrophotometric determination of the total antioxidant activity. *Talanta* 68:207–213.

Rocha, F. R. P., B. F. Reis, E. A. G. Zagatto, J. L. F. C. Lima, R. A. S. Lapa, and J. L. M. Santos. 2002. Multicommutation in flow analysis: Concepts, applications and trends. *Anal. Chim. Acta* 468:119–131.

Ruiz-Medina, A. and Llorent-Martínez, E. J. 2010. Recent progress of flow-through optosensing in clinical and pharmaceutical analysis. *J. Pharm. Biomed. Anal.* 53:250–261.

Ruzicka, J. 2000. Lab-on-valve: universal microflow analyzer based on sequential and bead injection. *Analyst* 125:1053–1060.

Ruzicka, J. 2014. Flow Injection Analysis. http://www.flowinjectiontutorial.com (accessed July 28, 2014).

Ruzicka, J. and E. H. Hansen. 1975. Flow injection analyses. Part I. A new concept of fast continuous flow analysis. *Anal. Chim. Acta* 78:145–157.

Ruzicka, J. and E. H. Hansen. 1988. *Flow Injection Analysis.* New York, NY: Wiley.

Ruzicka, J. and G. D. Marshall. 1990. Sequential injection: A new concept for chemical sensors, process analysis and laboratory assays. *Anal. Chim. Acta* 237:329–343.

Ruzicka, J., C. H. Pollema, and. K. M. Scudder. 1993. Jet ring cell: A tool for flow injection spectroscopy and microscopy on a renewable solid support. *Anal. Chem.* 65:3566–3570.

Šatínský, D., P. Solich, P. Chocholouš, and R. Karlícek. 2003. Monolithic columns—A new concept of separation in the sequential injection technique. *Anal. Chim. Acta* 499:205–214.

Skeggs, Jr., L. T. 2000. Persistence... and prayer: from the artificial kidney to the AutoAnalyzer. *Clin. Chem.* 46:1425–1436.

Solich, P. 2008. Chromatographic separations. In *Advances in Flow Injection Analysis and Related Techniques*, eds. S. D. Kolev, and I. D. McKelvie, pp. 265–285. Amsterdam: Elsevier.

Stojkovic, M., T. D. Mai, and P. C. Hauser. 2013. Determination of artificial sweeteners by capillary electrophoresis with contactless conductivity detection optimized by hydrodynamic pumping. *Anal. Chim. Acta* 787:254–259.

Theodoridis, G. A., C. K. Zacharis, and A. N. Voulgaropoulos. 2007. Automated sample treatment by flow techniques prior to liquid-phase separations. *J. Biochem. Biophys. Methods* 70:243–252.

Trojanowicz, M., ed. 2008. *Advances in Flow Analysis.* Weinheim: Wiley-VCH.

Tzanavaras, P. D., E. Thiakouli, and D. G. Themelis. 2009. Hybrid sequential injection–flow injection manifold for the spectrophotometric determination of total sulfite in wines using *o*-phthalaldehyde and gas-diffusion. *Talanta* 77:1614–1619.

van Staden, J. F., M. G. Mashamba, and R. S. Stefan. 2002. An on-line potentiometric sequential injection titration process analyser for the determination of acetic acid. *Anal. Bioanal. Chem.* 374:141–144.

Vermeir, S., K. Beullens, P. Mészáros, E. Polshin, B. M. Nicolaï, and J. Lammertyn. 2009. Sequential injection ATR-FTIR spectroscopy for taste analysis in tomato. *Sens. Actuators B Chem.* 137:715–721.

Vishnikin, A. B, T. Y. Svinarenko, H. Sklenářová, P. Solich, Y. R. Bazel, and V. Andruch. 2010. 11-Molybdobismuthophosphate—A new reagent for the determination of ascorbic acid in batch and sequential injection systems. *Talanta* 80:1838–1845.

Vishnikin, A. B., H. Sklenářová, P. Solich, G. A. Petrushina, and L. P. Tsiganok. 2011. Determination of ascorbic acid with Wells–Dawson type molybdophosphate in sequential injection system. *Anal. Lett.* 44:514–527.

Wang, Y., M. L. Chen, and J. H. Wang. 2007 New developments in flow injection/ sequential injection on-line separation and preconcentration coupled with electro- thermal atomic absorption spectrometry for trace metal analysis. *Appl. Spectrosc. Rev.* 42:103–118.

Zárate, N., M. P. Ruiz, R. Pérez-Olmos, A. N. Araújo, M. Conceiçao, and B. S. M. Montenegro. 2009. Development of a sequential injection analysis system for the potentiometric determination of nitrite in meat products by using a Gran's plot method. *Microchim. Acta* 165:117–122.

CHAPTER **3**

Miniaturization and Automation
Nanotechnology

María Pedrero

CONTENTS

3.1 INTRODUCTION

Food quality, food safety, and food traceability are issues of major importance nowadays due to the increasing number of threats, such as the increase in prices and in scarcity of raw ingredients, which lead to fraud and adulteration, or the hard competition in the food market that can find unscrupulous companies using cheaper illegal raw materials. It should never be forgotten that the quality of the food we eat is of great importance in the correct functioning of our body and, in fact, in the possibility of enjoying a healthy life.

Current trends in food analysis include the development of new faster devices able to provide a high sensitivity and selectivity. Moreover, automation, multiplexing, and miniaturization are nowadays basic in the development of new instrumentation where the demand for highly parallel screening of thousands of samples in a short times can be fulfilled together with such characteristics as versatility, precision, and low cost. In this sense, microfluidic devices are designed to miniaturize, multiplex, and automate serial sample analysis using submicroliter quantities of reagents, thus leading to significant savings in reagent and sample consumption, these devices being compatible with commercially available optical and electrochemical detectors.

One of the fields where microfluidic technologies are acquiring a higher impact is point of care testing (POCT) with applications in the fields of emergency testing, home tests, doctor's office screening, decentralized hospital tests, and environmental, forensic, military or agro-food analysis. The increasing interest in this technology is mainly due to its ability to allow miniaturization and integration of laboratory protocols into portable devices.

Nowadays, there is high interest in nanoscience because, due to the wide range of new technologies, its potential is evident to contribute to considerable improvement in industrial products and processes and introduce new functionalities. The integration of complex electronics systems into millimeter-sized chips should allow, in the near future, the implementation of complex and sophisticated instrumentation in cheap and portable devices for rapid detection of, for example, analytes of relevance in food monitoring. In this sense, nanotechnology is of great importance in the development of lab-on-a-chip systems where basic physical and chemical fundamentals cannot be applied and laws at the nanometer scale still need to be established (Ríos et al., 2009). Despite the great interest these devices arouse nowadays, very few applications can be found in the literature for the field of food analysis.

3.2 MINIATURIZATION OF ANALYTICAL SYSTEMS AND NANOTECHNOLOGY

Nanotechnology offers valuable tools for miniaturized analytical devices. Three levels of miniaturization can be considered, defined by the use of the prefixes mini- (in the range from 1 mm to the centimeter scale), micro- (in the 1 μm to 1 mm range), and nano- (below 1 μm). The term "nanosystem" thus refers to a system whose main structures are in the 1–100 nm range, nano-electromechanical systems (NEMS) constituting representative devices used at nanoscale, and nano-fluidic structures (Ríos et al., 2009). Recently, microfluidic systems and detection devices have been applied to 10^1–10^3 nm scale fluidic systems, which are called extended nanospace to distinguish them from those at the 10^0–10^1 nm scale space which belong to conventional nanotechnology (Mawatari et al., 2012).

Nanotechnology is finding increasing use in microfluidics in solving challenges related to selectivity and sensitivity. Nanotechnology tools integrated into microfluidic devices offer the opportunity to modify surfaces with nanomaterials such as carbon nanotubes (CNTs) or metallic nanoparticles (NPs), both of which are significant in the improvement of sensitivity and selectivity.

These analytical nanosystems are built with atomic precision using nanotechnology elements. These materials and devices can, in fact, be obtained from molecular components that assemble themselves chemically by principles of molecular recognition or from larger entities without atomic-level control (Ríos et al., 2009).

As an important feature of miniaturization at the nano-level is that the development of miniaturized analytical systems entails the use of extremely low volumes of both sample and reagents, in the picoliter (pL) to nanoliter (nL) range, together with the presence of laminar flows. This was already achieved about 10 years ago without any mechanical valves or external pumps by using electrokinetic phenomena (Szekely and Freitag, 2005), the main objective then being the integration of functional components within monolithic systems using both lithography and micromolding technologies. At present, however, very little is known about fluid dynamics in extended nanospace, so research in this field is of major importance to further development of nanofluidic control methods (Mawatari et al., 2012). These methods must take into account the unique fluid properties at this scale, including higher viscosity, lower permittivity, and high proton mobility. As described in detail by Mawatari et al., because fluidic control by pressure-driven flow is difficult at the nanospace scale, nowadays electroosmotic flow control and electrophoretic control, generated by an external electric field, are mainly used for extended nanofluidic systems, and also because of their ease of operation (Mawatari et al., 2012). Nevertheless, other methodologies—such as fluidic control by pressure-driven flow or shear-driven flow, where the flow is induced by moving a surface toward a fixed plate—may also be envisaged as key technologies for deriving full advantage form nanofluidic properties in the future.

On the other hand, integration of chemical processes in small spaces requires sophisticated detection technologies due to the small volumes of extended nanospace channels. Considering that the time-averaged amount of analyte in 1 fL of a 1 nmol L^{-1} fluid is less than a single molecule, if the dimensions of the detection volume are decreased to 100 nm, single-molecule detection is required for even 1 μmol L^{-1} solutions; this means that the detection method to be used on microchips requires single molecule sensitivity. The main detection methods used nowadays in these systems are optical (where the relatively short pathlengths also have to be considered) and electrochemical. A description of actual approaches in this context has recently been given elsewhere (Mawatari et al., 2012).

One of the ultimate objectives in the field of miniaturization field is the development of integrated microsystems wherein the entire analytical process can be performed. Here, it is important to mention the development of nanosensors based, for example, on the use of nanoelectrodes in electrochemical transduction systems, which allow integration of the detection system at the nanoscale level. Thus, the design of miniaturized analytical systems at the nanoscale level should take into account the following issues as compared to the macro-world (Ríos et al., 2009; Mawatari et al., 2012):

1. The ratio of surface to volume increases drastically.
2. In nanofluidic systems, diffusion is the dominant mass transport process.
3. Interfacial forces determined by fluid and channel properties drive the fluids' motion when it is not governed by an external field (electric, magnetic, thermal, etc.).
4. Electroosmotic flow is nowadays usually chosen to transport liquid through nanofluidic devices owing to the ease of operation.
5. Detection and imaging methods are required to monitor the introduction of sample and reagents into the channels and their transportation along them.
6. The viscosity of water is increased and proton mobility is enhanced in these small nanospaces, which can influence the kinetics of a chemical reaction.

Ríos et al. (2009) gave the first thorough description of the tools used in the design of miniaturized analytical systems. Some examples of nanodevices stand out among them:

1. *Nanowires for the fabrication of switchable microchip devices.* Piccin et al. (2007) used nickel nanowires to switch on demand the detection and the separation processes on microfluidic devices through their controlled reversible magnetic positioning and orientation, respectively.
2. For mixing solutions and initiating chemical reactions in microfluidic systems, Helman et al. (2007) used highly focused nanosecond laser pulses to generate cavitation bubbles that expand and collapse within the channel, disrupting the laminar flow and producing a localized region of mixed fluid.
3. *Nanopippetors as dispensing devices.* Byun et al. (2007) developed an electro-osmosis-based nanopipettor with no moving parts consisting of a microfabricated electroosmotic flow pump, a polyacrylamide grounding interface, and

a nanoliter-to-picoliter pipette tip characterized by its good levels of accuracy and precision.

4. Another example of nanoscale dispensing unit is that described by Kovaric and Jacobson (2007) for microfluidic channels. It dispenses volumes at the femtoliter scale, the transport being produced electrokinetically by applying up to 10 V directly from an analogue output board.

5. *Nanochannels or nanojunctions to change the flow rates.* Mellors et al. (2010) developed a microfluidic chip for the automated real-time analysis of individual cells using capillary electrophoresis (CE) and electrospray ionization mass spectrometry (ESI-MS). The described microchip incorporated a nanochannel segment (nanojunction, ~100 nm deep) connecting the end of the separation channel to the electroosmotic pump side channel, in order to reduce the electroosmotic flow in the side channel. Thus, through greatly increasing the hydrodynamic flow resistance of the side channel, the nanojunction allowed effective convective transport of the CE eluent to the electrospray emitter.

A major issue where nanotechnology is involved is the method of nanochannel fabrication. As reviewed by Perry and Kandlikar, this is done by using one of four methods (Perry and Kandlikar, 2006): (1) bulk nanomachining and wafer bonding, (2) surface nanomachining, (3) buried channel technology, or (4) nanoimprinting technology. As stated by these authors, bulk nanomachining and wafer bonding is the simplest method and also requires the least processing time with the lowest costs. Buried channel technology is at the intermediate level for the number of processing steps and operational costs, but it shows the largest freedom of design among all the considered methods.

In general, the fabrication methods for achieving well-defined nanostructures on a substrate are classified into two types: (1) top-down processes based on the nanostructure and directly fabricated on a bulk material using lithography, etching, and/or direct milling; and (2) bottom-up processes where the nanostructures are designed and built up on a material by manipulating and self-assembling atoms and molecules. The basics of the technologies employed in nanofabrication methods have been thoroughly described by Mawatari et al. (2012).

On the other hand, NEMS have been defined as integration of mechanical elements, sensors and actuators, and electronics on a common silicon substrate through nanofabrication technology (Schmitt et al., 2006). The fabrication of NEMS is nowadays mainly devoted to the field of biosensors and gas sensors, the field of food analysis not yet having many applications in this sense.

3.3 NANOTECHNOLOGY IN CHROMATOGRAPHIC AND ELECTROPHORETIC SYSTEMS: EXAMPLES OF APPLICATIONS TO FOOD ANALYSIS

Although quite a deal of work has been done on the miniaturization of the components and devices integrated in chromatographic and electrophoretic equipment, miniaturization of the whole apparatus is not yet a reality. In fact, miniaturization is only achieved when chromatographic or electrophoretic separations are carried out in chips where all the principles of micro- and nanofluidics are involved, these systems being connected to the lab-on-a-chip approaches. The common characteristics associated with nano–liquid chromatography (nano-LC) techniques include column internal diameters between 10 and 150 µm, flow rates in the range from 10 to 10^3 mL min^{-1}, and injection volumes at the nanoliter level (Ríos et al., 2009).

In spite of the work and development devoted to microchip fabrication, the use of nanotechnology is seen mainly only in the constituents of the overall devices. Thus, examples can be found of the use of CNTs as stationary phases in gas chromatography (GC) (Kartsova and Makarov, 2002) in both packed and open-tubular approaches, showing a high surface-to-volume ratio as well as better thermal and mechanical stability than traditional stationary phases (Li and Yuan, 2003; Saridara and Miltra, 2005). Taking advantage of the possibility to easily deposit CNTs by lithography, an ultrafast GC methodology based on single-walled CNT stationary phases in microfabricated channels for the continuous monitoring of the gas composition of atmospheric environments has been described (Stadermann et al., 2006). Also, gold nanoparticles (AuNPs) have been used as stationary phase for capillary liquid chromatography (CLC) taking advantage of their stability at high pH, rigidity, ease of manipulation, high area-to-volume ratio, and ease of chemical modification.

Nanotechnology can also be used in the detection systems, as was shown by Zellers et al. (2007) who described a wireless integrated micro-GC for complex vapor mixtures which included among its parts an integrated chemiresistor array coated with thin films of different Au-thiolated monolayer-protected nanoparticles whose responses varied with the nature of the analyte vapor.

CLC and MS detectors can be interfaced with nano-ESI, which improves the sample ionization efficiency in the MS system. These CLC–ESI–MS systems usually require online preconcentration steps in precolumns also used for cleanup and to avoid blocking of the capillaries in the inlet when real samples are to be analyzed (Ríos et al., 2009).

Nanoflow ESI interfaces are ideal for direct coupling of fluidic devices with the mass spectrometer as they cover flow ranges down to a few nL min^{-1}. Also, the use of nanospray, where small droplets are generated from emitters with an inner diameter of several micrometers, brings several advantages, such as (Ríos et al., 2009): (1) lower electrospray voltages, (2) an increase of the number of analyte ions that enter the MS, (3) lower consumption of samples, (4) the possibility of coupling on line with nano-LC techniques due to the low flow rate, (5) pure water solutions can be sprayed without discharges, (6) a higher ionization efficiency due to the smaller initial ratio of droplets generated, (7) the diameter of an emitter can be miniaturized and the flow rate reduced without losing mass spectrometric sensitivity, (8) ion suppression effects are reduced as compared to electrospray, (9) each droplet formed will contain on average a single or a few analyte molecules, eliminating ion suppression effects and making nanospray a very attractive technique, and (10) the number of charges available per analyte molecule is much higher than in electrospray, which enhances the probability of ionizing an analyte. However, there are also some drawbacks to be considered when implementing nanospray, including the fact that the influence of very small dead volumes, which normally have a minor impact on micro-LC separations, becomes significant in nano-LC, or that the quality of the nanoscale connections depends a great deal on the experience of the user. Also, variations in tip diameter will have a significant influence on the appearance of the mass spectrum.

Aqay et al. (2011) have described a method based on nano-LC-quadrupole time-of-flight-MS (nano-LC–QTOF-MS) for the screening of ochratoxins in wheat and cereal samples. The mycotoxin ochratoxin A (OTA) and cross-reacting mycotoxin analogues were analyzed in these samples after extraction using a novel direct inhibition flow cytometric immunoassay with superparamagnetic microbeads coated with monoclonal antibodies. The limit of detection (LOD) for OTA was 0.15 ng g^{-1}, below the lowest maximum level of 3 ng g^{-1} established by the European Union. Apart from the high sensitivity observed, the use of nano-LC with MS detection allowed the use of small injection volumes and ultralow flow rates, resulting in a low consumption of hazardous chemicals.

A nano-ESI interface combined with a microliter flow rate has recently been used by Oellig and Schwack (2014) in a microflow injection analysis TOF-MS (µL-FIA–TOF-MS) for the determination of high-throughput planar solid phase extraction (HTpSPE) extracts for pesticide residues in vegetable matrices (apples, red grapes, tomatoes, cucumbers, bananas, blackberries currants, and savoy cabbage). The extraction system allowed the omission of the LC separation step, while nano-ESI technology was chosen for signal intensity, low solvent consumption, and flow rate–dependent reduced matrix effects. HTpSPE required only a few minutes per sample, which made it possible to clean in parallel various samples at minimal cost with low sample and solvent consumption. After that, for the µL-FIA-TOF-MS, a further 6 minutes per sample were needed. The developed methodology is envisaged as being applicable as routine screening method.

Nanotechnology has also been employed in the field of CE, the main applications of this technology being devoted to the fields of proteomic, pharmaceutical, and environmental analysis. Examples can be found in literature of CE microchips applied to food analysis that integrate tools from nanotechnology in the detection stage (Escarpa et al., 2008). The most common detection systems are electrochemical devices followed by laser-induced fluorescence, although chemiluminescence and ultraviolet (UV) detection systems have also been employed (Martín et al., 2012). The main reason for the predominance of electrochemical techniques is their high compatibility with micro- and nanotechnologies, the large number of electroactive analytes, and the fact that other detection techniques require more sophisticated and expensive instrumentation. González-Crevillén et al. (2007) described for the first time the use of CNT materials in CE microchips for the analysis of dietary antioxidants, water-soluble vitamins, vanilla flavors, and isoflavones in food samples. The use of CNTs allowed direct detection of analytes in complex natural samples and the detection of fraud without the integration of complex preconcentration steps on the microdevices. Well-defined and resolved peaks with enhanced voltammetric currents were obtained in ultrafast separations, thus also demonstrating the applicability of CNTs as ideal material for electrochemical sensing in food analysis.

Nickel(II) oxide NPs have been used as modifiers in carbon paste electrodes employed in capillary zone electrophoresis with amperometric detection for the determination of glucose, sucrose, and fructose in honey samples (Cheng et al., 2008). Under the optimized conditions, the three carbohydrates were separated within 20 min with LODs ranging from 3.0×10^{-7} to 6.0×10^{-7} mol L^{-1}, thus demonstrating that nickel oxide NPs can avoid the usual passivation of the carbon electrodes in the presence of carbohydrates.

As a more recent example, microchip electrophoresis has been coupled to a copper-nanowire electrochemical detector that exhibited electrocatalysis toward carbohydrates, constituting a selective and sensitive detector that was applied in the fast and reliable analysis of monosaccharides in honey samples (García and Escarpa, 2014). Glucose and fructose were separated in less than 250 s, the obtained results agreeing, with errors lower than 10%, with those obtained by high-pressure LC (HPLC)–refractive index.

Microchip electrophoresis with label-free detection using deep UV fluorescence detection with excitation at 266 nm was applied by Ohla et al. (2011) for the determination of biologically active compounds (dopamine, serotonin, tryptophan, tyrosine, and the isoquinoline alkaloid salsolinol) in bananas, in less than 1 min. For comparison purposes, microchip electrophoresis was also coupled with MS detection using microfluidic

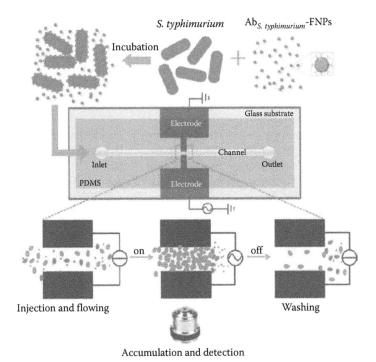

FIGURE 3.1   Schematic representation of the principle of detection of *S. typhimurium* using positive dielectrophoresis-driven online enrichment with fluorescent NP label. (Reprinted with permission from He, X. et al. 2013. *Biosens. Bioelectron.* 42:460–466.)

nanoelectrospray glass chips where a monolithically integrated nanospray tip was placed in front of a quadrupole MS. In this approach, the high viscosity of the sample solutions could easily block the nanospray emitter of the glass chip, which was only about 10 μm in inner diameter. Moreover, high signals were obtained for monosaccharides and disaccharides causing ion suppression for some analytes, but the presence of serotonin and salsolinol in the extracted banana sample was proven.

He et al. (2013) described a method for the rapid and sensitive detection of *Salmonella typhimurium*, a foodborne infectious pathogen, in microfluidic channels using positive dielectrophoresis-driven online enrichment and fluorescent NP label. The bacteria were labeled with antibody-conjugated Rubpy-doped fluorescent NPs and enriched by positive dielectrophoresis, then being detected continuously in a microfluidic chip (Figure 3.1) by fluorescence microscopy. The assay of the bacteria at low concentration levels was achieved in less than 2 h with a LOD of 110 cfu mL^{-1} in artificially contaminated mineral water samples.

3.4 MINIATURIZATION OF THE ENTIRE PROCESS: ON THE WAY TO NANOSENSORS

In the last two decades, the sensors field has seen considerable changes due to the incorporation of nanomaterials such as NPs and CNTs. Among NPs, quantum dots (QDs) are by far the most used in optical sensing. In addition to increasing or decreasing fluorescence, they can be exploited in sensor applications based on Förster (photochemically induced) fluorescence resonance energy transfer mechanisms (FRET) and also in the measurement of surface plasmon resonance (SPR) (Ríos et al., 2009). Physically, CNTs, metallic NPs, and semiconductor nanowires contribute to the fabrication of robust, sensitive and selective electrochemical sensors. Regarding mass sensors, the use of nanocantilevers allows limits of detection in the femtomole to attomole range with the possibility to detect at the single-molecule level in real time.

Although nanotechnology has been applied in the development of novel biosensors for food analysis, very few of them are used in automated systems, no examples having been found showing their application in nanofluidic systems.

A rapid, sensitive and easy-to-use biosensing device for rapid field-based diagnosis to aid the protection of the food supply chain was described by Pal et al. (2008). The system consisted of a sandwich-based flow conductimetric immunosensor for the detection of *Bacillus cereus* in alfalfa sprouts, strawberries, lettuce, tomatoes, fried rice, and cooked corn after homogenization in a stomacher, filtration, and dilution. They fabricated polyaniline nanowire–tagged antibodies which were bound to the pathogen before further antibody capture on the sensor surface. The conductive polyaniline nanowires formed a bridge between the two electrodes, generating a concentration-dependent signal and giving LODs ranging from 35.3 to 88.4 cfu mL^{-1} for the six food types. Although, owing to its rapid detection time (6 min), the method could not be used to enumerate the number of cells in a sample, it could be employed to specifically identify contaminated food samples.

AuNPs were used by Chen et al. (2008) as signal amplification system in a piezoelectric flow-based DNA sensor for *Escherichia coli* O157:H7. Apple juice, milk, and ground beef samples were centrifuged to isolate the bacteria, then genomic DNA was extracted and the specific gene was amplified by polymerase chain reaction (PCR) using synthetic primers including a tag sequence. The amplicons were denatured and allowed to hybridize with previously immobilized synthesized thiolated probes specific to the bacterial *eaeA* gene fragment before the AuNP-labeled secondary oligonucleotide complementary to the tag was introduced. As can be seen in Figure 3.2, the use of AuNPs effectively amplified the signals in frequency change. Thus in 3 h, the sample preparation and analysis were finished with a LOD of 5.3×10^2 cfu mL^{-1} for PCR products without enrichment of the culture. The work demonstrated the usefulness of the AuNPs as effective amplifiers in the quartz crystal microbalance (QCM) DNA sensor due to their relatively large mass compared to the DNA targets. When using this amplification system, only a short period of pre-enrichment may be necessary for bacteria detection in food products, given that a higher LOD was found in the food samples tested which was attributed either to a few bacterial cells being lost when the *E. coli* O157:H7 cells were isolated from the samples or to inhibitors coming from the food.

A microfluidic electrochemical immunosensor coupled with a flow injection system was described by Fernández-Baldo et al. (2009) for the determination of low levels of *Botrytis cinerea*, a plant-pathogenic fungus producing grey mold, in commercial apple fruit tissues samples. The immunosensor was coupled to carbon-based screen-printed electrodes (SPCEs) modified with multiwalled CNTs, these producing an enhancement of the electrochemical responses obtained for the reduction of the redox mediator, 4-tertbutyl-*o*-benzoquinone, to 4-tertbutylcatechol. The authors found the method promising for the analysis of symptomless plants, due to the short analysis time (30 min), minimization of the waste of expensive reagents, good accuracy, reproducibility, specificity and sensitivity with a LOD of 0.02 µg mL^{-1}, well below the LOD achieved with an ELISA (enzyme-linked immunosorbent assay) kit (10 µg mL^{-1}).

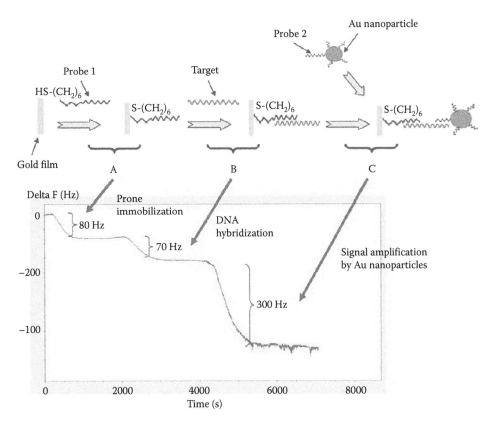

FIGURE 3.2   Time-dependent frequency changes of the circulating-flow QCM sensor described for the determination of *E. coli* O157:H7 in food samples. (Reprinted with permission from Chen, S. et al. 2008. *J Microbiol. Methods* 73:7–17.)

Norouzi et al. (2010) used a glassy carbon electrode (GCE) modified with both AuNPs and MWCNTs in the electrooxidation of thiocoline for a Fourier transform continuous cyclic voltammetric biosensor used in a flow-based system for the detection of the organophosphorus pesticide monocrotophos. The CNTs contained chitosan microspheres to increase the immobilization level and improve the stability of acetylcholinesterase (AChE). The combination of mixed-walled CNTs and AuNPs responded for a promotion of the electron transfer and catalyzed the electrooxidation of thiocoline, amplifying the detection sensitivity. The response time of the biosensor was below 70 s, reaching a LOD of 10 nmol L^{-1} and also showing a long-term stability of 50 days. Figure 3.3 shows the electrochemical cell used in the flow-injection system.

Yang et al. (2008) synthesized a hydrogen titanate ($H_2Ti_3O_7$) nanotube network substrate giving a 3-dimensional porous matrix that enabled a high loading of the amount of the enzyme L-lactate oxidase (Lox), while maintaining the enzyme catalytic activity, and provided a pathway for direct electron transfer between the redox center of the enzyme and the electrode surface. A linear response for lactic acid was found in the 0.5–12 mM concentration range. This constitutes an example of the application of nanotechnology to increase the sensitivity of electrochemical enzymatic biosensors by increasing the electrode surface area, which has been applied in the field of food analysis. Sometimes the increase of sensitivity through nanotechnology also occurs with the use of bienzymatic

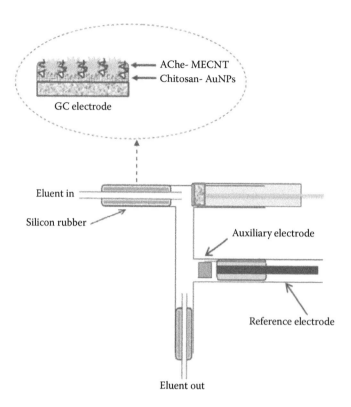

FIGURE 3.3   Scheme showing the monocrotophos biosensor and the electrochemical cell used in flow injection analysis. (Reprinted with permission from Norouzi, P. et al. 2010. *Int. J. Electrochem. Sci.* 5:1434–1446.)

systems. As an example, Pérez and Fàbregas (2012) described an amperometric bienzymatic biosensor based on the incorporation of Lox and horseradish peroxidase (HRP) into a CNT/polysulfone membrane used in the modification of screen-printed electrodes (SPEs). The reliability of the developed biosensor was demonstrated through its application to the determination of L-lactate in wine and beer samples by comparing the results obtained with those obtained using a spectrophotometric kit. Although the developed biosensor showed high accuracy and sensitivity with a low LOD of 0.05 mg L^{-1}, and reduced the assay time from 23 min in the spectrophotometric kit to 5 min, its stability was lower than with other biosensors (less than 2 weeks).

An electrical percolation-based label-free biosensor for real-time detection has been described by Yang et al. (2010) based on electrical percolation through a single-walled carbon nanotube (SWCNT)–antibody complex that formed a network functioning as a "biological semiconductor" (BSC). The conductivity of a BSC is directly related to the number of contacts facilitated by the antibody–antigen "connectors" within the SWCNT network. BSCs were fabricated by immobilizing a prefunctionalized SWCNT–antibody complex directly on a poly(methylmethacrylate) (PMMA) and polycarbonate (PC) surface. As can be seen in Figure 3.4, each BSC was connected via silver electrodes to a computerized ohmmeter, thereby enabling continuous electronic measurement of molecular interactions (e.g., antibody–antigen binding) via the change in resistance. Using anti-staphylococcal enterotoxin B (SEB) IgG (immunoglobulin G) to functionalize the BSC,

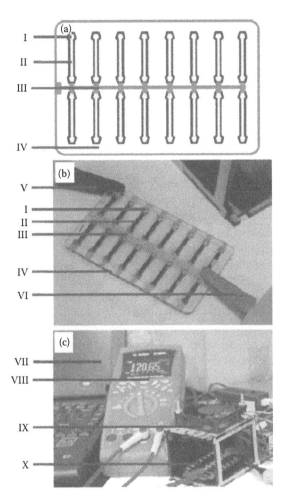

FIGURE 3.4 Electrical percolation-based biosensor. (a) A scheme of a sensor with 16 BSC specific electrodes (I), a ground connector common electrode (III), and space for the SWNT–antibody bionanocomposite gate (II) all printed on a PMMA board (IV). (b) A photograph of the actual biosensor with the electrode specific connector to the ohmmeter (V), the BSC specific electrode (I), the SWNT–antibody complex gate (II), the common ground electrode (III) printed on PMMA board (IV) and the positive ground ohmmeter connector (VI). (c) The electrical percolation–based biosensor setup. The laptop computer (VII) is connected to a digital ohmmeter (VIII) via USB port. The BSC dryer (IX) and the BSC connected to the ohmmeter (X). (Reprinted with permission from Yang, M. et al. 2010. *Biosens. Bioelectron.* 25:2573–2578.)

the biosensor was able to detect SEB at concentrations as low as 5 ng mL^{-1}. The measurements were performed on the chip in wet conditions. Direct measurement of the toxin in several food matrices was tested without sample preparation. The BSC's sensitivity with samples of SEB in milk was about 20% lower than for samples of SEB in buffer, while for samples of SEB in baby food the sensitivity was about 30% lower. Thus, a sample purification step is needed to improve the method's sensitivity and make it comparable to traditional ELISA assays.

Recent years have seen a large boost in the use of magnetic beads (MBs) as bioimmo-bilization platforms, biomolecular conveyors, and separators and concentrators of various analytes, as well as to control electrochemical processes that occur at the electrode surfaces. MBs constitute a versatile tool in the development of biosensors, as they provide a large surface area for attachment of various recognition elements and analytes, which can then easily be separated from the liquid phase by a magnet, and spread immediately to zoom out. The use of MBs greatly facilitates the preparation steps of the biosensor, and allows a reduction in the assay time, resulting in designs that generally achieve better LODs and lower matrix effects. Given their interesting properties and large commercial availability, different functionalized MBs have been used as immobilization surfaces in the development of magnetobiosensors in the food analysis field as well. As an example, Tang et al. (2009) used two types of nanomaterials in a microfluidic immunoassay applied to the determination of aflatoxin B_1 (AFB$_1$) in red paprika samples where the fungi that grow induce the formation of aflatoxins, their concentration increasing with the storage time. Thus, multifunctional MBs with $CoFe_2O_4$ NPs as the core and Prussian blue NP-doped silica as the shell were synthesized to immobilize AFB$_1$–bovine serum albumin conjugate. The modified beads were magnetically captured on the surface of an indium–tin–oxide (ITO) electrode and AuNPs modified with HRP-labeled antibodies were used for the amplification of the analytical signal in a competitive format. The LOD obtained of 6.0 pg mL^{-1} was 80 times lower than that obtained in the absence of AuNPs. Due to the presence of the doped Prussian blue NPs, the electron mediator addition in the detection solution was avoided. Also, Paniel et al. (2010) developed an electrochemical immunosensor for the detection of ultratrace amounts of aflatoxin M_1 (AFM$_1$) in milk samples. The sensor used magnetic nNPs coated with a specific antibody and a competitive immunoassay between the toxin and an AFM$_1$–HRP conjugate. Then, the MBs were deposited on the surface of SPEs and the enzymatic response was measured amperometrically, achieving a low LOD of 0.01 ppb, which was under the recommended level for this toxin (0.05 µg L^{-1}). The system allowed the measurement of AFM$_1$ directly in milk samples after a single centrifugation step without dilution or other pretreatment steps.

Fluorescent NPs, known as QDs, can be used in multiplex detection by conjugating these semiconductor nanocrystals with different specific biorecognition molecules. Their broad absorption spectra and narrow emission spectra make them ideal for optical multiplexing. Other properties of QDs include negligible photobleaching and fairly high quantum yields. As an example of their use in the field of food analysis, Peng et al. (2009) described the simultaneous multicolor determination of five low–molecular weight chemical drug residues in pork muscle tissue samples, based on the use of biotinylated denatured bovine serum albumin-coated cadmium telluride QD conjugates. An immune complex was formed between specific antibodies and the antigens and an indirect competitive fluorescent linked immunosorbent assay (ic-FLISA) was established for the simultaneous drug detection (Figure 3.5), where the corresponding antibodies had been coupled to QDs with different emission maximums. The LOD of the simultaneous array was 0.06 µg kg^{-1}, which is low enough to verify compliance of products with the legislation tolerance, and the procedure could be used as a simple and inexpensive screening method.

Although QDs offer many advantages over conventional fluorophores, their synthesis is considered difficult, and commercial variants are still limited in number and expensive.

Colloidal carbon NPs have been used as labels in a rapid screening method based on an indirect competitive immunostrip test for detection of the pesticide thiabendazole in fruit juices after sample dilution (Blazcová et al., 2010). The strip-assay was based on

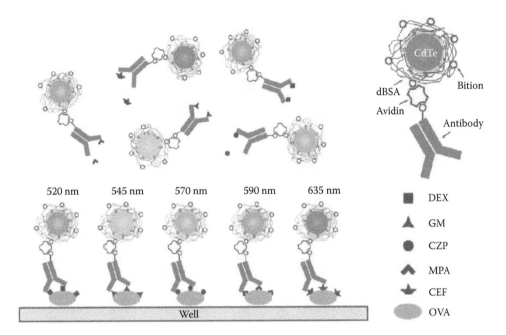

520 nm   545 nm   570 nm   590 nm   635 nm

Well

dBSA   Bition
Avidin
Antibody
CdTe

■ DEX
▲ GM
● CZP
◆ MPA
⚘ CEF
⬭ OVA

FIGURE 3.5   Sketch of the indirect competition FLISA protocol used for the multiplex determination of drugs in pork tissue samples. DEX, dexamethasone; MPA, medroxy-progesterone acetate; GM, gentamicin; CZP, clonazepam; CEF, ceftiofur; and dBSA, denatured bovine serum albumin. (Reprinted with permission from Peng, C. et al. 2009. *Biosens. Bioelectron.* 24:3657–3662.)

the migration of the sample and a secondary antibody labeled with carbon NPs along a reagent-coated membrane strip where the affinity reactions took place. The intensity of the black color formed in the test strip decreased with increasing thiabendazole concentrations, a LOD of 0.08 ng mL^{-1} being obtained in buffer solutions. Also based on this scanning densitometry, the strip assay was able to determine thiabendazole in fruit juices at a 50 ng mL^{-1} concentration level, which is below the maximum residual levels for this compound in fruits (50 mg kg^{-1}). A good agreement was found between the results given by the strip and those obtained by ELISA. The carbon NPs used as detection label provided higher color intensity than colloidal AuNPs.

An example of the opportunities offered by nanoscience and nanotechnology in the development of multiplexed devices, in particular for food analysis, is found in the biosensing system described by Adrian et al. (2009) for the multianalyte screening of antibiotics in milk samples. Following a sample dilution, they used a flow-based wavelength-interrogated optical biosensor with class-selective bioreceptors for the simultaneous screening of sulfonamides, fluoroquinolones, β-lactams, and tetracyclines, the most used antibiotics in the veterinary field. The label-free sensor, shown in Figure 3.6, detected changes in the refraction index close to the modified chip surface by scanning the resonance condition at which a light wave was coupled in the waveguide through a conveniently designed grating. The system could detect 34 antimicrobials in milk samples below the legislated maximum residue limits without the need of complex sample treatments.

The simultaneous detection of food-borne mycotoxins has been described by Mak et al. (2010) based on the combination of the specificity of immunoassays with the

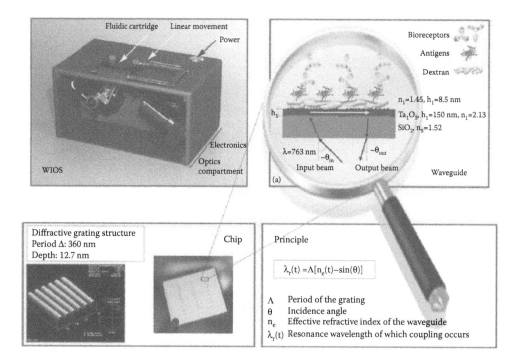

FIGURE 3.6 Pictures showing the wavelength-interrogated optical system (WIOS). The chip is a customized platform that contains 24 sensing sites. Each site is waveguide grating. Light from a vertical cavity-surface-emitting laser emitting at around 763 nm is incident on the first grating. The waveguide mode is excited and propagates into the waveguiding layer. The second grating sends out the guided light (at a different angle), which is collected by large plastic optical fibers. Electronics controls the laser-wavelength modulation and amplifies the signals detected. A computer acquires and processes the data. (Reprinted with permission from Adrian, J. et al. 2009. *Trends Anal. Chem.* 28:769–777.)

sensitivity and simplicity of magnetic detection. The developed system consisted of a multiplex magnetic nanotag-based detection platform for AFB_1, zearalenone, and HT-2 that functioned on a sub-picomolar concentration level achieving a LOD of 50 pg mL^{-1}. Unlike fluorescent labels, magnetic nanotags can be detected with inexpensive giant magnetoresistive sensors (e.g., spin-valve sensors). Thus, the samples were added to the antibody-immobilized sensor array prior to the addition of a biotinylated detection antibody. The sensor response was recorded in real time upon the addition of streptavidin-linked magnetic nanotags on the chip. With the right antibody–analyte pairing, the platform could differentiate multiple mycotoxins in the same run. Still, the system should be validated with real samples before its adaptation to point-of-use testing.

3.5 LAB-ON-A-CHIP SYSTEMS APPLIED TO FOOD ANALYSIS

The main reason why nanotechnology has still not reached a great impact in food analysis as opposed, for example, to clinical analysis is that sample preparation involving complex protocols is usually needed in food analysis together with analyte separation using advanced detection schemes. Still, lab-on-a-chip technology is continuously developing in a search for advanced sample handling and treatment, so it should go from actual application in medicine, biology, etc., to find more uses in food analysis.

Nevertheless, in recent years food analysis has seen the development of microfluidic devices including the integration of nanomaterials. This can be related to the fact that the analytical challenges inherent to these samples—poor selectivity and sensitivity together with the sample complexity—can potentially be solved through the increase in sensitivity and selectivity offered by nanotechnologies when applied to lab-on-a-chip systems.

CNTs have been used as detectors for the determination of total isoflavones and also in the fast detection of antioxidant profiles, being applied both as flow with integrated calibration and as separation systems (Crevillén et al., 2009). An enhancement in sensitivity for the detection of dietary antioxidants in apple samples, independent of the CNT material used, was observed in comparison to the detector without CNTs, together with sharp and well-resolved peaks for all compounds, in contrast with tailing unresolved peaks for the same analytes using unmodified detectors. More recently, CNTs have been press-transferred onto poly(methylmethacrylate) substrates for electrochemical microfluidic sensing, thus constituting new disposable electrodes where CNTs act as exclusive electrochemical transducers. This system has been applied in the fast and reliable qualitative and quantitative class-selective determination of isoflavones (Vilela et al., 2014). The electrochemical isoflavone index determination was based on the comigration of the total glycosides and total aglycones in less than 250 s with good repeatability and reproducibility. The advantages of this coupling of nanoscaled detectors to microfluidic systems, apart from the fact that CNTs are the exclusive transducer, include the facts that these electrodes can be fabricated from commercial sources using a simple protocol that can be afforded in any laboratory, and that they are well-suited to mass production, disposability, and other nanomaterials and/or biological materials (Escarpa, 2014).

An example of the employment of NPs in the detection step within microfluidic systems to improve sensitivity and selectivity was the amperometric detection of five sulfonamide drugs after electrokinetic separation in beef, pork, chicken, and fish samples (Won et al., 2013). An Al_2O_3-AuNP-modified carbon paste electrode was used, placed at the end of the separation microchannel. Al_2O_3 was used to preconcentrate the sulfonamides through an adsorption process while, as aluminum oxide is nonconductive, AuNPs were used to enhance the performance of the modified electrode. The LODs obtained were in the 0.91–2.21 fmol L^{-1} range, 10^6 times lower than the values given by LC with fluorescence detection, and the modified electrode had a lifetime of 2 months.

Recently, Brod et al. (2014) have described a high throughput method for multidetection of genetically modified organisms (GMOs) using a microfluidic dynamic array. Analysis of genetically modified crops by DNA-based methods was undertaken in a Fluidigm system from BioMark consisting of a nanoscale network of fluid lines, valves, and chambers under pressure control (Figure 3.7) (Spurgeon et al., 2008). The samples and arrays were individually pipetted into the corresponding inlets and immediately loaded and mixed in the chip. A PCR cycling step was undertaken and, at the end of the thermal cycling, the analysis software generated amplification curves for each of the reaction chambers in the chip. In this way, the chip allowed a high number of reactions taking place, each in a low volume (8 nL), its sensitivity being increased through the use of a preamplification step to ensure an adequate amount of input DNA. In this paper, six test samples with an increasing degree of complexity were prepared, preamplified, and analyzed in the microfluidic system, with elements present at concentrations as low as 0.06% being successfully detected. Based on the data they obtained, the authors concluded that

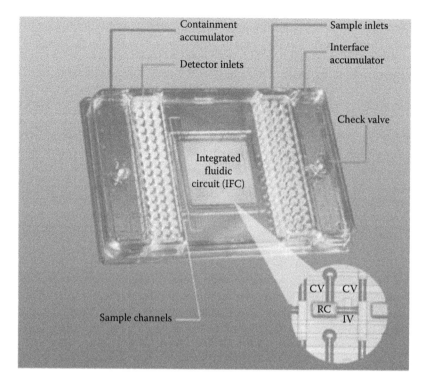

FIGURE 3.7 Photograph of a 48 samples × 48 assays dynamic array chip showing the position of the sample inlets and the detector inlets in which the gene expression assay reagents are added. (Reprinted with permission from Spurgeon, S. L., R. C. Jones, and R. Ramakrishnan. 2008. *PLoS One* 3(2):e1662.)

the system may prove a useful tool in the screening methods for detection of authorized and unauthorized GMOs in food and feed, especially for laboratories with a large number of samples that would benefit from reduced analysis times and reduced amounts of chemicals per sample. However, real samples should be tested to check for matrix effects and fully validate the system as a high-throughput method for GMO multidetection.

3.6 CONCLUSIONS

As has been described throughout this chapter, advances in nanotechnology and, in particular, in the synthesis of nanomaterials together with the availability of methods to functionalize their surfaces are contributing to increasing the selectivity and sensitivity of microfluidic methods, including in the field of food analysis. For example, functionalized magnetic NPs allow the preconcentration of analytes together with their separation from the sample matrix, thus contributing to an increase in selectivity; and the wide variety of optical and electrochemical properties of such nanomaterials as metallic NPs, QDs, or CNTs are exploited in improving the sensitivity of microfluidic methodologies, this sometimes based only on their large surface area-to-volume ratio. These nano-tools also increase multiplexing capabilities and reduce costs. Nanomaterials-based systems offer simplicity of optical and electrochemical configurations, ease of fabrication and a great potential for miniaturization, simple handling, and short assay times. Moreover, the use of nanotechnology in chromatographic and electrophoretic systems should not be forgotten, although in these cases it is mainly seen in the individual constituents of the whole device.

Finally, nanotechnology can for sure provide novel microfluidic solutions for food applications still to be explored. It can be used in pumping and flow devices but nanomaterials are already also employed as detectors, improving analytic performance and opening new avenues for future implementation of applications in the field of food analysis.

ACKNOWLEDGMENTS

The financial support of the Spanish Ministerio de Economía y Competitividad Research Projects, CTQ2012-34238, and the AVANSENS Program from the Comunidad de Madrid (S2009PPQ-1642) are gratefully acknowledged.

REFERENCES

Adrian, J., S. Pasche, D. G. Pinacho et al. 2009. Wavelength-interrogated optical biosensor for multi-analyte screening of sulfonamide, fluoroquinolone, β-lactam and tetracycline antibiotics in milk. *Trends Anal. Chem.* 28:769–777.

Aqay, P., J. Peters, A. Gerssen, W. Haasnoot, and M. W. F. Nielen. 2011. Immunomagnetic microbeads for screening with flow cytometry and identification with nano-liquid chromatography mass spectrometry of ochratoxins in wheat and cereal. *Anal. Bioanal. Chem.* 400:3085–3096.

Blazková, M., P. Rauch, and L. Fukal. 2010. Strip-based immunoassay for rapid detection of thiabendazole. *Biosens. Bioelectron.* 25:2122–2128.

Brod, F. C. A., J. P. van Dijk, M. M. Voorhuijzen et al. 2014. A high-throughput method for GMO multi-detection using a microfluidic dynamic array. *Anal. Bioanal. Chem.* 406:1397–1410.

Byun, C. K., X. Wang, Q. Pu, and S. Liu. 2007. Electroosmosis-based nanopipettor. *Anal. Chem.* 79:3862–3866.

Chen, S., V. C. H. Wu, Y. Chuang, and C. Lin. 2008. Using oligonucleotide-functionalized Au nanoparticles to rapidly detect foodborne pathogens on a piezoelectric biosensor. *J Microbiol. Methods* 73:7–17.

Cheng, X., S. Zhang, H. Zhang, Q. Wang, P. He, and Y. Fang. 2008. Determination of carbohydrates by capillary zone electrophoresis with amperometric detection at a nano-nickel oxide modified carbon paste electrode. *Food Chem.* 106:830–835.

Crevillén, A. G., M. Pumera, M. C. González, and A. Escarpa. 2009. Towards lab-on-a-chip approaches in real analytical domains based on microfluidic chips/electrochemical multi-walled carbon nanotube platforms. *Lab Chip* 9:346–353.

Escarpa, A. 2014. Lights and shadows on food microfluidics. *Lab Chip* 14:3213–3224. DOI: 10.1039/C4LC00172.

Escarpa, A., M. C. González, M. A. López-Gil, A. G. Crevillén, M. Hervás, M. García. 2008. Microchips for CE: Breakthroughs in real-world food analysis. *Electrophoresis* 29:4852–4861.

Fernández-Baldo, M., G. Messina, M. Sanz, and J. Raba. 2009. Screenprinted immunosensor modified with carbon nanotubes in a continuous-flow system for the *Botrytis cinerea* determination in apple tissues. *Talanta* 79:681–686.

García, M. and A. Escarpa. 2014. Microchip electrophoresis–copper nanowires for fast and reliable determination of monossacharides in honey samples. *Electrophoresis* 35:425–432.

González-Crevillén, A., M. Ávila, M. Pumera, M. C. González, and A. Escarpa. 2007. Food analysis on microfluidic devices using ultrasensitive carbon nanotubes detectors. *Anal. Chem.* 79:7408–7415.

He, X., C. Hu, Q. Guo, K. Wan, Y. Li, and J. Shangguan. 2013. Rapid and ultrasensitive *Salmonella typhimurium* quantification using positive dielectrophoresis driven on-line enrichment and fluorescent nanoparticles label. *Biosens. Bioelectron.* 42:460–466.

Hellman, A. N., K. R. Rau, H. H. Yoon et al. 2007. Laser-induced mixing in microfluidic Channels. *Anal. Chem.* 79:4484–4492.

Kartsova, L. A. and A. A. Makarov, 2002. Properties of carbon materials and their use in chromatography. *Russ, J. Appl. Chem.* 75:1725–1731.

Kovarik, M. L. and S. C. Jacobson. 2007. Attoliter-scale dispensing in nanofluidic channels. *Anal. Chem.* 79:1655–1660.

Li, Q. and D. Yuan. 2003. Evaluation of multi-walled carbon nanotubes as gas chromatographic column packing. *J. Chromatogr.* A 1003:203–209.

Mak, A. C., S. J. Osterfeld, H. Yu et al. 2010. Sensitive giant magnetoresistive-based immunoassay for multiplex mycotoxin detection. *Biosens. Bioelectron.* 25:1635–1639.

Martín, A., D. Vilela, and A. Escarpa. 2012. Food analysis on microchip electrophoresis: An updated review. *Electrophoresis* 33:2212–2227.

Mawatari, K., T. Tsukahara, Y. Tanaka, Y. Kazoe, P. Dextras, and T. Kitamori. 2012. *Extended-Nanofluidic Systems for Chemistry and Biotechnology.* London: Imperial College Press.

Mellors, J. S., K. Jorabchi, L. M. Smith, and J. M. Ramsey. 2010. Integrated microfluidic device for automated single cell analysis using electrophoretic separation and electrospray ionization mass spectrometry. *Anal. Chem.* 82:967–973.

Norouzi, P., M. Pirali-Hamedani, M. R. Ganjali, and F. Faridbod. 2010. A novel acetylcholinesterase biosensor based on chitosan-gold nanoparticles film for determination of monocrotophos using FFT continuous cyclic voltammetry. *Int. J. Electrochem. Sci.* 5:1434–1446.

Oellig, C. and W. Schwack. 2014. Planar solid phase extraction clean-up and microliter-flow injection analysis–time-of-flight mass spectrometry for multi-residue screening of pesticides in food. *J. Chromatogr.* A 1351:1–11.

Ohla, S., P. Schulze, S. Fritzsche, and D. Belder. 2011. Chip electrophoresis of active banana ingredients with label-free detection utilizing deep UV native fluorescence and mass spectrometry. *Anal. Bioanal. Chem.* 399:1853–1857.

Pal, S., W. Ying, E. C. Alocilja, and F. P. Downes. 2008. Sensitivity and specificity performance of a direct-charge transfer biosensor for detecting *Bacillus cereus* in selected food matrices. *Biosyst. Eng.* 99:461–468.

Paniel, N., A. Radoi, and J. L. Marty. 2010. Development of an electrochemical biosensor for the detection of aflatoxin M1 in milk. *Sensors* 10:9439–9448.

Peng, C., Z. Li, Y. Zhu et al. 2009. Simultaneous and sensitive determination of multiplex chemical residues based on multicolor quantum dot probes. *Biosens. Bioelectron.* 24:3657–3662.

Pérez, S. and E. Fàbregas. 2012. Amperometric bienzymatic biosensor for L-lactate analysis in wine and beer samples. *Analyst* 137(16):3854–3861.

Perry, J. L. and S. G. Kandlikar. 2006. Review of fabrication of nanochannels for single phase liquid flow. *Microfluid. Nanofluid.* 2:185–193.

Piccin, E., R. Laocharoensuk, J. Burdick, E. Carrilho, and J. Wang. 2007. Adaptive nanowires for switchable microchip devices. *Anal. Chem.* 79:4720–4723.

Ríos, A., A. Escarpa, and B. Simonet. 2009. *Miniaturization of Analytical Systems.* Chichester: John Wiley & Sons Ltd.

Saridara, C. and S. Miltra. 2005. Chromatography on self-assembled carbon nanotubes. *Anal. Chem.* 77:7094–7097.

Schmitt, R., S. Driessen, and B. Engelmann. 2006. Controlling the assembly of micro systems by image processing. *Microsyst. Technol.* 12:640–645.

Spurgeon, S. L., R. C. Jones, and R. Ramakrishnan. 2008. High throughput gene expression measurement with real time PCR in a microfluidic dynamic array. *PLoS One* 3(2):e1662.

Stadermann, M., A. D. McBrady, B. Dick et al. 2006. Ultrafast gas chromatography on single-wall carbon nanotube stationary phases in microfabricated channels. *Anal. Chem.* 78:5639–5644.

Szekely, L. and R. Freitag. 2005. Study of the electroosmotic flow as a means to propel the mobile phase in capillary electrochromatography in view of further miniaturization of capillary electrochromatography systems. *Electrophoresis* 26:1928–1939.

Tang, D., Z. Zhong, R. Niessner, and D. Knopp. 2009. Multifunctional magnetic bead-based electrochemical immunoassay for the detection of aflatoxin B1 in food. *Analyst* 134:1554–1560.

Vilela, D., A. Martín, M. C. González, and A. Escarpa. 2014. Fast and reliable class-selective isoflavone index determination on carbon nanotube press-transferred electrodes using microfluidic chips. *Analyst* 139:2342–2347.

Won, S. Y., P. Chandra, T. S. Hee, and Y. B. Shim. 2013. Simultaneous detection of antibacterial sulfonamides in a microfluidic device with amperometry. *Biosens. Bioelectron.* 39:204–209.

Yang, M. L., J. Wang, H. Q. Li, J. G. Zheng, and N. Q. N. Wu. 2008. A lactate electrochemical biosensor with a titanate nanotube as direct electron transfer promoter. *Nanotechnology* 19(7):075502 (6pp).

Yang, M., S. Sun, H. A. Bruck, Y. Kostova, and A. Rasooly. 2010. Electrical percolation-based biosensor for real-time direct detection of staphylococcal enterotoxin B (SEB). *Biosens. Bioelectron.* 25:2573–2578.

Zellers, E. T., S. Reidy, R. A. Veeneman et al. 2007. An integrated micro-analytical system for complex vapor mixtures. *Transducers Eurosens.* '07 3A31:1491–1496.

CHAPTER 4

Multicommutated Flow Injection Analysis, Multisyringe Flow Injection Analysis, and Multipumping Flow Systems
Theory and Trends

Víctor Cerdà Martín, Jose Manuel Estela Ripoll, and Jessica Avivar Cerezo

CONTENTS

4.1 INTRODUCTION

In the 1950s laboratory tests began to be used as diagnosis tools, which led to spectacular increase in demand for them. The huge number of analyses required and the need to have results in a short time could only be met by the advent of automatic analytical methods such as flow methods. Flow techniques arose in 1957 with segmented flow analysis (SFA) [1], which afforded not only substantially increased throughput but also substantial savings in samples and reagents. SFA laid the foundations for modern flow techniques. Since then flow techniques have been in continuous evolution toward new developments in the search of low-cost highly reproducible fast determinations.

Several flow techniques have been developed and used for analytical and monitoring applications. These have proved to be excellent tools for handling of solutions and thus for wet chemical analysis. In addition, flow systems can be regarded as closed laboratories that avoid cross-contamination. They have gained importance for clinical, industrial, and environmental purposes, since automation and miniaturization of fluidic-based analysis are essential to make them rapid and efficient for routine and research tasks [2]. Moreover by minimizing reagents and sample volumes, not only costs are reduced but also waste generation and thus environmental impact. Nowadays, the amount and toxicity of the wastes generated are as important as any other analytical features when developing a new analytical method. Moreover, the progress of flow techniques toward miniaturization and saving of reagents has contributed to greener analytical chemistry [3].

Multicommutated flow techniques are especially well suited to accomplish these aims since they have shown great potential in comparison with previous flow techniques in minimizing reagents consumption and waste production and in providing more environmentally friendly methodologies, since liquids are propelled to the system only when required and returned to their reservoirs when not required—that is, reagents are only dispensed when required and only the amounts of reagent strictly required are injected. As an example, a method for carbaryl pesticide determination reduced the amount of the main reagent (p-aminophenol) employed from 140 to 5 μg exploiting a multicommutated flow system [4].

Flow techniques can be classified in two main groups: those with computer-less working capability that can be operated by hand, and those requiring computerized control. Initially flow systems were operated exclusively by hand, for example, SFA [1] and flow injection analysis (FIA) [5]. Later, computers facilitated the development of computerized controlled flow techniques such as sequential injection analysis (SIA) [6], multicommutated flow injection analysis (MCFIA) [7], multisyringe flow injection analysis (MSFIA) [8], and multipumping flow systems (MPFS) [9], based on multicommutation operation. All of them have common components such as impulsion pumps, which act as liquid drivers, and a series of plastic tubes or a manifold intended to carry liquid streams to the detector.

Initially further development of the computer-controlled flow techniques was hindered owing to unavailability of suitable commercial software and general lack of experience in coupling personal computers to instruments. However, most of the advantages of current flow techniques are in part consequences of the incorporation of computers. The earliest automatic methods used devices suited to particular applications, which restricted their scope to very specific uses such as the control of manufacturing processes or to situations where the number of samples to be analyzed was large enough to justify the initial effort and investment required. Nowadays, computerized flow techniques allow the implementation of the same analytical method (hardware), with little or no alteration, on different types of samples simply by software modification.

The foundations and comparison of main flow techniques based in multicommutation are presented in this chapter.

4.2 MULTICOMMUTATION

The introduction of the arm used in the earlier air-segmented flow analyzers to select either the sample or the carrier solution to be aspirated toward the analytical path has been considered as the beginning of commutation in flow analysis [10]. In general, multi-commutation refers to flow systems designed with discrete computer-controlled devices resulting in flow networks in which all the steps involved in sample processing can be independently implemented. In short, multicommutation is inherent to flow systems that may present several stages. The simplest mechanical commutation involves only redirecting the streams, which can be accomplished by using three-way solenoid valves. Commutation is also responsible for additions of components (including sampling loops) to the analytical path and/or stream redirecting.

Several contributions involving different individually operated discrete devices [11], a number of linked three-way valves [12], several commutators [13], and establishment of tandem streams [11,14] were proposed before the term multicommutation was coined. Sometimes the presence of several valves in the system has been erroneously taken as an indicator of a multicommutated system. There are multicommutated systems with a single valve [15] and flow systems with several commutators where the aspect of multi-commutation is not highlighted [16,17].

Multicommutation permits different tasks to be accomplished simultaneously, so improving system performance and analytical characteristics. The flow system can be reconfigured by the control software, presenting increased versatility, potential for auto-mation, and minimization of both reagent consumption and waste generation.

One possibility with multicommutation is the establishment of a tandem stream, that is, a number of aliquots of different solutions can be introduced into the manifold by rapidly and sequentially switching the commutators, usually computer-controlled valves. This unique stream can be seen as a set of neighboring solution slugs that undergo rapid mixing while being transported through the analytical path. In con-trast to most flow systems, sample/reagent interaction starts in the sampling step, thus increasing the mean residence time without affecting sampling rate. The tandem stream idea was initially exploited for the development of an improved single-line system for spectrophotometric multiparametric analysis of natural waters into which several plugs of different reagent solutions were sequentially introduced [11]. Multicommutated flow systems with tandem streams have also been proposed for improving sample/reagent mixing in the spectrophotometric determination of glycerol in alcoholic fermentation juices [18].

Multicommutation also allows implementation of procedures for assessment of accu-racy. For example, the same analyte can be determined by two different methods with the same flow network. In this way, a multicommutated flow manifold was proposed for chloride determination in natural waters with different sample matrices based on the spectrophotometric procedure using mercury(II) thiocyanate and the turbidimetric method using silver nitrate [19]. Moreover, in-line addition/recovery tests were imple-mented for every assayed sample in order to detect matrix effects.

Potentialities of stopped-flow approaches are also expanded with multicommuta-tion [20]. Procedures involving in-line concentration are usually time-consuming. This drawback can be circumvented by using a multicommutated flow manifold, as illustrated in the determination of cadmium, nickel and lead in foodstuffs and plant materials by inductively coupled plasma optical emission spectroscopy (ICP-OES) [21]. In-line sorp-tion and elution were simultaneously carried out by using three ion-exchange columns managed by a four-way valve, allowing a sampling rate of 90 determinations/h.

As stated before, the low reagent consumption together with the high injection throughput achieved by using multicommutated techniques needs to be highlighted [22]. In this sense, Semenova et al. [23] compared the results obtained using an MSFIA system for total arsenic determination by hydride generation–atomic fluorescence spectrometry (HG-AFS) with those using SIA and FIA. In comparison with FIA, MSFIA consumed 15–18 times less reagents and increased the injection throughput by a factor of 2.5. In comparison with SIA, MSFIA increased the injection throughput by a factor of 3 and had similar reagent consumption. This is also illustrated in the enzymatic determination of glucose in soft drinks and sugarcane juices exploiting multicommutated flow systems where the required amounts of peroxidase and glucose oxidase were 85% lower than in the analogous batch procedure [24].

Furthermore, the incorporation of feedback mechanisms in the control software expands the performance of multicommutated flow analyzers. With active devices strategically placed in the manifold, sample processing can be modified according to preliminary measurements. In this way, parameters such as sample residence time, sample dispersion or sample volume can easily be modified, as demonstrated in the wide range determination of uranium and thorium in environmental samples [25]. In the last years, smart systems based on flow analyzers, such as FIA, SIA, MSFIA, or MPFS, have been developed for the determination of environmental parameters [26–29], control of industrial processes [30], and quality control of food [31,32].

There is a clear linkage between the development of flow analysis and that of the commutation concept [33]. Evolution of the commutation concept has led to the proposal and development of different generations of flow analyzers. The main multicommutated flow techniques are described in following sections.

4.3 MULTICOMMUTATED FLOW ANALYSIS

4.3.1 Theoretical Foundations and Instrumentation

In MCFIA, conventional six-port injection valves used in FIA are replaced by three-way solenoid commutation valves (Figure 4.1). Each valve acts as an independent commutator and its operation is controlled by software. The use of solenoid valves instead of injection valves provides flexibility and saves reagents, since these are injected into the system only when necessary.

The earliest MCFIA systems used a single-channel propulsion system to aspirate the liquids via individual valves with stoppers made by melting the end of a small piece of polyethylene tubing (Figure 4.2) [7]. Since aspiration devices tend to insert air bubbles or degas liquids in the system, it is preferable to use liquid propulsion devices such as peristaltic or multipiston pumps instead. Peristaltic pumps are the most common liquid drivers used in MCFIA. However, as the flexible tubing of peristaltic pumps needs to be replaced periodically, this factor is one of the main shortcomings of MCFIA [34].

Another major shortcoming of solenoid valves is the unfavorable effect of the heat released by the solenoid coil when the valves are activated for a long time. The resulting increase in temperature can deform the poly(tetrafluoroethylene) (PTFE) inner membranes of the valves and render them unusable. Thus, when solenoid valves are used, a solenoid-protection system (Figure 4.3) is necessary in order to minimize heat generation enlarging the lifetime of the valves. This system provides a hit-and-hold circuit for solenoid valves by stepping down DC voltage to one-third of input voltage (Figure 4.4).

MCFIA shares some advantages of FIA, such as a high throughput (higher than that for SIA) and a reduced consumption of reagents as a result of the return of unused sample

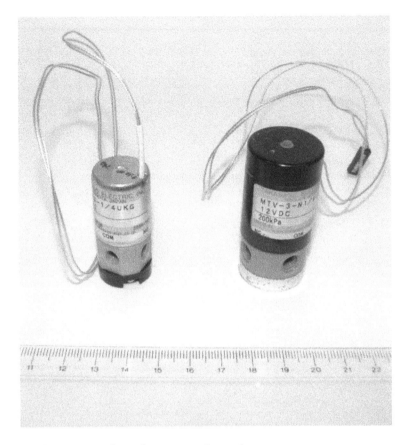

FIGURE 4.1   Three-way solenoid commutation valves.

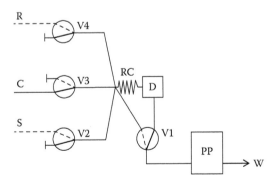

FIGURE 4.2   Schematic depiction of an MCFIA system for the spectrophotometric determination of iron. C: carrier (perchloric acid solution); D: detector (480 nm); PP: peristaltic pump; R: reagent (thiocyanate solution); RC: reaction coil; S: sample; V: three-way
solenoid valve; W: waste.

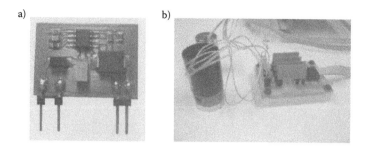

FIGURE 4.3   (a) Valve protector and (b) valve protection platform.

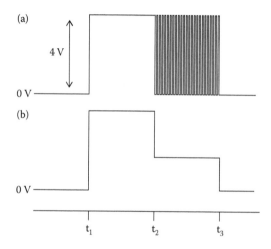

FIGURE 4.4   (a) Voltages used to switch and keep the valve activated. (b) Equivalent DC voltages.

and reagents to their respective reservoirs. However, MCFIA also shares one disadvantage of FIA, namely: the vulnerability of peristaltic pump tubing to aggressive reagents and, especially, solvents.

MCFIA has not had great success as a novel flow technique in itself. There are few publications of MCFIA methods in comparison with other flow techniques. However, three-way solenoid valves have been widely implemented in other flow systems. Thus, three-way solenoid valves have been one of the major contributions of MCFIA.

4.3.2 MCFIA Applications

By way of example, Figure 4.5 shows a schematic depiction of an MCFIA system for determination of mercury in fish by cold vapor atomic absorption spectrophotometry (CVAAS) [35]. Independently controlled solenoid valves were used for reagents and sample injection. First, V2 and V3 were switched on at the same time to mix an equal volume of sample or standard and reducing reagent. This volume was determined by the flow

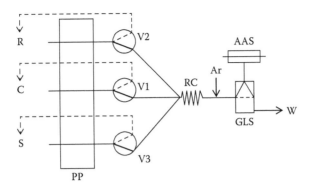

FIGURE 4.5 MCFIA system for mercury determination in fish by cold vapor absorption spectrophotometry. AAS: atomic absorption spectrometer quartz cell; Ar: argon carrier gas; C: carrier (HCl); GLS: gas–liquid separator; PP: peristaltic pump; R: reducing reagent ($NaBH_4$ in NaOH); RC: reaction coil; V: three-way solenoid valve; and W: waste.

rate and the time program of these two valves. Subsequently the mixture was transported to a gas–liquid separator by the carrier (V1 On). The HCl consumption and the waste production were reduced to one-half and one-third, respectively, in comparison with the reference flow injection CVAAS procedure.

Another example of an MCFIA system is shown in Figure 4.6. This MCFIA system was designed to implement an automated titration procedure [36]. The system was applied to samples with a wide range of acid concentration such as vinegar and lemon, orange, pineapple, maracock, and acajou juices.

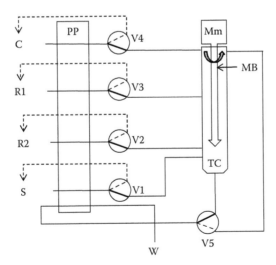

FIGURE 4.6 Schematic diagram of an MCFIA system for photometric titration procedures. C: carrier (H_2O); MB: magnetic stirring bar; Mm: DC mini-motor; PP: peristaltic pump; R1: dye indicator solution; R2: titrant solution; S: sample; TC: titration chamber; V: three-way solenoid valve; and W: waste.

4.4 MULTISYRINGE FLOW INJECTION ANALYSIS

4.4.1 Theoretical Foundations and Instrumentation

MSFIA [8,37] was developed in 1999. This technique arose with the aim of combining the advantages of previous flow techniques while avoiding their disadvantages. It was designed as a novel multichannel technique combining the multichannel operation and high injection throughput of FIA with the robustness and the versatility of SIA, together with the use of three-way solenoid valves (MCFIA).

Figure 4.7 shows a typical multisyringe burette for use in MSFIA. A multisyringe burette consists of a conventional automatic titration burette that can be equipped with up to four syringes, which are used as liquid drivers. Pistons are mounted on a common steel bar driven by a single stepper motor. Thus, all pistons are moved simultaneously and unidirectionally for either liquid delivery (dispense) or aspiration (pick up); this is equivalent to multichannel peristaltic pumps in FIA but avoids the disadvantages of fragile tubing and frequent recalibrations. The ratio of flow rates between channels can be modified by using syringes of appropriate cross-sectional dimensions similarly to tubing diameters in

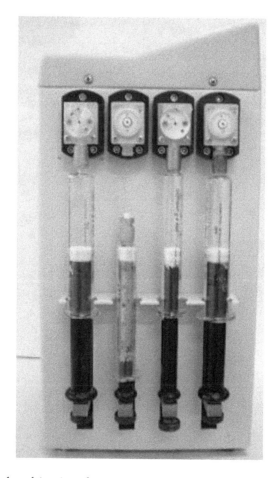

FIGURE 4.7   Typical multisyringe burette.

FIA. Each syringe has a three-way solenoid valve (N-Research, Caldwell, NJ, USA) at the head, which facilitates the application of multicommutation schemes allowing four kinds of liquid displacement: on-dispense, off-dispense, on-pick-up, and off-pick-up. High chemical robustness is provided by the use of resistant polymers, for example, poly(ethylene-co-tetrafluoroethylene) (ETFE) (head valves) and PTFE (piston heads, poppet flaps).

The evolution of multisyringe modules since the start of MSFIA has been noteworthy; currently, multisyringe modules with stepper motors with a resolution of 40,000 steps per full stroke are available. These modules can dispense volumes with higher precision than the first prototypes, whose stepper motors had a resolution of 5000 steps per full stroke. Thus, the multisyringe module allows precise handling of microliter amounts and a wide flow rate range (0.057–30 mL min^{-1}, depending on the syringe volume 1–10 mL). Syringes of 0.5, 1, 2.5, 5, 10, and 25 mL are available, enabling a wide flow rate range and a great range of combinations.

Each multisyringe burette has four backside ports (1–4, Figure 4.8) which enable the power of additional external multicommutation valves, micropumps, or other instruments to be exploited either directly or via a relay allowing remote software control. This increases the possibilities for constructing sophisticated flow networks. For example, the versatility of MSFIA has been demonstrated by coupling it to inductively coupled plasma mass spectrometry (ICP-MS) [38], high-performance liquid chromatography (HPLC) [39], gas chromatography–mass spectrometry (GC/MS) [40,41] and capillary electrophoresis [42,43], among others.

Thus, MSFIA systems combine some of the advantages of other flow techniques, namely: the high throughput of FIA, since the sample and reagents are incorporated in parallel, what leads to high mixing efficiency; the robustness of SIA since liquids only come into contact with the walls of the glass syringes and Teflon tubing, the high flexibility and low sample and reagent consumption of SIA, since flow rates and propelled volumes are precisely known and defined by software-based remote control and liquids are delivered to the system only when required; together with the ability to use MCFIA solenoid valves, which can be actuated without the need to stop the MSFIA piston. Switching between valves is so fast that no overpressure arises in the operation.

FIGURE 4.8 Backside connections of the multisyringe burette.

By using parallel moving syringes as liquid drivers, the shortcomings of peristaltic pumping such as pulsation, required recalibration of flow rates and limitations regarding applicable reagents, are overcome. Thus, MSFIA is an ideal multichannel technique for challenging analytical procedures, which require precise flow rates, and high pressure stability such as those with sorbent columns implementation [44], enabling at the same time, the handling of aggressive and volatile solutions. The only disadvantage of MSFIA in front of other flow techniques is the need of periodical syringe refilling which causes a lower injection frequency than using a FIA approach.

4.4.2 MSFIA Applications

Figure 4.9 shows an MSFIA system for the determination of total antioxidant capacity using the 2,2-diphenyl-1-picrylhydrazyl (DPPH*) assay applied to several food products [45]. The method was based on consumption of DPPH* by antioxidant species present in the sample by monitoring absorbance at 517 nm in stopped-flow approach. First the absorbance of radical solution in the absence of antioxidant compounds was measured. Then 50 μL of sample is aspirated into the holding coil (HC) (S1 and V5 in On position), and then the flow direction is changed and the sample, carrier, and reagent are sent toward the detector (S1, S2, and S3 On position). The flow is stopped for a duration of 180 s while the reaction is monitored. Finally, the flow cell is washed with carrier and reagent solution. The DPPH* assay was performed under strictly controlled reaction conditions with reduced handling of samples, control of temperature, reduced contact of radical and antioxidant species with oxygen and other substances, reduced solvent loss due to evaporation, and excellent repeatability.

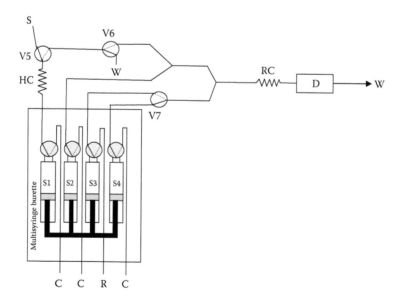

FIGURE 4.9   MSFIA system for the determination of total antioxidant capacity using the DPPH* assay. C: carrier (ethanol solution); D: detector; HC: holding coil; R: 2,2-diphenyl-1-pirylhydrazyl reagent in carrier; RC: reaction coil; S1–S4: syringes; S: sample or antioxidant standard solution; V: three-way solenoid valve; and W: waste.

As an example of the high versatility of the MSFIA technique, Figure 4.10 shows a multisyringe ion chromatography system with chemiluminescence detection for oxalate determination in beer and urine samples [46]. It is based in the use of a short surfactant-coated silica monolithic column functionalized with octadecyl (C_{18}) groups. Column coating, uncoating, ion chromatography and chemiluminescence detection are performed quickly and automatically, achieving a throughput of 48 h^{-1}. Syringe 1 (S1) performs sample loading and the oxalate separation. The headvalve of S1 was replaced by a two-way connector made from polyoxymethylene (Delrin) connected to an external three-way solenoid valve that bears higher pressure than conventional solenoid valves. Syringes 2 and 3 were used for the postcolumn injection of the two chemiluminometric reagents. An eight-port selection valve was used to introduce samples, coating solution, and uncoating solution into the flow network. Thus, this system operates as a SIA system for sample loading and column coating, and as an MCFIA system for the oxalate clean-up and chemiluminescence detection, being a useful tool for rapid oxalate analysis.

Figure 4.11 shows an MSFIA system coupled to cold vapor atomic fluorescence spectrometry (CV-AFS) for mercury determination in rice [47]. As can be seen for sample pick-up an extra three-way solenoid valve (V) was connected to syringe 1 (S1), syringe 2 (S2) was used to dispense the reducing solution and a solenoid micropump was used to remove the liquid-excess from the gas–liquid separator, keeping a constant liquid volume in it.

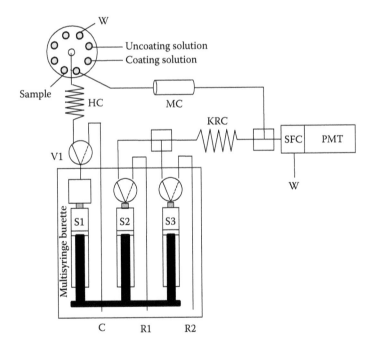

FIGURE 4.10 Schematic depiction of an multisyringe chromatography system applied to oxalate determination in beer and urine with chemiluminescence detection. HC: holding coil; KRC: knotted reaction coil; MC: monolithic column—10 mm × 4.6 mm C_{18} coated with cetyltrimethyl ammonium bromide (CTAB); PMT: photomultiplier; R1: tris(2,2′-bipyridyl)dichlororuthenium(II) solution; R2: cerium(IV) solution; S: syringe; SFC: spiral flow cell; V: three-way solenoid valve; and W: waste.

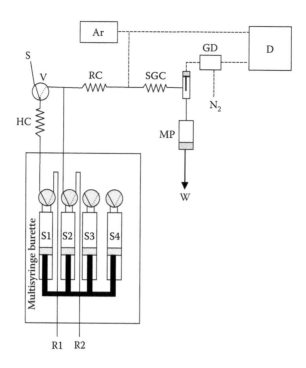

FIGURE 4.11 MSFIA CV-AFS system for the determination of mercury. D: Fluorescence detector; GD: gas dryer; GL: gas–liquid separator; HC: holding coil; MP: micropump; R1: nitric acid; R2: SnCl$_2$ + HCl; RC: reaction coil; S1–S4: syringe; S: sample; SGC: spray generator coil; V: three-way solenoid valve; and W: waste.

4.5 MULTIPUMPING FLOW SYSTEMS

4.5.1 Theoretical Foundations and Instrumentation

MPFS, which were developed in 2002 [9], are based on the use of solenoid piston micropumps (Figure 4.12) where each stroke propels a preset volume of liquid (8, 10, 20, 25, and 50 µL) the flow rate of which is determined by the stroke frequency, allowing continuous flow methodologies. Micropumps are the only active components of the flow manifold in MPFSs, providing operational simplicity and assuring a straight forward run-time control of important analytical variables [48].

A micropump is a solenoid-operated device designed to provide a discrete volume of fluid [49]. Figure 4.13 schematically depicts the parts of a micropump. The body of the micropumps is made of steel at the top (solenoid part) and Teflon at the bottom where the pumping system (diaphragm) is located. The flow path is isolated from the operating mechanism by a flexible diaphragm. Thus, when the solenoid is energized, the diaphragm is retracted, creating a partial vacuum within the pump body, pulling liquid through the inlet check valve (A) and simultaneously closing the outlet check valve (B). When the solenoid is deenergized a spring pushes the diaphragm down, expelling a discrete volume of liquid through check valve B while simultaneously closing check valve A. A complete On–Off cycle is required for each discrete dispense. A pulsed flow is created by repeatedly cycling of the solenoid micropumps. Micropumps when activated use a voltage source of 12 or 24 V.

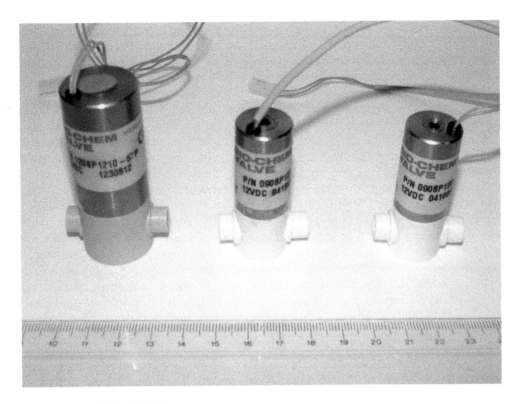

FIGURE 4.12 Solenoid micropumps.

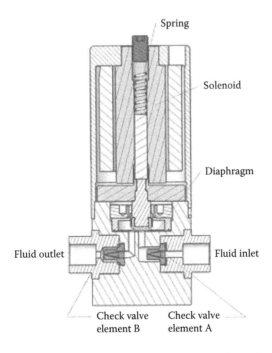

FIGURE 4.13 Schematic depiction of the inner parts of a solenoid micropump.

These micropumps operate as both liquid propeller and valve, which endows MPFS with high flexibility, robustness and low cost. MPFS offer numerous possibilities, such as the application to preconcentration procedures, avoiding time-consuming steps such as sample loading, replacement of the injection coil, or the periodic refill of the liquid driver. Minimal reagent consumption is also achieved, since each micropump is operated individually when inserting the solutions. Another advantage in comparison with other flow techniques is the pulsed flow of the micropumps, which is more effective and faster at homogenizing reaction zones [50], leading to higher analyte peaks. Thus, the main features of MPFSs to be highlighted are simplicity and low cost, favoring economical, portable, and miniaturized flow analyzers [51,52] that facilitate field measurements.

Great versatility and flexibility is provided in terms of flow system networking, especially when combining MPFS with multicommutation solenoid valves (V). Actually, a typical multipumping system is similar to an MCFIA system; the former can also be used to implement MCFIA as it affords control of any combination of valves and solenoid micropumps. Primary differences between the two are that to ensure reproducible flow-rates multipumping systems require control of not only valve switching but also the stroke frequency, and that solenoid micropumps do not require protection. Furthermore, when solenoid micropumps and valves are combined the set-up is simpler. A schematic depiction of a controller system (Sciware Systems S.L., Bunyola, Balearic Islands, Spain) is shown in Figure 4.14. This module can be controlled through the interface RS232C. This system can independently regulate the function of up to eight solenoid micropumps and/or valves. The solenoid devices are set on a mother board connected to a protection interface, which in turn is connected to the relay outputs. The power source, the Rs serial interface, and the I/O cards are integrated in the module.

Nevertheless, some disadvantages of micropumps are the susceptibility to blockage by particles and to back-pressure, requiring recalibration of the dispensed volume. On the

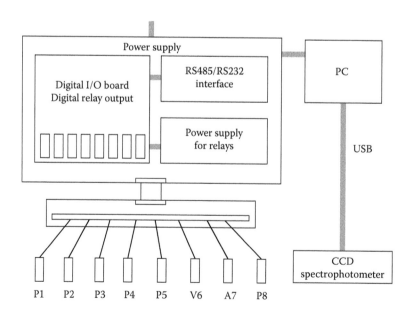

FIGURE 4.14   Schematic depiction of a micropump, solenoid valve and agitator controller system. (Adapted from Sciware Systems S.L., Bunyola, Balearic Island, Spain.)

one hand, dispensing precision of the larger pumps is better, for example, 20 μL stroke volume micropumps are recommended. On the other hand, a higher mixing efficiency is achieved using micropumps with lower stroke volumes, for example, 8 μL [48].

4.5.2 MPFS Applications

An MPFS–MCFIA set-up for the spectrophotometric determination, solid-phase extraction (SPE), and speciation analysis of iron is shown in Figure 4.15 [53]. In this system, the use of a solenoid valve allows the flow deviation toward the chelating disk to carry out SPE procedures when necessary. Thus, the determination of iron is feasible in a wide range of concentrations. However, micropumps cannot overcome the back-pressure from the SPE disk. For example, deviations as high as 2 μL were found in the indicated stroke volume of 8 μL/stroke micropumps requiring periodic recalibration.

An MPFS system for the spectrophotometric determination of phytic acid (PA) in foodstuffs is shown in Figure 4.16 [54]. The procedure involved a batch extraction step that was carried out in 40 parallel ion-exchange minicolumns. Then the extracts were analyzed with the proposed MPFS system. This system allowed different procedures for sample and reagent mixing, that is, merging zone configuration and tandem stream, by simultaneous and sequential operation of MP2 and MP3, respectively. MP1 contained the carrier solution. Low reagent consumption and waste generation were accomplished together with a high sample throughput of 150 samples/h.

Figure 4.17 shows a spectrophotometric MPFS system for determination of total tannins in beverages with minimized reagent consumption [55]. As can be seen, the flow system consisted of four solenoid micropumps (MP) and a three-way solenoid valve (V). MP2–MP4 were simultaneously actuated to insert sample and reagent aliquots into the analytical path. The sampling zone was then transported to the detector by MP1. The solenoid valve was used to carry out the sample replacement, by activating both MP2 and V, directing the liquid to the waste. The stopped-flow strategy could be implemented in the

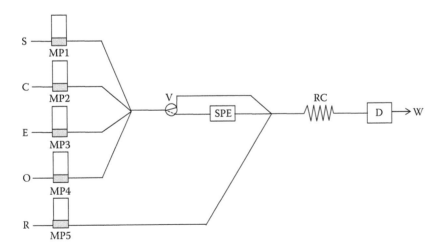

FIGURE 4.15   MPFS for iron determination. C: carrier; D: detector; E: eluent; MP: micropump; O: oxidizing agent; R: chromogenic reagent; RC: reaction coil; S: sample; SPE: column containing the SPE disk; V: three-way solenoid valve; and W: waste.

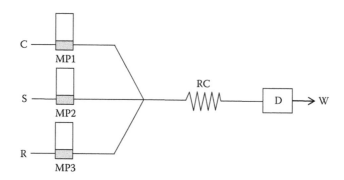

FIGURE 4.16 MPFS for PA spectrophotometric determination in plant extracts. C: carrier (NaCl); D: detector; E: eluent; MP: micropump; R: reagent; RC: reaction coil; S: sample; and W: waste.

proposed system. Moreover, interferences due to reducing species were minimized, sampling throughput was improved (75 samples/h) and reagent consumption and waste generation were reduced up to 83-fold and 60-fold, respectively, in comparison with the batch procedure. Thus, this MPFS system is a more environmentally friendly, low cost, fast, and reliable alternative for total tannin determination in beverages.

An automatic and fast screening miniaturized fluorometric MPFS system for glibenclamide determination in alcoholic beverages and pharmaceutical formulations is shown in Figure 4.18 [56]. Glibenclamide was monitored in acidic medium and in the presence of an anionic surfactant (sodium dodecyl sulfate [SDS]) to enhance the fluorometric measurements by the formation of an organized micellar medium. The analytical procedure started by intercalating a volume of sample (MP2) between two plugs of SDS (MP3). Subsequently, the reaction zone was pushed to the detector by MP1 containing a sulfuric acid solution that acted as carrier. Main advantages of this system are simplicity, versatility, portability, reduced reagent consumption, and waste generation. The sensitivity of the

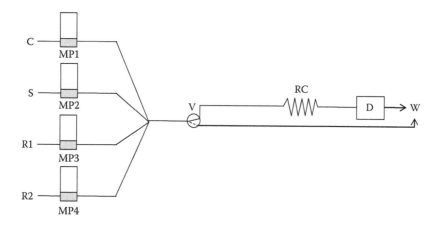

FIGURE 4.17 MPFS for tannin determination in beverages. C: carrier (H_2O); D: spectrophotometric detector; MP: micropump; R1: Folin–Denis reagent; R2: NaOH; RC: reaction coil; S: sample; V: three-way solenoid valve; and W: waste.

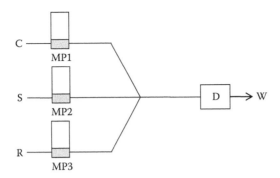

FIGURE 4.18 MPFS for glibenclamide chemical and toxicological control in beverages and biological matrices. C: carrier (H_2SO_4); D: fluorometer detector; MP: micropump; R: reagent (SDS); S: sample; and W: waste.

system could be easily improved by implementing preconcentration procedures due to the modular structure of MPFSs.

4.6 FUTURE TRENDS AND CRITICAL COMPARISON OF MULTICOMMUTATED FLOW TECHNIQUES

Nowadays, the most challenging trend in analytical chemistry is the development of more compact, smarter, faster, simpler, and more environmentally friendly analyzers. The evolution of flow techniques attempts to meet this challenge.

In MPFSs the number of system parts are reduced, which leads to a reduction of requirements for control or the occurrence of malfunctions. Micropumps are the only active components of the flow manifold in MPFSs, providing operational simplicity and assuring straightforward run-time control of important analytical variables. Furthermore, MPFSs provide continuous-flow methodologies, avoiding the repeated syringe refilling required in MSFIA or SIA, and the use of peristaltic tubes to circumvent the chemical resistance limitations of FIA systems. Moreover, the degree of mixing in MPFSs is superior to that reported in laminar flow–based techniques due to its inherently pulsed flow stream [57]. Thus, MPFSs are very attractive tools for developing low-cost miniaturized portable analyzers. However, one of the major shortcomings of MPFSs is the low back-pressure capacity of micropumps. In this sense, MSFIA provides higher back-pressure resistance and more precise and accurate flow rates and dispensed volumes. Therefore, more simple flow systems can be obtained by exploiting MPFS but MSFIA provides higher versatility, allowing the insertion of parts causing back-pressure such as SPE columns and monolithic columns in the manifold.

One of the major trends of automation based on the use of flow techniques is hyphenation. Hyphenation of flow techniques improves their individual advantages. For example, MSFIA–MPFS hyphenation allows sample treatment and separation to be carried out in a shorter time than with conventional MSFIA procedures [58,59]. In general, the combination of multicommutated flow techniques can offer several advantages, such as higher injection throughput and flexibility. MCFIA itself as it was originally presented has not had great success; however, solenoid valves that typify this technique have been widely incorporated in many flow manifolds, extending multicommutation to other flow techniques. Very simple analyzers have been developed by hyphenating MCFIA and MPFS [53,55].

In conclusion, hyphenation of multicommutated flow techniques provides fast, environmentally friendly, portable, low-cost analyzers with more favorable analytical features than the individual techniques alone. Multicommutated flow techniques also permit the development of highly sophisticated systems by facilitating the coupling of common laboratory detectors such as ICP-MS, ICP-OES, GC, and HPLC. Thus, multicommutation plays a major role in the automation that is essential in the present and for the future basis of development of efficient automated flow analyzers.

REFERENCES

1. Skeggs, L. 1957. An automatic method for colorimetric analysis. *Am. J. Clin. Pathol.* 28:311–322.
2. Trojanowicz, M. 2008. *Advances in Flow Analysis.* Germany: Wyley-VHC, Weinheim.
3. Rocha, F. R. P., J. A. Nóbrega, and O. F. Filho. 2001. Flow strategies to greener analytical chemistry. An overview. *Green Chem.* 3:216–220.
4. Reis, B. F., A. Morales-Rubio, and M. de la Guardia. 1999. Environmentally friendly analytical chemistry through automation: Comparative study of strategies for carbaryl determination with *p*-aminophenol. *Anal. Chim. Acta* 392:265–272.
5. Ruzicka, J. and E. H. Hansen. 1975. Flow injection analyses. I. New concept of fast continuous flow analysis. *Anal. Chim. Acta* 78:145–157.
6. Ruzicka, J. and G. D. Marshall. 1990. Sequential injection: A new concept for chemical sensors, process analysis and laboratory assays. *Anal. Chim. Acta* 237:329–343.
7. Reis, B. F., M. F. Giné, E. A. G. Zagatto, J. L. F. C. Lima, and R. A. Lapa. 1994. Multicommutation in flow analysis. Part 1. Binary sampling: Concepts, instrumentation and spectrophotometric determination of iron in plant digests. *Anal. Chim. Acta* 293:129–138.
8. Cerdà, V., J. M. Estela, R. Forteza et al. 1999. Flow techniques in water analysis. *Talanta* 50:695–705.
9. Lapa, R. A. S., J. L. F. C. Lima, B. F. Reis, J. L. M. Santos, and E. A. G. Zagatto. 2002. Multi-pumping in flow analysis: Concepts, instrumentation, potentialities. *Anal. Chim. Acta* 466:125–132.
10. Rocha, F. R. P., B. F. Reis, E. A. G. Zagatto, J. L. F. C. Lima, R. A. S. Lapa, and J. L. M. Santos. 2002. Multicommutation in flow analysis: Concepts, applications and trends. *Anal. Chim. Acta* 468:119–131.
11. Malcome-Lawes, D. J. and C. Pasquini. 1988. A novel approach to non-segmented flow analysis: Part 3. Nitrate, nitrite and ammonium in waters. *J. Autom. Chem.* 10:192–197.
12. Pasquini, C. and L. C. Faria. 1991. Operator-free flow injection analyzer. *J. Autom. Chem.* 13:143–146.
13. Reis, B. F., M. F. Giné, F. J. Krug, and H. Bergamin. 1992. Multipurpose flow injection system. Part 1. Programmable dilutions and standard additions for plant digests analysis by inductively coupled plasma atomic emission spectrometry. *J. Anal. At. Spectrom.* 7:865–868.
14. Israel, Y., A. Lásztity, and R. M. Barnes. 1989. On-line dilution, steady-state concentrations for inductively coupled plasma spectrometry achieved by tandem injection and merging-stream flow injection. *Analyst* 114:1259–1265.

15. Almeida, C. M. N. V., R. A. S. Lapa, J. L. F. C. Lima, E. A. G. Zagatto, and M. C. U. Araújo. 2000. An automatic titrator based on a multicommutated unsegmented flow system its application to acid–base titrations. *Anal. Chim. Acta* 407:213–223.

16. Zagatto, E. A. G., B. F. Reis, M. Martinelli, F. J. Krug, H. Bergamin, and M. F. Giné. 1987. Confluent streams in flow injection analysis. *Anal. Chim. Acta* 198:153–163.

17. Stewart, K. K., J. F. Brown, and B. M. Golden. 1980. A microprocessor control system for automated multiple flow injection analysis. *Anal. Chim. Acta* 114:119–127.

18. Kronka, E. A. M., P. R. Borges, R. Latanze, A. P. S., Paim, and B. F. Reis. 2001. Multicommutated flow system for glycerol determination in alcoholic fermentation juice using enzymatic reaction and spectrophotometry. *J. Flow Injection Anal.* 18:132–138.

19. Oliveira, C. C., R. P. Sartini, B. F. Reis, E. A. G. Zagatto, and J. L. F. C. Lima. 1997. Flow analysis with accuracy assessment. *Anal. Chim. Acta* 350:31–36.

20. Ruzicka, J. and E. H. Hansen. 1979. Stopped flow and merging zones—A new approach to enzymic assay by flow injection analysis. *Anal. Chim. Acta* 106:207–224.

21. Miranda, C. E. S., B. F. Reis, N. Baccan, A. P. Packer, and M. F. Giné. 2002. Automated flow analysis system based on multicommutation for Cd, Ni and Pb on-line pre-concentration in a cationic exchange resin with determination by inductively coupled plasma atomic emission spectrometry. *Anal. Chim. Acta* 453:301–310.

22. Zagatto, E. A. G., B. F. Reis, C. C. Oliveira, R. P. Sartini, and M. A. Z. Arruda. 1999. Evolution of the commutation concept associated with the development of flow analysis. *Anal. Chim. Acta* 400:249–256.

23. Semenova, N. V., L. O. Leal, R. Forteza, and V. Cerdà. 2002. Multisyringe flow-injection system for total inorganic arsenic determination by hydride generation-atomic fluorescence spectrometry. *Anal. Chim. Acta* 455:277–285.

24. Kronka, E. A. M., A. P. S. Paim, B. F. Reis, J. L. F. C. Lima, and R. A. Lapa. 1999. Determination of glucose in soft drink and sugar-cane juice employing a multicommutation approach in flow system and enzymic reaction. *Fresenius J. Anal. Chem.* 364:358–361.

25. Avivar, J., L. Ferrer, M. Casas, and V. Cerdà. 2011. Smart thorium and uranium determination exploiting renewable solid-phase extraction applied to environmental samples in a wide concentration range. *Anal. Bioanal. Chem.* 400:3585–3594.

26. Rius, A., M. P. Callao, and F. X. Rius. 1995. Self-configuration of sequential injection analytical systems. *Anal. Chim. Acta* 316:27–37.

27. Chow, C. W. K., D. E. Davey, and D. E. Mulcahy. 1997. An intelligent sensor system for the determination of ammonia using flow injection analysis. *Lab. Autom. Inf. Manage.* 33:17–27.

28. Grassi, V., A. C. B. Dias, and E. A. G. Zagatto. 2004. Flow systems exploiting in-line prior assays. *Talanta* 64:1114–1118.

29. Ferrer, L., J. M. Estela, and V. Cerdà. 2006. A smart multisyringe flow injection system for analysis of sample batches with high variability in sulfide concentration. *Anal. Chim. Acta* 573–574:391–398.

30. Bonastre, A., R. Ors, and M. Peris. 2000. Monitoring of a wort fermentation process by means of a distributed expert system. *Chemom. Intell. Lab. Syst.* 50:235–242.

31. Peris, M. 2002. Present and future of expert systems in food analysis. *Anal. Chim. Acta* 454:1–11.

32. Bonastre, A., R. Ors, and M. Peris. 2004. Advanced automation of a flow injection analysis system for quality control of olive oil through the use of a distributed expert system. *Anal. Chim. Acta* 506:189–195.

33. Krug, F. J., H. Bergamin, and E. A. G. Zagatto. 1986. Commutation in flow injection analysis. *Anal. Chim. Acta* 179:103–118.
34. Cerdà, V. and C. Pons. 2006. Multicommutated flow techniques for developing analytical methods. *Trends Anal. Chem.* 25:236–242.
35. Silva, M. F., I. V. Tóth, and A. O. S. S. Rangel. 2006. Determination of mercury in fish by cold vapor atomic absorption spectrophotometry using a multicommutated flow injection analysis system. *Anal. Sci.* 22:861–864.
36. da Silva, M. B., C. C. Crispino, and B. F. Reis. 2010. Automatic photometric titration procedure based on multicommutation and flow-batch approaches employing a photometer based on twin LEDs. *J. Braz. Chem. Soc.* 21:1854–1860.
37. Horstkotte, B., O. Elsholz, and V. Cerdà. 2005. Review on automation using multisyringe flow injection analysis. *J. Flow Injection Anal.* 22:99–109.
38. Avivar, J., L. Ferrer, M. Casas, and V. Cerdà. 2012. Fully automated lab-on-valve-multisyringe flow injection analysis-ICP-MS system: An effective tool for fast, sensitive and selective determination of thorium and uranium at environmental levels exploiting solid phase extraction. *J. Anal. At. Spectrom.* 27:327–334.
39. Quintana, J. B., M. Miró, J. M. Estela, and V. Cerdà. 2006. Automated on-line renewable solid-phase extraction liquid chromatography exploiting multisyringe flow injection-bead injection lab-on-valve analysis. *Anal. Chem.* 78:2832–2840.
40. Brunetto, M. R., Y. Delgado, S. Clavijo et al. 2010. In-syringe-assisted dispersive liquid–liquid microextraction coupled to gas chromatography with mass spectrometry for the determination of six phthalates in water samples. *J. Sep. Sci.* 33:1779–1786.
41. Clavijo, S., M. R. Brunetto, and V. Cerdà. 2014. In-syringe-assisted dispersive liquid–liquid microextraction coupled to gas chromatography with mass spectrometry for the determination of six phthalates in water samples. *J. Sep. Sci.* 37:974–981.
42. Horstkotte, B., O. Elsholz, and V. Cerdà. 2007. Development of a capillary electrophoresis system coupled to sequential injection analysis and evaluation by the analysis of nitrophenols. *Int. J. Environ. Anal. Chem.* 87:797–811.
43. Horstkotte, B., O. Elsholz, and V. Cerdà. 2008. Multisyringe flow injection analysis coupled to capillary electrophoresis (MSFIA–CE) as a novel analytical tool applied to the pre-concentration, separation and determination of nitrophenols. *Talanta* 76:72–79.
44. Almeida, M. I. G. S., J. M. Estela, and V. Cerdà. 2011. Multisyringe flow Injection potentialities for hyphenation with different types of separation techniques. *Anal. Lett.* 44:360–373.
45. Magalhaes, L. M., M. A. Segundo, S. Reis, and J. L. F. C. Lima. 2006. Automatic method for determination of total antioxidant capacity using 2,2-diphenyl-1-picrylhydrazyl assay. *Anal. Chim. Acta* 558:310–318.
46. Maya, F., J. M. Estela, and V. Cerdà. 2011. Multisyringe ion chromatography with chemiluminescence detection for the determination of oxalate in beer and urine. *Microchim. Acta* 173:33–41.
47. da Silva, D. G., L. A. Portugal, A. M. Serra, S. L. C. Ferreira, and V. Cerdà. 2013. Determination of mercury in rice by MSFIA and cold vapour fluorescence spectrometry. *Food Chem.* 137:159–163.
48. Santos, J. L. M., M. F. T. Ribeiro, A. C. B. Dias, J. L. F. C. Lima, and E. E. A. Zagatto. 2007. Multi-pumping flow systems: The potential of simplicity. *Anal. Chim. Acta* 600:21–28.
49. http://www.biochemvalve.com/micropumps.pdf.

50. Pons, C., R. Forteza, A. O. S. S. Rangel, and V. Cerdà. 2006. The application of multicommutated flow techniques to the determination of iron. *Trends Anal. Chem.* 25:583–588.
51. Horstkotte, B., C. M. Duarte, and V. Cerdà. 2011. A miniature and field-applicable multipumping flow analyzer for ammonium monitoring in seawater with fluorescence detection. *Talanta* 85:380–385.
52. Horstkotte, B., C. M. Duarte, and V. Cerdà. 2012. Multipumping flow systems devoid of computer control for process and environmental monitoring. *Int. J. Environ. Anal. Chem.* 92:344–354.
53. Pons, C., R. Forteza, and V. Cerdà. 2005. Multi-pumping flow system for the determination, solid-phase extraction and speciation analysis of iron. *Anal. Chim. Acta* 550:33–39.
54. Carneiro, J. M. T., E. A. G. Zagatto, J. L. M. Santos, and J. L. F. C. Lima. 2002. Spectrophotometric determination of phytic acid in plant extracts using a multipumping flow system. *Anal. Chim. Acta* 474:161–166.
55. Infante, C. M. C., V. R. B. Soares, M. Korn, and F. R. P. Rocha. 2008. An improved flow-based procedure for microdetermination of total tannins in beverages with minimized reagent consumption. *Microchim. Acta* 161:279–283.
56. Ribeiro, D. S. M., J. A. V. Prior, C. J. M. Taveira, J. M. A. F. S. Mendes, and J. L. M. Santos. 2011. Automatic miniaturized fluorometric flow system for chemical and toxicological control of glibenclamide. *Talanta* 84:1329–1335.
57. Lima, J. L. C., J. L. M. Santos, A. C. B. Dias, M. F. T. Ribeiro, and E. A. G. Zagatto. 2004. Multi-pumping flow systems: An automation tool. *Talanta* 64:1091–1098.
58. Avivar, J., L. Ferrer, M. Casas, and V. Cerdà. 2010. Automated determination of uranium(VI) at ultra trace levels exploiting flow techniques and spectrophotometric detection using a liquid waveguide capillary cell. *Anal. Bioanal. Chem.* 397:871–878.
59. Fajardo, Y., L. Ferrer, E. Gómez, F. Garcías, M. Casas, and V. Cerdà. 2008. Development of an automatic method for americium and plutonium separation and preconcentration using an multisyringe flow injection analysis-multipumping flow system. *Anal. Chem.* 80:195–202.

CHAPTER 5

Combination of Flow Injection Analysis and Other Analytical Systems
Multianalyte Approaches

Susana Campuzano, Felipe Conzuelo,
A. Julio Reviejo, and José M. Pingarrón

CONTENTS

Multicomponent analysis by flow injection and related flow techniques is mainly focused on the simultaneous determination of a few target compounds, generally two or three (Saurina, 2008, 2010). For more analytes, the effort required to develop an adequate method may not be matched by the features normally offered by flow systems; in such circumstances, separation techniques seem to be a more reasonable choice. Nevertheless, the interest of flow methods in multicomponent analysis cannot be underestimated. In general, FI methods do not physically separate components that flow together through the system and reach the detector(s) at the same time. Hence, in the absence of separation, alternative mechanisms are needed to ensure detection of each analyte under selective conditions. For this purpose, various physicochemical approaches, including the use of specific reagents, multiway detectors and multichannel manifolds, can be exploited (Saurina, 2008). The design of the flow manifold, including configuration of channels as well as injection, reaction, and detection elements, and optimization of experimental conditions, such as flow rates, reactor dimensions, injection volume(s), and chemical (reaction) conditions (Bosque-Sendrá et al., 2003), are fundamental to reaching the desired selectivity. Beyond these physicochemically based mechanisms, application of chemometric methods can be successful when selectivity has not been achieved experimentally (Saurina, 2010).

Several approaches for multiplexed detection of target analytes have been described by coupling of flow injection analysis (FIA) with different separation, sample pretreatment, and detection techniques (Hartwell and Grudpan, 2012). As will be illustrated by the examples selected, among different target analytes the flow-based analytical methods have been applied mainly to the detection of food additives, pesticides, clinical biomarkers, and drugs.

Various sample pretreatment processes may be coupled with the FIA system. Examples are solvent–solvent extraction (Fujiwara et al., 1994; Kyaw et al., 1998), sample clean-up such as filter and dialysis (Maeder et al., 1990; Ogbomo et al., 1991; Miró and Frenzel, 2004; Kritsunankul et al., 2009; Cabal et al., 2010; Kritsunankul and Jakmunee, 2011), digestion (Calleri et al., 2004), dilution (Economou et al., 2006), and preconcentration (Di and Davey, 1994; Purohit and Devi, 1997; Beauchemin and Specht, 1998). An interesting approach is the so-called bead injection technique, in which active microbeads are used as a solid surface to accommodate chemical reactions or to selectively trap/separate and accumulate the analyte of interest from sample matrices (Hartwell et al., 2004; Miró et al., 2008; Oliveira et al., 2010). Beads with suitable sizes, that is, small enough to be moved with the flow of solution but big enough not to clog the valve and tubing, have been used as solid phase separation media. These beads can be discarded after each analysis to prevent memory effect.

According to the detection system, special flow-through cells are designed for continuous flow of solution into and out of the detector. Many of these cells are commercially available for coupling to various detectors (Smith et al., 1977; Di Benedetto and Dimitrakopoulos, 1997; Abate et al., 2002; Wang and Huang, 2011; Hellma, web page; BASi LCEC Flowcells, web page), enabling flow injection analysis-based systems with different detection techniques (e.g., electrochemical, mass spectrometry [MS], chemiluminescence) (Caccuri et al., 1999; Zhang and Cass, 2006; Ogończyk and Koncki, 2007; Vahl et al., 2008; Hartwell and Grudpan, 2012).

FIA systems present several advantages, especially when combined with biosensors, such as greater accuracy and precision, the use of small volumes of sample and the ability to manipulate samples online and to perform determinations based on transient species (Felix and Angnes, 2010; Worsfold et al., 2013). The kinetic nature of FIA systems allows an easy automation and a high sample frequency. It also offers better option for standardization and optimization (Bhadekar et al., 2011). The design and development of an automated flow assembly for a biosensor was described by Saini and Suri (2010). This automated flow assembly offers the advantage of rapid sample throughput, small sample volumes, simplicity, and minimal handling. Because of inherent sensitivity of certain detectors, FIA procedures hold great potential biological and biotechnological applications (Saini and Suri, 2010).

This chapter summarizes work on multianalyte detection approaches using various formats of flow-based analysis methods, classified according to the type of analytical technique to which the FIA system has been coupled. An overview of relevant efforts in this area is discussed, and some examples of relevant approaches are highlighted, along with future prospects and challenges.

5.1 COMBINATION OF FIA WITH DIFFERENT DETECTION TECHNIQUES

5.1.1 Optical Detection Systems

The majority of flow methods coupled to optical detection are based on spectrophotometry, fluorescence, and chemiluminescence (CL) (Llorent-Martinez et al., 2011).

A FIA-based method with spectrophotometric detection was developed for the simultaneous determination of nitrites and nitrates (Ahmed et al., 1996). By the use of a selector valve to split the sample into two streams, nitrite was detected after the formation of a stable purple azo dye, while nitrate was reduced online to nitrite and its concentration was determined from the difference in absorbance values. The method was applied to the determination in beer, meat, flour, and cheese, with a sample throughput of up to 30 h^{-1}.

A similar approach was used by Ferreira et al. (1996) for the simultaneous assay of nitrite and nitrate in meat products. The system was based on splitting the flow after injection and the subsequent confluence of it before reaching the detector, allowing the reduction of nitrate to nitrite online in part of the sample plug. Each channel had a different residence time, therefore separate peaks where obtained for nitrite and nitrite + nitrate. The same system was later used by Pinho et al. (1998) for the evaluation of nitrite and nitrate contents in 15 different brands of pâté.

Similar procedures for the simultaneous determination of nitrate and nitrite based on the use of sequential injection analysis (SIA) have also been reported. The system developed by Kazemzadeh and Ensafi (2001) was based on the reaction of nitrite with safranine-O and allowed a sampling rate of 20 h^{-1}, with limits of detection (LODs) of 0.5 and 3 ng mL^{-1} for nitrite and nitrate, respectively. Oliveira et al. (2004) reported a system for the determination in cured meat samples based on the Shinn reaction, with detection limits of 9 μg L^{-1} of N for both nitrite and nitrate, and a sampling rate of 9 h^{-1}.

An optical fiber biosensor for the determination of the pesticides propoxur and carbaryl was described based on the inhibition of acetylcholinesterase (AChE) by these analytes and using chlorophenol red as pH indicator (Xavier et al., 2000). The AChE was covalently immobilized onto controlled-pore glass (CPG) beads and packed in a thermostated bioreactor connected to a flow-through cell that contains CPG-immobilized chlorophenol red, placed at the common end of a bifurcated fiberoptic bundle. In the presence of a constant acetylcholine concentration, the color of the pH-sensitive layer changes and the measured reflectance signal can be related to the carbamate concentration in the sample solution. The linear dynamic range obtained was from 0.8 to 3.0 mg L^{-1} and from 0.03 to 0.50 mg L^{-1}, for the determination of carbaryl and propoxur, respectively. The biosensor allowed LODs of 0.4 and 25 ng for propoxur and carbaryl, respectively, and has been applied to the determination of propoxur in spiked vegetables (onion and lettuce) using ultrasound extraction, achieving good recoveries rates.

A channel-resolved multianalyte immunosensing system for flow-through chemiluminescence (CL) detection of α-fetoprotein (AFP) and carcinoembryonic antigen (CEA) was developed (Fu et al., 2008). In this method, two polyethersulfone membranes modified with the corresponding capture antibodies were set in two channels of a flow cell, then two mixtures of the sample and corresponding alkaline phosphatase labeled antibodies were introduced into the channels for online incubation (Figure 5.1). Upon injection of the substrate, the catalyzed CL signals from the two channels were sequentially collected with the aid of an optical shutter for CL detection of the two analytes. Under optimal conditions AFP and CEA could be assayed in the ranges of 5.0–150 and 0.50–80 ng mL^{-1} with LODs of 1.5 and 0.25 ng mL^{-1}, respectively. The assay results of clinical serum samples with the proposed system were in acceptable agreement with those of a reference method in single-analyte test mode. This flow-through two-channel immunosensing system based on the channel-resolved technique implemented provided an automated, reusable, simple, sensitive, easily automated, and low-cost approach for multianalyte immunoassay without requiring an expensive array detector.

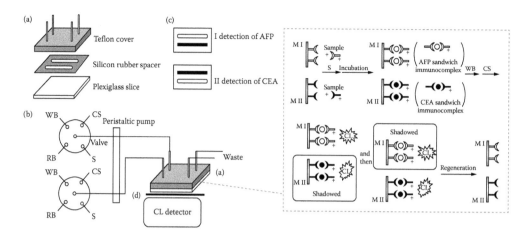

FIGURE 5.1 Scheme of the flow-through channel-resolved immunosensing system for the determination of AFP and CEA: (a) flow cell; (b) flow-through CL immunosensing system and process of immunoassay; (c) transect of flow cell for immunoassay; and (d) optical shutter. WB: wash buffer; RB: regeneration buffer; CS: CL substrate; S: mixture of sample and tracer antibody; M I: anti-AFP immobilized membrane; M II: anti-CEA immobilized membrane. (Adapted from Fu, Z. et al. 2008. *Biosens. Bioelectron.* 23:1063–1069. With permission.)

A novel automated CL read-out system for analytical flow-through microarrays based on multiplexed immunoassays has been developed (Kloth et al., 2009). This new fully automated microarray chip reader combines a fluidic system with a mechanical reagent supply for at least 47 successive measurements, a disposable compact microarray chip, a temperature-controlled microfluidic flow cell unit, cooled antibody syringes, and a sensitive CCD (charge-coupled device) camera for the CL chip imaging (Figure 5.2). The microarray chip, which consists of two channels for parallel measurement and regeneration (allowing up to 47 measurement cycles per channel) and contains all reagents

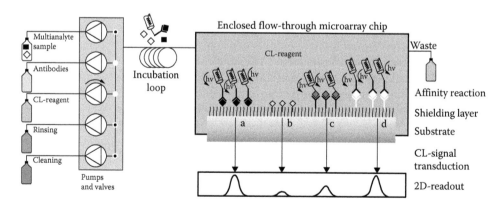

FIGURE 5.2 Schematic setup of an analytical CL flow-through microarray platform for quantification of analytes with indirect (a and b), direct (c), and sandwich (d) assay formats. (Reprinted from Kloth, K., R. Niessner, and M. Seidel, 2009. *Biosens. Bioelectron.* 24:2106–2112. With permission.)

needed for a working day, was designed for the analysis of up to 13 different antibiotics in milk applying an indirect competitive microarray immunoassay. As an initial example, penicillin G was quantified in milk on this microarray chip with an LOD of 1.1 μg L^{-1} in an assay time of 6 min.

The photolytic degradation of organophosphorus pesticides in the presence of light has been utilized for screening food samples for the presence of these analytes. A sensitive CL immunoassay technique based on a FIA format was developed for atrazine detection by immobilizing the antibody on a protein-A–Sepharose matrix packed in a glass capillary column, which functioned as an immunoreactor. The CL signals generated during the biochemical reactions, after the competitive binding of antigen and antibody occurred, were correlated with atrazine concentrations in the analyzed samples and allowed the detection of atrazine down to 0.01 ng mL^{-1} (Chouhan et al., 2010). Waseem et al. (2009) developed a simple and rapid FI method for the determination of dithiocarbamate fungicides (maneb, nabam, and thiram) also based on CL detection but involving the photodegradation of dithiocarbamate fungicides as a result of UV irradiation in an alkaline medium and reaction of the photoproducts with luminol in the absence of an oxidant. Linear calibration graphs were obtained between 0.01 and 4.0 mg L^{-1} for maneb and nabam and between 0.05 and 1.0 mg L^{-1} for thiram. The method exhibited LODs of 10, 8.0, and 5.0 ng mL^{-1} for maneb, nabam, and thiram, respectively, with a sample throughput of 100 h^{-1}.

FIA combined with CL has been also used for detection of carbofuran and promecarb (Figure 5.3) (Pérez-Ruiz et al., 2002), atrazine, and other triazines (Beale et al., 2009) by making use of the property of the pesticides to be converted into methylamine upon exposure to UV radiation. The methylamine generated is made to react with tris-ruthenium, allowing the CL detection of triazines.

Apart from spectrophotometric detection, Franko et al. (2009) demonstrated the suitability of FIA systems, based on AChE inhibition and thermal lens spectrometric (TLS) detection for the rapid and sensitive screening of pesticides. Owing to the high

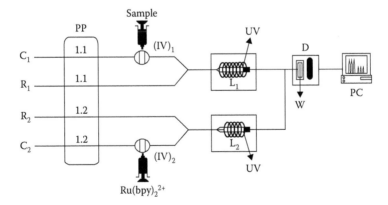

FIGURE 5.3   FI manifold for the determination of carbofuran and promecarb. PP: peristaltic pump (with flow rates given in mL min^{-1}); C_1, C_2: carrier (water); R_1: 0.10 mol L^{-1} phosphate buffer of pH 6.5; R_2: 1.5×10^{-3} mol L^{-1} potassium peroxydisulfate and 0.05 mol L^{-1} phosphate buffer of pH 5.7; $(IV)_1$, $(IV)_2$: injection valves; L_1, L_2: photoreactors; D: luminometer; W: waste. (Reprinted from Pérez-Ruiz T. et al. 2002. *Analyst* 127:1526–1530. With permission.)

sensitivity of TLS, several steps of sample preparation can be avoided (e.g., preconcentration, purification, isolation) and the incubation times can be reduced. High sample throughputs (10 h^{-1}) are also achievable for toxic oxo organophosphorus pesticides using spectrophotometric detection, providing LODs at the level of 10 ppb malaoxon equivalents, which is still about 50 times below maximal residue levels. Testing of the method for practical application on a set of 60 samples gave no false negative results and a level of 1.7% of false positives.

Wutz et al. (2011) developed for the first time a multianalyte immunoassay based on an automated flow-through CL microarray technique for identification and quantification of antibiotic derivatives in honey samples using regenerable antigen microarrays, an indirect competitive immunoassay format using horseradish peroxidase (HRP)-labeled antibodies and CL read-out with a CCD camera. The method allows the analysis of four analytes (enrofloxacin, sulfadiazine, sulfamethazine, and streptomycin) simultaneously in 8 min with adequate recoveries and without purification or extraction. Due to the regenerability of the microarray each chip could be individually calibrated before the analysis and allowed more than 40 assays, which reduces the costs per analysis and permits an automated work flow in routine laboratories.

Zhang et al. (2013) developed a novel electrochemiluminescence (ECL) immunosensor array using flow injection for the sequential detection of multiple tumor markers by site-selectively immobilizing multiple antigens on different electrodes. A disposable indium tin oxide (ITO) glass array was employed as detection platform and low-toxicity carbon dots coated with silica (SiO_2 at C-dots) were used for the labeling of antibodies for signal amplification. Such immunosensor array had attractive advantages such as excellent precision, high sensitivity (LODs between 0.003 and 0.006 ng mL^{-1}), simple operation, negligible cross-talk, fine reproducibility, good storage stability, and successful application to determine CEA, prostate-specific antigen (PSA), and AFP in serum samples. The results for real sample analysis demonstrated that the newly constructed immunosensor array provided a promising, addressable, and simple strategy for the multidetection of tumor markers with high throughput, cost-effectiveness, and sufficiently low detection limits for clinical applications.

5.1.2 Electrochemical Detection

Taking advantage of the relative simplicity, ease of miniaturization, possibility of *in situ* measurements, low cost, and high sensitivity of electroanalytical techniques, various electrochemical detection approaches have been coupled to FIA for the multiplexed determination of target analytes, with either direct or indirect detection. Commonly used electroanalytical techniques include potentiometry, conductometry, voltammetry, and amperometry, among others. Although amperometry has been the preferred option in most applications, potentiometry (Lee et al., 2001, 2002; Suwansa-Ard et al., 2005) and conductometry (Suwansa-Ard et al., 2005) have also been employed (Llorent-Martinez et al., 2011).

In comparison with other instrumental methods of chemical analysis, electroanalytical instrumentation is relatively easy to miniaturize. The main trends in modern electroanalysis include development of chemical and biochemical sensors based on progress in chemical and biochemical methods of molecular recognition, and development of measuring devices of a large integration scale, including sensor arrays. This is accompanied by the use of new materials, including nanomaterials and nanostructures, along with their coupling with flow systems, which is the aspect considered in this review.

Most electrochemical detectors, such as amperometric and potentiometric detectors, are surface detectors. They respond to substances that are either oxidizable or reducible and the electrical output results from an electron flow caused by the chemical reaction that takes place at the surface of the electrodes (Rao et al., 2002; Mehrvar and Abdi, 2004; Trojanowicz, 2009). Successful operation of a surface detector requires a reproducible radial concentration distribution. There are several types of flow-through detection cells, each type being characterized by parameters such as the length, diameter, and shape of its detection channel, which determine the laminar character of the liquid flow under the given experimental conditions and the predominant mode of the mass transport within the cell.

Electrochemical detectors have special features for flow analysis, such as the following: (1) The flowing liquid continuously cleans the surface of the indicator or working electrode. Consequently, electrochemical and/or mechanical regeneration of the working electrode surface with intensive washing, solution or solvent switching, potential cycling, and so on is generally less critical than in batch electroanalytical techniques. (2) The continuously streaming carrier solution removes reaction products and impurities leached from the electrode. (3) The convective transport of an analyte and/or a reactant reduces response time and improves the LOD compared with batch-type measurements with pure diffusion transport. (4) Differences in the response rate for the primary and interfering ions, in case of potentiometric detectors, can improve selectivity (Llorent-Martinez et al., 2011).

The simultaneous determination of sulfite and phosphate in wine was proposed by Yao et al. (1994), where a FIA system was coupled with amperometric detection. In this approach two different reactors with immobilized sulfite oxidase and co-immobilized purine nucleoside phosphorylase-xanthine oxidase were incorporated in a parallel configuration. The enzymatically generated hydrogen peroxide was selectively detected on a poly(1,2-diaminobenzene)-coated platinum electrode. Because of the different residence times at each channel, two different peaks could be obtained, the first corresponding to sulfite and the second to phosphate. Using this system, the analytes could be simultaneously analyzed with a linear range of 1×10^{-5}–2×10^{-3} M (sulfite) and 2×10^{-5}–5×10^{-3} M (phosphate) and a sample throughput of 30 h^{-1}.

Cardwell and Christophersen reported a dual channel FI system with amperometric detection for the determination of ascorbic acid and sulfur dioxide in wines and fruit juices (Cardwell and Christophersen, 2000). Here, the ascorbic acid was detected at a glassy carbon electrode polarized at 0.42 V (vs. Ag/AgCl), whereas the sulfur dioxide was detected at a Pt electrode polarized at 0.90 V (vs. Ag/AgCl) after separation of the analytes by a gas diffusion unit. The determination of ascorbic acid showed a linear range between 3 and 50 mg L^{-1} with an LOD of 1.5 mg L^{-1}; for sulfur dioxide the linear range was between 0.25 and 15 mg L^{-1} and an LOD of 0.05 mg L^{-1} was obtained. The sample frequency achieved with the system was 30 h^{-1}. The proposed method showed a good agreement with a reference method in the results obtained for white wines and juice samples, while for red wines and sweet wines an extraction procedure of the analytes by solid-phase extraction was required.

The FI determination of acesulfame-K, cyclamate, and saccharin in wines, yogurts, diet soft drinks, and sweetener tablets was reported by Nikolelis and Pantoulias (2001). The detection was performed with filter-supported bilayer membranes from lyophilized egg phosphatidylcholine, using two Ag/AgCl reference electrodes biased by an external power supply at 25 mV. Transient electrochemical signals were recorded with a different time of appearance for each artificial sweetener after the injection of the sample into the flow stream, thus allowing their selective detection in mixtures.

A multianalyte flow electrochemical cell was developed and integrated into a mono-channel FIA system for simultaneous bioelectrocatalytic detection of five sugars (glucose, sucrose, fructose, galactose, and lactose) in food samples by amperometry, using dehydrogenase-based carbon paste electrodes (Maestre et al., 2005). The applicability of the system was demonstrated by comparing the glucose, sucrose, fructose, galactose, and lactose contents determined in fruit juices and milk derivatives with those provided using commercial enzymatic kits.

Biosensor-based FI systems using a semidisposable enzyme reactor have been developed to determine carbamate pesticides in water samples (Suwansa-Ard et al., 2005). AChE was immobilized on silica gel by covalent binding. pH and conductivity electrodes were used to detect the ionic change of the sample solution due to hydrolysis of acetylcholine. Carbamate pesticides inhibited AChE and the decrease in the enzyme activity was used to determine these pesticides. Detection limits for the potentiometric and conductometric systems, both at 10% inhibition, corresponding to 0.02 and 0.3 ppm were obtained for carbofuran and carbaryl, respectively. Both systems also provided the same linear ranges, 0.02–8.0 ppm for carbofuran, and 0.3–10 ppm for carbaryl, and were used to analyze carbaryl in water samples from six wells in a vegetable growing without requiring any sample preconcentration.

Wei et al. (2009) described for the first time the application of a biosensor based on photoelectro-synergistic catalysis—a combination approach of photocatalysis and electrocatalysis—together with FIA for the highly sensitive detection of organophosphorus pesticides. This amperometric biosensor was based on the immobilization of AChE by adsorption into nanostructured $PbO_2/TiO_2/Ti$. A wide linear inhibition response for trichlorfon was observed in the range of 0.01–20 µM with an LOD of 0.1 nM. The results obtained for real samples were in excellent correspondence with those determined by GC.

Medeiros and coworkers reported a simple FIA system with multiple pulsed amperometric (MPA) detection for the simultaneous determination of the phenolic antioxidants butylated hydroxyanisole (BHA) and butylated hydroxytoluene (BHT) (Medeiros et al., 2010). A dual-potential waveform was applied to a boron-doped diamond (BDD) electrode for the determination of both analytes, first $E_{det1} = 850$ mV (vs. Ag/AgCl; 200 ms) was applied for the oxidation of BHA, then $E_{det2} = 1150$ mV (vs. Ag/AgCl; 200 ms) was applied for the oxidation of BHA and BHT. The determination of both species was done by concentration subtraction. The methodology developed exhibited a linear range of 0.050–3.0 µM for BHA and 0.70–70 µM for BHT, with LODs of 0.030 µM (BHA) and 0.40 µM (BHT). This method, which is simple, quick and presents good precision and accuracy, was successfully applied to the determination in commercial mayonnaise samples, with results in agreement with a reference HPLC (high-performance liquid chromatography) method. The same authors also reported a similar FI method for the simultaneous determination of two pairs of synthetic colorants in food samples, including juices, gelatins, and sports drinks. The pairs tartrazine–sunset yellow (TT–SY) and brilliant blue–sunset yellow (BB–SY) were determined by a dual potential waveform with $E_{det1} = -150$ mV (vs. Ag/AgCl; 400 ms) for the reduction of SY and $E_{det2} = -450$ mV (vs. Ag/AgCl; 100 ms) for the reduction of the pairs TT–SY and BB–SY at a cathodically pretreated BDD electrode (Medeiros et al., 2012). The pairs of colorants could be determined between the concentration ranges 5.0–60.0 µM (TT)/1.0–50.0 µM (SY) and 5.0–60.0 µM (BB)/1.0–50.0 µM (SY) with LODs of 2.5 µM (TT)/0.80 µM (SY) and 3.5 µM (BB)/0.85 µM (SY), respectively.

Another simple FIA-MPA strategy was developed for the simultaneous determination of paracetamol (PA) and caffeine (CA) (Silva et al., 2011). A sequence of potential pulses (waveform) was selected in such a way that PA is selectively oxidized at E_1 (+1.20 V/50 ms) and both compounds (PA + CA) are simultaneously oxidized at E_2 (+1.55 V/50 ms) (Figure 5.4); hence, current subtraction (using a correction factor) can be used for the selective determination of CA. This method is simple, cheap, fast (140 injections h^{-1}), and presents selectivity for the determination of both compounds in pharmaceutical samples, with results similar to those obtained by HPLC at a 95% confidence level.

Three amperometric biosensors were implemented in a continuous flow system (Figure 5.5) to perform the multiplexed quantification of sucrose, fructose, and glucose in a single experiment (Vargas et al., 2013). The good performance and stability of the biosensors, as well as the applicability of the developed continuous system, were demonstrated through the simultaneous determination of the three sugars in an infant food reference material, showing great promise as an affordable and useful analytical tool for quality control in the food industry.

Otieno et al. (2014) fabricated and validated a modular microfluidic immunoarray featuring novel online capture of analyte proteins on magnetic beads to detect multiple cancer biomarker proteins. The magnetic bead–protein bioconjugates were captured in an online chamber on eight antibody-decorated gold nanoparticle-film sensors and detected amperometrically. In simultaneous assays, the microfluidic system gave ultralow LODs of 5 fg mL^{-1} for interleukin (IL)-6 and 7 fg mL^{-1} for IL-8 in serum. Accuracy was demonstrated by measuring both analytes in conditioned media from oral cancer cell lines and showing good correlations with standard ELISAs. The online capture chamber facilitates rapid, sensitive and repetitive protein separation and measurement in 30 min in a semiautomated system adaptable to multiplexed protein detection based on inexpensive commercial components and hence accessible to virtually any biomedical laboratory.

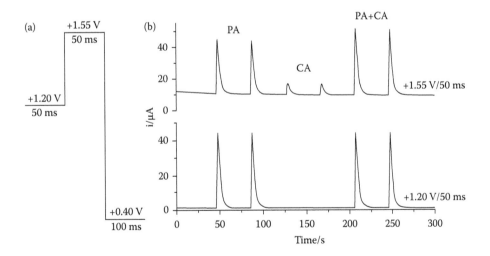

FIGURE 5.4 (a) MPA waveform (cyclic form) applied to the BDD working electrode as a function of time; (b) FIA-MPA responses (n = 2) for solutions containing only PA (60.0 mg L^{-1}), only CA (8.0 mg L^{-1}), and PA + CA (60.0 + 8.0 mg L^{-1}). Cleaning potential pulse: +0.40 V (amperogram not shown); supporting electrolyte: acetic acid/acetate buffer (0.1 mol L^{-1}; pH 4.7); flow rate: 4.5 mL min^{-1}; sample injection volume: 100 μL. (Reprinted from Silva, W. C. et al. 2011. *Electroanalysis* 23:2764–2770. With permission.)

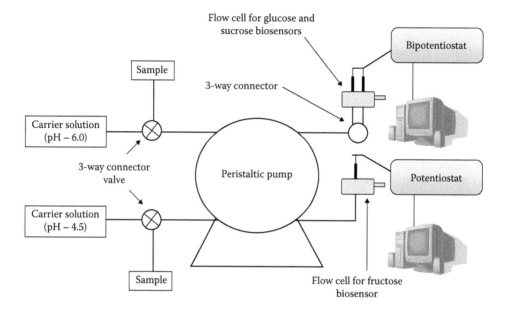

FIGURE 5.5 Schematic diagram of the flow system developed for the multiplexed detection of glucose, fructose, and sucrose using enzyme bioelectrodes as amperometric biosensors. (Reprinted from Vargas, E. et al. 2013. *Talanta* 105:93–100. With permission.)

5.1.3 Other Detection Techniques

In order to avoid limitations found when analyzing real samples, such as interferences from electroactive species in electrochemical sensors or spectral overlapping from other compounds in optical ones, other detection techniques have been coupled with FI systems such as calorimetric (Zheng et al., 2006) and piezoelectric (Halámek et al., 2005) biosensors and MS (Nanita et al., 2009).

A FI calorimetric biosensor was developed for the determination of dichlorvos (Zheng et al., 2006). The enzyme chicken liver esterase was used as the biorecognition element and acetyl-1-naphthol as the substrate. This enzyme was immobilized on an ion exchange resin, which was then packed in the enzyme reaction cell. The reference cell was filled with the same batch of the resin, but with a completely inactivated enzyme. As a result, the enzymatic reaction occurred in the enzyme reaction cell, but not in the reference cell and there was a temperature difference at the outlets of the two cells. The detection was based on the inhibition of the enzyme by the analyte, measuring the difference of the signal obtained with and without inhibition.

A piezoelectric biosensor, for the detection of several organophosphorus pesticides was developed (Halámek et al., 2005). The sensor was based on the immobilization of a reversible inhibitor of cholinesterase on the surface of the sensor. The binding of AChE to this inhibitor was monitored with a mass-sensitive piezoelectric quartz crystal. In the presence of an inhibiting substance in the sample, the binding of the enzyme to the immobilized compound was reduced, and the decrease of mass change was proportional to the concentration of the analyte in the sample. This sensor was applied to the determination of pesticides in river water samples.

Nanita et al. (2009) developed a FI approach for high-throughput pesticide residue quantitation without chromatographic separation prior to MS/MS. The method allowed the fast quantitation of agrochemicals (sulfonylureas and carbamates) in food and water samples. Samples were injected directly into a triple quadrupole instrument and data were obtained at the rate of 15 s/injection, allowing sample injection every 65 s, representing a significant improvement from the 15 – 30 min needed in typical HPLC-MS/MS methods, with accuracy limit of 0.01 ng mL^{-1} in food samples and 0.1 ng mL^{-1} of water samples with LODs of 0.003 mg mL^{-1} for food and 0.03 ng mL^{-1} for water samples.

5.2 INCORPORATION OF SAMPLE PRETREATMENT AND SEPARATION SYSTEMS

Process monitoring requires techniques that are selective and sensitive while providing rapid feedback. Several types of detectors can offer a rapid response but are limited by relatively poor selectivity, sensitivity, and versatility. The combination of FIA-based systems with sample pretreatment and/or separation techniques can overcome these limitations. Several sample pretreatment techniques such as dialysis, filtration, gas diffusion, derivatization, and ion exchange have already been successfully applied to FIA systems (Karlberg and Kuban, 2000).

Although separation techniques such as gas chromatography (GC), HPLC, and capillary electrophoresis (CE) are complex and tedious, they are extremely versatile since multianalyte determinations are possible when incorporated into flow systems. In this sense, the coupling with FIA is a simple concept that combines all the advantages of two mature techniques: the automated and versatile sample pretreatment offered by FIA, together with the high resolution and multicomponent options provided by the separation system coupled to it. Simpler approaches for analyte separation with FIA setups have been also suggested, a review of flowing stream systems comprising membrane separation was presented by Miró and Frenzel (2004).

A microdialysis-coupled flow injection amperometric sensor (μFIAS) (Figure 5.6) was used to determine glucose, galactose, and lactose in raw milk samples (Rajendran and Irudayaraj, 2002). With the multianalyte sensor it was possible to detect glucose and galactose by sequential injection of their corresponding oxidase enzymes (glucose oxidase and galactose oxidase), while lactose was determined by injection of a mixture of β-galactosidase and glucose oxidase enzymes. The sensor showed linear responses between 0.05 and 10 mM (glucose), 0.1 and 20 mM (galactose), and 0.2 and 20 mM (lactose). The relative standard deviation (RSD) values of the sensor measurements for glucose, galactose, and lactose were 3%–4% (n = 3). The sensor measurements for lactose content in milk were in agreement when compared with a standard method using an infrared spectrophotometer.

Successful applications of FIA-CE approaches have been described for drug analysis (dissolution rate analysis), paper and pulp liquors (anions), and food analysis (beverages) (Karlberg and Kuban, 2000). A novel interface for coupling microchip-based CE system with amperometric detection using FI sample introduction (Figure 5.7) was developed by Fu and Fang (2000). An H-channel configuration was used with a separation capillary positioned between two tubular side-arms, and a falling-drop interface connected to one side-arm, was developed to achieve electrical isolation between the FI and CE systems. End-column amperometric detection was accomplished with separation voltage decoupled from the detection system, employing a microdisk working electrode positioned

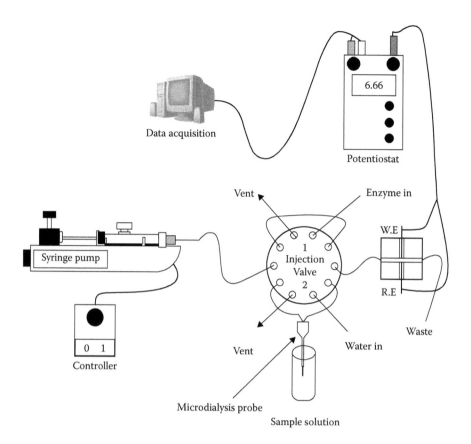

FIGURE 5.6   Schematic representation of the microanalysis-coupled FIAS. W.E: working electrode; R.E: reference electrode. (Reprinted from Rajendran, V. and J. Irudayaraj, 2002. *J. Dairy Sci.* 85:1357–1361. With permission.)

immediately outside the capillary outlet in the other side-arm, which functioned as a large-volume reservoir. In this work, dispersion in the FI system and FI–CE interface was minimized by intercalating the sample zone between two air segments, and by using the shortest possible length of transport conduit between the FI equipment and the FI–CE falling-drop interface. Performance of the FI–CE amperometric system was demonstrated by separating sucrose and glucose in about 60 s, achieving a sampling frequency of over 65 h⁻¹. Responses for sucrose and glucose were linear in the range of 10–1000 μM with sensitivities of 0.011 and 0.025 nA μM⁻¹, respectively. LODs of 2 and 1 μM were obtained for sucrose and glucose, respectively.

Dantan et al. (2001) developed a FIA system coupled with HPLC or CE for monitoring chemical and pharmaceutical production processes. In this system, a derivatization step, automated using a FIA-system (see Figure 5.8), was introduced to transform the activated ester involved in the reaction—suitable for the detection by HPLC but less so for amine quantification—to a nonreactive amide. The FIA-CE hyphenation method developed yielded good results appropriate for near real-time process monitoring.

The FIA–HPLC coupling is commonly used for analyte derivatization in order to improve the sensitivity of the methodology or reduce the assay time. As an example, an HPLC system coupled online with colorimetric detection of antioxidant activity using the

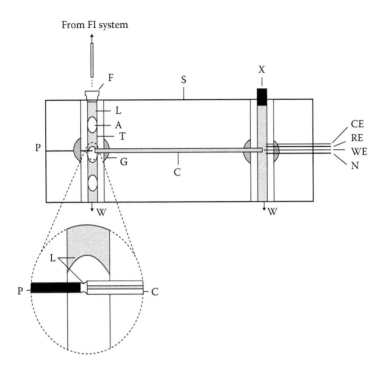

From FI system

FIGURE 5.7   Schematic diagram of the microchip FI–CE amperometric detection device used. C: fused-silica capillary; S: planar glass base; G: epoxy; F: conical inlet; L: liquid phase; X: glass plug; A: air bubble; T: Tygon® tube; P: platinum anode; N: platinum cathode. (Reprinted from Fu, C.-G. and Z.-L. Fang, 2000. *Anal. Chim. Acta* 422:71–79. With permission.)

free radicals DPPH (2,2-diphenyl-1-picrylhydrazyl) and DOPH (2,2-di(tert-octylphenil)-1-picrylhydrazyl) as color reagents in the FIA system was especially developed in order to separate antioxidants and to determine their activities in a single step (Nuengchamnong et al., 2004). This method, which can be applied for the separation and selective detection of flavonoids with antioxidant activity, improves the sensitivity in comparison with the method that used HPLC alone, allowing LODs for rutin and quercetin of 500 and 200 ng, corresponding to concentrations of 82 and 60 µM, respectively, in the injected sample. Moreover, this method can be used during lead-finding procedures for dereplication of these compounds in complex mixtures. The results showed that the developed system coupling HPLC online with antioxidant activity detection could separate the flavonoids rutin and quercetin in plant extracts such as *Sophora japonica* and *Morus alba* and simultaneously determine their antioxidant activity.

Gonzalez-San Miguel et al. (2007) coupled a multisyringe flow injection analysis system with chromatographic separation using a monolithic column and a diode array spectrophotometer (Figure 5.9). This multisyringe liquid chromatography (MSC) system has been successfully applied to the determination of amoxicillin, ampicillin, and cephalexin. Given the simplicity, versatility, and low-cost equipment of MSC, it can be regarded as an alternative to HPLC using low pressure and it could be a very attractive technique for improving selectivity in flow analytical systems (FIA and SIA). Back pressure was not a limiting factor in the MSC system, which allowed the use of similar flow rates to those used in HPLC under the same conditions.

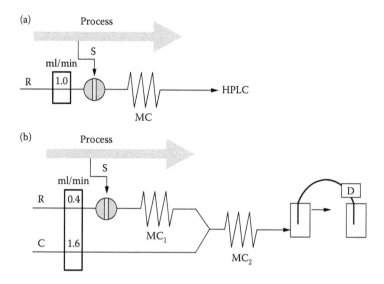

FIGURE 5.8 FIA manifold configurations for sampling and derivatization prior to (a) HPLC and (b) CE separation. R: reagent solution; C: carrier solution; S: sample injection; MC: mixing coil; D: UV–Vis detector. (a) R: octylamine/DMSO; S: 50 µL: MC: 100 cm/0.5 mm (b) R: propylamine/DMSO; C: water; S: 100 µL; MC1: 40 cm/0.5 mm; MC2: 200 cm/0.5 ram-knotted. (Reprinted from Dantan, N., W. Frenzel, and S. Küppers, 2001. *Chromatographia* 54:187–190. With permission.)

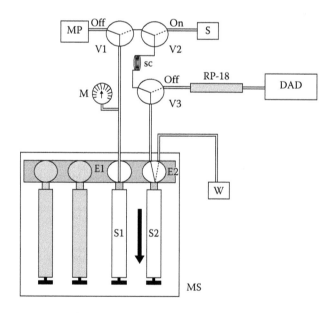

FIGURE 5.9 MSC system for isocratic separation. MS: multisyringe burette; M: manometer; V1–V3: solenoid valves; E1: two-way connector; E2: three-way solenoid valve; MP: mobile phase; S: sample; W: waste; RP-18: monolithic column; DAD: diode array detector; and SC: sample coil. (Reprinted from Gonzalez-San Miguel, H. M., J. M. Alpizar-Lorenzo, V. Cerda-Martin, 2007. *Talanta* 72:296–300. With permission.)

A monolithic minicolumn was incorporated in a FIA manifold for the simultaneous analysis of eight analytes, including sweeteners (aspartame, acesulfame-K, saccharin) and a few preservatives and antioxidants (García-Jiménez et al., 2007). The single-channel FIA system with a short monolithic C18 column, which allowed the separation of analytes according to their retention time, was used for quantification by measuring the intrinsic UV absorption of the analytes. The system was applied to the detection in different foodstuffs and the results obtained were in agreement with a reference HPLC method.

The same authors also reported the use of the system coupling an ultrashort C18 monolithic column with a FIA scheme for the determination of three antioxidants (propylgallate, butylhydroxyanisole, and butylhydroxytoluene) (García-Jiménez et al., 2009) and four preservatives (methylparaben, ethylparaben, propylparaben, and butylparaben) (García-Jiménez et al., 2010) at μg mL^{-1} level. The proposed approach, able to separate the analytes in only 85 or 150 s, respectively, was satisfactorily validated for the determination of the selected analytes in commercial food and cosmetics samples without any prior derivatization reaction.

A hybrid FIA/HPLC system incorporating monolithic column chromatography was developed for the determination of six opiate alkaloids (morphine, pseudomorphine, codeine, oripavine, ethylmorphine, and thebaine) and four biogenic amines (vanilmandelic acid, serotonin, 5-hydroxyindole-3-acetic acid, and homovanillic acid) in human urine, using tris(2,2'-bipyridyl)ruthenium(III) and acidic potassium permanganate CL detection (Adcock et al., 2007). This hybrid system approaches the automation and separation efficiency of HPLC while maintaining the positive attributes of FIA, such as manifold versatility, speed of analysis, and portability.

5.3 OUTLOOK AND FUTURE TRENDS IN THE DEVELOPMENT OF FLOW-BASED SYSTEMS FOR MULTI-ANALYTE DETECTION

On-site analysis is in demand for many areas of work including industrial quality control, agricultural and environmental studies, food analysis, and health care. All these areas of applications may require multianalyte detection and determinations. The high-throughput feature of FIA systems has shown to be very welcome in all these fields.

Although there are a large number of examples reported about multianalyte detection approaches using different technological platforms based on the coupling of FIA with different analytical techniques, only a few of them have been thoroughly studied with regard to application to real-life samples and even fewer have been validated.

Although great progress has been made, further research is required in two directions: (1) improving the selectivity of these systems, in order to quantify not only a family of analytes, but also distinguish between individual compounds; and (2) widening the range of applications of the developed methods to real samples. In general, the contribution of chemometrics, especially multivariate calibration methods, seems to be crucial to extending the applicability of multicomponent FI determinations to solve more complex analytical problems.

Future research will be focused in the development of small portable and automated systems that can be applied to a large number of species simultaneously in real samples. Downscaling of the analytical system toward a micro, nanofluidic format has become the trend for the development of future devices. Similarly to the larger-scale flow analysis, microfluidics systems are coupled with various methods to enhance the sensitivity of the measurements. The integration of sensing systems with nanoscience, however, involves a fundamental challenge. As the sensing systems scale to the micro/nanodomain, sensitivity and performance are compromised, because there are simply too few active sensing sites to produce a sensitive signal. According to the microchip-based FIA systems, the main challenges remain in the area of making an effective tiny pump and detector to be placed on the chip and the utilization of forces or methods of solution delivery that do not require external power. In addition, the introduction of new materials in electrochemical biosensors would increase their range or applications when coupling with FIA systems.

In summary, although from the point of view of the authors the coupling of FIA systems with different analytical techniques, including separation, sample pretreatment and detection, offers a very attractive, promising, and fruitful research field on the development and applications of analytical methods to multianalyte detection, there are significant challenges to be met before the wide dissemination and application of these technologies to routine analysis. The effort devoted by the scientific community in this research field is unprecedented, allowing optimism for the years to come, in which the introduction of the couplings that have been described and the identification of novel ones in clinical, food, and environmental practice seem about to become reality.

REFERENCES

Abate, G., J. Lichtig, and J. C. Masini, 2002. Construction and evaluation of a flow-through cell adapted to a commercial static mercury drop electrode (SMDE) to study the adsorption of Cd(II) and Pb(II) on vermiculite. *Talanta* 58:433–443.

Adcock, J. L., P. S. Francis, K. M. Agg et al. 2007. A hybrid FIA/HPLC system incorporating monolithic column chromatography. *Anal. Chim. Acta* 600:136–141.

Ahmed, M. J., C. D. Stalikas, S. M. Tzouwara-Karayanni, and M. I. Karayannis, 1996. Simultaneous spectrophotometric determination of nitrite and nitrate by flow-injection analysis. *Talanta* 43:1009–1018.

BASi LCEC Flowcells, web page. http://www.basinc.com/products/ec/flowcells.php.

Beale, D. J., N. A. Porter, and F. A. Roddick, 2009. A fast screening method for the presence of atrazine and other triazines in waterusing flow injection with chemiluminescent detection. *Talanta* 78:342–347.

Beauchemin, D., and A. A. Specht, 1998. Analysis of river water by ICP-MS with on-line preconcentration using flow injection. *Can. J. Anal. Sci. Spectrosc.* 43:43–48.

Bhadekar, R., S. Pote, V. Tale, and B. Nirichan. 2011. Developments in analytical methods for detection of pesticides in environmental samples. *Am. J. Anal. Chem.* 2:1–15.

Bosque-Sendrá, J. M., L. Gámiz-Gracia, and A. M. García-Campana, 2003. An overview of qualimetric strategies for optimisation and calibration in pharmaceutical analysis using flow injection techniques. *Anal. Bioanal. Chem.* 377:863–874.

Cabal, J., J. Bajgar, and J. Kassa, 2010. Evaluation of flow injection analysis for determination of cholinesterase activities in biological material. *Chem.-Biol. Interact.* 187:225–228.

Caccuri, A. M., G. Antonini, P. G. Board et al. 1999. Proton release on binding of glutathione to alpha, mu and delta class glutathione transferases. *Biochem. J.* 344:419–425.

Calleri, E., C. Temporini, E. Perani et al. 2004. Development of a bioreactor based on trypsin immobilized on monolithic support for the on-line digestion and identification of proteins. *J. Chromatogr. A* 1045:99–109.

Cardwell, T. J. and M. J. Christophersen, 2000. Determination of sulfur dioxide and ascorbic acid in beverages using a dual channel flow injection electrochemical detection system. *Anal. Chim. Acta* 416:105–110.

Chouhan, R. S., K. V. Rana, C. R. Suri et al. 2010. Trace-level detection of atrazine using immuno-chemiluminescence: Dipstick and automated flow injection analyses formats. *J. AOAC Int.* 93:28–35.

Dantan, N., W. Frenzel, and S. Küppers, 2001. Flow injection analysis coupled with HPLC and CE for monitoring chemical production processes. *Chromatographia* 54:187–190.

Di, P. and D. E. Davey, 1994. Trace gold determination by on-line preconcentration with flow injection atomic absorption spectrometry. *Talanta* 41:565–571.

Di Benedetto, L. T., and T. Dimitrakopoulos, 1997. Evaluation of a new wall-jet flow-through cell for commercial ion-selective electrodes in flow injection potentiometry. *Electroanalysis* 9:179–182.

Economou, A., P. Panoutsou, and D. G. Themelis, 2006. Enzymatic chemiluminescent assay of glucose by sequential-injection analysis with soluble enzyme and on-line sample dilution. *Anal. Chim. Acta* 572:140–147.

Felix, F. S. and L. Angnes, 2010. Fast and accurate analysis of drugs using amperometry associated with flow injection analysis. *J. Pharm. Sci.* 99:4784–4804.

Ferreira, I. M. P. L. V. O., J. L. F. C. Lima, M. C. B. S. M. Montenegro, R. P. Olmos, and A. Rios, 1996. Simultaneous assay of nitrite, nitrate and chloride in meat products by flow injection. *Analyst* 121:1393–1396.

Franko, M., M. Sarakha, A. Čibej et al. 2009. Photodegradation of pesticides and application of bioanalytical methods for their detection. *Pure Appl. Chem.* 77: 1727–1736.

Fu, C.-G. and Z.-L. Fang, 2000. Combination of flow injection with capillary electrophoresis: Part 7. Microchip capillary electrophoresis system with flow injection sample introduction and amperometric detection. *Anal. Chim. Acta* 422:71–79.

Fu, Z., F. Yan, H. Liu et al. 2008. Channel-resolved multianalyte immunosensing system for flow-through chemiluminescent detection of α-fetoprotein and carcinoembryonic antigen. *Biosens. Bioelectron.* 23:1063–1069.

Fujiwara, T., K. Murayama, and T. Imdadullah Kumamaru, 1994. Automated method for the selective determination of gold by online solvent extraction and reversed micellar-mediated luminal chemiluminescence detection. *Microchem. J.* 49:183–193.

García-Jiménez, J. F., M. C. Valencia, and L. F. Capitán-Vallvey, 2007. Simultaneous determination of antioxidants, preservatives and sweetener additives in food and cosmetics by flow injection analysis coupled to a monolithic column. *Anal. Chim. Acta* 594:226–233.

García-Jiménez J. F., M. C. Valencia, and L. F. Capitán-Vallvey, 2009. Simultaneous determination of three antioxidants in food and cosmetics by flow injection coupled to an ultra-short monolithic column. *J. Chromatogr. Sci.* 47:485–491.

García-Jiménez J. F., M. C. Valencia, and L. F. Capitán-Vallvey, 2010. Parabens determination with a hybrid FIA/HPLC system with ultra-short monolithic column. *J. Anal. Chem.* 65:188–194.

Gonzalez-San Miguel, H. M., J. M. Alpizar-Lorenzo, and V. Cerda-Martin, 2007. Development of a new high performance low pressure chromatographic system using a multisyringe burette coupled to a chromatographic monolithic column. *Talanta* 72:296–300.

Halámek, J., J. Pribyl, A. Makower et al. 2005. Sensitive detection of organophos-phates in river water by means of a piezoelectric biosensor. *Anal. Bioanal. Chem.* 382:1904–1911.

Hartwell, S. K. and K. Grudpan, 2012. Flow-based systems for rapid and high-precision enzyme kinetics studies. *J. Anal. Methods Chem.* 2012, Article ID 450716, 10pp. doi: 10.1155/2012/450716.

Hartwell, S. K., K. Grudpan, and G. D. Christian, 2004. Bead injection with a simple flow-injection system: An economical alternative for trace analysis. *Trends Anal. Chem.* 23:619–623.

Hellma, web page. Product cells for flow through measurements, http://www.hellma-analytics.com/text/191/en/flow-through-cell.html.

Karlberg, B. and P. Kuban, 2000. FIA coupled to capillary electrophoresis—A tool for process monitoring?. *J. Flow Injection Anal.* 17:5–9.

Kazemzadeh, A. and A. A. Ensafi, 2001. Sequential flow injection spectrophotometric deter-mination of nitrita and nitrate in various samples. *Anal. Chim. Acta* 442:319–326.

Kloth, K., R. Niessner, and M. Seidel, 2009. Development of an open stand-alone plat-form for regenerable automated microarrays. *Biosens. Bioelectron.* 24:2106–2112.

Kritsunankul, O. and J. Jakmunee, 2011. Simultaneous determination of some food addi-tives in soft drinks and other liquid foods by flow injection on-line dialysis coupled to high performance liquid chromatography. *Talanta* 84:1342–1349.

Kritsunankul, O., B. Pramote, and J. Jakmunee, 2009. Flow injection on-line dialysis coupled to high performance liquid chromatography for the determination of some organic acids in wine. *Talanta* 79:1042–1049.

Kyaw, T., T. Fujiwara, H. Inoue et al. 1998. Reversed micellar mediated luminol chemilu-minescence detection of iron(II, III) combined with on-line solvent extraction using 8-quinolinol. *Anal. Sci.* 14:203–207.

Lee, H. S., Y. A. Kim, Y. A. Cho et al. 2002. Oxidation of organophosphorus pesti-cides for the sensitive detection by a cholinesterase-based biosensor. *Chemosphere* 46:571–576.

Lee, H. S., Y. A. Kim, D. H. Chung et al. 2001. Determination of carbamate pesti-cides by a cholinesterase-based flow injection biosensor. *Int. J. Food Sci. Technol.* 36:263–269.

Llorent-Martinez, E. J., P. Ortega-Barrales, M. L. Fernández-de Cordova et al. 2011. Trends in flow-based analytical methods applied to pesticide detection: A review. *Anal. Chim. Acta* 684:30–39.

Maeder, G., J. L. Veuthey, M. Pelletier et al. 1990. Spectrophotometric determination of ethanol in blood using a flow injection system with an immobilized enzyme (alcohol dehydrogenase) reactor coupled to an on-line dialyser. *Anal. Chim. Acta* 231:115–119.

Maestre, E., I. Katakis, A. Narváez et al. 2005. A multianalyte flow electrochemical cell: Application to the simultaneous determination of carbohydrates based on bioelec-trocatalytic detection. *Biosens. Bioelectron.* 21:774–781.

Medeiros, R. A., B. C. Lourenção, R. C. Rocha-Filho et al. 2010. Simple flow injec-tion analysis system for simultaneous determination of phenolic antioxidants with multiple pulse amperometric detection at a boron-doped diamond electrode. *Anal. Chem.* 82:8658–8663.

Medeiros, R. A., B. C. Lourenção, R. C. Rocha-Filho et al. 2012. Flow injection simulta-neous determination of synthetic colorants in food using multiple pulse amperomet-ric detection with a boron-doped diamond electrode. *Talanta* 99:883–889.

Mehrvar, M. and M. Abdi, 2004. Recent developments, characteristics, and potential applications of electrochemical biosensors. *Anal. Sci.* 20:1113–1126.

Miró, M. and W. Frenzel, 2004. Automated membrane-based sampling and sample preparation exploiting flow-injection analysis. *Trends Anal. Chem.* 23:624–636.

Miró, M., K. Hartwell, J. Jakmunee et al. 2008. Recent developments in automatic solid-phase extraction with renewable surfaces exploiting flow-based approaches. *Trends Anal. Chem.* 27:749–761.

Nanita, S. C., A. M. Pentz, and F. Q. Bramble, 2009. High-throughput pesticide residue quantitative analysis achieved by Tandem mass spectrometry with automated flow injection. *Anal. Chem.* 81:3134–3142.

Nikolelis, D. P. and S. Pantoulias, 2001. Selective continuous monitoring and analysis of mixtures of acesulfame-K, cyclamate, and saccharin in artificial sweetener tablets, diet soft drinks, yogurts, and wines using filter-supported bilayer lipid membranes. *Anal. Chem.* 73:5945–5952.

Nuengchamnong, N., A. Hermans Lokkerbol, and K. Ingkaninan, 2004. Separation and detection of the antioxidant flavonoids, rutin and quercetin, using HPLC coupled on-line with colorimetric detection of antioxidant activity. *Naresuan Univ. J.* 12:25–37.

Ogbomo, I., R. Kittsteiner-Eberle, U. Englbrecht et al. 1991. Flow-injection systems for the determination of oxidoreductase substrates: Applications in food quality control and process monitoring. *Anal. Chim. Acta* 249:137–143.

Ogończyk, D. and R. Koncki, 2007. Potentiometric flow-injection system for determination of alkaline phosphatase in human serum. *Anal. Chim. Acta* 600:194–198.

Oliveira, H. M., M. A. Segundo, J. L.F.C. Lima et al. 2010. On-line renewable solid-phase extraction hyphenated to liquid chromatography for the determination of UV filters using bead injection and multisyringe-lab-on-valve approach. *J. Chromatogr. A* 1217:3575–3582.

Oliveira, S. M., T. I. M. S. Lopes, and A. O. S. S. Rangel, 2004. Spectrophotometric determination of nitrite and nitrate in cured meat by sequential injection analysis. *J. Food Sci.* 69:C690–C695.

Otieno, B. A., C. E. Krause, A. Latus et al. 2014. On-line protein capture on magnetic beads for ultrasensitive microfluidic immunoassays of cancer biomarkers. *Biosens. Bioelectron.* 53:268–274.

Pérez-Ruiz T., C. Martínez-Lozano, V. Tomás et al. 2002. Chemiluminescence determination of carbofuran and promecarb by flow injection analysis using two photochemical reactions. *Analyst* 127:1526–1530.

Pinho, O., I. M. P. L. V. O. Ferreira, M. B. P. P. Oliveira, and M. A. Ferreira, 1998. FIA evaluation of nitrite and nitrate contents of liver pâtés. *Food Chem.* 62:359–362.

Purohit, R. and S. Devi, 1997. Determination of nanogram levels of zirconium by chelating ion exchange and on-line preconcentration in flow injection UV–visible spectrophotometry. *Talanta* 44:319–326.

Rajendran, V. and J. Irudayaraj, 2002. Detection of glucose, galactose, and lactose in milk with a microdialysis-coupled flow injection amperometric sensor. *J. Dairy Sci.* 85:1357–1361.

Rao, T. N., B. H. Loo, B. V. Sarada et al. 2002. Electrochemical detection of carbamate pesticides at conductive diamond electrodes. *Anal. Chem.* 74:1578–156.

Saini, S. and C. R. Suri, 2010. Design and development of microcontroller based auto-flow assembly for biosensor application. *Int. J. Comput. Appl.* 6:21–27.

Saurina, J. 2008. Multicomponent flow injection analysis. In *Advances in Flow Analysis*, ed. M. Trojanowicz. Weinheim, Germany: Wiley-VCH.

Saurina, J. 2010. Flow-injection analysis for multi-component determinations of drugs based on chemometric approaches. *Trends Anal. Chem.* 29:1027–1037.

Silva, W. C., P. F. Pereira, M. C. Marra et al. 2011. A simple strategy for simultaneous determination of paracetamol and caffeine using flow injection analysis with multiple pulse amperometric detection. *Electroanalysis* 23:2764–2770.

Smith, R. M., K. W. Jackson, and K. M. Aldous, 1977. Design and evaluation of a fiber optic fluorometric flow cell. *Anal. Chem.*, 49:2051–2053.

Suwansa-Ard, S., P. Kanatharana, P. Asawatreratanakul, C. Limsakul, B. Wongkittisuksa, and P. Thavarungkul, 2005. Semi disposable reactor biosensors for detecting carbamate pesticides in water. *Biosens. Bioelectron.* 21:445–454.

Trojanowicz, M. 2009. Recent developments in electrochemical flow detections—A review: Part I. Flow analysis and capillary electrophoresis. *Anal. Chim. Acta* 653:36–58.

Vahl, K., H. Kahlert, D. Böttcher et al. 2008. A potential high-throughput method for the determination of lipase activity by potentiometric flow injection titrations. *Anal. Chim. Acta* 610:44–49.

Vargas, E., M. Gamella, S. Campuzano et al. 2013. Development of an integrated electrochemical biosensor for sucrose and its implementation in a continuous flow system for the simultaneous monitoring of sucrose, fructose and glucose. *Talanta* 105:93–100.

Wang, L. and S. Huang, 2011. Design of a flow-through voltammetric sensor based on an antimony-modified silver electrode for determining litholrubine B in cosmetics. *J. Anal. Methods Chem.* Article ID 896978, 7pp.

Waseem, A., M. Yaqoob, and A. Nabi, 2009. Photodegradation and flow-injection determination of dithiocarbamate fungicides in natural water with chemiluminescence detection. *Anal. Sci.* 25:395–400.

Wei, Y., Y. Li, Y. Qu et al. 2009. A novel biosensor based on photoelectro-synergistic catalysis for flow-injection analysis system/amperometric detection of organophosphorous pesticides. *Anal. Chim. Acta* 643:13–18.

Worsfold, P. J., R. Clough, M. C. Lohan et al.. 2013. Flow injection analysis as a tool for enhancing oceanographic nutrient measurements—A review. *Anal. Chim. Acta* 803:15–40.

Wutz, K., R. Niessner, and M. Seidel, 2011. Simultaneous determination of four different antibiotic residues in honey by chemiluminescence multianalyte chip immunoassays. *Microchim. Acta* 173:1–9.

Xavier, M. P., B. Vallejo, M. D. Marazuela et al. 2000. Fiber optic monitoring of carbamate pesticides using porous glass with covalently bound chlorophenol red. *Biosens. Bioelectron.* 14:895–905.

Yao, T., M. Satomura, and T. Nakahara, 1994. Simultaneous determination of sulfite and phosphate in wine by means of immobilized enzyme reactions and amperometric detection in a flow-injection system. *Talanta* 41:2113–2119.

Zhang, J. and A. E. G. Cass, 2006. Kinetic study of site directed and randomly immobilized his-tag alkaline phosphatase by flow injection chemiluminescence. *J. Mol. Recognit.* 19:243–246.

Zhang, Y., W. Liu, S. Ge et al. 2013. Multiplexed sandwich immunoassays using flow-injection electrochemiluminescence with designed substrate spatial-resolved technique for detection of tumor markers. *Biosens. Bioelectron.* 41:684–690.

Zheng, Y. H., T. C. Hua, D. W. Sun et al. 2006. Detection of dichlorvos residue by flow injection calorimetric biosensor based on immobilized chicken liver. *Food Eng.* 74:24–29.

CHAPTER 6

Flow Injection Analysis
Recent Developments and Future Trends

Marcela A. Segundo, Luísa Barreiros,
Ildikó V. Tóth, and Luís M. Magalhães

CONTENTS

Flow injection analysis (FIA) was proposed almost 40 years ago. It is now reaching its maturity and, despite the unfulfilled promise of large-scale implementation in routine analysis, this technique has contributed to research advance and formation of many analytical chemists and originated more than 21,500 papers in peer-reviewed international journals (Flow Injection Analysis Database, 2014). Nevertheless, recent reviews (Vidigal et al., 2013; Worsfold et al., 2013; Batista et al., 2014) indicate that the flame is still burning and scientists believe in FIA as a suitable tool to face new analytical challenges. This chapter aims to provide a personal view of what are likely to be the future topics of research with special impact in food analysis. This view is based on literature survey, on the papers presented at the 18th International Conference on Flow Injection Analysis (ICFIA), held in Porto, in September 2013, on papers collected in a special issue of Talanta (2015), and also on the most recent general books about flow analysis (Kolev and McKelvie, 2008; Trojanowicz, 2008).

6.1 RECENT DEVELOPMENTS

FIA systems evolved to give rise to sequential injection analysis (SIA) (Ruzicka and Marshall, 1990) and multicommutated approaches, including multicommutation (Rocha et al., 2002), multisyringe flow injection analysis (MSFIA) (Cerdà et al., 1999; Segundo and Magalhães, 2006), and multipumping systems (Lapa et al., 2002) as a result of the decreased cost of personal computers in the 1990s, which made them affordable as routine laboratory equipment. Nowadays, computer control is essential not only for data acquisition and processing but also for implementation of programmable flow and control of actuator devices. Recently, novel approaches relying on computer control have been proposed and their main features are summarized in the following sections.

6.1.1 Simultaneous Injection-Effective Mixing Analysis

The flow configuration of simultaneous injection-effective mixing analysis (SIEMA) is a hybrid format of FIA, SIA, and multicommutation in flow-based analysis that was proposed by Japanese researchers (Teshima et al., 2010). Sample and reagent solutions are aspirated into each holding coil through each solenoid valve by a syringe pump, and then the zones are simultaneously dispensed (injected) into a mixing coil by reversed flow toward a detector through a confluence point (Figure 6.1). This results in effective mixing and rapid detection with low reagent consumption.

The main difference from multicommutation systems is that flow reversal occurs in SIEMA while the main difference from MSFIA is that only one channel is present in the propulsion element, sending all holding coil contents simultaneously into a reaction coil. Despite its clear applicability to biological samples for determination of urinary biomarkers (Ponhong et al., 2011; Ratanawimarnwong et al., 2012; Vichapong et al., 2013; Ponhong et al., 2015), so far no application to food analysis has been reported, opening new opportunities in this field.

6.1.2 Cross Injection Analysis

Proposed by Nacapricha et al. (2013), cross injection analysis (CIA) has passed the proof-of-concept phase. It is based on an analytical platform drilled in an acrylic rectangular block where a single channel (designated as analytical flow path) is placed in the X-axis

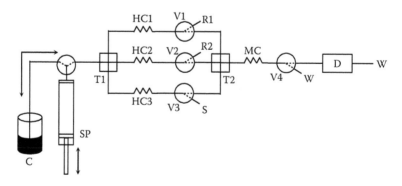

FIGURE 6.1 Schematic representation of SIEMA manifold. SP: syringe pump; C: carrier solution; T_i: confluence points (four-way cross connectors); HC_i: holding coils; V_i: three way commutation valve (position on and off represented as dashed lines); MC: mixing coil; R_i: reagents; S: sample; D: detector; and W: waste.

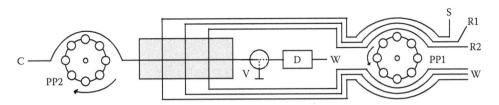

FIGURE 6.2 Schematic representation of CIA manifold. PP_i: peristaltic pump; C: carrier solution; V: three-way commutation valve (position on and off represented as dashed lines); R_i: reagents; S: sample; D: detector; and W: waste. The cross injection chip is represented in light gray, containing in this case three inlets for sample and reagents and one inlet for carrier.

direction and up to four channels (Y-channels) are placed perpendicular to the main channel. Each flow direction is associated with a propulsion element and, for the Y-direction channels, both outlet and inlet are connected to the peristaltic pump (Figure 6.2).

Hence, CIA accommodates two types of flow operation: alternate and simultaneous flow (Figure 6.3). In the first case, the carrier flow in the X-axis is operated in tandem with the Y-axis flow. Hence, sample and reagent loading in the Y-direction takes place without carrier flow, providing confined zones at the confluence points (Figure 6.3a). In the simultaneous approach, both X- and Y-axis flows are activated (Figure 6.3b). In this case, sample and reagent zones are driven by the carrier as long as they are introduced in the confluence points, providing larger interaction between zones and the introduction of more sample/reagent compared to the alternate approach. Compared to previous multicommutation and multipumping techniques, the CIA manifold is simpler, requiring fewer actuator elements (valves or micropumps) to introduce small slugs of sample and reagents. This feature makes CIA systems simpler and cheaper, requiring less maintenance. Despite having been applied so far only to iron determination in multivitamin tablets, CIA presents high potential for application to food samples.

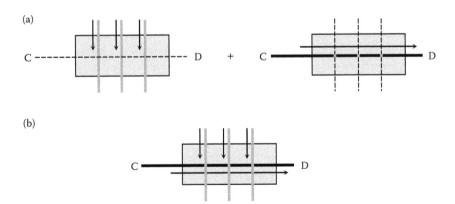

FIGURE 6.3 CIA performed in two modes: (a) reagents and sample are placed in CIA chip and later transported by carrier toward the detector; or (b) carrier, reagents, and sample flow simultaneously. C: carrier solution; D: detector.

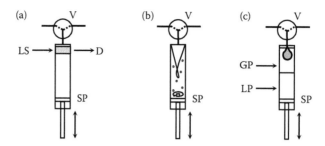

FIGURE 6.4 Schematic representation of lab-in-syringe comprising a detailed view of (a) implementation of analyte detection within the syringe; (b) stir bar assisted in-syringe dispersive liquid–liquid microextraction; and (c) in-syringe headspace single-drop micro-extraction. SP: syringe pump; V: three-way commutation valve (position on and off represented as dashed lines); LS: light source; D: detector; GP: gas phase; and LP: liquid phase.

6.1.3 Lab-in-Syringe

Lab-in-syringe systems were proposed in 2012 by researchers working under the leadership of Víctor Cerdà at the University of the Balearic Islands, Spain (Horstkotte et al., 2012a,b; Maya et al., 2012a,b; Suarez et al., 2012). It is based on the integration of several unitary operations taking place inside one syringe attached to a piston pump. As depicted in Figure 6.4, several configurations have been reported, allowing automation of solvent extraction and simultaneous detection of derivatized analytes by placing optical fibers connected to a light source and to a CCD (charge-coupled device) spectrometer at the top part of the syringe, close to its outlet. Hence, when organic solvents lighter than water are applied, they coalesce at the top of the syringe, where the concentrated analyte is detected (Figure 6.4a) (Maya et al., 2012a). In fact, more than 12 applications have been proposed, mainly featuring dispersive liquid–liquid microextraction (DLLME, Figure 6.4b) (Maya et al., 2014) and also headspace single-drop microextraction (Figure 6.4c) (Sramkova et al., 2014). Both extraction modes were performed in-syringe, featuring as novelties the possibility of stirred assisted dispersion by placing a small magnet inside the syringe for DLLME and the enhancement of volatile analyte extraction into the reagent solution drop by pressure decrease inside the syringe from aspiration against a closed port. In comparison with previous approaches, in-syringe DLLME uses a simpler instrumental setup, where mixing coils, phase separators, and other pumps are not required.

Several applications have been reported for food analysis, including the determination of ethanol in wines (Sramkova et al., 2014), trace inorganic mercury in mussel and tuna fish tissue (Mitani et al., 2014), phthalates (Clavijo et al., 2014), inorganic mercury (Mitani et al., 2014) in drinking waters, and the illegal food additive rhodamine B (Maya et al., 2012a).

6.2 CURRENT TRENDS

6.2.1 Automation of Liquid–Liquid Extraction

The automation of liquid–liquid extraction (LLE) using FIA and derived techniques offers several advantages, including higher analysis throughputs and reproducibility, and safer handling of volatile organic solvents along with minimization of sample loss and contamination. This is a topic of active research, with more than 20 papers published in the last 3 years, exploiting different approaches to implementing LLE in computer-controlled systems.

Initially, before computer control was implemented, LLE was carried out in flow injection systems by mixing two immiscible phases (flowing together at a confluence point, for instance), where the aqueous phase was generally the analyte-containing sample and the extractant was an organic solvent. After a given contact time, the phases were separated using some device (a small flow-through ampoule or membrane with affinity for organic compounds). Finally, a reaction color would take place in the organic phase or this would be directed to an atomic absorption detector. Nowadays, using programmable-flow strategies that allow flow reversal, no devices are required for phase separation, especially when DLLME or headspace single-drop microextraction are implemented.

For DLLME, where a small volume of extracting solvent is dispersed by the action of a second solvent, increasing the effective extraction area, the recovery of extractant is not easy due to its small amount, which introduces poor reproducibility. Automation by flow injection techniques grants improvement in both aspects, by providing repeatable conditions for solvent mixing and controlled fluid handling when dealing with small volumes. Moreover, considering the features of lab-in-syringe systems, working with organic solvents lighter than water is possible, as the recovery of liquids placed at the top of the syringe occurs first. For other systems, namely recent proposals for SIA, a microcolumn packed with a hydrophobic material is used for the separation of phases, with application of an additional eluent to further elute the extractant placed on the solid phase (Anthemidis and Ioannou, 2009).

As well as the achievements described in Section 6.1.3 using lab-in-syringe systems, automation of LLE has also been proposed by other groups, namely by Russian researchers resorting to the stepwise injection concept (Mozzhukhin et al., 2007). Recently, the stepwise injection approach was applied to evaluation of ammonia in concrete with headspace single-drop microextraction (Timofeeva et al., 2015), and to evaluation of antipyrine in saliva using DLLME (Bulatov et al., 2015). It has also been used to monitor the concentration of L-cysteine in biologically active supplements and animal feed, with full potential for implementation of food analysis (Bulatov et al., 2013).

6.2.2 Hyphenation to Chromatography for Sample Treatment Using Lab-on-Valve

Chromatography is often employed in food analysis because separation of analytes is required from interfering matrix components prior to quantification. However, most of the biological molecules present in food matrices (proteins, for instance) are not compatible with organic solvents present in chromatographic mobile phases, which hinders direct sample analysis. Prior sample treatment is needed, with purification and preconcentration of analytes.

Several proposals dealing with automation of sample treatment coupled to chromatography have been made in the last few years (Miró et al., 2011). The connection between the flow-based system and the chromatograph can be accomplished at-line, inline, or online. At-line connection is the simplest as no physical attachment exists between the automatic flow system and the chromatograph, requiring the transport of treated sample container to the chromatograph autosampler, for instance. Online hyphenation requires that the treated sample is automatically directed to the separative column, via an injection valve or other device. Inline connection means that the separative operation takes place within the flow system, using a monolithic column or entrapped beads within the lab-on-valve device (Ruzicka, 2000).

Another important feature of lab-on-valve systems is the possibility of implementing the bead injection concept. Also proposed by Ruzicka and coworkers (Ruzicka and Scampavia, 1999), bead injection consists of manipulating solid suspensions within the flow injection systems, promoting the retention of particles in different segments, namely a preconcentration/extraction channel or the detector itself. The bead injection concept allows the implementation of solid-phase extraction procedures with surface renewal for each sample. Using a flow programming strategy, a fresh portion of sorbent can be packed for each analysis and discarded after it, circumventing the problems or surface fouling reported for online sample treatment.

This technique has been applied to different samples, including several types of waters and biological samples (cell lysates, urine). Concerning food additives, an automatic system comprising online automatic renewable molecularly imprinted solid-phase extraction of riboflavin has been proposed (Oliveira et al., 2010) that allowed a strict control of the different steps within the extraction protocol, promoting selective interactions in the cavities of the molecularly imprinted sorbent (Figure 6.5). Sample determination throughput of six per hour was attained, including sample preparation (1 mL) and analysis, with successful application to pig liver extract (certified reference material BCR 487) and also to infant milk formula (certified reference material NIST 1846) and an energy drink in which riboflavin was an additive.

6.2.3 Low-Pressure Chromatography

Proposed in 2003 by the Czech group headed by Petr Solich, low-pressure chromatography in the SIA format (Satinsky et al., 2003) was made possible by the commercial availability of C_{18} monolithic columns. These columns consist of a monolith (single cylindrical block) of silica derivatized with octadecyl moieties (as in conventional HPLC [high-performance liquid chromatography] particles); they have meso- and micropores that allow flow through without the pressure increase observed for conventional packed columns. This feature allowed their integration to low pressure systems, including FIA and those based on derived techniques. Recently, the state-of-the-art has been reviewed both for systems using the SIA approach, giving sequential injection chromatography (SIC) (Idris, 2014) and for their integration into MSFIA systems (Fernandez et al., 2009).

Although most SIC and MSFIA-LC applications are focused on analysis of pharmaceutical formulations, three preservatives—two authorized (benzoic acid, sorbic acid) and one illegal (salicylic acid)—were successfully separated and quantified in several food products, including mayonnaise, jams, fruit juices, syrups, and soft drinks (Jangbai et al., 2012). Separation within less than 3 min was performed on a C_{18} monolithic column (5×4.6 mm) integrated into a SIA system, using 1% acetonitrile:ammonium acetate buffer pH 4.5 as a mobile phase. Recently, melamine, a toxic triazine added illegally to milk to increase its nitrogen content, was determined by micellar chromatography implemented in a SIC system, using a similar column and aqueous sodium dodecyl sulfate:propanol (92.5:7.5) as mobile phase (Batista et al., 2014).

MSFIA chromatographic analysis was also applied to food analysis (Fernandez et al., 2012), including orange juice, strawberry milkshake, and malt, for simultaneous determination of six water-soluble vitamins (thiamine, riboflavin, ascorbic acid, nicotinic

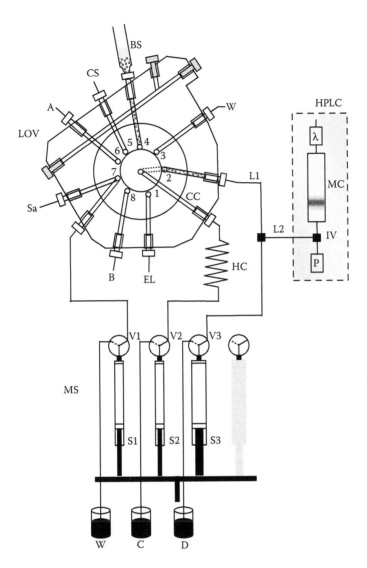

FIGURE 6.5 Schematic representation of the manifold for determination of riboflavin in foodstuff. LOV: lab-on-valve; MS: multisyringe; HPLC: liquid chromatograph; S_i: syringe; V_i: three-way commutation valve (position off: dashed line, position on: solid line); A: air; CS: conditioning solvent (50% (v/v) MeOH/H_2O); BS: bead suspension in conditioning solvent; C: carrier solution (H_2O); D: diluent (H_2O); W: waste; CC: central channel; EL: eluent (50% (v/v) MeOH/H_2O + 1% (v/v) CH_3COOH); B: channel for bead discarding; Sa: sample/standard solution; HC: holding coil; L1: connection tubing (8 cm); L2: connection tubing (44 cm); P: chromatographic pump; IV: injection valve; MC: monolithic chromatographic column; λ: diode array detector. (Reproduced with kind permission from Springer Science+Business Media: *Anal. Bioanal. Chem.*, Exploiting automatic on-line renewable molecularly imprinted solid-phase extraction in lab-on-valve format as front end to liquid chromatography: Application to the determination of riboflavin in foodstuffs, 397, 2010, 77–86, Oliveira, H. M. et al.)

acid, nicotinamide, and pyridoxine) using a dual isocratic elution with 5 mM sodium 1-hexanesulfonate (pH 7) and with 5 mM sodium 1-hexanesulfonate: methanol, 80:20 (v/v). Finally, a 10 mm long monolithic column has been successfully implemented in a low-pressure FIA system for the analysis of methylxanthines (theobromine, theophylline, and caffeine) in brewed coffee samples (Santos and Rangel, 2012), exhibiting an enhanced resolution between the target compounds.

6.3 CONCLUSIONS AND FUTURE PERSPECTIVES

FIA and, in particular, computer-controlled derived techniques still have an important contribution to make to quality control aspects in food analysis. Future trends will accompany the changes observed in analytical laboratories, where mass spectrometry-based detectors are replacing molecular spectrophotometry (e.g., diode array detectors [DAD]). Hence, it is expected the hyphenation of flow injection techniques to mass spectrometry, particularly for sample treatment (extraction, sample matrix removal) using FIA. Other less exploited feature, designated as reversed FIA (Mansour and Danielson, 2012), may also have an important role in future years. In this case, sample is applied as carrier, which allows an enhancement of detection limits. The only constraints are possible sample scarcity or high cost and multiplication of artifacts due to interferences. The future application of FIA is left to the imagination and ingenuity of future food analysts.

ACKNOWLEDGMENTS

I V. Tóth thanks FSE and Ministério da Ciência, Tecnologia e Ensino Superior (MCTES) for the financial support through the POPH-QREN program. This work received financial support from the European Union (FEDER funds through COMPETE) and National Funds (FCT, Fundação para a Ciência e Tecnologia) through project UID/Multi/04378/2013 and also from EU FEDER funds under the framework of QREN (NORTE-07-0162-FEDER-000124). L. Barreiros and L. M. Magalhães thank FCT and POPH (Programa Operacional Potencial Humano) for their post-doc grants (SFRH/BPD/89668/2012 and SFRH/BD/97540/2013).

REFERENCES

Anthemidis, A. N. and K. I. G. Ioannou. 2009. On-line sequential injection dispersive liquid–liquid microextraction system for flame atomic absorption spectrometric determination of copper and lead in water samples. *Talanta* 79:86–91.

Batista, A. D., C. F. Nascimento, W. R. Melchert, and F. R. P. Rocha. 2014. Expanding the separation capability of sequential injection chromatography: Determination of melamine in milk exploiting micellar medium and on-line sample preparation. *Microchem. J.* 117:106–110.

Batista, A. D., M. K. Sasaki, F. R. P. Rocha, and E. A. G. Zagatto. 2014. Flow analysis in Brazil: Contributions over the last four decades. *Analyst* 139:3666–3682.

Bulatov, A., K. Medinskaia, D. Aseeva, S. Garmonov, and L. Moskvin. 2015. Determination of antipyrine in saliva using the dispersive liquid–liquid microextraction based on a stepwise injection system. *Talanta* 133:66–70.

Bulatov, A. V., A. V. Petrova, A. B. Vishnikin, and L. N. Moskvin. 2013. Stepwise injection spectrophotometric determination of cysteine in biologically active supplements and fodders. *Microchem. J.* 110:369–373.

Cerdà, V., J. M. Estela, R. Forteza et al. 1999. Flow techniques in water analysis. *Talanta* 50:695–705.

Clavijo, S., M. D. Brunetto, and V. Cerdà. 2014. In-syringe-assisted dispersive liquid-liquid microextraction coupled to gas chromatography with mass spectrometry for the determination of six phthalates in water samples. *J. Sep. Sci.* 37:974–981.

Fernandez, M., R. Forteza, and V. Cerdà. 2012. Multisyringe chromatography (MSC): An effective and low cost tool for water-soluble vitamin separation. *Anal. Lett.* 45:2637–2647.

Fernandez, M., H. M. G. S. Miguel, J. M. Estela, and V. Cerdà. 2009. Contribution of multi-commuted flow analysis combined with monolithic columns to low-pressure, high-performance chromatography. *TrAC-Trends Anal. Chem.* 28:336–346.

Flow Injection Analysis Database by E. H. Hansen. http://www.flowinjectiontutorial.com/Database.html (accessed December 9, 2014).

Horstkotte, B., M. Alexovic, F. Maya, C. M. Duarte, V. Andruch, and V. Cerdà. 2012a. Automatic determination of copper by in-syringe dispersive liquid-liquid microextraction of its bathocuproine-complex using long path-length spectrophotometric detection. *Talanta* 99:349–356.

Horstkotte, B., F. Maya, C. M. Duarte, and V. Cerdà. 2012b. Determination of ppb-level phenol index using in-syringe dispersive liquid-liquid microextraction and liquid waveguide capillary cell spectrophotometry. *Microchim. Acta* 179:91–98.

Idris, A. M. 2014. The second five years of sequential injection chromatography: Significant developments in the technology and methodologies. *Crit. Rev. Anal. Chem.* 44:220–232.

Jangbai, W., W. Wongwilai, K. Grudpan, and S. Lapanantnoppakhun. 2012. Sequential injection chromatography as alternative procedure for the determination of some food preservatives. *Food Anal. Methods* 5:631–636.

Kolev, S. D. and I. D. McKelvie. 2008. *Advances in Flow Injection Analysis and Related Techniques*. Amsterdam: Elsevier.

Lapa, R. A. S., J. L. F. C. Lima, B. F. Reis, J. L. M. Santos, and E. A. G. Zagatto. 2002. Multi-pumping in flow analysis: Concepts, instrumentation, potentialities. *Anal. Chim. Acta* 466:125–132.

Mansour, F. R. and N. D. Danielson. 2012. Reverse flow-injection analysis. *TrAC-Trends Anal. Chem.* 40:1–14.

Maya, F., J. M. Estela, and V. Cerdà. 2012a. Completely automated in-syringe dispersive liquid–liquid microextraction using solvents lighter than water. *Anal. Bioanal. Chem.* 402:1383–1388.

Maya, F., B. Horstkotte, J. M. Estela, and V. Cerdà. 2012b. Lab in a syringe: Fully automated dispersive liquid–liquid microextraction with integrated spectrophotometric detection. *Anal. Bioanal. Chem.* 404:909–917.

Maya, F., B. Horstkotte, J. M. Estela, and V. Cerdà. 2014. Automated in-syringe dispersive liquid–liquid microextraction. *TrAC-Trends Anal. Chem.* 59:1–8.

Miró, M., H. M. Oliveira, and M. A. Segundo. 2011. Analytical potential of mesofluidic lab-on-a-valve as a front end to column-separation systems. *TrAC-Trends Anal. Chem.* 30:153–164.

Mitani, C., A. Kotzamanidou, and A. N. Anthemidis. 2014. Automated headspace single-drop microextraction via a lab-in-syringe platform for mercury electrothermal atomic absorption spectrometric determination after *in situ* vapor generation. *J. Anal. At. Spectrom.* 29:1491–1498.

Mozzhukhin, A. V., A. L. Moskvin, and L. N. Moskvin. 2007. Stepwise injection analysis as a new method of flow analysis. *J. Anal. Chem.* 62:475–478.

Nacapricha, D., P. Sastranurak, T. Mantim et al. 2013. Cross injection analysis: Concept and operation for simultaneous injection of sample and reagents in flow analysis. *Talanta* 110:89–95.

Oliveira, H. M., M. A. Segundo, J. L. F. C. Lima, M. Miró, and V. Cerdà. 2010. Exploiting automatic on-line renewable molecularly imprinted solid-phase extraction in lab-on-valve format as front end to liquid chromatography: Application to the determination of riboflavin in foodstuffs. *Anal. Bioanal. Chem.* 397:77–86.

Ponhong, K., N. Teshima, K. Grudpan, S. Motomizu, and T. Sakai. 2011. Simultaneous injection effective mixing analysis system for the determination of direct bilirubin in urinary samples. *Talanta* 87:113–117.

Ponhong, K., N. Teshima, K. Grudpan, J. Vichapong, S. Motomizu, and T. Sakai. 2015. Successive determination of urinary bilirubin and creatinine employing simultaneous injection effective mixing flow analysis. *Talanta* 133:71–76.

Ratanawimarnwong, N., K. Ponhong, N. Teshima et al. 2012. Simultaneous injection effective mixing flow analysis of urinary albumin using dye-binding reaction. *Talanta* 96:50–54.

Rocha, F. R. P., B. F. Reis, E. A. G. Zagatto, J. L. F. C. Lima, R. A. S. Lapa, and J. L. M. Santos. 2002. Multicommutation in flow analysis: Concepts, applications and trends. *Anal. Chim. Acta* 468:119–131.

Ruzicka, J. 2000. Lab-on-valve: Universal microflow analyzer based on sequential and bead injection. *Analyst* 125:1053–1060.

Ruzicka, J. and G. D. Marshall. 1990. Sequential injection—A new concept for chemical sensors, process analysis and laboratory assays. *Anal. Chim. Acta* 237:329–343.

Ruzicka, J. and L. Scampavia. 1999. From flow injection to bead injection. *Anal. Chem.* 71:257A–263A.

Santos, J. R. and A. O. S. S. Rangel. 2012. Development of a chromatographic low pressure flow injection system: Application to the analysis of methylxanthines in coffee. *Anal. Chim. Acta* 715:57–63.

Satinsky, D., P. Solich, P. Chocholous, and R. Karlicek. 2003. Monolithic columns— A new concept of separation in the sequential injection technique. *Anal. Chim. Acta* 499:205–214.

Segundo, M. A. and L. M. Magalhães. 2006. Multisyringe flow injection analysis: State-of-the-art and perspectives. *Anal. Sci.* 22:3–8.

Sramkova, I., B. Horstkotte, P. Solich, and H. Sklenarova. 2014. Automated in-syringe single-drop head-space micro-extraction applied to the determination of ethanol in wine samples. *Anal. Chim. Acta* 828:53–60.

Suarez, R., B. Horstkotte, C. M. Duarte, and V. Cerdà. 2012. Fully-automated fluorimetric determination of aluminum in seawater by in-syringe dispersive liquid–liquid microextraction using lumogallion. *Anal. Chem.* 84:9462–9469.

Talanta Special Issue dedicated to 18th ICFIA. 2015. Talanta 133:1–170. http://www.sciencedirect.com/science/journal/00399140/133 (accessed December 9, 2014).

Teshima, N., D. Noguchi, Y. Joichi et al. 2010. Simultaneous injection-effective mixing analysis of palladium. *Anal. Sci.* 26:143–144.

Timofeeva, I., I. Khubaibullin, M. Kamencev, A. Moskvin, and A. Bulatov. 2015. Automated procedure for determination of ammonia in concrete with headspace single-drop micro-extraction by stepwise injection spectrophotometric analysis. *Talanta* 133:34–37.

Trojanowicz, M. 2008. *Advances in Flow Analysis*. Weinheim: Wiley-VCH Verlag.

Vichapong, J., R. Burakham, N. Teshima, S. Srijaranai, and T. Sakai. 2013. Alternative spectrophotometric method for determination of bilirubin and urobilinogen in urine samples using simultaneous injection effective mixing flow analysis. *Anal. Methods* 5:2419–2426.

Vidigal, S. S. M. P., I. V. Toth, and A. O. S. S. Rangel. 2013. Sequential injection lab-on-valve platform as a miniaturisation tool for solid phase extraction. *Anal. Methods* 5:585–597.

Worsfold, P. J., R. Clough, M. C. Lohan et al. 2013. Flow injection analysis as a tool for enhancing oceanographic nutrient measurements—A review. *Anal. Chim. Acta* 803:15–40.

SECTION II

Preservatives

CHAPTER **7**

Determination of Nitrates and Nitrites

Claudia Ruiz-Capillas, Ana M. Herrero,
and Francisco Jiménez-Colmenero

CONTENTS

7.1 INTRODUCTION

Nitrates and nitrites are chemical compounds naturally present in the environment that form part of the nitrogen cycle. Major sources of nitrate include mineral deposits, soil, seawater, drinking water or freshwater systems, and the atmosphere. Nitrate plays an important role in plant nutrition and functions and is therefore used widely in fertilizers. Nitrate can accumulate in vegetables (Mensinga et al., 2003; Lundberg et al., 2004; Camargo and Alonso, 2006; EFSA, 2008; Lundberg et al., 2008). Levels tend to be higher in leaves and lower in seeds or tubers, and thus leaf crops generally have higher nitrate concentrations; for example, levels of 2.797 and 1.496 ppm have been recorded in spinach and celery respectively. On the other hand, the concentration of nitrite in fresh vegetables is generally low, depending on the method of cultivation, the rate and timing of fertilizer application, light intensity, daytime temperature, and soil characteristics. Human exposure to nitrate is mainly exogenous, from the consumption of vegetables, and to a lesser extent of water and other foods. Vegetables account for approximately 80% of the nitrate consumed via food and 10%–15% of the nitrate consumed via water (EFSA, 2008). Nitrate and nitrite are also formed endogenously in mammals, including humans. Nitrate is secreted in saliva and then converted to nitrite by oral microflora. In contrast, exposure to its metabolite nitrite is mainly from endogenous nitrate conversion.

Considering above all the potential health effects of nitrate exposure in infants and young children consuming lettuce or spinach, the European Commission has established a longer-term strategy to manage any risks from dietary nitrate exposure, based on the scientific opinion of the European Food Safety Authority (EFSA) on "Nitrate in vegetables" issued by the Panel on Contaminants in the Food Chain (CONTAM Panel) in 2008 (EFSA, 2008).

Then again, nitrate and nitrite are commonly used as preservatives in the food industry (Table 7.1). Preservatives are chemical compounds that when added to foods inhibit, retard, or prevent the activity and growth of spoilage and pathogenic microorganisms. Their chief purposes are to prolong the shelf life of foodstuffs by protecting them against deterioration caused by microorganisms, and to enhance their safety. Cheese and cheese products usually have only nitrates added, but both nitrites and nitrates are used in meat curing because they stabilize red meat color, inhibit some anaerobic spoilage and food poisoning microorganisms, delay the development of oxidative rancidity and contribute to flavor development. Nitrate is added as a source of nitrite, and in most cases is added along with nitrite to serve as a reserve source by slowly converting to nitrite. Nitrate is the stable form of oxidized nitrogen structures. Despite its low chemical reactivity, it can be reduced by microbiological action. Nitrite is much more reactive and can readily be oxidized to nitrate chemically or biologically, or else reduced to various compounds.

The exact mode of action by which inhibition occurs is unclear, but nitrites affect the growth of microorganisms in food through several reactions including (a) reacting with alpha-amino acid groups at low pH levels and (b) blocking sulfhydryl groups, which interferes with sulfur nutrition of the organism; (c) reacting with iron-containing compounds, which restricts the use of iron by bacteria; and (d) interfering with membrane permeability, which limits transport across cells (Urbain, 1971; Cassens, 1994; EFSA, 2003; Ray, 2004; Dave and Ghaly, 2011). Woods et al. (1989) reported that sodium nitrite at 200 mg/kg and a pH of 6.0 retarded the growth of *Achromobacter*, *Aerobacter*, *Escherichia*, *Flavobacterium*, *Micrococcus*, and *Pseudomonas* species in meat. Sofos et al. (1979) reported that *Clostridium botulinum* toxin formation in chicken frankfurter-type emulsions was delayed fivefold when 156 μg nitrite/g of meat was combined with 0.2% sorbic acid. In some countries, especially in Europe (EFSA, 2003, 2008; Directive 2006/52/EC), nitrate is added, as a salt of potassium or sodium or both, to cheese milk to prevent the growth of gas-producing bacteria. These cause blowing (swelling) of the cheese as a result of the proliferation of coliforms at the beginning of the maturation period, and of *Clostridium butyricum* and *Clostridium tyrobutyricum*, which cause late blowing. The effects of nitrite are enhanced by factors such as pH, salt, thermal processing and the presence of reducing agents/microorganisms. When nitrite is added to biological materials such as meat products, it reacts with various components occurring naturally in the matrix (myoglobin, nonheme proteins,

TABLE 7.1   Nitrite and Nitrate Additives Group with the E-Number

Groups	Names	E-Numbers
Nitrite (NO_2)	Potassium nitrite	E-249
	Sodium nitrite	E-250
Nitrates (NO_3)	Sodium nitrate	E-251
	Potassium nitrate	E-252

lipids), which means that the levels of nitrite detectable analytically (known as residual) necessarily vary. As a result, levels of residual nitrite fall during the different stages of processing, storage, preparation, and consumption depending on various factors relating to the type of product and the conditions applied in each of the phases mentioned above. Analysis of nitrite and/or nitrate content therefore detects only about 10%–20% of the original amount of nitrite added (Cassens et al., 1979; Cassens, 1997; Ruiz-Capillas et al., 2006, 2007; Ruiz-Capillas and Jiménez-Colmenero, 2008a).

7.2 POTENTIAL IMPLICATIONS OF NITRATE AND NITRITE FOR HUMAN HEALTH

The use of nitrates and nitrites as food additives has generally aroused controversy because of their potential effects on human health. Nitrate and nitrite are profoundly implicated in physiological systems needed for survival. Endogenous production of nitrate and nitrite in tissues can occur as a result of the activity of their most closely related components such as nitric oxide (Palmer et al., 1987; McKnight et al., 1999; Shiva, 2013). Nitric oxide reacts with oxygen to produce nitrite (Lundberg and Weitzberg, 2005; Lundberg et al., 2008), and it is through the greater stability of this ion that the action of NO can be detected. Nitric oxide, which is synthesized in humans, plays an important role in many metabolic functions: inhibition of platelet aggregation, as a neurotransmitter, in thrombosis and inflammation, in immune response, in control of blood pressure (vasoregulation), and in brain function. As a natural metabolite, it is necessary for the survival and functioning of the human biological system. It is widely accepted that nitrite is reduced to nitric oxide in the stomach and plays an important antimicrobial role there, destroying swallowed pathogens that can cause gastroenteritis in humans (Palmer et al., 1987; McKnight et al., 1999; Lundberg and Weitzberg, 2005; Lundberg et al., 2008; Shiva, 2013). The measurement of nitrite can therefore provide a reliable indicator of NO action within the body, and as such it can be used as a biomarker for physicians to gauge the health of an individual (Moorcroft et al., 2001).

The positive effect of nitrite on human health is thus known (protective role of nitric oxide); however, the interest in nitrite and its reaction is focused mainly on the potential formation of some compounds that are highly dangerous for human health. There are two main concerns associated with nitrate and nitrite: infant methemoglobinemia and formation of nitrosamines (on reaction with secondary amines), compounds with teratogenic, mutagenic, and carcinogenic effects (Cassens et al., 1979; Cassens, 1997; EFSA, 2003, 2008; Flores, 2008; De Mey et al., 2014). Nitrate per se is relatively nontoxic to humans, and its toxicity is attributed mainly to its reduction to nitrite through the mechanism of bacterial reduction as biotransformation in the saliva and digestive system. Excessive intake of nitrate and nitrite in the diet may cause toxic effects associated with the formation of methemoglobinemia, commonly known as "blue baby syndrome" because the infant's skin turns a bluish-gray color. Methemoglobinemia is a blood disorder in which the nitrite changes the normal form of hemoglobin, which carries oxygen in the blood to the rest of the body, into a form called methemoglobin that cannot carry oxygen (FAO/WHO, 2003a,b,c; EFSA, 2008).

The use of nitrites can also lead to the formation of carcinogenic nitrosamines (Warthesen et al., 1975; Gray and Randall, 1979; De Mey et al., 2014). Nitrosamine formation may occur either in food or in the human organism. Nitrite formed in the stomach from nitrate ingested in food can react with secondary and tertiary amines and amides to form *N*-nitroso compounds (NOCs), including nitrosamines and nitroamides. NOCs are potential carcinogens. The risk of nitrosamine formation in food is confined to products that become very hot in cooking (e.g., bacon) or are rich in nitrosable amines (fish and fermented products). However, nitrosamines have been found in cheeses even without the addition of nitrate (Klein et al., 1980). Nitrosamine formation is inhibited by antioxidants present in some foods, such as vitamin C and vitamin E. One possible means of reducing the risk of formation of these undesirable compounds is to use anti-oxidants in combination with potassium sorbate or phosphates (Shahidi, 1989; Pegg and Shahidi, 2000; Pourazrang et al., 2002; Ruiz-Capillas et al., 2015a,b). To avoid the formation of unhealthy compounds as far as possible and to reduce the risk that these entail, in recent decades the meat industry has modified the technologies used for cured meat production, resulting in lower levels of residual nitrite in products and beverages, including nitrite from fertilizers (Cassens, 1997; EFSA, 2003). Industries, researchers, and official bodies are making a considerable effort to reduce or eliminate the amount of nitrite present in foods (particularly in meat products) to minimize these adverse effects, with the least possible changes in the foods' sensory characteristics (color, taste, etc.). On the one hand, an effort has been made to reduce the amount of added nitrate and nitrite permitted in products as far as possible, and on the other hand various strategies have been assayed to replace added nitrites and so preserve the characteristics that these additives confer on the end product while maintaining its quality. These strategies include combination with ascorbic acid (E-330) and derivatives, and tocopherols (E-306 and following), which are especially effective in aqueous or fatty media, respectively (Mirvish et al., 1972; Pegg and Shahidi, 2000; Pourazrang et al., 2002; Sammet et al., 2006). Natural ingredients such as celery and beet sugar have also been tried as natural sources of nitrites and nitrates, mainly those that are reduced to nitrites by microorganisms, for example, *Staphylococcus* (Sindelar et al., 2007a,b; Viuda-Martos et al., 2010; Ruiz-Capillas et al., 2015a,b). Combinations of ingredients (celery, sodium lactate, orange dietary fiber, cochineal, vitamin C and vitamin E) have been used in a global strategy as substitutes for added chemical nitrites to formulate products that are healthier for consumers (Ruiz-Capillas et al., 2015a,b).

7.3 LEGAL CONSIDERATIONS REGARDING NITRATE AND NITRITE

For the reasons outlined earlier, effective control and regulation of the use of these preservatives in foods is very important to ensure consumer safety. The EFSA has established an acceptable daily intake (ADI) of nitrite, that is, the amount that can be ingested without adverse effects on human health (EFSA, 2003). The ADI of nitrite in a daily diet is up to 0.07 mg of nitrite per kg of body weight per day (equivalent to 3.6 mg/day for a person weighing 60 kg). Because some nitrate from food is converted to nitrite in our body, there is also an ADI for the amount of nitrate we eat. The acceptable daily limit of nitrate is up to 3.7 mg/kg of body weight per day (equivalent to 219 mg/day for a person weighing 60 kg), (FAO/WHO, 2003a,b,c; IARC/WHO, 2010).

The use of nitrates and nitrites in food (such as vegetables, meat products, and cheese) is regulated in the European Union (Spanish Royal Decree 1118/2007 approving the positive list of additives other than colorings and sweeteners for use in the manufacture of food products and conditions of use). This statute establishes and differentiates between the maximum amount that may be added during manufacture (expressed in mg/kg as $NaNO_2$) and the maximum residual levels according to the food group. In recent years, the permissible added amounts and the residual amounts of nitrates and nitrites have provoked a certain amount of controversy, especially in the case of meat products, and this has given rise to some amendments of the regulations (Directives 95/2/EC and 2006/52/EC). The legislation authorizing nitrites thus establishes a fine balance between these two risks, taking into account the large variety of foods (vegetables, meat products, and cheese) processed and traded throughout the European Union.

7.4 DETERMINATION OF NITRATE AND NITRITE

In view of the importance of nitrate and nitrite for purposes of legal regulation, food safety, and technological efficacy, it is essential to have appropriate and accurate analytical methods to determine the levels of these additives in different foods and beverages and be able to gauge their possible consequences. The literature contains references to numerous methodologies for the detection, determination and monitoring of nitrate and/or nitrite in various types of matrix, for example, environmental (water, atmosphere), food (meat and fish products, cheese, vegetables, etc.), and physiological samples (serum, plasma etc.) (Fazio and Warner, 1990; Karovicová and Simko, 2000; Moorcroft et al., 2001; Wood et al., 2004; Ruiter and Bergwerff, 2005). Interest in the simultaneous detection of these ions has grown over the years for their chemical implications; the chemical reduction of nitrate to the more reactive nitrite in particular features prominently in many reports and is often the only means of detecting the relatively inert nitrate ion. Most methodologies for determining nitrates and nitrites involve the use of ion chromatography, gas and liquid chromatography, capillary electrophoresis, enzymatic reduction, or electrochemistry (biosensors with amperometric monitoring among others). Different detection systems have been used, based on UV, mass spectroscopy, spectrophotometry, fluorescence, chemiluminescence or electrochemical detection (Usher and Telling, 1975; Wood et al., 2004; Ruiter and Bergwerff, 2005; Jiménez-Colmenero and Blazquez, 2007). Most of these methods require complex instrumentation to work successfully, and this limits them in practice to a laboratory-based environment. Determination of nitrates normally entails prior chemical reduction to nitrite (which is more reactive) before the detection stage. Various reduction agents (vanadium, cadmium, and others) have been assayed, and also several enzymes (nitrate reductase, nitrate ion-selective electrode, and others) (Senn and Carr, 1976; Davison and Woof, 1978; Guevara et al., 1998; Mori, 2000, 2001; Miranda, 2001; Patton et al., 2002; Gal et al., 2004; Pinto et al., 2005; Campbell et al., 2006; Colman and Schimel, 2010a,b). Of particular interest among the many existing methodologies for determination of nitrates and nitrites are those based on flow injection analysis (FIA), which have seen considerable development in recent years owing to the numerous advantages that they offer. FIA is a major alternative to conventional control methods, both in laboratories devoted to routine analyses and in official and industrial laboratories. The system offers major advantages, ensuring precise, simple, and fast determination of small amounts of nitrates and nitrites with little consumption of reactants and small sample volumes (Ruiz-Capillas and Jiménez-Colmenero, 2008a,b). This chapter considers the current status of the available published FIA methodologies for determination of nitrate and nitrite in foods and beverages. The review analyses different aspects relating to extraction, separation, detection, and quantification procedures in complex matrices.

7.5 FIA METHODOLOGIES FOR NITRATE AND NITRITE DETERMINATION

FIA, which was first described by Ruzicka and Hansen (1975), is the most widely used method for both simultaneous and sequential determination of nitrates and nitrites in various matrices (Valcarcel and Luque de Castro, 1987; Moorcroft et al., 2001). Early studies of FIA focused on the determination of nitrates and nitrites in water, but by the mid-1980s the applications of FIA had been extended to other fields and to more complex matrices (food, biomedicine, control of industrial processes, etc.) (Ahmed et al., 1996; Ensafi and Kazemzadeh, 1999; Higuchi and Motomizu, 1999; Brabcová et al., 2003; Reis-Lima et al., 2006). The system can be installed online and is compatible with a large number of detectors (potentiometric, amperometric, spectrophotometric, fluorometric, etc.) (Hulanicki et al., 1987; Lima et al., 1995; Moorcroft et al., 2001). Generally speaking, depending on the matrix, FIA nitrate and nitrite determination falls into two phases. The first phase is extraction of these compounds from the foods or beverages. The second is the FIA procedure as such. The two phases are considered below.

7.5.1 Extraction and Preparation of an Extract Containing Nitrate and Nitrite

Extraction is a crucial phase in FIA determination of nitrates and nitrites, as it affects nitrite and nitrate recoveries, especially in the case of complex matrices, for instance in meat products, fish, cheese or vegetables. It is the most critical phase in terms of achieving adequate nitrate and nitrite recoveries and reducing possible interferences. As noted earlier, when nitrite is added to a biological material, it reacts with various components in the matrix, which means that the levels of nitrite that are detectable analytically (known as residual) necessarily vary. The nitrate ion is relatively inert, but long-term storage of the sample prior to analysis should be discouraged in view of the propensity for bacterial spoilage. This is also true of nitrite, but in this case there are a number of additional precautions that need to be considered. Nitrite is stable in neutral or alkaline solutions but will decompose on standing in acidic conditions, and the decomposition is accelerated when the solutions are heated. The redox properties of nitrite are such that many matrix constituents will act to either reduce it (e.g., ascorbic acid, sulfamic acid, urea, iodide) or oxidize it [e.g., $Cr(III)$, $Fe(III)$, $Cu(II)$, MnO_4^-, CrO_4^{2-}, BrO_3^- and $Ce(IV)$], resulting in poor recoveries. A number of reviews have covered the effects of such interferences within food and environmental matrices (Adams, 1997; Fanning, 2000; Moorcroft et al., 2001).

Most nitrite and nitrate extraction processes for FIA determination use the AOAC method (AOAC, 1990). This process requires homogenization of the samples followed by heating and filtration to produce an extract ready to inject into the FIA manifold (van Staden and Makhafola, 1996; Ruiz-Capillas et al., 2007; Ruiz-Capillas and Jiménez-Colmenero, 2008a).

Over the years numerous changes to the extraction process have been proposed to improve the efficiency of nitrite extraction and eliminate interferences in the color-developing reaction of the various exogenous or endogenous compounds (ascorbic acid, fat, protein, etc.) present in biological samples and beverages. Most purification processes are based on deproteination and clarification of the extract, for which several types of agents have been used, mainly borax (sodium tetraborate anhydride), potassium aluminum sulfate, mercuric chloride, zinc sulfate, or potassium hexacyanoferrate. One of the most commonly used is borax buffer, which is intended to counteract the effect of low pH on the stability of nitrite/nitrate (Merino et al., 2000). These reagents are known as Carrez I (consisting mainly of borax) and Carrez II which includes zinc acetate (in case of meat products) or zinc sulfate (in case of cheese). Some authors use only Carrez I in the extraction, or else a combination of the two—first Carrez I and then Carrez II on the extract from the first (ISO, 1985, 2000; Pinho et al., 1998; Higuchi and Motomizu, 1999; AOAC, 2000; Andrade et al., 2003). In some cases 24% trichloroacetic acid (MSDA, 1973) and hydrochloric acid (Fox and Suhre, 1985; Fox, 1989) have been used to precipitate the protein.

One very important factor in the extraction process is the pH of both sample and extract. Some authors have reported that the higher the pH of the matrix, the greater is the recovery achieved (Merino et al., 2000). In this regard it has been proposed to treat samples with NaOH or a drop of chloroform and keep them in the refrigerator (Gine et al., 1980). In those cases the reagents mentioned above, as well as precipitating proteins from meat extracts, serve to fix the pH between 6.0 and 6.5 so as to avoid nitrate and nitrite loss. One of the major criticisms leveled at the use of Carrez is that the pH of the final solution (approximately pH 5) tends to promote nitrite loss. However, this sample deproteination procedure is further complicated by the fact that it produces cloudy filtrates resulting in serious experimental errors (Usher and Telling, 1975). For that reason it is often necessary to apply clearing techniques such as centrifugation, active carbon or gel filtration on Sephadex G25—this last only used for beverages (Torro et al., 1998).

For determination of nitrites in clean aqueous solutions such as water, there is no need for extraction; it is generally enough to filter the sample before it is injected into the FIA system. In some cases preconcentration of the sample may be necessary for determination of nitrite by FIA or sequential injection analysis (SIA). One such example would be direct nitrite determination in nonpolluted water samples where colorimetry is not sensitive enough to determine NO_2 (Miró et al., 2000, 2001). Preconcentration processes have been based on anionic exchange resins (Dowex 1-X8) (Wada and Hattori, 1971) or on ion pairs coprecipitated with microcrystalline biphenyl (Puri et al., 1998), or with naphthalene (Wang et al., 1998). Dialysis-FIA systems have been proposed for determination of nitrite in residual waters, and even a solid phase (monofunctional C_{18}) which is held inside a glass column for an automatic SIA has been tried (Miró et al., 2000).

7.5.2 Measurements of Nitrate and Nitrite Content

The most common approach to nitrite detection by FIA is undoubtedly the modification of the Shinn reaction (Shinn, 1941) better known as the Griess colorimetric indicator reaction (Liang et al., 1994; AOAC, 1997; Butt et al., 2001; Ötzekin et al., 2002; Ruiz-Capillas et al., 2006, 2007; Ruiz-Capillas and Jiménez-Colmenero, 2008a,b; Colman and Schimel, 2010a,b). The Griess assay typically relies on the diazotization of a suitable aromatic amine by acidified nitrite with the subsequent coupling reaction providing a highly colored azo chromophore from which the concentration of nitrite can be assessed (Figure 7.1). The absorption maximum for the azo product is generally in the range 500–600 nm (depending on the reagents used) and can be detected using conventional visible spectrometry. The most common arrangement utilizes sulfanilamide and N-(1-naphthyl) ethylenediamine (NED) (Moorcroft et al., 2001; Ruiz-Capillas and Jiménez-Colmenero, 2008a).

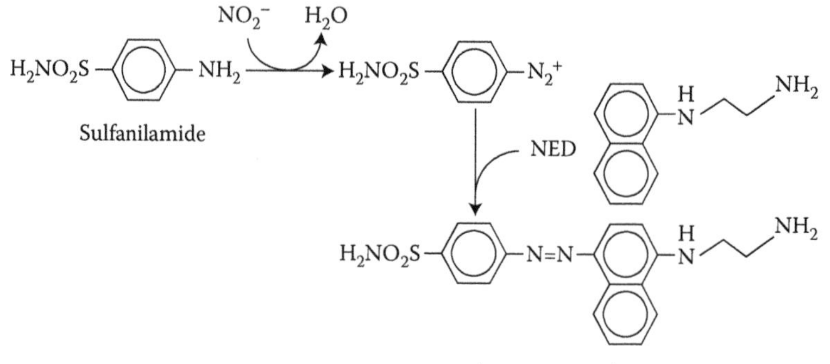

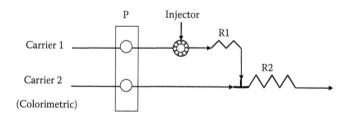

FIGURE 7.1 Griess reactions implicated in the determination of nitrate and nitrate by FIA. (Adapted from Promega. 2009. Griess Reagent System. Tecnical Bulletin. Promega Corporation: Madison, USA. Available at URL: http://www.promega.com/tbs/tb229/tb229.pdf.)

The Griess reaction take places automatically in the FIA system, where the sample extracts are generally injected through the injection system valve into the corresponding carrier streams (Figure 7.2). These masking and buffering reagents draw the injected extract into coils of different lengths; it is then mixed directly with the color-forming reagent in a second reaction coil where the color reaction takes place. The compounds that have been formed then enter the detection system (spectrophotometric, amperometric, etc.) to yield the FIA response curve (Figure 7.2).

The reagent carrier 1 (Figure 7.2) is generally composed of ammonium chloride and EDTA (ethylenediaminetetraacetic acid) in varying concentrations and pH levels (van Staden and Makhafola, 1996; Oliveira et al., 2004; Ruiz-Capillas et al., 2006, 2007) or ammonium chloride, sodium tetraborate, and EDTA (Ferreira et al., 1996; Pinho et al., 1998) or a buffer 7.5 (Ruiz-Capillas et al., 2007). The reagent or carrier 2 (Figure 7.2), which is responsible for the color, is mainly one of those used in the traditional diazo-coupling reaction (the Griess reaction), generally composed of sulfanilamide, (NED) and phosphoric acid (Ferreira et al., 1996; van Staden and Makhafola, 1996; Pinho et al., 1998; Ruiz-Capillas et al., 2006, 2007) or other acids such as HCl (Oliveira et al., 2004; Frenzel et al., 2004). The acid is used to promote the color reaction. This color reaction produces a response in the detector, which is connected to a chart recorder where the

FIGURE 7.2 FIA diagram for determination of nitrite (P: pump; R1, R2: reactors). (Adapted from Ruiz-Capillas, C. and F. Jiménez-Colmenero. 2008a. *Food Addit. Contam., Part A.* 25:1167–1178.)

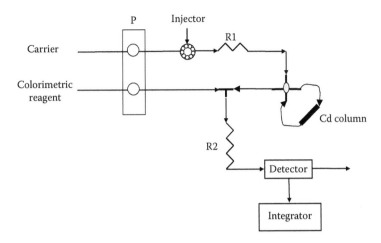

FIGURE 7.3 FIA diagram for determination of nitrate (P: pump; R1, R2: reactors). (From Ruiz-Capillas, C. and F. Jiménez-Colmenero. 2008a. *Food Addit. Contam., Part A.* 25:1167–1178.)

response is recorded in the form of a peak. The response is proportional to the concentration of nitrite in the sample or in the standard solutions. The peaks produced on the chart recorder by known concentrations of standard nitrite solutions are used to calibrate the scale.

Although initially FIA was used only for nitrite determination, nitrate determination is also possible (Figure 7.3). This requires prior reduction of nitrate to nitrite with reduction agents such as vanadium (Miranda, 2001) or cadmium, in the form of filings or granules, or washing with solutions of mercury(II), silver(I), or copper(II) ions (Nydahl, 1976; Davison and Woof, 1978; Gal et al., 2004; Colman and Schimel, 2010a,b). Of these, the reducing agent of choice for colorimetric nitrate determinations has long been copperized (copper-washed) cadmium granules packed into small columns. The pH required for this reaction is in the range 7.0–8.5 (Nydahl, 1976). Despite their long predominance as reducing agents of choice for colorimetric nitrate determinations in water, flow-through cadmium reactors are difficult to prepare and activate, pose health risks to analysts and waste stream processors, increase waste stream disposal costs, and are incompatible with discrete analyzers. To get around these drawbacks, several nitrate reductase–based nitrate assays have been done (Senn and Carr, 1976; Guevara et al., 1998; Mori, 2000, 2001; Patton et al., 2002; MacKown and Weik, 2004; Pinto et al., 2005; Campbell et al., 2006). These nitrate reductase enzymes offer a good alternative as soluble, nontoxic replacements for cadmium (Patton et al., 2002). However, they also have a drawback in that the level of enzymatic reduction in colorimetric nitrate + nitrite determination assays is low.

Most FIA methods include the online reduction step (nitrate to nitrite) prior to the color reaction (Griess or other reaction), following the same procedure as in the case of nitrite determination. The sample is then treated with azo dye reagent and the absorbance due to the sum of nitrite and nitrate is measured: nitrate is determined from the difference in absorbance values (Figure 7.3) (Ruiz-Capillas et al., 2006, 2007). In this way, nitrates and nitrites can be determined in the same system. In fact, FIA systems have been developed for simultaneous determination of nitrates and nitrites, with the saving in time that this affords. Some authors (Anderson, 1979; Ahmed et al., 1996) have determined nitrites and nitrates simultaneously using a selector valve to split the sample into two streams (one for nitrate and the other for nitrite) after it was injected into the carrier stream. Ferreira et al. (1996) and Pinho et al. (1998) also determined nitrate and nitrite simultaneously with FIA by splitting of the flow after injection and subsequent confluence of the flow before it reached the detector, allowing the reduction of nitrate to nitrite in part of the sample plug in an online copper–cadmium reactor column. Since each channel has a different residence time, two peaks were obtained: for nitrite and for nitrite plus nitrate. Two sequential columns have also been tried: copper and copper-coated cadmium have been used for simultaneous determination of nitrite and nitrate in a flow injection system (Ensafi and Kazemzadeh, 1999).

Although the Griess reaction is the one most commonly used for the determination of nitrates and nitrites, other amines and solutions have also been proposed as potential reactants for nitrite because they offer important advantages, including excellent sensitivity. However, it also has certain limitations as regards control requirements for the reaction conditions, stability of the reactants (affected by numerous oxidation reactions), interferences (such as SO_2) or the use of some reactants with carcinogenic effects (Ahmed et al., 1996; van Staden and Makhafola, 1996; Haghighi and Tavassoli, 2002). The amines and solutions that have been proposed as potential reactants for nitrite and potential coupling agents applied to the analysis of food extracts by FIA spectrophotometric determination include 3-nitroaniline, safranin, ferrous and thiocyanate solutions, and H_2SO_4/iodine (Ahmed et al., 1996; Kazemzadeh and Ensafi, 2001; Andrade et al., 2003; Penteado et al., 2005; Ruiz-Capillas and Jimenez Colmenero, 2008a). For example, safranin O (3,7-diamino-2,8-dimethyl-5-phenylphenazinium chloride) has also been reported to form diazonium salts in an acidic solution, which causes the reddish-orange dye color of the solution to change to blue in acidic media and which, absorbing at 520 nm, has been used for flow injection spectrophotometric determination of nitrite in beef and commercial sausages. The reaction of safranin with the nitrite weakens the color of the dye rather than reducing its concentration as in the case of diazo dye (Kazemzadeh and Ensafi, 2001; Penteado et al., 2005). 3-Nitroaniline, with NED, has been reported as a spectrophotometric (color-forming) reagent for nitrite and nitrate determination in agent in beef and pork sausages (Ahmed et al., 1996). Andrade et al. (2003) also reported the formation of $FeSCNNO^+$ for spectrophotometric determination at 460 nm. The method is based on the reduction of nitrite to NO in a sulfuric acid medium and the reaction of nitric oxide with ferrous and thiocyanate ions forming $FeSCNNO^+$. The reduction of nitrite with potassium iodine followed by chemiluminescence detection of the NO released in a FIA system has been tested for the determination of nitrites in food and other biological materials (Sen et al., 1994). Ensafi and Kazemzadeh (1999) also proposed the use of gallocyanine in a flow injection method with spectrophotometric detection for nitrite determination. The reaction is based on the catalytic effect of nitrite on the oxidation of gallocyanine by bromate in acidic media and the decrease in absorbance of the system at 530 nm.

A FIA gas diffusion (FIGD) system has also been used for determination of nitrites in food products and aqueous solutions (Haghighi and Tavassoli, 2002). The sample solution is injected into a stream of water which then reacts with a stream of hydrochloric acid. The gaseous products (NO, NO_2, HNO_2, and NOCl) are separated from the liquid stream by the home-made gas–liquid separator and are swept by the carrier O_2 gas into a home-made flow-through cell that has been positioned in the cell compartment of a UV–Vis spectrophotometer. The transient absorbance of the gaseous phase is measured at 205 nm. This method is less sensitive than the Griess methods but there are also fewer limitations. Some of these FIA systems have further been combined with sensors; for example, there is a gas diffusion membrane system that is combined with an optical sensor for the determination of nitrite in wastewaters and meat extracts (Frenzel et al., 2004). The color reagent is prepared by dissolving sulfanilamide, NED, and concentrated hydrochloric acid in water. Sulfuric acid is used as a releasing agent to set HNO_2 and NO_2 free from the nitrite solutions. This sensor is more selective because the analyte is separated by the gas-diffusion process; however, only a few analytes are sufficiently volatile to pass through the membrane and they are effectively separated from the sample matrix. In this way, compounds known to interfere in the Griess reaction are almost entirely excluded. In particular, colored and turbid samples can be analyzed without problems and interferences since extreme pH values or the presence of high salt contents in the samples are eliminated, so that the proposed sensor is almost nitrite specific. However, this sensor has a drawback in that its sensitivity is relatively low compared to the common spectrophotometric method (ISO, 1996), which limits its applicability (Frenzel et al., 2004). Torro et al. (1998) also developed a sensor–FIA combination that involved photochemical reduction of nitrate in a photoreactor consisting of a low-pressure mercury lamp. The nitrite produced is determined indirectly with the I_3^-/I^- system using a flow cell with two platinum (Pt) electrodes polarized at 100 mV.

Microelectronic technologies have also been used in nitrite biosensors and electrochemical sensors (Suzuki and Taura, 2001; Adhikari and Majumdar, 2004). Ameida et al. (2013) have also developed a method of electrochemical nitrite measurement by means of a gold working electrode covered with 1,2-diaminobenzene (DAB) integrated in a FIA system. This sensor helps improve selectivity, repeatability, stability, and sensitivity. A Nafion/lead-ruthenate pyrochlore electrode chemically modified for determination of NO_2 oxidation and NO reduction based on AC-impedance spectroscopy and FIA has also been tested (Zen et al., 2000). Quan and Shin (2010) also tested an electrochemical nitrite biosensor based on co-immobilization of copper-containing nitrite reductase and viologen-modified chitosan (CHIT-V) on a glassy carbon electrode (GCE).

7.6 CONCLUSION

The use of FIA to determine both nitrate and nitrite in foods and beverages is an important alternative and presents several advantages over traditional control methods in routine laboratory analysis, official bodies, and industry due to the many advantages that it offers. FIA is a highly versatile, flexible, and economical technique, suitable for fast, routine analysis of a large number of samples with high precision, simple protocols, and instrumentation (detector, sensors, etc.) that is widely available. Affordably priced and easy to use, it requires minimum intervention of the operator and lower consumption of samples and reactants. This is especially important when working with toxic or expensive reactants. It can be installed online and is compatible with a large number of detectors (potentiometric, amperometric, spectrophotometric, fluorometric, etc.). This methodology responds to growing demands for the mechanization and automation of analysis from many laboratories faced with requirements for increasing numbers of routine analyses. This, along with its advantageous cost/efficiency ratio, favors the use of the method over the numerous alternative analytical methods for determining nitrates and nitrites. Increasing demand for rapid online analysis will ensure the continued development and combination of this FIA system for determination of these compounds.

Advances in the automation of extraction and cleaning processes could help to improve recoveries of nitrates and nitrites and conditions for the application of FIA in complex matrices such as muscle-based foods, in order to reduce possible interference associated with natural matrix components.

ACKNOWLEDGMENTS

This research was supported by projects AGL 2010-1915, AGL 2011-29644-C02-01 of the Plan Nacional de Investigación Científica, Desarrollo e Innovación Tecnológica (I + D + I) Ministerio de Ciencia y Tecnología and Intramural projects CSIC: 201470E073. The author thanks to Aleix Unzaga Nuño and Josu for the support.

REFERENCES

Adams, J. B. 1997. Food additive-additive interactions involving sulphur dioxide and ascorbic and nitrous acids: A review. *Food Chem.* 59:401–409.

Adhikari, B. and S. Majumdar 2004. Polymers in sensor applications. *Prog. Polym. Sci.* 29:699–766.

Ahmed, M. J., C. D. Stalikas, S. M. Tzouwara-Karayanni, and M. I. Karayannis. 1996. Simultaneous spectrophotometric determination of nitrite nitrate by flow-injection analysis. *Talanta* 43:1009–1018.

Ameida, F. L., S. G. dos Santos Filho, and M. B. A. Fontes. 2013. FIA-automated system used to electrochemically measure nitrite and interfering chemicals through a 1-2 DAB/Au electrode: Gain of sensitivity at upper potentials. *J. Phys.: Conf. Ser.* 421:1–8.

Anderson, L. 1979. Simultaneous spectrophotometric determination of nitrite and nitrate by flow injection analysis. *Anal. Chim Acta* 110:123–128.

Andrade, R., C. O. Viana, S. G. Guadagnin, F. G. R. Reyes, and S. Rath. 2003. A flow-injection spectrophotometric method for nitrate and nitrite determination through nitric oxide generation. *Food Chem.* 80:597–602.

AOAC. 1990. *Official Methods of Analysis. Association of Official Analytical Chemistry* (15th ed.). Washington, DC: Association of Official Analytical Chemists, Official method 973.31.

AOAC. 1997. *Official Methods of Analysis* (14th ed.), Vol. II. Arlington, VA: Association of Official Analytical Chemists.

AOAC. 2000. *Official Method of Analysis 33.7. Nitrate and Nitrite in Cheese, Modified Jones Reduction Method*, 16. (p. 73). Rockville, Maryland, USA: Association of Official Analytical Chemists, Official method 976.14.

Brabcova, M., P. Rychlovsky, and I. Nemcova. 2003. Determination of nitrites, nitrates and their mixtures using flow injection analysis with spectrophotometric detection. *Anal. Lett.* 36:2303–2316.

Butt, S. B., M. Riaz, and M. Z. Iqbaz. 2001. Simultaneous determination of nitrite and nitrate by normal phase ion-pair liquid chromatography. *Talanta* 55:789–797.

Camargo, J. A. and A. Alonso. 2006. Ecological and toxicological effects of inorganic nitrogen pollution in aquatic ecosystems: A global assessment. *Environ. Int.* 32:831–849.

Campbell, W. H., P. Song, and G. G. Barbier. 2006. Nitrate reductase for nitrate analysis in water. *Environ. Chem. Lett.* 4:69–73.

Cassens, R. G. 1994. *Meat Preservation, Preventing Losses And Assuring Safety* (1st ed.). Trumbull, CT: Food and Nutrition Press, Inc.

Cassens, R. G. 1997. Residual nitrite in cured meat. *Food Technol.* 51:53–55.

Cassens, R. G., M. L. Greaser, T. Ito, and M. Lee. 1979. Reactions of nitrite in meat. *Food Technol.* 33:46–57.

Colman, B. P. and J. P. Schimel. 2010a. Understanding and eliminating iron interference in colorimetric nitrate and nitrite analysis. *Environ. Monit. Assess.* 165:633–641.

Colman, B. P. and J. P. Schimel. 2010b. Erratum to: Understanding and eliminating iron interference in colorimetric nitrate and nitrite analysis. *Environ. Monit. Assess.* 165:693.

Dave, D. and A. E. Ghaly. 2011. Meat spoilage mechanisms and preservation techniques: A critical review. *Am. J. Agric. Biol. Sci.* 6:486–510.

Davison, W. and C. Woof. 1978. Comparison of different forms of cadmium as reducing agents for the batch determination of nitrate. *Analyst* 103:403–406.

De Mey, E., K. De Klerck, H. De Maere et al. 2014. The occurrence of N-nitrosamines, residual nitrite and biogenic amines in commercial dry fermented sausages and evaluation of their occasional relation. *Meat Sci.* 96(2 Part A):821–828.

Directive 1995/2/EC of the European Parlament and of the Council of 20 February 1995 on food additives other than colours and sweeteners.

Directive 2006/52/EC of the European Parliament and of the Council of 5 July amending Directive 95/2/EC on food additives other than colours and sweeteners and Directive 94/35/EC on sweeteners for use in foodstuffs.

EFSA. 2003. The effects of nitrites/nitrates on the microbiological safety of meat products. *EFSA J.* 14:1–31.

EFSA. 2008. Nitrate in vegetables Scientific Opinion of the Panel on Contaminants in the Food chain 1 (Question No EFSA-Q-2006–071). *EFSA J.* 689:1–79.

Ensafi, A. A. and A. Kazemzadeh. 1999. Simultaneous determination of nitrite and nitrate in various samples using flow injection with spectrophotometric detection. *Anal. Chim. Acta* 382:15–21.

Fanning, J. C. 2000. The chemical reduction of nitrate in aqueous solution. *Coord. Chem. Rev.* 199:159–179.

FAO/WHO. 2003a. Food and Agriculture Organisation of the United Nations/World Health Organization. *Nitrate (and Potential Endogenous Formation of N-Nitroso Compounds)*. WHO Food Additive series 50. Geneva: World Health Organisation. Available at URL: http://www.inchem.org/documents/jecfa/jecmono/v50je06.htm

FAO/WHO. 2003b. Food and Agriculture Organisation of the United Nations/World Health Organization. *Nitrite (and Potential Endogenous Formation of N-Nitroso Compounds)*. WHO Food Additive series 50. Geneva: World Health Organisation. Available at URL: http://www.inchem.org/documents/jecfa/jecmono/v50je05.htm.

FAO/WHO. 2003c. Food and Agriculture Organisation of the United Nations/World Health Organization. *Nitrate and Nitrite (Intake Assessment)*. WHO Food Additive series 50, Geneva: World Health Organisation. Available at URL: http://www.inchem.org/documents/jecfa/jecmono/v50je07.htm.

Fazio, T. and C. R. Warner. 1990. A review of sulfites in foods—Analytical methodology and reported findings. *Food Addit. Contam.* 7:433–454.

Ferreira, I. M. P. L. V. O., J. L. F. C. Lima, and M. C. B. S. M. Montenegro. 1996. Simultaneous assay of nitrite, nitrate and chloride in meat products by flow injection. *Analyst* 121:1393–1396.

Flores, J. 2008. Necesidad tecnológica de los nitritos y/o nitratos en los productos cárnicos crudos adobados. *Boletín AICE* junio:40–43.

Fox, J. B. and F. B. Suhre. 1985. The determination of nitrite: A critical review. *Crit. Rev. Anal. Chem.* 15:283–313.

Fox, P. F. 1989. *Developments in Dairy Chemistry*, Vol. 4. New York: Elsevier Applied Science.

Frenzel, W., J. Schulz-Brussel, and B. Zinvirt. 2004. Characterisation of a gas-diffusion membrane-based optical flow-through sensor exemplified by the determination of nitrite. *Talanta* 64:278–282.

Gal, C., W. Frenzel, and J. Möller. 2004. Re-examination of the cadmium reduction method and optimisation of conditions for the determination of nitrate by flow injection analysis. *Microchim. Acta* 146:155–164.

Gine, M. F., H. Bergamin, E. A. G. Zagatto, and B. F. Reis. 1980. Simultaneous determination of nitrate and nitrite by flow-injection analysis. *Anal. Chim. Acta* 114:191–197.

Gray, J. I. and C. J. Randall. 1979. Nitrite-*N*-nitrosamine problem in meats-update. *J. Food Prot.* 42:168–179.

Guevara, I., J. Iwanejko, A. Dembińska-Kieć et al. 1998. Determination of nitrite/nitrate in human biological material by the simple Griess reaction. *Clin. Chim. Acta* 274:177–188.

Haghighi, B. and A. Tavassoli. 2002. Flow injection analysis of nitrite by gas phase molecular absorption UV spectrophotometry. *Talanta* 56:137–144.

Higuchi, K. and S. Motomizu. 1999. Flow injection spectrophotometric determination of nitrite and nitrate in biological samples. *Anal. Sci.* 15:129–134.

Hulanicki, A., W. Matuszewski, and M. Trojanowicz. 1987. Flow injection determination of nitrite and nitrate with biamperometric detection at two platinum wire electrodes. *Anal. Chim. Acta* 194:119–127.

IARC. 2010. WHO (World Health Organization)/IARC (International Agency for Research on Cancer). *Ingested Nitrate and Nitrite, and Cyanobacterial Peptide Toxins*, Vol. 94. Lyon, France: International Agency for Research on Cancer.

ISO. 1985. *Lactosérum en poudre: Détermination des teneurs en nitrates et en nitrites— méthode par réduction au cadmium et spectrometrie*. Réf. No. ISO 6740-1985 (F). Geneva, Switzerland: International Organization for Standardization.

ISO. 1996. Water quality—Determination of nitrite nitrogen and nitrate nitrogen and the sum of both by flow analysis (continuous flow analysis and flow injection analysis). Geneva: International Organization for Standardization (ISO 13395:1996 (E)).

ISO. 2000. Determination of nitrate and nitrite contents—Method by enzymatic reduction and molecular absorption spectrometry after Griess reaction. Ref. No. ISP/TC 34/SC5 N546. Delft, The Netherlands: International Organization for Standardization.

Jiménez Colmenero, F. and J. Blazquez. 2007. Additives: Preservatives. In *Handbook of Processed Meats and Poultry Analysis*, eds. L. Nollet, and F. Toldrá. Boca Raton, FL: Taylor & Francis, LLC, pp. 91–108.

Karovicová, J. and P. Simko. 2000. Determination of synthetic phenolic antioxidants in food by high-performance liquid chromatography. *J. Chromatogr.* A882:271–281.

Kazemzadeh, A. and A. A. Ensafi. 2001. Sequential flow injection spectrophotometric determination of nitrite and nitrate in various samples. *Anal. Chim. Acta* 442:319–326.

Klein, D., A. Keshavarz, P. Lafont, J. Hardy, and G. Debry. 1980. Formation of nitrosamines in cheese products. *Ann. Nutr. Aliment.* 34:1077–1088.

Liang, B., M. Iwatsuki, and T. Fukasawa. 1994. Catalytic spectrophotometric determination of nitrite using the chlorpromazine hydrogen-peroxide redox reaction in acetic-acid medium. *Analyst* 119:2113–2117.

Lima, J. L. F. C., A. O. S. S. Rangel, and M. R. S. Souto. 1995. Flow injection determination of nitrate in vegetables using a tubular potentiometric detector. *J. Agric. Food Chem.* 43:704–707.

Lundberg, J. O. and E. Weitzberg. 2005. NO generation from nitrite and its role in vascular control. *Arterioscler. Thromb. Vasc. Biol.* 25:915–922.

Lundberg, J. O., E. Weitzberg, J. A. Cole, and N. Benjamin. 2004. Nitrate, bacteria and human health. *Nat. Rev. Microbiol.* 2:593–602.

Lundberg, J. O., E. Weitzberg, and M. T. Gladwin. 2008. The nitrate–nitrite–nitric oxide pathway in physiology and therapeutics. *Nat. Rev. Drug Discovery* 7:156–167.

MacKown, C. T. and J. C. Weik. 2004. Comparison of laboratory and quick-test methods for forage nitrate. *Crop Sci.* 44:218–226.

McKnight, G. M., C. W. Duncan, C. Leifert, and M. H. Golden. 1999. Dietary nitrate in man: Friend or foe? *Br. J. Nutr.* 81:349–358.

Mensinga, T. T., G. J. Speijers, and J. Meulenbelt. 2003. Health implications of exposure to environmental nitrogenous compounds. *Toxicol. Rev.* 22:41–51.

Merino, L., U. Edberg, G. Fuchs, and P. Åman. 2000. Liquid chromatographic determination of residual nitrite/nitrate in foods. *J. AOAC Int.* 83:365–375.

Miranda, K. M., M. G. Espey, and D. A. Wink. 2001. A rapid, simple spectrophotometric method for simultaneous detection of nitrate and nitrite. *Nitric Oxide: Biol. Chem.* 5:62–71.

Miró, M., A. Cladera, J. M. Estela, and V. Cerdà. 2000. Sequential injection spectrophotometric analysis of nitrite in natural waters using an on-line solid-phase extraction and preconcentration method. *Analyst* 125(5):943–948.

Miró, M., W. Frenzel, V. Cerdà, and J. M. Estela. 2001. Determination of ultratraces of nitrite by solid-phase preconcentration using a novel flow-through spectrophotometric optrode. *Anal. Chim. Acta* 437:55–65.

Mirvish, S. S., L. Wallcave, M. Eagen, and P. Shubik. 1972. Ascorbate-nitrite reaction: Possible means of blocking the formation of carcinogenic N-nitroso compounds. *Science* 7:65–68.

Moorcroft, M. J., D. James, and R. G. Compton. 2001. Detection and determination of nitrate and nitrite—A review. *Talanta* 54:785–803.

Mori, H. 2000. Direct determination of nitrate using nitrate reductase in a flow system. *J. Health Sci.* 46:385–388.

Mori, H. 2001. Determination of nitrate in biological fluids using nitrate reductase in a flow system. *J. Health Sci.* 47:65–67.

MSDA 1973. *Office Central Féderal des Imprimis et du Matérial Berne* (5éme ed.), Vol. 2, Chapter 1. Bern, Switzerland: Manuel Suisse des Denrées Alimentaires.

Nydahl, F. 1976. Optimum conditions for reduction of nitrate to nitrite by cadmium. *Talanta* 23:349–357.

Oliveira, S. M., T. I. M. S. Lopes, and A. O. S. S. Rangel. 2004. Spectrophotometric determination of nitrite and nitrate in cured meat by sequential injection analysis. *J. Food Sci.* 69:C690–C695.

Ötzekin, N., M. S. Nutku, and F. B. Erim. 2002. Simultaneous determination of nitrite and nitrate in meat products and vegetables by capillary electrophoresis. *Food Chem.* 76:103–106.

Palmer, R. M. J., A. G. Ferrige, and S. Moncada. 1987. Nitric oxide release accounts for the biological activity of endothelium-derived relaxing factor. *Nature* 327:524–526.

Patton, C. J., A. E. Fischer, W. H. Campbell, and E. R. Campbell. 2002. Corn leaf nitrate reductase: A nontoxic alternative to cadmium for photometric nitrate determinations in water samples by air-segmented continuous-flow analysis. *Environ. Sci. Technol.* 36:729–735.

Pegg, R. B. and F. Shahidi. 2000. *Nitrite Curing of Meat. The N-Nitrosamine Problem and Nitrite Alternatives.* Trumbull, CT: Food and Nutrition Press Inc.

Penteado, J. C., L. Angnes, J. C. Masini, and P. C. C. Oliveira. 2005. FIA-spectrophotometric method for determination of nitrite in meat products: An experiment exploring color reduction of an azo-compound. *J. Chem. Educ.* 82:1074–1078.

Pinho, O., I. M. P. L. V. O. Ferreira, M. B. P. P. Oliveira, and M. A. Ferreira. 1998. FIA evaluation of nitrite and nitrate contents of liver pates. *Food Chem.* 62:359–362.

Pinto, P. C. A. G., J. J. F. C. Lima, and M. L. M. F. S. Saraiva. 2005. An enzymatic analysis methodology for the determination of nitrates and nitrites in water. *Int. J. Environ. Anal. Chem.* 85:29–40.

Pourazrang, H., A. A. Moazzami, and B. S. Fazly Bazzaz. 2002. Inhibition of mutagenic N nitroso compound formation in sausage samples by using L-ascorbic acid and α-tocopherol. *Meat Sci.* 62:479–483.

Promega. 2009. Griess Reagent System. Tecnical Bulletin. Promega Corporation: Madison, USA. Available at URL: http://www.promega.com/tbs/tb229/tb229.pdf.

Puri, S., M. Satake, and G. F. Wang. 1998. New method for the simultaneous spectrophotometric determination of nitrate and nitrite in water samples using column preconcentration. *Ann. Chim.* 88:685–695.

Quan, D. and W. Shin. 2010. A nitrite biosensor based on co-immobilization of nitrite reductase and viologen-modified chitosan on a glassy carbon electrode. *Sensors (Basel)* 10:6241–6256.

Ray, B. 2004. Control by antimicrobial preservatives. In *Fundamental Food Microbiology* (3rd ed.). Boca Raton, FL: CRC Press, pp. 439–506.

Reis-Lima, M. J., S. M. V. Fernandes, and A. O. S. S. Rangel. 2006. Determination of nitrate and nitrite in dairy samples by sequential injection using an in-line cadmium-reducing column. *Int. Dairy J.* 16:1442–1447.

Ruiter, A. and A. A. Bergwerff. 2005. Analysis of chemical preservatives in foods. In *Methods of Analysis of Food Components and Additives*, ed. Ötles S, Chapter 14. Boca Raton, FL: Taylor & Francis, pp. 423–444.

Ruiz-Capillas, C., P. Aller-Guiote, J. Carballo, and F. Jiménez-Colmenero. 2006. Biogenic amine formation and nitrite reactions in meat batter as affected by high-pressure processing and chilled storage. *J. Agric. Food Chem.* 54:9959–9965.

Ruiz-Capillas, C., P. Aller-Guiote, and F. Jiménez-Colmenero. 2007. Application of flow injection analysis to determine protein-bound nitrite in meat products. *Food Chem.* 101:812–816.

Ruiz-Capillas, C., A. M. Herrero, S. Tahmouzi et al. 2015b. Properties of reformulated hot dog sausage without added nitrites during chilled storage. *Food. Sci. Technol. Int.* doi: 10.1177/1082013214562919.

Ruiz-Capillas, C. and F. Jiménez-Colmenero. 2008a. Determination of preservatives in meat products by flow injection analysis (FIA). *Food Addit. Contam., Part A.* 25:1167–1178.

Ruiz-Capillas, C. and F. Jiménez-Colemenero, 2008b. Aplicación del análisis de inyección de flujo (FIA) a la determinación de nitratos y nitritos en productos cárnicos. *Eurocarne* 171:51–58.

Ruiz-Capillas, C., S. Tahmouzi, M. Triki, L. Rodriguez-Salas, F. Jiménez-Colmenero, and A. M. Herrero. 2015a. Free-nitrite added Asian hot dog sausages reformulated with nitrite replacers. *J. Food Sci. Technol.* 52:4333–4341.

Ruzicka, J. and E. H. Hansen. 1975. Flow injection analyses. 1. New concept of fast continuous-flow analysis. *Anal. Chim. Acta* 78:145–157.

Sammet, K., R. Duehlmeier, H. P. Sallmann, C. Canstein, T. Mueffling, and B. Nowak. 2006. Assessment of the antioxidative potential of dietary supplementation with α-tocopherol in low-nitrite salami-type sausages. *Meat Sci.* 72:270–279.

Sen, N. P., P. A. Baddoo, and S. W. Seaman. 1994. Rapid and sensitive determination of nitrite in foods and biological-materials by flow-injection or high-performance liquid-chromatography with chemiluminescence detection. *J. Chromatogr. A* 673: 77–84.

Senn, D. R. and P. W. Carr. 1976. Determination of nitrate at the part per billion level in environmental samples with a continuous-flow immobilized enzyme reactor. *Anal. Chem.* 46(7):954–958.

Shahidi, F. 1989. Current status of nitrite-free meat curing systems. In *Proceedings of the 35th International Congress of Meat Science and Technology,* Vol. 3, pp. 897–902, August 20–25, 1989, Copenhagen, Denmark.

Shinn, M. B. 1941. Colorimetric method for the determination of nitrite. *Ind. Eng. Chem. Anal. Ed.* 13:33–35.

Shiva, S. 2013. Nitrite: A physiological store of nitric oxide and modulator of mitochondrial function. *Redox Biol.* 1:40–44.

Sindelar, J. J., J. C. Cordray, J. G. Sebranek, J. A. Love, and D. U. Ahn. 2007a. Effects of varying levels of vegetable juice powder and incubation time on color, residual nitrate and nitrite, pigment, pH, and trained sensory attributes of ready-to-eat uncured ham. *J. Food Sci.* 72:388–395.

Sindelar, J. J., J. C. Cordray, D. G. Olson, J. G. Sebranek, and J. A. Love. 2007b. Investigating quality attributes and consumer acceptance of uncured, no-nitrate. *J. Food Sci.* 72:551–559.

Sofos, N., F. F. Busta, K. Bhothipaksa, and C. E. Allen. 1979. Sodium nitrite and sorbic acid effects on *Clostridium botulinum* toxin formation in chicken frankfurter-type emulsions. *J. Food Sci.* 44:668–675.

Suzuki, H. and T. Taura. 2001. Thin-film Ag/AgCl structure and operational modes to realize long-term storage. *J. Electrochem. Soc.* 148:E468–E474.

Torro, I. G., J. V. G. Mateo, and J. M. Calatayud. 1998. Flow-injection biamperometric determination of nitrate (by photoreduction) and nitrite with the NO_2-/I-reaction. *Anal. Chim. Acta* 366:241–249.

Urbain, W. M. 1971. Meat preservation. In *The Science of Meat and Meat Products* (2nd ed.), eds. J.F. Price, and B.S. Schweigert, pp. 402–451. San Francisco, CA: W.H. Freeman & Company. ISBN: 0-7167-0820-5.

Usher, C. D. and G. M. Telling. 1975. Analysis of nitrate and nitrite in foodstuffs—Critical review. *J. Sci. Food Agric.* 26:1793–1805.

Valcarcel, M. and M. D. Luque de Castro. 1987. *Flow-Injection Analysis: Principles and Applications.* Chichester, England: Ellis Horwood.

van Staden, J. F. and M. A. Makhafola. 1996. Spectrophotometric determination of nitrite in foodstuffs by flow injection analysis. *Fresenius' J. Anal. Chem.* 356:70–74.

Viuda-Martos, M., M. Ruiz-Navajas, Y. López-Fernandez, and J. A. Pérez-Álvarez. 2010. Effect of added citrus fibre and spice essential oils on quality characteristics and shelf-life of mortadella. *Meat Sci.* 85:568–576.

Wada, E. and A. Hattori. 1971. Spectrophotometric determination of traces of nitrite by concentration of azo dye on an anion-exchange resin: Application to sea waters. *Anal. Chim. Acta* 56:233–240.

Wang, G. F., K. Horita, and M. Satake. 1998. Simultaneous spectrophotometric determination of nitrate and nitrite in water and some vegetable samples by column preconcentration. *Microchem. J.* 58:162–174.

Warthesen, J. J., R. A. Scanlan, D. D. Bills, and I. M. Libbey. 1975. Formation of heterocyclic n-nitrosamines from reaction of nitrite and selected primary diamines and amino-acids. *J. Agric. Food Chem.* 23:898–902.

Wood, R., L. Foster, A. Damant, and P. Key. 2004. *Analytical Methods for Food Additives*, 253pp. Cambridge, UK: Woodhead.

Woods, L. F. J., J. M. Wood, and P. A. Gibb. 1989. Nitrite. In *Mechanisms of Action of Food Preservation Procedure*, ed. G.W. Gould, 225pp. Essex, UK: Elsevier Science Publisher.

Zen, J. M., A. S. Kumar, and H. F. Wang. 2000. A dual electrochemical sensor for nitrite and nitric oxide. *Analyst* 125:2169–2172.

CHAPTER 8

Determination of Sulfites

Claudia Ruiz-Capillas, Ana M. Herrero,
and Francisco Jiménez-Colmenero

CONTENTS

8.1 INTRODUCTION

Sulfites occur naturally in very small concentrations in the earth and in the atmosphere, and they are a natural ingredient in certain vegetables, foods and beverages as a result of endogenous fermentation by the yeasts in beer and wine (Taylor et al., 1986). Sulfites also occur naturally in the human body (Taylor et al., 1986; Adams, 1997; Food and Drug Administration [FDA], 2000), and they can be found in greater concentrations in natural waters or wastewaters as a result of industrial pollution, and in treatment plant effluents dechlorinated with sulfur dioxide, or as a consequence of volcanic activity, etc.

Sulfites are inorganic salts that have long been used in foods, beverages, and pharmaceuticals for the different functions they perform. Sulfur dioxide was used by the Romans in winemaking to keep the wine good for longer. It inhibits the growth of bacteria and molds, prevents oxidation of wine, and preserves its aroma and freshness, thus guaranteeing its quality by preventing further oxidation of polyphenols. An excess of sulfites impairs the quality of wine, causing it to lose color, giving it a sharp smell, and altering its flavor. Sulfites are also used on fruits and vegetables to prevent enzymatic and nonenzymatic browning, on shrimp and lobster to prevent melanosis or "black spot," in meat and meat products as antimicrobial and antioxidant agents, in dough as a conditioner, and to bleach certain food starches (e.g., potato) and cherries. They are also used in the production of some food packaging materials (e.g., cellophane) (Taylor et al.,

1986; Fazio and Warner, 1990; Sapers, 1993; FDA, 2000; Wood et al., 2004; Ruiter and Bergwerff, 2005). In addition, sulfites are used in pharmaceuticals to maintain the stability and potency of some medications (Papazian, 1996; Knodel, 1997). In some cases sulfites are used along with other additives such as nitrites or ascorbic acid, as this has been shown to enhance antimicrobial activity (Mirvish et al., 1972; Baird-Parker and Baillie, 1974; Raevuori, 1975).

The sulfites or sulfiting agents that can be added to foods or beverages are potassium bisulfite, potassium metabisulfite, sodium bisulfite, sodium dithionite, sodium metabisulfite, sodium sulfite, sulfur dioxide, and sulfurous acid. These can also be declared under the common names sulfites (sulphites) or sulfiting agents (sulphiting agents) (Table 8.1). Each is assigned an E number according to the rules and regulations of use laid down by the European Union (EU) (Table 8.1), with specific references to various foods such as wines, beers, fresh vegetable produce, canned vegetables, dried fruits, nuts, crustaceans, condiments, spices and meat preparations, and other processed foods (Ruiz-Capillas and Jiménez-Colmenero, 2009). The World Health Organization (WHO) recognizes nuts, fruit preserves, fruit juices, and wine as the main sources of dietary sulfite intake (FDA, 2000; WHO, 2000).

One essential aspect in evaluating the presence of these compounds in different matrices is that when added to foods they often disappear as a result of reversible and/or irreversible chemical reactions. When sulfites are added to foods, they come into contact with an aqueous medium there and undergo a process of dissociation in which the oxyanions may be separated from their cations depending on the pH, ionic strength and temperature of the medium. This process produces a dynamic chemical balance among species (sulfur dioxide, sulfurous acid, and sulfite and bisulfite anions) that tends toward the formation of one or another depending on the conditions in the medium so that all the forms coexist but in different proportions in different conditions (Wedzicha, 1992; Margarete et al., 2006).

Part of the dissociated sulfite (largely in the form of bisulfite or HSO_3^-) can bind reversibly or irreversibly to certain components of the food. This fraction is known as combined sulfite; the fraction not bound to the food is called free sulfite and remains dissociated in dynamic equilibrium. The sum of the two fractions is called total sulfite. Thus, it is often important to measure both free and bound forms of sulfite that are present in foods.

TABLE 8.1   Sulphiting Agents (Sulphite) with the Chemical Formula and E-Number

Sulphiting Agents (Sulphite)	Chemical Formula	E-Number
Sulphur dioxide	SO_2	E-220
Sodium sulphite	Na_2SO_3, $Na_2SO_3.7H_2O$	E-221
Sodium hydrogen sulphite	$NaHSO_3$	E-222
Sodium metabisulphite	$Na_2S_2O_5$	E-223
Potassium metabisulphite	$K_2S_2O_5$	E-224
Calcium sulphite	$CaSO_3$, $CaSO_3-2H_2O$	E-226
Calcium hydrogen sulphite	$Ca(HSO_3)_2$	E-227
Potassium hydrogen sulphite	$KHSO_3$	E-228

Note: An SO_2 content of not more than 10 mg/kg (mg/L) is not considered to be present.

8.2 POTENTIAL HEALTH IMPLICATIONS OF SULFITES

In general, for purposes of toxicology the use of sulfites in normal conditions or in permitted levels is not a problem for consumers, although the compounds are believed to exert various adverse effects on humans, chiefly in population groups that are sensitive or vulnerable to them, for instance persons suffering from sulfite oxidase deficiency and asthmatics. In the human organism the sulfite ingested with food is generally metabolized by an enzyme called sulfite oxidase. In persons with normal enzymatic activity there is no problem, but in persons who are highly sensitive to sulfites, the levels in some foods are enough to produce adverse effects associated with allergic reactions and food intolerance symptoms. Although the precise mechanisms of the sensitivity responses to sulfites are not fully understood, three have been identified: inhalation of sulfur dioxide (SO_2) generated in the stomach as a result of ingestion of sulfite-containing foods or beverages; deficiency in a mitochondrial enzyme; or an IgE-mediated immune response (Taylor et al., 1986; FDA, 1988a; Lester, 1995). The principal reactions to consumption of sulfites are asthmatic and allergic, with dermatological, pulmonary, gastrointestinal, and cardiovascular symptoms such as dermatitis, itching, rashes, urticaria, irritation of the gastrointestinal tract, colic, nausea, diarrhea, wheezing, coughing, difficulty in breathing, tightness in the chest, aggravation of asthma and even anaphylactic shock. In extreme cases it can produce a potentially deadly allergic reaction (Lester, 1995; Knodel, 1997). Adverse reactions to sulfites in nonasthmatics are extremely rare. The Food and Drug Administration (FDA) estimates that one out of a hundred people is sulfite-sensitive, and of that group 5% have asthma. Another source states that 5% of asthmatics are sulfite-sensitive, compared to only 1% of the nonasthmatic population (Knodel, 1997).

Another aspect to be considered regarding sulfites is the loss of nutritional value in some foods when sulfites are added. This is due to the ability of sulfites to break down thiamine (vitamin B1) into their components, thiazol and pyrimidine (Walker, 1985; Cassens, 1994). For this reason they need to be controlled and their use should be restricted to the technologically necessary minimum, especially in foods that are rich in thiamine, such as meat and meat derivatives.

8.3 REGULATORY STATUS OF SULFITES

The control and regulation of the use of sulfites in food and beverages is important for effectiveness and basically to ensure consumer safety. In this regard various national and international institutions (U.S. FDA, United States Department of Agriculture [USDA], Bureau of Alcohol, Tobacco, and Firearms [BATF], Environmental Protection Agency [EPA]) have laid down a number of rules to regulate sulfite levels in foods. Sulfites or sulfiting agents have been used as food additives in the United States since 1664 and have been approved for use in the United States since the 1800s (Lester, 1995). In 1974 the FAO and WAO (World Allergy Organization) International Scientific Conference established an acceptable daily intake (ADI) of sulfite (expressed as SO_2) of 0.7 mg/kg body weight; however, since then studies have been conducted to determine whether the intake in some population groups exceeds the ADI, mainly due to the presence of higher levels than permitted in the foods that they consume and to adverse health reactions to sulfites (Papazian, 1996; Timbo et al., 2004). These compounds are "generally regarded as safe" (GRAS) by the FDA (Papazian, 1996) in certain foods and in certain concentrations. The FDA requires that the presence of sulfites be declared on food labels when used as an ingredient in the food and also when used as a processing aid or when present in an ingredient used in the food (e.g., dried fruit pieces). Sulfites must be declared in these cases when the concentration in the food is $\geq$10 ppm total SO_2. Sulfiting agents are not currently considered GRAS for use in meats, foods recognized as major sources of vitamin B1 (sulfites have been found to destroy thiamine), or "fruits or vegetables intended to be served raw to consumers or to be presented to consumers as fresh" (FDA, 1988b, 2000; FAO/WHO, 2007).

The European Community has introduced various directives (Directives 95/2/EC, 2006/52/EC, and Real Decreto (RD) 1118/2007) to try to regulate the use of food additives, including sulfites, setting a maximum level (mg/kg or mg/L as appropriate), expressed as SO_2, for different foods and beverages. These levels vary widely depending on the products. Thus for example, the maximum levels for crustaceans and cephalopods range between 50 and 300 mg/kg (of the edible part); in meat products, the limit for burger meat, for instance, is 450 mg/kg (Triki et al., 2013); for dry biscuit it is between 30 and 50 mg/kg; for vegetables it is between 50 and 2000 mg/kg; and for beverages between 20 and 2000 mg/kg. Similarly, ceilings are set on total sulfurous anhydride in wines depending on the type—white, red, etc. (Council regulation (EC) No. 1493/1999, May 17, 1999 and Council regulation (EC) No. 1622/2000, July 24, 2000). Both the FDA and the EU regulations (Directive of 2004) on allergens require that wine labels bear specific messages advising of the presence of sulfites ("contains sulfites"), whenever the level exceeds 10 ppm.

In order to assess whether the use of sulfites in the specified food is appropriate and within the permitted limits, it is essential to maintain rigorous controls of concentrations in many products. The use of fast, sensitive, selective, and low-cost methods for determining sulfite is therefore important for food assurance and quality control (Ruiz-Capillas and Jiménez-Colmenero, 2009).

8.4 ANALYTICAL METHODS FOR SULFITE DETERMINATION

Numerous methods have been developed for determining sulfites in foods and beverages. These procedures involve titrimetry, electrochemistry, fluorometry, chemiluminescence spectrometry, colorimetry, gas–liquid chromatography, liquid chromatography, and others (Fazio and Warner, 1990; Wood et al., 2004; Ruiter and Bergwerff, 2005; Jiménez-Colmenero and Blazquez-Solana, 2007; Ruiz-Capillas and Jiménez-Colmenero, 2008, 2009). Most of the procedures are based on conversion of the various forms of sulfites to sulfur dioxide and range from the classic method of Monier and Williams—consisting basically of distillation in an acid medium followed by iodometric evaluation of the sulfur dioxide produced, as optimized over the years—to (a) spectrophotometric methods (e.g., based on a colorimetric reaction with p-rosaniline after reaction with mercuric extractant) (Fazio and Warner, 1990; AOAC, 2005a,b); (b) a combination of distillation/spectrophotometry using 5,5'-dithiobis-(2-nitrobenzoic acid) (DTNB), etc. (Banks and Board, 1982; Wedzicha and Mountfort, 1991; Li and Zhao, 2006); (c) differential pulse polarography (DPP) (Holak and Patel, 1987; Stonys, 1987; AOAC, 2005c); (d) chromatography: anion exclusion chromatography with electrochemical detection is the most commonly used system (Banks and Board, 1982; Anderson, et al., 1986; Kim and Kim, 1986; Paino-Campa, et al., 1991), although conductivity detection (Sullivan and Smith, 1985; Ruiz et al., 1994) and direct UV detection have also been reported (Pizzoferrato et al., 1998); (e) capillary electrophoresis (to determine the sulfate formed from sulfite by Monier–Williams distillation) (Trenerry, 1996); (f) enzymatic methods using sulfite oxidase to catalyze oxidation from sulfite to sulfate, releasing hydrogen peroxide that is measured spectrophotometrically by linking it to the oxidation of NADH (nicotinamide adenine dinucleotide) in the presence of NADH peroxidase (Edberg, 1993). Many of these methods have limitations in that, like the conventional Monier–Williams procedure for instance, they are tedious and time consuming (which limits the number of samples that can be analyzed in a working day), the level of detection is high, they cannot be used on certain products containing volatile compounds, and so on (Sullivan et al., 1986, 1990; Fazio and Warner, 1990). Chromatographic methods have lower detection limits and analyses take less time, but determination is difficult in products where sulfur dioxide is strongly bound, and moreover the equipment they require is costly and sometimes difficult to operate for routine laboratories. For their part, enzymatic methods are hard to standardize and less sensitive. On the other hand, FIA offers major advantages for determination of sulfites in foods and beverages, and as a result has become increasingly important (Ruiz-Capillas and Jiménez-Colmenero, 2009).

This chapter provides an overview of the various procedures currently in use, with specific reference to those based on flow injection analysis (FIA), as this offers advantages such as sensitivity, selectivity, rapidity, low cost and automation, among others. FIA is reviewed here as an alternative to conventional methods for determining sulfites in foods and beverages, looking in detail at aspects of extraction, separation, detection and quantification procedures in different matrices.

8.5 FLOW INJECTION METHODOLOGIES FOR SULFITE DETERMINATION

FIA offers a fast analytical response in real time. It is a highly versatile and flexible, low-cost technique suitable for routine determination of large numbers of samples. It is also very accurate and easy to handle, with low reagent consumption. It uses small sample volumes and less of toxic substances and is compatible with almost any detection principle. The instrumentation is relatively simple and can be miniaturized (Ruzicka and Hansen, 1975; Osborne and Tyson, 1988; Ruiz-Capillas and Jiménez-Colmenero, 2008, 2009). FIA procedures have been used extensively to determine sulfite in foods and beverages. From its beginnings (Ruzicka and Hansen, 1975), FIA has been widely used and developed for routine analysis of various compounds, and particularly for the determination of sulfites in foods and beverages in place of traditional methods. As with most analytical methods, determination of additives by FIA generally consists of two phases: extraction of a sulfiting agent from the food, and separation/evaluation. Especially in complex matrices like foods (meat, fish, etc.), because of the instability of the analyte and the possibility of binding with other compounds in the food or beverage concerned, extraction and subsequent determination are crucial stages in all processes of FIA analysis.

8.5.1 Sulfite Extraction Procedures for Flow Injection Analysis

Various different extraction conditions (distillation, acid extractant, alkaline extractant, etc.) have been tried for determination of sulfites in foods and beverages. These processes require some procedure to remove and recover the total sulfites (free and reversibly bound) and avoid loss of sulfites during extraction, which can be substantial. This is especially important in complex food matrices where extraction is a crucial step in sulfite determination.

Sulfites may be present in foods and beverages as sulfurous acid, inorganic sulfites and a variety of reversibly and irreversibly combined forms. When sulfiting agents are added to foods they react rapidly with a variety of food components (sugars, aldehydes, ketones, proteins, etc.), some reversibly bound (probably as hydrosulfonate adducts) and some irreversibly bound (mainly by reaction with alkanes or aromatic compounds). Sulfites may even be present, in a dynamic equilibrium, in free (nonbound) form (mainly sulfur dioxide, bisulfite, and sulfite ions) (Wedzicha and Mountfort, 1991; Wedzicha, 1992; Pizzoferrato et al., 1998). Sulfites themselves can undergo chemical reactions such as volatilization of sulfur dioxide, oxidation to sulfates, and so on. (Fazio and Warner, 1990). The oxidation of sulfite in solution has been a major problem in the determination of this compound. The presence of iron and other metal ions in trace amounts in analytical grade Na_2SO_3 (e.g., in the preparation of standard solutions) may be responsible for the fast degradation of unstabilized standard sulfite solutions.

The amount or presence of sulfite in each of these states (freely, reversibly or irreversibly bound) depends on a number of factors such as the level of addition, the type and composition of the food or beverage, and the processing and storage conditions. For example, reversibly bound sulfite may dissociate into free sulfite when a food is treated with acidic solutions and heated to boiling (Fazio and Warner, 1990; Lück and Jager, 1997). On the other hand, the free sulfite fraction is rapidly converted to molecular sulfur dioxide when the sulfited food is acidified (Wedzicha, 1992). Analytical determination of sulfite does not therefore reflect the preservatives that were initially added.

In this connection, various different types of extraction process have been tried, depending on the type of sample but above all in complex food matrices, to avoid loss of labile forms of sulfite during extraction and thus remove all the sulfite in the sample, which can then be transferred to the FIA system (Linares et al., 1989; MacLaurin et al., 1990; Sullivan et al., 1990; Bendtsen and Jorgensen, 1994; Frenzel and Hillmann, 1995; Atanassov et al., 2000; León et al., 2004; AOAC, 2005a,b,c).

The extraction procedures most commonly performed for determination of sulfites by FIA include the use of solutions containing acetone, formaldehyde, ethanol, tetrachloromercurate (TCM), etc.; alkaline extractions have also been assayed but they have been found to present problems of loss of sulfites (Sullivan et al., 1986, 1990; Frenzel and Hillmann, 1995; León et al., 2004; AOAC, 2005a,b,c; Ruiz-Capillas and Jiménez-Colmenero, 2009). The procedure for extraction with formaldehyde works through the formation of a stable derivative, hydroxymethyl-sulfonate (Frenzel and Hillmann, 1995). Extraction with 10% ethanol and KOH has been performed in wines (Decnopweever and Kraak, 1997). The use of ethanol for extraction has also been combined with a process of degassing in an ultrasonic bath with an antifoaming agent (1-octanol) in beer (Bendtsen and Jorgensen, 1994). In some cases, the process has been simplified by degassing the sample and equilibrating with nitrogen (Mana and Spohn, 2001).

Of all extraction processes, the one with sodium TCM solutions is the most widely used for its efficiency in both solid and liquid samples. In liquid or semisolid solutions such as water, wines, fruit juices, etc., the sulfites are extracted with TCM and injected directly into the FIA system, without any prior treatment or following adjustment of the pH (Granados et al., 1986; Linares et al., 1989; Bartroli et al., 1991; Huang et al., 1992; Safavi and Haghighi, 1997; Araujo et al., 1998; Su and Wei, 1998; Mana and Spohn, 2001; Corbo and Bertotti, 2002; AOAC, 2005a,b,c; Araujo et al., 2005). TCM solution helps to release the sulfite from solid foods and form a stable sulfite–mercuric complex, providing excellent sulfite recovery and preventing oxidation of solutions (Sullivan et al., 1986; Ruiter and Bergwerff, 2005). Bartroli et al. (1991) reported that a 0.01 M Hg solution was the one that provided the best sensitivity in a FIA system. TCM has also been used to stabilize standard sulfite solutions, as these pose problems of instability (mainly oxidation) over time in storage (Massom and Townshend, 1986). Other stabilizers have also been proposed, and some have been used for extraction, including EDTA (ethylenediaminetetraacetic acid), formaldehyde, glycerol, and isopropanol (Massom and Townshend, 1986; Bendtsen and Jorgensen, 1994; Safavi and Haghighi, 1997; León et al., 2004).

In any case sample preparation and analysis should be as fast as possible to avoid loss of labile forms of sulfite.

8.5.2 Sulfite Release/Separation/Detection Systems

The second phase of FIA procedures for determining sulfites in foods and beverages involves injection of the liquid sample extract (containing sulfur dioxide) into the FIA system (Figure 8.1) for determination and quantification. As noted, extraction processes generally entail the formation of stable complexes to minimize loss of labile forms of sulfite. Depending on the particular FIA methodology, release processes may be necessary before the evaluation as such. These processes are intended to promote rupturing of the bonds linking the sulfites to aldehydes and TCM that form during extraction. They can be performed outside the FIA system during sample preparation or in a prehydrolysis unit incorporated in the FIA system to automate this reaction (Figure 8.2). Briefly then, such processes can be accompanied by hydrolysis in a basic medium (mainly with NaOH, or KOH and hydroxymethanesulfonate) to release most of the bound sulfites from the extract solution (Sullivan et al., 1990; Bartroli et al., 1991; Huang et al., 1992; Frenzel and Hillmann, 1995; Decnopweever and Kraak, 1997; Atanassov et al., 2000; León et al., 2004; AOAC 2005a, b, c; Ruiz-Capillas and Jiménez-Colmenero, 2009). The

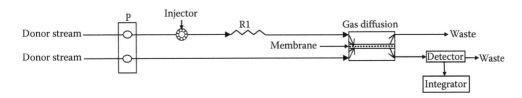

FIGURE 8.1 FIA diagram for sulfite determination (P: pump; R1, R2: reactors). (Adapted from Ruiz-Capillas, C. and F. Jiménez-Colmenero. 2009. *Food Chem.* 112:487–493.)

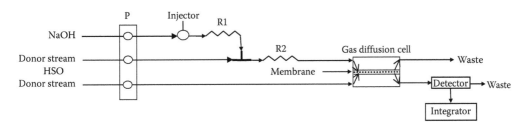

FIGURE 8.2 FIA diagram with hydrolysis (pre-hydrolysis) treatment for sulfite determination (P: pump; R1, R2: reactors). (Adapted from Ruiz-Capillas, C. and F. Jiménez-Colmenero. 2009. *Food Chem.* 112:487–493.)

release of most of the bound sulfites from the extract solution is complete above pH 8 (Kim and Kim, 1986; Sullivan et al., 1990; AOAC, 2005c; Ruiz-Capillas and Jiménez-Colmenero, 2009). Following is a description of the different sulfite separation and determination procedures.

8.5.2.1 Separation Step

Sulfite determination can be performed by a simple flow system analysis without any separation system, based only on a color reaction in a reaction coil; however, the process generally involves a sulfite separation step (depending on the complexity of the sample), for which there are a number of possible procedures: gas diffusion, gas–liquid system, dialysis, or ion exchange (Ruiz-Capillas and Jiménez-Colmenero, 2009). Of the different sulfite separation procedures that have been assayed, the most widely used is the one based on a gas diffusion cell process (Granados et al., 1986; Sullivan et al., 1986, 1990; Linares et al., 1989; Bartroli et al., 1991; Huang et al., 1992; Bendtsen and Jorgensen, 1994; Frenzel and Hillmann, 1995; Decnopweever and Kraak, 1997; Araujo et al., 1998, 2005; Su and Wei, 1998; Atanassov et al., 2000; Mana and Spohn 2001; Corbo and Bertotti, 2002; León et al., 2004; AOAC, 2005a,b,c; Dvorak et al., 2006). This is similar to the gas diffusion system used for other determinations, for example, of trimethylamines (Ruiz-Capillas and Horner, 1999). In general, the gas diffusion unit (Figure 8.3) consists of two liquid streams, a strong acidic donor solution (where the SO_2 gas is released) and an acceptor solution (which collects the released SO_2) containing an indicator, separated by a membrane (generally Teflon) that is permeable only to gases. The membrane is not only necessary to separate the strong acidic donor solution from the acceptor solution, but also forms a barrier against potential interferences in the sample.

The donor stream transports a strong acidic donor solution intended to promote the release of the SO_2 present in the test solution. This solution is generally composed of sulfuric acid, but citric acid, HCL, HNO_3, and other substances have also been used (Sullivan et al., 1986, 1990; Bendtsen and Jorgensen, 1994; León et al., 2004; AOAC 2005a,b,c; Araujo et al., 2005; Ruiz-Capillas and Jiménez-Colmenero, 2009) (Figures 8.1 and 8.2).

The acceptor stream, where the released SO_2 is collected, is usually a solution containing an acid–base indicator. The SO_2 gas released from the donor diffuses through the membrane and is dissolved into the acceptor solution. Which acceptor is used (iodine, bromocresol green, malachite green, etc.) in sulfite determination depends on the detection system concerned (spectrophotometry, fluorometry, amperometry, etc.) (Figures 8.1 and 8.2).

Determination of sulfite can be done without any separation system, for instance in the case of brine or stock solution, but the use of gas diffusion has the advantage that it can be used for food extracts, even strongly colored ones such as red wines, so that they can be processed without sample pretreatment. Another inherent advantage

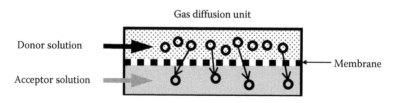

FIGURE 8.3   Diagrams of the gas diffusion unit for FIA.

of this separation system is the possibility of optimizing the sample treatment conditions in the donor stream without this affecting the detection chemistry in the acceptor stream (Frenzel and Hillmann, 1995). A separation process based on a minicolumn of ion exchange resin in thiocyanate form has also been used for spectrophotometric sulfite determination at 455 nm (Yaqoob et al., 1991).

There have been studies on interferences in this system of SO_2 determination with gas diffusion, chiefly in connection with the presence of compounds in beverages that are able to pass through the permeable membrane, such as nitrite, sulfide and cyanide (Frenzel and Hillmann, 1995). No interferences were detected with concentrations of nitrite up to 3 times higher than those of sulfite. Cyanine interference became apparent only when present at concentrations above 10 mg/L (Frenzel and Hillmann, 1995). In some cases, EDTA has been assayed as a means of preventing interference from some cations and anions; however, it was found that EDTA practically doubled the sensitivity of the sulfite determination (Safavi and Haghighi, 1997).

8.5.2.2 Detection Procedures

A number of different sulfite detection procedures have been reported, with or without separation processes, including spectrophotometry, potentiometry, amperometry, fluorescence, enzymes, sensors, electrodes, and others.

The color reaction in the FIA system is achieved with various different colorimetric reagents such as iodine, bromocresol green, p-rosaniline–formaldehyde solution, p-aminoazobenzene, luminol, and malachite green. Malachite green is the most commonly used solution in the determination of SO_2 by FIA in solid matrices (muscle-based foods) (Sullivan et al., 1986, 1990; Linares et al., 1989; Atanassov et al., 2000; AOAC, 2005a,b,c; Ruiz-Capillas and Jiménez-Colmenero, 2009). These color reagents lose color in proportion to the amount of sulfite present in the sample, which is generally measured spectrophotometrically at 615 nm in the case of malachite green. Iodine is measured at 620 nm, p-aminoazobenzene is detected at 520 nm, p-rosaniline-formaldyde with a maximum absorption at 578 nm, or bromocresol green an optimal absorption at 620 nm, and luminol with chemiluminescence detection. However, there are drawbacks to some of these reagents. For example, the complexation of p-rosaniline and sulfur dioxide is rather slow, reproducibility has been found to be poor, the reagent is toxic and carcinogenic, and the detection limit of sulfur dioxide with this technique is lower. The kinetics of the sulfite/p-aminoazobenzene reaction is very slow and extremely long reaction coils are needed to achieve sufficient sensitivity. Bromocresol green is a less toxic reagent and the kinetics is faster; however, this method cannot be used for sparkling wines and beers because carbon dioxide interferes with photometric detection (Decnopweever and Kraak, 1997). To get round some of these problems, other acceptors have been used, such as 4,4-dithiodipyridine, which reacts rapidly with sulfite to yield a thiol anion complex that is detected at 324 nm (Frenzel and Hillmann, 1995). MacLaurin et al. (1990) also used 5,5′-dithiobis-(2-nitrobenzoic acid) (DTNB) in brine. The sulfite reacted with the DTNB to produce the chromophoric species 2-nitro-5-mercaptobenzoic acid, which was monitored spectrophotometrically at 412 nm. More recently, o-phthalaldehyde (OPA)/ammonium reagent and fluorescent detection have been used. The highest sensitivity was achieved at an excitation wavelength of 330 nm and an emission wavelength of 390 nm at pH 6.5 (Mana and Spohn, 2001).

A copper electrode in an alkaline medium with amperometric detection has been used for sulfite determination in beverages (Corbo and Bertotti, 2002). Other systems have been reported such as a potentiometric sensor with a hydrated titanium oxide anion exchanger and an epoxy resin matrix membrane (Hassan et al., 2001); enzymes (sulfite

oxidase, etc.) immobilized in different systems and coupled to a FIA system, with amperometric or chemiluminescence detection (Massom and Townshend, 1986; Yaqoob et al.,
2004); chip-based micro-FIA for the reaction between a solution of Ce(IV) and sulfite
sensed by Rh6G for chemiluminescence determination of sulfite (He et al., 2005); and an
electrochemical device, based on a carbon paste electrode modified with *p*-rosaniline-
formaldehyde (PRA) in a phosphate buffer at pH 7 (Méndez et al., 2013).

8.6 CONCLUSIONS

Sulfites in foods or beverages need to be quantified to assure quality and compliance with
legal regulations and to minimize the health risk to consumers. Because of its noticeable advantages, FIA has become one of the most widely used methods for determining
sulfites. FIA gives a fast, simple and precise analytical response in real time, and small
amounts of sulfite can be determined with low reagent consumption when small volumes
of samples are available. The instrumentation is relatively simple and miniaturization is
possible. It offers low interference and can be easily applied for routine analyses in control laboratories or industrial applications.

ACKNOWLEDGMENTS

This research was supported by projects AGL2010-19515 and AGL 2011-29644-C02-01
of the Plan Nacional de Investigación Científica, Desarrollo e Innovación Tecnológica
(I + D + I) Ministerio de Ciencia y Tecnología. The author thanks Mª Carmen Pérez for
everything.

REFERENCES

Adams, J. B. 1997. Food additive–additive interactions involving sulphur dioxide and
ascorbic and nitrous acids: A review. *Food Chem.* 59:401–409.
Anderson, C., C. R. Warner, D. H. Daniels, and K. L. Padgett. 1986. Ion chromato-
graphic determination of sulfites in foods. *J. AOAC* 69:14–19.
AOAC. 2005a. Sulfites in meats. Qualitative test. In *Official Methods of Analysis of
AOAC International* (18th ed.) eds. W. Horwtz and G. W. Latimer, Chapter 47.
Gaithersburg, MD: AOAC International, Official Method 961.09.
AOAC. 2005b. Sulfurous acid (total) in food. Modified Monier-Williams method. In
Official Methods of Analysis of AOAC International (18th ed.) eds. W. Horwtz
and G. W. Latimer, Chapter 47. Gaithersburg, MD: AOAC International, Official
Method 962.16.
AOAC. 2005c. Sulfites (total) in foods. Differential pulse polarographic (DPP) method.
In *Official Methods of Analysis of AOAC International* (18th ed.) eds. W. Horwtz
and G. W. Latimer, Chapter 47. Gaithersburg, MD: AOAC International; Official
Method 987.04.
Araujo, A. N., C. M. C. M. Couto, J. L. F. C. Lima, and M. C. B. S. M. Montenegro.
1998. Determination of SO_2 in wines using a flow injection analysis system with
potentiometric detection. *J. Agric. Food Chem.* 46:168–172.

Araujo, C. S. T., J. L. de Carvalho, D. R. Mota, C. L. de Araujo, and N. M. M. Coelho. 2005. Determination of sulphite and acetic acid in foods by gas permeation flow injection analysis. *Food Chem.* 92:765–770.

Atanassov, G., R. C. Lima, R. B. R. Mesquita, A. O. S. S. Rangel, and I. V. Toth. 2000. Spectrophotometric determination of carbon dioxide and sulphur dioxide in wines by flow injection. *Analusis* 28:77–82.

Baird-Parker, A. C. and M. A. H. Baillie. 1974. The inhibition of *Clostridium botulinum* by nitrite and sodium chloride. In: *Proceedings of the International Symposium on Nitrite in Meat Products*, eds. Tinbergen, B. J., and B. Krol, p. 77. Zeist, The Netherlands: PUDOC, Wageningen.

Banks, J. G. and R. G. Board. 1982. Comparison of methods for the determination of free and bound sulfur-dioxide in stored British fresh sausage. *J. Sci. Food Agric.* 33:197–203.

Bartroli, J., M. Escalada, C. J. Jorquera, and J. Alonso. 1991. Determination of total and free sulfur-dioxide in wine by flow-injection analysis and gas-diffusion using para-aminoazobenzene as the colorimetric-reagent. *Anal. Chem.* 63:2532–2535.

Bendtsen, A. B. and S. S. Jorgensen. 1994. Determination of total and free sulfite in unstabilized beer by flow injection analysis. *J. AOAC Int.* 77:948–951.

Cassens, R. G. 1994. *Meat Preservation, Preventing Losses and Assuring Safety* (1st ed.), pp 79–92. Trumbull, CT: Food and Nutrition Press, Inc., ISBN: 0917678346.

Corbo, D. and M. Bertotti. 2002. Use of a copper electrode in alkaline medium as an amperometric sensor for sulphite in a flow-through configuration. *Anal. Bioanal. Chem.* 374:416–420.

Council regulation (EC) No. 1493/1999 of 17 May 1999 on the common organization of the market in wine.

Council regulation (EC) No 1622/2000 of 24 July 2000 laying down certain detailed rules for implementing Regulation (EC) No 1493/1999 on the common organisation of the market in wine and establishing a Community code of oenological practices and processes.

Decnopweever, L. G. and J. C. Kraak. 1997. Determination of sulphite in wines by gas-diffusion flow injection analysis utilizing spectrophotometric pH-detection. *Anal. Chim. Acta* 337:125–131.

Directive 2006/52/EC of the European Parliament and of the Council of 5 July amending Directive 95/2/EC on food additives other than colours and sweeteners and Directive 94/35/EC on sweeteners for use in foodstuffs.

Directive No. 95/2/EC of the European Parliament and of the Council of 20 February 1995 on food additives other than colours and sweeteners.

Dvorak, J., P. Dostalek, K. Sterba et al. 2006. Determination of total sulphur dioxide in beer samples by flow-through chronopotentiometry. *J. Inst. Brew.* 112:308–313.

Edberg, U. 1993. Enzymatic determination of sulfite in foods: NMKL interlaboratory study. *J. AOAC Int.* 76:53–58.

FAO/WHO. 2007. Expert committee of food additives sulfur dioxide. http:// www.inchem.org/documents/jecfa/jeceval/jec_2215.htm (accessed 12.11.07).

Fazio, T. and C. R. Warner. 1990. A review of sulphites in foods: Analytical methodology and reported finding. *Food Addit. Contam.* 7:433–454.

Food and Drug Administration (FDA). 1988a. Sulfiting agents in standardized foods: Labeling requirements. *Food Drug Admin.* (*Code of Federal Regulations*), 53:51062–51084.

Food and Drug Administration (FDA). 1988b. Sulfiting agents: Affirmation of GRAS status. *Food Drug Admin. (Code of Federal Regulations)* 53:51065–51084.

Food and Drug Administration (FDA). 2000. Sulfites: An important food safety issue, an update on regulatory status and methodologies. http://www.cfsan.fda.gov/~dms/fssulfit.html.

Frenzel, W. and B. Hillmann. 1995. Gas-diffusion flow-injection analysis to the determination of sulfite and sulfur-dioxide in environmental-samples. *Chem. Anal. Warsaw* 40:619–630.

Granados, M., S. Maspoch, and M. Blanco. 1986. Determination of sulphur dioxide by flow injection analysis with amperometric detection. *Anal. Chim. Acta* 179: 445–451.

Hassan, S. S. M., S. A. Marei, I. H. Badr, and H. A. Arida. 2001. Flow injection analysis of sulfite ion with a potentiometric titanium phosphate-epoxy based membrane sensor. *Talanta* 54:773–782.

He, D. Y., Z. J. Zhang, and Y. Huang. 2005. Chemiluminescence microllow injection analysis system on a chip for the determination of sulfite in food. *Anal. Lett.* 38:563–571.

Holak, W. and B. Patel, 1987. Differential pulse polarographic-determination of sulfites in foods—Collaborative study. *J. AOAC* 70:572.

Huang, Y. L., J. M. Kim, and R. D. Schmid. 1992. Determination of sulfite in wine through flow-injection analysis based on the suppression of luminol chemiluminescence. *Anal. Chim. Acta* 266:317–323.

Jiménez-Colmenero, F. and J. Blazquez-Solana. 2007. Additives: Preservatives. In: *Handbook of Processed Meats and Poultry Analysis*, eds. L. Nollet and F. Toldrá. Boca Raton, FL: Taylor & Francis, LLC.

Kim, H. J. and Y. K. Kim. 1986. Analysis of free and total sulfites in food by ion chromatography with electrochemical detection. *J. Food Sci.* 51:1360–1361.

Knodel, L. C. 1997. Current issues in drug toxicity; potential health hazards of sulfites. *Toxic Subst. Mech.* 16:309–311.

León, J. A., C. Santín, J. M. Pich, and F. Centrich. 2004. Determinación de dióxido de azufre en ajo en polvo mediante análisis por inyección de flujo (FIA). *I Congreso Nacional de Laboratorios Agroalimentarios*, Lugo, October.

Lester, M. R. 1995. Sulfite sensitivity: Significance in human health. *J. Am. Coll. Nutr.* 14:229–232.

Li, Y. and M. Zhao. 2006. Simple methods for rapid determination of sulfite in food products. *Food Control* 17:975–980.

Linares, P., M. D. L. DeCastro, and M. Valcarcel. 1989. Simultaneous determination of carbon-dioxide and sulfur-dioxide in wine by gas-diffusion flow-injection analysis. *Anal. Chim. Acta* 225:443–448.

Lück, E. and M. Jager. 1997. Sulfur dioxide. In *Antimicrobial Food Additives—Characteristics, Uses, Effects* (2 ed.), ed. G. F. Edwards, p. 262. Berlin: Springer.

MacLaurin, P., K. S. Parker, A. Townshends, P. J. Worsforld, N. W. Barnett, and M. Crane. 1990. Online determination of sulfite in brine by flow-injection analysis. *Anal. Chim. Acta* 238:171–175.

Mana, H. and U. Spohn. 2001. Sensitive and selective flow injection analysis of hydrogen sulfite/sulfur dioxide by fluorescence detection with and without membrane separation by gas diffusion. *Anal. Chem.* 73:3187–3192.

Margarete, R., D. Machado, M. C. F. Toledo, and E. Vicente. 2006. Sulfitos em alimentos. *Braz. J. Food Technol.* 9:265–275.

Massom, M. and A. Townshend. 1986. Flow-injection determination of sulphite and assay of sulphite oxidase. *Anal, Chim Acta* 179:399–405.

Méndez, L., L. M. Ortega, and M. Choy. 2013. Sensor electroquímico para la determinación de sulfito. *QuimicaViva* 12:247–261.

Mirvish, S. S., L. Wallcave, M. Eagen, and P. Shubik. 1972. Ascorbate-nitrite reaction: Possible means of blocking the formation of carcinogenic N-nitroso compounds. *Science* 7:65–68.

Osborne, B. G. and J. F. Tyson. 1988. Review: Flow injection analysis—A new technique for food and beverage analysis. *Int. J. Food Sci. Technol.* 23:541–554.

Paino-Campa, G., M. J. Penaegido, and C. García-Moreno. 1991. Liquid-chromatographic determination of free and total sulfites in fresh sausages. *J. Sci. Food Agric.* 56:85–93.

Papazian, R. 1996. Sulfites: Safe for most, dangerous for some. *FDA Consum. Mag.* 30.

Pizzoferrato, L., G. Di Lullo, and E. Quattrucci. 1998. Determination of free, bound and total sulphites in foods by indirect photometry-HPLC. *Food Chem.* 63:275–279.

Raevuori, M. 1975. Effect of nitrite and erythorbate on growth of *Bacillus cereus* in cooked sausage and laboratory media. *Zentralbl. Bakteriol. Orig. B* 161:280–287.

Real Decreto (RD) 1118/2007, de 24 de agosto, por el que se modifica el Real Decreto 142/2002, de 1 de febrero, por el que se aprueba la lista positiva de aditivos distintos de colorantes y edulcorantes para su uso en la elaboración de productos alimenticios, así como sus condiciones de utilización.

Ruiter, A. and A. A. Bergwerff. 2005. Analysis of chemical preservatives in foods. In *Methods of Analysis of Food Components and Additives*, ed. S. Ötles, Chapter 14. Boca Raton, FL: Taylor & Francis.

Ruiz, E., M. I. Santillana, M. De Alba, M. T. Neeto, and S. García-Castellanoa. 1994. High-performance ion chromatography determination of total sulfites in foodstuffs. *J. Liq. Chromatogr.* 17:447–456.

Ruiz-Capillas, C. and W. F. A. Horner. 1999. Determination of trimethylamine-nitrogen and total volatile basic-nitrogen in fresh fish by flow injection analysis. *J. Sci. Food Agric.* 79:1982–1986.

Ruiz-Capillas, C. and F. Jiménez-Colmenero. 2008. Determination of preservatives in meat products by flow injection analysis (FIA). *Food Addit. Contam.: Part A* 25:1167–1178.

Ruiz-Capillas, C. and F. Jiménez-Colmenero. 2009. Application of flow injection analysis for determining sulphites in food and beverages: A review. *Food Chem.* 112:487–493.

Ruzicka, J. and E. H. Hansen. 1975. Flow injection analyses. 1. New concept of fast continuous-flow analysis. *Anal. Chim. Acta* 78:145–157.

Safavi, A. and B. Haghighi. 1997. Flow injection analysis of sulphite by gas-phase molecular absorption UV/VIS spectrophotometry. *Talanta* 44:1009–1016.

Sapers, G. M. 1993. Browning of foods: Control by sulfites, antioxidants, and other means. *Food Tech.* 47:75–84.

Stonys, D. B. 1987. Determination of sulfur-dioxide in foods by modified Monier–Williams distillation and polarographic detection. *J. AOAC* 70:114–117.

Su, X. L. and W. Wei. 1998. Flow injection determination of sulfite in wines and fruit juices by using a bulk acoustic wave impedance sensor coupled to a membrane separation technique. *Analyst* 123:221–224.

Sullivan, D. M. and R. L. Smith, 1985. Determination of sulfite in foods by ion chromatography. *Food Tech.* 39:45.

Sullivan, J. J., T. A. Hollingworth, M. M. Wekell, R. T. Newton, and J. E. Larose. 1986. Determination of sulfite in food by flow-injection analysis. *J. AOAC* 69:542–546.

Sullivan, J. J., T. A. Hollingworth, M. M. Wekell et al. 1990. Determination of total sulfite in shrimp, potatoes, dried pineapple, and white wine by flow-injection analysis—Collaborative study. *J. AOAC* 73:35–42.

Taylor, S. L., N. A. Higley, and R. K. Bush. 1986. Sulfites in foods: Uses, analytical methods, residues, fate, exposure assessment, metabolism, toxicity, and hypersensitivity. *Adv. Food Res.* 30:1–76.

Timbo, B., K. M. Koehler, C. Wolyniak, and K. C. Klontz. 2004. Sulfites a food and drug administration review of recalls and reported adverse events. *J. Food Prot.* 67:1806–1811.

Trenerry, V. C. 1996. The determination of the sulphite content of some foods and beverages by capillary electrophoresis. *Food Chem.* 55:299–303.

Triki, M., A. M. Herrero, F. Jimenez-Colmenero, and C. Ruiz-Capillas. 2013. Storage stability of low-fat sodium reduced fresh merguez sausage prepared with olive oil in konjac gel matrix. *Meat Sci.* 94:438–446.

Walker, R. 1985. Sulphiting agents in foods: Some risk/benefit considerations. *Food Addit. Contam.*, Part A, 2:5–24.

Wedzicha, B. L. 1992. Chemistry of sulphiting agents in food. *Food Addit. Contam.* 9:449–459.

Wedzicha, B. L. and K. A. Mountfort. 1991. Reactivity of sulfur-dioxide in comminuted meat. *Food Chem.* 39:281–297.

WHO. 2000. WHO—World Health Organisation. Evaluation of certain food additives (Fifty-first report of the Joint FAO/WHO Expert Committee on Food Additives). *WHO Technical Report Series*, No. 891, Geneva.

Wood, R., L. Foster, A. Damant, and P. Key. 2004. *Analytical Methods for Food Additives*. Boca Raton, Cambridge: CRC Press Woodhead Publishing Limited.

Yaqoob, M., A. Nabi, A. Weseem, and M. Massom-Yasinzai. 2004. Determination of sulphite using an immobilized enzyme with flow injection chemiluminescence detection. *Luminescence* 19:26–30.

Yaqoob, M., M. A. Siddiqui, and M. Masoom. 1991. Spectrophotometric flow-injection determination of nitrate and nitrite in potable water using 8-hydroxyquinoline. *J. Chem. Soc. Pak.* 13:248–252.

CHAPTER 9

Determination of Sorbic Acid and Sorbates, Benzoic Acid, and Benzoates

José Manuel Cano Pavón, Amparo García de Torres, and Elisa Isabel Vereda Alonso

CONTENTS

9.1 INTRODUCTION

Preservation is usually defined as a method used to maintain an existing condition or to prevent damage likely to be brought about by chemical (oxidation), physical (temperature, light), or biological (microorganism) factors. The main function of food preservation is to delay the spoiling of foodstuffs and to prevent any alterations in their taste or, in some cases, their appearance. This can be done in different ways, through processing methods including canning, dehydration (drying), smoking and freezing, the use of packaging, and the use of food additives.

A food additive is defined as a substance or mixture of substances that is present in food as a result of production, processing, storage, or packaging. Food additives may be divided into six categories [1]: nutritional additives (which include vitamins, minerals, amino acids, and fibers); flavoring agents (which comprise sweeteners, natural flavors, synthetic flavors, and flavor enhancers); coloring agents; texturizing agents (such as emulsifiers and stabilizers); miscellaneous additives (such as enzymes, various solvents, chelating agents, and catalyst); and preservatives.

There are three types of food preservatives [1]: antimicrobials, which inhibit the activity or growth of microorganisms and molds; antioxidants, which are used to prevent the oxidation of vitamins, minerals, and lipids of foods; and antibrowning agents which prevent both enzymatic and nonenzymatic browning of foodstuffs. The main purpose of preservatives is to maintain the safety of food for human consumption, and retain its nutritional value and its overall quality. The factors that affect microbial growth in a food product are the water activity, the pH, the presence or absence of oxygen, the availability of nutrients, and the temperature.

Since foods are an excellent source of nutrients for the attraction and growth of microorganisms, the primary reason for using preservatives is to make foods safer by eliminating the influence of biological factors. The greatest threat to consumers is that of food being spoiled, or becoming toxic through the effect of microorganisms (e.g., bacteria, yeasts, molds) occurring in them. Some of these organisms can secrete poisonous substances (toxins), which are dangerous to human health and can even be fatal.

Preservatives are one of the 26 major additives categories that are used in the food processing. To ensure that preservatives really do help make foodstuffs safer, their use is subject to premarket safety assessment and authorization procedures. At the European level, the bodies responsible for the safety assessment, authorization, control, and labeling of preservatives and other additives are the European Food Safety Authority (EFSA) and the European Commission Parliament and Council. At the international level, there is a Joint Expert Committee from the Food and Agriculture Organization (FAO) and the World Health Organization (WHO) Joint Expert Committee for Food Additives (JECFA) [2].

Safety assessment of preservatives, as for other food additives, is based on reviews of all available toxicological data, including observations in humans and in animal models. From the available data, a maximum level of an additive that has no demonstrable toxic effect is determined. This is called the "no observed adverse effect level" (NOAEL) and is used to determine the "acceptable daily intake" (ADI). The ADI refers to the amount of a food additive that can be taken daily in the diet, over a lifetime span, without any negative effect on health. Food additive legislation adopted by the European Union is included in several European Parliament and Council Directives (Directives 95/2/EC and 2006/52/EC, which have been replaced by the Regulation (EC) 1333/2008 in 2011).

Preservative food additives are used to prevent the proliferation of bacteria, yeast, and molds [3]; sorbic acid, benzoic acid, and their salts are used extensively in this context. They are commonly used in a great variety of foods, namely, fruit products, jams, relishes, beverages, dressings, salads, pie and pastry fillings, icings, olives, and sauerkraut [4] to prevent alteration and degradation by microorganisms during storage. These additives are widely regarded as most active against yeast and molds, and least active against bacteria; however, it is difficult to obtain substantive evidence on their relative activity from available studies. Excessive addition of these preservatives may be harmful to consumers because of their tendency to induce allergic contact dermatitis, convulsions, and hives, among others effects [5,6]. Therefore, the development of convenient and inexpensive analysis methods of these preservatives is of great importance for food safety.

As for other food additives, the ADI for preservatives has been calculated because, as already mentioned, if they are consumed in inappropriate quantities they may have adverse effects on human health. Legislation on food additives is necessary to establish specific conditions for their use; thus, food additive legislation has been developed by the European Union and included in several European Parliament and Council Directives [7,8]. The ADI is 25 mg/kg for sorbic acid [9] and 5 mg/kg body weight for benzoic acid [10].

9.1.1 Sorbic Acid

Sorbic acid (2,4-hexadienoic acid) (CAS No. 110-44-1) is a straight-chain unsaturated fatty acid with the formula:

$$\text{\raisebox{0pt}{\Large $\diagup\!\diagup\!\diagup\!\diagup$}}\;\text{COOH}$$

A.W. Van Hoffman was the first to isolate sorbic acid from the berries of the mountain ash tree in the year 1859. The antimicrobial (preservative) properties of sorbic acid were recognized in the 1940s. In the late 1940s and 1950s it became commercially available. Since then, sorbic acid has been extensively tested and used as a preservative in many foods. Its use in wine was legalized in France in 1959 and in Germany in 1971 [11]. Today, sorbic acid or its salts (sodium sorbate, potassium sorbate, and calcium sorbate) are effective antifungal agents in baked products, cheeses, beverages (fruit juices, wines), marmalades, jellies, dried fruits, and margarine. Their E numbers are E200 Sorbic acid; E201 Sodium sorbate; E202 Potassium sorbate, and E203 Calcium sorbate (EU No. 1129/2011 Regulation on labelling of foodstuffs). Concentrations of sorbic acid used in food preservatives are in the range of 0.02%–0.3%; higher levels may cause undesirable changes in the taste of foods [12]. In general, the salts are preferred over the acid form because they are more soluble in water, but it is the acid form that is active. The major food groups contributing to dietary intake of sorbic acid constitute a wide variety permitted at the following levels: various foods 200–2000 mg/kg (liquid egg 5000 mg/kg, cooked seafood 6000 mg/kg) and soft drinks, wine, etc. 200–300 mg/kg (sacramental grape juice 2000 mg/kg, liquid tea concentrates 600 mg/kg) [13].

Sorbic acid (pKa = 4.76) has the advantages that it is odorless and tasteless at the level of use ($\leq$0.3%) and that it has a low cost. The antimicrobial activity of sorbic acid is primarily against fungi and yeast, less so against bacteria; it must be used in sufficient amounts to inhibit growth fully, or it is ineffective [3]. Alcohol has a synergistic action with sorbic acid. The fewer yeast cells present, the more effective the treatment [14]. The activity of sorbic acid depends on the pH; the antimicrobial action increases as the pH value decreases below 4.75. In other words, the proportion of undissociated form of sorbic acid increases (above 50%) as the pH drops below 4.75, this can lead to increased antimicrobial action. Its utilization is possible up to pH 6.5, at which the proportion of undissociated acid is still 1.8%, so it is relatively ineffective as a preservative of grape juice at allowable levels.

Sorbic acid was fed to rats at 0%, 1.5%, and 10% of their diet for 2 years without carcinogenic effects being found. At a feeding level of 10% of the diet, some liver and kidney enlargement occurred, but no toxic or physiological symptoms were seen. The no-effect level was established at 1.5% of the diet [15]. The oral median lethal dose (LD_{50}) for rats for potassium sorbate is 4.0–7.16 g/kg of body weight; for the acid, investigators found 7.36–10.5 g/kg [16]. This, along with other published results, indicates the relative harmlessness of this additive.

Sorbic acid is degraded biochemically like a fatty acid. Some microorganisms, such as *Penicillium roqueforti*, have the ability to decarboxylate sorbic acid and thus convert it into 1,3-pentadiene, which has no antimicrobial activity and in addition may contribute to an off-flavor in cheeses [3]. In wines, it is generally used as the potassium salt, since the acid is only sparingly soluble in wine. Studies have shown that sorbic acid will react with SO_2 by addition to the C-4 double bond to form:

$$\underset{\underset{HSO_3}{|}}{CH_3-CH-CH_2-CH} = CH-COOH$$

It was suggested this reaction proceeds in a 1:1 ratio. Thus the amount of free SO_2 would be reduced, allowing for bacterial growth [14]. Its use as preservative in wines is avoided unless some sulfurous acid is also used, because it gives the resulting wine a geranium after-taste, bottles are expeditiously sealed, and their contents promptly consumed [17].

9.1.2 Benzoic Acid

Benzoic acid, $C_7H_6O_2$ or C_6H_5COOH (CAS No. 65-85-0; E210 [EU No. 1129/2011 Regulation on labelling of foodstuffs]), molecular weight 122.13, is a white solid that starts to sublime at 100°C, with a melting point of 122°C and a boiling point of 249°C. Its solubility in water is low (2.9 g/L at 20°C), and its solution in water is weakly acid (it is a weak acid, pKa = 4.19). It is soluble in ethanol and very slightly soluble in benzene and acetone.

The earliest mention of benzoic acid appears in the sixteenth century. The substance received its name from "gum bezoin," the plant from whose resin it was first derived. This plant was for a long time the only source for this acid. In the nineteenth century, benzoic acid was synthesized from coal tar. Today it is manufactured by treating molten phthalic anhydride with steam in the presence of a zinc oxide catalyst, by the hydrolysis of benzo-trichloride, or by the oxidation of toluene with nitric acid or sodium dichromate or with air in the presence of a transition metal salt catalyst. The estimated global production capacity for this acid is about 600,000 tonnes per year.

Benzoic acid occurs naturally free and bound as benzoic esters in many plant and animal species and thus occurs in food at varying levels. High levels are found in most berries, especially strawberries and raspberries; prunes; tea; species such as cinnamon, nutmeg, clove, and anise; and cherry bark and cassia bark. Concentrations of naturally occurring benzoic acid in several foods did not exceed average values of 40 mg/kg of food [18]. Maximum concentrations, reported for benzoic acid or benzoate added to food for preservation purposes were in the range of 2000 mg/kg of food.

Benzoic acid and its sodium, potassium, or calcium salts represented by the E numbers E210, E211, E212, and E213, respectively (EU No. 1129/2011 Regulation on labelling of foodstuffs) are commonly used as food preservatives to prevent alteration and degradation by microorganisms during storage. Its seems that although benzoic acid has been given an E number and has been approved for use as a food additive in many countries, its salts (sodium, calcium, and potassium benzoate) are more commonly used. This is presumably because the salts are more easily dissolved. The effectiveness of benzoic acid and its salts depends on the acidity (pH) of the food. It may be found in acidic foods and beverages; items such as fruit juices (citric acid), sparkling drinks (carbon dioxide), soft drinks (phosphoric acid), and pickles (vinegar) or other acidified food are preserved with benzoic acid and benzoates. The mechanism starts with the absorption of benzoic acid into the cell. If the intracellular pH changes to 5 or lower, the anaerobic fermentation of glucose through phosphofructokinase is decreased by 95%. After oral uptake, benzoic acid and benzoates are rapidly absorbed from the gastrointestinal tract and metabolized in the liver by conjugation with glycine, resulting in the formation of hippuric acid, which is rapidly excreted via the urine. Owing to rapid metabolism and excretion, an accumulation of the benzoates or their metabolites is not to be expected. For humans, the WHO's International Programme on Chemical Safety (IPCS) suggests a provisional tolerable intake would be 5 mg/kg body weight per day. There have been reports that people suffering from asthma, acetylsalicylic acid (ASA, or aspirin) sensitivity, or urticarial may have allergic reactions and/or find that their symptoms become worse after eating foods containing benzoic acid. This may be particularly true of the foods that also contain tartrazine (E102) [19]. The mechanism of adverse reactions is unknown, but evidence suggests that the cyclooxygenase pathway of arachidonic acid metabolism may be affected.

There has also been some concern that benzoic acid and its salts may react with ascorbic acid (present as an ingredient in beverages) to form small quantities of benzene [20]. This is a worry because benzene is toxic and linked to many forms of cancer. Formation of benzene is exacerbated in soft drinks if they are stored for extended periods at elevated temperatures. Although the levels and frequency at which such benzene formation has occurred in the past have not been considered to pose a public health risk, the soft drinks industry has developed methods to prevent or minimize its occurrence. In recent years, the use of benzoates has been reduced because of new processing techniques but it is still necessary to use these preservatives in some beverages to maintain their quality.

In processed foods, benzoic and its salts are used up to a level of 0.1%. The effectiveness of these preservatives increases with decreasing pH. The composition of the food therefore has an effect on their efficiency and suitability for use. This is because the ratio of undissociated (i.e., free) benzoic acid to ionized benzoic acid increases as the pH decreases. It is generally accepted that the undissociated benzoic acid is the active antimicrobial agent. Although no definitive theory has been yet proposed to explain this antimicrobial effects, it is believed to be related to the high lipid solubility of the undissociated benzoic acid, which allows it to accumulate on the cell membranes or on various structures and surfaces of the bacterial cell, effectively inhibiting its cellular activity.

9.2 DETERMINATION OF SORBIC ACID, BENZOIC ACID, AND THEIR SALTS

Because sorbic acid and benzoic acid and their salts exhibit strong UV absorption, the technique most commonly used to determine these preservatives in foods is spectrophotometric. Other methods, based on chromatographic techniques, have also been reported, including gas chromatography (GC), high-performance liquid chromatography (HPLC), and capillary electrophoresis (CE) [21–23]. Generally, all these methods require sample derivatization or extraction, making the analytical procedures complex and time consuming. The flow injection analysis (FIA) methodology is a versatile, flexible, and economical technique that can be installed online as it is compatible with a large number of detectors. With this technique automation of the complex procedures of derivatization, extraction, etc., is possible, responding to the growing demand for routine analysis in food analysis laboratories.

9.2.1 FIA with Spectrophotometric Detection

As stated above, spectrophotometric methods are the most commonly used to determine sorbic or benzoic acid. However, only two reports have been found in the literature that combine flow injection (FI) methodology with spectrophotometric measurements to determine these preservatives. This scarcity might be explained by the advantages of the traditional spectrophotometric methods in terms of simplicity and specificity. There is no methodological reason that limits the application of FIA [24].

The most widely used photometric method for determining sorbic acid in wine samples relies on its oxidation with a mixture of $K_2Cr_2O_7$ and H_2SO_4 in a water bath at 100°C, followed by reaction of the resulting malonaldehyde with thiobarbituric acid, also at 100°C, to give a red product, the absorbance of which is measured at 532 nm [25]. For this colorimetric method to be effective, the reaction mixture with thiobarbituric acid must be cooled in an ice bath. This method is time consuming as it entails the prior distillation of the wine in order to isolate the sorbic acid from the potentially interfering sample matrix [21]. A simple, rapid, and accurate spectrophotometric FIA method based on this method was developed for the determination of sorbic acid in wine samples. This FIA-spectrophotometric method requires no prior distillation of the wine, which allows a sampling rate of 40 samples per hour, being more sensitive and selective, and uses less volume of sample and reagents compared with the method in batch [26]. A schematic diagram of the FIA system is shown in Figure 9.1.

The FIA system operates as follows. First the carrier solution (deionized water) is pumped into the system as a blank to establish a baseline. Samples (diluted wine 1:25/wine:deionized water) or standard solutions are pumped at 0.7 mL/min, and 225 μL is injected through a six-port valve into the system, where it is merged with a stream of a 1:1 oxidizing mixture (0.01 N $K_2Cr_2O_7$ and 0.3 N H_2SO_4). Both streams are mixed in the reactor R_1 (200 cm × 0.5 mm i.d.) where the oxidation of the sorbic acid is achieved. The resulting malonaldehyde is mixed in the reactor R_2 (500 cm × 0.5 mm i.d.) with a stream of thiobarbituric acid solution (0.5%). Both reactors are immersed in a water bath at 100°C. After this, the mixture is cooled rapidly by passing through the reactor R_3 (200 cm × 0.5 mm i.d.), which is immersed in an ice bath. The absorbance of the red product formed is measured at 532 nm. The ice bath is very important because it provides a better baseline and improves the reproducibility.

Multivariable calibration permits the simultaneous determination of multicomponent mixtures and it is mainly based on spectroscopy data. Full-spectrum multivariate calibration methods offer the advantage of speed in the determination of the analytes, avoiding separation steps in the analytical procedures. Partial least squares (PLS) has become the usual first-order multivariate tool because of the quality of the calibration models obtained, the ease of its implementation, and the availability of software [27]. However, all first-order methods, of which PLS is no exception, are sensitive to the presence of unmodeled interferents, that is, compounds occurring in new samples that have not been included during the training step of the multivariate model. This situation is encountered

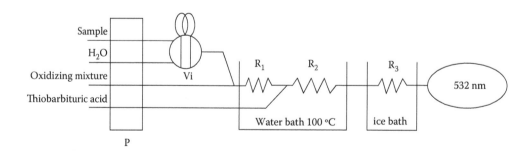

FIGURE 9.1 Schematic diagram of the spectrophotometric FIA system for determination of sorbic acid in wines. P, peristaltic pump; Vi, injection valve; R_1, R_2, and R_3, reactors (200, 500, 200 cm × 0.5 mm i.d., respectively).

when natural samples of complex composition, such as food samples, are examined. The problem can be alleviated, in part, thanks to the implementation of wavelength selection procedures, which are able, to some extent, to "filter" the spectral regions where serious interferences occur, leaving adequate spectral windows for successful application of multivariate methods [28]. In general, however, no first-order multivariate model can be made immune to unexpected interferents with spectral overlap over the whole useful spectral region where the analyte absorbs.

A good alternative to the above-discussed problem is to move to high-order data, which are particularly useful for the quantitative analysis of complex multicomponent samples. In the work of Marsili et al. [4], a method using second-order data provided by diode array spectrophotometric detection in a flow injection system with an imposed double-pH gradient, analyzed by both parallel factor analysis (PARAFAC) and the multivariate curve resolution–alternating least squares (MCR-ALS) model, was employed in the simultaneous determination of benzoic and sorbic acids in real juice samples. pH was selected as modulation for the second data dimension, given the acid–base properties of the analytes, assuming that an observable change in spectra with the pH takes place. FIA systems are well-suited for generating reproducible pH gradients and, combined with diode array UV–Vis detection, they provide valuable second-order data. The FIA system was very simple; a single channel was designed to generate the pH gradient (Figure 9.2a), with a carrier stream of acid Britton–Robinson solution (BRH) pumped continuously into the system. When an appropriate volume of the basic Britton–Robinson solution (BROH) is injected, the FIA signal recorded at 270 nm has the shape pictorially shown in Figure 9.2b. The baseline pH before the peak appears is acid, and the injected volume of BROH is responsible for the generation of a double pH gradient. The pH increases from the front of the flow injection peak, reaching its maximum value at the center of the peak, and then progressively decreases to reach its baseline value. Samples diluted 100-fold of

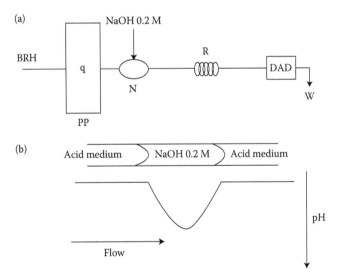

FIGURE 9.2 (a) Flow injection system used for the generation of a double-pH gradient; BRH, acid Britton–Robinson solution; IV, injection valve; R, reactor; DAD, diode array detector; W, waste; q, 0.56 mL/min; PP, peristaltic pump. (b) Scheme of the pH gradient profile.

commercial juice prepared in BRH were introduced into the FI system as a carrier stream, then BROH was injected, creating the necessary pH gradient close to the center of the sample. For each FI peak registered, spectra were collected every 2 nm in the range of 200–300 nm, and the pH gradient produced 33 equally spaced data points, making a total of $51 \times 33 = 1683$ data points for each sample matrix. The construction of five-component models proved to be satisfactory for all samples: these five components were sorbic acid, sorbate, benzoic acid, benzoate, and an average background contribution from the fruit juice. PARAFAC and MCR-ALS were applied to a set of calibration samples and each of the test samples using the five-component model. The results obtained were close to those obtained by the official method according to the paired t test [29].

9.2.2 FIA Combination with HPLC

Analytical methods that simultaneously determine these additives are of interest for controlling the maximum regulation-permitted levels of individual additives for maintaining food quality and characteristic as well as promoting food safety. The most popular method used for the simultaneous determination of food additives in food samples is HPLC [30], which needs a proper sample pretreatment to homogenize, extract, clean up, and concentrate the analytes from the complex matrix interferences in food samples. For food analysis, the conventional dialysis sample pretreatment has been applied for removing or reducing the high-molecular-weight molecules (such as proteins and fatty matters), suspended particulate matters and other matrices as food additive molecules. However, the conventional dialysis procedure is usually tedious and time consuming and consumes large amounts of sample, reagent, and materials. Online dialysis seems to be a good choice for HPLC, which is simple, quick, and inexpensive. This technique is operated by continuously fed liquid food sample on the donor side of the dialysis membrane while the solution in the acceptor side is flowed or stopped. A FIA online dialysis (FIAD) system for determining of five food additives (acesulfame-K, saccharin, caffeine, benzoic acid, and sorbic acid) in soft drinks and other liquid samples was described by Kritsunankul and Jakmunee [31]; the FIAD–HPLC system offers a simple, convenient, and low-consumption sample pretreatment system. A schematic diagram of the manifold configuration is shown in Figure 9.3.

First, the HPLC system (P_3) starts to operate. Eluent is pumped through the column to clean the system. The donor (P_1) and acceptor (P_2) solutions, at flow rates of 0.2 mL/min, are flowed to fill all tubings and channels. Three consecutive operation times of one operation cycle (14 min) comprise a dialysis time of a standard or sample, a travel time of the dialysate to the HPLC loop, and a cleaning time of the FIAD–HPLC system. For the first injection, a mixed standard or sample solution (900 µL) at V_1 is injected into the donor stream and is then propelled to the dialysis cell (DC), where small solute molecules in the donor solution are dialyzed through the dialysis membrane into the acceptor solution for a dialysis time of 6 min 10 s. During the dialysis period the donor stream is continuously flowed while the acceptor stream is stopped, in order to enhance the sensitivity. The dialysate zone containing food additives is filled into a sample loop (20 µL) of the HPLC valve (V_2), with a suitable travel time of 2 min 30 s. Then it is further injected into the HPLC system and analyzed under normal HPLC conditions, using an isocratic mobile phase, a reversed-phase (C_{18}) analytical column, and a photodiode array UV detector. The chromatogram is recorded for 13 min 18 s, which is within one operation cycle of the FIAD system. While the dialysate of the first injection is injected

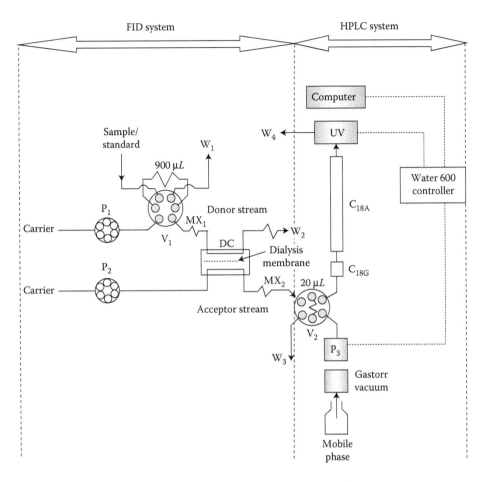

FIGURE 9.3   Manifold of the FIAD–HPLC system used for the determination of some food additives. Flow rate of carrier of donor and acceptor streams: 0.2 mL/min. P_1 and P_2: peristaltic pumps 1 and 2; P_3, HPLC pump; V_1: manual-rotary injection valve; V_2: HPLC manual-rotary injection valve; DC: dialysis cell; MX_1: mixing coil 1 (12 cm × 0.8 mm i.d.); MX_2: mixing coil 2 (7.5 cm × 0.8 mm i.d.); C_{18A}: C_{18} analytical column; C_{18G}: C_{18} guard column; UV: photodiode array detector; and W_1, W_2, W_3, and W_4: wastes 1, 2, 3, and 4.

into the HPLC system, the second injection is loaded into the sample loop of the FIAD valve (V_1). After a period of 5 min 20 s for cleaning of the donor and acceptor lines, the second injection is started. When a chromatographic separation of the first injection is ended (13 min 18 s of analysis time), the dialysate of the second injection is injected into the HPLC system, and the third injection is loaded into the FIAD system. In this way, the sample throughput is increased, resulting in an injection throughput of approximately 4.3 chromatograms per hour.

This system, like other FIA systems, has advantages of the high degree of automation in sample pretreatment, online sample separation and dilution, good sample clean-up for prolongation the lifetime of the expensive HPLC column, low consumption of chemicals and materials, and short analysis time.

9.2.3 FIA Combination with GC

The advantages of FIA-HPLC systems for the simultaneous determination of these preservatives have been clearly described above. However, the simultaneous separation of additives by HPLC is hindered by polarity differences. GC, with or without derivatization, is also employed for the selective determination of food preservatives [32]. The Association of Official Analytical Chemists (AOAC) official GC method [33] for preservatives in foods involves several extractions, evaporation, derivatization to a trimethylsilyl ester, and flame ionization detection (FID). Mass spectrometry (MS) is a sensitive and selective technique that is gradually gaining ground in the determination of these additives by GC, but involves sample pretreatments similar to that of the AOAC method.

Solid-phase extraction (SPE), introduced in the early 1970s, has superseded other sample preparation alternatives, aided by the variety of polar and nonpolar sorbents, and exchangers, that have been made commercially available. Thus, SPE has been used as extraction technique for determining food preservatives, including sorbic and benzoic acids [34,35]. A drawback of offline SPE procedures is that they can be time consuming and cumbersome to perform, often requiring many steps (conditioning and flushing the cartridge, and eluting the analytes by hand) before achieving a concentrated extract suitable for instrumental analysis, of which only a small portion is actually injected onto the chromatographic column. As compared to offline SPE, online SPE offers a series of advantages [36]. The use of online SPE techniques has made possible the development of faster methods by reducing the sample preparation time and thus increasing the sample throughput. Conditioning, washing, and elution steps can be performed automatically and some systems also permit extraction of one sample while another one is being analyzed by chromatography. Other important advantages of online coupling are decreased risk of contamination of the sample or sample extract, elimination of analyte losses by evaporation or by degradation during sample preconcentration, and improved precision and accuracy. Higher sensitivity is also achieved in online configurations due to the transfer and analysis of the totality of the extracted species to the analytical system, in contrast to offline SPE procedures where only an aliquot of the extract is injected into the column. The analysis of the whole sample leads to lower limits of detection or, alternatively, smaller sample volumes may be used to obtain enough sensitivity for a large variety of compounds. In addition, online SPE has low solvent consumption requirements, thereby decreasing the costs of disposal of waste organic solvents. Two online (via a FI system) SPE-GC methods have been found in the literature for the simultaneous determination of food preservatives (including sorbic and benzoic acids) [32,37]. In these works, liquid foods, in a nitric acid medium containing an internal standard, were directly loaded onto the sorbent column for simultaneous enrichment and matrix removal. Solid samples required some pretreatment before the residue was dissolved in 0.1 M HNO_3. Both methods used an Amberlite XAD-2 sorbent column inside the loop of the injection valve of FIA system, where 5 mL of treated sample or standard solution plus internal standard was continuously pumped through the sorbent column, then the column was flushed with n-hexane carried by N_2 in order to remove residual aqueous phase from the column and connections. Finally, the eluent carried by the N_2 stream was passed through the minicolumn, and the eluate was collected in glass vials containing anhydrous sodium sulfate, and 1 or 2 μL aliquots were manually injected into the gas chromatograph. After each determination, the sorbent column was cleaned with acetone to remove residual triglycerides and other adsorbed organic compounds; finally, the column was conditioned with 0.1 M HNO_3.

9.2.4 FIA Combination with CE

Nowadays, capillary electrophoresis (CE) has become an attractive analytical technique for preservatives owing to its high separation efficiency, low sample consumption, and high analysis speed [38]. The advantages of SPE have been discussed in Section 9.2.3.

Based on the flow analysis techniques of sequential injection analysis (SIA), introduced by Ruzicka and Marshall [39], multicommutation [40] and electroosmotic pump [41,42], electrokinetic flow analysis (EFA) was proposed [43–45].

Multicommutation [40] provides flow analysis technique with flexibility and controllability. As a fluid delivery device, the electroosmotic pump has proved to be suitable for flow analysis with its large flow range (10 μL/min to 5.0 mL/min), stable flow rate (<4%, 4 h), pulseless driving, simplified apparatus, convenient control operation, and proper back pressure (≤1.1 MPa) [41,42], etc. However, electrolyte solutions cannot be introduced into the porous core of the pump because of the influence on the surface charge density of the porous core and the electroosmotic flow (EOF). This problem can be resolved by aspirating sample and reagent solutions into a holding coil, as performed in SIA.

A simple and controllable EFA system, consisting of one electroosmotic pump, five solenoid valves, and one online homemade SPE unit, combined with capillary zone electrophoresis (CZE) has been proposed for determination of benzoic and sorbic acid in food products. Tetrabutylammonium bromide (TBAB) was adopted as an ion pair reagent to improve the retention of the preservatives on a C_8 column with bonded silica sorbent, which was also used to remove sample matrices. By using the SPE unit, the EFA-SPE-CZE system was able to perform the SPE operation and CZE separation simultaneously. The preservatives were separated and determined under optimized conditions with p-hydroxybenzoic acid as an internal standard. The EFA-SPE-CZE system was verified to improve sample throughput and analytical sensitivity and selectivity through the analysis of benzoic and sorbic acid in three kinds of food products: milk beverage, fruit jam, and soy sauce [46].

9.2.5 Other Methods

Electrochemical methods for benzoic acid determination, especially those employing amperometric biosensors, have been regarded as promising, because of their effectiveness, simplicity, and selectivity. They are based on tyrosinase, a copper-containing polyphenol oxidase, which possesses the ability to oxidize mono- and diphenols to corresponding o-quinones. Quinones can be reduced at an electrode surface to enable low-potential detection of phenolic compounds. The addition of benzoic acid causes enzyme inhibition and a decrease of the steady-state current signal proportional to the inhibitor concentration can be observed. Several biosensors based on tyrosinase inhibition have been developed for detection of benzoic acid. Morales et al. [47] employed a graphite–Teflon–tyrosinase composite biosensor for analysis of foodstuffs, such as mayonnaise sauce and cola soft drinks. Other authors have reported a highly reversible biosensor based on immobilization of tyrosinase by calcium carbonate nanomaterials for monitoring of benzoic acid in beverages [48]. Regarding flow methods, an approach based on sequential injection (SI) combined with monosegmented flow (which involves locating a sample zone between two gas bubbles), and an amperometric tyrosinase-based biosensor for inhibitive determination of benzoic acid have been reported [49]. In the biosensor, tyrosinase was entrapped in titania gel modified with multiwalled carbon nanotubes (MWCNT) and Nafion. In this work, the determinations were performed in electrochemical cell of capacity ca. 5 mL. The first stage included introduction of phosphate buffer (pH 6) and catechol solutions into the electrochemical cell. After establishing the baseline, an initial current signal was recorded. Subsequently, a sample was added and the current response in the presence of benzoic acid was registered. The sensor displayed a response time about 30 s. Afterwards, the standard was added and the corresponding current response was measured. The procedure of standard addition was repeated four times. In order to keep the catechol concentration at a constant level during the analysis, appropriate portions of diluted catechol solution were added additionally along with sample and standard additions. The method was applied to determination of benzoic acid in cola reference material and other beverages. No special sample pretreatment is necessary; samples were only diluted 50-fold with 0.1 M phosphate buffer solution (pH 6.0). Acceptable repeatability and reproducibility were obtained. Nevertheless, the lifetime of the biosensor appeared not to be satisfactory. After 5 days of storage in buffer solution (pH 6) at 4°C, the biosensor retained 89% of its initial current response and after an additional week only 34%. This response decrease may be attributed to enzyme leaching from the matrix and the inherent instability of tyrosinase [50].

REFERENCES

1. Branen, A. L., P. M. Davidson, S. Salminen, and J. H. Thorngate III, eds. 2002. *Food Additives* (2nd ed.). New York: Marcel Dekker, Inc., ISBN: 0-8247-9343-9.
2. http://www.eufic.org/article/en/expid/basics-food-additives/
3. Belitz, H.-D., W. Grosch, and P. Schieberle. 2009. *Food Chemistry* (4th ed.). Berlin, Heidelberg: Springer-Verlag, ISBN: 978-3-540-69934-7.
4. Marsili, N. R., A. Lista, B. S. Fernández Band, H. C. Goicoechea, and A. C. Olivieri. 2004. New method for the determination of benzoic and sorbic acids in commercial orange juices based on second-order spectrophotometric data generated by a pH gradient flow injection technique. *J. Agric. Food Chem.* 52(9):2479–845.
5. Soni, M. G., G. A. Burdock, S. L. Taylor, and N. A. Greenberg. 2001. Safety assessment of propyl paraben: A review of the published literature. *Food Chem. Toxicol.* 39:513–532.
6. Darbre, P. D. 2003. Underarm cosmetics and breast cancer. *J. Appl. Toxicol.* 23:89–95.
7. Directive No 95/2/EC of the European Parliament and of the Council of 20 February 1995 on food additives other than colours and sweeteners.
8. Directive 2006/52/EC of the European Parliament and of the Council of 5 July amending Directive 95/2/EC on food additives other than colours and sweeteners and Directive 94/35/EC on sweeteners for use in foodstuffs.
9. FAO/WHO. 2007a. Expert Committee of Food Additives: Sorbic Acid. Available: www.inchem.org/documents/jecfa/jeceval/jec_2181.htm (accessed June 27, 2007).
10. FAO/WHO. 2007b. Expert Committee of Food Additives: Benzoic Acid. Available: www.inchem.org/documents/jecfa/jeceval/jec_184.htm (accessed June 27, 2007).
11. Dharmadhikari, M. 1992. Sorbic acid. *Vineyard Vintage View* 7(6):1–5.
12. Sofos, J. N. 2000. Sorbic acid. In *Natural Food Antimicrobial Systems*, ed. A. S. Naidu, pp. 637–660. Boca Ratón, FL: CRC Press LLC, ISBN: 0-8493-2047-X.
13. Wood, R., L. Foster, A. Damant, and P. Key. 2004. *6 – E200–03: Sorbic Acid and Its Salts, Analytical Methods for Food Additives*. Cambridge, UK: Woodhead Publishing Lim. ISBN: 978-1-85573-722-8.

14. Ough, C. S. and M. A. Amerine. 1988. *Methods for Analysis of Musts and Wines* (2nd ed.). New York: John Wiley and Sons, ISBN: 0-471-62757-7.
15. Gaunt, I. F., K. R. Butterworth, J. Hardy, and S. D. Gangolli. 1975. Long-term toxicity of sorbic acid in the rat. *Food Cosmet. Toxicol.* 13:31–45.
16. Deuel, H. J., R. Alfin-Slater, C. S. Weil, and H. F. Smyth, Jr. 1954. Sorbic acid as a fungistatic agent for foods. I. Harmlessness of sorbic acid as a dietary component. *Food Res.* 19:1–12.
17. Vogt, E. and L. Jakob. 1984. *Der wein: Bereitung, Behandlung, Untersuchugn (The Wine: Preparation, Treatment, Investigation).* Verlag Eugen Ulmer, Stuttgart, Germany: Eugen Ulmer Gmblt & Co. ISBN: 3800112140.
18. International Programme on Chemical Safety. Concise International Chemical Assessment Document No. 26. Benzoic Acid and Sodium Benzoate. http://www.inchem.org/documents/cicads/cicads/cicad26.htm.
19. *UK Food Guide*—Benzoic acid, www.ukfoodguide.net.
20. German Federal Institute for Risk Assessment, www.bfr.bund.de/en/home.html.
21. Vereda Alonso, E., A. García de Torres, A. Rivero Molina, and J. M. Cano Pavón. 1998. Determination of organic acids in wines. A review. *Quim. Anal.* 17:167–175.
22. Wood, R., L. Foster, A. Damant, and P. Key. 2004. *Analytical Methods for Food Additives.* Cambridge, UK: Woodhead. ISBN: 1-85573-722-1.
23. Ruiter, A. and A. A. Bergwerff. 2005. Analysis of chemical preservatives in food. In *Methods of Analysis of Food Components and Additives*, ed. S. Ötles, Chapter 14. Boca Raton, FL: CRC Press, pp. 379–402.
24. Ruiz-Capillas, C. and F. Jiménez-Colmenero. 2008. Determination of preservatives in meat products by flow injection analysis (FIA). *Food Addit. Contam.* 25:1167–1178.
25. Caputi, A. and P. A. Stafford. 1977. Ruggedness of official colorimetric method for sorbic acid in wine. *J. AOAC* 60:1044–1047.
26. Rivero Molina, A., E. Vereda Alonso, M. T. Siles Cordero, A. García de Torres, and J. M. Cano Pavón. 1999. Spectrophotometric flow-injection method for determination of sorbic acid in wines. *Lab. Rob. Autom.* 11:299–303.
27. Haaland, D. M. and E. V. Thomas. 1988. Partial least-squares methods for spectral analysis. 1. Relation to other quantitative calibration methods and the extraction of qualitative information. *Anal. Chem.* 60:1193–1202.
28. Marsili, N. R., M. S. Sobrero, and H. C. Goicoechea. 2003. Spectrophotometric determination of sorbic and benzoic acids in fruit juices by a net analyte signal based method with selection of the wavelength range to avoid non modelled interferences. *Anal. Bioanal. Chem.* 376:126–133.
29. Miller, J. N. and J. C. Miller. 2005. *Statistics and Chemometrics for Analytical Chemistry.* Edinburgh: Pearson Education Limited, ISBN: 0-131-29192-O.
30. Zygler, A., A. Wasik, and J. Namiesnik. 2009. Analytical methodologies for determination of artificial sweeteners in foodstuffs. *Trends Anal. Chem.* 28:1082–1102.
31. Kritsunankul, O. and J. Jakmunee. 2011. Simultaneous determination of some food additives in soft drinks and other liquids foods by flow injection on-line dialysis coupled to high performance liquid chromatography. *Talanta* 84:1342–1349.
32. González, M., M. Gallego, and M. Valcárcel. 1998. Simultaneous gas chromatographic determination of food preservatives following solid-phase extraction. *J. Chromatogr. A* 823:321–329.
33. Helrich, K., ed. 1990. *Official Methods of Analysis of the Association of Official Analytical Chemist* (15th ed.), pp. 1143–1144. Arlington, VA: Association of Official Analytical Chemist (AOAC).

34. Moors, M., C. R. R. R. Teixeira, M. Jimidar, and D. L. Massart. 1991. Solid phase extraction of the preservatives sorbic acid and benzoic acid, and the artificial sweeteners aspartame and saccharine. *Anal. Chim. Acta* 255:177–186.
35. Chen, B. H., and S. C. Fu. 1995. Simultaneous determination of preservatives, sweeteners and antioxidants by paired-ion liquid chromatography. *Chromatographia* 41:43–50.
36. Rodríguez-Mozaz, S., M. J. López de Alda, and D. Barceló. 2007. Advantages and limitations of on-line solid phase extraction coupled to liquid chromatography–mass spectrometry technologies versus biosensors for monitoring of emerging contaminants in water. *J. Chromatogr. A* 1152:97–115.
37. González, M., M. Gallego, and M. Valcárcel. 1999. Gas chromatographic flow method for the preconcentration and simultaneous determination of antioxidant and preservative additives in fatty foods. *J. Chromatogr. A* 848:529–536.
38. Huang, H. Y., C. L. Chuang, C. W. Chiu, and J. M. Yeh. 2005. Application of microemulsion electrokinetic chromatography for the detection of preservatives in foods. *Food Chem.* 89:315–322.
39. Ruzicka, J., and G. D. Marshall. 1990. Sequential injection: A new concept for chemical sensors, process analysis and laboratory assays. *Anal. Chim. Acta* 237:329–343.
40. Reis, B. F., M. F. Giné, E. A. G. Zagatto, J. L. F. C. Lima, and R. A. Lapa. 1994. Multiconmutation in flow analysis. Part 1. Binary sampling: concepts, instrumentation and spectrophotometric determination of iron in plant digests. *Anal. Chim. Acta* 293:129–138.
41. Gan, W. E., L. Yang, Y. Z. He, R. H. Zeng, M. L. Cervera, and M. de la Guardia. 2000. Mechanism of porous core electroosmotic pump flow injection system and its application to determination of chromium(VI) in waste-water. *Talanta* 51:667–675.
42. Wang, L., Y. Z. He, G. N. Fu, Y. Y. Hu, and X. K. Wang. 2006. Study on pressurizing electroosmosis pump for chromatographic separation. *Talanta* 70:358–363.
43. Yang, L., Y. Z. He, W. E. Gan, M. Li, Q. S. Qu, and X. Q. Lin. 2001. Determination of chromium(VI) and lead(II) in drinking water by electrokinetic flow analysis system and graphite furnace atomic absorption spectrometry. *Talanta* 55:271–279.
44. Zhao, Y. Q., Y. Z. He, W. E. Gan, and L. Yang. 2002. Determination of nitrite by sequential injection analysis of electrokinetic flow analysis system. *Talanta* 56:619–625.
45. Hu, Y. Y., Y. Z. He, and L. L. Qian. 2005. On-line ion pair solid-phase extraction of electrokinetic multicommutation for determination of trace anion surfactants in pond water. *Anal. Chim. Acta* 536:251–257.
46. Han, F., Y. Z. He, L. Li, G. N. Fu, H. Y. Xie, and W. E. Gan. 2008. Determination of benzoic acid and sorbic acid in food products using electrokinetic flow analysis-ion pair solid phase extraction-capillary zone electrophoresis. *Anal. Chim. Acta* 618:79–85.
47. Morales, M. D., S. Morante, A. Escarpa, M. C. González, A. J. Riviejo, and J. M. Pingarron. 2002. Design of a composite amperometric enzyme electrode for the control of the benzoic acid content in food. *Talanta* 57:1189–1198.
48. Shan, D., Q. Li, H. Xue, and S. Cosnier. 2008. A highly reversible and sensitive tyrosinase inhibition-based amperometric biosensor for benzoic acid monitoring. *Sens. Actuators B* 134:1016–1021.
49. Kochana, J., J. Kozak, A. Skrobisz, and M. Wozniakiewicz. 2012. Tyrosinase biosensor for benzoic acid inhibition-based determination with the use of a flow-batch monosegmented sequential injection system. *Talanta* 96:147–152.
50. Bregg, R. K., ed. 2006. *Frontiers in Polymer Research.* New York: Nova Science Publishers, Inc., ISBN: 1-59454-824-2.

CHAPTER **10**

Determination of Parabens

Basil K. Munjanja

CONTENTS

10.1 INTRODUCTION

Preservatives are added to increase the shelf life and improve the safety of many consumer goods. Examples of these comprise nitrates, sulfites, sorbates, benzoates, and parabens. The parabens, which include compounds such methylparaben, ethylparaben, propylparaben, and butylparaben, are widely used as preservatives in many foodstuffs, cosmetics, and pharmaceutical products to prevent attack from bacteria and fungi. Figure 10.1 shows the structure of the respective parabens under discussion. Their power as microbial agents increases with increasing length of the alkyl chain of the ester group; however, shorter esters are commonly used because of their high solubility in water. Generally, they have low toxicity, due to their rapid metabolism by conjugation, but nowadays attention has shifted toward their occurrence in consumer products because of supposed possible side effects on human health such as irritation of the skin and contact dermatitis [1]. Furthermore, their presence has also been detected in human breast tumors [2]. Thus to ensure the continued safety of consumers, monitoring of parabens needs to be carried out in consumer products such as food, cosmetics, and pharmaceutical products, and the information obtained should be compared with acceptable intake values set by various regulatory bodies. For example, the Joint FAO/WHO Expert Committee on Food Additives (JECFA) in 1974 recommended that the acceptable daily intake (ADI) for methyl-, ethyl-, and propylparaben was 0–10 mg/kg body weight daily. After a critical review of the potential impacts of parabens on human health, the ADI for methylparaben and propylparaben was proposed as 55 mg/kg based on the no-adverse-effect level of 5500 mg/kg per day from a chronic study in rats [3].

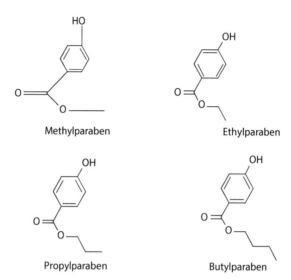

Methylparaben   Ethylparaben

Propylparaben   Butylparaben

FIGURE 10.1   Structures of various parabens.

Other regulatory bodies such as the U.S. FDA (Food and Drug Administration) have set the limit of methylparaben and propylparaben at 0.1% w/w. In cosmetics, the EEC and Danish cosmetic regulations give a combined maximum limit of 0.8% w/w for all the parabens. Various regulatory bodies set the maximum levels of all the parabens in pharmaceutical products at 1% w/w [3]. For this reason, robust analytical methods are required for the determination of parabens in a variety of matrices.

Many methods have been reported in the literature for the determination of parabens in different compounds based on gas chromatography [4], capillary electrophoresis (CE), and high-performance liquid chromatography (HPLC) [5]. However, a recent advance in their analysis relies on the use of flow injection analysis (FIA)–based techniques for their determination in various matrices such as foodstuffs, cosmetics, and pharmaceuticals. Flow injection–based methods exhibit advantages that include ease of automation, precision, and reduced contamination of the analyte [6].

The basic instrumentation of flow injection systems consists of the manifold, a pump for the transport of the reagent, rotary sample injection valves, a flow-through cell, and a detector [6], which can be chemiluminescence-, mass spectrometric-, or electrochemical-based, among other possibilities. For example Figure 10.2 shows the block diagram of a flow injection system coupled to chemiluminescence detection. The sample is introduced into the flow injection analysis system via the injection valve or via an autosampler, while the carrier and reagent streams are pushed forward to the detector by the pumps.

In terms of solution delivery, the peristaltic pump in flow injection systems delivers at levels of milliliters, but the sequential injection systems deliver solutions precisely down to microliter levels using a syringe pump. Thus, it can be appreciated that the sequential injection technique was developed to address the major drawback of flow injection systems, namely, the inability to deliver solutions at microliter levels. However, the main inconvenience associated with the syringe pump used in sequential, injection analysis is the limited volume capacity of the syringe barrel, which results in reduced flow rate of the solution due to its bidirectional operation. For this reason, multiport valves and

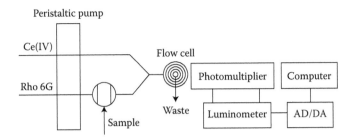

FIGURE 10.2   Block diagram of the flow injection chemiluminescence systems (AD/DA: analog to digital/digital to analog to converter). (Reprinted from *Anal. Chim. Acta*, 517, Myint, A. et al., Flow injection-chemiluminescence determination of paraben preservative in food safety, 119–124, Copyright 2004, with permission from Elsevier.)

multiport pumps have been added to accommodate more reagents and samples to create multicommutated systems [7,8]. As a result, these notable developments in flow injection and sequential injection analysis, which include multicommutated (developed in 1994), multisyringe flow injection analysis (MSFIA) (developed in 1999), and multipumping flow systems (MPFS), share the merits of flow injection and sequential injection, but they overcome the setbacks of the former. For this reason they offer better analytical performance overall [9].

This chapter reviews the automated sample processing methods and separation and detection methods used with flow injection analysis in the determination of parabens in various consumer products. The advantages and limitations of the analytical methods are also highlighted. In addition, the applications of the different flow-based methods in the determination of parabens are highlighted in Table 10.1.

10.2   AUTOMATED SAMPLE PREPARATION METHODS

Sample preparation is an important stage in the analytical process, and the ultimate results depend on this stage to a great extent. Good sample preparation technique should be able to separate the analyte from the target matrix with high selectivity and with high recoveries. Many of the sample preparation techniques have been based on the offline mode; examples of these include liquid-phase techniques (liquid–liquid extraction), analyte trapping (solid-phase extraction), and membrane extraction techniques. However, all these techniques have major drawbacks: they are tedious and consume large amounts of solvent, especially when the number of samples involved is large [10].

For this reason, recent developments in sample preparation include miniaturization and automation. Sample preparation methods such as liquid–liquid extraction and solid-phase extraction can be automated to flow injection analysis equipment without much difficulty. However, since the incorporation into the flow system of most sample pretreatment steps results in lack of selectivity, complex manifolds with a number of flow channels and detectors are required for the analysis of multianalytes [11].

Automated sample preparation methods offer advantages that include improved precision, reproducibility and reduced number of personnel. The disadvantages are high cost of instrumentation, systematic errors, and limited flexibility of the system.

TABLE 10.1   Applications of Flow-Based Methods in Analysis of Consumer Products

Analyte	Matrix	Column Specifications	Mobile Phase	System Specifications	Detector	LOD	References
Methyl, ethyl, and propyl paraben	Cosmetics	5 mm i.d., C8 bonded silica (50 μm particle size),	40% v/v, 10 mmol/L, sodium tetraborate buffer, and 60% v/v ethanol at 0.75 mL/min	FIA–SPE–MEKC	HPCE analyzer at 254 nm	0.07–0.1 μg/mL	[11]
Methyl, ethyl, propyl, and butyl paraben	Soy sauces			FI–CH	Chemiluminescence detector, based on cerium IV rhodamine 6G reaction in sulphuric acid	3.4×10^{-10}, 2.0×10^{-9}, 2.0×10^{-9}, 5.0×10^{-9} g/mL respectively	[28]
Methyl, ethyl, propyl, and butyl paraben	Cosmetics	C18 monolithic column (5 mm × 4.6 mm i.d.)	12% Acetonitrile: Water, 27% Acetonitrile: water at 2.5 mL/min	FIC	Chemiluminescence detector, based on cerium IV rhodamine 6G reaction in sulphuric acid	1.9×10^{-8} -9.9×10^{-5} M	[12]
Methyl, ethyl, propyl, and butyl paraben	Cosmetic	Ultrashort C18 monolithic column (5 mm × 4.6 mm i.d.)	12% Acetonitrile: Water, 27 % Acetonitrile: water at 2.6 mL/min	FIC	UV-Vis spectrophotometer at 254 nm	1.2×10^{-5} -2.0×10^{-3} M	[22]
Methyl paraben, propyl paraben, and sodium diclofenac	Pharmaceutical samples	Chromolith (25 mm C × 4.6 mm), C18 monolithic column	Acetonitrile: water (40:70 v/v) + 0.05% triethylamine, pH 2.5, gradient flow rate 8–20 μL/s	SIC with an eight port selection valve and 10mL syringe	UV-Vis spectrophotometer at 275 nm	0.25–0.50 μg/mL	[13]
Methyl, ethyl, propyl, and butyl paraben	Cosmetics and food	XDB C8 column	60% methanol:40% water at 1.0 mL/min	HPLC–FI–CH with a peristaltic pump, mixing tee and photomultiplier at −800 V	Chemiluminescence detector, based on cerium IV rhodamine 6G reaction in sulphuric acid	1.9×10^{-9} -5.3×10^{-9} g/mL	[20]

(Continued)

TABLE 10.1 (*Continued*)   Applications of Flow-Based Methods in Analysis of Consumer Products

Analyte	Matrix	Column Specifications	Mobile Phase	System Specifications	Detector	LOD	References
Methyl, ethyl, propyl, and butyl paraben + antioxidant and food sweeteners	Food and cosmetics	Silica C18 monolithic column (5 mm × 4.6 mm i.d.)	4% acetonitrile, 10 mM phosphate buffer at pH 6.0 at 3.5 mL/min	FIC with a peristaltic pump	DAD spectrophotometer with 10 mm flow glass cell light path		[14]
Methyl paraben, propyl paraben, and triamcinolone acetonitride	Pharmaceutical cream	25 × 4.6 mm i.d. C18 monolithic column with 10 mm pre column	Acetonitrile–methanol–water (35:5:65 v/v/v) + 0.05% nonylamine at pH 2.5, 0.6 mL/min	SIC with an 8 port selection valve and 10 mL syringe	UV–Vis Spectrophotometer, with a 10 mm flow cell	0.25–0.50 μg/mL	[17]
Naphazoline nitrate, methyl paraben	Pharmaceuticals	25 × 4.6 mm i.d. C18 column and 5 × 4.6 mm i.d. precolumn	Methanol: water (45:65 v/v) at pH 5.2 at 0.9 mL/min 3.5 mL of mobile phase/run Ethylparaben as internal standard	SIC with a six port selection valve and a 5.0 mL syringe pump	UV–Vis spectrophotometry fiber optic with a 10 mm Z-flow cell At 256 nm	0.02 μg/mL	[16]
Methylparaben, ambroxol hydrochloride, and benzoic acid	Pharmaceuticals	Chromolith SpeedROD RP-18e, 50–4.6 mm column with 10 mm precolumn	Acetonitrile–tetrahydrofuran–0.05 M acetic acid (10:10:90 v/v/v), pH 3.75, adjusted with triethylamine, at 0.48 mL/min	SIC with a six port selection valve and a 5.0 mL syringe pump	UV–Vis detector at 245 nm	2 μg/mL for methyl paraben	[21]

Abbreviations: DAD, diode array detector; FIC, flow injection chromatography; FIA–SPE–MEKC, flow injection analysis–solid phase extraction–micellar electro kinetic chromatography; HPLC–FI–CH, high performance liquid chromatography–flow injection–chemiluminescence; FI–CH, flow injection–chemiluminescence detection; HPCE, high performance capillary electrophoresis; LOD, limit of detection; LOQ, limit of quantitation; SIC, sequential injection chromatography.

10.2.1 Online Solid-Phase Extraction

While the extraction of parabens in various consumer products was previously carried out by liquid–liquid extraction, this technique suffered from major drawbacks such as being time consuming and matrix effects. For this reason solid-phase extraction is used as the technique of choice to minimize the shortcomings of this method.

Online solid-phase extraction coupled to flow injection analysis–micellar electrokinetic chromatography (MEKC) was used to concentrate the parabens and remove the matrix components. The breakthrough volume with three parabens solutions at 50 µg/mL was found to be 2.0 mL. Moreover, the enrichment factors obtained for methylparaben, ethylparaben, and propylparaben were 65, 68, and 72, respectively, compared with flow injection analysis MEKC without solid-phase extraction. Furthermore, matrix effects were not observed because the technique was highly selective (Table 10.1) [12]. Thus, the method could be successfully used to extract parabens in water-based lotions, oil-based lotions, gels, and creams.

10.3 SEPARATION TECHNIQUES

10.3.1 Flow Injection Chromatography

Flow injection analysis and sequential injection analysis can both be interfaced with chromatographic systems. Usually these are liquid chromatography systems that are coupled to solid-phase extraction columns that may contain reversed-phase sorbents, molecularly imprinted polymers, or ion exchange material [10]. The coupling of liquid chromatography to flow techniques offers interesting features: the high selectivity and separation efficiency of liquid chromatography is coupled with the simplicity and rapidity of the flow injection–based techniques. Thus, the combination of flow-based techniques and liquid chromatography systems makes possible the use of a cheaper instrumentation compared to the conventional liquid chromatography instrumentation. This is due to the use of monolithic columns that have high separation efficiency, at higher flow rates, and at a lower pressure [13]. For this reason they can be incorporated into flow injection and sequential injection systems without any problems [14].

Flow injection analysis equipment coupled to a short C_{18} monolithic column (5 mm length) was used to analyze a mixture of eight food additives consisting of four parabens, antioxidants, and other sweeteners in beverages and soups, and some cosmetics. The monolithic column included in the flow injection analysis manifold improved the separation efficiency by reducing the problem of large pressure drop commonly encountered with flow cells with packed solid supports such as silica C_{18} and Sephadex DEAE (Table 10.1). Compared to HPLC, flow injection analysis to a monolithic column has the advantages of being faster, cheaper, and simpler. However, the major disadvantage of the technique is that the resolution obtained is less than that obtained with a conventional HPLC method, although it is still good enough [15].

10.3.2 Sequential Injection Chromatography

Sequential injection chromatography involves the coupling of sequential injection analysis and liquid chromatography. It capitalizes on the good separating power of liquid chromatography, and the efficient sample processing ability of sequential injection. The technique can be carried out on monolithic columns, using a medium pressure pump, or on micro pumps using a conventional low-pressure pump. Monolithic columns are advantageous in the sense that they allow higher flow rates due to shorter diffusion paths. However, their limitations include large external porosity, which leads to low retention factors, and low radial dispersion coefficient [16]. On the other hand the advantages of monolithic columns outweigh their disadvantages, and for this reason they are widely coupled with flow-based methods.

In an experiment, monolithic columns were coupled to sequential injection analysis in the simultaneous determination of methylparaben, propylparaben, sodium diclofenac, and internal standard butylparaben in a pharmaceutical formulation. A Chromolith RP-18e column of 2.5 mm × 4.6 mm length coupled to a sequential injection system with an eight-port injection valve was used for this purpose (Table 10.1). The advantage of this technique is that separation can be carried out without the conventional HPLC instrumentation at a lower cost without consuming much solvent. The disadvantages observed included the lack of software for adequate separation analysis: the software was challenged in evaluating the four peaks of the parabens in one cycle, and as a result only the highest peak could be evaluated. Other disadvantages noted include limited flow rates for longer columns and hence limited separation due to limitations to do with column length and flow rate. Furthermore, in comparison with other conventional separation methods such as HPLC and gas chromatography, the technique is less sensitive and less reliable [14]. Another study by the same research group confirmed the advantages of sequential injection chromatography against conventional HPLC in the analysis of parabens in pharmaceutical formulations [17].

The same research group used the technique to separate methylparaben, propylparaben, and triamcinolone acetonide in a pharmaceutical formulation. The experiment was important in highlighting the advantages of coupling monolithic columns (Chromolith Flash RP-18e) with sequential injection analysis. The major drawbacks of the technique highlighted were the limited flow rates of the syringe pump due to maximum back-pressure (about 2.5 MPa) and limited volume of the syringe pump (not exceeding 10 mL) (Table 10.1). Because the column is short and the flow rate is limited, the number of compounds that can be separated does not exceed five [18].

10.3.3 Flow Injection Methods Based on Capillary Electrophoresis

CE is another versatile technique that can be applied in the determination of parabens in consumer products. CE operates in several modes including capillary zone electrophoresis (CZE), MEKC, nonaqueous capillary electrophoresis (NACE), and capillary electrochromatography (CEC) as well as others. Advantages shared by all these techniques are that the instrumentation is simple to operate and that they are cheaper than other separation techniques. Hence, CE is widely used in the determination of parabens in consumer products by different laboratories [19,20].

Several techniques for interfacing flow injection analysis to CE have been reported in the literature [11]. In most of the systems designed, buffer contamination took place as a result of the sample solution passing through the buffer reservoir, and hence sample consumption increased as the polluted buffer was removed. Han et al. [12] developed an improved split-flow interface to protect the running buffer from contamination and thus reduce its consumption. The technique involved coupling online solid-phase extraction with MEKC, for the determination of three parabens (methyl-, ethyl-, and propylparaben) in cosmetics (Table 10.1). A fourth paraben was used as an internal standard because it was not detected in the cosmetic samples. The advantage of this method is that improved limits of detection were obtained (0.07–0.1 µg/mL) [12].

10.4 DETECTION

Flow injection methods can be coupled to any detection system available. The most common detection systems include a chemiluminescence detector, fluorometer, ultraviolet–visible (UV–Vis) spectrophotometers, and a sophisticated mass spectrometer. Flow injection methods coupled to these detectors have been applied in the analysis of parabens in consumer products such as food, cosmetics, and pharmaceutical formulations. However, parabens have been determined mostly by UV–Vis spectrophotometers and chemiluminescence detectors as reported in the literature. Other detectors such as electrochemical and fluorescence detectors have disadvantages of poor reproducibility and requirements for additional derivatization [21].

10.4.1 Flow Injection Methods Based on UV Detection

Ultraviolet–visible spectrophotometry is used for the determination of species that absorb in a certain region of the electromagnetic spectrum (chromophores). They are usually substances that contain pi electrons or are aromatic. As parabens contain an aromatic ring, and they could be analyzed using a UV–Vis spectrophotometer.

Flow injection coupled to a UV–Vis spectrophotometer has been widely used in the determination of parabens in pharmaceutical products [18,22], cosmetics [23], and food [15] (Table 10.1). It offers the advantages of low cost and less complexity as a single-channel manifold is needed. However, the major drawback of the technique is that the detector has poor selectivity [24]. Therefore, methods have to be developed using highly selective detectors such as the mass spectrometer and the chemiluminescence detector. Nowadays the use of the mass spectrometer is preferred in most analytical laboratories because of its ability to quantify analytes in complex matrices such as food with minimal matrix effects. Moreover, it is also capable of quantifying many analytes in a single run. Hence, its coupling with flow-based techniques will further improve the analysis of parabens in various consumer products by combining the advantages of flow injection analysis such as speed and high throughput with those of the mass spectrometric detector. On the other hand, the use of the chemiluminescence detector coupled to flow injection analysis can be considered a technique of choice when low detection limits and wider dynamic linear ranges are the main concerns of the analysis, as will be shown in Section 10.4.2.

10.4.2 Flow Injection Based on Chemiluminescence Detection

Chemiluminescence is light emission as the result of a chemical reaction. The process is described as a "dark field" technique because it takes place in the absence of an external light source. This phenomenon gives rise to such advantages as the absence of background signals or interference, which is common for other spectroscopic techniques such as spectrophotometry and fluorometry. Furthermore, the achievement of very low detection limits and wide dynamic linear ranges is also made possible [25]. The major disadvantage of the technique is its very low selectivity because the chemiluminescence reagent responds to a family of compounds and not to a single compound [26]. Therefore in the analysis of a family of compounds such as parabens the selectivity observed will be very low as all the parabens give the same response to the chemiluminescence reagent, which is usually cerium(IV) in sulfuric acid medium (Table 10.1). However, it is also important to note that selectivity also depends on the chemiluminescence reaction selected, which is usually luminol oxidation, direct oxidation using potassium permanganate, or oxidation by tris(2,2 bipyridine) ruthenium(III) complex [27].

Chemiluminescence methods can be performed in batch mode, where the sample or reagent is injected directly into the chemiluminescence cell using a syringe. Another mode is represented by the flow-based methods in which the reagents flow continuously through manifolds to the detector. This technique has the merits of lower reagent consumption, higher sample throughput, and the capability for automation [26].

Most flow-based techniques coupled to chemiluminescence do not require separation before detection, However, for complex matrices, flow-based techniques are first coupled to a separation technique such as HPLC or CE (Table 10.1). The major drawback of HPLC is the additional need for pumps to deliver the postcolumn chemiluminescence reagent or for additional devices to mix the column eluate and chemiluminescence reagent. Furthermore, the analysis time is increased because of the necessary optimization of the separation parameters (mobile phase composition, flow rate) and chemiluminescence parameters (pH, temperature, catalyst, type of enhancer). On the other hand, CE coupled to chemiluminescence has the advantages of high resolution, reduced analysis time, and lower expense [28].

An important reaction in chemiluminescence detection is the cerium(IV)–rhodamine reaction in sulfuric acid medium. In the presence of parabens, this process is enhanced, as the cerium(IV) is reduced to cerium(III) and the energy is then transferred from cerium(III)* to rhodamine 6G to form the excited rhodamine 6G which emits radiation at 556 nm (Table 10.1).

Different parabens at the same concentration give decreasing chemiluminescence intensity as the length of the alkyl group increases (Figure 10.3) because, as the length of the chain increases, the parabens become more resistant to hydrolysis [13]. For this reason each peak in Figure 10.3 corresponds to the p-hydroxybenzoic acid that is generated from the acidic hydrolysis of parabens, which chemically are esters.

According to this reaction, the first use of flow injection analysis coupled to chemiluminescence detection was to determine methylparaben, ethylparaben, propylparaben, and butylparaben in soy sauces (Table 10.1). Satisfactory detection limits were obtained, and for methylparaben, ethylparaben, propylparaben, and butylparaben were 3.4×10^{-10}, 2.0×10^{-9}, 2.0×10^{-9}, and 5.0×10^{-9} g/mL, respectively (Table 10.1). The method showed higher sensitivity than the other methods, which required preconcentration (HPLC with UV detection) or derivatization (HPLC with fluorescence detection) [29].

Zhang et al. [21] used the HPLC–flow injection chemiluminescence method to determine methylparaben, ethylparaben, propylparaben, and butylparaben, based on the same chemiluminescence enhancement reaction by parabens of the cerium(IV)–rhodamine 6G system in sulfuric acid. The method was applied to orange juice, soy sauce, vinegar, and cola soda (Table 10.1). The advantage of this technique is that it showed greater sensitivity and wider dynamic linear ranges than the electrochemical and fluorometric detectors [21].

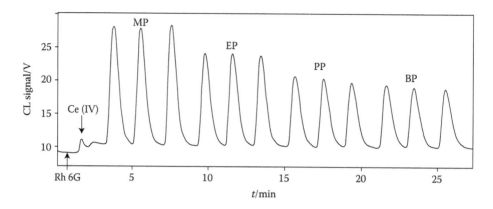

FIGURE 10.3   Reduction of the chemiluminescence signal for 1 mg/L solutions of parabens as the length of carbon chain increases. (Reprinted from *Talanta*, 79, Ballesta-
Claver, J., M. Valencia, and L. Capitan-Vallvey, Analysis of parabens in cosmetics by low
pressure liquid chromatography with monolithic column and chemiluminescent detection, 499–506, Copyright 2009, with permission from Elsevier.)

Liquid chromatography using an ultrashort monolithic column (5 mm) and chemiluminescence detection in a flow injection analysis type instrument manifold was applied
to determine four parabens in cosmetics products such as cleaning wipes and cleaning
mousse (Table 10.1). The detection was based on the oxidation of the hydrolysis products
of the parabens with cerium(IV) in the presence of rhodamine 6G, which gives chemiluminescence emission. While it shared the advantages of other flow injection analysis systems coupled to chemiluminescence detection such as improved speed and better
detection limits, the major disadvantage of the technique was its reduced precision and
limited ability for separation due to the short length of the column [13].

10.5 CONCLUSIONS AND OUTLOOK

While the benefits of flow-based methods such as the reduction of the analysis time and
higher sample throughput are greatly appreciated in the analytical sciences, their applications in the determination of parabens are greatly limited. For comparison, flow-based
techniques are widely applied in the analysis of other preservatives such as sulfites and
nitrites [30,31]. Thus, more investigations and publications on this subject are called
for. The applications of this technique in the analysis of foodstuffs are limited to simple
techniques such as flow injection and sequential injection analysis. However, the use of
novel hyphenated flow-based techniques would bring about a breakthrough in the determination of parabens in various consumer products. Moreover, there should be a shift to
the use of mass spectrometric detectors from the selective detectors such as chemiluminescence and UV–Vis.

    In conclusion, the advent of flow injection analysis has brought some advantages
in the analysis of consumer products. However, the application of the technique for the
determination of parabens should move in step with current developments in the instrumentation that include new detectors and hybrid systems, and more information should
be available on this subject.

REFERENCES

1. Eriksson, E., H. Andersen, and A. Ledin. 2008. Substance flow analysis of parabens in denmark complemented with a survey of presence and frequency in various commodities. *J. Hazard. Mater.* 156:240–259.
2. Darbre, P., A. Aljarrah, W. Miller, N. Coldham, M. J. Sauer, and G. Pope. 2004. Concentrations of parabens in human breast tumors. *J. Appl. Toxicol.* 24:5–13.
3. Soni, M., S. Taylor, N. Greenberg, and G. Burdock. 2002. Evaluation of the health aspects of methylparaben: A review of the published literature. *Food Chem. Toxicol.* 40:1335–1373.
4. Jain, R., M. K. Mudiam, A. Chauhan, C. Ratnasekhar, R. Murthy, and H. Khan. 2013. Simultaneous derivatization and preconcentration of parabens in food and other matrices by isobutyl chloroformate and dispersive liquid–liquid microextraction followed by gas chromatographic analysis. *Food Chem.* 141(1):436–443.
5. Saad, B., M. Bari, M. Saleh, K. Ahmad, and M. K. Talib. 2005. Simultaneous determination of preservatives (benzoic acid, sorbic acid, methylparaben and propylparaben) in foodstuffs using high performance liquid chromatography. *J. Chromatogr. A* 1073:393–397.
6. Xu, W., R. Sandford, P. Worsfold, A. Carlton, and G. Hanrahan. 2005. Flow injection techniques in aquatic environmental analysis: Recent applications and technological advances. *Crit. Rev. Anal. Chem.* 35:237–246
7. Hartwell, S., A. Kehling, S. Lapanantnoppakhun, and K. Grudpan. 2013. Flow injection/sequential injection chromatography: A review of recent developments in low pressure with high performance chemical separation. *Anal. Lett.* 46(11):1640–1671.
8. Mesquita, R. B. and A. O. S. S. Rangel. 2009. A review on sequential injection methods for water analysis. *Anal. Chim. Acta* 648:7–22.
9. Pinto, P. C. A., M. Lucia, M. F. Saraiva, and J. L. F. Lima. 2011. Sequential injection analysis hyphenated with other flow techniques: A review. *Anal. Lett.* 44(1–3): 374–397.
10. Theodoridis, G., C. Zacharis, and A. Voulgaropoulos. 2007. Automated sample treatment by flow techniques prior to liquid-phase separations. *J. Biochem. Biophys. Methods* 70:243–252.
11. Kuban, P. and B. Karlberg. 2009. Flow/sequential injection sample treatment coupled to capillary electrophoresis.a review. *Anal. Chim. Acta* 648:129–145.
12. Han, F., Y. He, and C. Yu. 2008. On-line pretreatment and determination of parabens in cosmetic products by combination of flow injection analysis, solid phase extraction and micellar electrokinetic chroatography. *Talanta* 74:1371–1377.
13. Ballesta-Claver, J., M. Valencia, and L. Capitan-Vallvey. 2009. Analysis of parabens in cosmetics by low pressure liquid chromatography with monolithic column and chemiluminescent detection. *Talanta* 79:499–506.
14. Satinsky, D., P. Solich, P. Chocholous, and R. Karlicek. 2003. Monolithic columns—A new concept of separation in the sequential injection technique. *Anal. Chim. Acta* 499:205–214.
15. Garcia-Jimenez, J., M. Valencia, and L. Capitan-Vallvey. 2007. Simultaneous determination of antioxidants, preservatives and sweetener additives in food and cosmetics by flow injection analysis coupled to a monolithic column. *Anal. Chim. Acta* 594:226–233.
16. Gritti, F. and G. Guiochon. 2011. Measurement of the eddy diffusion term in chromatographic columns. i. Applications to the first generation of 4.6 mm i.d monolithic columns. *J. Chromatogr. A* 1218:5216–5227.

17. Chocholous, P., D. Satinsky, and P. Solich. 2006. Fast simultaneous spectrophoto-metric determination of naphazoline nitrate and methylparaben by sequential injection chromatography. *Talanta* 70:408–413.
18. Satinsky, D., J. Huclova, P. Solich, and R. Karlicek. 2003. Reversed-phase porous silica rods, an alternative approach to high performance liquid chromatographic separation using the sequential injection chromatography technique. *J. Chromatogr. A* 1015:239–244.
19. Chu, Q., J. Wang, D. Zhang, and J. Ye. 2010. Sensitive determination of parabens by capillary zone electrophoresis with amperometric detection. *Eur. Food Res. Technol.* 231:891–897.
20. De Rossi, A. and C. Desiderio. 2002. Fast capillary electrochromatographic analysis of parabens and 4-hydroxybenzoic acid in drugs and cosmectics. *Electrophoresis* 23(19):3410–3417.
21. Zhang, Q., M. Lian, L. Liu, and H. Cui. 2005. High-performance liquid chroma-tography assay of parabens in wash-off cosmetic products and foods using chemilu-minescence detection. *Anal. Chim. Acta* 537:31–39.
22. Satinsky, D., J. Huclova, R. Ferreira, M. Montenegro, and P. Solich. 2006. Detrmination of ambroxol hydrochloride, methylparaben and benzoic acid in pharmaceutical preparations based on sequential injection technique coupled with monolithic column. *J. Pharm. Biomed. Anal.* 40:287–293.
23. Garcia-Jimenez, J., M. Valencia, and L.F. Capitán-vallvey. 2010. Parabens determi-nation with a hybrid FIA/HPLC system with utra-short monolithic column. *J. Anal. Chem.* 65(2):188–194.
24. Tzanavaras, P. and D. Themelis. 2007. Review of recent applications of flow injec-tion spectrophotometry to pharmaceutical analysis. *Anal. Chim. Acta* 588:1–9.
25. Liu, M., Z. Lin, and J.-M. Lin. 2010. A review on applications of chemilumines-cence detection in food analysis. *Anal. Chim. Acta* 670(1–2):1–10.
26. Mervatova, K., M. Polasek, and J. Calatayud. 2007. Recent applications of flow injection and sequential-injection analysis techniques to chemiluminescence deter-mination of pharmaceuticals. *J. Pharm. Biomed. Anal.* 45:367–381.
27. Fletcher, P., K. Andrew, A. C. Calokerinos, S. Forbes, and P. Worsfold. 2001. Analytical applications of flow injection with chemiluminescence detection—A review. *Luminescence* 16:1–23.
28. Christodouleas, D., C. Fotakis, A. Economou, K. Papadopoulos, M. Timotheou-Potamia, and A. Calokerinos. 2011. Flow-based methods with chemilumines-cence detection for food and environmental analysis: A review. *Anal. Lett.* 44(1–3):176–215.
29. Myint, A., Q. Zhang, L. Liu, and H. Cui. 2004. Flow injection-chemiluminescence determination of paraben preservative in food safety. *Anal. Chim. Acta* 517:119–124.
30. Ruiz-Capillas, C. and J.-C. Francisco. 2009. Application of flow injection analysis for determining sulphites in food and beverages: A review. *Food Chem.* 112:487–493.
31. Ruiz-Capillas, C. and J.-C. Francisco. 2008. Determining preservatives in meat products by flow injection analysis (FIA): A review. *Food Addit. Contam. Part A* 25:1167–1178.

CHAPTER 11

Determination of Acetic Acid and Acetates

Dayene do Carmo Carvalho, Helen Cristine de Rezende, Luciana M. Coelho, and Nívia M.M. Coelho

CONTENTS

11.1 CONSIDERATIONS

The use of additives in food production is mandatory since it can improve the taste, odor, and color of the products. Food additives, such as preservatives, are classified on the basis of their functional use. A preservative is defined as "a substance which when added to food is capable to inhibiting, retarding or arresting the process of fermentation, acidification or other decomposition of food" (Manual of Methods of Analysis of Foods–Food Additives, 2012). Notable preservatives are acetic acid and acetate salts, which are used in food production as a taste improver.

The control of preservatives in foods is important due to potential problems; for example, a high concentration of acetic acid in wine results in spoilage of the product. Due to the potential undesirable effect of acetic acid and acetate salts, their content in foods and beverages should be limited. Hence, the monitoring of these preservatives is desirable.

The quantitative techniques currently available to determine acetate concentrations include the use of ion suppression reversed-phase liquid chromatography (Skelly, 1982), high-pressure liquid chromatography (Hordijk, 1983), nonsuppressed ion chromatography (Haddad and Croft, 1986), gas chromatography (GC) (Rocchiccioli et al., 1989; Wittmann et al., 2000), and capillary electrophoresis (CE) (Wiliams et al., 1997; Hettiarachchi and Ridge, 1998). The accepted method for the determination of acetic acid in vinegar, for example, is based on the acid–base titration of the substrate with sodium hydroxide using phenolphthalein as an indicator. The decrease in the color intensity of the indicator is monitored spectrophotometrically at 480 nm and is proportional to the acid content. However, this method cannot be used for monitoring acetate in wine samples since the content of acetate in wine is almost 100 times lower than that in vinegar (Lenghor et al., 2002).

Considering the need for them in the food industry, reliable methods for the determination of acetic acid and acetate, mainly for continuous monitoring, such as online methods, are highly desirable. In this context, flow injection analysis (FIA) has been used extensively for various applications because this technique offers a number of important benefits, including reduced disposal costs and high sample throughput.

This chapter details the most recent strategies that employ the FIA technique for the determination of acetic acid and acetate salts used as additives in the food industry. We highlight the most important advantages and disadvantages of the different methodologies and briefly outline new trends. Specific considerations required for certain sample matrices, including techniques to clean up the interfering matrix of a real sample also merit and receive further discussion.

11.2 ACETIC ACID AND ACETATE SALTS AS ADDITIVES

Additives are required to guarantee the quality and durability of food products. Food additives are intentionally added to food and must be safe for a lifetime of consumption based on current toxicological evaluations (Manual of Methods and Analysis of Foods, 2012). Compounds used as preservatives added to maintain or improve the food quality act by preventing and/or retarding the microbial spoilage of food.

There are several sources of food contamination by microorganisms that, on contact with food, may promote its deterioration, changing its appearance, flavor, odor, and texture. These changes occur due to the intake of chemical substances, such as carbohydrates, proteins, and lipids, by microorganisms for use as an energy source.

Preservative agents are used to reduce the rate of food deterioration caused by microorganisms and consequently preserve the taste and appearance of the products. They act by increasing the latency phase, that is, the step during which the organism adapts to a new environment. This class of compounds includes acetic acid (E260) and acetates (potassium [E261], sodium [E262], and calcium [E263]), which can effectively inhibit the growth of *Bacillus, Clostridium, Listeria, Salmonella, Staphylococcus aureus, Pseudomonas, Lactobacillus pentosus, Saccharomyces cerevisiae, Escherichia coli, Campylobacter*, etc. (Levine and Fellers, 1940; Adams and Hall, 1988). Some of the microorganisms mentioned above are characteristic of fermented vegetables (Bautista-Gallego et al., 2008). These agents are also used to stabilize the acidity of food products and as sequestering agents and flavorings.

Acetic acid, calcium acetate, and sodium diacetate are widely used in human and veterinary medicine, cosmetics, and a variety of household products as buffering agents or antimicrobial agents. Considering the complete and rapid metabolism of acetic acid and its salts, their use in animal nutrition is not expected to contribute to human exposure.

Chemical preservatives such as E260–E263 are used in a number of food products including fish products of all types, fruit juices, soft drinks, pastries, salads, margarine, sauces, wines, dried fruits, citric fruits, bananas, desiccated vegetables, sugar, meat, etc.

The level of acetic acid or its salts added in the final food will typically be in the range of 200–2500 mg/kg complete feeding stuffs. The actual amount used will be dependent on the pH, moisture content, predicted storage time, and estimated potential number of spoilage microorganisms (Manual of Methods and Analysis of Foods, 2012).

European Union Register of Feed Additives describes that acetic acid (E260) and its calcium (E263) and sodium salts (sodium diacetate E262) are preservatives for use in all animal species and categories without a time limit and without maximum levels. The European Food Safety Authority (EFSA) has issued an opinion on the safety of calcium acetate added for nutritional purposes to food supplements (EFSA, 2009).

11.3 FLOW METHODS FOR DETERMINATION OF ACETIC ACID AND ACETATES

The topic of detection techniques for the determination of acetic acid and acetate salts has been approached from various perspectives.

Figure 11.1 shows the relevant analytical methods and instrumental techniques, particularly those that involve a separation step, such as chromatography and gaseous diffusion. Methods that feature electrochemical detection are covered, as well as those involving spectrometric detection. In general, the most commonly used analytical strategies for no-reference methods are those that involve electrochemical detection in combination with enzymatic methods. As has been noted in published papers on electrochemical developments, the majority of publications are concerned with quantification employing amperometric methods. Several papers have appeared on separation techniques of interest, in particular gaseous diffusion (Araújo et al., 2005; Coelho et al., 2010). Section 11.3.1 addresses the main analytical methods for the continuous monitoring of acetic and acetate salts, which offer attractive features, such as sensitivity and selectivity.

Optical detectors are by far the most commonly used of the analytical methods available for continuous monitoring. In fact, in around two-thirds of all FIA applications reported the use of this type of detector. However, several other types of detection methods are available for the determination of ions acetate and acetic acid involving, for instance, potentiometric, enzymatic, spectrophotometric, fluorometric, and chemiluminescence techniques. However, molecular absorption spectrophotometers continue to be the preferred choice in FIA, and in analytical chemistry in general, on account of their high versatility (Calatayud, 1996). Studies employing FIA and other methods for the determination of acetic acid and acetate ions are described in the following sections.

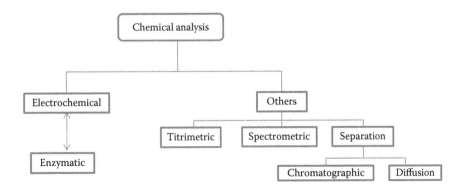

FIGURE 11.1   Chemical methods for the quantification of acetic acid and acetate salts.

11.3.1 Spectrometric Methods

FIA in spectroscopic methods is the most commonly used approach in studies on acetic acid and acetate determination. The determination of acetic acid/acetate employing the FIA methodology generally requires a combination of techniques to achieve good results. However, the methods used suffer severe interference from CO_2 and some acids (i.e., lactate) present in the sample.

One such method is the combination of titration with spectrophotometric determination, which is well established (Poppi and Pasquini, 1993). These authors used multivariate calibration, optimizing the FIA titration system for the selective determination of a mixture of acetic acid, benzoic acid, and hydrochloric acid, using NaOH as the titrant. The mixture containing the indicator–sample–alkaline gradient was analyzed using a spectrophotometer at 340–680 nm. This methodology allows the analysis of mixtures of acids in synthetic solutions within 1–5 min, although it is important to have data on the differences between the pKa values of the acids concerned.

Alternatively, spectrophotometric techniques can be used for the determination of acetic acid based on the UV absorbance at 205 nm (Tippkötter et al., 2008). The system was applied for the determination of acetic acid in *E. coli* fermentation broth.

In another study in which acetic acid concentrations were determined using spectrophotometric detection (Lenghor et al., 2002), the ease of analysis and flexibility of the method were noted; however, a sequential injection analysis (SIA) system was used. Compared with FIA, this system has a high sampling time (around 60 samples/h) and it is automatically controlled by a personal computer, which was not the case in the study by Poppi and Pasquini (1993).

Another approach is to combine spectroscopic techniques with photometric determination through gas diffusion, which has been used for the determination of acetate and glucose (Rocha and Ferreira, 2002). Those authors applied a proposed two-channel FIA system to the online determination of glucose and acetate during the fed-batch recombinant fermentation of *E. coli*. The results were compared with those obtained offline by high-performance liquid chromatography (HPLC) and using an enzymatic kit, employing simple reactions with inexpensive and widely available chemicals. It was reported that the consumption of chemicals was lower, leading to a reduction in waste products and more economical determinations.

Although the techniques of HPLC, GC, and mass spectrometry (MS) (Christiaens et al., 2004; Sanhueza et al., 1996) are very efficient, mainly when coupled with other methods, the FIA online system has been used successfully to determine acetic acid concentrations. Teeter et al. (1994) adapted a direct insertion membrane probe in mass spectrometer and the analytes of interest pass through the membrane and are ionized by electron implant ionization. The technique is cost effective and simple to implement, especially when used with an ion trap mass spectrometer. The high cost of acquisition of the equipment required explains the small number of studies employing online FIA and the preference for the use of photometric techniques together with an FIA system.

11.3.2 Electrochemical and Enzymatic Processes

The detection of acetic acid through enzymatic processes is based on a three-reaction mechanism sequence. Acetate reacts with adenosine triphosphate (ATP) to produce adenosine diphosphate (ADP) in a step catalyzed by acetate kinase (AK). In the presence of pyruvate kinase (PK), ADP reacts further with phosphoenolpyruvate (PEP) to give pyruvate, which is then reduced to lactate by NADH (nicotinamide adenine dinucleotide hydride) in the presence of lactate dehydrogenase (LDH). The amperometric biosensor measures the current decrease caused by the NADH decrease at the modified graphite electrode (Mieliauskiene et al., 2006).

Based on this enzymatic process a simplified system for acetate monitoring was developed. The enzymes (AK, PK, and LDH) were co-immobilized on the surface of a graphite electrode, using the electrochemical mediator brilliant cresyl blue (BCB) and a cross-linking agent poly(ethyleneglycol) diglycidyl ether (PEGDGE). Biosensors were tested in wine and vinegar samples. The results for the acetate concentration obtained using this biosensor and employing the enzymatic kit showed good agreement, verifying that the tri-enzyme electrode is suitable for the simple and accurate determination of acetic acid concentrations in food and beverages samples (Mizutani et al., 2003; Mieliauskiene et al., 2006).

Mizutani et al. (2003) determined the concentration of acetic acid in a manner similar to that of Mieliauskiene et al. (2006), the detection being performed using a combination of FIA with amperometric tri-enzyme biosensor detection. The biosensor was prepared by immobilizing AK, PK, and PyOx on a poly(dimethylsiloxane) (PDMS)-coated electrode. The oxygen consumption was monitored using the PDMS-coated electrode without interference from the PyOx reaction product (hydrogen peroxide). Thus, the biosensor-based system could be used for the determination of acetic acid from 0.05 to 20 mM with a sampling rate of 20 h^{-1} and it remained stable for one month. This system was applied in the analysis of wine samples (Mizutani et al., 2003).

In the process of the cultivation of recombinant *E. coli*, low yields are observed due to the formation and accumulation of acetic acid resulting from overflow metabolism. Thus, it is necessary to prevent and monitor the formation of acetic acid, which can be avoided by the introduction of glucose to the medium. However, the rate of glucose addition must be monitored in order to maintain the growth below a critical value. In this context, a flow injection system was developed for the determination of acetic acid in *E. coli* cultivations with electrochemical detection based on immobilized AK, PK, and LDH, with amperometric detection of NADH. A limitation associated with this system was that the enzyme AK lost 90% of its activity after only one fermentation. The authors thus suggested that, due to the low stability of immobilized AK in the fermentation measurements, alternative enzymes needed to be found (Tang et al., 1997).

11.3.3 Other Methods

Studies using the flow analysis technique for the determination of acetate ions and acetic acid are not very common compared with the development of methodologies for the determination of other ions, such as phosphate and sulfate ions. Details regarding methods not previously mentioned carried out on the determination of acetic acid and acetate ions using FIA can be seen in Table 11.1.

In the gaseous diffusion method, the determination of acetic acid is based on the fact that acetic acid diffuses through the membrane and dissolves into the acceptor solution.

Araújo et al. (2005) used a flow system with gaseous diffusion and conductometric detection for the determination of acetic acid in vinegar samples (Table 11.1). The FIA manifold for the determination of acetic acid is shown schematically in Figure 11.2. Due to the equilibrium $CH_3COOH + H_2O \leftrightarrow CH_3COO^- + H_3O^+$, the water conductivity

TABLE 11.1   Studies using FIA Systems in the Determination of Acetic Acid and Acetate

Sample	Analyte	Strategy	Detection	Reference
Wine, juice, fruits, and vinegar	Acetic acid	Gas diffusion	Conductometry	Araújo et al. (2005)
Ion standards	Acetate, citrate, L-lactate	Indirect immobilization of enzyme	Fluorometry	Tsukatani and Matsumoto (2006)
Pesticide	3-Indolylacetic acid	–	Fluorometry	Calatayud et al. (2006)
Outdoor air	Formaldehyde, formic and acetic acids, ammonia	Diffusion and electrophoresis	Conductometry	Coelho et al. (2010)

increases. The detection of acetic acid is based on the increase in the conductivity. A sample throughput of 80 h^{-1} was achieved. For acetic acid determination, the limits of detection and quantification were 5.0×10^{-6} and 1.7×10^{-5} mol/L, respectively. The vinegar samples were also analyzed to determine the acetic acid concentration using the standard titrimetric method (AOAC International, 1990) for comparison purposes. According AOAC (1990), a diluted 10 mL sample recently boiled and cooled was titrated with 0.5 mol/L NaOH using phenolphthalein as indicator.

Of the methods detailed in Table 11.1, FIA with gaseous diffusion and conductometric detection (Araújo et al., 2005) is the cheapest, and it consumes low amounts of reagents, uses simple materials, and allows rapid detection compared with the fluorometric detection methods.

11.4 CONCLUSIONS AND FUTURE PROSPECTS

The food industry is a field in which analytical chemistry plays an important role, contributing with new procedures of analysis and instrumentation. Methods for the determination and monitoring of preservatives, such as acetic acid and its acetate salts, are still scarce.

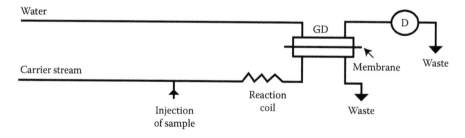

FIGURE 11.2   Schematic diagram of the flow injection system used for the determination of gaseous species. GD: gas diffusion; D: detector; the carrier stream was deionized water and the flow rate for acetic acid determination was 1.0 mL/min. (Adapted from Araújo, C. S. T. et al. 2005. *Food Chem.* 92:765–770.)

However, there are inherent difficulties associated with the types of samples involved. For example, in the medical area the main matrices are blood, serum, and urine, while in the food industry there are more types of samples and greater variations in their compositions. This hinders the application of electrochemical techniques, such as processes that involve biosensor design. Nevertheless, biosensors have been found to be versatile tools for the fast and accurate monitoring of acetate in model and real samples.

As noted by Kittsteiner-Eberle (1990), unfortunately, methods for the determination of preservatives in food are rare, especially those suitable for continuous run mode. Commonly used methods such as HPLC and GC do not provide satisfactory results due to the high level of interferences. A proposed procedure control for continuous monitoring based on the determination of the pH value (Forman et al., 1991) is attractive, but this is adversely affected by the presence of CO_2 and some acids (e.g., lactate) in the sample. Spectrometric techniques require appropriate derivatization to introduce a detectable chromophoric group.

The analytical capabilities of miniaturized separation techniques, such as capillary liquid chromatography (LC), nano-LC, and CE, are of interest in some fields of application, but these techniques have not yet reached the food industry, in particular for the determination of acetic acid and its salts. Thus, analytical chemists need to direct their attention toward these trends with the aim of narrowing the gap between science and the food industry.

ACKNOWLEDGMENTS

The authors are grateful for the financial support received from the Brazilian government agencies Conselho Nacional de Desenvolvimento Científico e Tecnológico (CNPq) and Coordenação de Aperfeiçoamento de Pessoal de Nível Superior (CAPES), from the MG state government agency Fundação de Amparo à Pesquisa do Estado de Minas Gerais (FAPEMIG), and GO state government agency Fundação de Amparo á Pesquisa do Estado de Goiás (FAPEG).

REFERENCES

Adams, M. R. and C. J. Hall. 1988. Growth inhibition of food-borne pathogens by lactic and acetic acids and their mixtures. *Int. J. Food Sci. Technol.* 23:287–292.

AOAC International. 1990. *Official Methods of Analysis* (15th ed.). Washington, DC: Association of Official Analytical Chemists.

Araújo, C. S. T., Carvalho, J. L., Mota D. R., Araújo, C. L., and N.N.M. Coelho. 2005. Determination of sulphite and acetic acid in foods by gas permeation flow injection analysis. *Food Chem.* 92:765–770.

Bautista-Gallego, J., Arroyo-López, F. N., Durán-Quintana, M. C., and A. Garrido-Fernández. 2008. Individual effects of sodium, potassium, calcium, and magnesium chloride salts on *Lactobacillus pentosus* and *Saccharomyces cerevisiae* Growth. *J. Food Prot.* 10:1412–1421.

Calatayud, J. M. 1996. *Flow Injection Analysis of Pharmaceuticals*. London: Taylor & Francis, 394pp.

Calatayud, J. M., Ascenção, J. G., and J. R. Albert-Garcia. 2006. FIA-fluorimetric determination of the pesticide 3-indolyl acetic acid. *J. Fluoresc.* 16:61–67.

Christiaens, B., Fillet, M., Chiap, P. et al. 2004. Fully automated method for the liquid chromatographic–tandem mass spectrometric determination of cyproterone acetate in human plasma using restricted access material for on-line sample clean-up. *J. Chromatogr. A* 1056:105–110.

Coelho, L. H. G., Melchert, W. R., Rocha, F. R., Rocha, F. R. P., and I. G. R. Gutz. 2010. Versatile microanalytical system with porous polypropylene capillary membrane for calibration gas generation and trace gaseous pollutants sampling applied to the analysis of formaldehyde, formic acid, acetic acid and ammonia in outdoor air. *Talanta* 83:84–92.

EFSA (European Food Safety Authority), 2009. Scientific opinion of the panel on food additives and nutrient sources added to food on calcium acetate, calcium pyruvate, calcium succinate, magnesium pyruvate magnesium succinate and potassium malate added for nutritional purposes to food supplements. *EFSA J.* 1088:1–25.

Forman, L. W., Thomas, B. D., and F. S. Jacobson. 1991. On-line monitoring and control of fermentation processes by flow injection analysis. *Anal. Chim. Acta* 249:101–111.

Haddad, P. R. and M. Y. Croft. 1986. The determination of lactate, acetate, chloride and phosphate in an intravenous solution by non-suppressed ion chromatography. *Chromatographia* 21:648–650.

Hettiarachchi, K., and S. Ridge. 1998. Capillary electrophoretic determination of acetic acid and trifluoroacetic acid in synthetic peptide samples. *J. Chromatogr. A* 817:153–158.

Hordijk, K. A. and T. E. Cappenberg. 1983. Quantitative high-pressure liquid chromatography-fluorescence determination of some important lower fatty acids in lake sediments. *Appl. Environ. Microbiol.* 46:361–369.

Kittsteiner-Eberle, R. 1990. Quantifizierung von Glycerin und Essigsiure. In *Jahresbericht, Lehrstuhl für Allgemeine Chemie und Biochemie*, pp. 23–44. Deutschland: TU München-Weihenstephan.

Lenghor, N., Jakmunee, J., Vilen, M., Sara, R., Christian, G. D., and K. Grudpan. 2002. Sequential injection redox or acid–base titration for determination of ascorbic acid or acetic acid. *Talanta* 58:1139–1142.

Levine, A. S. and C. R. Fellers. 1940. Action of acetic acid on food spoilage microorganisms. *J. Bacteriol.* 39: 499–514.

Manual of Methods and Analysis of Foods. 2012. Food safety and standards authority of India, Ministry of Health and Family Welfare Government of India, New Delhi. http://www.fssai.gov.in/Portals/0/Pdf/15Manuals/FOOD%20ADDITIVES.pdf.

Mieliauskiene, R., Nistor, M., Laurinavicius, V., and E. Csoregi. 2006. Amperometric determination of acetate with a tri-enzyme based sensor. *Sens. Actuators B* 113:671–676.

Mizutani, F., Hirata, Y., Yabuki, S., and S. Iijima. 2003. Flow injection analysis of acetic acid in food samples by using trienzyme/poly(dimethylsiloxane)-bilayer membrane-based electrode as the detector. *Sens. Actuators B* 91:195–198.

Poppi, R. J. and C. Pasquini. 1993. Spectrophotometric determination of a mixture of weak acids using multivariate calibration and flow injection analysis titration. *Chemom. Intell. Lab. Syst.* 19:243–254.

Rocha, I. and E. C. Ferreira. 2002. On-line simultaneous monitoring of glucose and acetate with FIA during high cell density fermentation of recombinant E. coli. *Anal. Chim. Acta* 462:293–304.

Rocchiccioli, F., Lepetit, N., and P. F. Bougneres. 1989. Capillary gas–liquid chromatographic/mass spectrometric measurement of plasma acetate content and (2–13C) acetate enrichment. *Biomed. Environ. Mass. Spectrom.* 18:816–822.

Sanhueza, E., Figuero, L., and M. Santana. 1996. Atmospheric formic and acetic acids in Venezuela. *Atmos. Environ.* 30:1861–1873.

Skelly, N. E. 1982. Ion-suppression reversed-phased liquid chromatographic determination of acetate in brine. *J. Chromatogr.* 250:134–140.

Teeter, B. K., Dejarme, L. E., Choudhury, T. K., Cooks, R. G., and R. E. Kaiser, R. E. 1994. Determination of ammonia, ethanol and acetic acid in solution using membrane introduction mass spectrometry. *Talanta* 41:1237–1245.

Tippkötter, N., Deterding A., and R. Ulber, R. 2008. Determination of acetic acid in fermentation broth by gas-diffusion technique. *Eng. Life Sci.* 8:62–67.

Tang, X. J., Tocaj, A., Holst, O., and G. Johansson. 1997. Process monitoring of acetic acid in *Escherichia coli* cultivation using electrochemical detection in a flow injection system. *Biotechnol. Tech.* 11:683–687.

Tsukatani, T. and K. Matsumoto. 2006. Flow-injection fluorometric quantification of pyruvate using co-immobilized pyruvate decarboxylase and aldehyde dehydrogenase reactor: Application to measurement of acetate, citrate and L-lactate. *Talanta* 69:637–642.

Wiliams, C. R., Boucher, R., Brown J., Scull J. R., Walker J., and D. Paolini. 1997. Analysis of acetate counter ion and inorganic impurities in pharmaceutical drug substances by capillary ion electrophoresis with conductivity detection. *J. Pharm. Biomed. Anal.* 16:469–475.

Wittmann, G., Van Langenhove, H., and J. Dewulf. 2000. Determination of acetic acid in aqueous samples, by water-phase derivatisation, solidphase microextraction and gas chromatography. *J. Chromatogr. A* 874:225–232.

CHAPTER 12

Determination of Lactic Acid

A.C.B. Dias and L.R. Braga

CONTENTS

12.1 INTRODUCTION

12.1.1 History and Consumption of Lactic Acid

The importance of lactic acid (LA) as a food preservative dates back to the beginning of the eighteenth century when scientists tried to understand the preservation and fermentation of sour milk. In 1857, Pasteur discovered that LA was present not as a milk component but as a metabolite produced during fermentation by a living microorganism [1]. Its presence as a natural fermentation metabolite in dairy products promoted preservation of the organoleptic properties for longer periods of time, making it an effective and natural food preservative. This finding initiated wide-ranging application and production of LA, with the first industrial production by microbial fermentation dating from 1881 [1]. Since then, research has considered new mechanisms of production, alternative microbiological routes, and utilization of various raw materials in order to guarantee efficiency, purity, and stability in LA production. The wide application of LA, as a food preservative and other industrial input requires about 130,000–150,000 metric tons per year [2]. According to a Global Industry Analyst announcement in January 2011, this amount is expected to reach approximately 329,000 metric tons by 2015 [3]. Ninety percent of today's world production of LA is obtained naturally by microbial fermentation of carbohydrates, such as glucose, sucrose, or lactose by homolactic organisms (bacteria or fungi) [1,3]. The rest is obtained synthetically by the hydrolysis of lactonitrile [4] or alternative routes involving catalyzed degradation of sugars, oxidation of propylene glycol or reaction of acetaldehyde [5]. Microbial fermentation is more advantageous than the chemical process in terms of experimental procedure (method), economic aspects, and environmental aspects because lactonitrile is a subproduct of petrochemical resources [1,6]. Furthermore, chemical synthesis produces only racemic mixtures of LA, which are impracticable as a food preservative [2].

Lactic acid occurs in two natural isomeric forms: L(+)-LA and D(−)-LA; the D(−) isomer is more harmful to humans than the L(+) isomer. From a nutritional point of view, the use of D(−)-LA as a preservative in food and beverages is undesirable because it is not easily metabolized by humans [7]. In addition, the World Health Organization recommends against its use for lactating infants and children [8]. The excessive consumption of the D(−) isomer can lead to metabolic damage, resulting in premature decalcification and acidosis [1,7]. As a result, the L(+) isomer is preferred for use as a food preservative due to its similarity with the human body's natural product, minimizing metabolic, allergic, and intolerance problems.

Microbial fermentation with some genera of LA bacteria or some specific fungi as the fermentation agent can produce L(+)-LA with high optical purity [1,3]. *Lactobacillus* (*L.*) produces both isomeric forms, depending on the species considered and culture conditions [7]. Species such as *Lactobacillus amylophilus* [9], *Lactobacillus casei* [10], *Lactobacillus rhamnosus* [11], and others [3,7] have been employed for production of the L(+) isomer. Filamentous fungi species also can be exploited, such as *Rhizopus* with the species *Rhizopus oryzae* and *Rhizopus arrhizus*, which are only able to convert starch into L(+)-LA [3]. As well as molasses, syrups, and dextrose, several new carbohydrate sources have been found for microbial fermentation, including novel technologies with renewable resources, such as starch, hemicellulose, and cellulose [4]. In commercial production, the choice of substrate depends on price, viability, purity, and energetic parameters [12]. The type of living organism also must be considered to avoid excessive or total production of D(−) isomer if the product is to be used as a preservative in food and beverages.

12.1.2 Preservative Properties of Lactic Acid

As the simplest hydroxycarboxylic acid with an asymmetrical carbon atom, LA (2-hydroxypropanoic acid) is a weak organic acid (pK_a 3.85) that is widely distributed in nature [13]. Its wide application as a preservative compound (E270) is associated with its physical–chemical properties, such as acid dissociation in aqueous media and bifunctional reactivity due to the presence of two functional groups (carboxyl and hydroxyl), which allows several kinds of chemical interactions [3]. These functions allow LA to be used as an acidulant, favoring, and pH-buffering agent. Its primary use is as an inhibitor of bacterial spoilage by pathogens, retarding proliferation and in certain conditions acting as a potent bactericide. Its preservative property is associated with impeding bacterial growth by decreasing the pH values of internal cellular membranes, causing inhibition of membrane transport and/or permeability functions [14]. In its nondissociated form, LA is soluble in the cell membrane and is transported into the cell by simple diffusion. Inside the cell, LA dissociates, promotes increased proton concentration, leading to higher acidity and causing a possible disruption of the membrane that damages the ability of nutrients to be transported and stops cell growth [15]. Then, the presence of LA acts as a bactericide impeding proliferation of pathogenic bacteria rather than causing direct extinction.

Lactic acid is present naturally as a preservative in dairy products, such as cheese, yogurt, butter, ice cream, and fermented milks. Lactic acid bacteria are present in the flora that converts carbohydrates in LA during the fermentation process [16]. Other fermented products, such as wine, pickles, and soy sauce, also naturally contain LA through alternative fermentative routes and microorganisms. In processed and industrialized food, LA can be found as a preservative in breads and bakery products, candy, soft drinks, soups, sherbets, beer, jams and jellies, mayonnaise, processed eggs, and others products [15]. The antimicrobial action associated with the mild flavor of LA is its most positive characteristic and the main reason for its use in the food and beverage industry. It acts as an acidity regulator in soft drinks and fruit juices and as a preservative in salads and dressings. With its mild flavor, it does not interfere with the flavor of the final product while it maintains microbial safety. Another valuable property is its enhancement of flavor, mainly savory flavors, with wide application in meat and dairy products. A recent application of LA is as a washing solution after harvest to increase the shelf life of fresh vegetables, fruits, and greenery [17,18]. Regarding food safety and human consumption, LA is classified as GRAS (generally recognized as safe) for use as a food additive by the U.S. FDA (United States Food and Drug Administration) [19], mainly due to its well-established natural origin in comparison with added synthetic chemicals and additives [12,20]. Natural food protection is a favored trend compared to the use of artificial preservatives that sometimes have considerable impact on nutritional aspects and sensory properties [20]. Lactic acid can be considered a natural preservative, easily produced by microbial fermentation at a low cost with high efficiency. Because of these characteristics, legislation based on the Codex Alimentarius Commission on food additives and contaminants (March, 2011) does not limit the acceptable daily intake (ADI) of LA, but recommends that it be controlled by good manufacturing practice (GMP).

12.1.3 Determination of Lactic Acid in Food and Beverages

Slight natural variations of LA concentration in food and beverages may occur in dairy products depending on certain fermentative parameters, such as kind and quantity of raw material and microorganism as well as biochemical and manufacturing conditions (temperature, agitation, energy, aeration, etc.) [3]. The amount of LA to be added to some processed food and beverages varies according to the purpose (enhance flavor, acidulate, or control bacteria). The amount must be controlled to avoid excessive acidification and/or alteration of the product's organoleptic properties. Thus, any modification in parameters during production of dairy products or erroneous calculations of LA additions can produce variability in taste and, more importantly, in the microbiological safety of the commercial product [21]. Then, determination of LA in commercial food and beverages is an important quality control tool for the organoleptic characteristics and estimation of shelf life.

Among the proposed methods for quantification of LA, high-performance liquid chromatography (HPLC) coupled with UV detection has been the technique most applied to analysis of fermented dairy and beverage products [22–24]. With this strategy, it is possible to quantify LA in the presence of other organic acids with a similar structure; however, there are some limitations and drawbacks, including problems associated with matrix color, excessive sample preparation and analysis time, separation efficiency, complexity, and cost. Some improvements in resolution and simplicity, reduction of consumption of chemicals, and reduction of sample preparation have been reported in electrophoretic methods coupled with UV detection [25,26]. HPLC and electropho-resis methods also allow isomeric separation and determination of D(−) and L(+)-LA in racemic mixtures by exploiting chiral columns [27] and complexation with cyclodextrins [28], respectively. However, most of these methods involve time-consuming steps due to the multiple analysis and complex sample preparations. Alternative methods involving individual determination of LA also can be exploited by spectrophotometry and enzy-matic reactions, with the enzymatic method being considered more specific for enantio-mer separation [26,29].

Accordingly, a wide range of methods can be employed for LA determination, but only a few can be considered simple, economical, and fast enough for use in any analyti-cal laboratory regardless of levels of experience. An effective response to laboratories' requirements for LA determination is the development of an analytical system based on automation and flow management solutions, chemical processes, and sample prepara-tion. Analytical systems based on principles of flow injection analysis (FIA) are simple to construct and easy to operate, thus allowing the accommodation of different protocols for LA determination in food and beverages.

12.2 FLOW SYSTEMS FOR LACTIC ACID DETERMINATION

Analytical chemistry laboratories regard FIA as the most useful tool for managing chemi-cal reactions and processes. With their large variety of flow systems, sample preparation units, and capacity for coupling in different kinds of detection systems, these manifolds have been widely employed for LA determination in varied samples. FIA is used in moni-toring LA production [30] for direct analysis in food and beverages, including dairy prod-ucts and wine [31,32]. Analysis of these samples generally demands complex procedures that can be carried out in FIA systems, promoting the minimization of sample handling, analyst contact and external contamination. Highly versatile, these systems are able to control sample preparation through insertion of individual or sequential units, such as solid-phase extraction, liquid–liquid extraction, gas diffusion, dialysis, UV reactor, and microwave radiation [33]. A noteworthy characteristic of flow systems for LA determina-tion is the realization of all analytical steps in one channel in a closed environment, from sample preparation through to detection.

To date, enzymatic methods have been the most widely applied technique for LA determination in flow systems [34]. With the development of technology and miniatur-ization, biosensors and electrochemical sensors have been proposed for quick and sensi-tive *in situ* analysis. Alternative methods have been noted in the literature, with a few applications to food and beverage analysis by spectrophotometric detectors. In chapter, an overview of enzymatic and nonenzymatic methods for determination of LA in flow systems is presented.

12.2.1 Enzymatic Methods

Enzymes have been used widely as analytical reagents due to their characteristics of high selectivity (occasionally specificity) and capacity for self-regeneration as biological cata-lysts [35]. Originating from biological systems, enzymes have been used as reagents in analytical determinations since the early 1950s, which development occurred simultane-ously with the origin of Skeggs's segmented flow analysis [36]. Their use became more widespread with the emergence of continuous-flow systems and FIA [37]. These systems developed chemical reactions in controlled flowing streams adjusted in terms of the length of analytical path, mixing conditions, residence time, flow rate, and limited dispersion. The combination of these conditions can circumvent the main problems in utilization of enzymes as analytical reagents: their high cost, limited stability when dissolved, restricted conditions of preparation, and low reusability [37]. The automation of enzymatic methods provides minimal consumption of enzyme reagent, and improvement of stability and of reutilization capacity, in addition to control of reaction conditions in a closed environment with a defined time and volume. Furthermore, flow systems permit, in a very simple way, the insertion of solid reagents by several approaches, minimizing the use of dissolved enzymes while increasing stability and reusability. Employing such approaches, these characteristics become very attractive for the immobilization of enzymes in filled minicolumns, open-tube wall reactors, solid sensors, and so on [35]. Figure 12.1 shows the main possibilities for insertion of a packed minicolumn with enzyme in a basic flow injection system: (a) before sample injection, (b) between the sample injection and detector, and (c) inside the detector (biosensor). These configurations can be adopted in any kind of flow system, but FIA is the example chosen to illustrate this chapter. By controlling the solution's flow rate, sample volume, and reaction time, enzyme immobilization on suitable supports increases stability and maintains biological activity.

Several flow strategies involving enzymatic reactions have been applied for determination of LA in food and beverages. Most of the work to be explored in this chapter involves the catalytic action of lactate dehydrogenase (LDH) and lactate oxidase (LOD) in the conversion of LA into pyruvic acid (Figure 12.2). LDH requires a biological redox

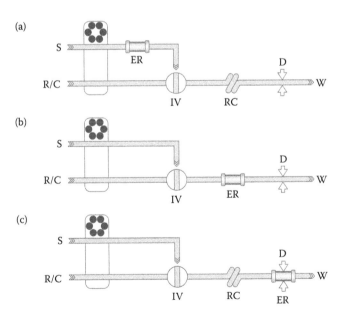

FIGURE 12.1   Flow injection manifolds for accommodation of minicolumns packed with immobilized enzyme. (a) ER before sample injection; (b) ER in the analytical path; and (c) ER inside the detection system. ER = enzyme reactor (solid support with immobilized enzyme), S = sample, R = reagent, C = carrier stream, IV = injection valve (or a similar injection device) RC = reaction coil, D = detector, and W = waste.

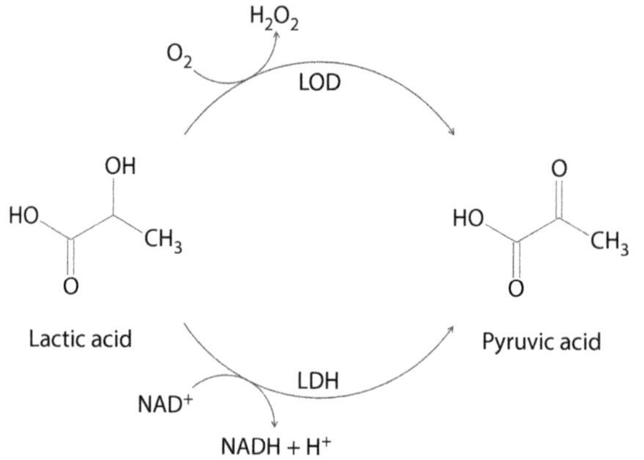

FIGURE 12.2 Mechanism of enzymatic reaction of conversion of LA to pyruvic acid. LDH = lactate dehydrogenase, LOD = lactate oxidase, and NAD⁺/NADH = nicotinamide adenine dinucleotide oxidized form/reduced form.

cofactor for enzyme regeneration in the form of nicotinamide adenine dinucleotide (NAD⁺ in oxidized form, and NADH in reduced form) for substrates of the dehydrogenase reaction [38]. The oxidation of LA is catalyzed in the presence of LDH and NAD⁺, producing pyruvic acid and the reduced species NADH. The enzyme LOD catalyzes the oxidation of LA to pyruvic acid plus H_2O_2, in an oxygenated medium. For quantification of LA, NADH and H_2O_2 can be monitored by spectrophotometry, fluorometry, chemiluminescence, or amperometry [26,29,39]. Both reaction mechanisms are illustrated in Figure 12.2.

Isomeric separation can be attained when enzymes are selected according to their specificity. For example, L(+)-LDH is specific to converting L(+)-LA and D(−)-LDH to D(−)-LA, and the same applies for LOD. In several approaches, D and L isomers can be determined individually or simultaneously with slight modifications in flow manifolds. Considering the specificity of enzymes toward LA, purchasing commercial enzymes with high purity is still rather expensive, and once dissolved, many of them exhibit limited stability [40]. Here, applications of enzymatic methods in FIA for LA determinations in food and beverages will be highlighted.

12.2.1.1 Applications of Lactate Dehydrogenase

For the catalytic action of LDH, a coenzyme is necessary to give adequate kinetic parameters and equilibrium products; the cofactor nicotinamide adenine dinucleotide (NAD⁺) is the one most used for LA determination in flow systems [41]. However, this reagent is quite expensive, and to obtain an adequate analytical sensitivity a large amount must be used, especially when the product is monitored by optical detectors. Shu et al. [42] reported the concern about high consumption of NAD⁺ in enzymatic determination of LA by spectrophotometry. The strategy for reduction of this reagent was to carry out the enzymatic reaction inside a packed minicolumn immobilized with LDH coupled to one port of a rotatory valve in a sequential injection analysis (SIA) system. The system permitted the selection of small volumes of samples and reagents with easy handling under operating conditions such as stopped flow and flow direction reversion. In addition, with the multiport valve coupled to a piston pump, it was possible to accomplish a sequence of aspirations with well-defined volumes and an appropriate sequence of buffer solutions, coenzyme solutions, and samples.

This system was employed for D-lactic determination as an indicator of reduced freshness in vacuum-packed chilled raw pork. Modified silica was immobilized with D-LDH mixed with L-alanine aminotransferase (ALT) to attain equilibrium and packed into a minicolumn (30 × 4.0 mm i.d.). The enzyme reactor was coupled between the sample injection device (multiport valve) and the detector, similarly to the FIA configuration shown in Figure 12.1b. To achieve suitable mixing conditions for the cofactor and sample inside the enzyme reactor, a stopped flow of 90 min with several flow reversal steps was applied. This repetitive movement of the fluids inside the minicolumn resulted in higher analytical signals due to the increased contact between sample, cofactor, and enzyme, with a lower volume of reagents. The consumption of cofactor NAD$^+$ was reduced, just 0.16 μmol per sample being used. This condition represented a decrease of 60 times compared to the reference batch method. NADH detection was performed at 340 nm in a spectrophotometer. Interference of matrix color was suffered even after sample preparation. To resolve this problem, the treated sample was passed through a minicolumn packed with blank silica without any enzymes, and the value was corrected. However, the authors could have taken advantage of the versatility of SIA and put the blank minicolumn in one port of a selective valve to maximize automation of this determination.

Because colored matrices absorb a large range of UV–VIS wavelengths, matrix color interference is very noticeable in oenological analysis involving spectrophotometric determinations. Red and white wines are the most complex matrices in terms of colored compound interferences. One strategy to circumvent this problem is to insert an online sample preparation unit to provide a discolored sample; for example, a dialysis unit (DU) could be built with a semipermeable membrane that prevents the passage of colored compound into the analytical path. Several configurations of DU have been employed in flow systems (Figure 12.3).

Depending on the objective and the application, the DUs can be implemented in different positions of flow manifold, which defines the mode of selecting the sample volume. Figure 12.4 shows possible configurations of a DU inserted in a basic flow injection

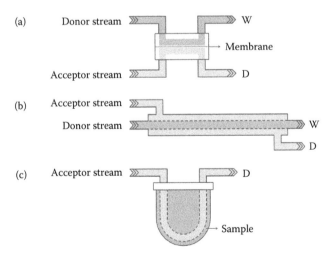

FIGURE 12.3 DU geometries for accommodation in flow systems: (a) sandwich type; (b) tubular concentric; and (c) microdialysis probe. D = detector, W = waste.

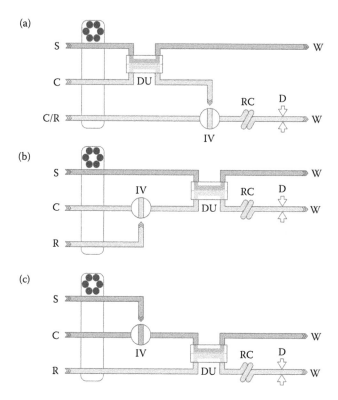

FIGURE 12.4   Flow injection manifolds for accommodation of a DU. (a) DU before injection valve; (b) DU after insertion of a defined volume of reagent by an IV; and (c) DU after injection valve. DU = dialysis unit, S = sample, R = reagent, C = carrier stream, IV = injection valve (or a similar injection device), RC = reaction coil, D = detector, and W = waste.

system, where two manifolds (Figure 12.4a,b) process the sample continuously through a DU, and after, a selected volume of sample or reagent solution are inserted by an injection valve defining the sample zone. The third manifold (Figure 12.4c) defines the sample volume before passage through the DU, reducing sample consumption.

Lima et al. [43] developed a flow injection system that could directly inject red and white wine samples continuously in a DU coupled online to a spectrophotometric detector. L(+)-Lactic acid was determined by inserting LDH/NAD$^+$ reagent solution in a defined loop (25 µL) that was carried through the acceptor stream in the DU "sandwich" (Figure 12.3a). A simplified manifold of this approach is illustrated in Figure 12.4b. After the DU, the flow stream carried the dialyzed sample and the LDH/NAD$^+$ plug toward a 75 cm reactor coil that was immersed in a temperature-controlled bath (37°C) to reach enzymatic equilibrium and form the product. Determination of the main organic acid is a very important parameter for the taste of wine [44]. During malolactic fermentation, the proportion of lactic and malic acids may change, altering the wine's characteristics [45]. Therefore, sequential determination of malic acid was performed in this system by adding a parallel loop with a malate dehydrogenase (MDH)/NAD$^+$ solution. With the difference in residence time between the loops, it was possible to quantify LA and malic acid sequentially. Several types of port as well as red and white wine were analyzed without interferences from the color matrix.

In spite of the positive results obtained by dissolved enzymes used as reagents solutions, it remains impossible to guarantee the efficacy of the catalyst action beyond 6 h of analysis. Thus, immobilized enzymes packed into minicolumn reactors or into detection systems, as a biosensor, have been applied in flow injection procedures [40]. Immobilization strategies in flow systems allow the use of high concentrations of enzymes within small volumes, resulting in a high conversion rate (high activity) with minimum dilution of the sample, and therefore, a lower detection limit [37]. The versatility of flow systems allows the insertion of solid reactors immobilized with enzymes, resulting in lower consumption of reagents, high sampling rates, and low costs. Mechanisms and methods of enzyme immobilization on silica or glass beads are beyond the scope of this chapter but can be found in several articles in the literature [34–41].

In flow systems, the strategy most applied to enzyme immobilization is a minicolumn or continuous-flow reactor packed with the enzyme immobilized on modified silica. The main concern for any solid-phase column is avoiding back-pressure inside the system through homogeneous arrangement of the solid phase [33]. Puchades et al. [45] developed a flow injection system with different packed columns for simultaneous determination of L(+)-LA and D(−)-malic acid. LDH and MDH were immobilized in controlled-pore glass beads of silica packed into columns of 8.0 cm in length and 1.2 mm in inner diameter. The enzyme reactors were coupled online after the injection valve (Figure 12.1b) and filled with the dialyzed sample after the online DU as arranged in Figure 12.4a, in order to minimize the sample's color. However, some species continued to interfere in the analytical signal once the polyphenols also dialyzed. Then, a third packed column filled only with silica beads without enzymes was inserted into the flow manifold. This procedure was necessary because the detection system was based on the NADH fluorescence, and some compounds in the wine matrix show a strong emission at 450 nm after excitation at 340 nm. After the split of the dialyzed sample volume at a confluence point, one half was carried toward the blank reactor and the other half to another confluence point that split both enzymatic reactors. By adjusting the length of reaction coils between the column and detector, the residence times of the three portions of the sample were sized so as to obtain three distinguishable analytical signals without overlap. After optimizing chemical (concentrations, pH, buffer composition) and physical conditions (flow rate, reaction coil length, and sample volume), samples of red, rose, and white wine were analyzed, achieving a sampling frequency of 30 determinations per hour, free from interferences. It is important to note that the minicolumns were used for 6 months in a laboratory environment with temperature control (20°C) without any significant decrease in activity. An important drawback of this work was that NAD$^+$ solution was used as a carrier stream to transport the injected sample volume toward the three mini-columns, over-consuming an expensive reagent.

Mataix et al. [44] adopted a similar approach intended to reduce the use of NAD$^+$. It also exploited an online sample DU prior to enzymatic reaction developed inside two different minicolumns packed with immobilized LDH and MDH for lactic and malic acid determination, respectively. Better control of NAD$^+$ consumption was achieved by inserting a syringe with a hypodermic needle into the loop of the injection valve to spare this reagent. Sample volume was defined by an injection valve inserted before the DU, as exemplified in the manifold shown in Figure 12.4c. For sequential determination of the organic acids, the dialysate was introduced into the analytical path coupled to an additional injection valve, which was operated to fill a loop of 80 μL. The rest of the solution was transported to the MDH minicolumn for malic determination. After an established length of time, the additional valve was switched and the sample loop of 80 μL was inserted and carried through the LDH column for LA determination. The authors evaluated spectrophotometric and fluorometric detection systems; both presented acceptable analytical parameters and are appropriate for implementation in routine laboratory wine analysis. Twenty samples of wine were analyzed without sample preparation, with the exception of sweet wines, which, due to their high viscosity, were diluted 1:2 in distilled water.

Apart from its optical properties, many reports cite NADH as a potent electroactive species with oxidation capacity in a variety of solid electrodes [38,41]. As such, amperometric detection becomes an elegant alternative for LA determination in flow injection systems. For detection, the amperometric flow cell must be specially built or adapted to purpose. Commercial electrochemical thin-layer flow cells can be used to support different kinds of working electrode, including those based on enzymatic reactions [46–48]. Figure 12.5 shows a typical amperometric flow cell, with a flow inlet that passes through a confined space where the reference, auxiliary, and working electrodes are connected, and continues to the flow outlet. Generally, this flow cell is very simple and can be built with Teflon or acrylic materials. The dimensions and distance between the reference and working electrodes must be as small as possible to minimize the uncompensated resistance. During the sample passage through the flow cell, the NADH oxidation current produced at the electrode is directly proportional to the LA concentration [47]. However, studies have demonstrated that direct oxidation of NADH demands high potential (up to 1.0 V), causing high background currents and coupled side reactions with inevitable poisoning of the electrode surface, which degrades analytical sensibility and selectivity [45,49]. Efforts have been made to develop new redox materials capable of reducing the oxidation overpotential and to improve the electron transfer rate. Alternatively, another electroactive species can be added for oxidation at a lower potential during the oxidation of NADH.

Yao and Wasa [41] used hexacyanoferrate(III) as a mediator for oxidation of NADH with an oxidation potential of 0.4 V versus Ag/AgCl. To catalyze this reaction, a well-known diaphorase compound was immobilized on a platinum electrode with bovine serum albumin. The current produced by hexacyanoferrate oxidation was proportional to the LA concentration. For the LA determination, a flow injection system was used with a sample splitting zone after injection (10 µL) in a carrier stream containing $NAD^+/Fe(CN)_6^{3-}$. After the split, each sample plug passed through one of the immobilized enzyme reactors (L-LDH and D-LDH). After the enzymatic reactor, each part of the sample containing of NADH, moved by different residence time, were proportioned by coil

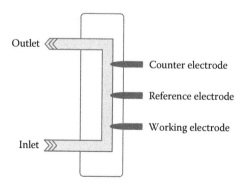

FIGURE 12.5 Flow cell for amperometric detection of LA.

lengths of 300 cm for D-LDH and 20 cm for L-LDH. With these conditions, two signals arrived at different times in the amperometric detector, resulting in the determination of L- and D-LA.

More benefits are observed when enzymes are immobilized in a solid support and coupled as a working electrode in an amperometric cell system (Figure 12.5). The strategy of using enzymes as biological supports for working electrodes in electroanalytical detectors gives what are known as sensing electrodes or biosensors. Most LA determinations in dairy food are based on biosensors coupled to flow systems and amperometric detection, presenting slight variations in instrumentation and procedure. Some specific differences can be found in electrode manufacturing, surface coating and treatment of enzyme immobilization. The use of biosensors in food and beverages analysis is limited primarily because concomitant species may be oxidized or reduced at the electrode, and the parallel oxidation–reduction reaction would damage the electrode surface. For this reason, online DUs usually are employed to resolve this problem.

Nanjo et al. [47] described a dual biosensor for simultaneous amperometric determination of D and L-LA in alcoholic beverages after online microdialysis sampling. A microdialysis probe was used as sampling device to collect the dialyzed sample during the passage of buffer solution inside the hollow fiber (Figure 12.3c). After the DU, a loop of sample (5 μL) was filled and propelled in the direction of the dual enzyme sensor by an NAD$^+$ solution. To obtain isomeric optical purity, the working electrode was protected with a thin film of poly(1,2-diaminobenzene) after immobilization of L-LDH/diaphorase and D-LDH/diaphorase on a platinum disk. This approach was effective in protecting the dual biosensor, avoiding the nonspecific reactions at the electrode, such as those with ascorbic acid, urate, and cysteine. The current obtained for individual quantification of D- and L-LA resulted in H$_2$O$_2$ being generated enzymatically at the platinum electrode surface. The oxidation of NADH is catalyzed by diaphorase, which mechanism incorporated an oxygen molecule, producing electroactive H$_2$O$_2$ that was monitored by applying 0.5 V versus Ag/AgCl. Analysis of beer, wine, and sake diluted with distilled water was assessed with good recoveries. Studies of stability showed that during 3 h of use per day, there was decrease in activity of the dual biosensor, and after 3 weeks the activity decreased by 58%–68%. It is important to stress that the microdialysis probe was immersed in sample solution, resulting in a nondestructive methodology for LA determination in beverages.

Radoi et al. [49] developed a similar flow injection procedure for amperometric determination of L-LA in probiotic yogurts after an online dialysis process. Compared to the work discussed above, the main modification was the composition of the working electrode. A screen-printed electrode modified with carbon nanotubes was treated to covalently link the redox mediator Variamine blue; subsequently, LDH was deposited on the surface. With this strategy, it was possible to work with a convenient value of potential for oxidation of NADH in the Variamine blue–modified electrode (200 mV vs. Ag/AgCl). A hollow fiber microdialyzer (Figure 12.3b), immersed in sample solution, was coupled in a single-line flow injection system to act as a barrier to large interfering molecules, allowing the passage of LA transported by an NAD$^+$ solution toward the amperometric flow cell. Samples of probiotic yogurts were diluted before dialysis to give LA concentrations in a linear range of the analytical calibration curve.

An efficient strategy for reducing cofactor NAD$^+$ consumption and increasing its stability involves incorporating it into the electrochemical sensor. Shu and Wu [48] developed a biosensor for D-LA determination based on a carbon paste electrode mixed with D-LDH, ALT, and cofactor (NAD$^+$). This kind of strategy simplifies the flow injection system, so that just a single line is necessary to inject an aliquot of sample by a pneumatic injection valve of 50 μL, and transport it in the direction of the amperometric biosensor. Under these conditions, NAD$^+$ solutions and sample treatment units were not used. ALT was used as a mediator for eletrocatalytic NADH-oxidation. In contact with the sodium glutamate present in the carrier solution, it completely eliminated the interference effect of pyruvate [42].

In terms of the main applications of LDH in flow systems for LA determination, it is important to recognize that there are some limitations in relation to the physical form of their utilization in the reaction process and sample analysis. As a reagent in solution, LDH has limited stability; and consumption of the reagent is high, which can be ameliorated by immobilization procedures and construction of a packed reactor for online reaction. However, the main limiting factor in use of this enzyme is its dependence on a cofactor (NAD$^+$), which further escalates costs already incurred on the enzyme reagent. Numerous studies have focused on alternative approaches to minimizing the consumption of this reagent in association with exploiting the positive characteristics of flow systems and biosensor technology.

12.2.1.2 Applications of Lactate Oxidase

Lactate oxidase (LOD) offers more advantages than LDH as it does not require a cofactor for its catalytic action. Lactic acid determination can be measured by the catalytic effect of LOD through monitoring the consumption of oxygen or the production of hydrogen peroxide (Figure 12.2). Its application in FIA involves the same concerns as those about LDH: achieving sufficient activity without loss of stability when dissolved solutions are used. It also involves procedures of immobilization in solid supports for packed minicolumns in a bioreactor (Figure 12.1). In fact, utilization of LOD in flow systems provides higher sampling frequency and less reagent consumption, making it an excellent analytical quality control strategy for food and beverages with LA as a preservative.

Mizutani et al. [50] proposed a typical rapid and accurate automated determination of L-LA that employed an amperometric biosensor using a LOD–polyion complex coating. The authors exploited the dispersion capacity of flow injection systems to obtain distinct ranges of analytical curves as function of LA concentration in samples of different origins, such as food and clinical samples, with only the sample volume varying. For this, a single-line manifold (Figure 12.1c) was constructed with a 60 cm distance between the injection valve (5 μL) and amperometric flow cell (detector). The enzyme electrode was based on glassy carbon immobilized with LOD followed by poly-L-lysine and poly(4-styrene sulfonate) in a buffer solution (pH 7). A typical amperometric flow cell was used (Figure 12.5). A potential of 1.0 V versus Ag/AgCl was sufficient for oxidation of hydrogen peroxide produced by the catalytic action of LOD in the LA oxidation. To achieve the required equilibrium and kinetic reaction parameters, the carrier flow stream was a phosphate buffer saturated with air. The enzyme electrode was coated with a polycomplex membrane showing permselectivity based on the solute size. This prevented molecules with higher molecular weights from passing through, making it selective for small molecules, such as LA and H_2O_2. Diluted sour milks were analyzed with good results in comparison to the official method. A sampling frequency of 120 h^{-1}, without complex sample preparation steps, indicates the selectivity and throughput of the proposed strategy.

Immobilization of enzymes on solid supports increases stability, but considerations of the rate of inactivation after immobilization sometimes are ignored. Bori et al. [51]

studied the stability of LOD produced by two different sources (*Aerococcus viridans* and *Pediococcus* species) in order to verify the lifetime and temperature stability for LA determination. The immobilization was done in a cell built by the authors with a U-shaped channel (120 mm × 2.5 mm i.d.) formed between two plates of Teflon. Inside the channel, a natural protein membrane (pig's intestine) was used as a chemical boundary for LOD immobilization. The U-shaped reactor was attached to the single-line flow instead of the enzyme reactor in the manifold configuration of Figure 12.1b, between the injection valve (50 µL) and detector. The amperometric flow cell detector was formed by a glassy carbon electrode, a solid-state palladium electrode, and a stainless steel electrode as working (590 mV), reference, and counter electrodes, respectively. Results showed that LOD from *Aerococcus viridans* immobilized in the reactor was more stable, maintaining activity during 1300 analyses for 6 weeks; meanwhile, for *Pediococcus species*, a decrease was observed after 200 analyses. Lactic acid was successfully determined in milk, yogurt, kefir, and cheese samples after dilution and fat removal, resulting in a robust and versatile system for quality control in industrial food.

Lactic acid can show variations in concentration depending on the particular food or beverage sample. It is therefore strongly recommended that analytical methods should be able to obtain a linear range variation and/or lower detection limits. Chemiluminescent methods are based on a chemical reaction that produces light as a function of the analyte concentration and generally exhibit lower detection limits than electrochemical reactions. This approach is considered highly selective and demonstrates excellent figures of merit. It is recommended for use in LA determination in samples of yogurts and milk as it does not need complicated sample preparation. A classic reaction that produces chemiluminescence exploits the oxidation of luminol in a basic medium catalyzed by Fe^{2+}. It is employed widely in flow systems due to the facility of automation of luminescent conditions, its high analytical throughput, and its low reagent consumption. Martelli et al. [52] proposed a multicommutated flow system for LA determination based on the reaction of luminol oxidation by hydrogen peroxide provenient of LOD catalysis. Two manifolds were tested to determine which was the better configuration for maintaining LOD stability and activity for longer periods of analysis. One of them had LOD dissolved in a buffer solution and the other had LOD immobilized on porous silica beads packed into a minicolumn. As expected, immobilization proved more advantageous than solution, mainly in terms of enzyme consumption per determination (0.54 µg vs. ≪0.0005 µg). Solenoid valves were used for solution management and a peristaltic pump was used as a propeller solution in an aspiration mode. This carried the samples and reagents to the LOD reactor that was coupled as in Figure 12.1b, and further to the detection system. Yogurt samples were dissolved in water and analyzed following the proposed procedure, giving results in agreement with the conventional method (Boehringer UV-Kit), with a high sampling frequency and lower detection limits (1.3×10^{-5} mol L^{-1}) than electrochemical methods (Table 12.1).

Although LOD has a high specific activity for LA oxidation, other oxidases can be used for this conversion. Wu et al. [53] employed an oxidase enzyme present in natural animal tissue to convert LA in hydrogen peroxide and pyruvic acid. A porcine kidney tissue sample was reported with high activity due to the presence of α-hydroxyacid oxidase. A mixture of tissue, glass wool, and glutaraldehyde was macerated and inserted into a minicolumn (80 mm × 5 mm i.d.) that was coupled to a flow injection system before an injection valve (150 µL) as exemplified in Figure 12.1a. After the production of H_2O_2 in the tissue reactor, the plug was filled and injected into a flowing mixture

TABLE 12.1 Comparative Parameters of Some Flow Systems for Lactic Acid Determination in Food and Beverages Exploiting Enzymes as Reagents

Enzyme	Flow System/ Mode	Sample	Sample Preparation	Sampling Frequency (h^{-1})	DL (mol L^{-1})	Detection	Reference
LDH	FIA/ER	Wine	Online dialysis	15	–	FLU	[45]
	SIA/ER	Pork meat	Acid extraction	15	–	SPEC	[42]
	FIA/solution	Wine	No need	40	5.5×10^{-4} 5.5×10^{-4}	SPEC	[43]
	FIA/ER	Wine	Online dialysis	15	1.1×10^{-4}	SPEC/FLU	[44]
	FIA/BS	Alcoholic beverages	Online dialysis	40	–	AMP	[47]
	FIA/BS	Yogurt	Online dialysis	3	–	AMP	[49]
LOD	FIA/BS	Sour milk	Dilution	120	2.0×10^{-5}	AMP	[50]
	MCFS/ER	Yoghurt	Dissolution	55	1.3×10^{-5}	CL	[52]
	FIA/ER	Milk	Filtration and dilution	40	0.2×10^{-6}	CL	[53]
	FIA/ER	Milk, yogurt, and kefir	Filtration	40	5.0×10^{-5}	AMP	[51]

Note: BS = enzyme used as a biosensor (immobilized in detection system), ER = enzyme reactor (enzyme immobilized on solid support packed in a reactor), FIA = flow injection analysis, SIA = sequential injection analysis, MCFS = multicommutation flow system, AMP = amperometric, FLU = fluorometric, SPEC = spectrophotometric, CL = chemiluminescence, and DL = detection limit.

of luminol and potassium ferricyanide, producing a chemiluminescence signal proportional to the LA content in the sample. Milk samples were processed in the flow system with a prior dilution of 1:100 with water, giving good results compared to the titration method. Sensor stability was confirmed up to 240 determinations, and a lower detection limit (0.2×10^{-6} mol L^{-1}) was attained compared to other flow analysis strategies (Table 12.1).

One of the main requirements for implementation of a new analytical method to a routine analysis is the reduction of cost of one determination in relation to the actual/reference method. In the studies described above, the replacement of an expensive enzyme by a commercially available material in a flow injection system resulted in a cheaper and faster strategy that could be used in routine analytical laboratories to provide quality control and monitoring of LA in food and beverages.

12.2.2 Nonenzymatic Methods

The literature presents a few alternatives to enzymatic methods for determination of LA in food and beverages. Most are based on batch procedures and spectrophotometric detection, such as those employing the conversion of LA to acetaldehyde by strong oxidizing agents and subsequent reaction with chromophore reagents and metal ions [54–56]. With the use of concentrated acids, high temperature, stirring, and centrifugation steps, these methods have rather severe reaction conditions and complex procedures. Thus, the application of flow analysis systems can improve the control of the procedures and chemical reaction, allowing the use of milder conditions to obtain a simple, cheap, and robust alternative to enzymatic methods.

Gómez-Álvarez et al. [57] developed a flow injection system to monitor the amount of LA present in milk using a photochemical reaction. Lactic acid determination consisted of its oxidation by a solution containing Fe^{3+} and subsequent irradiation with a halogen lamp (500 W). After the reduction, Fe^{2+} reacted with o-phenanthroline, yielding a ferroin complex [$Fe(o\text{-phen})_3^{2+}$] that was monitored spectrophotometrically (512 nm). A halogen lamp was placed above the reaction, located inside a water bath to prevent heating of the tubing. With adjustment of the flow injection parameters it was possible to control the time and volume of the sample and reagent inside the reaction coil and to apply a stopped-flow approach, allowing more exposure to irradiation. The method was linear between 0.5 and 50 mg mL^{-1} LA, the detection limit was 0.16 mg mL^{-1}, and the sample throughput was 30 samples per hour. However, due to the high concentration of LA found in milk and other dairy products, the method was validated using a nonlinear mathematical model involving ANOVA (analysis of variance) treatment and good recovery results were obtained. Furthermore, this strategy provided acceptable selectivity in relation to the main interfering compounds found in milk samples without pretreatment. However, higher reagent consumption is still noted when continuous flow is used. Reduction of sample and reagent volumes can be achieved by computer-controlled systems, such as sequential injection analysis (SIA).

Dias et al. [58] exploited the versatility of the sequential injection system through the use of a chemical reaction to convert LA and a gas diffusion unit to separate the gaseous compound formed; the technique also used posterior detection by spectrophotometry. The reaction was based on a batch procedure [54] in which LA was oxidized by Ce^{4+} in a strong acidic medium (4.4 mol L^{-1} H_2SO_4) under high temperatures (90°C). With the flow conditions and operational procedure of SIA, the kinetic parameters were improved, allowing the determination of LA under mild conditions (0.36 mol L^{-1} H_2SO_4 and $45°C$). For the quantification, the carbon dioxide produced by oxidization was passed through a gas diffusion unit (Figure 12.3a) coupled into the manifold between the selecting valve and spectrophotometric detector. The CO_2, proportional to the LA concentration, was monitored by change in coloration of a bromothymol blue solution (pH 8.4), an acid–base indicator, measured at 619 nm. The system had a sample throughput of 22 samples per hour and a consumption of 0.04 g of Ce^{4+} per determination of LA. The analytical curve was linear up to 100.0 mg L^{-1} LA, with limits of detection and quantitation of 0.158 and 1.6 mg L^{-1}, respectively. Samples of yogurt and fermented sugarcane were analyzed after sample preparation involving filtration, centrifugation, dilution, and removal of dissolved CO_2 by N_2 purge. These results were in agreement with the HPLC reference method. Thus, alternatives to enzymatic reaction can become innovative when complex batch strategies are employed in flow systems, resulting in improvements in reaction conditions, simplification, and total automation of operational variables (sample/reagent volume, flow rate, sequence of analytical procedures, insertion of separation units, and temperature).

Sequential injection systems are considered one of the most flexible and robust flow strategies. They are highly recommended for implementation in industrial quality control and monitoring of fermentation/biological processes. In routine analysis, these systems promote rapid analysis and accuracy and precision of volumetric parameters, mainly when associated with fast detection systems. In this context, Fourier transform infrared (FTIR) spectroscopy has been considered as an efficient tool for online quantitative analysis. It has great scanning speed, operational simplicity and minimal sample preparation [59]. Schindler et al. [60] coupled a sequential injection system to a FTIR detector for simultaneous quantification of sugars, alcohol, and some organic acids, including LA, in wine samples. With a simple adaptation in the FTIR detector, it could accommodate a flow-through cell with CaF_2 and ZnSe windows (2 mm thick) in an optical path of 25 μm. A simple operational procedure involved aspiration of original sample or standard solutions toward the holding coil; subsequently, the flow was reversed and the mixture was propelled toward the FTIR detector. Wine samples were injected without dilution or other treatment, giving complex spectra with several overlaps. For spectral resolution, a chemometric tool—partial least squares (PLS)—was applied. The results were in agreement with the HPLC reference method. Sample analysis was possible in 3 min, making it an excellent option for screening purposes.

Separation methods based on HPLC have primarily been used as a reference method for LA determination and results validation; nonetheless, some flow strategies involving online sample preparation have been applied. HPLC is an analytical technique that demands efficient clean-up in order to eliminate potential interferences before injection onto the column, resulting in complex procedures, high costs, and long amounts of time for analysis. Automation of sample preparation can be considered an excellent alternative to minimize HPLC problems, producing adequate control of sample preparation with the possibility of a direct connection to the column. Removal of microparticles and macromolecules by DUs (Figure 12.3a) exploiting FIA has been the best option for cleaning up samples of yogurt, milk, soft drinks [61], and wine [62], improving the column stability and lifetime.

Although few studies have been published involving nonenzymatic methods for LA determination, the works presented here are very attractive in terms of their low reagents consumption, operational simplicity, and low cost.

12.3 FINAL REMARKS

This chapter has provided an overview of LA determination in food and beverages in flow systems with the objective of emphasizing the important participation of LA as a natural preservative. FIA was the most used strategy of automation with both enzymatic and nonenzymatic reactions in comparison with flow strategies based on computer device control. However, flow systems based on computer-controlled devices are more able to control all of the flow parameters, because they consume fewer reagents and minimize analyst involvement. In addition, these systems allow the insertion of packed reactors or strategies with bead injection analysis or fluidized beds depending on the manifold. In prospect, new approaches could be developed in multicommutated, multisyringe, and multipumping flow systems, exploiting the decreased enzyme consumption and increased stability for lower detection limits in LA determination.

REFERENCES

1. Young-Jung, W., K. Jin-Nam, and R. Hwa-Won. 2006. Biotechnological production of lactic acid and its recent applications. *Food Technol. Biotechnol.* 44:163–172.
2. Rojan, P. J., M. Nampoothiri, and A. Pandey. 2007. Fermentative production of lactic acid from biomass: An overview on process developments and future perspectives. *Appl. Microbiol. Biotechnol.* 74:524–534.
3. Martinez, F. A. C., E. M. Balciunas, J. M. Salgado, J. M. D. Gonzalez, A. Converti, and R. P. S. Oliveira, 2013. Lactic acid properties, applications and production: A review. *Trends Food Sci. Technol.* 30:70–83.
4. Hofvendahl, K. and B. Hahn–Hagerdal. 2000. Factors affecting the fermentative lactic acid production from renewable resources. *Enzyme Microb. Technol.* 26:87–107.
5. Narayanan, N., P. K. Roychoudhury, and A. Srivastava. 2004. L(+) lactic acid fermentation and its product polymerization. *Electron. J. Biotechnol.* 2:167–179.
6. Choudhury, B., A. Basha, and T. Swaminathan. 1998. Study of lactic acid extraction with higher molecular weight aliphatic amines. *J. Chem. Technol. Biotechnol.* 72:111–116.
7. Liu, S. Q. 2003. Practical implications of lactate and pyruvate metabolism by lactic acid bacteria in food and beverage fermentations. *Int. J. Food Microbiol.* 83:115–131.
8. WHO. 1974. Toxicological evaluation of some food additives including anticaking agents, antimicrobials, antioxidants, emulsifiers, and thickening agents. *WHO Food Addit. Ser.* 5:461–465.
9. Yumoto, I. and K. Ikeda. 1995. Direct fermentation of starch to L-(þ)-lactic acid using *Lactobacillus amylophilus*. *Biotechnol. Lett.* 17:543–546.
10. John, R. P. and K. M. Nampoothiri. 2008. Strain improvement of *Lactobacillus delbrueckii* using nitrous acid mutation for L—Lactic acid production. *Survival* 24:3105–3109.
11. Lu, Z., M. Wei, and L. Yu. 2012. Enhancement of pilot scale production of L-(þ)-lactic acid by fermentation coupled with separation using membrane bioreactor. *Process Biochem.* 47:410–415.
12. Datta, R. and M. Henry. 2006. Lactic acid: Recent advances in products, processes and technologies—A review. *J. Chem. Technol. Biotechnol.* 81:1119–1129.

13. Vijayakumar, J., R. Aravindan, and T. Viruthagiri. 2008. Recent trends in the production, purification and application of lactic acid. *Chem. Biochem. Eng. Q.* 22:245–264.

14. Benninga, H. 1990. *A History of Lactic Acid Making: A Chapter in the History of Biotechnology.* New Jersey: Springer.

15. FDA. *Methods to Reduce/Eliminate Pathogens from Produce and Fresh-Cut Produce*, Chapter V. http://www.fda.gov/Food/FoodScienceResearch/SafePractices forFoodProcesses/ucm091363.htm. (accessed May 20, 2014).

16. Leroy, F. and L. D. Vuyst. 2004. Lactic acid bacteria as functional starter cultures for the food fermentation industry. *Trends Food Sci. Technol.* 15:67–78.

17. Ramos, B., F. A. Miller, T. R. S. Brandão, P. Teixeira, and C. L. M. Silva. 2013. Fresh fruits and vegetables—An overview on applied methodologies to improve its quality and safety. *Innovative Food Sci. Emerging Technol.* 20:1–15.

18. Park, S.-H., M.-R. Choi, J.-W. Park et al. 2011. Use of organic acids to inactivate *Escherichia coli* O157:H7, *Salmonella typhimurium*, and *Listeria monocytogenes* on organic fresh apples and lettuce. *J. Food Sci.* 76:M293–M298.

19. FDA. *Generally Recognized as Safe (GRAS).* http://www.fda.gov/food/ingredientspackaginglabeling/gras/default.htm (accessed June 01, 2014).

20. Reis, J. A., A. T. Paula, S. N. Casarotti, and A. L. B. Penna, 2012. Lactic acid bacteria antimicrobial compounds: Characteristics and applications. *Food Eng. Rev.* 4:124–140.

21. Caplice, E. and G. F. Fitzgerald. 1999. Food fermentations: Role of microorganisms in food production and preservation. *Int. J. Food Microbiol.* 50:131–149.

22. Zotou, A., Z. Loukou, and O. Karava. 2004. Method development for the determination of seven organic acids in wines by reversed-phase high performance liquid chromatography. *Chromatographia* 60:39–44.

23. Tormo, M. and J. M. Izco. 2004. Alternative reversed-phase high-performance liquid chromatography method to analyse organic acids in dairy products. *J. Chromatogr. A.* 1033:305–310.

24. Shui, G. and L. P. Leong. 2002. Separation and determination of organic acids and phenolic compounds in fruit juices and drinks by high-performance liquid chromatography. *J. Chromatogr. A* 977:89–96.

25. Mato, I., S. Suárez-Luque, and J. F. Huidobro. 2007. Simple determination of main organic acids in grape juice and wine by using capillary zone electrophoresis with direct UV detection. *Food Chem.* 102:104–112.

26. Mato, I., S. Suarez-Luque, and J. F. Huidobro. 2005. A review of the analytical methods to determine organic acids in grape juices and wines. *Food Res. Int.* 38:1175–1188.

27. Buglass, A. J. and S. H. Lee. 2001. Sequential analysis of malic acid and both enantiomers of lactic acid in wine using a high-performance liquid chromatographic column-switching procedure. *J. Chromatogr. Sci.* 39:453–458.

28. Kodama, S., A. Yamamoto, A. Matsunaga, T. Soga, and K. Minoura. 2000. Direct chiral resolution of lactic acid in food products by capillary electrophoresis. *J. Chromatogr. A* 875:371–377.

29. Mazzei, F., F. Botrè, and G. Favero. 2007. Peroxidase based biosensors for the selective determination of D, L-lactic acid and L-malic acid in wines. *Microchem. J.* 87:81–86.

30. Shu, H., H. Hakanson, and B. Mattiasson. 1995. On-line monitoring of D-lactic acid during fermentation process using immobilized D-lactate dehydrogenase in a sequential injection analysis system. *Anal. Chim. Acta* 300:277–285.

31. Segundo, M. A., J. L. F. C. Lima, and A. O. S. S. Rangel. 2004. Automatic flow systems based on sequential injection analysis for routine determinations in wines. *Anal. Chim. Acta* 513:3–9.

32. Perez-Olmos, R., J. C. Soto, N. N., Zarate, A. N., Araujo, J. L. F. C. Lima, and M. L. M. F. S. Saraiva. 2005. Application of sequential injection analysis (SIA) to food analysis. *Food Chem.* 90:471–490.

33. Zagatto, E. A. G., and A. C. B. Dias. 2007. In-line sample preparation in flow analysis. In *Trends in Sample Preparation*, ed. M. A. Z. Arruda, pp. 197–232. New York: Nova Science Publishers, Inc.

34. Silvestre, C. I. C., P. C. A. G. Pinto, M. A. M. Segundo, L. M. F. S. Saraiva, and J. L. F. C. Lima. 2011. Enzyme based assays in a sequential injection format: A review. *Anal. Chim. Acta* 689:160–177.

35. Mottola, H. A. 1983. Enzymatic preparations in analytical continuous-flow systems. *Anal. Chim. Acta* 145:27–39.

36. Skeggs Jr., L. T. 1957. Automatic method for colorimetric analysis. *Am. J. Clin. Pathol.* 28:311–322.

37. Hansen, E. H. 1989. Flow-Injection enzymatic assays. *Anal. Chim. Acta* 216:257–273.

38. Yao, T., Y. Kobayashi, and S. Musha. 1982. Flow injection analysis for l-lactate with immobilized lactate dehydrogenase. *Anal. Chim. Acta* 138:81–85.

39. Luque de Castro, M. D., J. González-Rodríguez, and P. Pérez-Juan. 2005. Analytical methods in wineries: Is it time to change? *Food Rev. Int.* 21:231–265.

40. Hansen, E. H. 1996. Principles and applications of flow injection analysis in biosensors. *J. Mol. Recognit.* 9:316–325.

41. Yao, T. and T. Wasa. 1985. Simultaneous determination of l(+)- and d(–)-lactic acid by use of immobilized enzymes in a flow-injection system. *Anal. Chim. Acta* 175:301–304.

42. Shu, H. C., E. H. Hakanson, and B. Mattiason. 1993. D-lactic acid in pork as a freshness indicator monitored by immobilized D-lactate dehydrogenase using sequential injection analysis. *Anal. Chim. Acta* 283:727–737.

43. Lima, J. L. F. C., T. I. M. S. Lopes, and A. O. S. S. Rangel. 1998. Enzymatic determination of L(+) lactic and L(–) malic acids in wines by flow-injection spectrophotometry. *Anal. Chim. Acta* 366:187–191.

44. Mataix, E. and M. D. Luque de Castro. 2001. Determination of L-(–)-malic acid and L-(C)-lactic acid in wine by a flow injection-dialysis-enzymic derivatisation approach. *Anal. Chim. Acta* 428:7–14.

45. Puchades, R., M. A. Herrero, A. Maquieira, and J. Atienza. 1991. Simultaneous enzymatic determination of L(–) malic acidand L(+) lactic acid in wine by flow injection analysis. *Food Chem.* 42:167–182.

46. ALS Co., Ltd. *Electrochemistry & Spectroelectrochemistry*. http://www.als-japan. com/1401.html (accessed June 01, 2014).

47. Nanjo, Y., T. Yano, R. Hayashi, and T. Yao. 2006. Optically specific detection of D- and L-lactic acids by a flow-injection dual biosensor system with on-line microdialysis sampling. *Anal. Sci.* 22:1135–1138.

48. Shu, H-C. and N-P. Wu. 2001. A chemically modified carbon paste electrode with D-lactate dehydrogenase and alanine aminotranferase enzyme sequences for D-lactic acid analysis. *Talanta* 54:361–368.

49. Radoi, A., D. Moscone, and G. Palleschi. 2010. Sensing the lactic acid in probiotic yogurts using an l-lactate biosensor coupled with a microdialysis fiber inserted in a flow analysis system. *Anal. Lett.* 43:1301–1309.

50. Mizutani, F., S. Yabuki, and Y. Hirata. 1996. Flow injection analysis of L-lactic acid using an enzyme-polyion complex-coated electrode as the detector. *Talanta* 43:1815–1820.

51. Bori, Z., G. Csiffary, D. Virag, M. Toth-Markus, A. Kiss, and N. Adanyi. 2012. Determination of L-Lactic acid content in foods by enzyme-based amperometric bioreactor. *Electroanalysis* 24:158–164.

52. Martelli, P. B., B. F. Reis, A. N. Araujo, and M. C. B. S. M. Montenegro. 2001. A flow system with a conventional spectrophotometer for the chemiluminescent determination of lactic acid in yoghurt. *Talanta* 54:879–885.

53. Wu, F., Y. Huang, and C. Huang. 2005. Chemiluminescence biosensor system for lactic acid using natural animal tissue as recognition element. *Biosens. Bioelectron.* 21:518–522.

54. Barker, S. B. and W. H. Summerson. 1941. The colorimetric determination of lactic acid in biological material. *J. Biol. Chem.* 138:535–554.

55. Figenschou, D. L. and J. P. Marais. 1991. Spectrophotometric method for the determination of microquantities of acid lactic in biological material. *Anal. Biochem.* 195:308–312.

56. Madrid, J., A. Martinez-Teruel, F. Hernandez, and M. D. Megias. 1999. A comparative study on the determination of lactic acid in silage juice by colorimetric high-performance liquid chromatography and enzymatic methods. *J. Sci. Food Agric.* 79:1722–1726.

57. Gómes-Álvarez, E., E. Luque-Pérez, A. Ríos, and M. Valcárcel. 1999. Flow injection spectrophotometric determination of lactic acid in skimmed milk based on a photochemical reaction. *Talanta* 50:121–131.

58. Dias, A. C. B., R. A. O. Silva, and M. A. Z. Arruda. 2010. A sequential injection system for indirect spectrophotometric determination of lactic acid in yogurt and fermented mash samples. *Microchem. J.* 96:151–156.

59. Fairbrother, P., W. O. George, and J. M. Williams. 1991.Whey fermentation: On-line analysis of lactose and lactic acid by FTIR spectroscopy. *Appl. Microbiol. Biotechnol.* 35:301–305.

60. Schindler, R., R. Vonach, B. Lendl, and R. Kellner. 1998. A rapid automated method for wine analysis based upon sequential injection (SI)-FTIR spectrometry. *Fresenius' J. Anal. Chem.* 362:130–136.

61. Vérette, E., F. Qian, and F. Mangani. 1995. On-line dialysis with high-performance liquid chromatography for the automated preparation and analysis of sugars and organic acid in foods and beverages. *J. Chromatogr. A* 705:195–203.

62. Kritsunankul, O., B. Pramote, and J. Jakmunee. 2009. Flow injection on-line dialysis coupled to high performance liquid chromatography for the determination of some organic acids in wine. *Talanta* 79:1042–1049.

Synthetic Antioxidants

Determination of Butylhydroxytoluene, Butylhydroxyanisole, and *tert*-Butylhydroquinone

*Thiago Faria Tormin, Eduardo Santos Almeida,
Raquel Maria Ferreira Sousa, Eduardo Mathias Richter,
and Rodrigo Alejandro Abarza Munoz*

CONTENTS

13.1 DETERMINATION OF ANTIOXIDANTS IN FOOD

Various additives are commonly added to processed food to prevent spoilage and preserve their organoleptic and nutritional characteristics; to control pH, flavor, color, and texture; and to avoid any changes that may occur during processing, storage, and use, without causing any risk to the health of the consumer. There are many classes of food additives such as colorants, emulsifiers, stabilizers, flavor enhancers, antioxidants, and others.

Antioxidants are an important group of these food additives, as they are able to prevent or delay lipid oxidation, commonly known as rancidity, thereby maintaining the quality of the food within its shelf life. Antioxidants are low cost substances, used in small concentrations, easy to apply, and do not change the characteristics or stability of the food (Takemoto et al., 2009; André et al., 2010). Their functional mechanisms have been widely studied. Berthollet in 1797 and later Davy in 1817 were the first scientists to study the retardation of oxidative reactions by certain compounds (oxidation inhibitors); they proposed the theory of "catalyst poisoning" in oxidative reactions, which was complemented by the free radical theory of peroxidation. Later, Duclaux demonstrated that atmospheric oxygen is the main factor responsible for the oxidation of free fatty acids.

Years later, Tsujimoto showed that rancid odors in fish oils are caused by the oxidation of unsaturated triglycerides (Shahidi, 2005).

Unsaturated lipid substrates are the prime compounds of deterioration in food and cosmetics. They are constituted of mixtures of tri-, di-, and monoacylglycerols, free fatty acids, glycolipids, phospholipids, sterols, and other substances (Chen et al., 2011). Hydrolysis, polymerization, and isomerization reactions can cause the deterioration of food, and the resulting compounds are harmful to health (Chen et al., 2011; Ng et al., 2014). However, the most important reaction that occurs with lipids is oxidation, because lipids are easily oxidized, even by atmospheric oxygen. This process results in the destruction of essential triglycerides and fat-soluble vitamins (A, D, E, and K), which leads to the formation of aldehydes, ketones, and free fatty acids. This is undesirable because the palatability, odor, texture, consistency, appearance, and nutritional value of the food are altered (Takemoto et al., 2009; André et al., 2010).

Lipids can undergo oxidation due to the presence of oxygen and an initiator such as heat (thermolysis), light (photolysis), metal ions, and photosensitizers (natural dyes or pigments that absorb visible or UV light) (Kumarathasan et al., 1992; Laguerre et al., 2007). The autoxidation process occurs in three phases, which are represented in Figure 13.1a.

In the initiation phase, homolytic breakdown of the hydrogen in the alpha position of the double bond chain, occurs by the action of oxygen or by initiators, resulting in free radical formation. The radical lipid formed is stabilized by resonance structures (Figure 13.1b). The propagation phase starts with the addition of molecular oxygen to the radical lipid (L•) obtained in first phase to afford the peroxyl radical (LOO•). In sequence, the peroxyl radical abstracts a hydrogen atom from L–H to generate LOOH and another radical (L•). In this phase, fragmentation can also be observed, along with rearrangement and cyclization. In the last phase—termination—the peroxyl radical is transformed into

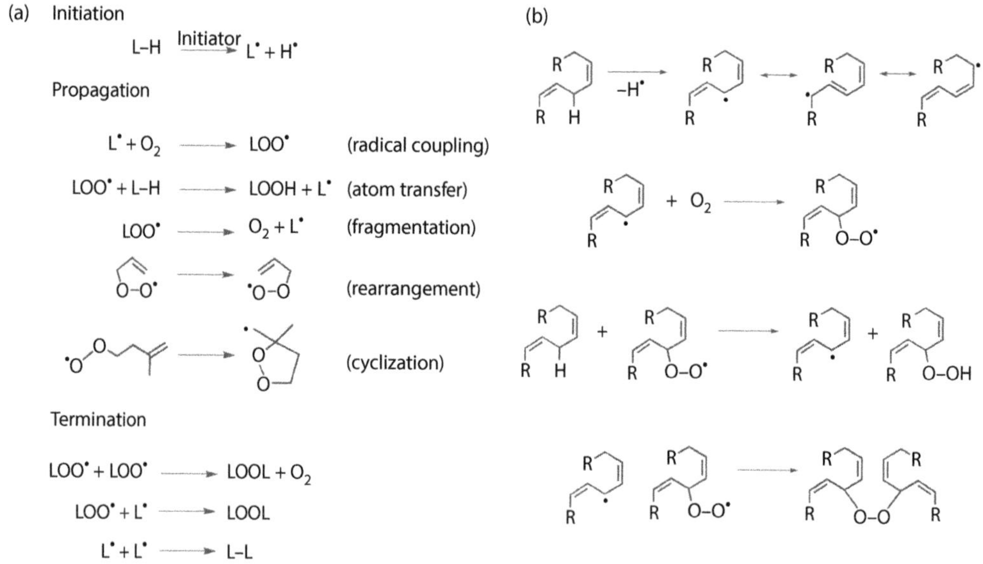

FIGURE 13.1 Autoxidation in lipids: (a) stages of the process; (b) process in a general lipid. (Adapted from Porter, N. A., S. E. Caldwell, and K. A. Mills. 1995. *Lipids* 30(4):277–290; Laguerre, M., J. Lecomte, and P. Villeneuve. 2007. *Prog. Lipid Res.* 46(5):244–282.)

nonradical oxidation compounds. This main mechanism involves scission of the double bond adjacent to the hydroperoxyl group, leading to the formation of hydrocarbons, aldehydes, ethers, alcohols and volatile ketones, oxidized aldehydes, triacylglycerols, and their polymers (Porter et al., 1995; Laguerre et al., 2007).

This oxidative degradation process of lipids in food can be inhibited by antioxidant compounds. Beyond the preservation of food, antioxidants absorb free radicals and help prevent cardiovascular, neurological, and carcinogenic diseases in biological systems (André et al., 2010).

Antioxidants are a heterogeneous group of substances including vitamins, minerals, pigments, and other chemical substances. They are classified as follows (André et al., 2010):

- Natural (or biological)—organic acids, organosulfur compounds, flavonoids, and terpenes.
- Synthetic—propyl gallates and quinone derivatives.
- Primary or long-term antioxidants (phenols and secondary amines) that are capable of donating hydrogen and stop the chain reaction.
- Secondary antioxidants (organic phosphites and thioesters) that are decomposers of peroxides.
- There are also chelating agents (EDTA [ethylenediaminetetraacetic acid] salts), which act by sequestering metallic ions (mainly copper and iron) that catalyze lipid oxidation. Enzymes are biological antioxidants (glucose oxidase and catalase) capable of removing oxygen from the system (Laguerre et al., 2007).

Despite the great number of antioxidant compounds available, only a few are used in the food industry and their use (especially of synthetic ones) is regulated by national or international authorities, with upper limits for daily intake levels (Takemoto et al., 2009; André et al., 2010). The most commonly used are the naturally occurring tocopherols and synthetic phenolic antioxidants, including propyl gallate (PG), butylhydroxyanisole (BHA), butylhydroxytoluene (BHT), and *tert*-butylhydroquinone (TBHQ).

Daily levels of acceptable intake (ADI) according to the Food and Agriculture Organization (FAO) of the United Nations are 0.3, 0.5, and 0.7 mg kg^{-1} body weight/day for BHT, BHA, and TBHQ, respectively. ADI values for various antioxidants and other food additives can be found in the FAO website (Food and Agriculture Organization of the United Nations, 2014).

BHA is a more effective antioxidant in suppressing oxidation in animal fats than in vegetable oils. Like most of the phenolic antioxidants, its efficiency is limited in unsaturated oils from seeds and vegetables. It has low stability at high temperatures, but is effective at controlling the oxidation of short-chain fatty acids. BHT has similar properties to BHA; however, while BHA is a synergist for PG, BHT is not. Both BHA and BHT may confer an odor to food when applied at high temperatures. BHA and BHT are synergists with each other. BHA acts as a scavenger of peroxide radicals, while BHT acts as a regenerator of BHA radicals. TBHQ is a white crystalline powder, shiny, moderately soluble in oils and fats, and does not form complexes with copper and iron ions. Generally, TBHQ is more effective in vegetable oils than BHA and BHT. On the other hand, TBHQ presents similar efficiency in animal fat in comparison to BHA and is more effective than BHT. TBHQ is considered the best antioxidant for frying oils because it resists heating and provides excellent stability in finished products (Ramalho and Jorge, 2006).

The structures of the antioxidants BHA, BHT, and TBHQ are shown in Figure 13.2.

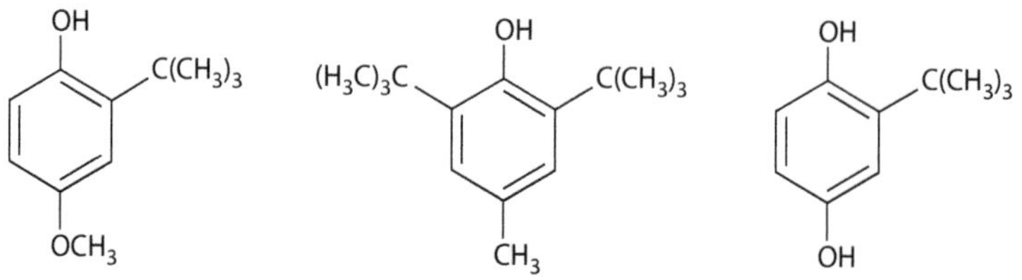

FIGURE 13.2   Structures of the synthetic phenolic antioxidants BHA, BHT, and TBHQ.

The phenolic structure of primary antioxidants, such as BHA, BHT, and TBHQ, allows the donation of a proton to a free radical molecule, which regenerates the acylglycerol molecule and interrupts the mechanism of oxidation by free radicals. Then, the phenolic derivatives are transformed into free radicals and these radicals are self-stabilized without promoting or propagating the oxidation reaction (Ramalho and Jorge, 2006). Reactions (13.1) and (13.2) indicate the action of a primary antioxidant

$$ROO^\bullet + AOH \rightarrow ROOH + AO^\bullet \tag{13.1}$$

$$R^\bullet + AOH \rightarrow RH + AO^\bullet \tag{13.2}$$

13.2 ANALYTICAL METHODS FOR THE DETERMINATION OF ANTIOXIDANTS IN FOOD

The main concern with the use of antioxidants is related to the possible toxicological effects associated with these compounds (Botterweck et al., 2000; de Campos and Toledo, 2000; Sasaki et al., 2002). Experiments on animals have demonstrated the possibility of certain antioxidants exerting a carcinogenic effect (Sasaki et al., 2002). Therefore, the analytical determination of the levels of antioxidants in food is essential to ensure that the quantities effectively present in the food at the time of consumption do not exceed the legal limits and, at the same time, to verify whether these quantities are sufficient to maintain the quality of the food. Moreover, analytical data of the concentrations of different antioxidants found in food can be used to estimate the potential for ingestion of these additives in order to ensure safe use (de Campos and Toledo, 2000). The official method of the Association of Official Agricultural Chemists (AOAC) for the determination of antioxidants uses a large amount of sample and solvents and involves several extraction steps (de Campos and Toledo, 2000; AOAC, 2003; Takemoto et al., 2009). To overcome this drawback, various methods have been proposed for analysis of antioxidants in food. Another point is the banning of some antioxidants in some countries, such as in Japan, where the use of TBHQ is not permitted (Juntachote et al., 2006); however, this antioxidant is added to food products exported to that country.

The antioxidants BHA, BHT, and TBHQ were mostly investigated in the literature. A review (Karovičová and Šimko, 2000) devoted to the determination of phenolic antioxidants in food by high-performance liquid chromatography (HPLC) presented 24 analytical methods (papers). Eighteen of them investigated at least one of the three antioxidants and nine of them showed the determination of the three antioxidants simultaneously. André et al. (2010) conducted a more complete review, and once again, articles containing at least one of the three antioxidants (TBHQ, BHA, or BHT) accounted for approximately 50% of the studies reviewed. They reported in the review various food matrices such as oils, cereals, breads, vegetables, fruits, dairy products, meats, fish, seafood, sugar, honey, beverages, supplements, and stimulants. Extraction modes were reported, including dynamic thermal desorption, supercritical fluid extraction, solid-phase extraction and micro-extraction, solid–liquid extraction, and liquid–liquid extraction, which was the most used sample treatment. HPLC was used in most studies, followed by electrophoresis and gas chromatography (GC) (André et al., 2010). These techniques were used due to the easy separation of the analytes. Initially, GC was more advantageous because it was faster than liquid chromatography (LC), even considering that the time of sample preparation for GC was larger. LC did not require a derivatization step; a single extraction step or even direct injection of the sample in the case of liquid samples such as vegetable oils was sufficient, which increased in analytical frequency of LC methods (Van Niekerk and Du Plessis, 1980). Likely for this reason, LC was the most used technique for the development of analytical methods for the determination of antioxidants in food. Recent attention has been given to capillary electrophoresis (CE), which is an analytical separation technique that offers several advantages including its high speed, efficiency, reproducibility, small sample volume, and ease of removing contaminants. Micellar electrokinetic capillary electrophoresis (MEKC) is the typical choice for the separation of neutral species such as phenolic antioxidants. An MEKC method using spectrophotometric detection in the UV region has been reported for the determination of BHA and BHT in edible oils after liquid–liquid extraction as sample preparation (Delgado-Zamarreño et al., 2007). Combined with electrochemical detection, CE holds great promise due to its high sensitivity and selectivity (Guan et al., 2006). Other techniques reported in the literature were thin-layer chromatography (Pyka and Sliwiok, 2001), voltammetry (Raymundo et al., 2007), and infrared spectroscopy (Blanco and Alcalá, 2006).

Molecular absorption spectrophotometry is probably the oldest instrumental technique used for the analysis of antioxidants in food samples; there are works that date from the late 1950s. However, even with its high robustness and low cost, there are few studies in the literature. Due to the complexity of samples, generally oils and fats, spectrophotometry presents little selectivity, thus necessitating laborious sample preparation, with steps of extraction, distillation, and derivatization (Hansen et al., 1959; Szalkowski and Garber, 1962; Dilli and Robards, 1977; Komaitis and Kapel, 1985).

13.3 FLOW INJECTION METHODS FOR FOOD ANTIOXIDANTS

Flow injection methods have excellent potential in almost all areas of analysis since they provide high-throughput analysis (increase of analytical frequency) and the possibility of automation. However, the application of analytical methods based on flow analysis for food analysis has still received little attention, due to the complexity of food matrices (Lopez-Fernandez et al., 1995). Therefore, the development of automated methods for the determination of additives, such as synthetic antioxidants (e.g., BHA, BHT, and TBHQ), is a promising area of investigation. Flow injection solid-phase spectrophotometry was used for the determination of BHA and PG in fatty food (dehydrated chicken soup, chicken cream, bull calf, and chicken broth) and cosmetics, providing greater selectivity to the conventional spectrophotometric technique. After extraction with petroleum either for food and with hexane for cosmetics, the two antioxidants were separated by injecting them into a C18 silica column, using 30:70 (v/v) methanol:water as the carrier solution. The flow rate was 1.4 mL min^{-1} and the detection limit was 0.7 mg mL^{-1} for BHA and 0.9 mg mL^{-1} for PG (Capitán-Vallvey et al., 2003). In a later study by the same authors, it was shown that binary mixtures of BHA/BHT and BHT/PG can be determined in similar food samples (cola drink and dehydrated soup). The extraction process was performed in two stages: the first one with hexane and the second with a mixture of 70% methanol in water. A C18 silica column under flow rates of 1.10 and 1.24 mL min^{-1} was used for the respective mixtures. The carrier mixture was composed of 50:50 (v/v) methanol:water with a detection limit of approximately 2 mg mL^{-1} for the three compounds (Capitán-Vallvey et al., 2003). A third study by this group demonstrated the possibility of determining BHA and seven additives simultaneously in various food and cosmetics. Similarly to previous articles, a monolithic C18 silica column was used to separate the eight analytes and a peristaltic pump was used to perform the propulsion. Sample preparation was performed in several extractors, each one appropriate for a certain sample, including hexane, methanol, acetonitrile, diethyl ether, and aqueous solutions containing sodium chloride and sodium bicarbonate. The additional aid of ultrasound, common filtration and/or solid-phase extraction was required for some samples. Two carrier solutions were used: a 10 mmol L^{-1} phosphate buffer pH 6.0 with 4% (v/v) acetonitrile and, subsequently, a 30:70 (v/v) methanol:water solution (changed by a rotary valve), to carry the more apolar analytes. The flow rate was 3.5 mL min^{-1} and the detection limit for BHA was 1.60 mg mL^{-1} (García-Jiménez et al., 2007). The same research group developed a method for the determination of BHA, BHT, and PG in seven different food (including chewing gum, dehydrated soup, one-minute soup, and bouillon cube), using the same separation column from the previous work. The antioxidants were extracted with hexane, methanol, and 2-propanol due to the specific nature of the matrix. Two compositions of methanol:water as carrier solutions were used, starting at 42% (v/v) methanol and finishing with 70% (v/v) methanol. The limit of detection was below 1 mg mL^{-1} for all three antioxidants with a flow rate of 2.4 mL min^{-1}, which allowed the separation in 85 s only. These four studies showed that flow injection analysis (FIA) in the solid phase allows for the determination of multianalytes with the same advantages of traditional FIA, including low cost, simplicity, versatility, low analysis time, and possible use in routine analysis (García-Jiménez et al., 2009).

13.3.1 Flow Injection Methods Using Electrochemical Detectors for Food Antioxidants

The electrochemical oxidation of phenolic antioxidants is widely reported in the literature and the mechanism of electrochemical oxidation of BHA, BHT, and TBHQ is well known, as described in Figure 13.3. In the three cases, a quinone derivative is formed after a single step involving a two-electron transfer. The formation of a quinone carbocation through electrochemical oxidation is more readily stabilized in the TBHQ molecule in comparison with BHA and BHT, respectively. Therefore, the electrochemical oxidation of TBHQ occurs at less positive potentials than that of BHA and BHT, respectively. Cyclic voltammetry of BHA reveals the interconversion of BHA into TBHQ in the second voltammetric cycle, as illustrated in Figure 13.3. The first cycle presents the electrochemical oxidation of BHA at 0.6 V (vs. Ag/AgCl) generating *tert*-butylquinone species (TBQ) after a chemical step as illustrated in the inset of Figure 13.3. The second cycle reveals the electrochemical oxidation of TBHQ into TBQ through the formation of the oxidation

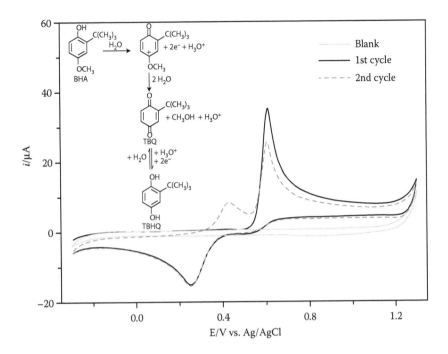

FIGURE 13.3   Cyclic voltammetry of 1 mmol L⁻¹ BHA in 0.1 mol L⁻¹ HCl on a glassy carbon electrode. Scan rate of 50 mV s⁻¹ and initial potential at −0.3 V. Inset: mechanism of oxidation of BHA and its conversion to TBHQ. (Adapted from de la Fuente, C. et al. 1999. *Talanta* 49(2):441–452.)

peak. The mechanism of electrochemical oxidation of BHT is presented in Figure 13.4. After the formation of a carbocation, a hydroquinone derivative is probably formed (Figure 13.4). Since the electrochemical oxidations of TBHQ, BHA, and BHT are easily attained at different electrodes, several electrochemical methods have been reported for the determination of antioxidants in various food matrices.

Differential-pulse polarography was reported to evaluate the content of THBQ in edible oils (Tonmanee and Archer, 1982; Cortés et al., 1993). Differential-pulse voltammetry using a carbon paste electrode modified with nickel phthalocyanine was applied to determine BHA in potato flakes (Ruiz et al., 1995). Square-wave voltammetric methods employing carbon-disk ultramicroelectrodes (UME) (Ceballos and Fernández, 2000), glassy carbon electrodes (Raymundo et al., 2007), platinum electrodes

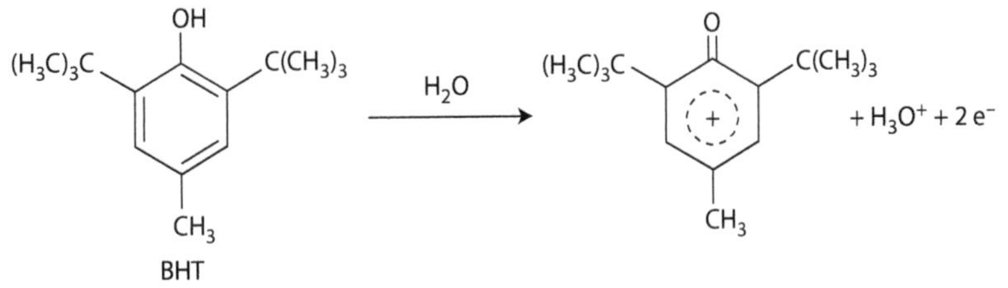

FIGURE 13.4   Oxidation mechanism of BHT. (Adapted from Medeiros, R. A., R. C. Rocha-Filho, and O. Fatibello-Filho. 2010b. *Food Chem.* 123(3):886–891.)

(Raymundo et al., 2007), and boron-doped diamond electrodes (BDDE) were reported for measurement of BHA and BHT in edible oils; BHA, BHT, and TBHQ in mayonnaise; and BHA and BHT mayonnaise and margarine (Ceballos and Fernández, 2000; Raymundo et al., 2007; Medeiros et al., 2010b). Linear-sweep voltammetry was employed to evaluate BHA, BHT, and TBHQ (and also pyrogallol) in peanut oil, salad oil, sesame oil, cake, biscuits, and milk candy using a glassy carbon electrode and chemometric approaches for data treatment, since the electrochemical oxidation of the four antioxidants occurs at similar potential values so that the individual determination of all molecules in a single run is complicated (Ni et al., 2000).

Table 13.1 lists the electrochemical methods developed for antioxidant analysis in several samples, including the details of sample preparation. The hardest challenge to adapt these methods to flow systems is the sample preparation step, especially for solid samples. The following paragraphs discuss some examples of electroanalytical methods associated with FIA applied to antioxidant (BHA, BHT, and TBHQ) analysis in food samples.

Yáñez-Sedeño et al. (1991) developed an analytical method using an FIA system coupled with amperometric detection at a glassy carbon electrode for the determination of BHA in corn oil. The application of the constant potential of 0.60 V versus Ag/AgCl/3 mol L^{-1} KCl (selected from hydrodynamic voltammetry studies) allowed for the selective determination of BHA in the presence of BHT. The adsorption of the phenolic compound or its oxidation product on the surface of the glassy carbon working electrode was verified after performing successive injections of standard solutions. This effect was minimized by adding 5% (v/v) methanol in the carrier solution and in the samples (RSD = 3.9%, $n = 15$). The optimized flow rate and injection volume were 3.5 mL min^{-1} and 200 μL, respectively. According to Yáñez-Sedeño et al. (1991), sample preparation was based on the dissolution of the sample in n-hexane, followed by extraction with acetonitrile (in three portions), solvent evaporation (rotating vacuum evaporator), dissolution of the dried residue in methanol, and subsequent dilution in the electrolyte (0.1 mol L^{-1} HClO$_4$). In this work, recovery tests for BHA conducted with samples in the presence or absence of BHT (as a possible interfering molecule) ranged between 88% and 100%. Other components present in commercial oil samples such as citric acid, propylene glycol, and propyl gallate were evaluated, and only propyl gallate was considered to be an interfering molecule.

The use of modified electrodes was explored by Ruiz et al. (1995), who applied a nickel-phthalocyanine polymer modified electrode. The peak current measured with the modified electrode was approximately twofold higher than that obtained with the glassy carbon electrode. With regards to the potential peak, this was slightly shifted to the less positive potential relative to the electrode polymer obtained at an unmodified electrode. Based on hydrodynamic voltammetry, a value of 0.70 V (versus Ag/AgCl/3 mol L^{-1} KCl) was selected as the best potential for the amperometric determinations of BHA in biscuits (Ruiz et al., 1995). Tests for possible interference in BHA amperometric determinations were performed using sodium bisulfite, citric acid, ascorbic acid, TBHQ, BHT, and PG. The optimized flow rate and injection volume were 3.3 mL min^{-1} and 200 μL, respectively. The samples were reduced to a fine powder with a mortar and pestle, subjected to extraction (with portions of 5 mL of 50% v/v methanol/water) under stirring and centrifugation, and diluted in the carrier electrolyte (0.2 mol L^{-1} HClO$_4$). The proposed method proved to be fairly precise (RSD = 3.5%, $n = 10$) and accurate (recovery values between 98% and 102%).

TABLE 13.1   Electrochemical Methods Developed for Antioxidant Analysis in Several Samples, Including the Details of Sample Preparation

Electrode	Technique	Sample	Sample Preparation	Analyte(s)	Reference
Hg	DPP	Edible oils	Dissolution in buffer solution (acetate buffer in 1:5 toluene/absolute ethanol solution).	TBHQ	Tonmanee and Archer (1982)
Hg	DPP	Edible (olive and corn) oils	Liquid–liquid extraction using hexane–ethyl acetate (99:1 v/v) with emulsion formation.	TBHQ	Cortés et al. (1993)
NiP-CPE	DPV	Potato flakes	Liquid–liquid extraction using methanol/water (50% v/v) followed by pre-concentration in a vacuum rotary evaporator.	BHA	Ruiz et al. (1994)
C-UME	SWV	Edible oils	Liquid–liquid extraction using acetonitrile.	BHA and BHT	Ceballos and Fernandez (2000)
GCE	LSV	Peanut oil, salad oil, sesame oil, cake, biscuits and milk candy	Liquid–liquid extraction in methanol (for oil samples) and dissolution in petroleum ether and subsequent liquid–liquid extraction in methanol (for solid food samples) with the aid of shaking followed by centrifugation.	BHA, BHT, PG, and TBHQ	Ni et al. (2000)
GCE and Pt disk	SWV	Mayonnaise	Liquid–liquid extraction using methanol followed by centrifugation.	BHA, BHT, and TBHQ	Raymundo et al. (2007)
BDDE	SWV	Mayonnaise and margarine	Liquid–liquid extraction using ethanol with the aid of agitation and centrifugation.	BHA and BHT	Medeiros et al. (2010b)

Note: NiP-CPE: carbon paste electrode modified with Ni-phthalocyanine; C-UME: carbon ultramicroelectrode; GCE: glassy carbon electrode; BDDE: boron-doped diamond electrode; DPP: differential-pulse polarography; DPV: differential-pulse voltammetry; LSV: linear-sweep voltammetry; and SWV: square-wave voltammetry.

Prabakar and Narayanan (2010) demonstrated the amperometric determination of BHA using a wax composite electrode modified with nickel hexacyanoferrate (NiHCF) coupled to a FIA system. The modified electrode, in comparison to the bare electrode, gave a considerable improvement in the sensitivity (approximately sevenfold) and decreased the oxidation potential of BHA. The potential chosen for amperometric determinations was 0.4 V versus Ag/AgCl/3 mol L^{-1} KCl (a decrease of 300 mV in the overpotential oxidation). The optimized flow rate and injection volume were 4.0 mL min^{-1} and 20 μL, respectively. The method was also applied to the determination of BHA in potato chips (Prabakar and Narayanan, 2010). A fine powder of samples was dissolved in a 10% (v/v) methanol solution and mechanically agitated for 20 min, using three portions of 5 mL. The supernatant was subsequently diluted in the carrier solution (0.1 mol L^{-1} NaNO$_3$; pH 7). To evaluate the accuracy of the method, recovery studies were performed (ranging from 96.8% to 100%) for three different samples of potato chips.

Garcia and Ortiz (1998) developed a FIA method with amperometric detection at a glassy carbon electrode modified with 4-hydroxybenzaldehyde–formaldehyde polymer for the determination of BHT in oil samples. Hydrodynamic voltammetry showed maximum current values at relatively high potential values (>1.0 V versus Ag/AgCl/3 mol L^{-1} NaCl), but the potential was set at 0.70 V due to the fact that this value resulted in a good response; moreover, a relatively low value prevents the oxidation of other compounds. The determinations were made using a carrier solution containing 50% (v/v) methanol and phosphate buffer (pH 7.2). The optimized flow rate was 1.1 mL min^{-1} and the injection volume was not reported. Liquid–liquid extractions of 5 mL of corn oil were carried out with five aliquots of 5 mL of a solution containing 60% (v/v) methanol and 40% (v/v) phosphate buffer (pH 7.2). The extract was obtained by separating the oil phase from the aqueous phase; the BHT determination was performed using the standard addition method. The samples were analyzed by the proposed method and HPLC-UV, with good agreement between them (error of 2%).

The simultaneous determination of two phenolic antioxidants (BHA and BHT) in an FIA system was demonstrated using multiple-pulse amperometric (MPA) detection by Medeiros et al. (2010a,b) (Table 13.1). This approach involved the constant application of a sequence of two potential pulses at a BDDE. BHA was detected by its electrochemical oxidation in the first potential pulse (0.85 V versus Ag/AgCl/KCl$_{sat}$) free of interference from BHT. In the second potential pulse of 1.15 V versus Ag/AgCl/KCl$_{sat}$, both BHA and BHT were oxidized at the BDDE surface. Current subtraction from the contribution of the BHA response at 1.15 V was necessary to accurately quantify BHT. The use of MPA for simultaneous determinations in an FIA system was presented earlier (Santos et al., 2008). In some applications of MPA detection, a third or fourth potential pulse is applied for electrode cleaning (Santos et al., 2009; Gimenes et al., 2010; Silva et al., 2011; Miranda et al., 2012; Gimenes et al., 2013, 2014). According to Medeiros et al. (2010a,b), a third potential pulse to clean the working electrode was not necessary, because the adsorption process was reduced at the BDDE, as indicated by the low relative standard deviation (ranging from 1.9% to 4.1%) obtained in repeatability studies (interday and intraday). The supporting electrolyte solution used in this work was an ethanol–water solution (30% ethanol v/v) with 10 mmol L^{-1} KNO$_3$ (pH adjusted to 1.5 with HNO$_3$). Unlike in the previously mentioned studies, Medeiros et al. (2010a,b) analyzed commercial mayonnaise samples. The sample preparation step involved extraction by dissolving about 1 g of the sample in ethanol, stirring for 5 min and centrifugation for 10 min. The extraction procedure was repeated twice and a portion of extract was taken for further dilution in the supporting electrolyte. The treated samples were analyzed by FIA with MPA detection (proposed method) and by HPLC, with consistent results (95% confidence according t-test) for the determination of BHA and BHT in mayonnaise.

Table 13.2 lists the FIA methods with amperometric detection developed for antioxidant analysis in food samples including the antioxidant type, sample, and limit of detection.

All electroanalytical methods, including the ones associated with FIA, devoted to antioxidant analysis in food samples require a prior sample treatment step such as

TABLE 13.2   Studies Reported in the Literature for Analysis of the Content of Antioxidants in Food Samples by Flow Injection Analysis with Amperometric Detection

Electrode	Sample	Analyte(s)	LOD	Reference
GCE	Corn oil	BHA and BHT	2.5 µg L^{-1} (1.4 × 10^{-8} and 1.1 × 10^{-8} mol L^{-1} for BHA and BHT)	Yáñez-Sedeño et al. (1991)
Ni-PP	Biscuits	BHA	2.7 µg L^{-1} (1.5 × 10^{-8} mol L^{-1})	Ruiz et al. (1995)
HBFP	Industrial and vegetable oil	BHT	0.001 g L^{-1} (4.5 × 10^{-6} mol L^{-1})	García and Ortiz (1998)
NiHCF	Potato chips	BHA	6 × 10^{-7} mol L^{-1}	Prabakar and Narayanan (2010)
BDDE	Commercial mayonnaise samples	BHA and BHT	0.030 µmol L^{-1} (BHA) 0.40 µmol L^{-1} (BHT)	Medeiros et al. (2010a,b)

Note: BDDE: boron-doped diamond electrode; GCE: glassy carbon electrode; LOD: limit of detection; HBFP: 4-hydroxybenzaldehyde–formaldehyde polymer; Ni-PP: nickel–phthalocyanine polymer; and NiHCF: nickel hexacyanoferrate.

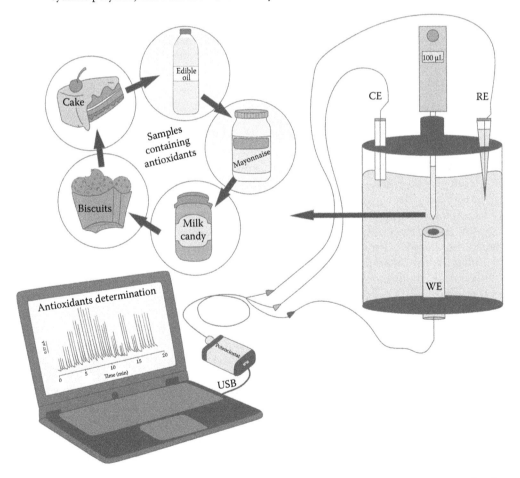

FIGURE 13.5   Scheme of a BIA cell designed for electroanalytical applications in food.

solid–liquid or liquid–liquid extraction. The adaptation of the sample treatment step to the FIA systems shows great promise for high-throughput analysis of antioxidants in food samples. This adaptation would be easier for liquid samples such as oil. For example, the online formation of oil–water emulsions could be an alternative sample preparation method for analysis of antioxidants by FIA with amperometric detection.

Another promising strategy not explored in antioxidant analysis is the batch injection analysis (BIA) technique, which is easily combined with electrochemical detectors (Wang and Taha, 1991). The BIA technique can be considered as an alternative to FIA for the development of analytical methods with high analytical frequency as required for routine purposes (Wang and Taha, 1991). BIA only requires an electronic micropipette that injects a sample plug directly onto the working electrode surface (wall-jet configuration), which is immersed in a large-volume blank solution (Quintino and Angnes, 2004). Figure 13.5 schematically depicts a BIA cell designed for electroanalytical applications based on schemes reported in the literature (Tormin et al., 2011; Silva et al., 2012). Disadvantages associated with the pump and valves of the FIA system (especially when organic solvents are used as the carrier) and the disposal of carrier solutions are eliminated (Quintino and Angnes, 2004). The direct determination of antioxidants in oil samples by an FIA method would require the use of organic solvents. Recent studies have been reported in the literature demonstrating the determination of BHA in biodiesel samples (Tormin et al., 2011). Additionally, the use of MPA detection for the simultaneous determination of BHA and TBHQ in biodiesel samples in the BIA system has been reported (Tormin et al., 2011, 2012). Due to the similar sample matrix of biodiesel samples, these methods can easily be adapted for the quantification of these antioxidants (and also BHT) in edible oil samples.

REFERENCES

André, C., I. Castanheira, J. M. Cruz, P. Paseiro, and A. Sanches-Silva. 2010. Analytical strategies to evaluate antioxidants in food: A review. *Trends Food Sci. Technol.* 21(5):229–246.

AOAC. 2003. *Official Methods of Analysis of AOAC International.* (17th ed.), Arlington, VA: Association of Official Analytical Chemists.

Blanco, M. and M. Alcalá. 2006. Simultaneous quantitation of five active principles in a pharmaceutical preparation: Development and validation of a near infrared spectroscopic method. *Eur. J. Pharm. Sci.* 27(2–3):280–286.

Botterweck, A. A. M., H. Verhagen, R. A. Goldbohm, J. Kleinjans, and P. A. van den Brandt. 2000. Intake of butylated hydroxyanisole and butylated hydroxytoluene and stomach cancer risk: Results from analyses in the Netherlands Cohort Study. *Food Chem. Toxicol.* 38(7):599–605.

Capitán-Vallvey, L. F., M. C. Valencia, and E. Arana Nicolás. 2003. Simple resolution of butylated hydroxyanisole and *n*-propyl gallate in fatty foods and cosmetics samples by flow-injection solid-phase spectrophotometry. *J. Food Sci.* 68(5):1595–1599.

Ceballos, C. and H. Fernández. 2000. Synthetic antioxidants in edible oils by square-wave voltammetry on ultramicroelectrodes. *J. Am. Oil Chem. Soc.* 77(7):731–735.

Chen, B., D. J. McClements, and E. A. Decker. 2011. Minor components in food oils: A critical review of their roles on lipid oxidation chemistry in bulk oils and emulsions. *Crit. Rev. Food Sci. Nutr.* 51(10):901–916.

Cortés, A. G., A. J. R. García, P. Yáñez-Sedeño, and J. M. Pingarrón. 1993. Polarographic determination of *tert*-butylhydroquinone in micellar and emulsified media. *Anal. Chim. Acta* 273(1–2):545–551.

de Campos, G. C. M. and M. C. F. Toledo. 2000. The determination of BHA, BHT and TBHQ in fats and oils by high performance liquid chromatography. *Braz. J. Food Technol.* 3:65–71.

de la Fuente, C., J. A. Acuña, M. D. Vázquez, M. L. Tascón, and P. S. Batanero. 1999. Voltammetric determination of the phenolic antioxidants 3-*tert*-butyl-4-hydroxyanisole and *tert*-butylhydroquinone at a polypyrrole electrode modified with a nickel phthalocyanine complex. *Talanta* 49(2):441–452.

Delgado-Zamarreño, M. M., I. González-Maza, A. Sánchez-Pérez, and R. Carabias Martínez. 2007. Analysis of synthetic phenolic antioxidants in edible oils by micellar electrokinetic capillary chromatography. *Food Chem.* 100(4):1722–1727.

Dilli, S. and K. Robards. 1977. Detection of the presence of BHA by a rapid spectrofluorimetric screening procedure. *Analyst* 102(1212):201–205.

Food and Agriculture Organization of the United Nations. 2014. In *Food Safety and Quality: Tertiary Butylhydroquinone, Butylated Hydroxyanisole and Butylated Hydroxytoluene.* Available from http://www.fao.org/ag/agn/jecfa-additives/specs/Monograph1/. Assessed in August 1, 2015.

García, C. D. and P. I. Ortiz. 1998. Determination of *tert*-butylhydroxytoluene by flow injection analysis at polymer modified glassy carbon electrodes. *Electroanalysis* 10(12):832–835.

García-Jiménez, J. F., M. C. Valencia, and L. F. Capitán-Vallvey. 2007. Simultaneous determination of antioxidants, preservatives and sweetener additives in food and cosmetics by flow injection analysis coupled to a monolithic column. *Anal. Chim. Acta* 594(2):226–233.

García-Jiménez, J. F., M. C. Valencia, and L. F. Capitán-Vallveyn. 2009. Simultaneous determination of three antioxidants in foods and cosmetics by flow injection coupled to an ultra-short monolithic column. *J. Chromatogr. Sci.* 47(6):485–491.

Gimenes, D. T., R. R. Cunha, M. M. A. de et al. 2013. Two new electrochemical methods for fast and simultaneous determination of codeine and diclofenac. *Talanta* 116(0):1026–1032.

Gimenes, D. T., W. T. P. dos Santo, T. F. Tormin, R. A. A. Munoz, and E. M. Richter. 2010. Flow-injection amperometric method for indirect determination of dopamine in the presence of a large excess of ascorbic acid. *Electroanalysis* 22(1):74–78.

Gimenes, D. T., M. C. Marra, R. A. A. Munoz, L. Angnes, and E. M. Richter. 2014. Determination of propranolol and hydrochlorothiazide by batch injection analysis with amperometric detection and capillary electrophoresis with capacitively coupled contactless conductivity detection. *Anal. Methods* 6(10):3261–3267.

Guan, Yueqing, Qingcui Chu, Liang Fu, Ting Wu, and Jiannong Ye. 2006. Determination of phenolic antioxidants by micellar electrokinetic capillary chromatography with electrochemical detection. *Food Chem.* 94(1):157–162.

Hansen, P. V., F. L. Kauffman, and L. H. Wiedermann. 1959. A direct spectrophotometric determination of butylated hydroxyanisole in lard and in hardened lard. *J. Am. Oil Chem. Soc.* 36(5):193–195.

Juntachote, T., E. Berghofer, F. Bauer, and S. Siebenhandl. 2006. The application of response surface methodology to the production of phenolic extracts of lemon grass, galangal, holy basil and rosemary. *Int. J. Food Sci. Technol.* 41(2):121–133.

Karovičová, J. and P. Šimko. 2000. Determination of synthetic phenolic antioxidants in food by high-performance liquid chromatography. *J. Chromatogr. A* 882(1–2):271–281.

Komaitis, M. E. and M. Kapel. 1985. Spectrophotometric determination of BHA in edible fats and oils. *J. Am. Oil Chem. Soc.* 62(9):1371–1372.

Kumarathasan, R., A. B. Rajkumar, N. R. Hunter, and H. D. Gesser. 1992. Autoxidation and yellowing of methyl linolenate. *Prog. Lipid Res.* 31(2):109–126.

Laguerre, M., J. Lecomte, and P. Villeneuve. 2007. Evaluation of the ability of antioxidants to counteract lipid oxidation: Existing methods, new trends and challenges. *Prog. Lipid Res.* 46(5):244–282.

Lopez-Fernandez, J. M., A. Rios, and M. Valcarcel. 1995. Assessment of quality of flow injection methods used in food analysis: A review. *Analyst* 120(9):2393–2400.

Medeiros, R. A., B. C. Lourenção, R. C. Rocha-Filho, and O. Fatibello-Filho. 2010a. Simple flow injection analysis system for simultaneous determination of phenolic antioxidants with multiple pulse amperometric detection at a boron-doped diamond electrode. *Anal. Chem.* 82(20):8658–8663.

Medeiros, R. A., R. C. Rocha-Filho, and O. Fatibello-Filho. 2010b. Simultaneous voltammetric determination of phenolic antioxidants in food using a boron-doped diamond electrode. *Food Chem.* 123(3):886–891.

Miranda, J. A. T. de, R. R. Cunha, D. T. Gimenes, R. A. A. Munoz, and E. M. Richter. 2012. Simultaneous determination of ascorbic acid and acetylsalicylic acid using flow injection analysis with multiple pulse amperometric detection. *Quim. Nova* 35:1459–1463.

Ng, C.-Y., X.-F. Leong, N. Masbah, S. K. Adam, Y. Kamisah, and K. Jaarin. 2014. Heated vegetable oils and cardiovascular disease risk factors. *Vasc. Pharmacol.* 61(1):1–9.

Ni, Y., L. Wang, and S. Kokot. 2000. Voltammetric determination of butylated hydroxyanisole, butylated hydroxytoluene, propyl gallate and *tert*-butylhydroquinone by use of chemometric approaches. *Anal. Chim. Acta* 412(1–2):185–193.

Porter, N. A., S. E. Caldwell, and K. A. Mills. 1995. Mechanisms of free radical oxidation of unsaturated lipids. *Lipids* 30(4):277–290.

Prabakar, S. J. R. and S. S. Narayanan. 2010. Flow injection analysis of BHA by NiHCF modified electrode. *Food Chem.* 118(2):449–455.

Pyka, A. and J. Sliwiok. 2001. Chromatographic separation of tocopherols. *J. Chromatogr. A* 935(1–2):71–76.

Quintino, M. S. M., and L. Angnes. 2004. Batch injection analysis: An almost unexplored powerful tool. *Electroanalysis* 16(7):513–523.

Ramalho, V. C. and N. Jorge. 2006. Antioxidants used in oils, fats and fatty foods. *Quim. Nova* 29(4):755–760.

Raymundo, M., dos Santos, M. M. da Silva Paula, C. Franco, and R. Fett. 2007. Quantitative determination of the phenolic antioxidants using voltammetric techniques. *LWT—Food Sci. Technol.* 40(7):1133–1139.

Ruiz, M. A., M. G. Blázquez, and J. M. Pingarrón. 1995. Electrocatalytic and flow-injection determination of the antioxidant *tert*-butylhydroxyanisole at a nickel phthalocyanine polymer modified electrode. *Anal. Chim. Acta* 305(1–3):49–56.

Ruiz, M. A., M. P. Calvo, and José M. Pingarrón. 1994. Catalytic-voltammetric determination of the antioxidant *tert*-butylhydroxyanisole (BHA) at a nickel phthalocyanine modified carbon paste electrode. *Talanta* 41(2):289–294.

Santos, W. T. P. dos, E. G. N. de Almeida, H. E. A. Ferreira, D. T. Gimenes, and E. M. Richter. 2008. Simultaneous flow injection analysis of paracetamol and ascorbic acid with multiple pulse amperometric detection. *Electroanalysis* 20(17):1878–1883.

Santos, W. T. P. dos, D. T. Gimenes, E. G. N. de Almeida, S. de P. Eiras, Y. D. T. Albuquerque, and E. M. Richter. 2009. Simple flow injection amperometric system for simultaneous determination of dipyrone and paracetamol in pharmaceutical formulations. *J. Braz. Chem. Soc.* 20:1249–1255.

Sasaki, Y. F., S. Kawaguchi, A. Kamaya et al. 2002. The comet assay with 8 mouse organs: Results with 39 currently used food additives. *Mutat. Res./Genet. Toxicol. Environ. Mutagen.* 519(1–2):103–119.

Shahidi, F. 2005. *Bailey's Industrial Oil & Fat Products* (6th ed.), Vol. 6. New Jersey: John Wiley & Sons.

Silva, R. A. B., R. H. O. Montes, E. M. Richter, and R. A. A. Munoz. 2012. Rapid and selective determination of hydrogen peroxide residues in milk by batch injection analysis with amperometric detection. *Food Chem.* 133(1):200–204.

Silva, W. C., P. F. Pereira, M. C. Marra et al. 2011. A simple strategy for simultaneous determination of paracetamol and caffeine using flow injection analysis with multiple pulse amperometric detection. *Electroanalysis* 23(12):2764–2770.

Szalkowski, C. R. and J. B. Garber. 1962. Antioxidant measurement, determination of 2,6-di-*tert*-butyl-4-hydroxytoluene (BHT): Application to edible fats and oils. *J. Agric. Food Chem.* 10(6):490–495.

Takemoto, E., J. Teixeira Filho, and H. T. Godoy. 2009. Validation of methodology for the simultaneous determination of synthetic antioxidants in vegetables oils, margarine and vegetables hydrogenated fats by HPLC/UV. *Quim. Nova* 32:1189–1194.

Tonmanee, N. T. and V. S. Archer. 1982. Determination of *tert*-butylhydroquinone in edible oils by differential-pulse polarography. *Talanta* 29(11, Supplement 1):905–909.

Tormin, T. F., R. R. Cunha, E. M. Richter, and R. A. A. Munoz. 2012. Fast simultaneous determination of BHA and TBHQ antioxidants in biodiesel by batch injection analysis using pulsed-amperometric detection. *Talanta* 99:527–531.

Tormin, T. F., D. T. Gimenes, E. M. Richter, and R. A. A. Munoz. 2011. Fast and direct determination of butylated hydroxyanisole in biodiesel by batch injection analysis with amperometric detection. *Talanta* 85(3):1274–1278.

Van Niekerk, P. J. and L. M. Du Plessis. 1980. High-performance liquid chromatographic determination of *tert*-butyl-hydroquinone in vegetable oils. *J. Chromatogr. A* 187(2):436–438.

Wang, J. and Z. Taha. 1991. Batch injection analysis. *Anal. Chem.* 63(10):1053–1056.

Yáñez-Sedeño, P., J. M. Pingarrón, and L. M. Polo Díez. 1991. Determination of *tert*-butylhydroxyanisole and *tert*-butylhydroxytoluene by flow injection with amperometric detection. *Anal. Chim. Acta* 252(1–2):153–159.

CHAPTER **14**

Determination of Propyl, Octyl, and Dodecyl Gallates

Agustina Gómez-Hens and Juan Godoy-Navajas

CONTENTS

14.1 INTRODUCTION

Propyl (PG, E 310), octyl (OG, E 311), and dodecyl (or lauryl) (DG, E 312) esters of gallic acid, usually named alkyl gallates, form a part of the synthetic phenolic antioxidants, mainly used as food additives, with PG being the most frequently used alkyl gallate. The molecular structure of these compounds is composed of a hydrophilic head (phenolic ring) and a hydrophobic alkyl chain (Figure 14.1). PG is slightly soluble in water but OG and DG are insoluble in water. The three compounds are soluble in ethanol, ethyl ether, and oils.

Alkyl gallates are mainly used to protect fats, oils, and fat-containing food from rancidity that results from the formation of peroxides. Their use as antioxidant additives is regulated by legal authorities in a limited number of foods, with maximum limits in each case. The acceptable PG daily intake (ADI) recommended by the joint FAO/WHO Expert Committee on Food Additives (JECFA) is 0–1.4 mg of additive per kilogram body weight, and the maximum level recommended in foodstuffs is 200 mg/kg [1]. A recent reevaluation by the European Food Safety Authority (EFSA) of PG as food additive has concluded that it is not of safety concern at the current uses and use levels [2].

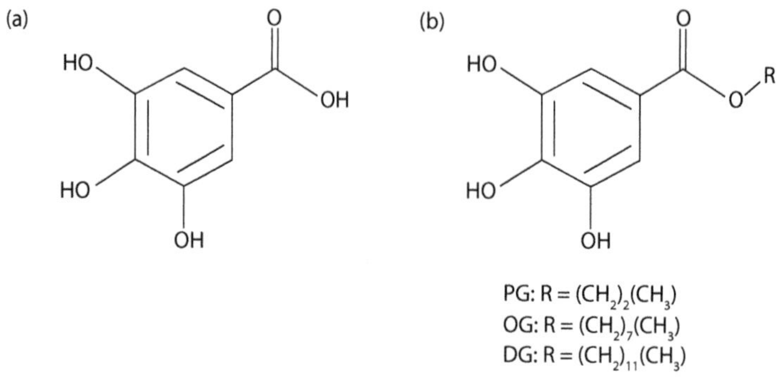

(a)

(b)

PG: R = (CH$_2$)$_2$(CH$_3$)
OG: R = (CH$_2$)$_7$(CH$_3$)
DG: R = (CH$_2$)$_{11}$(CH$_3$)

FIGURE 14.1 Chemical structures of gallic acid (a) and alkyl gallates (b).

PG is mainly absorbed in the body by the gastrointestinal tract. It is hydrolyzed to gallic acid and 4-O-methylgallic acid by cellular carboxylesterase. In addition, PG is converted to a dimer and ellagic acid via autoxidation [3]. Free gallic acid or a conjugated derivative of 4-O-methylgallic acid with glucuronic acid are excreted in the urine. It has been reported that OG and DG are absorbed and hydrolyzed to a lesser degree than PG [4].

PG is also used to prevent slow deterioration of cosmetics and personal care products caused by chemical reactions with oxygen, inhibiting the formation or accumulation of free radicals that may cause the deterioration of the product. The Cosmetic Ingredient Review (CIR) Expert Panel evaluated the scientific data and concluded that PG was safe in the practices of use in cosmetics and personal care products at concentrations less than or equal to 0.1% [5]. It is used in many product categories, including lipsticks, bath products, skin cleansing products, moisturizers, skin care products, makeup products, self-tanning products, and sunscreen and suntan products.

In addition to the use of alkyl gallates as food and cosmetic additives, the multifunctional action of these compounds and their potential application for other purposes has been widely investigated, as described below. Some of these applications, which are discussed below, involve their use as enzymatic inhibitors, as anticancer, antifungal and antimicrobial agents, and as additives in nonfood and noncosmetic industries (Figure 14.2). These applications are mainly based on the antioxidant and prooxidant properties of these compounds, which, together with their potential toxicity, are also discussed below.

14.2 FEATURES AND APPLICATIONS OF ALKYL GALLATES

14.2.1 Antioxidant and Prooxidant Activities of Alkyl Gallates and Their Potential Toxicity as Food Additives

The antioxidant activity of alkyl gallates in food is ascribed to their inhibitor effect on lipoxygenase activity [6–8]. Lipoxygenases are a family of nonheme iron-containing dioxygenases widely distributed in both animal and plant kingdoms. The main substrates are linoleic and linolenic acids and the primary products are fatty acid hydroperoxides, which are capable of producing membrane damage and promoting cell death. Analysis of

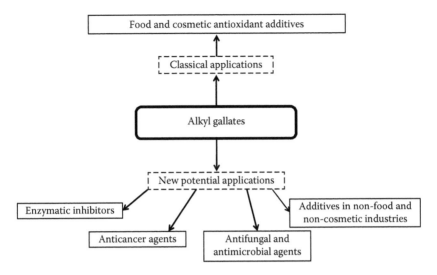

FIGURE 14.2   Classical and new applications of alkyl gallates.

inhibition kinetics by Lineweaver–Burk plots indicates that alkyl gallates are reversible competitive inhibitors, the alkyl chain length being significantly related to the inhibitory activity. Thus, it has been reported that DG is a more potent lipoxygenase inhibitor than OG. The DG inhibitory concentration leading 50% activity loss (IC_{50}) has been estimated to be 0.07 µM, whereas that for OG was 1.3 µM [7]. The ability of alkyl gallates to chelate transition metal ions, which are promoters of free radical formation, also contributes to their antioxidant activity [8].

A theoretical study on the antioxidant properties of PG in aqueous and lipid media has been carried out using several free radicals [9]. The results showed that in aqueous media at physiological pH, the neutral and deprotonated forms of PG are in equilibrium, the deprotonated form being the better antioxidant molecule. The reactions mainly involve the hydrogen transfer mechanism, with transfer of the hydrogen from the phenolic hydroxyl groups, rather than a single electron transfer mechanism. The antioxidant activity of PG in lipid media is also based on the hydrogen transfer mechanism, mainly by the abstraction of the hydrogen from the hydroxyl group at the 4(OH) position. PG is also a good antioxidant in lipid media, but the calculated rate constants are lower than those obtained in aqueous media.

Although the antioxidant capacity of alkyl gallates has been widely demonstrated, prooxidant properties have also been described, which calls their safety into question due to the controversial effects [10–12]. The antioxidant effect of alkyl gallates is closely related to their hydrogen donor activity [13], while the cytotoxic effects reported for these compounds, which are discussed below, are assumed to be due to the prooxidant action [14]. For instance, it has been reported that gallic acid, as a metabolite of PG, generates hydrogen peroxide in HL-60 human leukemia cells, increasing the amount of a deoxyguanosine derivative that is a characteristic biomarker of oxidative stress [10]. OG and DG have also been suggested to present both antioxidant and prooxidant properties depending on their concentration and cellular conditions [15].

A review published in 1986 on the toxicological effects of alkyl gallates [4] described different oral short- and long-term toxicity assays in several animals using large amounts of these compounds (5.0 g/kg). The results obtained showed that no effects were observed at these high doses. However, a more recent study on the genotoxicity and cytotoxicity of PG indicated that the consumption of PG as a food additive at a relatively high dose may induce toxicity and carcinogenicity [16]. This study evaluated the effect of a 1.0 mM (0.21 g/L) PG solution on the growth and death of A549 lung cancer cells, concluding that PG can lead to cell growth inhibition. It has also been reported that PG may contribute to mitochondrial impairment and, consequently, inhibition of cell respiration, which may lead to adenosine triphosphate depletion [3,17].

Although PG is recognized as safe, its absolute safety remains unknown [18]. The toxicological behavior of PG has been widely investigated, but the exact mechanism of toxicity remains unclear. In fact, there is a controversy about the effects of PG on carcinogenesis and mutagenesis as both enhancing and suppressing effects have been described. Antioxidative and cytoprotective properties of PG may change to prooxidative, cytotoxic, and genotoxic properties [19]. A study on the kinetics of the transformation of PG in the perfused rat liver has shown that PG exerts several harmful effects, especially inhibition of gluconeogenesis [20]. Alkyl gallates are extensively bound to cellular structures, which increases the residence time and increases toxicity and metabolic effects. The study of the cytotoxicity of alkyl gallates in rat hepatocytes [21] has shown that butyl gallate, OG, and DG, which have long alkyl side-chains, are more toxic than methyl, ethyl, and propyl gallates, and the potency of their toxicity is associated with impairment of mitochondrial respiration. A potential antifibrotic activity of PG has been suggested from the results obtained by studying its effect in activated hepatic stellate cells, which play a central role in liver fibrosis, demonstrating the apoptosis-inducer characteristics of PG [22].

An ecotoxicological study of the effects of PG in aquatic systems has shown that PG should be classified as toxic to aquatic organisms [23]. The study was carried out using several ecological systems, but the IC_{50} values obtained ranged from 10 to 1090 μM, which might be related to a variety of factors.

14.2.2 Inhibitory Activity of Alkyl Gallates

Some of the properties of alkyl gallates described above, such as their antioxidant activity, are closely related to their behavior as enzyme inhibitors. A study of the inhibitory effect of PG on tyrosinase has shown that it inhibits diphenolase activity of this enzyme, acting as a reversible and mixed-type inhibitor [24]. The inhibitor concentration leading to 50% loss of activity (IC_{50}) was determined to be 0.685 mM. This inhibitory property has been used to study its potential application for control of pericarp browning of harvested fruits. PG has been also described as an efficient inhibitor of tomato fruit ripening [25]. The investigation of the inhibitory activity of a series of alkyl gallates against hyaluronidase and collagenase has shown that OG is a potent inhibitor of these enzymes, which could justify its use as a cosmetic ingredient [26]. Thus, OG might be useful to improve inflammation and allergic reactions, and to prevent wrinkles and skin aging. Another study about the antioxidant activity of DG has shown that it prevents generation of superoxide radicals by xanthine oxidase, which comes from its ability to inhibit the enzyme [27]. According to Lineweaver–Burk plots, the inhibition is noncompetitive. Also, it has been shown that DG inhibits formation of uric acid acting as a competitive inhibitor.

Other inhibitory studies of OG and other antioxidants have demonstrated that they are highly effective in preventing reactions of free thiyl radicals with oleic acid [28]. Thiyl radicals, which are formed from thiol compounds by physical, chemical, and biochemical energy transfer, reversibly attack the double bonds of unsaturated fatty acids, leading to mixtures of *cis*- and *trans*-fatty acids as well as addition products (thioethers). The results obtained showed that OG is highly effective in preventing stereomutation and addition reactions of free tetradecanethiyl radicals with oleic acid.

The inhibitory effect of PG and DG on free radical–induced hemolysis and depletion of intracellular glutathione (GSH) in human erythrocytes has been studied [29]. The mechanism of erythrocyte hemolysis induced by thermolysis of 2,2′-azobis(2-amidinopropane)hydrochloride (AAPH) has been correlated with lipid peroxidation and oxidation of membrane proteins. This process also provokes the depletion of intracellular GSH. The results obtained showed that PG and DG inhibit hemolysis, the effect being dose dependent.

A study of the action of PG in rat liver has shown its inhibitory effect on hepatic gluconeogenesis at concentrations up to 200 mM, which is in part ascribed to the inhibition of pyruvate carboxylation [30]. Also, the potential use of OG as antidiabetic drug has been studied in streptozotocin-induced diabetic rats, indicating that OG reduced plasma glucose to near normal values, which might be due to potentiation of insulin [31]. Another study of the inhibitory effect of PG and other antioxidants has shown that these compounds are able to significantly inhibit the oxidation of β-carotene by hydroxyl free radicals, which is ascribed to the sequestering ability of the antioxidants against the free radicals [32].

14.2.3 Anticancer Activity of Alkyl Gallates

A widely studied potential application of the cytotoxicity of alkyl gallates is their use as anticancer agents [15,33]. As indicated above, the cytotoxic effects of alkyl gallates are assumed to be due to their prooxidant action, instead of their antioxidant capacity. These compounds have been shown to induce apoptosis through free radical generation and alterations in the energy potential of the cell; the mechanisms of cell death induced by alkyl gallates depends on the length of the alkyl chain [33]. The cytotoxic and antiproliferative effects described for OG and DG on tissues and cells are related to inhibitory action on protein kinase activity [13]. These alkyl gallates have shown an inhibitory potency on protein kinase activity that results in apoptosis induction 50–250 times greater than that of its precursor gallic acid for various human cell lines, indicating a selectivity for fast-growing cells, which supports the study of OG and DG as potential anticancer agents. It has been also reported that DG not only prevents the formation of chemically induced skin tumors in mice but is also able to selectively kill tumor cells in stabilized tumors [34]. Another study of the antiproliferative effect of DG in human breast cancer cells showed that it induced cell cycle alterations and apoptosis in all cell lines tested, with very low toxicity to normal cells and good selectivity to tumor cells [35].

Studies of the carcinogenicity of PG in a human leukemia cell line have suggested that gallic acid plays an important role in this effect since it is more easily oxidized than PG [11]. However, when rat hepatocytes were incubated with the esterase inhibitor diazinon, the cytotoxic effects of PG were enhanced. These results suggest that the hepatotoxicity of PG was not mediated by its metabolites [3].

PG has been reported to induce apoptosis in a large variety of cancer cells [36,37]. Recent research suggests that PG inhibits the growth of HeLa cells and leukemia cells by depleting intracellular GSH levels, inducing a G_1-phase arrest of the cell cycle and increasing the production of the superoxide anion [36,38]. The intracellular levels of reactive oxygen species (ROS) increased or decreased in PG-treated HeLa cells depending on the times of incubation (1 or 24 h) and doses (100–1600 μM). Thus, ROS levels were decreased in HeLa cells treated with 100–800 μM PG at 1 h, whereas the level was increased in 1600 μM PG at this time. At 24 h, 400–1600 μM PG significantly increased ROS levels in HeLa cells. Also, PG increased activities of superoxide dismutase and catalase in HeLa cells. The inhibitory effect of PG on the growth of HeLa cells via caspase-dependent apoptosis, with an IC_{50} of approximately 800 μM at 24 h, has been also described [36]. Another study reported that PG inhibits the growth rate of A549 lung cancer cell lines by inducing apoptosis through chromatin and DNA fragmentation [16].

The study of the toxicity of several polyphenolic antioxidants, including alkyl gallates, toward human promyelocytic leukemia cells (HL-60) has shown that the polyphenol cytotoxicity may be related to their prooxidant properties since, as indicated above, the same polyphenol compounds could behave as both antioxidants and prooxidants, depending on concentration and free radical source [39]. Polyphenols may act as substrates for peroxidases and other metalloenzymes, yielding quinone- or quinomethide-type prooxidant and/or alkylating products. Another characteristic of the action of polyphenols is that their cytotoxicity increases upon an increase in their lipophilicity. Thus, the concentration of gallic acid, BG and OG for 50% survival of HL-60 cells (IC_{50}) is 750, 110, and 30 μM, respectively.

Another comparative cytotoxicity study of alkyl gallates on mouse tumor cell lines has shown that gallates bearing long alkyl groups are toxic to rat hepatocytes, but at higher concentrations than those found for tumor cells [40]. When the alkyl chain length increased from four to five carbons, the respiratory rate of hepatocytes was strongly inhibited and the selectivity of these compounds toward tumor cells was strongly decreased. The results obtained in this study indicate that alkyl gallates are selectively cytotoxic to tumor cells, which may be due to the mitochondrial dysfunctions of these cells.

14.2.4 Antifungal and Antimicrobial Activities of Alkyl Gallates

Several studies have reported the potential antifungal and antimicrobial activities of alkyl gallates. Although these properties are of special interest in the food industry, the results described so far have been obtained in the absence of food matrices.

The antifungal activity of PG, OG, and DG has been studied using *Saccharomyces cerevisiae* and *Zygosaccharomyces bailii* [41]. OG was the only active compound, with minimum fungicidal concentration of 25 μg/mL. In contrast, neither PG nor DG showed any fungicidal activity up to 1600 μg/mL. The length of alkyl group in the gallates was associated with their antifungal activity, which disappeared after the alkyl length reached the maximum, the so-called "cutoff" phenomenon, and the dodecyl group in DG is beyond this point. On the other hand, the propyl group in PG is not long enough to elicit the activity. Another study, in which a high number of alkyl gallates were assayed showed that branched alkyl gallates have stronger activity than linear alkyl gallates [42].

Several investigations have reported the potential use of alkyl gallates as antimicrobial agents. These studies involve the determination of the minimal inhibitory concentration (MIC), which is defined as the minimal concentration of antimicrobial compound that inhibits visible growth of the strain tested. The study of the anti-Salmonella activity of several alkyl gallates has shown that the length of the alkyl group is not a major contributor but it can affect this activity in some extent [43]. MIC values of 12.5, 50, and 1600 µg/mL for OG, DG, and PG, respectively, were obtained in this study, which show that OG was the most effective. Another article on the effect of several phenolic antioxidants against three bacterial species, two Gram-positive (*Staphylococcus aureus*, *B. cereus*) and one Gram-negative (*Pseudomonas fluorescens*) bacteria, showed that OG would be effective as an antimicrobial agent [44]. This study was carried out by assaying different strains of each microorganism, reporting a MIC range of 5.5–29.2 mg/mL OG for *S. aureus* and 44–100 mg/mL OG for *P. fluorescens*. However, significant differences in MIC values of OG among strains of *B. cereus* were obtained, which ranged between 19 and >1600 mg/mL. This wide variation in the response to OG was related to the spore-forming character of this bacterial genus and differences between sporulation–germination patterns among the strains of *B. cereus*.

14.2.5 Alkyl Gallates as Additives in Nonfood and Noncosmetic Industries

In addition to the general use of alkyl gallates as additives in the food and cosmetic industries, their particular features have prompted the study of their potential application for other uses, such as in the polymer, biodiesel, and textile industries. Some examples are described below.

Inorganic oxide nanoparticles are added to organic polymers to modify the thermal stability, optical, mechanical, and/or electrical properties of these polymers. For instance, composites containing TiO_2 nanoparticles dispersed in a polymeric matrix have been used to obtain visible-transparent UV filters. However, a problem of these composites is that nanoparticles tend to aggregate, leading to a loss of optical clarity. The incorporation of hydrophobic organic modifiers as capping agents is a strategy to obtain stable dispersions in organic solvents and matrices. The usefulness of DG for this purpose has been illustrated by the synthesis of DG-capped TiO_2 nanoparticles, which facilitated the dispersion of the nanomaterial in organic solvents and the preparation of composite materials [45]. Several alkyl gallates were assayed as modifiers of the surface of TiO_2 nanoparticles, which were dispersed into a polystyrene matrix, obtained by *in situ* bulk radical polymerization of styrene [46]. The results obtained showed that the thermal and thermooxidative stability of the polystyrene matrix improved with the use of these nanoparticles, giving the best results when the nanoparticle surface was modified with OG. The use of OG has been also investigated to prepare biocompatible nanodispersions as delivery systems for food additives [47].

The presence of unsaturated fatty acids in biodiesel makes it susceptible to oxidation or autoxidation during long-term storage. Several studies carried out by assaying different antioxidants to improve the oxidative stability of biodiesel have shown the usefulness of PG for this purpose [48–50].

Polypropylene is one of the most widely consumed and produced polyalkenes in the plastics industry, widely applied in packaging, tubing, automobiles, and household appliances. However, this polymer readily suffers from uncontrolled oxidation during processing, storage, and use, which causes undesirable changes including discoloration and deterioration in mechanical properties. The addition of antioxidants is an effective way to prevent the degradation of polypropylene. Thus, the potential use of gallates as antioxidant additives for polypropylene has been investigated using OG, DG, and hexadecyl gallate (HG) [51]. The results showed that these compounds could provide long-term stabilization to polypropylene under conditions of oxidative degradation, with the antioxidant efficiency increasing as the length of alkyl chain increases. This behavior is ascribed to the fact that the compatibility of polypropylene with HG is better than with the other gallates owing to the longer alkyl chain.

The covalent grafting of alkyl gallates onto wool through a laccase-catalyzed reaction has been used to obtain a multifunctional textile material with antioxidant, antibacterial, and water-repellent properties [52]. The results obtained from this study show that the length of the alkyl chain of gallates plays an important role in the functional properties of wool fibers. The radical scavenging activity of gallate-modified wool, which is related to the gallates' antioxidant activity, was determined by measuring the decrease in absorbance of 1,1-diphenyl-2-picrylhydrazyl radical (DPPH*) at 515 nm. All the gallates assayed showed similar scavenging activity on DPPH in solution, indicating that the alkyl chain length was not directly related to this activity because the antioxidant efficiency of gallates depends upon the reactivity of the phenolic groups. However, after incorporation onto wool, the antioxidant activity of gallates increased with the increase of the length of gallate alkyl chain, which suggests that during the reaction grafting onto the wool fibers more phenolic groups remained active/nonoxidized for the compounds with longer alkyl chains, compared to the assays carried out in solution. Also, OG and DG showed excellent water repellence and the reduction of bacterial growth reached 37% and 50% for wool modified with OG and DG, respectively.

14.3 ANALYTICAL METHODS FOR THE DETERMINATION OF ALKYL GALLATES

A large number of methods have been described for the determination of alkyl gallates. Although this chapter is mainly devoted to methods using flow injection analysis, a brief description of other methods is also included. Tables 14.1 and 14.2 summarize some features of these methods. Most of these methods require the separation of alkyl gallates from the sample matrix, which is a critical step. Extraction from food samples with methanol and acetonitrile is one of the procedures most commonly used for this purpose [1].

14.3.1 Flow Injection Methods

Flow injection methods for the determination of alkyl gallates have been mainly focused on the determination of PG alone or together with other additives, principally other antioxidants such as butylated hydroxyanisole (BHA), butylated hydroxytoluene (BHT), and tertiary butylhydroquinone (TBHQ) (Table 14.1). Both optical and electrochemical detection systems have been used for this purpose.

Photometric detection has been used for the determination of PG and other antioxidants such as BHA [53,54] and BHT [55] using flow injection systems and a diode array detector (DAD), which allows the measurement of the intrinsic absorbance of each analyte at its individual wavelength. Thus, a flow sensor for the determination of PG and BHA has been described using the differential transient retention of the analytes in a solid phase packed in a flow-through cell [53]. The use of hydrophilic adsorbents, such as Sephadex G-15 and G-25, resulted in low retention of the antioxidants, and similar residence times. In contrast, the use of C_{18}-bonded silica as adsorbent with average particle sizes of 55–105 μm produced good retention and elution of both antioxidants. PG experienced a faster transient retention than BHA, which allowed the separation and determination of both analytes. PG was determined at 40 s from injection, measuring at 272 nm, and BHA

TABLE 14.1   FIA Methods for the Determination of Gallates

Gallate (and Other Antioxidant)	Detection System	Detection Limit	Application	Reference
PG (BHA)	Photometry (DAD)	PG: 0.9 µg/mL	Food and cosmetics	[53]
PG (BHA)	Photometry	PG: 0.06 µg/mL	Food and cosmetics	[54]
PG (BHT)	Photometry (DAD)	PG: 2.0 µg/mL	Food and cosmetics	[55]
PG (and others)	Photometry (DAD)	PG: 0.02 µg/mL	Food and cosmetics	[56]
PG	Chemiluminescence	PG: 36 ng/mL	Food	[57]
PG (and others)	Chemiluminescence	PG: 0.87 µg/mL	Cosmetics	[58]
PG	Amperometry	PG: 0.85 µg/mL and 1.4 µg/mL	Food	[59]
PG, OG (BHA)	Amperometry	PG: 0.23 µg/mL OG: 0.32 µg/mL	Food	[60]
PG (and others)	Amperometry	PG: 0.19 µg/mL	Food	[61]
PG (and others)	Conductometry		Kinetic parameters	[62]

PG: propyl gallate; BHA: butylated hydroxyanisole; BHT: butylated hydroxytoluene; OG: octyl gallate; and DAD: diode array detector.

at 200 s and measuring at 288 nm, after PG desorption by the carrier itself. The detection limit of the method was 0.7 µg/mL for BHA and 0.9 µg/mL for PG. The method allows the determination of both analytes in ratios between 1:10 and 10:1 and was applied to the analysis of several food and cosmetic samples. Recovery data ranged between 101% and 105% for both analytes. The same approach was also described for the individual determination of each antioxidant [54], in this instance reaching a detection limit for PG of 0.06 µg/mL. A similar photometric method has been described for the simultaneous determination of PG and BHT in ratios between 1:8 and 8:1 [55]. The detection limits obtained in this instance for BHT and PG were 2.5 and 2.0 µg/mL, respectively. The method was applied to the analysis of food (fats and edible oils) and cosmetic (bath oils and lotions) samples, and the results were validated using a liquid chromatographic (LC) method.

A flow injection system coupled to a monolithic column has been described for the simultaneous determination of antioxidants (PG and BHA), sweeteners (potassium acesulfame, sodium saccharin, and aspartame), and preservatives (methylparaben, ethylparaben, propylparaben, and butylparaben), using photometric detection [56]. The monolithic column used as separation system was a 5 mm commercial precolumn of silica C_{18}. The mixture was separated in only 400 s with resolution factors greater than 1.1 in all cases. Detection was accomplished by means of a DAD at the respective wavelength of each compound. The detection limit obtained for PG was 0.02 µg/mL. The method was applied to the analysis of food and cosmetic samples and the results were compared with those obtained using a conventional LC method.

TABLE 14.2 Other Methods for the Determination of Gallates

Gallate (and Other Antioxidants)	Method	Features	Application	Reference
PG, OG, and DG (and others)	LC with photometric detection (DAD)	PG: 0.85 µg/mL OG: 0.95 µg/mL DG: 5.72 µg/mL	Food	[65]
PG, OG, and DG (and others)	LC with fluorometric detection	PG: 0.19 µg/mL OG: 0.57 µg/mL DG: 0.18 µg/mL	Food	[67]
PG and OG (and other)	LC with TOF-MS detection	PG: 0.02 µg/g OG: 0.2 µg/g	Food	[68]
PG, OG, and DG (and others)	LC with quadrupole-orbitrap MS	PG: 5 µg/g OG: 1 µg/g DG: 1 µg/g	Food	[69]
PG (TBHQ)	CE with amperometric detection	PG: 75.1 µg/g	Cosmetics	[70]
PG (BHA)	CE with amperometric detection	PG: 1.1 µg/mL	Food	[71]
PG and OG (and others)	CE with photometric detection (DAD)	PG: 0.09 µg/mL OG: 0.31 µg/mL	Food	[72]
PG, OG and DG (and others)	CE with photometric detection	PG: 9 µg/mL OG: 15 µg/mL DG: 16 µg/mL	Food	[73]
DG (BHT and BHA)	CE with photometric detection	DG: 0.27 µg/mL	Food	[74]
PG (and others)	CE with electrochemical detection	PG: 0.064 µg/mL	Cosmetics	[75]
PG (and others)	CE with electrochemical detection	PG: 0.061 µg/mL	Food	[76]
PG (BHA)	Fluorometry	PG: 0.055 µg/mL	Food	[77]
PG	Fluorometry	PG: 0.02 µg/mL	Food and cosmetics	[78]
PG (BHA)	Fluorometry	PG: 0.03 µg/mL	Food	[79]
PG (BHA)	Fluorometry	PG: 2.2 ng/mL	Cosmetics	[80]
PG (BHA)	Photometry (DAD)	PG: 0.5 µg/mL	Food	[81]
PG, OG, and DG (and others)	Resonance light scattering	PG: 2.12 ng/mL OG: 1.69 ng/mL DG: 1.48 ng/mL	Food	[85]

PG: propyl gallate; OG: octyl gallate; DG: dodecyl gallate; TBHQ: tertiary butylhydroquinone; BHA: butylated hydroxyanisole; BHT: butylated hydroxytoluene; LC: liquid chromatography; CE: capillary electrophoresis; and DAD: diode array detector.

A chemiluminescence method has been described for the determination of PG in edible oils using the enhancing effect of this antioxidant on the luminescent signal of 2-phenyl-4,5-di(2-furyl)-1H-imidazole (PDFI) and potassium ferricyanide system in an alkaline medium [57]. The method involves the use of solid-phase extraction with a C_{18} cartridge as the stationary phase for sample treatment. The interferents were retained on the cartridge while PG was eluted. A flow system was used to introduce the reagents (PDFI, sodium hydroxide, and potassium ferricyanide) and the sample into the sample cell of the instrument and to measure the enhanced chemiluminescence signal, which was linearly related to the PG concentration. A detection limit of 36 ng/mL PG was obtained.

A method based on the combined use of low-pressure LC with a reversed-phase monolithic column and a flow injection system to perform chemiluminescence detection has been proposed for the determination of PG and other additives, such as phloroglucinol, 2,4-dihydroxybenzoic acid, salicylic acid, and methylparaben, in healthcare products [58]. The cerium(IV)–rhodamine 6G system was used to obtain the corresponding chemiluminescence signals from the analytes. The determination required only 280 s and the detection limit obtained for PG was 0.87 μg/mL. The method was applied to the analysis of healthcare products.

Amperometric detection has been the main electrochemical system used in the flow injection methods described for the determination of alkyl gallates. A graphite–Teflon composite amperometric tyrosinase biosensor has been developed for the determination of PG in foodstuffs [59]. The enzyme reaction involves the catalytic oxidation of PG to the corresponding *o*-quinone, which is electrochemically reduced. The detection limit was 0.85 μg/mL using an aqueous phosphate buffer (pH 6.5) solution, and 1.4 μg/mL using an acetonitrile–Tris buffer (pH 7.4). Investigation of the selectivity of the method showed that BHA, BHT, OG, and DG did not cause significant interference in aqueous medium, in which they are scarcely soluble. However, TBHQ interfered in the determination of PG in this medium, acting as a substrate of tyrosinase as it is a phenolic derivative with a free *ortho* position in the aromatic ring. All the antioxidants assayed as potential interferents in the organic medium affected the PG determination, except BHT, which it is not a tyrosinase substrate. The method was applied to the determination of PG in dehydrated broth bars [59] and olive oil samples.

Amperometric detection has also been used in a flow injection method for the determination of PG, OG, and BHA using a poly(vinyl chloride) graphite composite electrode [60]. Although the detection limits obtained are relatively high (>0.2 mg/L), the method is characterized for its simplicity and low cost of instrumentation. The applicability of the method was checked by determining the analytes in soup and oil samples. A polypyrrole electrode modified with a tetrasulfonated Ni(II) phthalocyanine complex has been described as an amperometric detector in a flow injection system for the determination of PG, BHA, and TBHQ, which had been previously separated using LC [61]. The detection limit obtained for PG was 0.19 μg/mL.

A stopped-flow manifold system has been developed to determine the activity and kinetics of immobilized tannase using PG as substrate and conductometric detection [62]. Tannase is a hydrolytic enzyme with important applications in the food industry. The enzyme was covalently immobilized onto the surface of aminopropyl controlled pore glass (CPG) using glutaraldehyde as the crosslinking reagent and the immobilized enzyme reactor was inserted within the tube-type electrode pair (cell constant = 103.2 cm⁻¹) for real-time conductometric measurements. The sample was introduced into the system by a switching valve and the conductometric signals were monitored after stopping the carrier flow. The Michaelis constant of immobilized tannase, using PG as substrate, was slightly higher (7.28 mM) than that of free enzyme (6.24 mM), which indicated the reduction of substrate affinity after immobilization.

A flow injection system with coulometric detection has been described to study the usefulness of PG, OG, DG, and other antioxidants to minimize the oxidative degradation of active pharmaceutical ingredients [63]. Flow injection analysis provided an oxidative profile from of a hydrodynamic voltammetry system in less than 1 min. The parameter used for this study was the relative oxidation potential, $E_{1/2}$, which is the applied potential corresponding to 50% of the total signal obtained for an electrochemical reaction. The lower the $E_{1/2}$ value, the more prone to oxidation is the compound. Thus, an effective antioxidant for a given drug will have a lower $E_{1/2}$ value than the drug. The hypothesis was that the optimal antioxidant would have a $E_{1/2}$ value approximately 100 mV less than that of the drug. The results obtained showed that this system enabled the efficient screening of drug samples and antioxidant candidates to assist with drug formulation studies.

14.3.2 Liquid Chromatographic Methods

Most methods for the determination of alkyl gallates are based on the use of liquid chromatography (LC). The features of these methods, which usually involve the use of reverse phases made of chemically bonded octadecyl silica and photometric detection, have been reviewed [1,64]. Thus, only the most recent methods are included in Table 14.2.

The photometric determination of ten antioxidants, including PG, OG, and DG, in edible vegetable oils has been described using LC and a DAD [65], together with a second-order calibration method based on the use of the alternating penalty trilinear decomposition algorithm [66]. This method mathematically decomposes the overlapping peaks into the pure profiles of each chemical species. Thus, the second-order data, generated by recording spectra during the development of the chromatogram, allow the quantification of the analytes even with overlapped peaks or in the presence of interferences and baseline drift. The detection limits obtained for PG, OG, and DG are 0.85, 0.95, and 5.72 μg/mL, respectively [65]. A similar method for the determination of these analytes was described using fluorescence detection instead of photometric detection [67]. The detection limits reported for PG, OG, and DG in this instance are 0.19, 0.57, and 0.18 μg/mL, respectively, which are better than those obtained using photometry.

The use of mass spectrometry (MS) as the detection system in LC methods for the determination of alkyl gallates has been relatively restricted. PG and OG, together with other antioxidants and preservatives, have been identified and quantified in edible vegetable oils using LC/time-of-flight MS (TOF-MS) [68]. The analytes were extracted from the samples with acetonitrile saturated with hexane. The detection limits obtained for PG and OG were 0.02 and 0.2 μg/g, respectively. A method combining QuEChERS (quick, easy, cheap, effective, rugged, and safe sample preparation method) with ultrahigh-performance liquid chromatography (UPLC) and electrospray ionization quadrupole orbitrap (ESI Q-Orbitrap) high-resolution MS has been described for the determination of 43 antioxidants, preservatives, and synthetic sweeteners in dairy products [69]. The results obtained show the capability of this approach for the separation and quantitation of a high number of analytes in only about 14 min.

14.3.3 Capillary Electrophoretic Methods

Different capillary electrophoretic (CE) techniques have been used for the determination of synthetic antioxidants, including alkyl gallates (Table 14.2). CE coupled with amperometric detection has been described for the determination of PG and TBHQ in cosmetic samples [70], and PG and BHA in food samples [71]. A porous etched joint was used in the first instance to eliminate the influence of high voltage on the electrochemical detection. The detection limit obtained for PG in cosmetic samples was 75 µg/mL and the recovery values were in the range of 93.6%–98.8%.

Capillary electrochromatography (CEC) using a mixed styrene- and methacrylate ester-based polymeric monolith as separation column and photometric detection (214 nm) has been described for the separation and determination of PG, OG, and other antioxidants [72]. The separation required a two-step gradient method varying the concentrations of acetonitrile in a phosphate solution. The detection limits obtained for PG and OG were 0.09 and 0.31 µg/m, respectively, and the method was applied to the analysis of edible oils.

Microemulsion electrokinetic chromatography (MEEKC) for PG, OG, DG, and other antioxidants, using photometric detection (214 nm), has been also described [73]. The microemulsion was constituted of octane, butan-1-ol, SDS, phosphate buffer, and propan-2-ol. The detection limits obtained in this instance for PG, OG, and DG were 9, 15, and 16 µg/mL, respectively. The method was applied to the determination of PG in noodles.

Micellar electrokinetic capillary chromatography (MEKC) has been used for the determination of DG, BHA, and BHT with photometric detection (280 nm) [74]. Bis-(2-ethylhexyl) sodium sulfosuccinate was used as the pseudostationary phase, reaching a detection limit for DG of 0.27 µg/mL. This compound was determined in edible oils using a previous liquid–liquid extraction with acetonitrile. MEKC has been also applied to the determination of PG, TBHQ, BHA, and BHT mixtures in cosmetic [75] and food [76] samples using sodium dodecyl sulfate and electrochemical detection.

14.3.4 Direct Fluorometric and Photometric Methods

Several fluorometric and photometric methods for the direct determination of PG that do not involve LC or CE separations have been reported (Table 14.2). These methods are usually based on the use of a derivatization reaction to improve the fluorescent or absorbent features of PG, achieving very low detection limits.

A fluorometric method has been described for the determination of PG in foodstuffs using 4-chloro-7-nitrobenzofurazan as reagent [77]. The reaction product was extracted with cyclohexane and the detection limit obtained was 0.055 µg/mL. BHA gave a similar reaction, but TBHQ and BHT did not interfere with the method.

The luminescence of PG with terbium(III), which is the result of an intramolecular energy transfer process, has been used to develop a kinetic method for the determination of this antioxidant in edible and cosmetic oils using the stopped-flow mixing technique [78]. This technique allowed the automation of the method and the measurement of kinetic data in only 0.2 s. A detection limit of 0.02 µg/mL PG was obtained.

The joint use of stopped-flow mixing technique and a T-format luminescence spectrometer has been described for the simultaneous determination of PG and BHA in food samples [79]. The method involves two different and independent reactions. The PG determination is based on the formation of a terbium(III) chelate in the presence of Triton X-100 and tri-*n*-octylphosphine oxide, while BHA is determined by monitoring its reaction product with the oxidized form of Nile blue. Both systems were excited at 310 nm, and the emission wavelengths were 545 and 665 nm for PG and BHA, respectively. The absence of overlap in the emission spectra allowed the determination of each analyte in each channel of the instrument. The detection limit obtained for PG was 0.03 μg/mL.

The simultaneous fluorometric determination of PG and BHA in cosmetics has also been described using excitation–emission fluorescence matrix data, which were processed by applying the second-order calibration method based on self-weighted alternating normalized residue fitting algorithm [80]. The detection limit obtained for PG was 2.2 ng/mL.

PG and BHA were simultaneously determined in food samples using stopped-flow mixing technique and a DAD [81]. The method was based on the different kinetic behavior of the analytes when reacted with 3-methylbenzothiazolin-2-one hydrazone in the presence of cerium(IV) and the measurement of the absorbance of each reaction product at a different wavelength. The determination was carried out using a system of two equations, which was resolved by multiple linear regression.

The cupric reducing antioxidant capacity (CUPRAC) method [82], which involves the use of copper(II)–neocuproine reagent, has been modified to assay both lipophilic and hydrophilic antioxidants simultaneously, including PG and DG [83]. The modification makes use of methyl-β-cyclodextrin, which forms inclusion complexes, increasing the solubility of lipophilic antioxidants and avoiding the turbidity of alcohol–water solutions.

The Folin–Ciocalteu method, which is also an antioxidant capacity assay based on the use of molybdotungstophosphate heteropolyanion reagent, has been also modified to enable the simultaneous determination of hydrophilic and lipophilic antioxidants, such as PG and DG [84]. The modification was performed by using an isobutanol-diluted version of the reagent and providing an alkaline medium with aqueous sodium hydroxide such that both organic and aqueous phases, necessary for lipophilic and hydrophilic antioxidants, respectively, were supplied simultaneously.

14.3.5 Other Methods for the Determination of Alkyl Gallates

Resonance light scattering (RLS) detection has been described for the kinetic determination of PG, OG, DG, and other antioxidant additives in foodstuffs using the capability of these compounds to reduce gold(III) to gold nanoparticles (AuNPs) in the presence of cetyltrimethylammonium bromide [85]. The formation of the NPs was followed by monitoring the variation of the RLS signal with time, using stopped-flow mixing technique, which allows the measurement of the initial rate of the system. The detection limits for the three alkyl gallates ranged between 7 and 10 nmol/L. The method does not distinguish among the different antioxidants, but it can be used as a very sensitive and fast screening method to detect the presence of these compounds in foodstuffs.

Raman spectroscopy has been used to study a series of structurally related antioxidant phenolic esters, including PG and OG [86]. Distinct vibrational patterns were observed for these compounds, which evidences the usefulness of this technique to obtain structural information.

ACKNOWLEDGMENTS

The authors gratefully acknowledge financial support from the Spanish Ministerio de Economía y Competitividad MINECO (Grant No. CTQ2012-32941), the Junta de Andalucía Program (Grant No. P09-FQM4933) and the FEDER program.

REFERENCES

1. André, C., I. Castanheira, J. M. Cruz, P. Paseiro, and A. Sanches-Silva. 2010. Analytical strategies to evaluate antioxidants in food: A review. *Trends Food Sci. Technol.* 21:229–246.
2. Aguilar, F., R. Crebelli, B. Dusemund et al. 2014. Scientific opinion on the re-evaluation of propyl gallate (E 310) as a food additive. *EFSA J.* 12:3642–3688.
3. Nakagawa, Y., K. Nakajima, S. Tayama, and P. Moldeus. 1995. Metabolism and cytotoxicity of propyl gallate in isolated rat hepatocytes: Effects of a thiol reductant and an esterase inhibitor. *Mol. Pharmacol.* 47:1021–1027.
4. Van der Heijden, C. A., P. J. C. M. Janssen, and J. J. T. W. A. Strik. 1986. Toxicology of gallates: A review and evaluation. *Food Chem. Toxicol.* 24:1067–1070.
5. Becker, L. Final report on the amended safety assessment of propyl gallate. *Int. J. Toxicol.* 26:89–118.
6. Ha, T. J. and K. Nihei, I. Kubo. 2004. Lipoxygenase inhibitory activity of octyl gallate. *J. Agric. Food Chem.* 52:3177–3181.
7. Ha, T. J. and I. Kubo. 2007. Slow-binding inhibition of soybean lipoxygenase-1 by dodecyl gallate. *J. Agric. Food Chem.* 55:446–451.
8. Kubo, I., N. Masuoka, T. J. Ha, K. Shimizu, and K. Nihei. 2010. Multifunctional antioxidant activities of alkyl gallates. *Open Bioact. Compd. J.* 3:1–11.
9. Medina, M. E., C. Iuga, and J. R. Alvarez-Idaboy. 2013. Antioxidant activity of propyl gallate in aqueous and lipid media: A theoretical study. *Phys. Chem. Chem. Phys.* 15:13137–13146.
10. Kawanishi, S., S. Oikawa, and M. Murata. 2005. Evaluation for safety of antioxidant chemopreventive agents. *Antioxid. Redox. Signaling* 7:1728–1739.
11. Kobayashi, H., S. Oikawa, K. Hirakawa, and S. Kawanishi. 2004. Metal-induced oxidative damage to cellular and isolated DNA by gallic acid, a metabolite of antioxidant propyl gallate. *Mutat. Res. Genet. Toxicol. Environ. Mutagen.* 558:111–120.
12. Aruoma, O. I., A. Murcia, J. Butler, and B. Halliwell. 1993. Evaluation of the antioxidant and prooxidant actions of gallic acid and its derivatives. *J. Agric. Food Chem.* 41:1880–1885.
13. Serrano, A., C. Palacios, G. Roy et al. 1998. Derivatives of gallic acid induce apoptosis in tumoral cell lines and inhibit lymphocyte proliferation. *Arch. Biochem. Biophys.* 350:49–54.
14. Sierra-Campos, E., M. A. Valdez-Solana, D. Matuz-Mares, I. Velazquez, and J. P. Pardo. 2009. Induction of morphological changes in Ustilago maydis cells by octyl gallate. *Microbiology* 155:604–611.
15. de Cordova, C. A. S., C. Locatelli, L. S. Assuncao et al. 2011. Octyl and dodecyl gallates induce oxidative stress and apoptosis in a melanoma cell line. *Toxicol. In Vitro* 25:2025–2034.
16. Hamishehkar, H., S. Khani, S. Kashanian, J. E. N. Dolatabadi, and M. Eskandani. 2014. Geno- and cytotoxicity of propyl gallate food additive. *Drug Chem. Toxicol.* 37:241–246.
17. Han, Y. H., H. J. Moon, B. R. You, and W. H. Park. 2010. Propyl gallate inhibits the growth of calf pulmonary arterial endothelial cells via glutathione depletion. *Toxicol. In Vitro* 24:1183–1189.
18. Dolatabadi, J. E. N. and S. Kashanian. 2010. A review on DNA interaction with synthetic phenolic food additives. *Food Res. Int.* 43:1223–1230.

19. Jacobi, H., B. Eicke, and L. Witte. 1998. DNA strand break induction and enhanced cytotoxicity of propyl gallate in the presence of copper(II). *Free Radicals Biol. Med.* 24:972–978.

20. Eler, G. J., I. S. Santos, A. G. de Moraes et al. 2013. Kinetics of the transformation of n-propyl gallate and structural analogs in the perfused rat liver. *Toxicol. Appl. Pharmacol.* 273:35–46.

21. Nakagawa, Y. and S. Tayama. 1995. Cytotoxicity of propyl gallate and related compounds in rat hepatocytes. *Arch. Toxicol.* 69:204–208.

22. Che, X. H., W. Y. Jiang, D. R. Parajuli, Y. Z. Zhao, S. H. Lee, and D. H. Sohn. 2012. Apoptotic effect of propyl gallate in activated rat hepatic stellate cells. *Arch. Pharmacal. Res.* 35:2205–2210.

23. Zurita, J. L., A. Jos, A. del Peso, M. Salguero, M. López-Artíguez, and G. Repetto. 2007. Ecotoxicological effects of the antioxidant additive propyl gallate in five aquatic systems. *Water Res.* 41:2599–2611.

24. Lin, Y. F., Y. H. Hu, H. T. Lin et al. 2013. Inhibitory effects of propyl gallate on tyrosinase and its application in controlling pericarp browning of harvested longan fruits. *J. Agric. Food Chem.* 61:2889–2895.

25. Xu, F., D. W. Zhang, J. H. Wang, et al. 2012. n-Propyl gallate is an inhibitor to tomato fruit ripening. *J. Food Biochem.* 36:657–666.

26. Barla, F., H. Hagashijima, S. Funai et al. 2009. Inhibitive effects of alkyl gallates on hyaluronidase and collagenase. *Biosci. Biotechnol. Biochem.* 73:2335–2337.

27. Kubo, I., N. Masuoka, P. Xiao, and H. Haraguchi. 2002. Antioxidant activity of dodecyl gallate. *J. Agric. Food Chem.* 50:3533–3539.

28. Klein, E. and N. Weber. 2001. *In vitro* test for the effectiveness of antioxidants as inhibitors of thiyl radical-induced reactions with unsaturated fatty acids. *J. Agric. Food Chem.* 49:1224–1227.

29. Ximenes, V. F., M. G. Lopes, M. S. Petronio, L. O. Regasini, D. H. S. Silva, and L. M. da Fonseca. 2010. Inhibitory effect of gallic acid and its esters on 2,2'-azobis(2-amidinopropane)hydrochloride (AAPH)-induced hemolysis and depletion of intracellular glutathione in erythrocites. *J. Agric. Food Chem.* 58:5355–5362.

30. Eler, G. J., R. M. Peralta, and A. Bracht. 2009. The action of n-propyl gallate on gluconeogenesis and oxygen uptake. *Chem. Biol. Interact.* 181:390–399.

31. Latha, R. C. R. and P. Daisy. 2013. Therapeutic potential of octyl gallate isolated from fruits of Terminalia bellerica in streptozotocin-induced diabetic rats. *Pharm. Biol.* 51:798–805.

32. Soares, D. G., A. C. Andreazza, and M. Salvador. 2003. Sequestering ability of butylated hydroxytoluene, propyl gallate, resveratrol, and vitamins C and E against ABTS, DPPH, and hydroxyl free radicals in chemical and biological systems. *J. Agric. Food Chem.* 51:1077–1080.

33. Locatelli, C., P. C. Leal, R. A. Yunes, R. J. Nunes, and T. B. Creczynski-Pasa. 2009. Gallic acid ester derivatives induce apoptosis and cell adhesion inhibition in melanoma cells: The relationship between free radical generation, glutathione depletion and cell death. *Chem. Biol. Interact.* 181:175–184.

34. Ortega, E., M. C. Sadabal, A. I. Ortiz et al. 2003. Tumoricidal activity of lauryl gallate towards chemically induced skin tumours in mice. *Br. J. Cancer* 88:940–943.

35. Calcabrini, A., J. M. García-Martínez, L. González et al. 2006. Inhibition of proliferation and induction of apoptosis in human breast cancer cells by lauryl gallate. *Carcinogenesis* 27:1699–1712.

36. Han, Y. H., H. J. Moon, B. R. You, S. Z. Kim, S. H. Kim, and W. H. Park. 2010. Propyl gallate inhibits the growth of HeLa cells via caspase-dependent apoptosis as well as a G1 phase arrest of the cell cycle. *Oncol. Rep.* 23:1153–1158.
37. Chen, C. H., W. C. Lin, C. N. Kuo, and F. J. Lu. 2011. Role of redox signaling regulation in propyl gallate-induced apoptosis of human leukemia cells. *Food Chem. Toxicol.* 49:494–501.
38. Han, Y. H. and W. H. Park. 2009. Propyl gallate inhibits the growth of HeLa cells via regulating intracellular GSH level. *Food Chem. Toxicol.* 47:2531–2538.
39. Sergediene, E., K. Jönsson, H. Szymusiak, B. Tyrakowska, I. M. C. M. Rietjens, and N. Cenas. 1999. Prooxidant toxicity of polyphenolic antioxidants to HL-60 cells: description and quantitative structure-activity relationships. *FEBS Lett.* 462:392–396.
40. Frey, C., M. Pavani, G. Cordano et al. 2007. Comparative cytotoxicity of alkyl gallates on mouse tumor cell lines and isolated rat hepatocytes. *Comp. Biochem. Physiol. Part A* 146:520–527.
41. Fujita, K. and I. Kubo. 2002. Antifungal activity of octyl gallate. *Int. J. Food Microbiol.* 79:193–201.
42. Ito, S., Y. Nakagawa, S. Yazawa, Y. Sasaki, and S. Yajima. 2014. Antifungal activity of alkyl gallates against plant pathogenic fungi. *Bioorg. Med. Chem. Lett.* 24:1812–1814.
43. Kubo, I., K. Fujita, and K. Nihei. 2002. Anti-Salmonella activity of alkyl gallates. *J. Agric. Food Chem.* 50:6692–6696.
44. Gutiérrez-Larraínzar, M., J. Rúa, D. de Arriaga, P. del Valle, and M. R. García-Armesto. 2013. *In vitro* assessment of synthetic phenolic antioxidants for inhibition of foodborne *Staphylococcus aureus, Bacillus cereus* and *Pseudomonas fluorescens. Food Control* 30:393–399.
45. Bosh-Jiménez, P., M. Lira-Cantu, C. Domingo, and J. A. Ayllón. 2012. Solution processable TiO$_2$ nanoparticles capped with lauryl gallate. *Mater. Lett.* 89:296–298.
46. Dzunuzovic, E. S., J. V. Dzunuzovic, T. S. Radoman et al. 2013. Characterization of in situ prepared nanocomposites of PS and TiO$_2$ nanoparticles surface modified with alkyl gallates: Effect of alkyl chain length. *Polym. Compos.* 34:399–407.
47. Kalaitzaki, A., M. Emo, M. J. Stebe, A. Xenakis, and V. Papadimitriou. 2013. Biocompatible nanodispersions as delivery systems of food additives: A structural study. *Food Res. Int.* 54:1448–1454.
48. Serrano, M., M. Martínez, and J. Aracil. 2013. Long term storage stability of biodiesel: influence of feedstock, commercial additives and purification step. *Fuel Process. Technol.* 116:135–141.
49. Rawat, D. S., G. Joshi, B. Y. Lamba, A. K. Tiwari, and S. Mallick. 2014. Impact of additives on storage stability of Karanja (Pongamia Pinnata) biodiesel blends with conventional diesel sold at retail outlets. *Fuel* 120:30–37.
50. Tang, H., A. Wang, S. O. Salley, and K. Y. S. Ng. 2008. The effect of natural and synthetic antioxidants on the oxidative stability of biodiesel. *J. Am. Oil Chem. Soc.* 85:373–382.
51. Xin, M., Y. Ma, K. Xu, and M. Chen. 2014. Gallate derivatives as antioxidant additives for polypropylene. *J. Appl. Polym. Sci.* 131:39850(1–7).
52. Hossain, K. M. G., M. Díaz González, J. M. Dagá Mobmany, and T. Tzanov. 2010. Effects of alkyl chain lengths of gallates upon enzymatic wool functionalization. *J. Mol. Catal. B: Enzym.* 67:231–235.

53. Capitán-Vallvey, L. F., M. C. Valencia, and E. Arana Nicolás. 2003. Simple resolution of butylated hydroxyanisole and n-propyl gallate in fatty foods and cosmetics samples by flow-injection solid-phase spectrophotometry. *Food Chem. Toxicol.* 68:1595–1599.

54. Capitán-Vallvey, L. F., M. C. Valencia, and E. Arana Nicolás. 2001. Monoparameter sensors for the determination of the antioxidants butylated hydroxyanisole and n-propyl gallate in foods and cosmetics by flow injection spectrophotometry. *Analyst* 126:897–902.

55. Capitán-Vallvey, L. F., M. C. Valencia, and E. Arana Nicolás. 2004. Solid-phase ultraviolet absorbance spectrophotometric multisensor for the simultaneous determination of butylated hydroxytoluene and co-existing antioxidants. *Anal. Chim. Acta* 503:179–186.

56. García-Jiménez, J. F., M. C. Valencia, and L. F. Capitán-Vallvey. 2007. Simultaneous determination of antioxidants, preservatives and sweetener additives in food and cosmetics by flow injection analysis coupled to a monolithic column. *Anal. Chim. Acta* 594:226–233.

57. Kang, J., L. Han, Z. Chen, J. Shen, J. Nan, and Y. Zhang. 2014. Sensitized chemiluminescence of 2-phenyl-4,5-di(2-furyl)-1H-imidazole/$K_3Fe(CN)_6$/propyl gallate system combining with solid-phase extraction for the determination of propyl gallate in edible oil. *Food Chem.* 159:445–450.

58. Ballesta-Claver, J., M. C. Valencia, and L. F. Capitán-Vallvey. 2011. Analysis of phenolic compounds in health care products by low-pressure liquid-chromatography with monolithic column and chemiluminescence detection. *Luminescence* 26:44–53.

59. Morales, M. D., M. C. González, A. J. Reviejo, and J. M. Pingarrón. 2005. A composite amperometric tyrosinase biosensor for the determination of the additive propyl gallate in foodstuffs. *Microchem. J.* 80:71–78.

60. Luque, M., A. Rios, and M. Valcárcel. 1999. A poly(vinyl chloride) graphite composite electrode for flow-injection amperometric determination of antioxidants. *Anal. Chim. Acta.* 395:217–223.

61. Riber, J., C. de la Fuente, M. D. Vázquez, M. L. Tascón, and P. Sánchez Batanero. 2000. Electrochemical study of antioxidants at a polypyrrole electrode modified by a nickel phthalocyanine complex. Application to their HPLC separation and to their FIA system detections. *Talanta* 52:241–252.

62. Chang, F. S., P. C. Chen, R. L. C. Chen, F. M. Lu, and T. J. Cheng. 2006. Real-time assay of immobilized tannase with a stopped-flow conductimetric device. *Bioelectrochemistry* 69:113–116.

63. Webster, G. K., R. A. Craig, C. A. Pommerening, and I. N. Acworth. 2012. Selection of pharmaceutical antioxidants by hydrodynamic voltammetry. *Electroanalysis* 24:1394–1400.

64. Karovicova, J. and P. Simko. 2000. Determination of synthetic phenolic antioxidants in food by high-performance liquid chromatography. *J. Chromatogr. A* 882:271–281.

65. Wang, J. Y., H. L. Wu, Y. Chen et al. 2012. Fast analysis of synthetic antioxidants in edible vegetable oil using trilinear component modeling of liquid chromatography-diode array detection data. *J. Chromatogr. A* 1264:63–71.

66. Xia, A. L., H. L. Wu, D. M. Fang, Y. J. Ding, L. Q. Hu, and R. Q. Yu. 2005. Alternating penalty trilinear decomposition algorithm for second-order calibration with application to interference-free analysis of excitation-emission matrix fluorescence data. *J. Chemom.* 19:65–76.

67. Wang, J. Y., H. L. Wu, Y. M. Sun et al. 2014. Simultaneous determination of phenolic antioxidants in edible vegetable oils by HPLC-FLD assisted with second-order calibration based on ATLD algorithm. *J. Chromatogr. B* 947–948:32–40.

68. Li, X. Q., C. Ji, Y. Y. Sun, M. L. Yang, and X. G. Chu. 2009. Analysis of synthetic antioxidants and preservatives in edible vegetable oil by HPLC/TOF-MS. *Food Chem.* 113:692–700.

69. Jia, W., Y. Ling, Y. Lin, J. Chang, and X. Chu. 2014. Analysis of additives in dairy products by liquid chromatography coupled to quadrupole-orbitrap mass spectrometry. *J. Chromatogr. A* 1336:67–75.

70. Sha, B. B., X. B. Yin, X. H. Zhang, X. W. He, and W. L. Yang. 2007. Capillary electrophoresis coupled with electrochemical detection using porous etched joint for determination of antioxidants. *J. Chromatogr. A* 1167:109–115.

71. Xiang, Q., Y. Gao, Y. Xu, and E. Wang. 2007. Capillary electrophoresis-amperometric determination of antioxidant propyl gallate and butyrated hydroxyanisole in foods. *Anal. Sci.* 23:713–717.

72. Huang, H. Y., Y. J. Cheng, and C. L. Lin. 2010. Analyses of synthetic antioxidants by capillary electrochromatography using poly(styrene-divinylbenzene-lauryl methacrylate) monolith. *Talanta* 82:1426–1433.

73. Darji, V., M. C. Boyce, I. Bennett, M. C. Breadmore, and J. Quirino. 2010. Determination of food grade antioxidants using microemulsion electrokinetic chromatography. *Electrophoresis* 31:2267–2271.

74. Delgado-Zamarreño, M. M., I. González-Maza, A. Sánchez-Pérez, and R. Carabias Martínez. 2007. Analysis of synthetic phenolic antioxidants in edible oils by micellar electrokinetic capillary chromatography. *Food Chem.* 100:1722–1727.

75. Guan, Y., Q. Chu, L. Fu, and J. Ye. 2005. Determination of antioxidants in cosmetics by micellar electrokinetic capillary chromatography with electrochemical detection. *J. Chromatogr. A.* 1074:201–204.

76. Guan, Y., Q. Chu, L. Fu, T. Wu, and J. Ye. 2006. Determination of phenolic antioxidants by micellar electrokinetic chromatography with electrochemical detection. *Food Chem.* 94:157–162.

77. Chen, M., Z. Tai, X. Hu, M. Liu, and Y. Yang. 2012. Utility of 4-chloro-7-nitrobenzofurazan for the spectrofluorimetric determination of butylated hydroxyanisole and propyl gallate in foodstuffs. *J. Food Sci.* 77: C401–C407.

78. Panadero, S., A. Gómez-Hens, and D. Pérez-Bendito. 1995. Kinetic determination of propyl gallate in edible and cosmetic oils with sensitized terbium(III) luminescence detection. *Analyst* 120:125–128.

79. Aguilar-Caballos, M. P., A. Gómez-Hens, and D. Pérez-Bendito. 2000. Simultaneous stopped-flow determination of butylated hydroxyanisole and propyl gallate using a T-format luminescence spectrometer. *J. Agric. Food Chem.* 48:312–317.

80. Wang, J. Y., H. L. Wu, Y. Chen, M. Zhai, X. D. Qing, and R. Q. Yu. 2013. Quantitative determination of butylated hydroxyanisole and n-propyl gallate in cosmetics using three-dimensional fluorescence coupled with second-order calibration. *Talanta* 116:347–353.

81. Aguilar-Caballos, M. P., A. Gómez-Hens, and D. Pérez-Bendito. 1997. Simultaneous kinetic determination of butylated hydroxyanisole and propyl gallate by coupling stopped-flow mixing technique and diode-array detection. *Anal. Chim. Acta* 354:173–179.

82. Apak, R., K. Güçlü, M. Özyürek, and S. E. Karademir. 2004. Novel total antioxidant capacity index for dietary polyphenols and vitamins C and E using their cupric

ion reducing capability in the presence of neocuproine: CUPRAC method. *J. Agric. Food Chem.* 52:7970–7981.

83. Celik, S. E., M. Özyürek, K. Güçlü, and R. Apak. 2007. CUPRAC total antioxidant capacity assay of lipophilic antioxidants in combination with hydrophilic antioxidants using the macrocyclic oligosaccharide methyl β-cyclodextrin as the solubility enhancer. *React. Funct. Polym.* 67:1548–1560.

84. Berker, K. I., F. A. O. Olgun, D. Ozyurt, B. Demirata, and R. Apak. 2013. Modified Folin–Ciocalteu antioxidant capacity assay for measuring lipophilic antioxidants. *J. Agric. Food Chem.* 61:4783–4791.

85. Andreu-Navarro, A., J. M. Fernández-Romero, and A. Gómez-Hens. 2011. Determination of antioxidant additives in foodstuffs by direct measurement of gold nanoparticle formation using resonance light scattering detection. *Anal. Chim. Acta* 695:11–17.

86. Calheiros, R., N. F. L. Machado, S. M. Fiuza et al. 2008. Antioxidant phenolic esters with potential anticancer activity: A Raman spectroscopy study. *J. Raman Spectrosc.* 39:95–107.

CHAPTER **15**

Determination of Phosphoric Acid and Phosphates

Yon Ju-Nam, Jesus J. Ojeda, and Hilda Ledo de Medina

CONTENTS

15.1 INTRODUCTION

A food additive is usually defined in the literature as an ingredient added to food as a part of its production (Furia, 1968; Branen et al., 1990). The use of food additives is normally regulated by the corresponding authorities and governmental organizations worldwide, through food-related framework directives. This set of laws concerns food additives authorized for use in foodstuffs intended for human intake. Food additives usually found in processed food are nutritional additives, preservatives, and flavoring, coloring, and texturing agents (Kritsunankul and Jakmunee, 2011). Some examples that can be mentioned include benzoic acid and its sodium salts, and sorbic acid and its potassium, calcium, and sodium salts, which are normally used as food preservatives with the objective of inhibiting microbial growth in foods (Branen et al., 1990). Phosphoric acid is added to noncarbonated and nonalcoholic carbonated beverages as an acidulant. Artificial sweeteners are also additives but are mostly used in low-calorie soft drinks and food (Branen et al., 1990). Several additives are also added to many ready-to-eat or instant food products.

The development of analytical methodologies for single and simultaneous determination of food additives in different food matrices is an area of great interest, especially to verify the maximum permitted levels of individual additives, and to maintain and ensure food quality and safety (Kritsunankul and Jakmunee, 2011). Although the research area of food additives and the analytical methodologies for their determination and monitoring is extensive, only phosphoric acid (E 338) and phosphates (E 339–343) will be covered in this chapter. The scope of this chapter will also be confined to the determination of these compounds from different food matrices by flow injection analysis (FIA) techniques.

15.1.1 Phosphoric Acid (E 338)

Food-grade phosphoric acid or orthophosphoric acid, E 338 (Table 15.1), is added to foods such as carbonated beverages, processed meat, chocolate, fats and oils, beer, jam, and sweets, to control acidity and also to enhance the antioxidant effects of other compounds present (Furia, 1973). Regarding the legislative framework for food additives in the European Union, each of the Member State is required to monitor their consumption and usage. The authorization and use of food additives in the European Union are based on the framework Directive 89/107/EEC (Report from the Commission on Dietary Food Additive Intake in the European Union, 1995). The objective is to ensure that their use does not exceed the acceptable daily intake (ADI) set for additives by the Scientific Committee on Food (SFC). On the basis of the framework directive, three specific directives are adopted by the Council and European Parliament. The Directive 94/35/EC corresponds to sweeteners, Directive 94/36/EC is for colors, and Directive 95/2/EC is for additives other than colors and sweeteners. According to the Report from the Commission on Dietary Food Additive Intake in the European Union, the ADI corresponding to E 338 is 70 mg/kg, for adults and young children (Report from the Commission on Dietary Food Additive Intake in the European Union, 1995). The ADI is based on 60 kg body weight for an adult and 15 kg for a young child.

Crystalline phosphoric acid has a hydrogen-bonded layer structure in which each molecule is attached to six others. Usually, impure phosphoric acid has its main application in fertilizer production, and in the synthesis of pure phosphoric acid (Van Wazer, 1953). However, the pure grade acid is used in food, detergents, pharmaceuticals, and metal treatment (e.g., pickling, cleaning, rust proofing, polishing). Phosphoric acid is the only inorganic acid widely used as food acidifier. This acid is the least expensive food-grade acidulant and is commercially supplied as 75%, 80%, and 85% aqueous solutions. These colorless solutions meet specifications required by the Food Chemical Codex (Nordic Council of Ministers, 2002).

Commercially available phosphoric acid is usually manufactured by one of two processes (Van Wazer, 1953). (i) The most cost-effective method consists of treating phosphorite with sulfuric acid. (ii) The other process involves the reduction of phosphorite to elementary phosphorus in an electric furnace or blast furnace. This consists of burning phosphorite in the presence of air in order to obtain phosphorus pentoxide, hydrating the oxide to produce up to 75%–85% phosphoric acid, and purifying the resulting product with hydrogen sulfide to eliminate arsenic.

As previously mentioned, in the food industry, phosphoric acid E 338 is widely used to acidify food products. E 338 is one of the strongest food-grade acidulants with the possibility of obtaining the lowest achievable pH (Gardner, 1973). This acid is mainly used in carbonated soft drinks, beer, and similar kinds of beverages, and in cheese production to adjust pH. E 338 is also employed as a yeast stimulant in the production of bread dough, in the clarification and acidification of collagen in the production of gelatin and jellies, and in the purification of vegetable oils (Gardner, 1973).

15.1.2 Phosphates (E 339–343)

Phosphates, E 339–343, are recognized as safe substances for their use in food production by the U.S. Food and Drug Administration (FDA) (Nordic Council of Ministers, 2002). Phosphates, such as monocalcium and dicalcium phosphate, and sodium phosphate, are

TABLE 15.1   Official Chemical Name, E Numbers, Synonyms, and Chemical Formulas of Food-Grade Phosphoric Acid and Phosphates

Chemical Name	E Number	Synonyms	Chemical Formula
Phosphoric acid	E 338	Orthophosphoric acid, monophosphoric acid	H_3PO_4
Monosodium phosphate	E 339 (i)	Sodium dihydrogen monophosphate, monosodium monophosphate, acid monosodium monophosphate, monobasic sodium phosphate	$NaH_2PO_4 \cdot nH_2O$ (n = 0, 1, or 2)
Disodium phosphate	E 339 (ii)	Disodium hydrogen monophosphate, disodium hydrogen orthophosphate, disodium monophosphate, secondary sodium phosphate, disodium orthophosphate, acid disodium phosphate	$Na_2HPO_4 \cdot nH_2O$ (n = 0, 2, 7, or 12)
Trisodium phosphate	E 339 (iii)	Trisodium monophosphate, trisodium phosphate, sodium phosphate, tribasic sodium phosphate, trisodium orthophosphate	$Na_3PO_4 \cdot nH_2O$ (n = 0, 0.5, 1, or 12)
Monopotassium phosphate	E 340 (i)	Potassium dihydrogen phosphate, monopotassium dihydrogen orthophosphate, monopotassium dihydrogen monophosphate, monobasic potassium phosphate, monopotassium monophosphate, potassium acid phosphate, potassium orthophosphate	KH_2PO_4
Dipotassium phosphate	E 340 (ii)	Dipotassium hydrogen monophosphate, dipotassium hydrogen phosphate, dipotassium hydrogen orthophosphate, dipotassium monophosphate, secondary potassium phosphate, dipotassium acid phosphate, dipotassium orthophosphate, dibasic potassium phosphate	K_2HPO_4
Tripotassium phosphate	E 340 (iii)	Tripotassium monophosphate, tripotassium phosphate, tripotassium orthophosphate, potassium phosphate, tribasic potassium phosphate	$K_3PO_4 \cdot nH_2O$ (n = 0, 1, or 3)
Monocalcium phosphate	E 341 (i)	Calcium dihydrogen phosphate, monobasic calcium phosphate, monocalcium orthophosphate	$Ca(H_2PO_4)_2 \cdot nH_2O$ (n = 0 or 1)
Dicalcium phosphate	E 341 (ii)	Calcium monohydrogen phosphate, calcium hydrogen orthophosphate, secondary calcium phosphate, dibasic calcium phosphate, dicalcium orthophosphate	$CaHPO_4 \cdot nH_2O$ (n = 0 or 2)
Tricalcium phosphate	E 341 (iii)	Tricalcium monophosphate, tribasic calcium phosphate, calcium orthophosphate	$Ca_3(H_2PO_4)_2$

(Continued)

TABLE 15.1 (*Continued*)   Official Chemical Name, E Numbers, Synonyms, and Chemical
Formulas of Food-Grade Phosphoric Acid and Phosphates

Chemical Name	E Number	Synonyms	Chemical Formula
Monomagnesium phosphate	E 343 (i)	Monomagnesium dihydrogen monophosphate, magnesium dihydrogen phosphate, monobasic magnesium phosphate, monomagnesium orthophosphate	$Mg(H_2PO_4)_2 \cdot nH_2O$ (n = 0–4)
Dimagnesium phosphate	E 343 (ii)	Dimagnesium hydrogen monophosphate, magnesium hydrogen phosphate, dibasic magnesium phosphate, dimagnesium orthophosphate	$MgHPO_4 \cdot nH_2O$ (n = 0–3)

Source: Data taken from Nordic Council of Ministers. 2002. *Food Additives in Europe 2000: Status of Safety Assessments of Food Additives Presently Permitted in the EU,* TemaNord 2002:560. Copenhagen: Ekspressen Kopi & Trykcenter.

phosphoric acid salts (Table 15.1). For instance, mono- and disodium phosphates are produced by treating phosphoric acid and sodium carbonate. The family of E 339–343 additives are used as food ingredients in baking powder and in self-rising flour production (Gardner, 1973). Calcium, magnesium, potassium, and sodium phosphates are employed as buffering agents and as optional emulsifying agents in the production of pasteurized cheese. Phosphates have been shown to prevent the formation of caries when added to foods in combination with fluorides and other calcium salts (Furia, 1973). According to the Report from the Commission on Dietary Food Additive Intake in the European Union, the ADI corresponding to E 339–343 is 70 mg/kg, for adults and young children (Report from the Commission on Dietary Food Additive Intake in the European Union, 1995).

Various uses of phosphates as food additives are linked to particular properties. Phosphates (E 339–343) are added to fresh and frozen meat muscle to retain moisture, prevent disintegration, enhance texture and flavor, help protein gelation, inhibit decay due to unwanted microbial growth, and avoid off-flavors and changes in color and odor (Dziezak, 1990; Strack, 1996; Chang and Regenstein, 1997; Weilmeier and Regenstein, 2004). The preservation of these qualities is linked to the property of phosphates to act as antioxidants. This property of phosphates and the mechanisms involved are still unclear; several published studies propose that this ability is based on their role as chelators or radical quenchers (Weilmeier and Regenstein, 2004).

Some studies have suggested that higher concentrations of sodium tripolyphosphate provide greater antioxidant effect in cooked minced beef and turkey (Trout and Dale, 1990; Craig et al., 1996). In contrast, in other reports it was suggested that there is no direct relationship between concentration of phosphates and oxidation when studies of phosphates antioxidant properties were done in frozen cooked mechanically recovered beef and pork (Huffman et al., 1987). These discrepancies in experimental observations might be due to differences in processing conditions and the various ways in which phosphates interact with different food matrices.

More recently, Weilmeier and Regenstein (2004) monitored lipid oxidation in raw mackerel and lake trout in order to acquire a better understanding of the antioxidant properties of phosphates. In this study, fish was treated with different antioxidants including several phosphates additives such as monosodium orthophosphate, disodium orthophosphate, disodium and tetrasodium pyrophosphate, sodium and potassium tripolyphosphate. In order to track the antioxidant activities of the phosphates, thiobarbituric acid-reactive substances (TBARS) were measured during different stages of the fish storage. Researchers observed that mackerel and lake trout, both fatty fish, behaved differently when treated with antioxidants (Weilmeier and Regenstein, 2004). Their work agrees with other reported studies, as they also observed that phosphates have the ability to inhibit oxidation in raw mackerel. However, this antioxidant effect was not as strong as that provided by other additives such as propyl gallate, ascorbic acid, or erythorbic acid. This experimental observation led Weilmeier and Regenstein (2004) to conclude that sodium tripolyphosphate would not be the antioxidant of choice to preserve mackerel. However, if this polyphosphate is employed for its antioxidant property and also for its other functions such as water retention in fish muscle, its use can be justified. It was also observed that smaller monovalent cation phosphates were better antioxidants (Weilmeier and Regenstein, 2004).

Food-grade phosphoric acid and phosphates have an important impact on various products' flavor and physical appearance. The measurement of their concentration, separately or as total phosphorus, in all phases of the food production can be used to track potential variations in pH, flavor, odor, and product appearance (Boyles, 1992).

In general, one of the most popular analytical methods used for the determination of food additives in food samples is high-performance liquid chromatography (HPLC) with various detection techniques (Tsang et al., 1985; Di Pietra et al., 1990; Pylypiw and Grether, 2000; Tfouni and Toledo, 2002; Garcia et al., 2003; Mota et al., 2003; Demiralay et al., 2006; Lino and Pena, 2010). These HPLC methods include UV (Ferreira et al., 2000; Wang et al., 2000; Saad et al., 2005; Dossi et al., 2006; Techakriengkrai and Surakarnkul, 2007), light scattering (Wasik et al., 2007), and spectrofluorimetric detection (Wrobel and Wrobel, 1997). Other analytical methods used to determine food additives and reported in the literature are ion chromatography with UV detection (Zhu et al., 2005), capillary electrophoresis with UV detection (Frazier et al., 2000), gas chromatography with flame ionization detection (Lin and Choong, 1999; Dong and Wang, 2006), and FIA coupled with a wide range of detection methods (Gonzalez et al., 1999; Lucena et al., 2005; Garcia-Jimenez et al., 2006, 2007; Han et al., 2008).

Several food sample pretreatment techniques for food additive analysis are required prior to the use of these analytical techniques. Filtration, dilution, centrifugation (Pylypiw and Grether, 2000; Demiralay et al., 2006; Wasik et al., 2007), offline dialysis (Kritsunankul and Jakmunee, 2011), liquid–liquid extraction (Saad et al., 2005), steam distillation (Ferreira et al., 2000) are some of the techniques used prior to the analyte analysis. These are used to remove particulate matter and minimize matrix interferences from food samples for the identification and quantification of the food additive (Kritsunankul and Jakmunee, 2011). As stated, the scope of the following sections will be the determination of phosphoric acid and phosphates from foodstuffs by the FIA technique.

15.2 FLOW INJECTION SYSTEMS

Constant development of analytical methods for measuring food additives to increase accuracy and minimize human interference in all analytical steps is the goal of a number of studies in food analysis. The flow injection system offers great advantages such as high sample throughput, low sample and reagent consumption, and the possibility to carry out a considerable number of chemical manipulations in a contamination-free system with satisfactory precision. The FIA technique is a type of continuous-flow analysis in which volumes of samples are injected in a precise manner using an analytical stream. This automatic analysis mode can also produce a fast, accurate, and exceptionally versatile system (Ranger, 1981).

In general, in a flow injection system, loading of standard and sample (sequential aliquots) into a sampler is assisted by the use of a pump. As the samples are aspirated, chemicals are simultaneously fed into the system by the pump. Then the samples are loaded by the use of an automatic injection valve. The sample and reagents are moved toward the reaction manifold where the chemical reaction occurs. The reaction product is then loaded into the flow and sent to a chosen/appropriate detector for generation of the analog signal (Ruzicka and Hansen, 1978).

Flow injection technique is based on injection or insertion of sample or standard solutions in volumes typically ranging from 20 to 100 μL into a flow cell (Karlberg and Pacey, 1989; Hansen and Miró, 2009). The technique of FIA is rigorous and depends on three factors: sample injection, reproducible timing, and controllable sample dispersion (Ranger, 1981). The creation of a concentration gradient of the injected sample and the precise timing of manipulation from the point of injection to the point of detection will result in reproducible signal. These peaks are the results of two kinetic processes such as dispersion (physical process) and reaction between the analyte and reagent species (chemical process) that occur at the same time (Ruzicka and Hansen, 1988; Hansen and Miró, 2009).

Correct design and downscaling of the normal flow injection system allow the incorporation of a variety of analytical techniques for the detection of different chemical species. Electrothermal atomic absorption spectrometry (Wang and Chen, 2008), inductively couple plasma atomic emission spectroscopy or mass spectrometry (Wang and Hansen, 2001), electrospray ionization mass spectrometry (Ogata et al., 2002, 2004), and chromatographic/electrophoretic column separation systems coupled to UV–Vis or mass spectrometric detection (Wu et al., 2003; Quintana et al., 2006) are among the detection and separation systems that have been coupled with flow injection devices.

In regard to phosphoric acid and phosphates, studies have shown that organic phosphate analysis is considered a difficult task because of their poor spectral characteristics and the lack of specific reagents (Woo and Maher, 1995; Benson et al., 1996; Parra et al., 2005). To resolve this problem, both direct and indirect methods based on complex analytical techniques have been used and reported in the literature. Direct methods include HPLC (Burbano et al., 1995; Sekiguchi et al., 2000), mid-infrared spectroscopy (Ishiguro et al., 2003), ^{31}P nuclear magnetic resonance spectroscopy (Turner et al., 2003), and inductively couple plasma mass spectrometry (Muñoz and Valiente, 2003). Indirect approaches are essentially based on the enzymatic hydrolysis of the organic phosphates. For instance, inositol phosphate, found in seeds and grains, undergoes enzymatic hydrolysis to generate *myo*-inositol and phosphate (Parra et al., 2005). These ions can be determined by gas chromatography (Dekoning, 1994) and UV–Vis spectrophotometry (Vieira and Nogueira, 2004), respectively.

There are few works reported in the literature regarding the use of flow injection systems to detect inorganic and organic phosphorus and total phosphorus from different food sources (Fernandes et al., 2000; Vieira and Nogueira, 2004; Parra et al., 2005). To improve flow injection methods, inorganic support materials have been used for enzymatic immobilization, which allows the use of smaller amounts of reagents and makes possible online enzymatic reaction (Lupetti et al., 2002). Considering the problems relating to phosphorus species determination, studies suggested that the use of an enzymatic reactor coupled to a flow injection system is suitable to determine the inorganic and total phosphorus. Therefore, flow injection based methodologies have been considered as good procedures for a simple, reliable, rapid, accurate and less time-consuming determination of phosphates in agricultural and food samples.

15.3 FLOW INJECTION ANALYSIS METHODOLOGIES FOR PHOSPHORIC ACID, PHOSPHATES, AND TOTAL PHOSPHORUS DETERMINATION

As mentioned in Section 15.2, few works can be found in the literature regarding the use of flow injection systems to detect inorganic and organic phosphorus and total phosphorus from different food sources, such as alcoholic drinks (Fernandes et al., 2000), cereals and grains (Vieira and Nogueira, 2004; Parra et al., 2005), vegetables (Lima et al., 1997), and mackerel and lake trout (Weilmeier and Regenstein, 2004). The following two sections are based on methodologies based on flow injection technique reported in the literature for the determination of total phosphorus (phosphoric acid and phosphates) in beer (Fernandes et al., 2000) and in vegetables (Lima et al., 1997). Although these methods are not directly related to the determination of E 338–343 as additives in food, they can be used as examples of flow injection methodologies adaptable to different food matrices.

15.3.1 Alcoholic Beverages

This section is based on the flow injection methodology developed for the analysis of total phosphorus in beer, reported by Fernandes et al. (2000) As those authors mentioned in their work, phosphorus-containing compounds (inorganic and organic phosphorus) from water used in the brewing process have a considerable impact on the final flavor and physical appearance of the product. The measurement of its concentration in all phases of beer production can be used to help track metabolic products of fermentation and correlate beer flavor trends. Prior to the analysis of total phosphorus, all phosphorus compounds present in the food matrix must be converted into an analyzable form such as orthophosphate (total phosphorus), and therefore, undergo a digestion procedure.

Fernandes et al. (2000) proposed an automatic method for continuous digestion with FIA to determine total phosphorus in beer. They developed a method that uses a two-stage photooxidation/thermal digestion procedure together with oxidizing and hydrolyzing reagents to convert all forms of phosphorus compounds (organically bound phosphates, condensed phosphates, and orthophosphates) into orthophosphates. In their procedure, organophosphates were digested by UV-catalyzed peroxodisulfate oxidation, and polyphosphates were hydrolyzed with a strong acid and heat. The resulting orthophosphate was reacted with phosphomolybdenum blue using stannous chloride as reducing agent and then analyzed by a UV–Vis spectrophotometric technique. According to the authors, the UV–thermal-induced digestion methodology gave recoveries greater than 90% for all compounds tested. Their results were reproducible and comparable to those of conventional digestion methods.

The following chemicals, reagents and samples are required to carry out the experiments designed by Fernandes et al. (2000):

1. 2 M sulfuric acid and 6 g/L potassium peroxodisulfate solutions are required for the online digestion.
2. Stannous chloride solution and a color reagent based on a 9.4 g/L ammonium molybdate solution (color reaction).
3. Sodium trimethylphosphate solutions (phosphorus standards) of 5.00–20.00 mg/L are required to be prepared daily for the calibration curve.
4. Beer samples are degassed in an ultrasonic bath for 10 min, and diluted 50 fold prior to their introduction into the flow injection system.

A flow injection manifold designed by Fernandes et al. (2000) for the analysis of total phosphorus in beer comprises of a digestion unit and a spectrophotometric determination unit (Figure 15.1). Digestion of the beer sample is carried out inside the flow tubes in the UV/thermal digestion unit. In this unit, condensed phosphates are hydrolyzed by acid and heat, and organophosphates are digested by UV-catalyzed peroxodisulfate oxidation. These are achieved by continuous online mixing of the phosphorus standards or

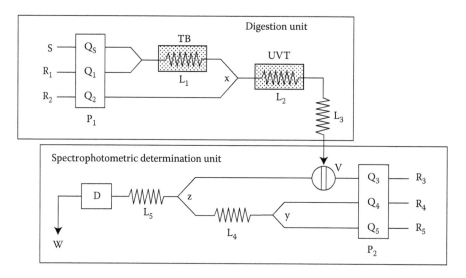

FIGURE 15.1 Diagram of the flow injection system developed by Fernandes et al. (2000) for the analysis of total phosphorus in beer samples. Conditions and components used by the authors are the following—flow rates: $Q_1 = Q_2 = 0.23$ mL/min and $Q_3 = Q_4 = Q_5 = 0.60$ mL/min; reagent solutions: $R_1 = 2$ M sulfuric acid, $R_2 = 6$ g/L potassium peroxodisulfate, $R_3 = 0.25$ M sulfuric acid, $R_4 = 9.4$ g/L ammonium molybdate $+ 0.65$ M sulfuric acid, $R_5 = 0.17$ g/L stannous chloride $+ 2$ g/L hydrazinium sulfate $+ 0.5$ M sulfuric acid; beer samples or standards: S; injection volume of 40 µL: V; tube lengths: $L_1 = 200$, $L_2 = 400$, $L_3 = L_4 = 100$, and $L_5 = 200$ cm; thermostatic bath at 90°C: TB; ultraviolet radiation tube: UVT; peristaltic pumps: P_1 and P_2; detector: D; confluence points: x, y, z; waste: W. (With kind permission from Springer Science+Business Media: *Fresenius J. Anal. Chem.*, Spectrophotometric flow injection determination of total phosphorus in beer using on-line UV/thermal induced digestion, 366, 2000, 112–115, Fernandes, S. M. V., J. L. F. C. Lima, and A. O. S. S. Rangel.)

beer diluted samples (S) with 2 M sulfuric acid (R_1). The resulting digested solution is streamed to a 2 m coil (L_1) immersed in a thermostatic bath set at 90°C. Then, the heated solution is combined with the peroxodisulfate solution (R_2) and streamed through a 4 m coil (L_2) that is wrapped around the UV tube. Then, the color reaction is carried out in the color development manifold. 40 μL of the digested solution is injected into a sulfuric acid carrier stream and mixed with the color (R_4) and reducing (R_5) reagents (already combined at confluence point y) at point z. The resulting phosphomolybdenum blue complex is finally monitored at a specific wavelength of 710 nm (D) (Fernandes et al., 2000).

Fernandes et al. (2000) concluded that flow injection online digestion is a reliable method for the rapid determination of total phosphorus in beer samples. Each sample analysis can be carried out in about 7 or 8 min. This efficient methodology can easily offer a reduction in the sample pretreatment time (many hours for manual digestion methodologies) and laboratory space (Fernandes et al., 2000). This work shows the versatility of the flow injection system for online analysis applications and suggests that this approach might be suitable for the determination of total phosphorus in other food matrices.

15.3.2 Vegetables

This section is based on one of the publications by Lima et al. (1997) in which they showed the possibility of using FIA system for the sequential determination of total nitrogen and phosphorus in digested vegetables using potentiometric and spectrophotometric detection systems, respectively. Lima et al. (1997) quantified total phosphorus by using the molybdenum blue method. The total phosphorus was determined spectrophotometrically, at the wavelength of 400 nm, as the yellow phosphovanadomolybdate complex formed in acidic solution from the reaction of orthophosphate with ammonium molybdate and ammonium vanadate. These authors opted for the potentiometric determination of nitrogen (in its ammonium form) instead of the spectrophotometric detection, as it requires less sample manipulation and allows nitrogen determination in a much wider concentration range. This was the main reason for the presence of two detection systems within this flow injection system, which allowed the simultaneous detection of total nitrogen and phosphorus from the same sample of vegetable.

The vegetables used for this study were lettuce, watercress, spinach, turnip, sprout, turnip leaf, and parsley. The sample preparation method used by the authors (Lima et al., 1997) was the following.

1. The vegetable samples were initially oven dried (at 80–100°C) and ground.
2. Acid digestions of the pulverized vegetable samples were carried out using sulfuric acid, salicylic acid, hydrogen peroxide, and selenium as the main reagents.
3. The samples were filtered through filter paper and then used for analysis or stored at 4°C.

Lima et al. (1997) also measured the concentration of phosphorus by a reference method in order to verify the reproducibility of the developed methodology and establish comparisons between them.

The flow injection configuration that Lima et al. (1997) used for this study is shown in Figure 15.2. Samples (or standards) are injected into a 0.8 mol/L H_2SO_4 carrier stream (R_2). A 2.5 mol/L hydroxide solution (R_3) is then streamed in the confluence point α in order to achieve the conversion of ammonium into volatile ammonia. This gas diffuses across the gas-permeable membrane in the gas-diffusion unit (GDU) and is reconverted to ammonium using a buffer solution (0.01 mol/L Tris-HCl pH 7.5), and then sent to the acceptor stream. After stream splitting and the addition of molybdate and stannous chloride solutions, the resulting phosphomolybdenum blue complex (measured at 710 nm) is quantified over the sample plug remaining in the donor stream. The total nitrogen determination is then carried out in the tubular ammonium electrode placed immediately after the GDU (Lima et al., 1997).

Lima et al. (1997) selected a volume of injection of 180 μL, and also found that increasing the volume produced only a slight increase of the magnitude of the potentiometric signals but had the disadvantage of lowering the upper limit of linear response for the colorimetric phosphate determination. However, decreasing the volume translated into poor reproducibility of the potentiometric measurements. In terms of the composition of the flowing solutions, the sensitivity of the analysis of total phosphorus by spectrophotometric detection depends on the acidity of the medium. Thus, a 0.8 mol/L H_2SO_4 solution was used as carrier stream (R_2) in order to ensure similar acidity in L_8. The acidity of the digests was taken into account in the selection of this concentration. The concentration of the sodium hydroxide solution (R_3) was dependent on the final

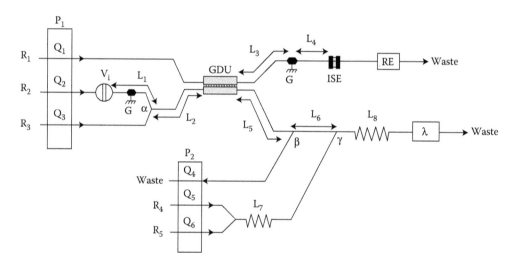

FIGURE 15.2 Diagram of the flow injection system developed by Lima et al. (1997) for the simultaneous determination of total nitrogen and phosphorus in vegetable digests. Conditions and components used by the authors are the following—flow rates: $Q_1 = 3.0$ mL/min, $Q_2 = Q_3 = 1.5$ mL/min, $Q_4 = 2.4$ mL/min, $Q_5 = Q_6 = 1.6$ mL/min; reagent solutions: $R_1 = 0.01$ mol/L tris-HCl pH $7.5 + 1 \times 10^{-6}$ mol/L NH_4^+, $R_2 = 0.8$ mol/L H_2SO_4, $R_3 = 2.5$ mol/L NaOH, $R_4 = 2.82$ g/L ammonium molybdate $+ 0.52$ mol/L H_2SO_4, $R_5 = 0.153$ g/L stannous chloride $+ 1.6$ g/L hydrazine sulfate $+ 0.38$ mol/L H_2SO_4; injection volume (180 μL): V_i; tube lengths: $L_1 = 16$, $L_2 = 50$, $L_3 = 19$, $L_4 = 13$, $L_5 = 20$, $L_6 = 22$, $L_7 = 105$, and $L_8 = 200$ cm; peristaltic pumps: P_1 and P_2; gas-diffusion unit: GDU; ground electrodes: G; tubular ammonium selective electrode: ISE; reference electrode: RE; spectrophotometer (set at 710 nm): λ; and confluence points: α, β, γ. (With kind permission from Springer Science+Business Media: *Fresenius J. Anal. Chem.*, Flow injection system with gas diffusion for the sequential determination of total nitrogen and phosphorus in vegetables, 358, 1997, 657–662, Lima, J. L. F. C., A. O. S. S. Rangel, and M. R. S. Souto.)

acidity of the digests. The hydroxide concentration was also considered, as this should be sufficient to increase the pH to a value that allowed quantitative conversion of ammonium into ammonia gas. A 2.5 mol/L NaOH solution was selected, as the sensitivity of the potentiometric measurement increased up to 2.5 mol/L of sodium hydroxide (Lima et al., 1997).

Adjustments of concentrations and ratios of the color development solutions (R_4 and R_5) for the phosphate determination were also made by Lima et al. (1997). Ammonium heptamolybdate 2.82 g/L with 0.52 mol/L H_2SO_4 (R_4) was chosen as a compromise between sensitivity and the linear working range. The concentration selected for the reducing reagent, stannous chloride, was 0.153 g/L (for maximum sensitivity). After several trials, maximum sensitivity and linearity were obtained for the 0.38 mol/L sulfuric acid. Under these optimized conditions of the flow injection system, the detection limit of the phosphate methodology, corresponding to three-times the standard deviation of the system background noise, was 4.1 mg/L of phosphate (Lima et al., 1997).

The results obtained in this work showed once again the versatility of flow injection systems, allowing the development of a flow injection manifold for simultaneous determination of total nitrogen and phosphorus in digested vegetable samples as Lima et al. (1997) achieved, and also to assume that this approach can be applied to other food matrices and materials for simultaneous quantification of these elements.

REFERENCES

Benson, R., I. McKelvie, B. Hart, Y. Truong, and I. Hamilton. 1996. Determination of total phosphorus in waters and wastewaters by on-line UV/thermal induced digestion and flow injection analysis. *Anal. Chim. Acta* 326:29–39.

Boyles, S. 1992. Method for the analysis of inorganic and organic acid anions in all phases of beer production using gradient ion chromatography. *J. ASBC* 50:61–63.

Branen, A. L., P. M. Davidson, and S. Salminen. 1990. *Food Additives*. New York: Marcel Dekker, Inc.

Burbano, C., M. Muzquiz, A. Osagie, G. Ayet, and C. Cuadrado. 1995. Determination of phytate and lower inositol phosphates in Spanish legumes by HPLC methodology. *Food Chem.* 52:321–325.

Chang, C. C. and J. M. Regenstein. 1997. Water uptake, protein solubility, and protein changes of cod mince stored on ice as affected by polyphosphates. *J. Food Sci.* 62:305–309.

Craig, J., J. A. Bowers, X. Y. Wang, and P. Seib. 1996. Inhibition of lipid oxidation in meats by inorganic phosphate and ascorbate salts. *J. Food Sci.* 61:1062–1067.

Dekoning, A. J. 1994. Determination of myo-inositol and phytic acid by gas cromatography using scillitol as internal standard. *Analyst* 119:1319–1323.

Demiralay, E. C., G. Ozkan, and Z. Guzel-Seydim. 2006. Isocratic separation of some food additives by reversed phase liquid chromatography. *Chromatographia* 63(1–2): 91–96.

Di Pietra, A. M., V. Cavrini, D. Bonazzi, and L. Benfenati. 1990. HPLC analysis of aspartame and saccharin in pharmaceutical and dietary formulations. *Chromatographia* 30:215–218.

Dong, C. and W. Wang. 2006. Headspace solid-phase microextraction applied to the simultaneous determination of sorbic and benzoic acids in beverages. *Anal. Chim. Acta* 562:23–29.

Dossi, N., R. Toniolo, S. Susmel, A. Pizzarello, and G. Bontempelli. 2006. Simultaneous RP-LC determination of additives in soft drinks. *Chromatographia* 63(11–12):557–562.

Dziezak, J. D. 1990. Phosphates improve many foods. *Food Technol.* 44(4):80–92.

Fernandes, S. M. V., J. L. F. C. Lima, and A. O. S. S. Rangel. 2000. Spectrophotometric flow injection determination of total phosphorus in beer using on-line UV/thermal induced digestion. *Fresenius J. Anal. Chem.* 366:112–115.

Ferreira, I. M. P. L. V. O., E. Mendes, P. Brito, and M. A. Ferreira. 2000. Simultaneous determination of benzoic and sorbic acids in quince jam by HPLC. *Food Res. Int.* 33(2):113–117.

Frazier, R. A., E. L. Inns, N. Dossi, J. M. Ames, and H. E. Nursten. 2000. Development of a capillary electrophoresis method for the simultaneous analysis of artificial sweeteners, preservatives and colours in soft drinks. *J. Chromatogr. A* 876(1):213–220.

Furia, T. E. 1973. *CRC Handbook of Food Additives.* (2nd ed.), Vol. I. Boca Raton, FL: CRC Press, Inc.

Garcia, I., Ortiz, M.C., Sarabia, L., Vilches, C., and Gredilla, E. 2003. Advances in methodology for the validation of methods according to the International Organization for Standardization. Application to the determination of benzoic and sorbic acids in soft drinks by high-performance liquid chromatography. *J. Chromatogr. A* 992(1–2):11–27.

Garcia-Jimenez, J. F., M. C. Valencia, and L. F. Capitan-Vallvey. 2006. Improved multianalyte determination of the intense sweeteners aspartame and acesulfame-K with a solid sensing zone implemented in an FIA scheme. *Anal. Lett.* 39(7):1333–1347.

Garcia-Jimenez, J. F., M. C. Valencia, and L. F. Capitan-Vallvey. 2007. Simultaneous determination of antioxidants, preservatives and sweetener additives in food and cosmetics by flow injection analysis coupled to a monolithic column. *Anal. Chim. Acta* 594(2):226–233.

Gardner, H. Wm. 1973. Acidulants in food processing. In *CRC Handbook of Food Additives* (2nd ed.), ed. T. E. Furia, Vol. I, pp. 225–270. Boca Raton, FL: CRC Press.

Gonzalez, M., M. Gallego, and M. Valcarcel. 1999. Gas chromatographic flow method for the preconcentration and simultaneous determination of antioxidant and preservative additives in fatty foods. *J. Chromatogr. A* 848(1–2):529–536.

Han, F., Y. Z. He, L. Li, G. N. Fu, H. Y. Xie, and W. E. Gan. 2008. Determination of benzoic acid and sorbic acid in food products using electrokinetic flow analysision pair solid phase extraction-capillary zone electrophoresis. *Anal. Chim. Acta* 618(1):79–85.

Hansen, E. H. and M. Miró. 2009. Flow injection analysis in industrial biotechnology. In *Encyclopedia of Industrial Biotechnology*, pp. 1–16. John Wiley & Sons, Inc.

Huffman, D. L., C. F. Ande, J. C. Cordray, M. H. Stanley, and W. R. Egbert. 1987. Influence of polyphosphate on storage stability of restructured beef and pork nuggets. *J. Food Sci.* 52:275–278.

Ishiguro, T., T. Ono, K. Nakasato, C. Tsukamoto, and S. Shimada. 2003. Rapid measurement of phytate in raw soymilk by mid-infrared spectrometry. *Biosci. Biotechnol. Biochem.* 67:752–757.

Karlberg, B. and G. E. Pacey. 1989. *Flow Injection Analysis—A Practical Guide.* BV Amsterdam: Elsevier.

Kritsunankul, O. and J. Jakmunee. 2011. Simultaneous determination of some food additives in soft drinks and other liquid foods by flow injection on-line dialysis coupled to high performance liquid chromatography. *Talanta* 84:1342–1349.

Lima, J. L. F. C., A. O. S. S. Rangel, and M. R. S. Souto. 1997. Flow injection system with gas diffusion for the sequential determination of total nitrogen and phosphorus in vegetables. *Fresenius J. Anal. Chem.* 358:657–662.

Lin, H. J. and Y. M. Choong. 1999. A simple method for the simultaneous determination of various preservatives in liquid foods. *J. Food Drug Anal.* 7(4):291–304.

Lino, C. M. and A. Pena. 2010. Occurrence of caffeine, saccharin, benzoic acid and sorbic acid in soft drinks and nectars in Portugal and subsequent exposure assessment. *Food Chem.* 121:503–508.

Lucena, R., S. Cardenas, M. Gallego, and M. Valcarcel. 2005. Continuous flow autoanalyzer for the sequential determination of total sugars, colorant and caffeine contents in soft drinks. *Anal. Chim. Acta* 530(2):283–289.

Lupetti, K. O., I. C. Vieira, and O. Fatibello-Filho. 2002. Flow injection spectrophotometric determination of isoproterenol using an avocado (*Persea americana*) crude extract immobilized on controlled-pore silica reactor. *Talanta* 57:135–143.

Mota, F. J. M., I. M. P. L. V. O. Ferreira, S. C. Cunha, M. Beatriz, and P. P. Oliveira. 2003. Optimisation of extraction procedures for analysis of benzoic and sorbic acids in foodstuffs. *Food Chem.* 82(3):469–473.

Muñoz, J. A. and M. Valiente. 2003. Determination of phytic acid in urine by inductively coupled plasma mass spectrometry. *Anal. Chem.* 75:6374–6378.

Nordic Council of Ministers. 2002. *Food Additives in Europe 2000: Status of Safety Assessments of Food Additives Presently Permitted in the EU, TemaNord 2002:560.* Copenhagen: Ekspressen Kopi & Trykcenter.

Ogata, Y., L. Scampavia, T. L. Carter, E. Fan, and F. Turecek. 2004. Automated affinity chromatography measurements of compound mixtures using a lab-on-valve apparatus coupled to electrospray ionization mass spectrometry. *Anal. Biochem.* 331:161–168.

Ogata, Y., L. Scampavia, J. Ruzicka, C. R. Scott, M. H. Gelb, and F. Turecek. 2002. Automated affinity capture-release of biotin-containing conjugates using a lab-on-valve apparatus coupled to UV/visible and electrospray ionization mass spectrometry. *Anal. Chem.* 74:4702–4708.

Parra, A., M. Ramon, J. Alonso, S. G. Lemos, E. C. Vieira, and A. R. A. Nogueira. 2005. Flow injection potentiometric system for the simultaneous determination of inositol phosphates and phosphate: Phosphorus nutritional evaluation on seeds and grains. *J. Agric. Food Chem.* 53:7644–7648.

Pylypiw, H. M. Jr. and M. T. Grether. 2000. Rapid high-performance liquid chromatography method for the analysis of sodium benzoate and potassium sorbate in foods. *J. Chromatogr. A* 883:299–304.

Quintana, J. B., M. Miró, J. M. Estela, and V. Cerdà. 2006. Automated on-line renewable solid-phase extraction-liquid chromatography exploiting multisyringe flow injection-bead injection-lab on valve (MSFI-BI-LOV) analysis. *Anal. Chem.* 78:2832–2840.

Ranger, C. 1981. Flow injection analysis: Principles, techniques, applications, design. *Anal. Chem.* 53(1):20–32.

Report from the Commission on Dietary Food Additive Intake in the European Union. 1995.

Ruzicka, J. and E. H. Hansen. 1978. Flow injection analysis part X, theory, technique and trends. *Anal. Chim. Acta.* 99:37–76.

Ruzicka, J. and E. H. Hansen. 1988. *Flow Injection Analysis* (2nd ed.). New York: Wiley-Interscience.

Saad, B., Md. F. Bari, M. I. Saleh, K. Ahmad, and M. K. M. Talib. 2005. Simultaneous determination of preservatives (benzoic acid, sorbic acid, methylparaben and propylparaben) in foodstuffs using high performance liquid chromatography. *J. Chromatogr. A* 1073:393–397.

Sekiguchi, Y., A. Matsunaga, A. Yamamoto, and Y. Inoue. 2000. Analysis of condensed phosphates in food products by ion chromatography with an on-line hydroxide eluent generator. *J. Chromatogr. A* 881:639–644.

Strack, H. J. 1996. Phosphates. Key ingredients in meat products. Food Technol. 26(3):92–95.

Techakriengkrai, I. and R. Surakarnkul. 2007. Analysis of benzoic acid and sorbic acid in Thai rice wines and distillates by solid-phase sorbent extraction and high-performance liquid chromatography. *J. Food Compos. Anal.* 20:220–225.

Tfouni, S. A. V. and M. C. F. Toledo. 2002. Determination of benzoic and sorbic acids in Brazilian food. *Food Control* 13:117–123.

Trout, G. R. and S. Dale. 1990. Prevention of warmed-over flavor in cooked beef: Effect of phosphate type, phosphate concentration, a lemon juice/phosphate blend, and beef extract. *J. Agric. Food Chem.* 38:665–669.

Tsang, W. S., M. A. Clarke, and F. W. Parrish. 1985. Determination of aspartame and its breakdown products in soft drinks by reverse-phase chromatography with UV detection. *J. Agric. Food Chem.* 33:734–738.

Turner, B. L., N. Mahieu, and L. M. Condron. 2003. Quantification of myo-inositol hexakisphosphate in alkaline soil extracts by solution 31P NMR spectrometry and spectral deconvolution. *Soil Sci.* 168:469–478.

Van Wazer, J. R. 1953. Phosphoric acid and phosphates. In *Encyclopedia of Chemical Technology* (1st ed.), eds. R. E. Kirk and D. F. Othner, Vol. 10. New York: Interscience Encyclopedia, Inc.

Vieira, E. C. and A. R. A. Nogueira. 2004. Orthophosphate, phytate, and total phosphorus determination in cereals by flow injection analysis. *J. Agric. Food Chem.* 52(7):1800–1803.

Wang, H., K. Helliwell, and X. You. 2000. Isocratic elution system for the determination of catechins, caffeine and gallic acid in green tea using HPLC. *Food Chem.* 68:115–121.

Wang, J.-H. and X.-W. Chen. 2008. Life sciences applications. *Comprehensive Analytical Chemistry, Vol. 54: Advances in Flow Injection Analysis and Related Techniques,* eds. S. D. Kolev and I. D. Mckelvie, pp. 559–590. Amsterdam: Elsevier.

Wang, J.-H. and E.H. Hansen. 2001. Interfacing sequential injection on-line preconcentration using a renewable micro-column incorporated in a "lab-on-valve" system with direct injection nebulization inductively coupled plasma mass spectrometry. *J. Anal. At. Spectrom.* 16:1349–1355.

Wasik, A., J. McCourt, and M. Buchgraber. 2007. Simultaneous determination of nine intense sweeteners in foodstuffs by high performance liquid chromatography and evaporative light scattering detection—Development and single-laboratory validation. *J. Chromatogr. A* 1157(1–2):187–196.

Weilmeier, D. M. and J. M. Regenstein. 2004. Antioxidant properties of phosphates and other additives during the storage of raw mackerel and lake trout. *Food Chem. Toxicol.* 69(2):FTC102–FTC108.

Woo, L. and W. Maher. 1995. Determination of phosphates in turbid waters using alkaline potassium peroxodisulphate digestion. *Anal. Chim. Acta* 315:123–135

Wrobel, K. and K. Wrobel. 1997. Determination of aspartame and phenylalanine in diet soft drinks by high- performance liquid chromatography with direct spectrofluorimetric detection. *J. Chromatogr. A* 773(1–2):163–168.

Wu, C.-H., L. Scampavia, and J. Ruzicka. 2003. Micro sequential injection: Automated insulin derivatization and separation using a lab-on-valve capillary electrophoresis system. *Analyst* 128:1123–1130.

Zhu, Y., Y. Guo, M. Ye, and F. S. James. 2005. Separation and simultaneous determination of four artificial sweeteners in food and beverages by ion chromatography. *J. Chromatogr. A* 1085(1):143–146.

CHAPTER **16**

Determination of Lactates

Wataru Yoshida and Isao Karube

CONTENTS

16.1 INTRODUCTION

Lactate is produced as a metabolic intermediate in humans and is a biomarker of human performance levels (Shah et al., 2007). In the food industry, sodium, potassium, and calcium lactates (E number: E325, E326, and E327, respectively) have been approved as food additives, and are used to regulate pH and as preservatives and flavoring agents. Lactate can also be used as a marker of fermentation (Avramescu et al., 2002) and a chemical indicator of flavor (Hemmi et al., 1995). Thus, determination of lactate concentrations is important not only for food process control but also for quality control of foods. In this chapter, we review flow injection analysis (FIA) for lactate determinations.

16.2 LACTATE BIOSENSORS

Biosensors are assembled using a combination of molecular recognition elements and transducers that detect recognition signals (Nakamura and Karube, 2003). Molecular recognition elements include enzymes, antibodies, peptides, aptamers, and cells. However, among these, enzymes are the most useful because they recognize targets and generate signals via enzymatic reactions such as redox reaction. In combination with transducers, oxidase and dehydrogenase enzymes find great utility in biosensor development because electrochemical signals from enzymatic reactions are easily detected using electrodes. Blood glucose sensors are particularly successful biosensors for the diagnosis of diabetes. These sensors are constructed by immobilizing glucose oxidase (GOD) or glucose dehydrogenase (GDH) on an electrode, allowing determination of blood glucose concentration within seconds of adding patient blood to the sensor chip. Similarly, lactate biosensors have been developed by immobilizing lactate oxidase (LOD) (Williams et al., 1970) or lactate dehydrogenase (LDH) on electrodes (Matsunaga et al., 1982), and a LOD-based biosensor has been commercialized. However, this lactate biosensor is not suitable for food analyses because it utilizes a disposable sensor chip. Analyses of food samples require a biosensing system that can be used to automatically analyze multiple food samples.

16.3 FLOW INJECTION ANALYSES FOR LACTATE

Flow injection analysis offers suitable systems for food sample analyses and can accommodate multiple samples. Moreover, sample pretreatment processes can be performed using flow systems in combination with FIA. Therefore, FIA systems have been developed for analyses of lactate in food using LOD and LDH. The enzymatic reactions of LOD and LDH are as follows:

$$\text{Lactate} + O_2 \xrightarrow{\ LOD\ } \text{Pyruvate} + H_2O_2$$

$$\text{Lactate} + NAD^+ \underset{\ }{\overset{LDH}{\rightleftharpoons}} \text{Pyruvate} + NADH + H^+$$

These reactions can be monitored by detecting hydrogen peroxide (H_2O_2) and NADH using electrochemical or optical methods, which have been incorporated into FIA systems.

Simple FIA for lactate sensing are based on incorporation of LOD- or LDH-modified electrodes into flow injection systems. FIA based on enzyme reactors that contain LOD or LDH have also been developed for electrochemical or optical lactate sensing. Other enzyme electrodes or reactors can also be incorporated into lactate FIA systems, leading to the development of FIA systems that simultaneously detect lactate and other target molecules (Renneberg et al., 1991). In addition, miniaturized flow cell that can be incorporated into FIA system (Nakamura et al., 2001) and enzyme thermistor–based FIA has been developed (Chen et al., 2011).

16.3.1 FIA Based on Enzyme Electrodes

Target molecules can be detected by incorporating enzyme electrodes into flow injection systems (Table 16.1). Pandey (1994) reported a graphite paste electrode that was modified with LDH and NAD+ using electroactive tetracyanoquinodimethane (TCNQ) as a mediator (Pandey, 1994). Direct oxidation of NADH on the electrode surface requires high operation potential, which induces side reactions. Hence, the mediator was coimmobilized on the electrode to reduce operation potential and the electrode was integrated into a flow system and was maintained at 0.2 V versus Ag/AgCl. Lactate was successfully detected using FIA, with linear range up to 4 mM and response times of 20–40 s, and 140 analyses were achieved without loss of response. A screen-printed carbon electrode modified with Meldola's Blue–Reinecke salt (MBRS–SPCE), LDH, and NAD+ was developed for lactate sensing in FIA (Piano et al., 2010). In this apparatus, MBRS–SPCE reduced operation potential to +0.05 V versus Ag/AgCl. In the corresponding FIA, the detection limit for lactate was 0.55 mM and the linear range was 0.55–10 mM.

Lactate oxidase–modified electrodes have also been used in FIA for lactate. In particular, a platinum electrode was modified with LOD using poly-o-phenylenediamine (PPD),

TABLE 16.1 Characteristics of Enzyme Electrode Based FIA Systems

Enzyme	Coimmobilized Reagent	Detection Limit	Linear Range	Sample Volume (μL)	Reference
LDH	TCNQ	N/A	up to 4 mM	70	Pandey (1994)
LDH	MBRS–SPCE	0.55 mM	0.55–10 mM	65	Piano et al. (2010)
LOD	PPD	2 μM	up to 0.2 mM	110	Palmisano et al. (1994)
LOD	2,6-DHN and AP-EA	0.01 mM	0.01–0.3 mM	50	Badea et al. (2003)
LOD	HRP and ferrocene	4 μM	4–40 μM	100	Zaydan et al. (2004)
LOD	Prussian Blue	0.5 μM	0.5 μM–1 mM	50	Yashina et al. (2010)

Note: LDH, lactate dehydrogenase; LOD, lactate oxidase; TCNQ, tetracyanoquinodimethane; MBRS–SPCE, Meldola's Blue–Reinecke salt; PPD, poly-*o*-phenylenediamine; 2,6-DHN, 2,6-dihydroxynaphthalene; AP-EA, 2-(4-aminophenyl)-ethylamine; HRP, horseradish peroxidase.

and was assembled using *in situ* electrochemical immobilization methods in a flow system (Palmisano et al., 1994). PPD reduces interference from ascorbate, urate, cysteine, and acetaminophen, leading to FIA detection limits for lactate of 2 μM and a linear range of up to 0.2 mM. Badea et al. (2003) reported LOD modification of a platinum electrode using 2,6-dihydroxynaphthalene (2,6-DHN) that was copolymerized with 2-(4-amino-phenyl)ethylamine (AP-EA) (Badea et al., 2003). This immobilization method was also used to produce Pt electrodes that were modified with GOD, L-amino acid oxidase, and alcohol oxidase, allowing detection of glucose, L-amino acid, and alcohol, respectively, using FIA. Using these apparatuses, the detection limit for lactate was 0.01 mM and the linear range was 0.01–0.3 mM. A carbon paste electrode for FIA was modified with LOD and horseradish peroxidase (HRP) using ferrocene as a mediator (Zaydan et al., 2004). Because HRP catalyzes the electrochemical reduction of hydrogen peroxide, hydrogen peroxide produced by LOD was detected with low operation potential. The detection limit was 4 μM and linear range was 4–40 μM in the corresponding FIA. In combination with FIA, a bioreactor containing *Streptococcus thermophilus* was used to assess fermentation in milk samples. Subsequently, highly sensitive lactate sensors in FIA were developed using LOD immobilized in gel membranes on Prussian blue–modified electrodes (Yashina et al., 2010). This FIA had a detection limit of 0.5 μM and linear range of 0.5 μM–1 mM.

16.3.2 FIA Based on Enzyme Reactors

In the presence of lactate, LOD and LDH produce hydrogen peroxide and NADH, respectively, and incorporation of enzyme reactors and detectors for hydrogen peroxide and NADH into flow injection systems has led to the development of FIA systems for lactate sensing (Table 16.2). Yao et al. (1982) reported FIA systems comprising LDH reactors and thin-layer electrochemical flow cells containing a glassy carbon working electrode, a silver–silver chloride reference electrode, and a platinum auxiliary electrode (Yao et al., 1982). In the presence of lactate, NADH was produced in the reactor from NAD^+, and was then amperometrically detected in the electrochemical flow cell. In FIA, the detection limit was 0.34 μM and the linear range was 1–80 μM. Direct oxidation of NADH on solid electrodes demands high potentials that cause high background currents. Thus, electrodes modified with 3-β-naphthoyl-Nile blue (Schelter-Graf et al., 1984) or with

TABLE 16.2 Characteristics of Enzyme Reactor Based FIA Systems

Enzyme	Detector	Detection Limit	Linear Range	Sample Volume (μL)	Reference
LDH	Electrochemical	0.34 μM	1.0–80 μM	100	Yao et al. (1982)
LDH	Electrochemical	10 μM	0.01–10 mM	50	Schelter-Graf et al. (1984)
LDH	Electrochemical	2.0 μM	0.01–2.0 mM	10	Yao and Wasa (1985)
LOD	Fluorescence	25 nM	25 nM–25 μM	20	Zaitsu et al. (1987)
LOD	Chemiluminescent	80 nM	100 nM to 1 mM	100	Hemmi et al. (1995)
LDH, LOD	Chemiluminescent	48 nM	up to 6 μM	50	Hansen et al. (1991)

Note: LDH, lactate dehydrogenase; LOD, lactate oxidase.

diaphorase (Yao and Wasa, 1985) as NADH sensors have been reported in FIA systems that can detect lactate with low potential.

Highly sensitive lactate-sensing FIA systems have been developed using combinations of LOD and peroxidase reactors. Zaitsu et al. (1987) reported a FIA system comprising a LOD reactor, HRP reactor, and fluorescence spectrophotometer (Zaitsu et al., 1987). In the presence of lactate, hydrogen peroxide was produced in the LOD reactor, and after reaction with the fluorogenic HRP substrate in the HRP reactor, the resulting fluorescent dye was detected using a fluorescence spectrophotometer. In FIA, the detection limit was 25 nM and the linear range was 25 nM–25 μM. We developed a chemiluminescence detector that was specifically designed for detecting weak chemiluminescence using a photodiode, and assembled a FIA system using a LOD reactor (Hemmi et al., 1995). In this system, hydrogen peroxide that was produced in the LOD reactor was mixed with chemiluminescence reagents in a flow cell, and the resulting light intensity was measured using the detector. This system produced a lactate detection limit of 80 nM and linear range of 100 nM–1 mM. Hansen et al. (1991) reported the use of LOD- and LDH-coimmobilized columns to amplify substrate-recycling signals. In this apparatus, LOD catalyzes the conversion of lactate and oxygen to pyruvate and hydrogen peroxide, and LDH converts pyruvate and NADH to lactate and NAD$^+$. Therefore, substrate recycling occurred in the LOD- and LDH-coimmobilized reactor. This recycling reaction consumed oxygen and NADH and generated NAD$^+$ and hydrogen peroxide. Thus, a FIA system for lactate sensing was developed using this reactor and a chemiluminescence detector, with a detection limit of 48 nM.

16.3.3 FIA for Sensing Multiple Metabolites

A major advantage of FIA is the ease with which multisensing systems can be assembled by incorporating several enzyme electrodes and/or enzyme reactors into one flow system (Table 16.3). Renneberg et al. (1991) developed a multisensing FIA system that simultaneously detected lactate, glucose, and glutamine through the incorporation of corresponding enzyme electrodes. In the FIA, LOD and GOD were separately immobilized on two electrodes, whereas for the glutamine sensor, glutaminase and glutamate oxidase were coimmobilized on an electrode. By connection of the FIA system to a bioreactor containing mammalian cell lines, the FIA system was used for online process control of cell culture (Renneberg et al., 1991). A dual enzyme electrode for lactate and pyruvate sensing

TABLE 16.3   Characteristics of FIA Systems for Sensing Multiple Metabolites

Enzyme	Detector	Metabolites	Detection Limit	Linear Range	Sample Volume (μL)	Reference
GOD, LOD, glutaminase/glutamate oxidase	Electrochemical	Lactate, glucose, glutamine	0.025 mM	up to 20 mM (lactate), up to 30 mM (glucose), up to 15 mM (lactate)	44	Renneberg et al. (1991)
LOD, pyruvate oxidase	Electrochemical	LOD, pyruvate	0.01 mM	0.01–5.0 mM	5	Yao and Yano (2004)
GOD, LOD	Fluorescence	Lactate, glucose	N/A	up to 30 mM	120	Dremel et al. (1992)
GOD, LOD	Electrochemical	Lactate, glucose	0.1 mM (lactate), 0.06 mM (glucose)	up to 12.6 mM (lactate), up to 7.7 mM	20	Marzouk et al. (2000)
GDH, LDH	Spectrometric	Lactate, glucose	11 mM (lactate), 2.8 mM (glucose)	11–33 mM (lactate), 2.8–22 mM (glucose)	N/A	Becker et al. (1993)

Note: LDH, lactate dehydrogenase; LOD, lactate oxidase; GOD, glucose oxidase; GDH, glucose dehydrogenase.

was also used in a FIA system and the electrode assembly contained dual electrodes with two platinum disks as a working electrode, which was separately modified with LOD and pyruvate oxidase (Yao and Yano, 2004). This FIA system was applied to simultaneous detection of lactate and pyruvate in rat serum and brain tissues.

Enzyme reactor–type multisensing FIA systems have also been developed. Dremel et al. (1992) reported a FIA system comprising a fiber-optic detector for oxygen consumption and two enzyme reactors containing GOD or LOD (Dremel et al., 1992). In this FIA system, the enzyme reactions consume oxygen, and thus glucose and lactate were simultaneously detected by measuring oxygen consumption. A similar FIA system was also developed using hydrogen peroxide production by GOD or LOD reactors as a proxy for glucose and lactate (Marzouk et al., 2000). Moreover, Becker et al. (1993) reported a FIA system for simultaneous detection of glucose and lactate using GDH and LDH reactors and a spectrophotometer. In these reactors, NADH was enzymatically produced from NAD^+ and the targets were simultaneously detected in photometric determinations of NADH.

16.3.4 Total Microanalysis System for FIA

Miniaturized flow cells for FIA systems offer advantages of reduced detection times, low cost of reagents, and small sample volumes. Thus, we developed an integrated flow cell as a total microanalysis system (μTAS) for lactate sensing (Nakamura et al., 2001). This integrated flow cell was constructed on a silicon chip (15×20 mm) that contained two hollows of a LOD reactor, a mixing cell, a spiral groove, and three pierced holes for chemiluminescence solutions, carrier, and waste. In this flow cell, LOD reacts with lactate to produce hydrogen peroxide at the LOD reactor. Subsequently, hydrogen peroxide reacts with the luminol reagent in the mixing cell and the resulting chemiluminescence is detected using a photodiode attached to the spiral groove. Incorporation of a flow cell into this FIA system allowed detection of lactate in 0.2-μL samples, with a response time of <30 s and a linear range of 0.5–5 mM.

16.3.5 FIA Based on Enzyme Thermistors

Enzyme thermistors can be used to detect enzyme substrates by measuring the heat produced from exothermic enzymatic reactions. Chen et al. (2011) incorporated an enzyme thermistor into a flow system for lactate sensing. In their system, the enzyme thermistor comprised an enzyme column containing LOD and catalase, and a reference column was incorporated into the flow system to correct for nonspecific signals. This FIA was used to detect lactate in milk samples without sample pretreatment, with a linear range of 25 nM–5.0 mM.

16.4 CONCLUSION

Numerous FIA systems for lactate sensing have been developed using LOD and LDH. These FIA systems are based on LOD- or LDH-modified electrodes or on LOD or LDH reactors. In the presence of lactate, LOD and LDH enzymatically produce hydrogen peroxide and NADH, respectively, which are electrochemically and optically detected.

Thus, FIA systems for lactate sensing have predominantly been developed using electrochemical and optical detectors. Incorporation of other enzyme electrodes and reactors has led to the development of multisensing FIA systems. Finally, FIA systems based on enzyme thermistors have been developed for lactate sensing using LOD. Together, these FIA systems have been used to detect lactate in food samples and human blood samples.

A major advantage of FIA systems is the ease of combination with other flow systems. In particular, FIA systems comprising sample dilution flow systems and lactate-sensing flow systems have been developed. Moreover, miniaturized flow cells such as μTAS have been developed for lactate sensing. Future incorporation of suitable flow systems into FIA will enable the development of miniaturized FIA systems that can automatically detect lactate in multiple samples without pretreatment.

REFERENCES

Avramescu, A., T. Noguerb, M. Avramescuc, and J. Marty. 2002. Screen-printed biosensors for the control of wine quality based on lactate and acetaldehyde determination. *Anal. Chim. Acta* 458:203–213.

Badea, M., A. Curulli, and G. Palleschi. 2003. Oxidase enzyme immobilisation through electropolymerised films to assemble biosensors for batch and flow injection analysis. *Biosens. Bioelectron.* 18:689–698.

Becker, T., W. Schuhmann, R. Betken, H. L. Schmidt, M. Leible, and A. Albrecht. 1993. An automatic dehydrogenase-based flow-injection system: Application for the continuous determination of glucose and lactate in mammalian cell-cultures. *J. Chem. Technol. Biotechnol.* 58:183–190.

Chen, Y., A. Andersson, M. Mecklenburg, B. Xie, and Y. Zhou. 2011. Dual-signal analysis eliminates requirement for milk sample pretreatment. *Biosens. Bioelectron.* 29:115–118.

Dremel, B. A., S. Y. Li, and R. D. Schmid. 1992. On-line determination of glucose and lactate concentrations in animal cell culture based on fibre optic detection of oxygen in flow-injection analysis. *Biosens. Bioelectron.* 7:133–139.

Hansen, E. H., L. Nørgaard, and M. Pedersen. 1991. Optimization of flow-injection systems for determination of substrates by means of enzyme amplification reactions and chemiluminescence detection. *Talanta* 38:275–282.

Hemmi, A., K. Yagiuda, N. Funazaki et al. 1995. Development of a chemiluminescence detector with photodiode detection for flow-injection analysis and its application to l-lactate analysis. *Anal. Chim. Acta* 316:323–327.

Marzouk, S. A. M., H. E. M. Sayour, A. M. Ragab, W. E. Cascio, and S. S. M. Hassan. 2000. A simple FIA-system for simultaneous measurements of glucose and lactate with amperometric detection. *Electroanalysis* 12:1304–1311.

Matsunaga, T., I. Karube, N. Teraoka, and S. Suzuki. 1982. Determination of cell numbers of lactic acid producing bacteria by lactate sensor. *Eur. J. Appl. Microbiol. Biotechnol.* 16:157–160.

Nakamura, H. and I. Karube. 2003. Current research activity in biosensors. *Anal. Bioanal. Chem.* 377:446–468.

Nakamura, H., Y. Murakami, K. Yokoyama, E. Tamiya, and I. Karube. 2001. A compactly integrated flow cell with a chemiluminescent FIA system for determining lactate concentration in serum. *Anal. Chem.* 73:373–378.

Palmisano, F., D. Centonze, and P. G. Zambonin, 1994. An *in situ* electrosynthesized amperometric biosensor based on lactate oxidase immobilized in a poly-o-phenylenediamine film: Determination of lactate in serum by flow injection analysis. *Biosens. Bioelectron.* 9:471–479.

Pandey, P. C. 1994. Tetracyanoquinodimethane-mediated flow injection analysis electrochemical sensor for NADH coupled with dehydrogenase enzymes. *Anal. Biochem.* 221:392–396.

Piano, M., S. Serban, R. Pittson, G. A. Drago, and J. P. Hart. 2010. Amperometric lactate biosensor for flow injection analysis based on a screen-printed carbon electrode containing Meldola's Blue-Reinecke salt, coated with lactate dehydrogenase and NAD+. *Talanta* 82:34–37.

Renneberg, R., G. Trott-Kriegeskorte, M. Lietz et al. 1991. Enzyme sensor-FIA-system for on-line monitoring of glucose, lactate and glutamine in animal cell cultures. *J. Biotechnol.* 21:173–185.

Schelter-Graf, A., H. L. Schmidt, and H. Huck 1984. Determination of the substrates of dehydrogenases in biological material in flow-injection systems with electrocatalytic NADH oxidation. *Anal. Chim. Acta* 163:299–303.

Shah, N. C., O. Lyandres, J. T. Walsh Jr., M. R. Glucksberg, and R. P. Van Duyne. 2007. Lactate and sequential lactate-glucose sensing using surface-enhanced Raman spectroscopy. *Anal. Chem.* 79:6927–6932.

Williams, D. L., A. R. Doig Jr., and A. Korosi. 1970. Electrochemical-enzymatic analysis of blood glucose and lactate. *Anal. Chem.* 42:118–121.

Yao, T., Y. Kobayashi, and S. Musha. 1982. Flow injection analysis for L-lactate with immobilized lactate dehydrogenase. *Anal. Chim. Acta* 138:81–85.

Yao, T. and T. Wasa. 1985. Simultaneous determination of l(+)- and d(−)-lactic acid by use of immobilized enzymes in a flow-injection system. *Anal. Chim. Acta* 175:301–304.

Yao, T. and T. Yano. 2004. On-line microdialysis assay of l-lactate and pyruvate *in vitro* and *in vivo* by a flow-injection system with a dual enzyme electrode. *Talanta* 63:771–775.

Yashina, E. I., A. V. Borisova, E. E. Karyakina et al. 2010. Sol-gel immobilization of lactate oxidase from organic solvent: Toward the advanced lactate biosensor. *Anal. Chem.* 82:1601–1604.

Zaitsu, K., M. Nakayama, and Y. Ohkura. 1987. Sensitive flow-injection determination of l-lactate in human blood with immobilized enzyme columns and fluorimetric detection. *Anal. Chim. Acta* 201:351–355.

Zaydan, R., M. Dion, and M. Boujtita. 2004. Development of a new method, based on a bioreactor coupled with an L-lactate biosensor, toward the determination of a non-specific inhibition of L-lactic acid production during milk fermentation. *J. Agric. Food Chem.* 52:8–14.

SECTION IV

Natural Antioxidants

CHAPTER 17

Determination of Tartaric Acid

*Ildikó V. Tóth, Sara S. Marques, Luís M. Magalhães,
and Marcela A. Segundo*

CONTENTS

Tartaric acid (2,3-dihydroxybutanedioic acid, molecular weight: 150.09 g/mol) is a white crystalline diprotic acid occurring naturally in plants (grapes, bananas, and tamarinds). The naturally occurring form of the acid, L-(+)-tartaric acid, has chiral properties. The mirror-image (enantiomeric) form, D-(−)-tartaric acid, can be obtained artificially. Being diprotic (pK_{a1} = 2.72; pK_{a2} = 4.79), tartaric acid provides an efficient buffer solution for pH values between 3 and 3.5, if present as major acid in solutions. Tartaric acid is also a ligand that forms complexes with diverse metal cations (e.g., iron, copper), thereby interfering with the catalytic activity of metal ions in various redox processes within the food matrix (Kreitman et al., 2013; Danilewicz, 2014).

Tartaric acid is the principal contributor of the acid fraction of wines. The role of tartaric acid in wines is based on its pH-controlling effect, which strongly influences the color, taste, and chemical and microbiological stability of the product. In wines with high alcoholic content, the formation of complexes with major ions of potassium and calcium can be significant to the extent that precipitation of tartrate crystals may occur in the bottle with time. Although this precipitation is harmless, it is undesirable in the view of the consumer.

Tartrates are commonly combined with baking soda to function as a leavening agent in the baking process. They are also added to foods to give a sour taste, as an antioxidant, and as an acidity regulator and/or sequestering agent. The incorporation of tartaric acid is permitted in most food products. Maximum limits have been established for some products including chocolate (5 g/kg), biscuits, and infant rusks (5 g/kg) (Bemrah et al., 2008); fruit juices, nectars, and their concentrates (4 g/kg) (CAC, 2013).

The acceptable daily intake (ADI) for the group of tartrates is set at 30 mg/kg bw/day based on the scientific literature, on the results of long-term studies on animals, on the compound's relative inertness and on the fact that it is a natural component of the human diet (JECFA, 1977; EU, 2001).

The required full assessment of food additives could not be conducted for tartaric acid by the time of the first assessment given the lack of usage data (Bemrah et al., 2008), so this additive was included in the second Total Diet Study in France (Bemrah et al., 2012). The ADI was not exceeded in the French population. The major contributions to tartaric acid dietary intake (as food additive) in adults were from pastries, cakes, mixed dishes, and vegetables. In the child population, the major contributors are biscuits, crackers, bars, pastries, cakes, and mixed dishes.

Regarding nutritional dietary assessment, tartaric acid has an additional interest. A recent study (Regueiro et al., 2014) indicates that urinary tartaric acid content can be used as a biomarker for moderate wine consumption. An efficient biomarker that can be used in the estimation of the intake of specific foods and dietary components, as an alternative or addition to self-reported dietary questionnaires, is of great interest for nutritional research. Actually, the authors of the study were able to differentiate between wine consumers and nonconsumers based on the urinary tartaric acid content.

17.1 FLOW INJECTION ANALYTICAL METHODS FOR DETERMINATION OF TARTRATES

The reference methods for tartaric acid determination are based either on the simultaneous determination of organic acids after high-performance liquid chromatography (HPLC) separation on an octyl-bonded silica column or on a determination of tartaric acid as calcium tartrate after a 12-h precipitation. In the gravimetric method, the precipitate is filtered, washed, and dried to constant weight, being finally titrated with EDTA (Wood et al., 2000; OIV, 2014). Routine analyses used to be performed in wines according to one recently withdrawn method of OIV, based on the Rebelein procedure, which consists of the separation of tartaric acid using an ion exchange resin before development of the color with vanadic acid (OIV, 2014). Both previous methods are tedious, laborious, and time consuming. In response to these limitations, flow methodologies have been developed to automate and thus improve the analytical features of this determination. Flow injection analysis (FIA) with spectrophotometric detection is the most common technique but fluorometric, potentiometric, and Fourier transform infrared spectroscopy (FTIR) detections were also employed. Applications in wine samples are overwhelming due to the importance of the analyte in this matrix. In the next section, various flow strategies grouped by detection technique are discussed and a summary of the analytical characteristics is presented in table format (Table 17.1). Figure 17.1 shows some basic configurations in schematic presentation for the applied flow injection configurations.

17.2 FLOW-BASED METHODS COUPLED TO SPECTROPHOTOMETRIC DETECTION

Spectrophotometric methods are, almost exclusively, based on the reaction of tartaric acid with vanadate. The first and benchmark application was published in 1989 by Lázaro et al. (1986). A detailed study on the time, temperature, and pH dependence of

TABLE 17.1   Application of Flow-Based Methods for the Determination of Tartaric Acid

Detection	Other Analytes	Sample Matrix	Flow Setup	LOD	Application Range	References
Spectrophotometry (490 nm)	N/a	Jerez wines	Flow injection, offline sample dilution, in the form of vanadate complex	0.02 g/L	0.02–0.4 g/L	Lázaro et al. (1986)
Spectrophotometry (490 nm)	Total acidity	Red and white wines, port wines	Flow injection, inline dialysis before sample injection, in the form of vanadate complex	0.07 g/L	0.5–4.0 g/L	Rangel and Tóth (1998)
Spectrophotometry (490 nm)	N/a	Table wines, port wines	Flow injection, continuous dialysis, in the form of vanadate complex	0.08 g/L	0.5–4.0 g/L	Silva and Alvares-Ribeiro (2002)
Spectrophotometry (490 nm)	N/a	Red and white table wines	Multicommutated flow setup, in the form of vanadate complex	–	0.5–10 g/L	Fernandes and Reis (2006)
Spectrophotometry (475 nm)	N/a	Wines	Flow injection, in the form of vanadate complex	–	Up to 1 g/L	Bastos et al. (2009)
Spectrophotometry (500 nm)	Potassium	Table and port wine samples	Multicommutated flow setup with inline dialysis unit, in the form of vanadate complex	0.1 g/L	1.0–5.0 g/L (table wines) 0.5–2.5 g/L (port wine)	Oliveira et al. (2010)
UV detection (185 nm) after capillary electrophoresis	Oxalic, formic, malic, succinic, maleic, glutaric, pyruvic, acetic, lactic, citric, butyric, benzoic, sorbic, ascorbic, and gluconic acids	Red and white wine, grape, apple, orange, and pineapple juice	Capillary ion analyzer and fixed-wavelength UV–Vis detector with 185 nm wavelength filter	0.38 mg/L	1.31–40 mg/L	Mato et al. (2006)

(Continued)

TABLE 17.1 (*Continued*) Application of Flow-Based Methods for the Determination of Tartaric Acid

Detection	Other Analytes	Sample Matrix	Flow Setup	LOD	Application Range	References
UV detection (210 nm) after HPLC separation	Malic, lactic, acetic, citric, and succinic acid	Thai wine samples	Flow injection inline dialysis sample pretreatment, and high-performance liquid chromatographic separation/detection	83 mg/L	250–7500 mg/L	Kritsunankul et al. (2009)
FTIR	Citric and malic acid, glucose, fructose, sucrose	Orange, grape, and apple juice, lemon drink, other soft drinks	Sequential injection, convective interaction media (CIM) disk separation, multivariate calibration for spectral data resolution, inline calibration standard preparation from a single standard solution	–	0.2–9.0 g/L	Lethanh and Lendl (2000)
FTIR (sheath-flow cell)	Citric and malic acid	Soft drinks	Multivariate calibration for spectral data resolution, inline calibration standard preparation from a single standard solution, stopped flow measurement	–	0.0–1.0 g/L	Ayora-Cañada and Lendl (2000)
Electrochemical (voltammetry)	Acetic, citric, lactic, malic, and succinic acid	Grape and orange juice, fermentation	HPLC separation, postcolumn reaction, electrochemical cell	–	0.75–300 mg/L	Kotani et al. (2004)
Electrochemical (potentiometry)	N/a	Red and white wines	Flow injection, single line setup, tartrate selective electrode	5.0×10^{-4} mol/L	7.5×10^{-4} to 6.0×10^{-2} mol/L	Sales et al. (2001)
Fluorescence ($\lambda_{Ex} = 340$ nm; $\lambda_{Em} = 460$ nm)	N/a	Red and white wines	Flow injection, immobilized enzyme reactors	–	1×10^{-4} to 1.2×10^{-3} mol/L	Tsukatani and Matsumoto (2000)

Note: LOD, limit of detection; HPLC, high-performance liquid chromatography; FTIR, Fourier transform infrared spectroscopy; N/a, not applicable.

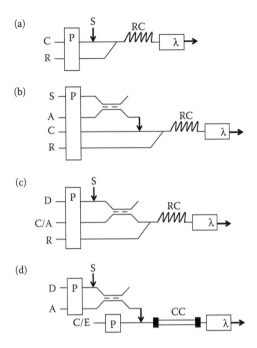

FIGURE 17.1 Schematic representation of the different flow strategies applied in the spectrophotometric determination of tartaric acid without (a) and with (b through d) inline dialysis unit. S, sample; P, pump; RC, reaction coil; CC, chromatography column; C, carrier; D, donor stream; A, acceptor stream; R, reagent; E, eluent; λ, detector; vertical arrows indicate injection valves.

the reaction was presented in this publication. A two-channel configuration (Figure 17.1a) was used to accomplish the inline formation of the acetic acid–acetate buffer. The samples were diluted in acetic acid and acetic acid was also used as a carrier solution. This is also the first reference to added low concentration (80 mg/L) tartaric acid in the reagent solution to improve the initial velocity of the reaction and therefore the overall sensitivity. Later, various modifications were introduced by other authors to improve the analytical characteristics of this spectrophotometric method.

A single flow injection system was proposed by Rangel and Tóth (1998) for carrying out the determinations without any previous sample treatment. The measurement of total acidity was also incorporated based on a pseudotitration strategy, while tartaric acid determination was based on colored vanadate complex formation. This methodology involved inline dialysis prior to injection (Figure 17.1b) to dilute the samples and to avoid interference from the sample matrix in the spectrophotometric detection. In order to be able use a single spectrophotometer, the color changes from both reactions were monitored using two flow cells aligned in the optical path.

Silva and Alvares-Ribeiro (2002) carried out a throughout optimization of the vanadate-based method. Experimental design was implemented to select the most important parameters (injection volume, flow rate, and vanadate concentration), followed by application of a modified simplex algorithm with a response function that included sensitivity, deviation from linearity at low concentrations, and residence time (used as an inverse measure of sampling rate) to optimize these variables. This system setup also comprised an inline dialysis unit (Figure 17.1c), however the variables of the dialysis process were not considered for optimization.

Fernandes and Reis (2006) developed a multicommutated flow method for this determination. Based on the use of solenoid valves, a flow network was set up for the spectrophotometric determination. A peristaltic pump aspirated the solutions into the flow manifold. No previous sample treatment was applied, and with the use of a low sample volume the intrinsic color of the red wine samples did not influence the accuracy of the method.

More recently, another multicommutated flow method was developed for the simultaneous determination of tartaric acid and potassium as a tool for evaluating the tartrate stability of wines (Oliveira et al., 2010). This system also resorted to inline dialysis of the samples in order to minimize matrix interferences for tartaric acid determination. A detailed study with different configurations of the dialysis unit was presented and various membrane materials were compared. The continuous-flow dialysis process was optimized for acceptor and donor channel flow rates, flow directions, and stop flow periods.

17.3 FLOW-BASED METHODS COUPLED TO FOURIER TRANSFORM INFRARED SPECTROSCOPY DETECTION

A sequential injection analysis method for the determination of organic acids (citric, malic, and tartaric acids) and sugars (glucose, fructose, and sucrose) in soft drinks by sequential injection FTIR spectroscopy has also been presented (Lethanh and Lendl, 2000). The method was based on the solid-phase extraction of the desired analyte groups and their separation from the sample matrix. A solid-phase extraction disk was incorporated into the manifold and, upon injection of a sample, the organic acids were completely retained on a convective interaction media (CIM) disk carrying quaternary amino moieties whereas sugars passed through to the flow cell. The organic acids were subsequently eluted by injection of an alkaline (pH 8.5) 1 M sodium chloride solution and recorded in their fully deprotonated form as a second flow injection peak.

The CIM disks, with low-volume monolith stationary phase, used in this application allowed the desired separation of the analyte under low-pressure conditions. The use of a mixing chamber and the precise control of timing allowed the inline preparation of the necessary high number of standard solution mixtures for the multivariate calibration. The mixing chamber also provided a homogeneous mixture before the transport to the detector. The large sampled volume (1 mL) allowed the recording of spectra without dispersion effects ($D = 1$). Samples were diluted and their pH was adjusted to 8.5 before injection. Nevertheless, validation for the tartaric acid concentration was not possible with the enzymatic kits used in this work.

The same research group (Ayora-Cañada and Lendl, 2000) presented a similar configuration, but without the separation of the analyte from the sample matrix. In this case, spectral matrix discrimination was used to reduce the sample matrix contribution to the analytical signal.

17.4 FLOW-BASED METHOD COUPLED TO FLUOROMETRIC DETECTION

Although no direct tartaric acid detection enzymatic assay is available commercially, tartaric acid can be accessed as a secondary substrate of D-malate dehydrogenase.

This reaction was exploited by Tsukatani and Matsumoto (2000) in a stopped-flow FIA method. An immobilized D-malate dehydrogenase enzyme reactor was employed and the reduced enzymatic cofactor NADH that was formed was monitored fluorometrically ($\lambda_{Ex} = 340$ nm; $\lambda_{Em} = 460$ nm). Due to the slow reaction rate, the flow was stopped with the sample in the reactor to increase reaction time. The intrinsic sample fluorescence was also assessed using a parallel blank reactor without immobilized enzyme. The method was validated through the analysis of red and white wine samples. The enzyme reactor stability was also evaluated and it was found that the sensitivity (evaluated as amplitude of response at a constant concentration of the analyte) gradually decreased to 60% within a week but then remained stable for a month. As D-malate cannot be present in naturally fermented wines (except for fraudulent addition), the interference of this primary substrate can be considered negligible.

17.5 FLOW-BASED METHODS COUPLED TO ELECTROCHEMICAL DETECTION

Sales et al. (2001) developed a potentiometric sensor for the determination of tartaric acid. Tartrate-sensitive electrodes were constructed, comprising poly(vinyl chloride) membranes. Various membrane compositions were evaluated and the one combining an ionic sensor (bis(triphenylphosphoranylidene)ammonium chloride), 4-*tert*-octylphenol, and *o*-nitrophenyl octyl ether was selected. The analytical characteristics of this electrode were evaluated using a single-line flow injection manifold. The accuracy of the developed method was demonstrated for the determination of tartaric acid in red and white wine samples.

In a different approach, voltammetry was applied as a postcolumn detection system (Kotani et al., 2004). A two-channel system was designed using an ion exclusion column where the column outlet merged downstream with the 2-methyl-1,4-naphthoquinone reagent. The electrochemical detector with a glassy carbon working electrode was used for the detection of acetic, citric, lactic, malic, succinic, and tartaric acids. Acids were detected by measuring the current height of a flow signal at the fixed potential. The variation in the current peak areas caused by the eluted acids was correlated with the acid concentration in the sample. The method was applied to the determination of organic acids in wines and during fruit juice fermentation; however, the separation of tartaric acid and citric acid was poor.

17.6 HYPHENATED FLOW METHODS

Flow injection methods are more and more frequently used as upfront inline sample preparation tools for chromatographic methods. These hyphenated methods have the advantage of automation of the troublesome sample cleanup procedures that are necessary for the efficient separation of the analytes and a "must have" for the high-sensitivity sophisticated detectors.

One example of a flow injection method coupled to an HPLC separation can be found for the determination of tartaric acid (Figure 17.1d). In this method, a dialysis unit was used for sample preparation, the injected sample was dialyzed, and a small portion (20 µL) of the dialyzed sample was injected into the high-pressure eluent stream followed by analyte separation on a C18 column and UV–Vis detection (photodiode array detector). The separation of six organic acids (tartaric, malic, lactic, acetic, citric, and succinic acids) was achieved within 8 min of analysis time. The timing sequence of injections and elution phases of the experimental protocol were optimized to guarantee good precision of the results.

17.7 CONCLUSIONS AND FUTURE TRENDS

In conclusion, from the existing literature, flow methods provide the various advantages of automation for tartaric acid determination; these include increased precision, increased sample throughput, and reduced reagent consumption and waste production. Membrane separation procedures using dialysis membranes for the dilution of the analyte and for separation of the small molecules from other higher-molecular-weight constituents of the matrix were frequent choices of researchers. In most of the applications, not only tartaric acid was targeted but multiparametric determinations were also performed. Biparametric systems resorted to a flow setup that provided the possibility of two distinct detection points. Multianalyte systems relied on the incorporation of chromatographic separation of analytes, and the detection was performed downstream to the column. With the increase in the number and type of the available low-pressure monolith analytical columns, it is expected that these determinations can be further improved.

In terms of the analytical merits of the flow systems, the LOD and application ranges are adequate for the determination of tartaric acid in wine samples. However, these methods were not evaluated in other matrixes. Considering that in the child population the largest intake is in the form of a food additive (Bemrah et al., 2012), and not as the natural constituent of wine, automated analytical methods directed to bakery products might be of interest.

Regarding chirality, it is a very challenging task to differentiate the natural L-(+)-tartaric acid form the artificial D-(−)-tartaric acid (Trojanowicz and Kaniewska, 2013). Until now mostly only ligand-exchange capillary electrophoresis (CE) systems were able to achieve separation (Schmid, 2012). Direct ligand-exchange resolution methods for the separation of tartrate enantiomers by CE using copper(II)–D-quinic acid as a chiral selector were investigated (Kodama et al., 2001; Knob et al., 2013). However, L-(+)-tartaric acid is present in large excess over the D-(−)-tartaric acid in food samples, causing analytical limitations related to the detection limit and the necessary extensive sample dilutions. In this context, flow-based methodologies providing reproducible automated sample treatment procedures may contribute to the further improvement of the CE methods.

From the current regulatory point of view, tartaric acid as a food additive seems not to have any direct health implications. Toxicity values indicate no potential danger for the population, providing a justification for the fact that the analytical methods developed are focused on the technological effects of the compound (such as cold stabilization in wineries). Antioxidant effects, direct or indirect (via copper and iron complexation or sequestration), might also be of interest in the future, and the developed flow methods presented here and in Chapter 38 have been shown to provide a reliable assessment of these effects.

ACKNOWLEDGMENTS

I. V. Tóth and L. M. Magalhães thank FSE and Ministério da Ciência, Tecnologia e Ensino Superior (MCTES) for the financial support through the POPH-QREN program.

This work received financial support from the European Union (FEDER funds through COMPETE) and National Funds (FCT, Fundação para a Ciência e Tecnologia) through project UID/Multi/04378/2013 and also from EU FEDER funds under the framework of QREN (Project NORTE-07-0124-FEDER-000067).

REFERENCES

Ayora-Cañada, M. J. and B. Lendl. 2000. Sheath-flow Fourier transform infrared spectrometry for the simultaneous determination of citric, malic and tartaric acids in soft drinks. *Anal. Chim. Acta* 417(1):41–50.

Bastos, S. S. T., P. A. R. Tafulo, R. B. Queirós, C. D. Matos, and M. G. F. Sales. 2009. Rapid determination of tartaric acid in wines. *Comb. Chem. High T. Scr.* 12(7):712–722.

Bemrah, N., J. C. Leblanc, and J. L. Volatier. 2008. Assessment of dietary exposure in the French population to 13 selected food colours, preservatives, antioxidants, stabilizers, emulsifiers and sweeteners. *Food Addit. Contam.: Part B Surveill.* 1(1):2–14.

Bemrah, N., K. Vin, V. Sirot et al. 2012. Assessment of dietary exposure to annatto (E160b), nitrites (E249–250), sulphites (E220–228) and tartaric acid (E334) in the French population: The second French total diet study. *Food Addit. Contam.: Part A* 29(6):875–885.

CAC, Codex Alimentarius Commission. 2013. *Codex General Standard for Food Additives.* Rome, Italy: World Health Organization, Food and Agriculture Organization of the United Nations.

Danilewicz, J. C. 2014. Role of tartaric and malic acids in wine oxidation. *J. Agr. Food Chem.* 62(22):5149–5155.

EU, European Commission. 2001. *Report from the Commission on Dietary Food Additive Intake in the European Union.* http://ec.europa.eu/food/fs/sfp/addit_flavor/flav15_en.pdf.

Fernandes, E. N. and B. F. Reis. 2006. Automatic spectrophotometric procedure for the determination of tartaric acid in wine employing multicommutation flow analysis process. *Anal. Chim. Acta* 557(1–2):380–386.

JECFA, Joint Expert Committee on Food Additives 1977. *431. Tartaric Acid and Monosodium Tartrate (WHO Food Additives Series 12).* Geneva, Switzerland: World Health Organization.

Knob, R., J. Petr, J. Ševčík, and V. Maier. 2013. Enantioseparation of tartaric acid by ligand-exchange capillary electrophoresis using contactless conductivity detection. *J. Sep. Sci.* 36(20):3426–3431.

Kodama, S., A. Yamamoto, A. Matsunaga, and K. Hayakawa. 2001. Direct chiral resolution of tartaric acid in food products by ligand exchange capillary electrophoresis using copper(II)-D-quinic acid as a chiral selector. *J. Chromatogr. A* 932(1–2):139–143.

Kotani, A., Y. Miyaguchi, E. Tomita, K. Takamura, and F. Kusu. 2004. Determination of organic acids by high-performance liquid chromatography with electrochemical detection during wine brewing. *J. Agr. Food Chem.* 52(6):1440–1444.

Kreitman, G. Y., A. Cantu, A. L. Waterhouse, and R. J. Elias. 2013. Effect of metal chelators on the oxidative stability of model wine. *J. Agri. Food Chem.* 61(39):9480–9487.

Kritsunankul, O., B. Pramote, and J. Jakmunee. 2009. Flow injection on-line dialysis coupled to high performance liquid chromatography for the determination of some organic acids in wine. *Talanta* 79(4):1042–1049.

Lázaro, F., M. D. Luque De Castro, and M. Valcárcel. 1986. Photometric determination of tartaric acid in wine by flow injection analysis. *Analyst* 111(7):729–732.

Lethanh, H. and B. Lendl. 2000. Sequential injection Fourier transform infrared spectroscopy for the simultaneous determination of organic acids and sugars in soft drinks employing automated solid phase extraction. *Anal. Chim. Acta* 422(1):63–69.

Mato, I., J. F. Huidobro, J. Simal-Lozano, and M. T. Sancho. 2006. Simultaneous determination of organic acids in beverages by capillary zone electrophoresis. *Anal. Chim. Acta* 565(2):190–197.

OIV, International Organisation of Vine and Wine. 2014. *Compendium of International Methods of Wine and Must Analysis.* Paris, France: International Organization of Vine and Wine.

Oliveira, S. M., T. I. M. S. Lopes, I. V. Tóth, and A. O. S. S. Rangel. 2010. Simultaneous determination of tartaric acid and potassium in wines using a multicommuted flow system with dialysis. *Talanta* 81(4–5):1735–1741.

Rangel, A. O. S. S., and I. V. Tóth. 1998. Sequential determination of titratable acidity and tartaric acid in wines by flow injection spectrophotometry. *Analyst* 123(4):661–664.

Regueiro, J., A. Vallverdu-Queralt, J. Simal-Gandara, R. Estruch, and R. M. Lamuela-Raventos. 2014. Urinary tartaric acid as a potential biomarker for the dietary assessment of moderate wine consumption: A randomised controlled trial. *Br. J. Nutr.* 111(9):1680–1685.

Sales, M. G. F., C. E. L. Amaral, and C. M. Delerue Matos. 2001. Determination of tartaric acid in wines by FIA with tubular tartrate-selective electrodes. *Anal. Bioanal. Chem.* 369(5):446–450.

Schmid, M. G. 2012. Chiral metal-ion complexes for enantioseparation by capillary electrophoresis and capillary electrochromatography: A selective review. *J. Chromatogr. A* 1267:10–16.

Silva, H. A. D. F. O. and L. M. B. C. Alvares-Ribeiro. 2002. Optimization of a flow injection analysis system for tartaric acid determination in wines. *Talanta* 58(6):1311–1318.

Trojanowicz, M. and M. Kaniewska. 2013. Flow methods in chiral analysis. *Anal. Chim. Acta* 801:59–69.

Tsukatani, T. and K. Matsumoto. 2000. Quantification of L-tartrate in wine by stopped-flow injection analysis using immobilized D-malate dehydrogenase and fluorescence detection. *Anal. Sci.* 16(3):265–268.

Wood, R., L. Foster, A. Damant, and P. Key. 2000. *Analytical Methods for Food Additives.* Cambridge, England: Woodhead Publishing Limited, CRC Press.

CHAPTER 18

Determination of Ascorbic Acid (Vitamin C)

M. Carmen Yebra-Biurrun

CONTENTS

18.1 INTRODUCTION

Ascorbic acid (AA, known as vitamin C) is a hydrosoluble, antioxidant vitamin. It is a hexanoic sugar acid with two dissociable protons (pK_a 4.17 and 11.57). The term vitamin C is used as the generic descriptor for all compounds exhibiting qualitatively the biological activity of AA. The principal natural compound with vitamin C activity is L-AA ($C_6H_8O_6$, molecular weight = 176.1). Vitamin C is designated as AA because of its ability to cure and prevent scurvy. The IUPAC–IUB Commission on Biochemical Nomenclature changed vitamin C (2-oxo-L-threo-hexono-1,4-lactone-2,3-enediol) to AA or L-AA in 1965. The chemical structure of AA is given in Figure 18.1. Dehydroascorbic acid (DHAA), the oxidized form of AA retains vitamin C activity and can exist as a hydrated hemiketal. Crystalline L-DHAA acid can exist as a dimer [1]. AA is rapidly oxidized to DHAA due to the presence of two hydroxyl groups in its structure. Oxidation reactions can be induced by exposure to increased temperatures, high pH, light, presence of oxygen or metals, and enzymatic action. This reaction is reversible and a principal step in the antioxidant activity of AA. Further oxidation generates diketogluconic acid, which has no biological function and the reaction is no longer reversible.

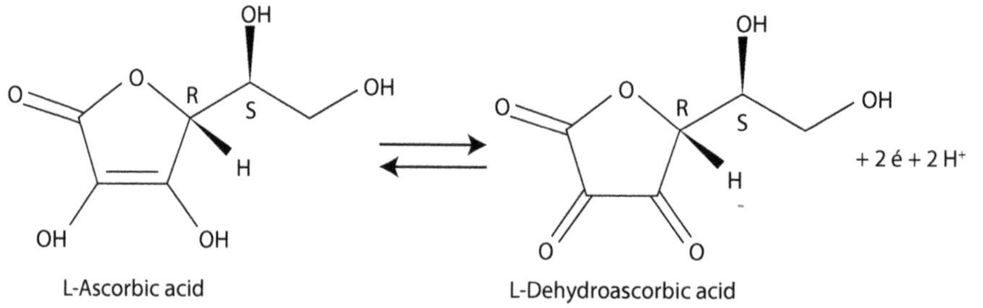

FIGURE 18.1   Structure of L-ascorbic acid (AA) and its oxidation to L-dehydroascorbic acid (DHAA).

Vitamin C is one of the most ubiquitous vitamins, which were first isolated by Nobel Prize winner, Dr. Albert Szent-Gyorgyi in 1928. With its antioxidant property, it plays a major role as a free radical scavenger and provides a defense system to the body against reactive oxygen species by preventing tissue damage [2]. Besides its role as antioxidant, vitamin C is also considered as an effective antiviral agent. It also takes part in a wide range of essential metabolic reactions for better functioning of the body. The necessity of AA for human health is firmly established because it is necessary to form collagen, for the healing of wounds, and for the repair and maintenance of cartilage, bones, and teeth. Furthermore, it is important for therapeutic purposes and biological metabolism. As an antioxidant, vitamin C plays an important role in protecting against heart disease, high cholesterol, high blood pressure, common cold, cancer, osteoarthritis, obesity and weight loss, cataracts, age-related macular degeneration, diabetes, Alzheimer disease, and other types of dementia [3]. Also, it is known for its use on a wide scale as an antioxidant agent in foods and drinks. It is present naturally in a wide range of foods, particularly vegetables and fruits (notably in members of the citrus family) and, as the human body cannot synthesize it, these are the major sources for the human diet. AA has limited stability and may be lost from foods during storage, preparation, and cooking [4].

AA concentration, or more accurately, the AA/DHAA ratio can be an indicator of the redox state of an organism, and the content of vitamin C in fruits is used as an index of the health-related quality of fruits.

Under regulation by European legislation [5], the food industry uses four AA forms as food additives, the first three of which are AA itself and their sodium and calcium salts (E300, E301, and E302, respectively). These compounds are water-soluble and are used in certain nonfatty foods to prevent oxidation. The fourth form groups fatty acid esters of AA (ascorbyl palmitate and ascorbyl stearate), which are hydrophobic and used to protect fats from oxidation (E304). Compounds added for their nutritional benefit are not normally regarded as food additives. At the same time, an additive used in a food for a technological purpose should not be declared as a nutrient, but with the category name of the function to which it relates. For example, if AA is being added as an antioxidant, it should be declared as an antioxidant and not as vitamin C. However, different rules may apply in different countries [6]. AA does not have a maximum limit of usage in foodstuff and it can be used on quantum satis basis (maximum numerical level is specified and substances shall be used in accordance with good manufacturing practice, at a level not higher than is necessary to achieve the intended purpose and provided that the consumer is not misled).

18.2 ANALYTICAL TECHNIQUES FOR ASCORBIC ACID DETERMINATION

In food analysis, a method for determining vitamin C should ideally account for AA, its salts, and its reversible oxidation product, DHAA, to give a total value for vitamin C. In addition, the ability to distinguish AA from its epimer D-isoascorbic acid (erythorbic acid) is useful in the analysis of processed foods.

Traditional methods for AA determination involve titration with an oxidant solution: dichlorophenol indophenol [7], potassium iodate [8], or bromate [9]. Thus, a titrimetric method is employed in the Association of Official Analytical Chemists (AOAC) Official Methods (method 985.33;vitamin C in ready-to-feed milk-based infant formula) [10]. Fluorometric [11,12] and ultraviolet-visible (UV–Vis) spectrophotometric [13–15] detection were also used for AA determination. In fact, AOAC Official Methods 967.22 and 984.26 utilize fluorometry to determine total vitamin C content in vitamin preparations and foods, respectively, by derivatization of DHAA using the o-phenylenediamine condensation reaction [10]. The AOAC Official Method 984.26 is a semiautomated method, being the adaptation of the manual 967.22 method to a Technicon Autoanalyzer system. Other methodologies for AA determination include chemiluminescence [16], electrochemical methods (amperometry, voltammetry, and potentiometry) [17–19], and chromatographic methods, above all high-performance liquid chromatography (HPLC) with electrochemical, fluorometric, UV–Vis or mass spectrometric (MS) detection [20–22]. In recent years two new official methods were proposed by the AOAC for the determination of vitamin C in infant formula and adult/pediatric nutritional formula based on HPLC and ultra-high-pressure liquid chromatography (UHPLC) both with UV detection [10]. It is important to notice that only HPLC procedures simultaneously separate L-AA, L-DHAA, and D-isoascorbic acid in foods [23,24]. Capillary electrophoresis (CE) offers an alternative to HPLC for AA determination and eliminates the need for organic mobile phases and expensive chromatography columns [25,26].

18.3 FLOW ANALYSIS TECHNIQUES FOR ASCORBIC ACID

Flow injection determination methods of AA were reviewed by Yebra-Biurrun in 2000 [27]. Since that time a great number of articles have been published involving AA determination based on flow analysis techniques. There are several identifiable generations of flow analysis techniques, the majority of them have been used in AA determination in a large number of samples, including foods, pharmaceutical, and biological samples.

The ScienceFinder database was searched up to March 2014 for flow analysis methods for the determination of AA. The keywords used as search criteria were "flow injection analysis," "sequential injection analysis," "lab-on-valve," "multicommutation," "multisyringe," "stopped-flow," "bead injection" as well as "ascorbic acid." As can be seen in Figure 18.2, flow injection analysis (FIA), sequential injection analysis (SIA), and lab-on-valve (LOV), and other flow analysis techniques such as multicommutated techniques (multicommutated flow injection analysis [MCFIA], multisyringe flow injection analysis [MSFIA], and multipumping flow systems [MPFS]), bead injection (BI), and stopped-flow techniques have been applied for AA determination. Figure 18.2 reveals that the most applied flow analysis technique for AA determination is FIA: more than 80% of the published articles are based on that methodology.

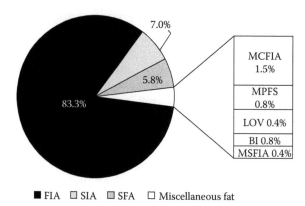

FIGURE 18.2　Distribution of ascorbic acid (AA) determinations according to the flow analysis technique used. BI: bead injection; FAT: flow analysis techniques; FIA: flow injection analysis; LOV: lab-on-valve; MCFIA: multicommutated flow injection analysis; MPFS: multi-pumping flow system; MSFIA: multisyringe flow injection analysis; SIA: sequential injection analysis; SFA: stopped-flow analysis.

Generally, AA is determined individually, and only about a 10% of the published methods determine AA simultaneously with other analytes such as uric acid, glucose, fructose, dopamine, iodate, bromate, hypochlorite, thiourea, glutathione, hydrogen peroxide, acetylsalicylic acid, kojic acid, ascorbyl glucoside, paracetamol, cysteine, and other water soluble vitamins (thiamine [vitamin B_1], folic acid [vitamin B_{12}], niacin [vitamin B_3], riboflavin [vitamin B_2], and pyridoxine [vitamin B_6]).

Also, it is important remark that the most of the flow analysis methodologies are proposed for L-AA determination, being only a minority those that determine total AA. The methods that determine total AA are based on oxidation of AA to DHAA by dissolved oxygen or mercuric chloride, or on reduction of DHAA to AA by dithiothreitol.

In the next sections will be described the main features and characteristics of flow analysis methodologies proposed for the determination of AA.

18.3.1 Flow Injection Analysis Methods for Ascorbic Acid Determination

Normal FIA (nFIA) and reversed FIA (rFIA) manifolds were used in AA determinations. The reported FIA procedures for the determination of AA have been grouped for discussion according to the detection technique employed. Most of the FIA methods currently available for the determination of AA are based on the oxidation of its 1,2-enediol group.

As can be seen in Figure 18.3, the FIA systems proposed are mainly based on molecular spectroscopic and electroanalytical detectors.

18.3.1.1 Molecular Spectroscopic Techniques

The most important details of the published FIA procedures for AA determination based on molecular spectroscopic detection (spectrophotometric, chemiluminescence, and fluorometric methods) following a chronological order, are contained in Tables 18.1 through 18.3.

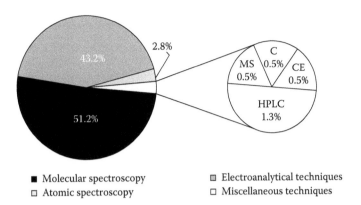

FIGURE 18.3 Distribution of flow injection analysis (FIA) determinations of ascorbic acid (AA) according to the detection technique used. C: calorimetry; CE: capillary electrophoresis; HPLC: high-performance liquid chromatography; MS: mass spectrometry.

Spectrophotometric methods for the determination of AA can be subdivided into two groups: methods based on measuring the intrinsic absorption of AA, and methods based on measuring the light absorption of products that result from the reduction of various regents by AA. These last are more frequently used. Thus, the method using chloramine T in the presence of starch–potassium iodide solution, the titration with 2,6-dichlorophenolindophenol [28–30], the formation of Fe(II)–phenanthroline [31–44], Fe(II)–ferrozine [45], Cu(I)–bathocuproine complexes [46], or the formation of copper thiocyanate precipitate [47,48] are used after previous metal ion reduction by AA. Discolorimetry was also employed by reduction of cerium(IV) [49], reduction of triiodide [50–56], reduction of Co(III)–EDTA complex [57], photochemical reduction of methylene blue [58–61], or the discoloring effect of AA on the reaction between Fe(III) and thiocyanate complex [42,62]. Solid-phase reactors using copper(II) phosphate and $Fe(OH)_3$ immobilized in a polyester resin have been employed, providing an alternative to obtain faster and more efficient sample conversion than is possible with conventional procedures, where the reagents are introduced into the flow system as solutions [37,46]. Direct spectrophotometry in the UV region is also used because AA absorbs in this region, presenting a maximum absorbance at 243 nm in strongly acid media and at 265 nm in neutral media. Since many compounds present in samples also absorb in the UV region and matrix effects need to be eliminated, several strategies have been proposed. This effect is generally overcome by making a second measurement. Furthermore, after AA has been decomposed, by means of sodium hydroxide [63–65], a column packed with modified silica that retains interferents by ionic pair formation between amine groups and interfering compounds is used [66,67]. An anion exchanger, placed in a suitable quartz flow cell, that retains the ascorbate anion in a flow-through solid-phase technique, where the carrier itself acts also as the eluting agent, is another alternative for eliminating interferences [66–72].

Chemiluminescent methods in FIA systems involved the decrease of the chemiluminescence signal, in presence of AA, produced by reaction between, among other combinations, luminol, hydrogen peroxide, and peroxidase [73,74]; luminol and ferricyanide [75]; hexacyanoferrate(III) and luminol [76]; luminol and permanganate [77–79]; luminol and periodate [80,81]; luminol, hydrogen peroxide and Fe(III) [73,74]; luminol

TABLE 18.1   FIA Methods for Ascorbic Acid (AA) Determination with Spectrophotometric Detection

Detection Technique	Description of the Method	LOD (μM)	SF (h⁻¹)	RSD (%)	Samples	Year	Reference
SP (Vis)	Titration with 2,6-dichlorophenolindophenol	No data	No data	0.3	Pharmaceuticals	1985	28
SP (Vis)	Titration with 2,6-dichlorophenolindophenol	No data	No data	0.3	Pharmaceuticals	1985	29
SP (Vis)	1. Chloramine T in presence of starch–potassium iodide solution as indicator. Normal FIA. $\lambda = 650$ nm	No data	90	0.97	Synthetic samples	1986	30
	2. Chloramine T in presence of methyl red-potassium bromide solution as indicator. FIA titration without a gradient chamber. $\lambda = 535$ nm	No data	30	0.82			
SP (Vis)	Chloramine T in presence of starch–potassium iodide solution as indicator. Normal FIA, $\lambda = 650$ nm	No data	90	2.5	Urine	1986	31
SP (Vis)	AA gave a negative peak after merging with a triiodide stream generated into the flow system. $\lambda = 580$ nm	No data	300	No data	Foods, pharmaceuticals	1986	32
SP (Vis)	Chloramine T in presence of starch-potassium iodide solution as indicator. $\lambda = 581$ nm	235	90	0.3–1.5	Soft drinks	1987	33
SP (Vis)	Reduction of Fe(III) to Fe(II) by using AA and subsequent formation of Fe(II) complex with 1,10-phenanthroline. $\lambda = 510$ nm	No data	No data	1.8	Fruit juices	1987	34
SP (Vis)	Reaction with ferric o-phenanthroline to form a colored complex that was measured at 510 nm	No data	No data	<2%	Vegetables	1990	35
SP (UV)	Reduction of vanadotungstophosphoric acid by AA. $\lambda = 360$ nm	5.7	80	1.5	Pharmaceuticals	1991	36
SP (Vis)	AA solution is colored and the absorbance of this colored solution is decreased	No data	No data	No data	No data	1991	37

(Continued)

TABLE 18.1 (*Continued*) FIA Methods for Ascorbic Acid (AA) Determination with Spectrophotometric Detection

Detection Technique	Description of the Method	LOD (μM)	SF (h⁻¹)	RSD (%)	Samples	Year	Reference
SP (Vis)	Reaction of AA with dichloroindophenol. $\lambda = 525$ nm	2.8	No data	0.5	Pharmaceuticals, soft drinks, urine	1991	38
SP (UV)	AA was determined in two portions of sample, one was analyzed directly and the other after treatment with NaOH to give a background correction. $\lambda = 245$ nm	1.13	180	0.6	Pharmaceuticals	1991	39
SP (Vis)	Reaction between AA and starch-potassium iodide. $\lambda = 580$ nm	No data	No data	0.9–2.6	Vegetables, honey, pharmaceuticals	1992	40

SP (Vis)	Photochemical reaction between AA and methylene blue. $\lambda = 650$ nm
	1. With irradiation of the reaction coil
	  a. reversed flow mode
	  b. merging zones approach
	2. With irradiation of the flow cell
		2.27, 3.97, 49.4	No data, 25, 15	2.4, 1.8, 2.0	Pharmaceuticals, soft drinks	1992	41

SP (Vis)	Ethylenediaminetetraacetic acid–Co(III) complex in a medium of 5% diethylamine. $\lambda = 540$ nm 1. Peak-height FI technique 2. Peak-width FI method	60 No data	60 60	0.94 2.02	Pharmaceuticals Urine	1992	42
SP (Vis)	Reduction of Fe(III) to Fe(II) by using AA and subsequent formation of Fe(II) complex with 1,10-phenanthroline. $\lambda = 508$ nm	No data	90	0.5	Pharmaceuticals	1992	43
SP (Vis)	Reduction of Fe(III) to Fe(II) by using AA and subsequent formation of Fe(II) complex with 1,10-phenanthroline. $\lambda = 510$ nm	No data	No data	0.27	Pharmaceuticals	1992	44

(*Continued*)

TABLE 18.1 (*Continued*) FIA Methods for Ascorbic Acid (AA) Determination with Spectrophotometric Detection

Detection Technique	Description of the Method	LOD (μM)	SF (h⁻¹)	RSD (%)	Samples	Year	Reference
SP (Vis)	Photochemical reaction between AA and methylene blue. $\lambda = 666$ nm	1.93	50–55	0.31–2.09	Pharmaceuticals	1993	45
SP (Vis)	FIA titration technique using Ce(IV) as a titrant in sulfuric acid media as a self-indicating system by monitoring the decrease in absorbance of cerium(IV)	No data	36	0.9	Pharmaceuticals	1993	46
SP (Vis)	AA was determined by passing the sample through a dialysis cell. Fe(III) was reduced to Fe(II) and subsequent formation of Fe(II) complex with 1,10-phenanthroline. $\lambda = 510$ nm	No data	60	0.5	Fruit juice	1993	47
SP (Vis)	Oxidation of AA with Fe(III) in presence of 2,2′-dipyridyl	No data	No data	No data	Pharmaceuticals, blood plasma	1994	48
SP (Vis) sulfuric	Fe(III) was reduced by AA and then mixed with 1,10-phenanthroline in sulfuric acid media. $\lambda = 510$ nm	No data	100	0.88	Pharmaceuticals	1994	49
SP (Vis)	The injection of the sample into the flux was done hydrodynamically (without using an injection valve). Iodide was employed as a carrier	No data	180	No data	Orange juice	1994	50
SP (Vis)	Photochemical reaction between AA and methylene blue. $\lambda = 666$ nm	No data	No data	<10	Vegetables Fruits	1994	51
SP (Vis)	Photochemical reaction between AA and methylene blue. $\lambda = 666$ nm	No data	55–60	No data	Pharmaceuticals	1995	52
SP (UV)	AA was treated with sodium hydroxide and a fraction was decomposed into substances that do not absorb in the UV region. $\lambda = 245$ nm	2.84 1.14	30 30	2.5 1.8	Soft drinks	1995	53
SP (Vis)	Based on the formation of ferroin. $\lambda = 510$ nm	No data	No data	No data	Pharmaceuticals	1995	54
SP (Vis)	Differential photometry at 510 nm	No data	No data	No data	Drugs	1996	55

(*Continued*)

TABLE 18.1 (*Continued*) FIA Methods for Ascorbic Acid (AA) Determination with Spectrophotometric Detection

Detection Technique	Description of the Method	LOD (μM)	SF (h⁻¹)	RSD (%)	Samples	Year	Reference
SP (UV)	AA reacts with I_3^- and excess I_3^- was determined. $\lambda = 350$ nm	No data	70	0.4	Foods	1996	56
SP (Vis)	Fe(III) was reduced by AA and then mixed with hexacyanoferrate(III). $\lambda = 700$ nm	No data	140	0.65	Pharmaceuticals	1997	57
SP (Vis)	Method based on the catalytic effect of AA on copper(II) porphyrin formation. $\lambda = 550$ nm	No data	No data	2.8	Urine, tea, blood	1997	58
SP (Vis)	Based on the accelerating effect of the online microwave FIA system on the reaction of AA with Na_2MoO_4	11.36	No data	No data	Pharmaceuticals	1997	59
SP (Vis)	Copper(II) contained on a solid-phase reactor was reduced by AA, and then detected as its chelate produced with bathocuproine. $\lambda = 480$ nm	0.3	80	0.75	Pharmaceuticals	1998	60
SP (Vis)	Fe(III) was reduced by AA, and then determined by using ferrozine as chromogenic reagent. $\lambda = 562$ nm	0.5	No data	0.56	Pharmaceuticals	1998	61
SP (Vis)	Fe(III) contained on a solid-phase reactor was reduced by AA, and then detected as its chelate produced with 1,10-phenanthroline. $\lambda = 510$ nm	0.16	90	0.19	Soft drinks, pharmaceuticals, urine	1998	62
SP (UV)	A column packed with modified silica was used for interfering retention. $\lambda = 265$ nm	2.0	30	1.5	Pharmaceuticals	1998	63
SP (UV)	AA was determined by a continuous flow-through solid-phase spectrophotometry. The system was based on the measurement of its intrinsic absorbance at 267 nm when retained on a Sephadex QAE A-25 anion exchanger gel layer	0.22 0.17 0.11	28 24 21	0.87 1.08 0.90	Pharmaceuticals Sweets Urine	1999	64
SP (Vis)	Based on the color change of a permanganate solution with AA due to the redox reaction. $\lambda = 525$ nm	No data	90	2.5	Pharmaceuticals	1999	65

(*Continued*)

TABLE 18.1 (*Continued*) FIA Methods for Ascorbic Acid (AA) Determination with Spectrophotometric Detection

Detection Technique	Description of the Method	LOD (μM)	SF (h^{-1})	RSD (%)	Samples	Year	Reference
SP (Vis)	The procedure is based on the inhibiting effect of AA on the enhancing effect of oxalate on the potassium dichromate–potassium iodide/rhodamine 6G system	0.45	No data	<2	Pharmaceuticals Tomatoes Oranges	1999	66
SP (Vis)	The method is based on the redox reaction that takes place between AA and Fe(III), yielding DHAA and Fe(II). Fe(II) reacts with 1,10-phenanthroline, producing the Fe(phen)$_3^{2+}$ complex (ferroin). $\lambda = 512$ nm	2.8–15.3	40–60	1.0–1.6	Soft drinks Beer	2000	67
SP (UV)	UV flow-through optosensor using sodium acetate/ acetic acid and 0.05 M NaCl as carrier/self-eluting solutions and Sephadex QAE A-25 anion exchanger gel as solid phase	0.11	18–23	<1.3	Pharmaceuticals	2000	68
SP (Vis)	The procedure is based on the oxidation. of AA with Fe(III) and 2,2'-dipyridyl	No data	40	1.2	Blood serum	2001	69
SP (Vis)	Fe(III) is quantitatively reduced to Fe(II) with AA in the presence of 1,10-phenanthroline (phen), producing the Fe(II)-phen complex. $\lambda = 510$ nm	2×10^{-4}	45	<0.7	Pharmaceuticals	2001	70
SP (Vis)	The method is based on the reduction of Fe(III) to Fe(II) by AA, and the subsequent reaction of the produced Fe(II) with 2,2'-dipyridyl-2-pyridylhydrazone (DPPH) in acidic medium to form a colored complex. $\lambda = 535$ nm	9.7	120	0.1	Pharmaceuticals	2001	71
SP (Vis)	The procedure is based on the oxidation of AA with Fe(III) and 2,2'-dipyrydyl	2.3	40	1.2	Blood serum	2001	72
SP (UV)	Flow-through solid-phase (Sephadex QAE A-25) spectroscopic determination. The absorbance was monitored directly on the solid phase with a double-beam spectrophotometer. $\lambda = 250$ nm	2.0	22	1.22	Pharmaceuticals	2002	73

(*Continued*)

TABLE 18.1 (*Continued*) FIA Methods for Ascorbic Acid (AA) Determination with Spectrophotometric Detection

Detection Technique	Description of the Method	LOD (µM)	SF (h⁻¹)	RSD (%)	Samples	Year	Reference
SP (Vis)	The method is based on the inhibitory effect of AA on the oxidation of pyrogallol red by iodate in sulfuric acid media	0.5	100	0.6–3.4	Pharmaceuticals, fruit juice	2002	74
SP (Vis)	The procedure is based on the oxidation of AA with Fe(III) and 2,2′-dipyridyl	1.1	40	1.2	Rat's tissues	2002	75
SP (UV)	Optosensor for the simultaneous determination of two species with different electric charges present at very different concentrations (AA [or acetylsalicylic acid] and thiamine). Selective sorption and determination of a cationic analyte on a cation exchange gel, while the other, anionic analyte is determined in the solution among the interstices of the cation-exchange gel in the same flow cell. λ = 253 nm	39.8	12	1.85	Pharmaceuticals	2002	76
SP (UV)	Method comprising two solid-phase reactors: a polyethylene minicolumn filled with solid iodine and other column filled only with silica gel, which are then incorporated in a flow system. λ = 267 nm	0.45	110	1.0	Pharmaceuticals, foods	2003	77
SP (Vis)	The method is based on the oxidation of AA by potassium chromate and on reaction of unreacted chromate with sym-diphenylcarbazide. λ = 548 nm	0.28	90	1.65	Pharmaceuticals, foods	2003	78
SP (Vis)	Fe(III) is reduced by AA in sodium dodecyl sulfate micellar medium. Reduced Fe(II) is chelated with 2,2′-bipyridine. λ = 520 nm	No data	54	0.18	Pharmaceuticals, foods	2003	79

(*Continued*)

TABLE 18.1 (*Continued*) FIA Methods for Ascorbic Acid (AA) Determination with Spectrophotometric Detection

Detection Technique	Description of the Method	LOD (μM)	SF (h⁻¹)	RSD (%)	Samples	Year	Reference
SP (Vis)	The method is based on the reduction of Fe(III) to Fe(II) by AA, the reaction of the produced Fe(II) with 1,10-phenanthroline (phen) in a weak acidic medium to form a colored complex, and the subsequent oxidation reaction of Fe(II) to Fe(III) by the coexisting peroxodisulfate. $\lambda = 510$ nm	0.23	300	0.3–1.2	Pharmaceuticals	2004	80
SP (Vis)	Indirect determination based on the redox reaction between Fe(III)–thiocyanate complex and L-AA in acidic medium. $\lambda = 462$ nm	2.0	No data	2.0	Pharmaceuticals, foods	2004	81
SP (Vis)	AA reduces Cu(II) to Cu(I), and Cu(I) reacts with 2,9-dimethyl-1,10-phenanothroline (neocuproine) to form Cu(neocuproine)⁺ complex, which is extracted with N-phenylbenzimidoylthiourea (PBITU) in chloroform. $\lambda = 460$ nm	No data	No data	No data	Fruits, beverages, pharmaceuticals	2005	82
SP (Vis)	AA reduces methyl viologen to form a stable blue colored free radical ion. $\lambda = 600$ nm	0.57	No data	1.7	Foods, pharmaceuticals, biological samples	2007	83
SP (Vis)	AA is determined by using its catalytic effect on the complexation reaction of Cu(II) with 5,10,15,20-tetrakis(4-N-trimethyl-aminophenyl) porphyrin. $\lambda = 400$ nm	0.03	35	0.11	Soft drinks	2007	84
SP (Vis)	Discoloring spectrophotometry based on the discoloring effect of AA on the reaction between Fe(III) and SCN⁻. $\lambda = 480$ nm	No data	72	No data	Pharmaceuticals	2007	85
SP (UV)	Multivariate curve resolution using alternating least squares is used to quantify AA and acetylsalicylic acid	No data	No data	<5.0	Pharmaceuticals	2008	86

(*Continued*)

TABLE 18.1 (*Continued*)   FIA Methods for Ascorbic Acid (AA) Determination with Spectrophotometric Detection

Detection Technique	Description of the Method	LOD (µM)	SF (h⁻¹)	RSD (%)	Samples	Year	Reference
SP (Vis)	The method is based on the reaction kinetics of reduction of Fe(III)–1,10-phenanthroline complex to a red Fe(II)–phen complex. $\lambda = 510$ nm	No data	45	No data	Pharmaceuticals	2008	88
SP (Vis)	Based on reaction of AA with 1,10-phenanthroline Fe(III) complex. $\lambda = 510$ nm	No data	No data	<3.0	Pharmaceuticals	2008	89
SP (Vis)	Method based on the reaction of AA with iodine, and photometric measurement of the residual iodine as triiodide at 350 nm	No data	No data	No data	Soft drinks	2009	90
SP (UV)	AA and DHAA are resolved by a reversed-phase column, and DHAA is reduced to AA by an online postcolumn reaction with dithiothreitol. $\lambda = 260$ nm	1.7	No data	No data	Beverages	2009	91
SP (UV)	Determination of AA at 300 nm before a photodegradation step	12.8	257	3.4	Fruit juices	2010	92
SP (Vis)	Indirect determination of AA based on reduction of Cu(II) to Cu(I), which combines with SCN⁻ to form copper thiocyanate precipitate	4.8	No data	No data	Beverages	2010	93
SP (Vis)	Indirect determination of AA based on reduction of Cu(II) to Cu(I), which combines with SCN⁻ to form copper thiocyanate precipitate	4.8	No data	No data	Beverages	2011	94
SP (Vis)	Based on the reduction of Cu(II) by AA followed by absorbance measurement of the complex of Cu(I) with 2,2′-biquinoline-4,4′-dicarboxylic acid at 558 nm	2.8	60	<1.7	Pharmaceuticals	2011	95
SP (Vis)	AA accelerates dediazoniation of diazonium ions. The derivatization product is monitoring from coupling unreacted diazonium ion with phenol to give an azo dye	1.4	No data	No data	Pharmaceuticals	2011	96

(Continued)

TABLE 18.1 (*Continued*) FIA Methods for Ascorbic Acid (AA) Determination with Spectrophotometric Detection

Detection Technique	Description of the Method	LOD (µM)	SF (h⁻¹)	RSD (%)	Samples	Year	Reference
SP (Vis)	In acidic medium AA is oxidized by Fe(III). Fe(II) forms a complex with *o*-phenanthroline hydrate. $\lambda = 510$ nm	No data	56	No data	Tomatoes	2011	97
SP (Vis)	Based on reduction of Fe(III) to Fe(II) by the AA, and by the subsequent reaction of the Fe(II) with 2,4,6-tripyridyl-s-triazine in buffered medium. $\lambda = 593$ nm	2.4×10^{-10}	180	0.8	Pharmaceuticals	2012	98
SP (Vis)	A polyester resin that adsorbs Fe^{3+} is packed in a glass tubing as a mini reaction column. When a stream of AA merged with the mini reaction column, the AA reacts with the Fe^{3+}, which is reduced to Fe^{2+} (no adsorbed by the resin). Fe^{2+} reacts with phenanthroline to form a complex	1.6	60	1.8	No data	2012	99

Note: LOD: detection limit; RSD: relative standard deviation; SF: sampling frequency; SP: spectrophotometry; DHAA: dehydroascorbic acid.

TABLE 18.2　FIA Methods for Ascorbic Acid (AA) Determination with Chemiluminescence Detection

Description of the Method	LOD (µM)	SF (h⁻¹)	RSD (%)	Samples	Year	Reference
Based on the decrease of chemiluminescence signal produced by the reaction between luminol, H_2O_2 and peroxidase in basic medium in presence of AA	No data	No data	2.0	Juices, wines	1990	100
Based on a luminol–Fe(III)–hydrogen peroxide system	1×10^{-6}	2–56	1.4	Pharmaceuticals, juices	1993	101
Based on the chemiluminescent reaction in basic medium of lucigenin with the products from the photooxidation of AA sensitized by toluidine	1×10^{-4}	80	0.4–1.2	Pharmaceuticals, juices, soft drinks, blood serum	1995	102
Based on the lucigenin chemiluminescence emitted by AA in basic medium	1×10^{-3}	30	1.9	No data	1996	103
Sensor based on the decrease of chemiluminescence signal produced by the reaction between luminol and ferricyanide in presence of AA	0.031	60	<5.0	Pharmaceuticals, vegetables	1996	104
Based on the suppression of the chemiluminescence reaction of potassium hexacyanoferrate(III) with luminol by AA	No data	No data	0.5	Pharmaceuticals, blood	1996	105
Suppression of the chemiluminescence reaction of $K_7Cu(IO_6)_2$ with luminol in basic medium by AA	0.015	No data	2.7	Pharmaceuticals	1996	106
Sensor based on the decrease of chemiluminescence signal produced by the reaction between luminol and permanganate in presence of AA	No data	No data	2.3	Beverages	1997	107
Direct chemiluminescence detection from the oxidation of AA with permanganate	No data	No data	No data	Pharmaceuticals	1997	108
Based on a luminol–Fe(III)–hydrogen peroxide system	No data	No data	No data	Pharmaceuticals	1997	109
Based on the inhibition of AA on the chemiluminescence reaction of the luminol/Fe(II)/O_2 system	1.1×10^{-3}	No data	No data	Pharmaceuticals, vegetables	1997	110
Chemiluminescence sensor containing an anion-exchange resin for immobilization of luminol and a cation exchange resin for immobilization of Fe(II)	2×10^{-3}	No data	2.8–4.2	Pharmaceuticals, vegetables	1997	111

(Continued)

TABLE 18.2 (*Continued*)　FIA Methods for Ascorbic Acid (AA) Determination with Chemiluminescence Detection

Description of the Method	LOD (μM)	SF (h⁻¹)	RSD (%)	Samples	Year	Reference
Chemiluminescence was directly produced by the reaction of AA with luminol in the presence of a catalyst (hexacyanoferrate(III) and hexacyanoferrate(II)) in an alkaline solution	6×10^{-6}	No data	No data	No data	1997	112
Based on the decrease of chemiluminescence signal produced by the reaction between luminol and permanganate in presence of AA	No data	No data	3.2	Vegetables	1998	113
Based on the suppression of the chemiluminescence of the luminol/H_2O_2/KIO_4 system by AA	0.06	No data	1–2.1	Pharmaceuticals	1999	114
Indirect determination of AA based on the suppression of luminol/KIO_4 chemiluminescence system	0.06	No data	1.0	Pharmaceuticals	1999	115
A reactor with immobilized ascorbate oxidase is used for selective decomposition of the AA. The signal is registered after the passage of the sample through the enzymatic reactor and is compared with that obtained for the sample that did not flow through the reactor. The difference between the two signals was dependent on the AA concentration	5	4	3.13	Fruit juices	2000	116
Chemiluminescence suppression method based on luminol–KIO_4^-–H_2O_2–AA system	0.06	No data	1.0	Pharmaceuticals	2000	117
Based on AA acidic reduction of potassium dichromate to generate Cr^{3+} and luminol–H_2O_2–Cr chemiluminescence reaction	1.3×10^{-3}	No data	<5.0	Pharmaceuticals	2000	118
Based on the inhibition effect of AA on the chemiluminescence reaction of luminol and superoxide anion radical	No data	No data	1.1	Pharmaceuticals	2001	119
The method is based on the chemiluminescent reaction of rhodamine B with Ce(IV) in sulfuric acid media	1.3×10^{-7}	No data	0.92	Fresh vegetables	2002	120
Based on the fast reduction reaction between Cr(VI) and AA in acid medium. Cr(III) generated couples with luminol–H_2O_2 CL chemiluminescent reaction in alkaline solution	1.2×10^{-3}	No data	1.2–2.3	No data	2003	121
Based on formaldehyde enhancing the chemiluminescence from the oxidation of AA by Mn(IV)	0.02	No data	2.3	Pharmaceuticals	2004	122

(*Continued*)

TABLE 18.2 (*Continued*)   FIA Methods for Ascorbic Acid (AA) Determination with Chemiluminescence Detection

Description of the Method	LOD (μM)	SF (h⁻¹)	RSD (%)	Samples	Year	Reference
Based on chemiluminescence reaction of AA with potassium permanganate in the presence of formaldehyde as enhancer	0.01	No data	2.5	Pharmaceuticals	2004	123
Based on the chemiluminescence emitted during the oxidation of AA by N-bromosuccinimide (NBS) in alkaline medium in the presence of fluorescein as energy transfer	0.05	No data	2.3	Pharmaceuticals	2004	124
Based on manganese-based chemiluminescence detection	0.05	No data	1.0	Pharmaceuticals	2004	125
Simultaneous determination of AA and L-cysteine (Cys) with partial least squares calibration. This method is based on the fact that both analytes can quantitatively reduce Fe^{3+} to Fe^{2+}, and that the reaction rates of AA and Cys with Fe^{3+} are different. The reduced product, Fe^{2+}, is detected with the luminol–Fe^{2+}–O_2 chemiluminescent system	0.17	No data	3.2	Pharmaceuticals, urine	2005	126
Based on the chemiluminescence emitted during the oxidation of AA by potassium permanganate in hydrochloric acid condition in the presence of formic acid as enhanced sensibility	0.02	No data	4.7	Pharmaceuticals	2007	127
Hydroxyl radical generated on line by the reaction between Fe^{3+} and H_2O_2 in HCl medium could oxidize rhodamine 6G to produce weak chemiluminescence. AA enhances the chemiluminescence	6×10^{-3}	60	3.1	Foods	2007	128
Inhibition of chemiluminescence of luminol–hypochlorite system. The hypochlorite is electrogenerated online by constant-current electrolysis	5×10^{-8}	No data	No data	Pharmaceuticals	2007	129
The poor chemiluminescence resulting from the oxidation of AA by Ce(IV) in acid media is enhanced by quinine	68.1	No data	2.5	Pharmaceuticals	2009	130
Indirect determination based on quenching of the titanium–luminol complex chemiluminescence caused by adding AA	5.7	515	No data	Pharmaceuticals	2009	131
The addition of trace L-AA caused significant quenching luminescence intensity in the mixing hydrogen peroxide, hydrogen carbonate, and CdSe/CdS quantum dots	6.7×10^{-3}	No data	<5.3	Human serum	2010	132

(*Continued*)

TABLE 18.2 (*Continued*)　FIA Methods for Ascorbic Acid (AA) Determination with Chemiluminescence Detection

Description of the Method	LOD (μM)	SF (h^{-1})	RSD (%)	Samples	Year	Reference
Based on the fact that the chemiluminescence signal from the reaction of luminol–$KBrO_3$ can be strengthened by AA	7×10^{-3}	No data	1.5	Pharmaceuticals, fruit juices	2010	133
Based on the chemiluminescence from the oxidation of AA by nanocolloidal manganese dioxide under acidic conditions	2.3	No data	2.1	Pharmaceuticals	2011	134
Evaluation of quenching effect of AA on peroxynitrite ($ONOO^-$)	2.8	No data	3.4–9.0	Fruit juices	2011	135
Based on a novel 3-(4′-nitryl)-5-(2′-sulfonic-phenylazo) rhodanine–HCl–7-$KMnO_4$ chemiluminescence reaction system	176	No data	No data	Foods	2011	136
Based on strong inhibition effect of AA on chemiluminescence emission produced from the luminol–H_2O_2-supported Co/SiO_2 system	4.2×10^{-4}	No data	No data	Pharmaceuticals	2011	137
Based on the inhibitory effect of AA on the chemiluminescence reaction between luminol and H_2O_2	6.0×10^{-3}	No data	6.7	Urine	2011	138
Based on the CdTe nanocrystals and potassium permanganate chemiluminescence system. AA is a sensitive enhancer of the above energy-transfer excitation process	2.9×10^{-4}	No data	2.3	Pharmaceuticals, fresh vegetables	2012	139
Based on that AA could inhibit the chemiluminescent reaction of glyoxal and potassium permanganate in hydrochloric acid	5.0×10^{-7}	No data	1.7	Fruits, fresh vegetables	2012	140
Simultaneous determination of AA and rutin (RT) using partial least squares calibration. Method based on the fact that AA and RT can reduce Fe^{3+} to Fe^{2+} at different reaction rates	0.05	No data	No data	Pharmaceuticals, urine	2013	141

Note: LOD: detection limit; RSD: relative standard deviation; SF: sampling frequency.

TABLE 18.3   FIA Methods for Ascorbic Acid (AA) Determination with Fluorimetric Detection

Description of the Method	LOD (µM)	SF (h^{-1})	RSD (%)	Samples	Year	Reference
The FIA system automatically buffers and dilutes the juice sample, which is then analyzed with a fiberoptic O optrode biosensor. The biosensor contains AA oxidase immobilized on the O optrode surface, and O consumption is measured by dye fluorescence quenching by molecular oxygen	No data	No data	No data	Fruit juices	1989	142
Total AA was quantitated by oxidizing to dehydroascorbic acid with HgCl$_2$. Then, the total DHAA reacts with o-phenylenediamine to form a fluorescent compound	No data	No data	No data	Foods	1989	143
Detection was achieved by monitoring the reduction of Ce(IV) to Ce(III) by AA in acidic conditions. The concentration of AA could be directly related to the concentration of Ce(III) produced, which was determined at 350 nm (excitation at 260 nm)	1.7	No data	0.86	Beverages	1991	144
AA samples were loaded into two serial sample injection valves and injected into a carrier system containing mercuric chloride and o-phenylenediamine. The formation of a fluorescent quinoxaline was monitored at 435 nm (excitation at 366 nm)	0.6	30	2.0	Pharmaceuticals	1991	145
Enzymatic kinetic spectrofluorometric procedure combining the merging zones based on the oxidation of AA to dehydroascorbic acid, which reacts with o-phenylenediamine to form a fluorescent quinoxaline	No data	30	2.0–4.1	Pharmaceuticals, beverages, urine	1995	146
The Leucothionine blue, formed during the photooxidation of AA sensitized by Thionine blue, is highly fluorescent	No data	80	1.6	Pharmaceuticals, juices, soft drinks	1997	147
Based on the oxidation of AA by thallium(III) and the parameter measured is the fluorescence of thallium(I) formed	No data	45 ± 5	1.3	Pharmaceuticals, juices	1998	148
Method for the determination of mixtures of thiamine and AA. The procedure is based on the oxidation with mercury(II) of the B1 and C vitamins to form thiochrome (TC) and quinoxaline derivative, respectively. Both reaction products exhibit fluorescence at the same wavelengths ($\lambda_{ex} = 356$ nm, $\lambda_{em} = 440$ nm)	7.4	25	0.6–2.3	Pharmaceuticals	2004	149

(Continued)

TABLE 18.3 (*Continued*) FIA Methods for Ascorbic Acid (AA) Determination with Fluorimetric Detection

Description of the Method	LOD (μM)	SF (h⁻¹)	RSD (%)	Samples	Year	Reference
This method is based on the reduction of Tl(III) with AA or cysteine in acidic media, producing fluorescence reagent, $TlCl_3^{2-}$ ($\lambda_{ex} = 227$ nm, $\lambda_{em} = 419$ nm)	0.8	16	2.1	Lemon juice, shampoo, water	2005	150
Perylene bisimide–linked nitroxide (PBILN) is used as a fluorescent reagent, which permits the selective determination of AA. The fluorescence of the perylene bisimide moiety in PBILN is quenched by the nitroxide moiety, which is linked to the perylene bisimide. When a stream of AA is merged with a stream of PBILN, the AA reacts with the nitroxide moiety of PBILN to form hydroxylamine, and the fluorescence properties of the perylene bisimide moiety are recovered	0.28	10	1.0	Fruit juices, soft drinks	2011	151

Note: LOD: detection limit; RSD: relative standard deviation; SF: sampling frequency.

and superoxide anion radical [82]; luminol and hypochlorite [82]; titanium–luminol complex [83], luminol, hydrogen peroxide and potassium periodate [84]; or glyoxal and potassium permanganate [85].

FIA-chemiluminescence sensors were made by immobilization of analytical reagents (luminol and ferricyanide, or luminol and permanganate) on an anion exchange resin column [75,86].

Other chemiluminescence FIA methodologies for AA determination are based on AA acidic reduction of potassium dichromate to generate Cr(III) and subsequent luminol–hydrogen peroxide-Cr(III) chemiluminescence reaction [87,88], and on chemiluminescence reaction of AA with potassium permanganate in the presence of formaldehyde [89,90], formic acid [91], or CdTe nanocrystals [92] as sensitivity enhancers. Quinine is also used as an enhancer of chemiluminescence produced by reaction between AA and Ce(IV) in acid media [93]. Pérez-Ruíz et al. [94] proposed a chemiluminescent reaction, in alkaline solution, of lucigenin with the products from the photooxidation of AA sensitized by toluidine blue.

Fluorometric methods are based on the formation of fluorescent species in redox reactions where oxidation of AA is produced. The fluorescent species can be o-phenylenediamine [95,96], quinoxaline [96–98], Leucothionine blue [99], and thallium(I) [100], where the violet fluorescence is attributed to the presence of the $TlCl_3^{2-}$ ion.

18.3.1.2 Electroanalytical Techniques

FIA methods with electroanalytical detectors are mainly based on the inherent redox chemistry of the analyte and reagents. The most important details of the published FIA determination based on electroanalytical detection (amperometry, coulometry, conductimetry, potentiometry, polarography, and voltammetry) following a chronological order are contained in Table 18.4.

Karlberg et al. [101] developed for first time a FIA method for AA determination using platinum electrodes for redox potential measurements of AA because such electrodes were used earlier in air-segmented flow systems. A high-sensitivity measurement of AA has been carried out amperometrically. Although this detector is not selective, because samples that contain redox compounds can cause interference, amperometric detection of AA achieves good results when the FIA manifold has a packed-bed immobilized-enzyme (ascorbate oxidase) reactor [102–108] or a carrier solution containing ascorbate oxidase [109,110], or an enzymic treatment of the sample is done [107,111]. Platinum or reticulated vitreous carbon flow-through electrodes and a wall-jet glassy electrode or a sessile mercury drop electrode without the need to deoxygenate the eluent or sample are used as detector cells. The photochemical reduction of methylene blue [112–114] and the redox reaction of iodate ion [115,116] are employed in amperometric determinations. The major advantage of coulometric methods is that the analyte is completely electrolyzed and can be determined without calibration graphs [117–119]. The direct and simultaneous determination of AA and glucose was carried out with a flow-through electrolysis cell for a coulometric determination of AA based on its quantitative electrolysis [120]. The major problem with the use of an electrochemical method as detector in FIA is the variability of the analytical current under different flow conditions, and the limited selectivity using the controlled-potential method. Square-wave voltammetry is a selective and sensitive electrochemical technique for use with FIA because the measuring time for each voltammetric run is only a few seconds [121]. The use of polymer-modified electrodes for sensor development has received considerable attention for AA determination. This is due to the advantages of these devices over

TABLE 18.4   FIA Methods for Ascorbic Acid (AA) Determination with Electroanalytical Detection

Detection Technique	Description of the Method	LOD (μM)	SF (h⁻¹)	RSD (%)	Samples	Year	Reference
P	The redox potential shift determined by the reduction of Ce(IV) to Ce(III) by AA was detected by platinum electrodes	No data	45–60	1.2	Pharmaceuticals	1978	152
C	Reticulated vitreous carbon flow-through electrode as detector	No data	24	1–2	Pharmaceuticals	1979	153
A	Reticulated vitreous carbon flow-through electrode as detector	No data	264	0.5	Pharmaceuticals	1979	153
A	An immobilized enzyme (AA oxidase) reactor was used	0.01	No data	1–3	Brain tissue	1983	154
A	Based on the redox reaction of iodate ion with AA	No data	No data	0.7	Aqueous solutions	1984	155
A	A sessile mercury drop electrode without the need to deoxygenate the eluent or sample was used	0.06	No data	1.0	Aqueous solutions	1985	156
Po	Ascorbate oxidase was immobilized on CH-Sepharose via carbodiimide	No data	60	5.0	Fruit juices	1985	157
V/A	Dual-electrode detector	No data	No data	No data	Aqueous solutions	1986	158
A	Thin-layer flow detector with a glassy carbon electrode coated with a film of protonated poly(4-vinylpyridine)	2.3×10^{-4}	No data	12	Aqueous solutions	1987	159
A	Cucumber juice with a high ascorbate oxidase activity was used as carrier solution	100	60	4.0	Aqueous solutions	1988	160
A	AA was determined on a capillary electrochemical detection cell	No data	No data	1.5–5	No data	1988	161
A	AA was oxidized by iodine and a decrease in the iodine signal was observed	1.0	No data	No data	Aqueous solutions	1989	162
V	Metalloporphyrin coated glassy carbon electrodes were used as electrocatalytic voltammetric sensor for AA	5.0–10.0	No data	No data	Aqueous solutions	1989	163

(Continued)

TABLE 18.4 (*Continued*) FIA Methods for Ascorbic Acid (AA) Determination with Electroanalytical Detection

Detection Technique	Description of the Method	LOD (µM)	SF (h⁻¹)	RSD (%)	Samples	Year	Reference
A	Ascorbate oxidase was immobilized on cyanogen bromide-activated sepharose 4B	0.022	30	1.0	Fruit juices	1990	164
A	Simultaneous enzymatic determination of glucose and AA	No data	No data	No data	Fruit juices	1990	165
O electrode	Enzymic treatment with cucumber juice. The O consumed in the enzymic reaction was detected	No data	No data	No data	Fruits	1990	166
A	Automated enzyme packed-bed system (ascorbate oxidase)	0.57	15	No data	Foods	1991	167
V, A	Electrodes modified by electrodeposition of poly(3-methylthiophene) were used as chemical sensors of AA	No data	No data	No data	No data	1991	168
P	Multichannel electrochemical detection system fabricated using a 16-channel microelectrode array	No data	No data	No data	Blood, urine	1992	169
A	1. Simultaneous determination of uric acid and AA: spectrophotometric determination of uric acid and amperometric determination of both analytes 2. Individual determination of uric acid and AA	0.85	80	0.8	Aqueous solutions	1992	170
V	Square-wave voltammetry. AA buffered at pH 2.87	20	No data	0.82–5.5	Soft drinks, fruit juices	1992	171
A	An enzyme reactor (ascorbate oxidase) was used	No data	No data	0.9	Pharmaceuticals	1993	172
A	Based on AA oxidation in acidic medium using bromate or iodate solutions as oxidizing reagents	No data	30	No data	Medicinal drugs	1994	173
P	The enzyme ascorbic oxidoreductase, extracted from *Cucurbita maxima*, is immobilized onto alkylamine glass beads using glutaraldehyde as a bifunctional agent. AA concentration is related to oxygen saturation. Fall in oxygen concentration., as a result of AA oxidation, is detected by a homemade oxygen electrode	0.02	90	1.0	Biological fluids, fruit juices	1994	174

(*Continued*)

TABLE 18.4 (*Continued*) FIA Methods for Ascorbic Acid (AA) Determination with Electroanalytical Detection

Detection Technique	Description of the Method	LOD (μM)	SF (h⁻¹)	RSD (%)	Samples	Year	Reference
V, A, Ch	Study of V, A, and Ch behavior of AA at polypyrrole-dodecyl sulfate film-coated electrodes	No data	No data	5.0	Aqueous solutions	1994	175
V, A	The voltammetric behavior of AA and a series of biologically important compounds were studied at a glassy carbon electrode pretreated electrochemically and at an electrode coated with a copolymer of maleic acid anhydride attached with Eastman-AQ55D	No data	No data	No data	Aqueous solutions	1994	176
P	Based on a reaction of AA with Cu(II) at pH 5.0 to form a readily oxidizable species	No data	No data	1.0	Soft drinks, fruit juices	1995	177
A	AA was detected in a wall-jet detection cell	No data	120	No data	Pharmaceuticals	1995	178
A	Electrocatalytic oxidation of AA at a redox polymer modified electrode	1.0	50	0.1	Aqueous solutions	1995	179
C	AA was determined with a coulometric detector using a porous carbon felt as working electrode	No data	No data	No data	No data	1995	180
A	Direct oxidation of L-ascorbate on a platinum electrode. The signal difference between the ascorbate oxidase-immobilized column and the blank column is measured as a net concentration of L-ascorbate	No data	No data	<2	Fruits	1995	181
A	Based on the photochemical reduction of Methylene Blue in 0.1 M phthalate buffer at pH 3.8	10.8	45–50	1.3–4.8	Pharmaceuticals	1996	182
A	Electrooxidation of AA on a dispersed-platinum glassy carbon electrode	No data	No data	No data	Fruit juices	1996	183
A	System using the diaphragm micropump equipped with a piezoelectric vibrator (bimorph oscillator)	No data	No data	<1	Soft drinks	1996	184

(*Continued*)

TABLE 18.4 (*Continued*) FIA Methods for Ascorbic Acid (AA) Determination with Electroanalytical Detection

Detection Technique	Description of the Method	LOD (μM)	SF (h⁻¹)	RSD (%)	Samples	Year	Reference
BA	Based on AA oxidation to dehydroascorbic acid in acidic medium using iodine–iodide solution as oxidizing reagent. The iodine amount consumed in the redox reaction was detected	No data	60	1.8	Fruit juices	1997	185
A	Electrocatalysis of AA on a glassy carbon electrode chemically modified with polyaniline films	1.0	No data	No data	Aqueous solutions	1997	186
A, V	AA was determined at a vitreous C electrode modified with 3,4-dihydroxybenzaldehyde	0.3, 10	No data	4.0, no data	Pharmaceuticals	1997	187
A	AA was determined with a chemically modified with methylene green (electron mediator) carbon paste electrode	0.01	No data	0.6	Aqueous solutions	1997	188
V	Electrodes modified by the electrodeposition of conducting organic polymers such as poly(3-methylthiophene), polypyrrole, and polyaniline are used as chemical sensors	10^{-2} to 10^{-3}	No data	No data	Aqueous solutions	1997	189
P	Clay-modified electrodes with ruthenium purple	5.7	No data	No data	Aqueous solutions	1997	190
A	Enzymic treatment with *Cucumis sativus* tissue which is a rich source of ascorbate oxidase	No data	No data	1.0	Pharmaceuticals, soft drinks/juices	1998	191
C	Based on the quantitative electrolysis of AA	No data	No data	No data	Soft drinks	1998	192
A	A four-channel multipotentiostat for simultaneous measurements with microelectrode arrays. To reduce the complexity of electrochemistry cell, only one reference, and one auxiliary electrode are used	No data	No data	No data	Synthetic samples	1998	193
A	Electrocatalytic determination of AA using tetraaza macrocycle-modified electrodes	0.47	No data	5.0	Beverages	1999	194
Co	Neutralization of AA in ammonia leading to a change in conductivity	No data	90	2.5	Pharmaceuticals	1999	65

(*Continued*)

TABLE 18.4 (*Continued*) FIA Methods for Ascorbic Acid (AA) Determination with Electroanalytical Detection

Detection Technique	Description of the Method	LOD (µM)	SF (h⁻¹)	RSD (%)	Samples	Year	Reference
A	Based on diamond films, fabricated by chemical vapor deposition	0.012	No data	No data	Aqueous solutions	1999	195
P	Indirect determination by using a tubular electrode with silver tube chemically pretreated with $HgCl_2$ and KI solutions	No data	40	No data	Pharmaceuticals	1999	196
A	An array of gold microelectrodes modified by electrochemical deposition of palladium is employed as working electrode	3	No data	<1	Urine	2000	197
V	Ascorbate sensor based on a glassy carbon electrode modified with a cellulose acetate polymeric film bearing 2,6-dichlorophenolindophenol	10	25	0.75–1.2	Pharmaceuticals, nonalcoholic beverages	2000	198
BA	Based on the oxidation of AA and reduction of dissolved oxygen and employs two identical activated platinum wire electrodes	8	No data	1.58	Pharmaceuticals, fruit juices	2000	199
P	Uses a tubular electrode based on the redox properties of copper(II) ions occluded in polyethylene–co-vinyl acetate (EVA) membrane	850	120	3.6	Pharmaceuticals	2000	200
A	Simultaneous determination of AA and SO_2 based on the oxidation of both analytes in a dual-channel amperometric detection system	8.5	30	1.0	Wines, fruit juices	2000	201
DEP	Two similar platinum electrodes are employed and polarized by a constant current	No data	No data	<0.5	Pharmaceuticals	2000	202
A	Quantification of AA, dopamine, epinephrine and dipyrone in mixtures is performed by using an array of microelectrodes with units modified by the electrodeposition of different noble metals	No data	No data	No data	Aqueous solutions	2000	203

(*Continued*)

TABLE 18.4 (*Continued*) FIA Methods for Ascorbic Acid (AA) Determination with Electroanalytical Detection

Detection Technique	Description of the Method	LOD (μM)	SF (h⁻¹)	RSD (%)	Samples	Year	Reference
P	Tubular electrode made by chemical pretreatment of a Ag tube with Hg(II) chloride and iodide used as a potentiometric sensor	No data	No data	No data	Pharmaceuticals	2000	204
A	Sensor based on electropolymerized aniline. Polyaniline is grown on both glassy carbon and screen-printed electrodes	2.45	No data	1.2–2.0	Pharmaceuticals, foods	2001	205
A	Utilization of Prussian blue film-modified electrode	2.49	No data	2.5	Pharmaceuticals	2001	206
A	Gold electrodes from compact discs modified with platinum	1.0	90	<1	Pharmaceuticals	2001	207
A	Cobalt(II) phthalocyanine (CoPc) modified graphite–epoxy electrode	13	No data	<1	Pharmaceuticals	2001	208
A	Movable tubular electrode, which allows standard addition to be performed automatically online	No data	55	3.8	Fruit juices	2002	209
A	Laccase covalently immobilized onto porous glass beads is used as a recognition element for L-AA. The immobilized enzymes are packed into a small polymer column, and then mounted in a water-jacketed holder. The biosensing system is assembled with the column unit and a flow-through type of an oxygen electrode for monitoring dissolved oxygen enzymatically consumed	15	No data	No data	Aqueous solutions	2002	210
A	Electrolytic device to extend determination range for biosensing with use of oxidase by using an immobilizing laccase column as a recognition element	No data	No data	No data	Aqueous solutions	2002	211

(*Continued*)

TABLE 18.4 (*Continued*) FIA Methods for Ascorbic Acid (AA) Determination with Electroanalytical Detection

Detection Technique	Description of the Method	LOD (μM)	SF (h^{-1})	RSD (%)	Samples	Year	Reference
A	Based on the photochemical reduction of methylene blue (MB) in 0.1 M phosphate buffer. MB is used as a redox mediator for the modification of a carbon paste electrode (CPE) due to its facile reducible-oxidizable behavior leached from the matrix. MB is reduced to nearly quasi-reversible at the modified carbon paste electrode (MCPE)	0.01	60	2.0	Pharmaceuticals	2003	212
A	Conductive boron-doped diamond thin-film electrodes	No data	No data	No data	Aqueous solutions	2003	213
P	Single stream manifold, consisting in a peristaltic pump, an injector with internal loop and two Pt electrodes connected to a PC by serial or parallel interface card	No data	No data	3.0–3.4	Aqueous solutions	2003	214
V	Polytoluidine blue (PTB) film is chemically modified on glassy carbon (GC) electrode by cyclic voltammetry in 0.1 M H$_3$PO$_4$ buffer	0.048	No data	4.2	Aqueous solutions	2003	215
V	Glucose oxidase, lactate oxidase, L-amino acid oxidase, and alkaline oxidase are immobilized on new films based on 2,6-dihydroxynaphthalene copolymerized with 2-(4-aminophenyl)-ethylamine onto the Pt electrodes	No data	No data	No data	Aqueous solutions	2003	216
V	Silica sol–gel glass-coated ferricyanide-doped Tosflex-modified screen-printed electrode used for the mediated oxidation of AA (neutral pH).The electrochemical mediation of AA is found to follow the Michaelis–Menten kinetic pathway	0.046	No data	2.68	Urine, pharmaceuticals, apple juice	2003	217
V	A polymer film of N,N-dimethylaniline having a positive charge on the quaternary ammonium group in its backbone is electrochemical deposited on a glassy carbon electrode surface resulting a film-coated GC electrode	No data	No data	No data	Aqueous solutions	2004	218

(*Continued*)

TABLE 18.4 (*Continued*)   FIA Methods for Ascorbic Acid (AA) Determination with Electroanalytical Detection

Detection Technique	Description of the Method	LOD (μM)	SF (h⁻¹)	RSD (%)	Samples	Year	Reference
A	The polymer film of *N,N*-dimethylaniline is deposited on the electrochemically pretreated glassy carbon electrode by continuous electrooxidation of the monomer. This film-coated electrode is be used as an amperometric sensor of AA	No data	No data	No data	Fruit juices	2004	219
A	Based on a L-AA biosensor based on ascorbate oxidase. The enzyme is extracted from the mesocarp of cucumber by using 0.05 mol L⁻¹ phosphate buffer, pH 5.8 containing 0.5 mol L⁻¹ NaCl	No data	No data	No data	Pharmaceuticals	2005	220
A	Methylene blue (MB) is incorporated into titanium phosphate (TiP) after pretreatment of TiP with the gas butylamine. The dye is strongly retained and not easily leached from the layered host matrix. The adsorbed MB on TiP is used to prepare modified carbon paste electrodes (MCPE)	No data	No data	No data	Pharmaceuticals	2005	221
V	Electrocatalytic determination of AA at the graphite electrode modified with a polyaniline film containing palladium particles	No data	No data	< 5.0	Pharmaceuticals	2006	222
A	Based on poly(vinyl chloride) tetrathiafulvalene–tetracyanoquinodimethane (TTF-TCNQ) composite electrode	15.1–31.8	No data	0.46–1.13	Aqueous solutions	2006	223
A	A glassy carbon electrode is modified with an electropolymerized film of 1-naphthylamine in aqueous solution	5.7	No data	3.4	Pharmaceuticals, nonalcoholic beverages	2007	224
A	A PVC/TTF-TCNQ composite electrode is employed as detector for the simultaneous detection of AA and uric acid	120	No data	0.7	Aqueous solutions	2007	225

(*Continued*)

TABLE 18.4 (*Continued*) FIA Methods for Ascorbic Acid (AA) Determination with Electroanalytical Detection

Detection Technique	Description of the Method	LOD (μM)	SF (h⁻¹)	RSD (%)	Samples	Year	Reference
V	The anodic oxidation of AA on a ruthenium oxide hexacyanoferrate–modified electrode is characterized by cyclic voltammetry	2.2	No data	2.0	Urine	2008	226
V	Thermoelectrochemistry at disposable screen-printed carbon electrodes	0.0287	No data	1.91	Orange juice	2008	227
A	Glassy carbon, carbon paste and modified carbon paste electrodes are used. All electrodes are inserted in a wall-jet device with an Ag/AgCl reference electrode and a platinum auxiliary electrode	No data	48	No data	Soft drinks	2008	228
P	The potentiometric FIA signal is based on the reaction of formation the sparingly soluble salts between thiols and Ag^+ ions. AA has no influence on potentiometric signal at any experimental concentration	No data	45	No data	Pharmaceuticals	2008	88
A	Dialysis sample pretreatment. AA is electrochemically oxidized at glassy carbon electrode giving an anodic peak current proportional to AA concentration	85.2	20	1.5	Pharmaceuticals, fruit juices	2008	229
A	Simultaneous determination of paracetamol (PC) and AA. The method allows the resolution of the mixtures without chemical pretreatment of the sample or electrode modification or the use of chemometric techniques for data analysis	1.1	60	0.59	Pharmaceuticals	2008	230
V	Poly(acriflavine), a compact thin film, is electropolymerized on the glassy carbon electrode by cyclic voltammetry for sensing AA and dopamine	1.5	No data	No data	Pharmaceuticals, fruit juices	2009	231
P	Platinum electrodes are used as an indicating system to follow the oxidation of vitamin C with potassium iodate, or potassium permanganate (acidic medium)	51.1–62.5	No data	11–12	Pharmaceuticals	2009	232

(*Continued*)

TABLE 18.4 (*Continued*) FIA Methods for Ascorbic Acid (AA) Determination with Electroanalytical Detection

Detection Technique	Description of the Method	LOD (μM)	SF (h⁻¹)	RSD (%)	Samples	Year	Reference
P	Novel periodate selective sensors based on newly synthesized Ni(II)–Schiff base complexes as neutral receptors	5.1	50	1.3	Pharmaceuticals, nonalcoholic beverages	2010	233
P	Based on the oxidation of AA by periodate. A new potentiometric periodate sensor is constructed to monitor this reaction. The selective membranes are of PVC with porphyrin-based sensing systems and a lipophilic cation as additive	30.1	96	0.2–3.5	Pharmaceuticals, nonalcoholic beverages	2010	234
A	An amperometric detector and an enzymic reaction are combined for the measurement of L-AA. The enzyme cell (containing immobilized ascorbate oxidase) is connected to a FIA system with a glassy carbon electrode as an amperometric detector	5	25	5.0	Foods	2010	235
V	Based on a thin layer dual-electrode detector developed for the selective and simultaneous determination of AA and hydrogen peroxide present in the same samples	1	No data	5.0	Soft drinks	2011	236
P	A polymeric membrane permanganate-selective electrode is used. By applying an external current, diffusion of permanganate ions across the polymeric membrane can be controlled precisely	0.078	No data	3.1	Pharmaceuticals, vegetables	2011	237
A	Uses a new and/or re-used graphite screen-printed electrode modified with silver hexacyanoferrate and a Nafion polymer layer	28.4	65–70	< 2	Orange juice, drugs	2011	238

(*Continued*)

TABLE 18.4 (*Continued*) FIA Methods for Ascorbic Acid (AA) Determination with Electroanalytical Detection

Detection Technique	Description of the Method	LOD (µM)	SF (h⁻¹)	RSD (%)	Samples	Year	Reference
A	A gold electrode modified with the self-assembled monolayer of a heterocyclic thiol, 3-mercapto-1,2,4-triazole (MTz), is used for the detection of uric acid (UA) and AA. MTz forms a less compact self-assembly on the gold electrode. The self-assembly of MTz on gold electrode favors the oxidation of both analytes at less positive potential	No data	No data	No data	Aqueous solutions	2012	239
A	Based on a tubular reactor containing the ascorbate oxidase enzyme immobilized. A gold electrode modified by electrochemical deposition of palladium is employed as working electrode	0.14	180	3.0	Honey	2012	240
A	Simultaneous determination of AA and acetylsalicylic acid. The procedure is based on application of two potential pulses	0.17	125	0.7–2.9	Pharmaceuticals	2012	241

Note: A: amperometry; BA: biamperometry; C: coulometry; Ch: chronoamperometry; Co: conductometry; DEP: differential electropotentiometry; LOD: detection limit; P: potentiometry; Po: polarography; RSD: relative standard deviation; SF: sampling frequency; V: voltammetry.

conventional electrode surfaces, such as sensitivity, selectivity, efficient electrocatalysis, and low limits of detection. Metalloporphyrin-coated glassy carbon electrodes are used as electrocatalytic voltammetric sensors for AA. Heterogeneous charge-transfer rates are often very slow at carbon electrodes, leading to poorly defined voltammetric responses. The metalloporphyrin-modified electrodes decrease by several hundred millivolts the potential required for the oxidation of AA and other organic compounds. The faster rates of electron transfer result in a well-defined voltammetric response and increased sensitivity [122]. Electrodes modified by electrodeposition of poly(3-methylthiophene) are used as chemical sensors of some organic and biological molecules such as AA because this modified surface catalyzes the oxidation of organic compounds [123,124]. Osmium-containing redox polymers as "molecular wires" in enzyme electrodes are also used to determine AA. These electrodes present low redox potentials and rapid charge transport characteristics [125]. A 3,4-dihydroxybenzaldehyde (DHB)-modified glassy carbon electrode has been used for AA determination. The oxidation of AA was facilitated by the DHB film on the electrode surface [126]. Other glassy carbon or screen-printed electrodes for the electrocatalytic determination of AA were modified with polyaniline films [127–130], a copolymer of maleic acid anhydride bonded with Eastman-AQ55D [131], a Ni(II) complex (1,5,8,12-tetraaza-2,4,9,11-tetramethylcyclo tetradecinatonickel(II)) [132], polytoluidine blue [133], ferricyanide [134], or silver ferricyanide [135].

18.3.1.3 Atomic Spectroscopic Techniques

The first contribution of a FIA determination of AA by an atomic spectroscopic technique is that proposed by Yebra et al. [136]. The most important details of the published FIA procedures for AA determination based on atomic spectroscopic techniques, in chronological order, are contained in Table 18.5.

All FIA atomic absorption methodologies developed for indirect AA determination involve the utilization of flame atomic absorption spectrometry (FAAS) as detector, and are based on oxidation of AA to DHAA and reduction of a metallic specie (Fe(III) to Fe(II), Cr(VI) to Cr(III), Mn(VII) to Mn(II) and Mn(IV) to Mn(II)). Most of the methods apply a microcolumn with a solid phase (polymeric adsorbent Amberlite XAD4 [136], cation-exchange resin Amberlite IR120 [137,138], or poly(aminophosphonic acid) chelating resin [139,140]) to retain the reduced metallic species. The other possibility is the utilization of a solid-phase reactor filled with the substance to be reduced. Thus, Noroozifar et al. [141] propose a reactor filled with MnO_2 suspended on silica gel beads.

18.3.1.4 Miscellaneous Techniques

Other detection techniques employed with FIA systems for AA determination are HPLC, CE, MS, and calorimetry. The most important details of these methodologies, in chronological order, are contained in Table 18.6.

HPLC coupled with UV/Vis [142], chemiluminescence [143] and electrochemical [144] detection is proposed to determine AA using a FIA system. Since CE in its modern form was first described by Jorgenson and Lukacs (1981), its application for the separation and determination of a variety of samples has been increasingly widespread because of its minimal sample volume requirement, short analysis time, and high separation efficiency. Nevertheless, only one method for AA determination involving a FIA system has been proposed [145]. Bhandari et al. [146] demonstrated the potential of automated FIA electrospray ionization tandem MS (ESI-MS/MS) for the mass-spectroscopic

TABLE 18.5   FIA Methods for Ascorbic Acid (AA) Determination with Atomic Spectroscopic Detection

Detection Technique	Description of the Method	LOD (μM)	SF (h⁻¹)	RSD (%)	Samples	Year	Reference
FAAS	Fe(III) is reduced by AA and then mixed with 1,10-phenanthroline in sulfuric acid media. The cationic complex forms an ion pair with picrate that is adsorbed online on Amberlite XAD-4. The unadsorbed iron is continuously monitored	1.1	90	2.9	Pharmaceuticals, soft drinks, sweets	1997	242
FAAS	Based on the redox reaction between chromate and AA in acid medium. The Cr(III), produced by reduction to Cr(VI) by AA, is adsorbed on a cation exchange resin microcolumn, then is eluted with diluted nitric acid	0.6	20	2.5	Pharmaceuticals, vegetables	2000	243
FAAS	Based on reduction of Fe(III). The Fe(II) produced is online is adsorbed on a minicolumn filled with a cation exchange resin, then is eluted reversely with diluted nitric acid	0.3	30	1.6	Pharmaceuticals, vegetables, fruit juices	2001	244
FAAS	Permanganate is reduced to Mn(II) in acid medium by AA. The Mn(II) formed is retained online on a poly(aminophosphonic acid) chelating resin, which is only selective for Mn(II). The nonreduced Mn(VII) is detected	0.3	90	2.2	Fruit juices	2001	245
FAAS	Cr(VI) is reduced to Cr(III) by AA. This Cr(III) formed is retained online on a poly(aminophosphonic acid) chelating resin, which is only selective for Cr(III). Nonreduced Cr(VI) is determined	0.6	90	2.0	Pharmaceuticals, fruit juices	2001	246
FAAS	Based on utilization of a solid-phase MnO₂ (30% m/m suspended on silica gel beads) reactor. The flow of the sample through the column reduces the MnO₂ to Mn(II) in an acidic carrier stream. Mn(II) is measured	1.1	95	<1.1	Pharmaceuticals, fruit juices, foods	2005	247

Note: LOD: detection limit; RSD: relative standard deviation; SF: sampling frequency; FAAS: flame atomic absorption spectrometry.

TABLE 18.6 FIA Methods for Ascorbic Acid (AA) Determination with Other Detectors

Detection Technique	Description of the Method	LOD (μM)	SF (h^{-1})	RSD (%)	Samples	Year	Reference
HPLC/UV	Application of a column containing silica gels modified with Co/tetrakis(carboxyphenyl)porphine that catalyzes most rapidly the oxidation reaction of AA that is accompanied by the formation of hydrogen peroxide. The resulting hydrogen peroxide is determined by using a FIA system equipped with the column containing glass beads modified with Mn/tetrakis(carboxyphenyl)porphine	No data	No data	4.1	Pharmaceuticals, soft drinks	2003	248
Calorimetry	An electrolytic device is developed and introduced into a calorimetric FIA system for L-ascorbate by using an immobilizing ascorbate oxidase column as a recognition element in combination with a substrate recycling method	No data	No data	No data	Aqueous solutions	2004	249
HPLC/CL	Based on AA inhibition of chemiluminescence from luminol-hypochlorite system. The hypochlorite is electrogenerated online by constant current electrolysis. This method is coupled with HPLC which filled with a C18 column for separation purposes	0.003	No data	No data	Fruits	2007	250
HPLC/A, HPLC/C	The system consisted of two solvent delivery pumps, a reaction coil and Metachem Polaris C18A reversed-phase column and a CoulArray electrochemical detector. The electrochemical detector includes two flow cells	0.09	No data	6.0	Pharmaceuticals, fruits, human blood serum	2008	251
CE/UV	Determination of L-AA, ascorbyl glucoside and kojic acid. The analysis is carried out using an unmodified fused-silica capillary	17.8	30	<5.0	Cosmetics	2008	252
MS	AA and folic acid are determined. Quantitative detection is achieved by electrospray ionization tandem mass spectrometry (ESI-MS/MS) in negative ion mode using the method of standard addition	5.9 ng g^{-1}	No data	0.2–3.1	Pharmaceuticals	2013	253

Note: A: amperometry; C: coulometry; CE: capillary electrophoresis; CL: chemiluminescence; HPLC: high-performance liquid chromatography; LOD: detection limit; MS: mass spectrometry; RSD: relative standard deviation; SF: sampling frequency.

quantification of two water-soluble vitamins (ascorbic and folic acids) using the method of standard addition. Calorimetry is another analytical technique proposed for the FIA determination of AA [147].

18.3.2 Sequential Injection Analysis Methods for Ascorbic Acid Determination

SIA is considered the second generation in the family of flow injection techniques. The principles upon which SIA is based are similar to those of FIA, namely controlled partial dispersion and reproducible sample handling. Unlike FIA, SIA comprises a computer-controlled multiport selection valve and a bidirectional syringe pump. Aliquots of sample and reagents are sequentially aspirated into a holding coil connected to the common port of the multiposition valve. The resultant stack of sample and reagent zones is propelled toward the detector. Flow reversal leads to mixing of the sample and reagent zones to create a zone of product whose properties are measured at the detector. In contrast to FIA, where the reagent flow is continuous, SIA can reduce the consumption of sample and reagents(s). Moreover, SIA is easily automated with a computer, which makes it possible to reduce human error and to obtain reproducible results. Characteristic advantages of SIA include its versatility, full compatibility with computer control, robustness, high sample throughput, and low sample and reagent consumption. The selection valve allows the coupling to the system of all detector types and sample treatment devices, through each of its inlets [148]. However, compared with FIA, few publications involving SIA have appeared in the literature so far for AA determination.

The most important details of the published SIA procedures for AA determination, in chronological order, are contained in Table 18.7. Most of them made us of the reducing property of AA. The first SIA method proposed for AA determination was published in 1998 by Sultan et al. [149]. This is a titrimetric method with spectrophotometric detection, which is the detector more coupled.

The use of solid-phase spectroscopy (SPS), based on using appropriate microbeads in the detection zone of the flow system, has been rarely described with SIA coupling. Nevertheless, Llorent-Martínez et al. [150] proposed a dual SIA optosensor in which two SPS luminescence detection techniques are used in the same manifold (fluorescence and chemiluminescence) for application to multivitamin determination (vitamins B_2, B_6, and C).

18.3.3 Lab-on-Valve Methods for Ascorbic Acid Determination

Lab-on-valve is similar in concept to sequential injection, but involves integration of all flow manifold elements and the detector flow cell into a microconduit attached to the back of the selection valve [151]. Only one method based on this methodology for AA determination has been published. Sorouraddin et al. [152] applied an on-chip microde-termination device, integrating the FIA system, to l-AA and DHAA determination in urine and pharmaceutical samples. The manifold components are used for flow, mixing, reaction, and detection. The reaction system is a coupled redox–complexation reaction between AA and a 1,10-phenanthroline-Fe(III) and a photothermal microscope is used for the ultrasensitive detection of the nonfluorescent reaction product (ferroin). For DHAA determination, dithiothreithol is used as reducing agent and the total AA is determined.

TABLE 18.7 Sequential Injection Analysis (SIA) Methods for Ascorbic Acid (AA) Determination

Detection Technique	Description of the Method	LOD (μM)	SF (h⁻¹)	RSD (%)	Samples	Year	Reference
SP (Vis)	Fe(III) in sulfuric acid media is used as an oxidant and 1,10-phenanthroline as an indicator and absorbance of the resulting tris(1,10-phenanthroline)-Fe(II) is measured. $\lambda = 510$ nm	No data	No data	<1.2	Pharmaceuticals	1998	254
SP (Vis)	Titrimetric spectrophotometric method based on the oxidation of vitamin C with cerium(IV) in sulfuric acid media. $\lambda = 510$ nm	No data	No data	<1.2	Pharmaceuticals	1999	255
SP (Vis)	The redox indicator, ferroin, tris(1,10-phenanthroline)Fe(II) is incorporated into the perfluorosulfonated cation exchange membrane Nafion and, together with optical fibers, a photodiode, and a light-emitting diode, is used for the construction of a redox optical sensor (optode)	No data	17	No data	Pharmaceuticals	2000	256
SP (Vis)	Determination of AA based on the redox reaction between AA and permanganate in an acidic medium and lead to a decrease in color intensity of permanganate. $\lambda = 525$ nm	No data	60	2.9	Pharmaceuticals	2002	257
SP (UV)	Simultaneous determination of AA and rutin trihydrate (RT). The methodology described is based on a solid-phase extraction microcolumn. The SPE microcolumn was used for retention of RT, while the AA was eluted with the solvent front phase. $\lambda = 262$ nm	5.7	26	0.70	Pharmaceuticals	2003	258
CL	Based on manganese-based chemiluminescence detection	0.05	No data	1.0	Pharmaceuticals	2004	125
SP (Vis)	A PC-controlled SIA system equipped with a spectrophotometric diode array detector for rapid monitoring and evaluation of antioxidation/radical scavenging activity of biological samples	0.5	45	1.8	Foods	2004	259
CL	The antioxidative activity is expressed as the percentage inhibition of luminol chemiluminescence due to the scavenging of hypochlorite ion by an antioxidant	No data	45	2.5	Aqueous solutions	2004	260

(Continued)

TABLE 18.7 (Continued)　Sequential Injection Analysis (SIA) Methods for Ascorbic Acid (AA) Determination

Detection Technique	Description of the Method	LOD (µM)	SF (h⁻¹)	RSD (%)	Samples	Year	Reference
SP (Vis)	The method for AA is based on the formation of a chromophore between guaiacol and hydrogen peroxide	No data	No data	No data	Pharmaceuticals, foods	2005	261
V	Voltammetric electronic tongue is designed and applied to the determination of oxidizable species. The use of artificial neural networks has solved the overlapped signal of AA, 4-aminophenol, and 4-acetamidophenol (paracetamol)	3.6	No data	No data	Aqueous solutions	2005	262
V	Determination of mixtures of AA, uric acid and acetaminophen (paracetamol). For their determination, a voltammetric electronic tongue is proposed coupled with artificial neural networks as chemometric tool for modelling the analytical system	No data	No data	No data	Aqueous solutions	2007	263
CL	Coupling of SIA and solid-phase spectroscopy with luminescence detection is carried out with a novel characteristic: the use of two different detection techniques in the same system enhancing the scope of applications. This approach allows the determination of compounds with dissimilar spectroscopic characteristics: vitamins B_2, B_6, and C	34.6	8	< 5	Pharmaceuticals	2008	264
SP (Vis)	Based on the oxidation reaction of vitamin C with potassium permanganate in sulfuric acid media. The titrimetric reaction is monitored by the decrease of absorbance of the permanganate. $\lambda = 525$ nm	No data	No data	0.2–1.3	Pharmaceuticals	2009	265
SP (Vis)	Synthesis and characterization of the guanidinium salt of 11-molybdobismuthophosphate (11-MBP) as specific reagent for AA. $\lambda = 720$ nm	2.0	15	1.4	Pharmaceuticals, fruit juices	2010	266
SP (UV)	Simultaneous determination of vitamin C and paracetamol based on the difference between absorption spectra of both substances at different pH	1.7	40	No data	Pharmaceuticals	2010	267

(Continued)

TABLE 18.7 (*Continued*) Sequential Injection Analysis (SIA) Methods for Ascorbic Acid (AA) Determination

Detection Technique	Description of the Method	LOD (μM)	SF (h⁻¹)	RSD (%)	Samples	Year	Reference
SP (Vis)	Based on an optoelectronic device constructed of two ordinary light emitting diodes compatible with optosensing films. This fiberless device containing chemoreceptor, semiconductor light source, and detector integrated in a miniaturized flow-through cell	20	No data	0.7–1.2	Aqueous solutions	2011	268
SP (Vis)	Two-electron reduced heteropolyblue is formed very fast at pH 3.75–4.75 in the reaction between Dawson-type molybdophosphate HPA and AA	1.0	60	No data	Pharmaceuticals, fruit juices	2011	269
CE/CC	Ascorbate and other organic anions are determined. The SIA manifold is based on a syringe pump and sample injection is carried out with a new design relying on a simple piece of capillary tubing to achieve the appropriate back pressure for the required split-injection procedure	20	No data	1.6	Aqueous solutions	2011	270

Note: CC: contactless conductivity; CE: capillary electrophoresis; CL: chemiluminescence; LOD: detection limit; RSD: relative standard deviation; SF: sampling frequency; SP: spectrophotometry; V: voltammetry.

18.3.4 Miscellaneous Flow Analysis Techniques for Ascorbic Acid Determination

Other flow techniques applied to AA determination are the multicommutated techniques, BI, and stopped-flow analysis. Multicommutated techniques include MCFIA, MSFIA, and MPFS [153].

The main important details of these procedures for AA determination are contained in Table 18.8. As can be seen in Table 18.8, among these flow analysis techniques, stopped-flow procedures are the most applied to AA determination, being mostly kinetic enzymatic determinations. BI methodologies for AA determination use spectrophotometric detection and a commercial flow cell, which is filled with appropriate solid beads, works as a flow-through chemical sensor integrating online reaction, retention, and detection on the solid-phase disposable beads.

It is important to note the method published by Fernández et al. [154] because they propose and validate a low-pressure chromatographic method based on monolithic column separation, and multicommutated flow programming for simultaneous determination of AA, thiamine, nicotinic acid, nicotinamide, pyridoxine, and riboflavin. This method is comparable to conventional HPLC methods, and the multisyringe chromatographic technique shows several advantages, between them the possibility of online use of reagents in all the steps of the determination. The production of waste and the consumption of solvents are lower than in HPLC methods (due to discontinuous flow), enabling the reduction of cost per analysis. Furthermore, dimensions of the system and its easy portability provide the opportunity for analysis "in the field." The main drawback of this methodology is the limited multianalyte resolution efficiency as a result of the impracticability of coupling monolith rods longer than 5.0 cm due to the increase of back pressure.

18.4 CONCLUSIONS

The development of flow analysis methods for AA is important because of the presence and significance of this analyte in foodstuffs, pharmaceuticals, and biological fluids, with implications in redox processes, human health, and food quality. However, so far there are few flow analysis methods for simultaneously determining AA and DHAA, or simultaneously determining several vitamins, including vitamin C (the sum of the contents of AA plus DHAA). Furthermore, most flow analysis methods do not include online sample preparation, except for analytical separations and derivatization reactions, and only two online sample dilution methods allow the fully automatic determination of AA.

FIA is still the most widely used flow analysis methodology for AA determination, and molecular spectroscopy is the most used detection technique. Flow analysis methodologies proposed for AA determination offer superb performances in terms of rapidity, precision, and accuracy. Therefore, in many instances chromatographic methods can be replaced by rapid flow analysis methods, and it is hoped that in future new flow analysis methodologies for determining AA will be developed, including the last generations of flow analysis techniques, which to date have been less exploited. This is important because they can be considered as progress of flow injection methods toward the concept of Green Analytical Chemistry, since, among other advantages, these strategies are very effective in minimizing reagent consumption and exploiting intermittent reagent addition.

TABLE 18.8 Miscellaneous Flow Analysis Techniques for Ascorbic Acid (AA) Determination

FAT	Detection Technique	Description of the Method	LOD (μM)	SF (h⁻¹)	RSD (%)	Samples	Year	Reference
MCFIA	SP	Titration procedure based on volumetric fraction variation and employing a reaction with 2,6-dichloroindopheno	No data	5–30	1.1	Fruit juices, soft drinks	2000	275
MCFIA	P	The method is based on the reduction of IO_3^- by AA and the detection is carried out employing a flow-through ion selective electrode for iodide	No data	15	1.0	Pharmaceuticals	2002	276
MCFIA	SP	Determination of hydrosoluble vitamins (AA, thiamine, riboflavin, and pyridoxine). The flow manifold is designed with computer-controlled three-way solenoid valves	0.5	60	1.0	Pharmaceuticals	2003	277
MCFIA	F	Based on the monitoring of quantum dots fluorescence quenching produced by AA	0.02	68	2.7	Pharmaceuticals, foods	2013	278
MPFS	CL	Relies on the inhibitory effect of AA on the oxidation of luminol by hydrogen peroxide in alkaline medium.	1.7	100	<1.0	Fruit juices	2006	279
MPFS	CL	AA acts as a radical scavenger, preventing the oxidation of luminol	0.3	200	0.62	Pharmaceuticals	2014	280
MSFIA	HPLC/ DAD	System equipped with monolithic column for simultaneous determination of six water-soluble vitamins (thiamine, riboflavin, AA, nicotinic acid, nicotinamide, and pyridoxine)	0.57	No data	0.2–2.5	Malt, orange juice, and infant milk	2012	281

(Continued)

TABLE 18.8 (*Continued*)   Miscellaneous Flow Analysis Techniques for Ascorbic Acid (AA) Determination

FAT	Detection Technique	Description of the Method	LOD (µM)	SF (h⁻¹)	RSD (%)	Samples	Year	Reference
BI	SP	The flow cell is filled by injecting in the flow system a homogeneous bead suspension of an appropriate solid support previously loaded with the chromogenic reagent. The solid beads work as a flow-through chemical sensing microzone integrating online separation/reaction/detection	0.02–0.14	13–16	4.02–4.19	Fruit juices, pharmaceuticals	2003	282
BI	SP	Simple BI spectroscopy-FIA system with spectrophotometric detection. The sensor is based on the decrease of absorbance obtained (720 nm) when Prussian blue is reduced by AA	0.45	13	5	Fruit juices, pharmaceuticals, sweets	2004	283
SFA	SP/Vis	Reaction of AA with 2,6-dichlorophenolindophenol	No data	No data	1.0–2.0	Aqueous solutions	1975	284
SFA	SP/Vis	Reaction of AA with 2,6-dichlorophenolindophenol	No data	No data	No data	Orange juice	1979	285
SFA	SP/Vis	Reaction of AA with 2,6-dichlorophenolindophenol	0.2	No data	No data	Aqueous solutions	1980	286
SFA	SP/Vis	Reaction of AA with 2,6-dichlorophenolindophenol	No data	No data	<1	Aqueous solutions	1980	287
SFA	SP/Vis	Reaction of AA with 2,6-dichlorophenolindophenol, determination of AA and DHAA	No data	No data	<1	Pharmaceuticals, human blood, plasma, and urine	1982	288
SFA	SP/Vis	Reaction of AA with 2,6-dichlorophenolindophenol	No data	No data	<2	Fruit juices, vegetables	1987	289

(*Continued*)

TABLE 18.8 (*Continued*) Miscellaneous Flow Analysis Techniques for Ascorbic Acid (AA) Determination

FAT	Detection Technique	Description of the Method	LOD (μM)	SF (h^{-1})	RSD (%)	Samples	Year	Reference
SFA	CL	Based on the decrease of CL signal produced by the reaction between luminol, H_2O_2, and peroxidase in presence of AA	No data	No data	<2	Fruit juices, Wines	1990	290
SFA	A	Biosensor comprising a rotating bioreactor and a stationary platinum ring for the simultaneous determination of fructose and ascorbate. Hexacyanoferrate(II) is the monitored species	7–8	No data	1.2	Foods	1993	291
SFA	F	Based on the interference of laccase by AA, causing a decrease in the rate of the laccasecatalyzed fluorescence reaction in the presence of Brij-35. The resulting change of fluorescence intensity (λex = 430 and λem = 530nm)	No data	No data	No data	Pharmaceuticals	1995	292
SFA	F	Kinetic procedure to determine total AA combining the merging zones principle with the stopped-flow technique. It is based on the oxidation of AA catalyzed by laccase to DHAA, which then reacts with o-phenylenediamine to form a fluorescent quinoxaline	No data	30	2.0–4.08	Wine, Beer, Urine Pharmaceuticals	1995	293
SFA	F	Based on the inhibition of L-AA on the formation of 2,3-diaminophenazine, which is an oxidant product of o-phenyl-enediamine catalyzed by laccase, in the presence of Brij-35, which shows strong enhancement on the fluorescence (λ_{ex} = 430 nm, λ_{em} = 530 nm)	0.08	No data	0.56–3.55	Pharmaceuticals	1997	294

(*Continued*)

TABLE 18.8 (*Continued*) Miscellaneous Flow Analysis Techniques for Ascorbic Acid (AA) Determination

FAT	Detection Technique	Description of the Method	LOD (μM)	SF (h⁻¹)	RSD (%)	Samples	Year	Reference
SFA	SP/Vis	Simultaneous determination of AA and L-cysteine	No data	No data	No data	Fruit juices	2000	295
SFA	SP/Vis	AA and L-cysteine reacts with Fe^{3+}-o-phenanthroline at in different reaction speeds to obtain Fe^{2+}-o-phenanthroline, $\lambda = 508$ nm	5.7	No data	<1	Fruit juices	2000	296
SFA	F	The method makes use of the stopped-flow mixing technique in order to achieve rapid oxidation of AA by dissolved oxygen to DHAA, which then reacts with o-phenyl-enediamine to form a fluorescent quinoxaline	0.11	No data	0.5	Pharmaceuticals, fruit juices, soft drinks, blood serum	2001	297
SFA	A	Online interfacing of a rotating bioreactor and continuous flow/stopped flow. Horseradish peroxidase, immobilized on a rotating disk, in presence of hydrogen peroxide catalyzes the oxidation of hydroquinone to p-benzoquinone, whose electrochemical reduction back to hydroquinone is detected on glassy carbon electrode	0.006	25	1.4–5.0	Pharmaceuticals	2004	298

Note: A: amperometry; BI: bead injection; CL: chemiluminescence; F: fluorescence; FAT: flow analysis technique; HPLC/DAD: high-performance liquid chromatography with diode array detector; LOD: detection limit; MCFIA: multicommutated flow injection analysis; MPFS: multipumping flow systems; MSFIA: multisyringe FIA; MPFS: multipumping flow systems; P: potentiometry; RSD: relative standard deviation; SF: sampling frequency; SFA: stopped-flow analysis; SP: spectrophotometry.

REFERENCES

1. Bartosz, G. 2014. *Food Oxidants and Antioxidants: Chemical, Biological, and Functional Properties.* Boca Raton: CRC Press.
2. Ball, G. F. M. 2006. *Vitamins in Foods Analysis, Bioavailability, and Stability.* Boca Raton: CRC Press.
3. Du, J., Cullen, J. J., and Buettner, G. R. 2012. Ascorbic acid: Chemistry, biology and the treatment of cancer. *Biochim Biophys Acta* 1826:443–457.
4. Johnston, C. S., F. M. Steinberg, and R. B. Rucker, 2007. Ascorbic acid. In *Handbook of Vitamin*, eds. J. Zempleni, R. B. Rucker, D. B. McCormick, and J. W. Suttie, pp. 489–520. Boca Raton: CRC Press.
5. Commission Regulation (EU) No 1129/2011 of 11 November 2011. http://eur-lex.europa.eu/LexUriServ/LexUriServ.do?uri=OJ:L:2011:295:0001:0177:En:PDF (accessed March 31, 2014).
6. Saltmarsh, M. 2013. *Essential Guide to Food Additives* (4th ed.). Cambridge: The Royal Society of Chemistry.
7. Svehla, G., L. Koltai, and L. Erdey. 1963. The use of 2,6-dichlorophenol-indophenol as indicator in iodometric titrations with ascorbic acid. *Anal. Chim. Acta* 29:442–447.
8. Deshmukh, G. S. and M. G. Bapat. 1955. Determination of ascorbic acid by potassium iodate. *Fresenius' Z. Anal. Chem.* 145:254–256.
9. Puzanowska-Tarasiewicz, H., M. Tarasiewicz, and N. Omieljaniuk. 1980. Titrimetric determination of ascorbic acid with bromate using perphenazine as redox indicator. *Fresenius' Z. Anal. Chem.* 303:412.
10. Association of Official Analytical Chemists (AOAC). 2012. *Official Methods of Analysis of AOAC International* (19th ed.), ed. G. W. Latimer. AOAC: Gaithersburg. http://www.eoma.aoac.org/ (accessed March 31, 2014).
11. Yang, J., Q. Ma, F. Huang, L. Sun, and J. Dong. 1998. A new fluorimetric method for the determination of ascorbic acid. *Anal. Lett.* 31:2757–2766.
12. Wang, L., L. Zhang, S. She, and F. Gao. 2005. Direct fluorimetric determination of ascorbic acid by the supramolecular system of AA with β-cyclodextrin derivative. *Spectrochim. Acta Part A* 61:2737–2740.
13. Arya, S. P., M. Mahajan, and P. Jain. 1998. Photometric methods for the determination of vitamin C. *Anal. Sci.* 14:889–895.
14. Zaporozhets, O. A. and E. A. Krushinskaya. 2002. Determination of ascorbic acid by molecular spectroscopic techniques. *J. Anal. Chem.* 57:286–297.
15. Güçlü, K., K. Sözgen, E. Tütem, M. Özyürek, and R. Apak. 2005. Spectrophotometric determination of ascorbic acid using copper(II)–neocuproine reagent in beverages and pharmaceuticals. *Talanta* 65:1226–1232.
16. Mi, C., T. Wang, P. Zeng, S. Zhao, N. Wanga, and S. Xu. 2013. Determination of ascorbic acid via luminescence quenching of LaF_3:Ce,Tb nanoparticles synthesized through a microwave-assisted solvothermal method. *Anal. Methods* 5:1463–1468.
17. Tonelli, D., B. Ballarin, L. Guadagnini, A. Mignani, and E. Scavetta. 2011. A novel potentiometric sensor for L-ascorbic acid based on molecularly imprinted polypyrrole. *Electrochim. Acta* 56:7149–7154.
18. Jo, A., M. Kang, A. Cha, H. S. Jang, J. H. Shim, N. S. Lee, M. H. Kim, Y. Lee, and C. Lee. 2014. Nonenzymatic amperometric sensor for ascorbic acid based on hollow gold/ruthenium nanoshells. *Anal. Chim. Acta* 819:94–101.

19. Pisoschi, A. M., A. Pop, A. I. Serban, and C. Fafaneata. 2014. Electrochemical methods for ascorbic acid determination. *Electrochim. Acta* 121:443–460.
20. Arya, S. P., M. Mahajan, and P. Jain. 2000. Non-spectrophotometric methods for the determination of Vitamin C. *Anal. Chim. Acta* 417:1–14.
21. Novakova, L., P. Solich, and D. Solichova. 2008. HPLC methods for simultaneous determination of ascorbic and dehydroascorbic acids. *Trends Anal. Chem.* 27:942–958.
22. Fenoll, J., A. Martínez, P. Hellín, and P. Flores. 2011. Simultaneous determination of ascorbic and dehydroascorbic acids in vegetables and fruits by liquid chromatography with tandem-mass spectrometry. *Food Chem.* 127:340–344.
23. Kutnink, M. A. and S. T. Omaye. 1987. Determination of ascorbic acid, erythorbic acid, and uric acid in cured meats by high performance liquid chromatography. *J. Food Sci.* 52:53–56.
24. Nisperos-Carriedo, M. O., B. S. Buslig, and P. E. Shaw. 1992. Simultaneous detection of dehydroascorbic, ascorbic, and some organic acids in fruits and vegetables by HPLC. *J. Agric. Food Chem.* 40:1127–1130.
25. Tang, Y. and M. Wu. 2005. A quick method for the simultaneous determination of ascorbic acid and sorbic acid in fruit juices by capillary zone electrophoresis. *Talanta* 65:794–798.
26. Široká, J., A. Martincová, M. Pospíšilová, and M. Polášek. 2013. Assay of citrus flavonoids, troxerutin, and ascorbic acid in food supplements and pharmaceuticals by capillary zone electrophoresis. *Food Anal. Method* 6:1561–1567.
27. Yebra-Biurrun, M. C. 2000. Flow injection determination methods of ascorbic acid. *Talanta* 52:367–383.
28. Koupparis, M. and P. Anagnostopoulou. 1984. An automated microprocessor-based spectrophotometric flow-injection analyzer. *J. Autom. Chem.* 6:186–191.
29. Koupparis, M. A., P. Anagnostopoulou, and H. V. Malmstadt. 1985. Automated flow-injection pseudotitration of strong and weak acids, ascorbic acid and calcium, and catalytic pseudotitrations of aminopolycarboxylic acids by use of a microcomputer-controlled analyzer. *Talanta* 32:411–417.
30. Ma, H., J. Feng, and B. Cao. 1991. Spectrophotometric determination of ascorbic acid in vitamin C tablets, beverages, orange and urine with 2,6-dichloroindophenol by flow injection analysis. *Fenxi Huaxue* 19:182–184.
31. Yamane, T. and T. Ogawa. 1987. Simple and rapid determination of L-ascorbic acid by flow injection analysis with spectrophotometric detection. *Bunseki Kagaku* 36:625–628.
32. Ma, Y. and J. Yang. 1990. Determination of ascorbic acid in food with flow injection spectrophotometry. *Huaxue Shijie* 31:505–507.
33. Fu, L. and Y. Ren. 1992. Determination of ascorbic acid in pharmaceutical products with iron(III)-1,10-phenanthroline system by flow injection spectrophotometry. *Fenxi Huaxue* 20:193–195.
34. Ma, Y. and J. Yang. 1992. Rapid determination of ascorbic acid in medicaments by flow injection spectrophotometry. *Yaowu Fenxi Zazhi* 12:26–28.
35. Alamo, J. M., A. Maquieira, R. Puchades, and S. Sagrado. 1993. Determination of titratable acidity and ascorbic acid in fruit juices in continuous-flow systems. *Fresenius' J. Anal. Chem.* 347:293–298.
36. Sultan, S. M., A. M. Abdennabi, and F. E. O. Suliman. 1994. Flow-injection colorimetric method for the assay of vitamin C in drug formulations using tris(1,10-phenanthroline)–iron(III) complex as an oxidant in sulfuric acid media. *Talanta* 41:125–130.

37. Molina-Diaz, A., I. Ortega-Carmona, and M. I. Pascual-Reguera. 1998. Indirect spectrophotometric determination of ascorbic acid with ferrozine by flow-injection analysis. *Talanta* 47:531–536.

38. Luque-Perez, E., A. Rios, and M. Valcarcel. 2000. Flow injection spectrophotometric determination of ascorbic acid in soft drinks and beer. *Fresenius' J. Anal. Chem.* 366:857–862.

39. Teshima, N., T. Nobuta, and T. Sakai. 2001. Simultaneous flow injection determination of ascorbic acid and cysteine using double flow cell. *Anal. Chim. Acta* 438:21–29.

40. Zenki, M., A. Tanishita, and T. Yokoyama. 2004. Repetitive determination of ascorbic acid using iron(III)-1.10-phenanthroline-peroxodisulfate system in a circulatory flow injection method. *Talanta* 64:1273–1277.

41. Martinovic, A., S. Cerjan-Stefenovic, and N. Radic. 2008. Flow injection analysis with two parallel detectors: Potentiometric and spectrophotometric determination of thiols and ascorbic acid in mixture. *J. Chem. Metrol.* 2:1–12.

42. Gong, A., X. Zhu, and W. Yu. 2007. Determination of vitamin C by in flow injection discoloring spectrophotometry. *Yaowu Fenxi Zazhi* 27:450–453.

43. Duan, Y. J., Z. J. Guo, and L. Liu. 2011. Determination of the contents of vitamin C in tomatoes by flow injection analysis. *Fenxi Shiyanshi* 30:24–26.

44. Wang, L. Y., Z. H. Xie, Y. Liu, Q. Cai, and H. L. Wang. 2012. Determination of ascorbic acid by flow-injection spectrophotometric method with on-line reaction column. *Fenxi Shiyanshi* 31:72–75.

45. Pereira, A. V. and O. Fatibello-Filho. 1998. Spectrophotometric flow injection determination of L-ascorbic acid with a packed reactor containing ferric hydroxide. *Talanta* 47:11–18.

46. Pereira, A. V. and O. Fatibello-Filho. 1998. Flow injection spectrophotometric determination of L-ascorbic acid in pharmaceutical formulations with online solid phase reactor containing copper(II) phosphate. *Anal. Chim. Acta* 366:55–62.

47. Zhong, M. H. and Y. Li. 2010. The determination for vitamin C by flow injection resonance light scattering method. *Jiangxi Shifan Daxue Xuebao* 34:564–566.

48. Zhong, M. H. and Y. Li. 2011. Determination of vitamin C by flow injection resonance light scattering method. *Guangpu Shiyanshi* 28:101–104.

49. Sultan, S. M. 1993. Flow injection titrimetric analysis of vitamin C in pharmaceutical products. *Talanta* 40:593–598.

50. Lazaro, F., A. Rios, M. D. Luque de Castro, and M. Valcarcel. 1986. Determination of vitamin C by flow injection analysis. *Analyst* 111:163–166.

51. Lazaro, F., A. Rios, M. D. Luque de Castro, and M. Valcarcel. 1986. Determination of vitamin C in urine by flow injection analysis. *Analyst* 111:167–169.

52. Hernandez-Mendez, J., A. Alonso Mateos, M. J. Almendral Parra, and C. Garcia de Maria. 1986. Spectrophotometric flow-injection determination of ascorbic acid by generation of triiodide. *Anal. Chim. Acta* 184:243–250.

53. Lazaro, F., M. D. Luque de Castro, and M. Valcarcel. 1987. Simultaneous determination of ascorbic acid and sulfite in soft drinks by flow injection analysis. *Analusis* 15:183–187.

54. Zhang, A. 1992. Spectrophotometric determination of ascorbic acid in medicines and foods by flow injection analysis. *Yingyang Xuebao* 14:307–310.

55. Danet, A., V. David, and M. Oancea. 1994. Online analytical device with hydrodynamic injection. Determination of vitamin C. *Rev. Chim. Bucharest* 45:1000–1006.

56. Miura, Y. and M. Akagi. 2009. Sensitive FIA of L-ascorbic acid in colored and turbid aqueous solution. *Bunseki Kagaku* 58:73–79.

57. Albero, M. I., M. S. Garcia, C. Sanchez-Pedreno, and Rodriguez, J. 1992. Determination of ascorbic acid in pharmaceuticals and urine by reverse flow injection. *Analyst* 117:1635–1638.

58. Sanz-Martínez, A., A. Ríos, and M. Valcárcel. 1992. Photochemical determination of ascorbic acid using unsegmented flow methods. *Analyst* 117:1761–1765.

59. Leon, L. E. and J. Catapano. 1993. Indirect flow-injection analysis of ascorbic acid by photochemical reduction of methylene blue. *Anal. Lett.* 26:1741–1750.

60. Zhao, Y., Wang, G., Xu, R., and Jia, Y. 1994. Determination of total vitamin C in foods by flow injection analysis. *Yingyang Xuebao* 16:186–191.

61. Liu, D. L., A. Sun, and G. Liu. 1995. Application of photochemical reactions in a flow injection analysis system. IV. Determination of ascorbic acid based on the photochemical reduction of methylene blueR. *Fenxi Huaxue* 23:187–190.

62. Noroozifar, M., M. Khorasani-Motlagh, and A. R. Farahmand. 2004. Automatic spectrophotometric procedure for determination of L-ascorbic acid based on reduction of iron(III)-thiocyanate complex. *Acta Chim. Slov.* 51:717–727.

63. Verma, K. K., A. Jain, A. Verma, and A. Chaurasia. 1991. Spectrophotometric determination of ascorbic acid in pharmaceuticals by background correction and flow injection. *Analyst* 116:641–645.

64. Jain, A., A. Chaurasia, and K. K. Verma. 1995. Determination of ascorbic acid in soft drinks, preserved fruit juices and pharmaceuticals by flow injection spectrophotometry: Matrix absorbance correction by treatment with sodium hydroxide. *Talanta* 42:779–787.

65. Llamas, N. E., M. S. DiNezio, and B. S. Fernandez Band. 2010. Flow-injection spectrophotometric method with on-line photodegradation for determination of ascorbic acid and total sugars in fruit juices. *J. Food Compos. Anal.* 24:127–130.

66. Fernandes, J. C. B., G. de Oliveira Neto, and L. T. Kubota. 1998. Use of column with modified silica for interfering retention in a FIA spectrophotometric method for direct determination of vitamin C in medicine. *Anal. Chim. Acta* 366:11–22.

67. Noroozifar, M. and M. Khorasani-Motlagh. 2003. Solid-phase iodine as an oxidant in flow injection analysis: Determination of ascorbic acid in pharmaceuticals and foods by background correction. *Talanta* 61:173–179.

68. Molina-Díaz, A., A. Ruiz-Medina, and M. L. Fernández-de Córdova. 1999. Determination of ascorbic acid by use of a flow-through solid phase UV spectrophotometric system. *Fresenius' J. Anal. Chem.* 363:92–97.

69. Ruiz-Medina, A., M. L. Fernández-de Córdova, M. J. Ayora-Cañada, M. I. Pascual-Reguera, and A. Molina-Díaz. 2000. A flow-through solid phase UV spectrophotometric biparameter sensor for the sequential determination of ascorbic acid and paracetamol. *Anal. Chim. Acta* 404:131–139.

70. Ruiz-Medina, A., P. Ortega-Barrales, M. L. Fernandez-De Cordova, and A. Molina-Diaz. 2002. Use of a continuous flow solid-phase spectroscopic sensor using two sensing zones: Determination of thiamine and ascorbic acid. *J. AOAC Int.* 85:369–374.

71. Ortega-Barrales, P., A. Ruiz-Medina, M. L. Fernandez-de Cordova, and A. A. Molina-Diaz. 2002. Flow-through solid-phase spectroscopic sensing device implemented with FIA solution measurements in the same flow cell: Determination of binary mixtures of thiamine with ascorbic acid or acetylsalicylic acid. *Anal. Bioanal. Chem.* 373:227–232.

72. Takayanagi, T., M. Nishiuchi, M. Ousaka, M. Oshima, and S. Motomizu. 2009. Monitoring of vitamin C species in aqueous solution by flow injection analysis coupled with an on-line separation with reversed-phase column. *Talanta* 79:1055–1060.

73. Kim, J. M., Y. Huang, and R. D. Schmid. 1990. Chemiluminescent determination of ascorbic acid in juices. *Anal. Lett.* 23:2273–2282.

74. Alwarthan, A. A. 1993. Determination of ascorbic acid by flow injection with chemiluminescence detection. *Analyst* 118:639–642.

75. Zhujun Z. and Q. Wei. 1996. Chemiluminescence flow sensor for the determination of ascorbic acid with immobilized reagents. *Talanta* 43:119–124.

76. Yang, W., B. Li, Z. Zhang, and G. Tian. 1996. Determination of trace ascorbic acid by flow injection chemiluminescence suppression method. *Fenxi Huaxue* 24:579–582.

77. Agater, I. B. and R. A. Jewsbury. 1997. Direct chemiluminescence determination of ascorbic acid using flow injection analysis. *Anal. Chim. Acta* 356:289–294.

78. Shen, A., B. Xu, M. Ji, and J. Wu. 1997. Determination of trace ascorbic acid by flow injection chemiluminescence method. *Fenxi Kexue Xuebao* 13:347.

79. Ye, Q. and Z. Wanf. 1998. Determination of trace ascorbic acid in vegetables by potassium permanganate - ascorbic acid- luminol chemiluminescence method. *Fenxi Huaxue* 26:6–13.

80. Wu, Y. Y., L. Q. Li, and Q. G. Liu. 1999. Study on the chemiluminescence reaction of luminol-KIO$_4$-ascorbic acid system and its application. *Fenxi Shiyanshi* 18:58–61.

81. Zhou, Y., T. Nagaoka, F. Li, and G. Zhu. 1999. Evaluation of luminol-H$_2$O$_2$-KIO$_4$ chemiluminescence system and its application to hydrogen peroxide, glucose and ascorbic acid assays. *Talanta* 48:461–467.

82. Li, Y. H. and J. R. Lu. 2001. Flow injection chemiluminescence determination of ascorbic acid. *Shaanxi Shifan Daxue Xuebao* 29:81–83.

83. Mahmoud, K. M. and A. H. Saheed. 2009. Indirect determination of ascorbic acid in pharmaceutical preparations using reversed FIA-CL method. *Dirasat: Pure Sciences* 36:91–100.

84. Li, F., W. Zhang, and G. Zhu. 2000. Determination of ascorbic acid by flow injection chemiluminescence suppression method. *Fenxi Huaxue* 28:1523–1526.

85. Wang, S. M., X. M. Fan, X. W Cui, C. Zhang, Z. Yang, and X. W. Zheng. 2012. Potassium permanganate-glyoxal chemiluminescence system for determination of ascorbic acid. *Fenxi Shiyanshi* 31:23–26.

86. Wang, F., W. Qin, and Z. Zhang. 1997. Flow-injection/chemiluminescence sensor for the determination of ascorbic acid. *Fenxi Huaxue* 25:1255–1258.

87. Du, J. X., Y. H. Li, and J. R. Lu. 2000. Determination of ascorbic acid by on-line coupling chemiluminescence. *Shaanxi Shifan Daxue Xuebao* 28:81–83.

88. Yang, W. P., Z. Wang, and Z. J. Zhang. 2003. Determination of trace-micro organic reductants by flow injection coupling chemiluminescence analysis. *Shaanxi Shifan Daxue Xuebao* 31:67–70.

89. Zhu, X., Y. He, M. Liu, J. Du, and J. Lu. 2004. Chemiluminescence determination of trace ascorbic acid with soluble manganese(IV)-formaldehyde system. *Fenxi Huaxue* 32:752–754.

90. Xie, C. G., C. F. Liu, W. G. Chang, and H. F. Li. 2004. New flow-injection chemiluminescence system for the determination of ascorbic acid. *Guangpu Shiyanshi* 21:439–441.

91. Fang, L. Q. and Xu, X. D. 2007. The determination of vitamin C with flow injection chemiluminescent system of potassium permanganate-formic acid. *Jiangxi Shifan Daxue Xuebao* 31:90–93.

92. Chen, H., B. Ling, F. Yuan, C. Zhou, J. Chen, and L. Wang. 2012. Chemiluminescence behaviour of CdTe-potassium permanganate enhanced by sodium hexametaphosphate and sensitized sensing of L-ascorbic acid. *Luminescence* 27:466–472.

93. Tang, Z. and H. Xiong. 2009. Determination of ascorbic acid by Ce(IV)-ascorbic acid-quinine chemiluminescence analytical system. *Shipin Kexue* 30:409–411.

94. Perez-Ruiz, T., C. Martinez-Lozano, and A. Sanz. 1995. Flow-injection chemiluminescence determination of ascorbic acid based on its sensitized photooxidation. *Anal. Chim. Acta* 308:299–307.

95. Vanderslice, J. T. and D. J. Higgs. 1989. Automated analysis of total vitamin C in foods. *J. Micronut. Anal.* 6:109–117.

96. Huang, H., R. D. Cai, and Z. Y. Yumin. 1995. Flow-injection stopped-flow spectrofluorometric kinetic determination of total ascorbic acid based on an enzymelinked coupled reaction. *Anal. Chim. Acta* 309:271–275.

97. Chung, H. K. and J. D. Ingle. 1991. Kinetic fluorometric FIA determination of total ascorbic acid, based on use of two serial injection valves. *Talanta* 38:355–357.

98. Perez-Ruiz, T., C. Martinez-Lozano, A. Sanz, and A. Guillen. 2004. Successive determination of thiamine and ascorbic acid in pharmaceuticals by flow injection analysis. *J. Pharmaceut. Biomed.* 34:551–557.

99. Perez-Ruiz, T., C. Martinez-Lozano, V. Tomas, and C. Sidrach. 1997. Flow injection fluorometric determination of ascorbic acid based on its photooxidation by Thionine Blue. *Analyst* 122:115–118.

100. Rezaei, B., A. A. Ensafi, and S. Nouroozi. 2005. Flow-injection determination of ascorbic acid and cysteine simultaneously with spectrofluorometric detection. *Anal. Sci.* 21:1067–1071.

101. Karlberg, B. and S. Thelander. 1978. Determination of readily oxidised compounds by flow injection analysis and redox potential detection. *Analyst* 103:1154–1159.

102. Bradberry, C. W. and R. N. Adams. 1983. Flow injection analysis with an enzyme reactor bed for determination of ascorbic acid in brain tissue. *Anal. Chem.* 55:2439–2440.

103. Greenway, G. M. and P. Ongomo. 1990. Determination of L-ascorbic acid in fruit and vegetable juices by flow injection with immobilized ascorbate oxidase. *Analyst* 115:1297–1299.

104. Daily, S., S. J. Armfield, B. G. D. Haggett, and M. E. A. Downs. 1991. Automated enzyme packed-bed system for the determination of vitamin C in foods. *Analyst* 116:569–572.

105. Friedrich, O., G. Sontag, and F. Pittner. 1993. Electrochemical FIA system combined with an immobilized enzyme reactor for the determination of ascorbic acid. *Ernaehrung* 17:605–608.

106. Matsumoto, K., T. Tsukatani, and S. Higuchi. 1995. Simultaneous sensing of five compounds in fruit by amperometric flow injection system with immobilized-enzyme reactors. *Sensor. Mater.* 7:167–177.

107. Vig, A., A. Igloi, N. Adanyi, G. Gyemant, C. Csutoras, and A. Kiss. 2010. Development and characterization of a FIA system for selective assay of L-ascorbic acid in food samples. *Bioprocess Biosyst. Eng.* 33:947–952.

108. da Silva, V. L., M. R. F. Cerqueira, D. Lowinsohn, M. A. C. Matos, R. C. Matos. 2012. Amperometric detection of ascorbic acid in honey using ascorbate oxidase immobilised on amberlite IRA-743. *Food Chem.* 133:1050–1054.

109. Uchiyama, S., Y. Tofuku, and S. Suzuki. 1988. Flow-injection determination of L-ascorbate with cucumber juice as carrier. *Anal. Chim. Acta* 208:291–294.

110. Cheregi, M. and A. F. Danet. 1997. Flow injection determination of L-ascorbic acid in natural juice with biamperometric detection. *Anal. Lett.* 30:2625–2640.

111. Matos, R. C., M. A. Augelli, J. J. Pedrotti, C. L. Lago, and L. Angnes. 1998. Amperometric differential determination of ascorbic acid in beverages and vitamin C tablets using a flow cell containing an array of gold microelectrodes modified with palladium. *Electroanalysis* 10:887–890.

112. Leon, L. E. 1996. Amperometric flow-injection method for the assay of L-ascorbic acid based on the photochemical reduction of Methylene Blue. *Talanta* 43:1275–1279.

113. Dilgin, Y., Z. Dursun, and G. Nisli, 203. Flow injection amperometric determination of ascorbic acid using a photoelectrochemical reaction after immobilization of methylene blue on muscovite. *Turk. J. Chem.* 27:167–180.

114. Dilgin, Y. and G. Nisli. 2006. Flow injection photoamperometric investigation of ascorbic acid using methylene blue immobilized on titanium phosphate. *Anal. Lett.* 39:451–465.

115. Ikeda, S., H. Satake, and Y. Kohri. 1984. Flow injection analysis with an amperometric detector utilizing the redox reaction of iodate ion. *Chem. Lett.* 6:873–876.

116. Goto, M., K. Akabori, and S. Maeda. 1994. Development of a flow injection method based on bromometry and iodometry for pharmaceutical analysis. *Bunseki Kagaku* 43:505–509.

117. Strohl, A. N. and D. J. Curran. 1979. Controlled potential coulometry with the flow-through reticulated vitreous carbon electrode. *Anal. Chem.* 51:1050–1053.

118. Chen, G. 1995. Study on coulometric detector based on a porous carbon felt as a working electrode for flow injection analysis and its application. *Fenxi Huaxue* 23:292–296.

119. Okawa, Y., H. Kobayashi, and T. Ohno. 1998. Direct and simultaneous determination of ascorbic acid and glucose in soft drinks with electrochemical filter/biosensor FIA system. *Bunseki Kagaku* 47:443–445.

120. Matuszewski, W., M. Trojanowicz, and L. Ilcheva. 1990. Simultaneous enzymatic determination of glucose and ascorbic acid using flow-injection amperometry. *Electroanalysis* 2:147–153.

121. Roy, P. R., T. Okajima, and T. Ohsaka. 2004. Simultaneous electrochemical detection of uric acid and ascorbic acid at a poly(N,N-dimethylaniline) film-coated GC electrode. *J. Electroanal. Chem.* 561:75–82.

122. Wang, J. and T. Golden. 1989. Metalloporphyrin chemically modified glassy carbon electrodes as catalytic voltammetric sensors. *Anal. Chim. Acta* 217:343–351.

123. Atta, N. F., A. Galal, A. E. Karagozler, G. C. Russell, H. Zimmer, and H. B. Mark. 1991. Electrochemistry and detection of some organic and biological molecules at conducting poly(3-methylthiophene) electrodes. *Biosens. Bioelectron.* 6:333–341.

124. Matos, R. C., L. Angnes, M. C. U. Araujo, and T. C. B. Saldanha. 2000. Modified microelectrodes and multivariate calibration for flow injection amperometric simultaneous determination of ascorbic acid, dopamine, epinephrine and dipyrone. *Analyst* 125:2011–2015.

125. Doherty, A. P., M. A. Stanley, and J. G. Vos. 1995. Electrocatalytic oxidation of ascorbic acid at [osmium(2,2'-bipyridyl)2-(poly-4-vinylpyridine)10Cl]Cl modified electrodes; implications for the development of biosensors based on osmium-containing redox relays. *Analyst* 120:2371–2376.

126. Gao, Z., K. Siow, N. A. Siong, and Y. Zhang. 1997. Determination of ascorbic acid in a mixture of ascorbic acid and uric acid at a chemically modified electrode. *Anal. Chim. Acta* 343:49–57.

127. Casella, I. G. and M. R. Guascito. 1997. Electrocatalysis of ascorbic acid on the glassy carbon electrode chemically modified with polyaniline films. *Electroanalysis* 9:1381–1386.

128. Erdogdu, G. and A. Karagozler. 1997. Ersin Investigation and comparison of the electrochemical behavior of some organic and biological molecules at various conducting polymer electrodes. *Talanta* 44:2011–2018.
129. O'Connell, P. J., C. Gormally, M. Pravda, and G. G. Guilbault. 2001. Development of an amperometric l-ascorbic acid (Vitamin C) sensor based on electropolymerised aniline for pharmaceutical and food analysis. *Anal. Chim. Acta* 431:239–247.
130. Shaidarova, L. G., A. V. Gedmina, I. A. Chelnokova, and G. K. Budnikov. 2006. Electrocatalytic oxidation and flow-injection determination of ascorbic acid at a graphite electrode modified with a polyaniline film containing electrodeposited palladium. *J. Anal. Chem.* 61:601–608.
131. Liu, A. and E. Wang. 1994. Determination of catechol derivatives on pretreated and copolymer coated glassy carbon electrode. *Talanta* 41:147–154.
132. Bae, Z. U., J. H. Park, S. H. Lee, and H. Y. Chang. 1999. Nickel(II) tetraaza macro-cycle modified electrodes for the electrocatalytic determination of l-ascorbic acid by the flow injection method. *J. Electroanal. Chem.* 468:85–90.
133. Yuan, J., Y. Chen, and L. Song. 2003. Electrocatalytic oxidation of ascorbic acid at the glassy carbon electrode modified with polytoluidine blue film and its application to the FI-determination of ascorbic acid. *Huaxue Fence* 39:691–693.
134. Zen, J. M., D. M. Tsai, and A. S. Kumar. 2003. Flow injection analysis of ascorbic acid in real samples using a highly stable chemically modified screen-printedelec-trode. *Electroanalysis* 15:1171–1176.
135. Mattos, I. L., F. Padilla, J. H. Zagal, E. H. L. Falcao, and R. Segura. 2011. Screen-printed electrode modified with silver hexacyanoferrate-nafion for ascorbic acid determination. *J. Chil. Chem. Soc.* 56:803–807.
136. Yebra-Biurrun, M. C. and R. M. Cespón-Romero. 1997. Indirect flow-injection determination of ascorbic acid by flame atomic absorption spectrometry. *Mikrochim. Acta* 126:53–58.
137. Zhang, Z. Q. and Y. C. Jiang. 2000. Flame atomic absorption spectrometry for the automatic indirect determination of ascorbic acid based on flow injection and reduc-tion of chromate. *At. Spectrosc.* 21:100–104.
138. Jiang, Y. C., Z. Q. Zhang, and J. Zhang. 2001. Flow-injection, on-line concentrat-ing and flame atomic absorption spectrometry for indirect determination of ascorbic acid based on the reduction of iron(III). *Anal. Chim. Acta* 435:351–355.
139. Yebra-Biurrun, M. C., R. M. Cespón-Romero, and A. Moreno-Cid. 2001. Flow injection flame AAS determination of ascorbic acid based on permanganate reduc-tion. *At. Spectrosc.* 22:346–349.
140. Yebra, M. C., R. M. Cespón, and A. Moreno-Cid. 2001. Automatic determina-tion of ascorbic acid by flame atomic absorption spectrometry. *Anal. Chim. Acta* 448:157–164.
141. Noroozifar, M., M. Khorasani-Motlagh, and K. Akhavan. 2005. Atomic absorp-tion spectrometry for the automatic indirect determination of ascorbic acid based on the reduction of manganese dioxide. *Anal. Sci.* 21:655–659.
142. Iwado, A., M. Mifune, H. Akizawa, N. Motohashi, and Y. Saito. 2003. Flow injection analysis of ascorbic acid using carriers modified with metal-porphine as oxidative solid catalysts. *J. Pharm. Biomed. Anal.* 30:1923–1928.
143. Xie, Z. P., H. M. Zhong, L. B. Luo, and J. W. Wang. 2007. HPLC-FI chemilu-minescence determination of VC in Gannan-orange with on-line electrogenerated hypochlorite. *Fenxi Shiyanshi* 26:95–97.

144. Gazdik, Z., O. Zitka, J. Petrlova et al. 2008. Determination of vitamin C (ascorbic acid) using high performance liquid chromatography coupled with electrochemical detection. *Sensors* 8:7097–7112.

145. Chen, X. G., J. S. Zhang, and X. M. Liu. 2008. Separation and determination of L-ascorbic acid, ascorbyl glucoside and kojic acid in commercial cosmetics by FI-CE. *Ziran Kexueban* 44:49–53.

146. Bhandari, D., V. Kertesz, and G. J. Van Berkel. 2013. Rapid quantitation of ascorbic and folic acids in SRM 3280 multivitamin/multielement tablets using flow-injection tandem mass spectrometry. *Rapid Commun. Mass. Spectrom.* 27:163–168.

147. Iida, Y., K. Akiba, and I. Satoh. 2004. Calorimetric determination of L-ascorbate using an FIA system armed with an electrolytic device. *Chem. Sensors* 20:109–111.

148. Ruzicka, J. and G. D. Marshall. 1990. Sequential injection: A new concept for chemical sensors, process analysis and laboratory assays. *Anal. Chim. Acta* 237:329–343.

149. Sultan, S. M. and N. I. Desai. 1998. Mechanistic study and kinetic determination of vitamin C employing the sequential injection technique. *Talanta* 45:1061–1071.

150. Llorent-Martínez, E. J., P. Ortega-Barrales, and A. Molina-Diaz. 2008. Sequential injection multi-optosensor based on a dual-luminescence system using two sensing zones: Application to multivitamin determination. *Microchim. Acta* 162:199–204.

151. Ruzicka, J. 2000. Lab-on-valve: Universal microflow analyzer based on sequential and bead injection. *Analyst* 125:1053–1060.

152. Sorouraddin, H. M., A. Hibara, M. A. Proskurnin, and T. Kitamori. 2000. Integrated FIA for the determination of ascorbic acid and dehydroascorbic acid in a microfabricated glass-channel by thermal-lens microscopy. *Anal. Sci.* 16:1033–1037.

153. Cerdá, V. and C. Pons. 2006. Multicommutated flow techniques for developing analytical methods. *Trends Anal. Chem.* 25:236–242.

154. Fernández, M., R. Forteza, and V. Cerdá. 2012. Multisyringe chromatography (MSC): An effective and low cost tool for water-soluble vitamin separation. *Anal. Lett.* 45:2637–2647.

155. Burns, D. T., N. Chimpalee, D. Chimpalee, and S. Rattanariderom. 1991. Flow-injection spectrophotometric determination of ascorbic acid by reduction of vanadotungstophosphoric acid. *Anal. Chim. Acta* 243:187–190.

156. Wang, Y. and M. Zhou. 1991. Determination of vitamin C content by negative-absorption flow injection analysis. *Shenyang Yaoxueyuan Xuebao* 8:102–104.

157. Kojlo, A., E. Kleszczewska, and H. Puzanowska-Tarasiewicz. 1994. Flow-injection spectrophotometric determination of ascorbic acid by oxidation with iron(III). *Acta Pol. Pharm.* 51:293–295.

158. He, H., Y. Shen, X. Qi, K. Ni, and Y. Jiang. 1995. Flow injection analysis of tablets and injections of ascorbic acid. *Zhongguo Yiyao Gongye Zazhi* 26:70–72.

159. Zhao, Z., D. Zhang, and S. Li. 1996. Determination of vitamin C content in drug preparations by FIA differential photometry. *Zhongguo Shenghua Yaowu Zazhi* 17:259–261.

160. Huang, Y., Q. Fang, J. Huang, and Y. Sun. 1996. Determination of vitamin C by FIA with reversed merging zone. *Fenxi Shiyanshi* 15:34–36.

161. Nobrega, J. A. and G. S. Lopes. 1995. Flow-injection spectrophotometric determination of ascorbic acid in pharmaceutical products with the Prussian Blue reaction. *Talanta* 43:971–976.

162. Tabata, M. and H. Morita. 1997. Spectrophotometric determination of a nanomolar amount of ascorbic acid using its catalytic effect on copper(II) porphyrin formation. *Talanta* 44:151–157.

163. Wu, X., X. Chen, G. Li, and Z. Hu. 1997. Rapid determination of vitamin C in pharmaceuticals by online microwave flow injection analysis. *Lanzhou Daxue Xuebao* 33:58–62.

164. Grudpan, K., K. Kamfoo, and J. Jakmunee. 1999. Flow injection spectrophotometric or conductometric determination of ascorbic acid in a vitamin C tablet using permanganate or ammonia. *Talanta* 49:1023–1026.

165. Fan, J., C. Ye, S. Feng, G. Zhang, and J. Wang. 1999. Flow injection kinetic spectrophotometric determination of ascorbic acid based on an inhibiting effect. *Talanta* 50:893–898.

166. Kleszczewski, T. and E. Kleszczewska. 2001. FIA of vitamin C in blood serum in humans at increasing ethanol concentration. *J. Pharm. Biomed. Anal.* 25:477–481.

167. Themelis, D. G., P. D. Tzanavaras, and F. S. Kika. 2001. On-line dilution flow injection manifold for the selective spectrophotometric determination of ascorbic acid based on the Fe(II)-2,2'-dipyridyl-2-pyridylhydrazone complex formation. *Talanta* 55:127–134.

168. Ensafi, A. A., B. Rezaei, and M. Beglari. 2002. Highly selective flow-injection spectrophotometric determination of ascorbic acid in fruit juices and pharmaceuticals using pyrogallol red-iodate system. *Anal. Lett.* 35:909–920.

169. Kleszczewski, T. and E. Kleszczewska. 2002. Flow injection spectrophotometric determination of L-ascorbic acid in biological matters. *J. Pharm. Biomed. Anal.* 29:755–759.

170. Noroozifar, M., M. Khorasani-Motlagh, and M. Estakhri. 2003. Indirect spectrophotometric determination of ascorbic acid by dissolution of diphenylcarbazide and chromate in a flow injection system. *Asian J. Spectrosc.* 7:19–25.

171. Memon, N., M. I. Bhanger, and M. A. Memon. 2003. Flow injection spectrophotometric determination of ascorbic acid with 2,2'-bipyridine in anionic surfactant micellar medium. *Pak. J. Anal. Chem.* 4:152–156.

172. Shrivas, K., K. Agrawal, and D. K. Patel. 2005. A spectrophotometric determination of ascorbic acid. *J. Chin. Chem. Soc.-Taip.* 52:503–506.

173. Janghel, E. K., V. K. Gupta, M. K. Rai, and J. K. Rai. 2007. Micro determination of ascorbic acid using methyl viologen. *Talanta* 72:1013–1016.

174. Liu, J. and J. I. Itoh. 2007. Kinetic method for determination of ascorbic acid on flow injection system by using its catalytic effect on the complexation reaction of an ultra sensitive colorimetric reagent of porphyrin with Cu(II). *Spectrochim. Acta* 67A:455–459.

175. Carneiro, R. L., J. W. B. Braga, R. J. Poppi, and R. Tauler. 2008. Multivariate curve resolution of pH gradient flow injection mixture analysis with correction of the Schlieren effect. *Analyst* 133:774–783.

176. N. Teshima, T. Nobuta, and T. Sakai. 2008. Simultaneous spectrophotometric determination of ascorbic acid and glutathione by kinetic-based flow injection analysis. *Bunseki Kagaku* 57:327–333.

177. Schmidt, E., W. R. Melchert, and F. R. P. Rocha. 2011. Polyvalent flow analysis system for spectrophotometric determination of pharmaceuticals. *Quim. Nova.* 34:1205–1210.

178. Hassan, R. O. and A. T. Faizullah. 2011. Reverse-FIA with spectrophotometric detection method for determination of Vitamin C. *J. Iran Chem. Soc.* 8:662–673.

179. Kukoc-Modun, L., M. Biocic, and N. Radic 2012. Indirect method for spectrophotometric determination of ascorbic acid in pharmaceutical preparations with 2,4,6-tripyridyl-*s*-triazine by flow-injection analysis. *Talanta* 96:174–179.

180. Hasebe, T. and T. Kawashima. 1996. Flow injection determination of ascorbic acid by iron(III)-catalyzed lucigenin chemiluminescence in a micellar system. *Anal. Sci.* 12:773–777.

181. Feng, M., J. Lu, X. Zhang, Z. Xu, and Z. Zhang. 1996. Flow-injection chemiluminescence analysis for determination of trace ascorbic acid. *Fenxi Huaxue* 24:1364.

182. Chen, H., W. Qin, and Z. Zhang. 1997. Highly sensitive chemiluminescence system for determination of ascorbic acid. *Fenxi Huaxue* 25:1079–1081.

183. Qin, W., Z. J. Zhang, and H. H. Chen. 1997. Highly sensitive chemiluminescence flow sensor for ascorbic acid. *Fresenius J. Anal. Chem.* 358:861–863.

184. Kubo, H. and A. Toriba. 1997. Chemiluminescence flow injection analysis of reducing agents based on the luminol reaction. *Anal. Chim. Acta* 353:345–349.

185. Danet, A. F., M. Badea, and H. Y. Aboul-Enein. 2000. Flow injection system with chemiluminometric detection for enzymatic determination of ascorbic acid. *Luminescence* 15:305–309.

186. Ma, Y., M. Zhou, X. Jin, B. Zhang, H. Chen, and N. Guo. 2002. Flow-injection chemiluminescence determination of ascorbic acid by use of the cerium(IV)-rhodamine B system. *Anal. Chim. Acta* 464:289–293.

187. Luo, W. F. and R. Q. Zeng. 2004. Flow injection chemiluminescence analysis of ascorbic acid. *Henan Shifan Daxue Xuebao* 32:69–72.

188. Anastos, N., N. W. Barnett, B. J. Hindson, C. E. Lenehan, and S. W. Lewis. 2004. Comparison of soluble manganese(IV) and acidic potassium permanganate chemiluminescence detection using flow injection and sequential injection analysis for the determination of ascorbic acid in Vitamin C tablets. *Talanta* 64:130–134.

189. B. Li, D. Wang, C. Xu, Z. Zhang. 2005. Flow-injection simultaneous chemiluminescence determination of ascorbic acid and L-cysteine with partial least squares calibration. *Microchim. Acta* 149(3–4):205–212.

190. He, D., S. He, and Z. A. Zhang. 2007. Chemiluminescence reaction between hydroxyl radical and rhodamine 6G and its applications. *Anal. Lett.* 40:2935–2945.

191. Xie, Z. P., Y. J. Xiao, and J. W. Wang. 2007. FI chemiluminescence determination of ascorbic acid with on-line electrogenerated hypochlorite. *Fenxi Shiyanshi* 26:64–66.

192. Chen, H., R. Li, L. Lin, G. Guo, and J. M. Lin. 2010. Determination of L-ascorbic acid in human serum by chemiluminescence based on hydrogen peroxide-sodium hydrogen carbonate-CdSe/CdS quantum dots system. *Talanta* 81:1688–1696.

193. Wu, X. Z., L. H. Wang, and X. H. Lu. 2010. Flow-injection enhancing chemiluminescence determination of ascorbic acid based on the reaction of luminolpotassium bromated. *Fenxi Shiyanshi* 29:101–104.

194. Wang, H. and J. X. Du. 2010. Investigation on the chemiluminescence reaction of nano-colloidal MnO_2 and ascorbic acid. *Fenxi Shiyanshi* 29:43–46.

195. Wada, M., M. Kira, H. Kido, R. Ikeda, N. Kuroda, T. Nishigaki, and K. Nakashima. 2011. Semi-micro flow injection analysis method for evaluation of quenching effect of health foods or food additive antioxidants on peroxynitrite. *Luminescence* 26:191–195.

196. Zhao, P., Y. Tan, W. Gao, and J. Yu. 2011. Determination of ascorbic acid by flow injection chemiluminescence and its mechanism. *Jinan Daxue Xuebao* 25:243–247.

197. Zhang, X. D., H. Y. Cao, and Y. M. Huang. 2011. Determination of ascorbic acid based on luminol chemiluminescence catalyzed by the supported Co/SiO$_2$. *Xinan Daxue Xuebao* 33:74–77.

198. Cai, L. and C. Xu. 2011. An automated method for determination of ascorbic acid in urine by flow injection chemiluminescence coupling with on-line removal of interference. *J. Chil. Chem. Soc.* 56:938–940.

199. Zeng, H. J., H. L. Liang, J. You, S. J. Li, G. H. Sui, and L. B. Qu. 2013. Simultaneous determination of ascorbic acid and rutin by flow-injection chemiluminescence method using partial least squares regression. *Faguang Xuebao* 34:369–374.

200. Schaffar, B. P. H., B. A. A. Dremel, and R. D. Schmid. 1989. Ascorbic acid determination in fruit juices based on a fiber optic ascorbic acid biosensor and flow injection analysis. In *Biosensors: Applications in Medicine, Environmental Protection and Process Control*, eds. R. D. Schmid, and F. Scheller, pp. 229–232. New York: VCH Publishers.

201. Elbashier, E. E. and G. M. Greenway. 1991. Determination of vitamin C in karkady using flow injection analysis. *J. Micronut. Anal.* 8:311–316.

202. Ensafi, A. A. and B. Rezaei. 1998. Flow injection analysis determination of ascorbic acid with spectrofluorometric detection. *Anal. Lett.* 31:333–342.

203. Maki, T., N. Soh, K. Nakano, and T. Imato. 2011. Flow injection fluorometric determination of ascorbic acid using perylenebisimide-linked nitroxide. *Talanta* 85:1730–1733.

204. Fogg, A. G., A. M. Summan, and M. A. Fernandez-Arciniega. 1985. Flow injection amperometric determination of ascorbic acid and dopamine at a sessile mercury drop electrode without deoxygenation. *Analyst* 110:341–343.

205. Stevanato, R., L. Avigliano, A. Finazzi-Agrò, and A. Rigo. 1985. Determination of ascorbic acid with immobilized green zucchini ascorbate oxidase. *Anal. Biochem.* 149:537–542.

206. Lunte, C. E., S. W. Wong, T. H. Ridgway, W. R. Heineman, and K. W. Chan. 1986. Voltammetric/amperometric detection for flow-injection systems. *Anal. Chim. Acta* 188:63–67.

207. Wang, J., T. Golden, and T. Peng. 1987. Poly(4-vinylpyridine)-coated glassy carbon flow detectors. *Anal. Chem.* 59:740–744.

208. Younghee, H. 1988. Amperometric determination of ascorbic acid at a thin layer flow cell. *Arch. Pharm. Res.* 11:56–60.

209. Abdalla, M. A. and H. M. Al-Swaidan. 1989. Iodimetric determination of iodate, bromate, hypochlorite, ascorbic acid, and thiourea using flow-injection amperometry. *Analyst* 114:583–586.

210. Uchiyama, S. and S. Suzuki. 1990. Flow-injection determination of total vitamin C using cucumber juice carrier. *Bunseki Kagaku* 39:793–795.

211. Aoki, A., T. Matsue, and I. Uchida. 1992. Multichannel electrochemical detection with a microelectrode array in flowing streams. *Anal. Chem.* 64:44–49.

212. Almuaibed, A. M. and A. Townshend. 1992. From Individual and simultaneous determination of uric acid and ascorbic acid by flow injection analysis. *Talanta* 39:1459–1462.

213. Fung, Y. S. and S. Y. Mo. 1992. Application of square-wave voltammetry for the determination of ascorbic acid in soft drinks and fruit juices using a flow-injection system. *Anal. Chim. Acta* 261:375–380.

214. Marques, I. D. H. C., E. T. A. Marques, A. C. Silva, W. M. Ledingham, E. H. M. Melo, V. L. da Silva, and J. L. Lima Filho. 1994. Ascorbic acid determination in biological fluids using ascorbate oxidase immobilized on alkylamine glass beads in a flow injection potentiometric system. *Appl. Biochem. Biotechnol.* 44:81–89.

215. Zhiqiang, G. 1994. Electrochemistry of ascorbic acid at polypyrrole/dodecyl sulphate film-coated electrodes and its application. *J. Electroanal. Chem.* 365:197–205.
216. Sano, A., T. Kuwayama, M. Furukawa, S. Takitani, and H. Nakamura. 1995. Determination of L-ascorbic acid by a flow injection analysis with copper(II)-mediated electrochemical detection. *Anal. Sci.* 11:405–409.
217. Garrido, E. M., J. L. F. C. Lima, and C. Delerue-Matos. 1995. Determination of ascorbic acid in pharmaceutical products by flow injection analysis using an amperometric detector. *Farmaco* 50:881–884.
218. Casella, I. G. 1996. Electrooxidation of ascorbic acid on the dispersed platinum glassy carbon electrode and its amperometric determination in flow injection analysis. *Electroanalysis* 8:128–134.
219. Kakizaki, T., K. Imai, and K. Hasebe. 1996. Flow injection with diaphragm pump and amperometric detector. *Anal. Commun.* 33:75–77.
220. Yu, A. M., C.-X. He, J. Zhou, and H. Y. Chen. 1997. Flow injection analysis of ascorbic acid at a methylene green chemically modified electrode. *Fresenius J. Anal. Chem.* 357:84–85.
221. Shyu, S. C. and C. M. Wang. 1997. Characterization of iron-containing clay modified electrodes and their applications for the detection of hydrogen peroxide and ascorbic acid. *J. Electrochem. Soc.* 144:3419–3425.
222. Matos, R. C., L. Angnes, and C. L. Do Lago. 1998. Development of a four-channel potentiostat for amperometric detection in flow injection analysis. *Instrum. Sci. Technol.* 26:451–459.
223. Granger, M. C., J. Xu, J. W. Strojek, and G. M. Swain. 1999. Polycrystalline diamond electrodes: Basic properties and applications as amperometric detectors in flow injection analysis and liquid chromatography. *Anal. Chim. Acta* 397:145–161.
224. Kolar, M. and D. Dobcnik. 1999. Preparation of iodide ion-selective electrode with chemically pretreated silver tube and its suitability in potentiometric flow injection analysis for the determination of vitamin C. *Slovenski Kemijski Dnevi* 1999:88–93.
225. Matos, R. C., M. A. Augelli, C. L. Lago, and L. Angnes. 2000. Flow injection analysis-amperometric determination of ascorbic and uric acids in urine using arrays of gold microelectrodes modified by electrodeposition of palladium. *Anal. Chim. Acta* 404:151–157.
226. Florou, A. B., M. I. Prodromidis, M. I. Karayannis, and S. M. Tzouwara-Karayanni. 2000. Flow electrochemical determination of ascorbic acid in real samples using a glassy carbon electrode modified with a cellulose acetate film bearing 2,6-dichlorophenolindophenol. *Anal. Chim. Acta* 409:113–121.
227. Song, J., C. Zhao, W. Guo, and X. Kang. 2000. Flow injection biamperometric determination of ascorbic acid. *Fenxi Huaxue* 28:38–41.
228. Fernandes, J. C. B., L. Rover, L. T. Kubota, and G. De Oliveira Neto. 2000. Potentiometric determination of L-ascorbic acid in pharmaceutical samples by FIA using a modified tubular electrode. *J. Braz. Chem. Soc.* 11:182–186.
229. Cardwell, T. J. and M. J. Christophersen. 2000. Determination of sulfur dioxide and ascorbic acid in beverages using a dual channel flow injection electrochemical detection system. *Anal. Chim. Acta* 416:105–110.
230. Abulkibash, A. M. S., M. E. Koken, M. M. Khaled, and S. M. Sultan. 2000. Differential electrolytic potentiometry, a detector in flow injection analysis for oxidation-reduction reactions. *Talanta* 52:1139–1142.

231. Kolar, M., D. Dobcnik, and N. J. Radic. 2000. Potentiometric flow-injection determination of vitamin C and glutathione with a chemically prepared tubular silver-electrode. *Pharmazie* 55:913–916.

232. Castro, S. S. L., V. R. Balbo, P. J. S. Barbeira, and N. R. Stradiotto. 2001. Flow injection amperometric detection of ascorbic acid using a Prussian Blue film-modified electrode. *Talanta* 55:249–254.

233. Muñoz, R. A. A., R. C. Matos, and L. Angnes. 2001. Gold electrodes from compact discs modified with platinum for amperometric determination of ascorbic acid in pharmaceutical formulations. *Talanta* 55:855–860.

234. Hosseinalizadeh-Khorasani, J., M. K. Amini, H. Ghanei, and S. Tangestaninejad. 2001. Flow-injection amperometric determination of ascorbic acid using a graphite-epoxy composite electrode modified with cobalt phthalocyanine. *Iran J. Chem. Chem. Eng.* 20:66–74.

235. Catarino, R. I. L., M. B. Q. Garcia, J. L. F. C. Lima, E. Barrado, and M. Vega. 2002. Relocation of a tubular voltammetric detector for standard addition in FIA. *Electroanalysis* 14:741–746.

236. Iida, Y., T. Satoh, and I. Satoh. 2002. Application of an electrolytic device to an FIA system for extension of the determination range of L-ascorbic acid. *Electrochem. Soc. Jap.* 70:515–517.

237. Iida, Y., T. Kikuchi, and I. Satoh. 2002. Determination of L-ascorbate by using an electrolytic device-FIA system with an oxidase column. *Chem. Sensors* 18:124–126.

238. Komatsu, M. and A. Fujishima. 2003. Detection of ascorbic acid in an ethanol-water mixed solution on a conductive diamond electrode. *Bull. Chem. Soc. Jpn.* 76:927–933.

239. Harjana, I. S., S. Muzakki, and, H. Miftahul. 2003. A bipotentiometric FIA system for determination of vitamin C. *Majalah Farmasi Indo* 14:276–283.

240. Badea, M., A. Curulli, and G. Palleschi. 2003. Oxidase enzyme immobilization through electropolymerized films to assemble biosensors for batch and flow injection analysis. *Biosens. Bioelectron.* 18:689–698.

241. Roy, P. R., M. S. Saha, T. Okajima, and T. Ohsaka. 2004. Electrooxidation and amperometric detection of ascorbic acid at GC electrode modified by electropolymerization of N,N-dimethylaniline. *Electroanalysis* 16:289–297.

242. Tomita, I. N., A. Manzoli, F. L. Fertonani, and H. Yamanaka. 2005. Amperometric biosensor for ascorbic acid. *Ecletica. Quim.* 30:37–43.

243. Cano, M., B. Palenzuela, and R. Rodriguez-Amaro. 2006. A PVC/TTF-TCNQ composite electrode for use as a detector in flow injection analysis. *Electroanalysis* 18:1727–1729.

244. D'Eramo, F., L. E. Sereno, and A. H. Arevalo. 2007. Electrocatalytic properties of a novel poly-1-naphthylamine-modified electrode using ascorbic acid as molecule probe. *Electroanalysis* 19:96–102.

245. Cano, M., B. Palenzuela, J. L. Avila, and R. Rodriguez-Amaro. 2007. Simultaneous determination of ascorbic acid and uric acid by using a PVC/TTF-TCNQ composite electrode as detector in a FIA system. *Electroanalysis* 19:973–977.

246. Paixao, T. R. L. C. and M. Bertotti. 2008. FIA determination of ascorbic acid at low potential using a ruthenium oxide hexacyanoferrate modified carbon electrode. *J. Pharm. Biomed. Anal.* 46:528–533.

247. Ke, J. H., H. J. Tseng, C. T. Hsu, J. C. Chen, G. Muthuraman, and J. M. Zen. 2008. Flow injection analysis of ascorbic acid based on its thermoelectrochemistry at disposable screen-printed carbon electrodes. *Sens. Actuat. B* 130:614–619.

248. Goreti, M., F. Sales, M. S. A. Castanheira, R. M. S. Ferreira, M. Carmo, V. G. Vaz, and C. Delerue-Matos. 2008. Chemically modified carbon paste electrodes for ascorbic acid determination in soft drinks by flow injection amperometric analysis Portugaliae. *Electrochim. Acta* 26:147–157.
249. Bunpeng, P., S. Lapanantnoppakhun, and J. Jakmunee. 2008. Flow injection amperometric method with dialysis sample pretreatment for determination of ascorbic acid. *Chiang Mai J. Sci.* 35:345–354.
250. Pio dos Santos, W. T., E. G. Nascimento de Almeida, H. E. A. Ferreira, D. T. Gimenes, and E. M. Richter. 2008. Simultaneous flow injection analysis of paracetamol and ascorbic acid with multiple pulse amperometric detection. *Electroanalysis* 20:1878–1883.
251. Nien, P. C., P. Y. Chen, and K. C. Ho. 2009. On the amperometric detection and electrocatalytic analysis of ascorbic acid and dopamine using a poly(acriflavine)-modified electrode. *Sens. Actuat. B* 140:58–64.
252. Abulkibash, A. M. S., S. Fraihat, and B. El Ali. 2009. Flow injection determination of vitamin C in pharmaceutical preparations by differential electrolytic potentiometry. *J. Flow Injection Anal.* 26:121–125.
253. Abdel, A., A. Ayman, and A. H. Kamel. 2010. Batch and hydrodynamic monitoring of vitamin C using novel periodate selective sensors based on a newly synthesized Ni(II)-Schiff bases complexes as a neutral receptors. *Talanta* 80:1356–1363.
254. Guerreiro, J. R. L., A. H. Kamel, and M. G. F. Sales. 2010. FIA potentiometric system based on periodate polymeric membrane sensors for the assessment of ascorbic acid in commercial drinks. *Food Chem.* 120:934–939.
255. Toniolo, R., N. Dossi, A. Pizzariello, S. Susmel, and G. Bontempelli. 2011. Simultaneous detection of ascorbic acid and hydrogen peroxide by flow-injection analysis with a thin layer dualelectrode detector. *Electroanalysis* 23:628–636.
256. Song, W., J. Ding, R. Liang, and W. Qin. 2011. Potentiometric flow injection system for determination of reductants using a polymeric membrane permanganate ion-selective electrode based on current-controlled reagent delivery. *Anal. Chim. Acta* 704:68–72.
257. Dey, R. S., S. Gupta, R. Paira, S. M. Chen, and C. R. Raj. 2012. Flow injection amperometric sensing of uric acid and ascorbic acid using the self-assembly of heterocyclic thiol on Au electrode. *J. Solid State Electrochem.* 16:173–178.
258. de Miranda, J. A. T., R. R. Cunha, D. T. Gimenes, R. A. A. Muñoz, and E. M. Richter. 2012. Simultaneous determination of ascorbic acid and acetylsalicylic acid using flow injection analysis with multiple pulse amperometric detection. *Quim. Nova* 35:1459–1463.
259. Sultan, S. M., Y. A. M. Hassan, and K. E. E. Ibrahim. 1999. Sequential injection technique for automated titration: Spectrophotometric assay of vitamin C in pharmaceutical products using cerium(IV) in sulfuric acid. *Analyst* 124:917–921.
260. Newcombe, D. T., T. J. Cardwell, R. W. Cattrall, and S. D. Kolev. 2000. An optical membrane redox chemical sensor for the determination of ascorbic acid. *Lab. Robot. Autom.* 12:200–204.
261. Lenghor, N., J. Jakmunee, M. Vilen, R. Sara, G. D. Christian, and K. Grudpan. 2002. Sequential injection redox or acid-base titration for determination of ascorbic acid or acetic acid. *Talanta* 58:1139–1144.
262. Legnerová, Z., D. Šatínský, and P. Solich. 2003. Using on-line solid phase extraction for simultaneous determination of ascorbic acid and rutin trihydrate by sequential injection analysis. *Anal. Chim. Acta* 497:165–174.

263. Polášek, M., P. Skála, L. Opletal, and L. Jahodář. 2004. Rapid automated assay of anti-oxidation/radical-scavenging activity of natural substances by sequential injection technique (SIA) using spectrophotometric detection. *Anal. Bioanal. Chem.* 379:754–758.

264. Nakamura, K., Y. Ohba, N. Kishikawa, and N. Kuroda. 2004. Measurement of antioxidative activity against hypochlorite ion by sequential injection analysis with luminol chemiluminescence detection. *Bunseki Kagaku* 53:925–930.

265. S. Y. Tham and C. H. Foo. 2005. Ascorbic acid assay using sequential injection analysis. *Malays. J. Biochem. Mol. Biol.* 12:8–13.

266. Gutés, A., F. Céspedes, S. Alegret, and M. del Valle. 2005. Sequential injection system with higher dimensional electrochemical sensor signals. Part 1. Voltammetric e-tongue for the determination of oxidizable compounds. *Talanta* 66:1187–1196.

267. Gutés, A., D. Calvo, F. Céspedes, and M. del Valle. 2007. Automatic sequential injection analysis electronic tongue with integrated reference electrode for the determination of ascorbic acid, uric acid and paracetamol. *Microchim. Acta* 157:1–6.

268. Sultan, S. M. and Y. A. M. Hassan. 2009. Sequential injection titrimetric analysis of vitamin C in drug formulation using potassium permanganate. *J. Flow Injection Anal.* 26:53–57.

269. Vishnikin, A. B., T. Ye. Svinarenko, H. Sklenárová, P. Solich, Y. R. Bazel, and V. Andruch. 2010. 11-Molybdobismuthophosphate—A new reagent for the determination of ascorbic acid in batch and sequential injection systems. *Talanta* 80:1838–1845.

270. Chu, N. and S. Fan. 2010. Simultaneous determination of vitamin C and paracetamol in pharmaceuticals using sequential injection pH gradient spectrophotometry. *Fenxi Yiqi* 2:46–51.

271. Pokrzywnicka, M., D. J. Cocovi-Solberg, M. Miro, V. Cerda, R. Koncki, and L. Tymecki. 2011. Miniaturized optical chemosensor for flow-based assays. *Anal. Bioanal. Chem.* 399:1381–1387.

272. Vishnikin, A. B., H. Sklenarova, P. Solich, G. A. Petrushina, and L. P. Tsiganok. 2011. Determination of ascorbic acid with Wells-Dawson type molybdophosphate in sequential injection system. *Anal. Lett.* 44:514–527.

273. Mai, T. D. and P. C. Hauser. 2011. Anion separations with pressure-assisted capillary electrophoresis using a sequential injection analysis manifold and contactless conductivity detection. *Electrophoresis* 32:3000–3007.

274. Paim, A. P. S., and B. F. Reis. 2000. An automatic spectrophotometric titration procedure for ascorbic acid determination in fruit juices and soft drinks based on volumetric fraction variation. *Anal. Sci.* 16:487–491.

275. Paim, A. P. S., C. M. N. V. Almeida, B. F. Reis, R. A. S. Lapa, E. A. G. Zagatto, and J. L. F. Costa Lima. 2002. Automatic potentiometric flow titration procedure for ascorbic acid determination in pharmaceutical formulations. *J. Pharm. Biomed. Anal.* 28:1221–1225.

276. Rocha, F. R. P., O. Fatibello Filho, and B. F. A. Reis. 2003. Multicommuted flow system for sequential spectrophotometric determination of hydrosoluble vitamins in pharmaceutical preparations. *Talanta* 59: 191–200.

277. Llorent-Martinez, E. J., L. Molina-Garcia, R. Kwiatkowski, and A. Ruiz-Medina. 2013. Application of quantum dots in clinical and alimentary fields using multicommutated flow injection analysis. *Talanta* 109:203–208.

278. Pires, C. K., A. F. Lavorante, L. M. T. Marconi, S. R. P. Meneses, and E. A. G. Zagatto. 2006. A multi-pumping flow system for chemiluminometric determination of ascorbic acid in powdered materials for preparation of fruit juices. *Microchem. J.* 83:70–74.

279. Sasaki, M. K., D. S. M. Ribeiro, C. Frigerio, J. A. V. Prior, J. L. M. Santos, and E. A. G. Zagatto. 2014. Chemiluminometric determination of ascorbic acid in pharmaceutical formulations exploiting photo-activation of GSHcapped CdTe quantum dots. *Luminescence* 29:901–907.

280. Ruedas Rama, M. J., A. Ruiz Medina, and A. Molina Diaz. 2003. Bead injection spectroscopic flow-through renewable surface sensors with commercial flow cells as an alternative to reusable flow-through sensors. *Anal. Chim. Acta* 482:209–217.

281. Ruedas Rama, M. J., A. Ruiz Medina, and A. A. Molina Diaz. 2004. Prussian blue-based flow-through renewable surface optosensor for analysis of ascorbic acid. *Microchem. J.* 78:157–162.

282. Karayannis, M. I. 1975. Kinetic determination of ascorbic acid by the 2,6-dichlorophenolindophenol reaction with a stopped-flow technique. *Anal. Chim. Acta* 76:121–130.

283. Obata, H., T. Tokuyama, Y. Nitta, M. Takagi, and K. Hiromi. 1979. A stopped-flow method for the determination of ascorbic acid in orange juice containing triose reductone. *Agric. Biol. Chem.* 43:2191–2192.

284. Hiromi, K., C. Kuwamoto, and M. Ohnishi. 1980. A rapid sensitive method for the determination of ascorbic acid in the excess of 2,6-dichlorophenolindophenol usinga stopped-flow apparatus. *Anal. Biochem.* 101:421–426.

285. Koupparis, M. A., K. M. Walczak, and H. V. Malmstadt. 1980. A compact automated microprocessor-based stopped-flow analyzer. *J. Autom. Chem.* 2:66–75.

286. Farasoglou, D. I. and M. I. Karayannis. 1982. Kinetic determination of ascorbic acid in pharmaceutical preparations and biological fluids applying a stopped-flow technique. *Chim. Chronika* 11:281–294.

287. Karayannis, M. I. and D. I. Farasoglou. 1987. Kinetic-spectrophotometric method for the determination of ascorbic acid in orange juice, parsley, and potatoes. *Analyst* 112:767–770.

288. Matsumoto, K., J. J. Baeza Baeza, and H. A. Mottola. 1993. Simultaneous kinetic-based determination of fructose and ascorbate with a rotating bioreactor and amperometric detection: Application to the analysis of food samples. *Anal. Chem.* 65:1658–1661.

289. Huang, H. P., R. X. Cai, Y. M. Du, and Y. E. Zeng. 1995. Laccase-based micellar enhanced spectrofluorimetric determination of L-ascorbic acid. *Chinese Chem. Lett.* 6:235–238.

290. Huang, H., R. Cai, T. Korenaga, X. Zhang, Y. Yang, Y. Du, and Y. Zeng. 1997. Micelle enhanced spectrofluorimetric determination of L-ascorbic acid based on laccase-linked coupling reaction using flow-injection stopped-flow technique. *Anal. Sci.* 13:67–70.

291. Zeng, S. and H. Tan. 2000. Simultaneous determination of ascorbic acid (Vc) and L-cysteine in fruit juice by stopped-flow kinetic spectrophotometry. *Shipin Kexue* 21:60–62.

292. Zeng, S. and H. Tan. 2000. Stopped-flow injection analysis method for simultaneous determination of ascorbic acid and L-cysteine. *Huaxue Fence* 36:439–441.

293. Perez-Ruiz, T., C. Martinez-Lozano, V. Tomas, and J. Fenol. 2001. Fluorimetric determination of total ascorbic acid by a stopped-flow mixing technique. *Analyst* 126:1436–1439.

294. Messina, G. A., A. A. J. Torriero, I. E. De Vito, and J. Raba. 2004. Continuous-flow/stopped-flow system for determination of ascorbic acid using an enzymatic rotating bioreactor. *Talanta* 64:1009–1017.

CHAPTER 19

Determination of Vitamin E and Similar Compounds

Mohammad Yaqoob and Abdul Nabi

CONTENTS

19.1 INTRODUCTION

19.1.1 Vitamins

Vitamins are well-known group of organic, low-molecular-weight substances that are essential in very small amounts for the normal function of human body. Vitamins have generally different chemical and physiological functions and are broadly distributed in natural food sources. Thirteen vitamins have been recognized in human nutrition and have been conveniently classified into two main groups: fat-soluble vitamins mainly include vitamin A (retinol), D (calciferol), E (tocopherol), and K (phylloquinone, menaquinone); and water-soluble vitamins include vitamin C (ascorbic acid) and members of the vitamin B group; B_1 (thiamine), B_2 (riboflavin), B_3 (niacin, nicotinic acid), B_5 (pantothenic acid), B_6 (pyridoxine, pyridoxal hydrochloride), B_7 (biotin), B_9 (folic acid), and B_{12} (cobalamin, cyanocobalamin) (Ball, 2006).

Vitamins play specific and vital functions in metabolism and their lack or excess can cause health problems. The supply of vitamins depends on diet; however, foods that contain the necessary vitamins can have reduced vitamin content after storage, processing, or cooking. Therefore, many people take multivitamin tablets and consume milk powder and vitamin-fortified beverages to supplement their diet (Garrett and Grisham, 1999).

19.1.1.1 Vitamin E

Vitamin E was discovered in 1922 as an important nutritional factor for reproduction in rats (Evans and Bishop, 1922). Vitamin E is a family of compounds comprising a 6-chromanol ring and an isoprenoid side-chain that includes both tocopherols and tocotrienols. Vitamin E is represented by eight different isomers of varying biological potency: four tocopherols and four tocotrienols differ in that the tocopherols have a saturated C_{16} isoprenoid side chain whereas the tocotrienols have a similar isoprenoid with an unsaturated side chain. The different tocopherols and tocotrienols (α, β, γ, and δ) differ in the methylation of the chromanol ring. These structural differences determine biological activity (Sure, 1942). α-Tocopherol is the most active and common vitamin as shown in Figure 19.1. Tocotrienols occur in foods as free alcohols and also as esters, tocopherols occur naturally as the free alcohols but acetate and succinate esters are used in pharmaceutical preparations because of their greater stability against oxidation (Holland et al., 1991). Tocopherol is destroyed rapidly by sunlight and artificial light containing wavelengths in the UV region. The vitamers are slowly oxidized by atmospheric oxygen to form mainly biologically inactive quinones; the oxidation is accelerated by light, heat, alkalinity, and certain trace metals. The presence of ascorbic acid completely prevents the catalytic effect of iron(III) and copper(II) on vitamin E oxidation by maintaining these metals in their lower oxidation states (Cort et al., 1978).

19.1.1.1.1 Functions

Vitamin E is an essential nutrient and helps the human body to function normally. It has many biological functions; the most important is its antioxidant function, protecting the vital phospholipids in cellular and subcellular membranes from peroxidative degeneration (Bell, 1987; Food and Nutrition Board, 2000). It has significant function in cell signaling (Zingg and Azzi, 2004; Azzi, 2007) and acts as a peroxyl radical scavenger to maintain the integrity of long-chain polyunsaturated fatty acids in the cell membranes for the maintenance of their bioactivity (Sattler, 2006).

As an enzymatic activity regulator, α-tocopherol inhibits protein kinase C activity, which is involved in cell proliferation and also plays a role in smooth muscle growth. α-Tocopherol has a stimulatory effect on the dephosphorylation enzyme, protein phosphatase 2A, which in turn cleaves phosphate groups from protein kinase C, leading to its deactivation, bringing the smooth muscle growth to a halt. This means that protein kinase

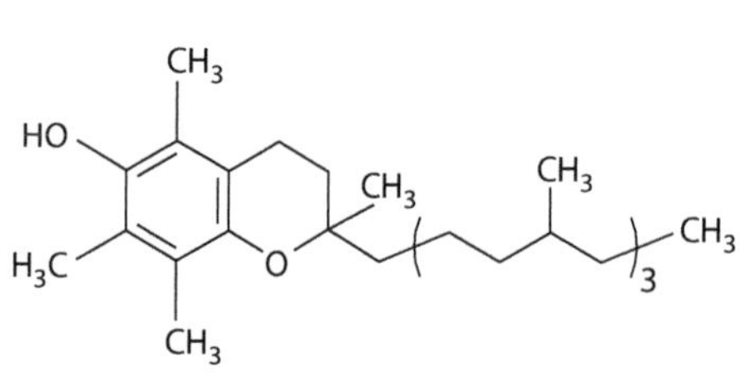

FIGURE 19.1   Chemical structure of α-tocopherol ($C_{29}H_{50}O_2$; MW 430.71).

C is a key signaling molecule in the regulation of growth (Schneider, 2005). α-Tocopherol has a role to regulate the expression of the CD36 (cluster of differentiation 36) scavenger receptor gene and the scavenger receptor class A that modulates expression of the connective tissue growth factor (Azzi and Stocker, 2000; Villacorta, 2003). The CTGF (connective tissue growth factor) gene is responsible for the repair of wounds and regeneration of the extracellular tissue that is lost or damaged during atherosclerosis (Villacorta, 2003).

Vitamin E also plays a role in neurological function (Muller, 2010), inhibition of platelet aggregation, and modification and stabilization of blood fats (Dowd and Zheng, 1995; Davies, 2007; Epand and Epand, 2008).

Vitamin E is a very active dietary antioxidant acting as a free radical and singlet state oxygen scavenger, playing a crucial role in the protection of the skin from free radical–generating factors such as UV radiation (Blasco et al., 2007). The term antioxidant refers to synthetic and natural compounds added to products to prevent deterioration by action of oxygen in air. The ability to utilize oxygen has provided humans with the benefit for metabolizing fats, proteins, and carbohydrates for energy. Oxygen as a free radical is highly reactive and can attack healthy cells of the body, causing them to lose structure and functions (Percival, 1998). The types of reactive oxygen species include hydroxyl radicals, superoxide anion radical, hydrogen peroxide, singlet oxygen, nitric oxide radical, hydrochloride radicals, and lipids peroxides. All are capable of reacting with membrane lipids, nucleic acid, proteins, and enzymes, resulting in cellular damage. There are number of pathways by which cells generate oxidants, for example, aerobic metabolism, oxidative burst from phagocytes (white blood cells), and xenobiotic metabolism. Vigorous exercise, chronic inflammation, infections, allergens, and exposure to drugs and toxins such as cigarette smoke, pollution, and pesticides also contribute to increasing the body's oxidants load. Reactive oxygen species can damage cells and generate disease of aging such as cancer, cardiovascular disease, cataracts, immune system decline, and brain dysfunction (Sies et al., 1992; Bourassa and Tardif, 2006).

Humans have evolved a highly sophisticated and complex antioxidant protection system involving components including nutrient-derived antioxidants such as vitamins C and E, antioxidant enzymes, and metal-binding proteins (Percival, 1998).

Several comprehensive reviews have been published on the existing chromatographic methods for the analysis of lipophilic antioxidants (tocopherols, tocotrienols, and carotenoids) in various sample matrices (Abidi, 2000; Aust et al., 2001); on electrochemical approaches in the sensing of natural or biological antioxidants and antioxidant capacity (mainly polyphenols and vitamins C and E) using cyclic voltammetry; on flow injection analysis (FIA) with amperometric detection in food and biological samples (Blasco et al., 2007); and on chemiluminescence (CL) and fluorescence (FL) methods for the analysis of oxidative stability, antioxidant activity, and lipid hydroperoxide content in edible oils (Christodouleas et al., 2012).

Since the mid-1980s, the European Union (EU) had devised full ingredient labeling standard codes (E numbers) that accurately describe additives used in the production of food (Food Labels, 2012) and commonly found on food labels through out the EU (European Directives 95/2/EC, 1995). Food additives are natural or artificial substances that may be added to food in small quantities to perform specific functions or to keep food fresh and consistent in quality. Antioxidants, for example, tocopherol, numeric code E306 (vitamin E, obtained from natural extracts, e.g., soya bean oil, wheat germ, rice germ, cottonseed, maize and green leaves, used as an antioxidant and nutrient) and synthetic α-, γ-, and δ-tocopherols obtained synthetically, added to fats and oils and having similar functions fall under numeric codes E307–E309, respectively.

19.1.1.1.2 Occurrence

Vitamin E in its tocopherol form is found in both animal foods and plants. The animal source of vitamin E is poor, with some being found in fish, meat, eggs, and milk. In mammalian muscle, the content of α-tocopherol is <1 mg/100 g. The concentration in different meat cuts of a given animal species increases with increasing fat content. In eggs, vitamin E is in the yolk; its concentration also depends on the amount of supplemental α-tocopheryl acetate in the chicken feed with different values of 0.46 and 1.10 (Bauernfeind, 1980), 0.70 (McLaughlin and Weihrauch, 1979), and 1.96 mg/100 g α-tocopherol (Syvaoja, 1985). Fish is a better source of vitamin E than meat. Tuna and salmon contain 0.53 and 0.7 mg/100 g α-tocopherol, respectively (Lehman et al., 1986) and sardines canned in tomato sauce contained 3.9 mg/100 g (Hogarty et al., 1989).

Vitamin E is an important constituent of chloroplast membranes in green plants. The best source of vitamin E is vegetables, seed nuts, and wheat germ oil. The oil components of all cereal grains, nuts, beans, and seeds are rich source of tocopherol. Major sources include margarine, mayonnaise, and salad dressings milk, tomato products, and apples (Sheppard et al., 1993). Vegetable oils are highly unsaturated and contain a high concentration of vitamin E to maintain the oxidative stability of their constituent poly saturated fatty acids. Fortified breakfast cereals, vegetable shortenings and cooking oils, peanut butter, eggs, and potato crisps are also the major sources of vitamin E in the Unites States (Bramley et al., 2000).

A huge activity of vitamin E is found in the germ fraction of cereal grains (Grams et al., 1970; Barnes and Taylor, 1981). Wheat germ and green leafy vegetables are the best sources of vitamin E. The dark green outer leaves of brassicas contain more vitamin E than the lighter green leaves. Vitamin E is not present in the innermost part of white cabbage or the florets of cauliflower and the α-tocopherol values determined for these vegetables depends on their green parts only (Booth and Bradford, 1963). Green apples contain more α-tocopherol than red or yellow apples (Booth and Frohock, 1961).

Vitamin E is added to whole milk powder and cereals to supplement dietary requirements. The vitamin E requirement increases with an increased intake of polyunsaturated fatty acid and several types of margarines are enriched with vitamin E. The acetate ester of α-tocopherol is used as a supplement on account of its greater stability (Ottaway, 2008).

19.1.1.1.3 Deficiency

Vitamin E deficiencies are rare in humans because its depletion takes a very long time. There are three possibilities when vitamin E deficiency occurs in humans: (i) insufficient absorption of dietary fat, (ii) very low birth weight in newborns, and (iii) genetic abnormalities (Traber and Sies, 1996). Symptoms of vitamin E deficiency include loss of hair, muscular weakness, and leg cramps as well as gastrointestinal disorders. Lack of vitamin E in the body causes red blood cells to become fragile. It may also lead to decrease in blood circulation within the body. Enlarged prostate, impotence, and muscular wasting are some other effects of vitamin E deficiency. Vitamin deficiency can lead to major complications in life because of its wide range of effects on the nervous, reproductive, muscular, and circulatory systems. However, biochemically, low levels of vitamin E can be measured in the blood and have been seen in such conditions as acne, anemia, infections, some cancers, periodontal disease, cholesterol gallstones, neuromuscular diseases, and dementias such as Alzheimer disease (Azzi and Stocker, 2000; Boothby and Doering, 2005; Litwack, 2007).

19.1.1.1.4 Requirements

The amount of vitamin E required by the body depends upon its size and the amount of polyunsaturated fats in the diet, as vitamin E is needed to protect these fats from oxidation. The requirement for vitamin E depends upon intake of refined oils, fried foods, or rancid oils. Supplemental estrogen or estrogen imbalance in women increases the need for vitamin E, as does air pollution. The recommended dietary allowance (RDA) for vitamin E is really quite low, many people do not consume this in their diet alone. Table 19.1 lists the RDAs and tolerable upper intake levels (ULs) for vitamin E. The new recommendations for vitamin E are expressed as milligrams of RRR-α-tocopherol equivalents. Dietary supplements of vitamin E are labeled in terms of international units (IU). 1 mg of synthetic vitamin E (*all-rac*-α-tocopheryl acetate is equivalent to 1 IU vitamin E, but only 0.45 mg RRR-α-tocopherol. 1 mg of natural vitamin E (RRR-α-tocopherol) provides 1.5 IU. For the UL, the Food and Nutrition Board recommended 1000 mg of any α-tocopherol form, which is equivalent to 1500 IU RRR- or 100 IU *all-rac*-α-tocopherol (Food and Nutrition Board, 2000; Hathcock et al., 2005; Combs, 2008).

19.1.1.1.5 Toxicity

Vitamin E is not stored in the body like other fat-soluble vitamins. As a result, toxicity is rare in the human body because of the wide variation in daily blood vitamin E levels (Karlsson, 1997). Excess intake is usually excreted through the urine and feces, and most doses are cleared by the body within few days; therefore, toxicity from vitamin E is unlikely.

19.1.2 Flow Injection Analysis

The term FIA was introduced in 1975 (Ruzicka and Hansen, 1975) to describe the use of injection into an unsegmented continuous flowing stream with subsequent detection of the analyte. FIA is now extensively used and well established and offers several advantages: it is simple, relatively require inexpensive equipment, ease of operation and great capacity for achieving results that are excellent in view of the rapidity, accuracy and precision. The concept of FIA is based on combination of four factors: (i) unsegmented flow, (ii) reproducible sample injection volumes, (iii) reproducible operational timing,

TABLE 19.1   RDA and ULs for Vitamin E (α-Tocopherol)

Age	RDA (mg/day)	UL (mg/day)
0–6 months	4 (6 IU)	–
7–12 months	5 (7.5 IU)	–
1–3 years	6 (9.1 IU)	200 (300 IU)
4–8 years	7 (10.4 IU)	300 (450 IU)
9–13 years	11 (16.4 IU)	600 (900 IU)
14–18 years	15 (22.4 IU)	800 (1200 IU)
19+ years	–	1000 (1500 IU)
Pregnancy[a]	15 (22.4 IU)	1000 (1500 IU)
Lactation[a]	19 (28.4 IU)	1000 (1500 IU)

Note: IU = international unit.

[a] 19+ years.

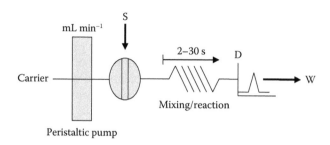

FIGURE 19.2   A single-channel FIA manifold using a carrier reagent stream: S = sample injection valve; D = detector; and W = waste.

and (iv) controlled sample dispersion (Ruzicka and Hansen, 1978, 1988; Hansen, 2004; Trojanowicz, 2008). A single-channel FIA system is shown in Figure 19.2.

In FIA, an accurate volume of liquid sample is introduced via an injection valve into an unsegmented, liquid carrier flowing through a narrow-bore tube or conduit. The injected sample forms a well-defined zone as it is transported through the conduit under laminar flow conditions. Reagents are mixed with the sample zone under the influence of radial dispersion, to produce reactive or detectable species that can be sensed by any one of a variety of flow-through detection devices. The analytical readout is in the form of a transient peak, the height of which is related to the concentration of the analyte. The whole process of sample injection, transport, reagent addition, reaction, and detection, can be accomplished rapidly, using minimum amounts of sample and reagents, with a high degree of precision (Ruzicka and Hansen, 1988; Barcelo, 2008).

19.1.3 Chemiluminescence

CL is a well-established spectrometric branch of analytical chemistry based on the production of electromagnetic radiation from a chemical reaction (Campana and Baeyens, 2001; Su et al., 2007). Due to its minimal instrumentation, no external source, simple optical system, and high sensitivity and selectivity, CL-based detection has become a useful tool in liquid phase coupled with flow injection (FI), sequential injection manifolds, liquid chromatography. and capillary electrophoresis systems in the field of environmental, pharmaceutical, clinical, biomedical, and food analysis (Mervartova et al., 2007; Campana et al., 2009; Fan et al., 2009; Gracia et al., 2009; Wang et al., 2009).

Different CL reactions in liquid phase have been applied with analytical purposes. These include luminol, tris(2,2′-bipyridyl)ruthenium(II), lucigenin, lophine, peroxyoxalate derivatives, acidic potassium permanganate, sulfite, gallic acid, morphine, codeine, pyrogallol acridinium esters, and others. Table 19.2 summarizes some of the commonly used CL reactions in liquid phase and their analytical application in different areas.

19.1.4 FI/Batch Methods for Determination of Vitamin E

19.1.4.1 FI–CL Methods

In the literature, a limited number of CL methods can be found for the determination of α-tocopherol and retinol in pharmaceutical preparations, infant-based milk formulas,

TABLE 19.2 Chemiluminescence (CL) Reactions in the Liquid Phase

S. No.	CL Reaction	λ(max)/Color	Quantum Yield[a]	Application Area	Reference
1.	Luminol (5-amino-2,3-dihydro-1,4-phthalazinedione) Luminol → 3-aminophthalate (3-APA) via H_2O_2, ^-OH, Catalyst → 3-APA + Light	425 nm/blue	0.01	Pharmaceuticals, hydroperoxides, drugs, neurotransmitters and their metabolites, environmental, food analysis, micronutrients, pesticides; forensic science, edible oils	Mervartova et al. (2007), Gracia et al. (2009), Campana et al. (2009), Lara et al. (2010), Iranifam (2013), Worsfold et al. (2013), Mestre et al. (2001), Fletcher et al. (2001), Gracia et al. (2005), Campana and Baeyens (2001), Barni et al. (2007), Liu et al. (2010), Christodouleas et al. (2012)
2.	Tris(2,2'-bipyridyl) ruthenium(II) (i) $Ru(bpy)_3^{2+} \longrightarrow Ru(bpu)_3^{3+} + e^-$ (Oxidation) (ii) $Ru(bpy)_3^{3+} \longrightarrow [Ru(bpu)_3^{2+}]^* +$ (Reduction with analyte) (iii) $[Ru(bpy)_3^{2+}]^* \longrightarrow Ru(bpy)_3^{2+} +$ Light Tris(2,2'-bipyridine)ruthenium(II)	680 nm/red	0.05	Oxalates, organic acids, amines, amino acids, proteins, alkaloids, pharmaceuticals, pesticides, inorganics, drugs, environmental, food and clinical analysis	Gorman et al. (2006), Gracia et al. (2009), Campana et al. (2009), Lara et al. (2010), Iranifam (2013), Mestre et al. (2001), Fletcher et al. (2001), Gracia et al. (2005), Liu et al. (2010)

(Continued)

TABLE 19.2 (*Continued*) Chemiluminescence (CL) Reactions in the Liquid Phase

S. No.	CL Reaction	λ(max)/Color	Quantum Yield[a]	Application Area	Reference
3.	Lucigenin (*N, N′*-dimethyl-9,9′-bisacridinium dinitrate) 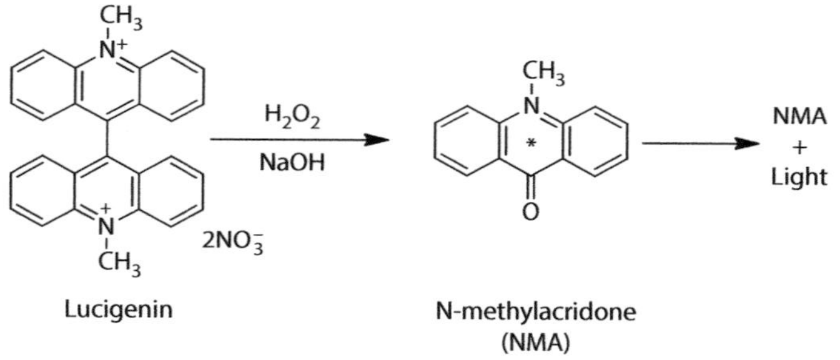	440 nm/ blue–green	0.016	Pharmaceuticals, environmental, inorganic species, drugs, edible oils	Mestre et al. (2001), Fletcher et al. (2001), Campana and Baeyens (2001), Christodouleas et al. (2012)
4.	Lophine (2,4,5-triphenylimidazole) 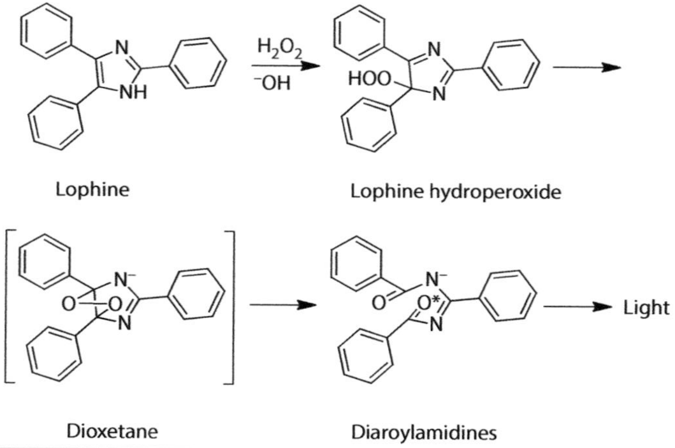	525 nm/yellow	0.01	Pharmaceuticals, environmental, inorganic, and organic compounds	Mestre et al. (2001), Fletcher et al. (2001), Campana and Baeyens (2001), Nakashima (2003)

Lucigenin

N-methylacridone (NMA)

Lophine

Lophine hydroperoxide

Dioxetane

Diaroylamidines

(*Continued*)

TABLE 19.2 (*Continued*) Chemiluminescence (CL) Reactions in the Liquid Phase

S. No.	CL Reaction	λ(max)/Color	Quantum Yield[a]	Application Area	Reference
5.	Peroxy-oxalate [bis(2,4,6-trichlorophenyl)oxalate] bis-(2,4,6-trichlorophenyl)oxalate Fluorophor + Light → Fluorophor* + CO_2 + 2 Cl	Blue	0.07–0.50	Catecholamines, drugs, phenols, environmental, food and beverages, clinical analysis, pharmaceuticals, pesticides, amino acids, catecholamines, amines	Gracia et al. (2009), Campana et al. (2009), Lara et al. (2010), Fletcher et al. (2001), Gracia et al. (2005), Campana and Baeyens (2001)
6.	Potassium permanganate ($KMnO_4$) $Mn(VII) \xrightarrow[H^+]{4e^-} Mn(III)$ $Mn(III) \xrightarrow[H^+]{e^-} Mn(II)^*$ $Mn(II)^* \xrightarrow[H^+]{e^-} Mn(II) + Light$	680 nm/red	0.05–0.5	Pharmaceuticals and clinical, forensic, food and consumer products, agricultural and environmental	Adcock et al. (2014), Lara et al. (2010), Iranifam (2013), Mestre et al. (2001), Barnett and Francis (2005), Adcock et al. (2007), Liu et al. (2010)
7.	Sulfite (SO_3^{2-}) $HSO_3^- + oxidant \longrightarrow HSO_3^\bullet$ $2HSO_3^\bullet \longrightarrow S_2O_6^{2-} + 2H^+$ $S_2O_6^{2-} \longrightarrow SO^{2-} + SO_2^*$ $SO_2^* \longrightarrow SO_2 + Light$	532 nm	NR	Pharmaceuticals, environmental, food and beverages, biomedical	Mervartova et al. (2007), Iranifam (2013), Mestre et al. (2001), Fletcher et al. (2001)

[a] Quantum yield = the fraction of reacting molecules that produce photons, that is, the number of light quanta emitted per reactant molecule; NR = not reported.

and blood serum. Waseem et al. (2009) have reported a FI–CL method for the determination of retinol and its derivatives and tocopherol in pharmaceuticals, based on their enhancement effect on the potassium permanganate–HCHO–CL reaction in an acidic medium. The optimized FI–CL manifold used is shown in Figure 19.3. It is well documented that molecules containing phenol or amine moieties efficiently generate CL upon oxidation with acidic potassium permanganate ($KMnO_4$) (Hindson and Barnett, 2001; Adcock et al., 2007). Tocopherol contains the phenol moiety (6-chromanol) which upon oxidation forms quinones and hydroquinones. It is feasible that Mn(II) may be complexed via the phenolic oxygen, and Mn(II) complexes can be oxidized to the Mn(III) species using acidic $KMnO_4$ (Cotton and Wilkinson, 1988). It was postulated (Barnett et al., 1993) that Mn(III) complexes can be reduced to an electronically excited Mn(II) complex, which then emits a photon. It is well known that dilute permanganate solution rapidly oxidizes olefins to diols via the transfer of two oxygen atoms from $KMnO_4$ to the olefins by complex formation (Wiberg and Saegebarth, 1957). It is assumed that retinol makes a complex with permanganate during which the reduction of permanganate occurs and generates CL via an electronically excited Mn(II) complex. The role of formaldehyde is to enhance the reduction of $KMnO_4$ to a lower excited-state Mn(II) species, which in turn gives CL as background, that can be observed when $KMnO_4$ alone reacts with formaldehyde.

The effects of common ions, excipients in pharmaceutical preparations, and related organic compounds were studied on the determination of retinol and α-tocopherol (1×10^{-7} mol/L) individually. The tolerance of each foreign species was taken as the largest concentration yielding less than ±5% of the error of the adoptive concentration of retinol and tocopherol. No clear interference could be found at 1000-fold for glycerol, Na^+, K^+, and SO_4^{2-}; at 500-fold for tocopherol acetate, methylcellulose, Ca^{2+}, Mg^{2+}, NH_4^+, vitamin D_3, cholesterol, and stearic acid; at 200-fold for sorbitol, mannitol, sucrose, glucose, starch, gum acacia, magnesium stearate, dextrin, Zn^{2+}, PO_4^{3-}, NO_3^-, HCO_3^-, and Cl^-; at 100-fold for oleic acid, triolein, eicosapentaenoic acid (EPA), and docosahexaenoic acid (DHA); and at 50-fold for poly(vinyl alcohol). The FI–CL method developed was successfully applied to

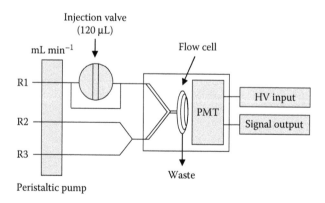

FIGURE 19.3 Flow injection CL manifold for the determination retinol and α-tocopherol. Optimum experimental conditions: R1 = 5% (v/v) aqueous ethanol solution; R2 = 2.5×10^{-5} mol/L $KMnO_4$ solution in 0.05 mol/L H_2SO_4; R3 = 0.5 mol/L formaldehyde solution in 0.05 mol/L H_2SO_4; 2 mL/min flow rate per channel; 120 μL sample loop volume; CR = chart recorder, HV input = negative high-voltage supply, and PMT = photomultiplier tube (900 V).

pharmaceuticals containing vitamin E (DL-α-tocopherol acetate) after transesterification (Ruperez, 2001; Yaqoob et al., 2009). The results obtained confirm that the developed method was not liable to interferences and was in good agreement with the amount labeled.

Asgher et al. (2011a) have reported another simple and sensitive two-channel FI method for the determination of retinol and α-tocopherol in pharmaceuticals and human blood serum. The method was based on the reduction of vanadium(V) by retinol and α-tocopherol and subsequent reaction of reduced vanadium with luminol generated CL emission. Vanadium(IV) has been shown to catalyze luminol oxidation using dissolved oxygen to produce light (Boyle et al., 1987; Qin et al., 1997; Li et al., 2002). Tocopherol and retinol reduces V(V) to V(IV), which catalyze the luminol reaction in the presence of dissolved oxygen to generate CL. It has been reported that vanadium, in its lower oxidation state, has very low redox potential in a strong alkaline solution; it can reduce dissolved oxygen to superoxide radical (Li et al., 2002), and the superoxide radical itself can react with alkaline luminol to generate CL or superoxide radical, which, in turn, dismutates to hydrogen peroxide (Murphy et al., 1993). Further reduction of H_2O_2 by vanadium possibly produces the highly reactive hydroxyl radical (analogous to Fe(II) reaction, Murphy et al., 1993; Yildiz and Demiryurek, 1998; Rose and Waite, 2001).

The effect of common excipients and other foreign ions were studied under the optimized conditions containing a standard solution of 100 μg/L retinol and 50 μg/L tocopherol (3 and 100 times lower than the lower concentration limits of these vitamins in blood serum). No interference could be found at 400-fold for Na^+, K^+, Ca^{2+}, Mg^{2+}, Zn^{2+}, NH_4^+, SO_4^{2-}, PO_4^{3-}, NO_3^-, HCO_3^-, Cl^-, glycerol, tocopheryl acetate, methyl cellulose, vitamin D, cholesterol, stearic acid, sucrose, glucose, starch, gum acacia, magnesium stearate, dextrin; at 200-fold for sorbitol, mannitol oleic acid, triolein, EPA, DHA, polyvinyl alcohol; at 50-fold for albumin and vitamin K; for 10-fold for β-carotene, and at 5-fold for ascorbic acid. The reported FI–CL method for the determination of retinol and α-tocopherol in pharmaceuticals and human blood serum samples was validated by comparing with a high-performance liquid chromatography (HPLC) method (Driskell et al., 1982). There was no statistical difference between the two methods at the 95% confidence level ($t_{calc} = 1.4 < t_{tab} = 3.2$ for retinol and tocopherol in pharmaceuticals and $t_{calc} = 0.8 < t_{tab} = 4.3$ for tocopherol in blood serum).

Asgher et al. (2011b) have also reported another FI–CL method to determine retinol and α-tocopherol in pharmaceuticals and human blood serum based on the enhancement effect of the lucigenin CL reaction in alkaline medium. Surfactants including polyoxyethylene lauryl ether (Brij-35), Triton X-100, cetyltrimethyl ammonium bromide (CTAB) and sodium dodecyl sulfate (SDS) enhance lucigenin CL intensity. With Brij-35, the enhancement was 67% for retinol and 58% for α-tocopherol. With CTAB, the enhancement was 16% for retinol whereas for α-tocopherol, the CL intensity was quenched by up to 95%. Retinol could be determined specifically in the presence of α-tocopherol using CTAB.

The CL spectrum of lucigenin under alkaline conditions in the presence of tocopherol and retinol was examined in the range of 400–650 nm using a spectrofluorimeter (RF-1501, Shimadzu, Japan) with the lamp off position. The spectrum gave a peak maximum at 500 nm, corresponding to the FL spectrum of lucigenin solution at an excitation wavelength of 366 nm. This indicated that the CL from the reaction of retinol and tocopherol in the presence of lucigenin was due to lucigenin CL. The quenching effect of CTAB was also investigated on tocopherol CL. The CL signal of tocopherol was almost quenched by CTAB, possibly due to strong ionic interaction between CTAB and tocopherol. The CL signal from retinol remained unchanged in the presence of CTAB, which means that CTAB can be used while determining retinol in the presence of tocopherol.

The lucigenin CL product N-methylacridone is insoluble in water and its deposits on the walls of the glass flow cell must be removed for maximum CL intensity (Kolpf and Nieman, 1985). The use of surfactants resolves this problem due to micelle formation. Investigations of the lucigenin CL reaction in the presence of surfactants and in micellar media has shown that sensitized and quenched CL could arise due to solubilization and alteration of the pH of the microenvironment (Hinze et al., 1984; Kamidate, 1991; Zhang and Chen, 2000).

The effects of some common excipients in drugs (lactose, sugars, or sugar polymers such as cellulose, starch, and mannitol), inorganic ions and related organic compounds on the determination of retinol 0.03 mg/L and tocopherol 0.04 mg/L were investigated under the optimum conditions. The tolerance limit was taken as the maximum concentration of the foreign substances, which caused an approximately ±5% relative error in the determination with respect to added retinol and tocopherol concentrations. No significant interference could be observed of these foreign substances with the determination of retinol and tocopherol.

The method was applied for the determination of retinol and tocopherol in commercially available tablets/capsules containing vitamins A and E. It could be seen from Table 19.3, that there were no significant differences between labeled contents of pharmaceutical formulations and those obtained by the FI–CL method and HPLC method (Driskell et al., 1982). Normal blood serum concentration of retinol and tocopherol are in the ranges 0.3–0.6 and 5.0–15 mg/L, respectively (Combs, 2008). Blood serum samples were also analyzed after appropriate dilution to fit in to the linear range and the results obtained were compared with an HPLC method. The results show that there were no significant differences obtained by these methods.

19.1.4.2 Batch CL Methods

Mambro et al. (2003) reported a batch CL inhibition method for the antioxidant activity measurements of different forms of vitamin E in pharmaceutical formulations and compared their CL inhibition on a luminol–hydrogen peroxide–horseradish peroxidase enzyme system for 10 min at 25°C in 10 μL samples. α-Tocopherol, mixed tocopherols

TABLE 19.3 Determination of Vitamins A and E in Different Pharmaceutical Formulations ($n = 6$)

| | | Retinol and α–Tocopherol Found | |
| | | Proposed FI–CL Method | HPLC Reference Method[a] |
Sample Matrix	Composition Labeled		
Vitamins A and E (capsule)	Retinyl palmitate 200 μg	195 ± 3.0 μg	198 ± 5.0 μg
	Tocopheryl acetate 200 μg	196 ± 3.5 μg	195 ± 2.7 μg
Vitamin E (capsule)	Tocopheryl acetate 200 mg	196 ± 1.8 mg	194 ± 2.2 mg
Vitamins A and E (tablet)	Retinyl palmitate 30,000 IU	29505 ± 25 IU	29304 ± 32 IU
	Tocopheryl acetate 70 mg	72 ± 1.0 mg	66 ± 1.6 mg
Vitamin E (tablet)	Tocopheryl acetate 100 mg	95 ± 1.2 mg	97 ± 1.5 mg
Vitamins A and E (injection)	Retinol 10,000 IU	9508 ± 10 IU	9354 ± 16 IU
	Tocopheryl acetate 3 mg	3.1 ± 0.1 mg	2.8 ± 0.2 mg

Source: Asgher, M. et al. 2011b. *Luminescence* 26:416–423. With permission.
[a] Driskell et al. (1982).

(containing α-, β-, γ-, and δ-tocopherol, D-tocopherols, and tocopherol excipients), and ronoxan MAP® (containing ascorbyl palimate, dry mixed tocopherols 30%, and malto-dextrin) inhibited CL intensity in an antioxidant concentration range of 1.0–50 μg/mL, above which these samples reached a plateau corresponding to a 99% inhibition of CL intensity. α-Tocopherol acetate did not inhibit the CL reaction even at above 50 μg/mL, probably due to the acetylated condition of the hydroxyl group of α-tocopherol respon-sible for its antioxidant activity. The formulation components did not interfere with the antioxidant measurements.

Bezzi et al. (2008) reported another batch CL inhibition method for the measurement of antioxidant activity of five common compounds (L-ascorbic acid, butylated hydroxy-toluene, α-tocopherol, β-carotene, and quercetin) found in olive oils using luminol CL with Co(II) as an EDTA (ethylenediaminetetraacetic acid) complex as catalyst at pH 9.0. Different concentrations of each antioxidant were mixed with optimized hydrogen per-oxide (1.3×10^{-4} mol/L), luminol (5.6×10^{-4} mol/L), Co(II) (7.3×10^{-4} mol/L), and EDTA (2.3×10^{-3} mol/L) and the emission intensity (I) was measured. The emission intensity (I_o) from 1.3×10^{-4} mol/L hydrogen peroxide without any antioxidant added was also mea-sured. By plotting I_o/I versus concentration of antioxidant, a linear regression equation was obtained. IC_{50}, defined as the concentration of antioxidant that reduces the emission intensity by 50%, was calculated from the regression equation for each antioxidant at $I_o/I = 2$. The scavenging activity of hydrogen peroxide (SAHP) defined as: SAHP = I/IC_{50} was also calculated. A calibration range of α-tocopherol 0.005–012 μM with slope of calibration line of 2.23, antioxidant activity (IC_{50}) of 0.31 μM, and SAPH of 3.22 μM^{-1} are reported. A mixture of acetone and ethanol (2:1 v/v) was used for preparation of homogeneous solutions of oils with luminol. It was obvious that L-ascorbic acid and butylated hydroxytoluene are the most powerful hydrogen peroxide scavengers, followed by α-tocopherol and β-carotene, and quercetin was the weakest scavenger under the experimental conditions described.

19.1.4.3 FI–Spectrophotometric Methods

Memon et al. (2004) have reported an indirect FI–spectrophotometric method for the determination of α-tocopherol in pharmaceuticals and facial oil samples in microemul-sion reaction medium using iron(III)–1,10-phenanthroline as an oxidant. Micelles and microemulsions have now found an important place in the field of analytical chemistry and replaced toxic organic solvents. The proposed method is an example of how micro-emulsion in conjunction with FIA can be useful in analytical problems that requires repro-ducible time control and extraction in toxic organic solvents in analytical procedures.

Jadoon et al. (2010) have reported a single-channel reverse FI–spectrophotometric method for the determination of vitamin E (α-tocopherol) in pharmaceuticals, infant milk powder, and blood serum samples (Figure 19.4). The chemical reaction was based on the reduction of iron(III) to iron(II) by vitamin E, then the *in situ* formed iron(II) reacts with potassium hexacyanoferrate(III) ($K_3[Fe(CN)_6]$) to produce soluble Prussian blue ($KFe^{III}[Fe^{II}(CN)_6]$ in ethanolic medium [12.5%, v/v with H_2O]), with an absorption wavelength of 735 nm.

The effects of common excipients and other foreign ions were checked under the optimized conditions containing a standard solution of 0.2 μg/mL of vitamin E. No interference was observed at 250-fold for glycerol, Na$^+$, K$^+$, Ca^{2+}, Mg^{2+}, Zn^{2+}, NH$_4^+$, SO$_4^{2-}$, PO$_4^{3-}$, NO$_3^-$, HCO$_3^-$, Cl$^-$, methylcellulose, cholesterol, stearic acid, sucrose, glucose, starch, gum acacia, magnesium stearate; at 100-fold for vitamin D, sorbitol, mannitol, oleic acid, triolein, EPA, DHA, polyvinyl alcohol, albumin, retinyl palmitate; at 10-fold

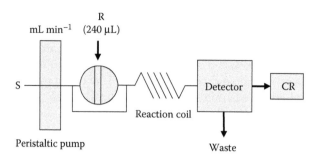

FIGURE 19.4 FI manifold for spectrophotometric determination of α-tocopherol. Optimum experimental conditions: S = sample/standard solutions in 12.5% (v/v) ethanol; R = mixture of 4×10^{-3} mol/L potassium ferricyanide and 2×10^{-3} mol/L iron(III) solutions; 250 cm reaction coil length; 0.6 mL/min flow rate; 240 μL injection loop volume; and CR = chart recorder.

for β-carotene and vitamin K; and at 5-fold for ascorbic acid and vitamin A (retinol). A selective extraction procedure for vitamin E, however, insures the absence of most of the interference described above and is demonstrated by analysis of certified reference material for method validation. Interference from retinol, carotene, and vitamin K is not very serious as their concentration in blood serum was <0.6 μg/mL retinol, whereas vitamin E was present in the range of 5–15 μg/mL (Combs, 2008); further, their influence can be subsided by dilution of the blood serum samples.

Recovery experiments were carried out with spiked blood serum samples and were in the range 94 ± 3 to 96 ± 5%. To determine the accuracy of the method, two level certified reference materials were analyzed: "fat-soluble vitamins, carotenoids and cholesterol in human serum" (SRM 968c, from NIST, Gaithersburg, MD). Values of 11.21 ± 0.47 and 18.15 ± 0.8 μg/mL (levels I and II, respectively) were obtained for vitamin E, which were in good agreement with certified values of 11.501 ± 0.618 and 18.887 ± 0.931 μg/mL.

Rishi et al. (2011) have reported two different FI–spectrophotometric methods for α-tocopherol determination in pharmaceuticals and infant milk powder. Method I, was based on the reduction of iron(III) to iron(II) in the presence of vitamin E, and reaction between iron(II) and 1,10-phenanthroline formed iron(II)-o-phenanthroline red complex with maximum wavelength of 510 nm. Method II was based on previously reported work (Prieto et al., 1999) involving the reduction of Mo(VI) to Mo(V) by vitamin E and the subsequent formation of a green phosphate/Mo(V) complex in aqueous sulfuric acid monitored at 695 nm. Triton X-100 was used as a sample carrier for solubility of vitamin E and for removal of the deposits of reaction product on the glass flow-through cell to achieve maximum peak height absorbance. The temperature of reaction coil I was kept at 85 ± 1°C and that of coil II at 20 ± 1°C.

The response of some common excipients in drugs—including sucrose, starch, and cellulose, inorganic ions, for example, sodium, potassium, calcium, magnesium, chloride, and sulfate ions 300- and 250-fold; nitrate, nitrite, zinc, copper, manganese, and chromium 100-fold; and organic compounds, for example, cholesterol, retinyl palmitate and acetate 75- and 80-fold; retinol, β-carotene vitamin D_3, and vitamin K 5- and 10-fold—on α-tocopherol determination were investigated under the conditions established for methods I and II, respectively. The tolerance limit was taken as the maximum concentration of the foreign substances that caused an approximately ±5% relative error in the determination with respect to α-tocopherol concentration (0.5 μg/mL). No significant interference could be observed for these foreign substances for α-tocopherol determination.

The results obtained for α-tocopherol were compared with an official method (Fabianek et al., 1968) and were in good agreement with no significant difference between these methods observed at 95% confidence level, and gave statistical (F and t tests) values of $F_{tab} = 1.01 < F_{cal} = 6.39$, $t_{cal} = 0.89 < t_{tab} = 2.78$ and $F_{tab} = 1.02 < F_{cal} = 6.39$, $t_{cal} = 2.6 < t_{tab} = 2.78$ for methods I and II, respectively.

Saima et al. (2013) reported another FI–spectrophotometric method for vitamins A and E analysis in pharmaceuticals, infant milk, and blood serum samples using a ferrozine–Fe(II) detection system. In the presence of vitamin A/E, iron(III) was reduced to iron(II), then the *in situ* iron(II) reacted with ferrozine to produce a magenta-colored complex with an absorption wavelength of 562 nm. The limits of detection (3s) were 0.06 and 0.03 μg/mL with relative standard deviations ($n = 4$) in the range of 0.8%–2.8%, respectively. The method was validated by comparing with HPLC reported method (Driskell et al., 1982) and the results have shown that there was no significant difference between the two methods at 95% confidence level.

The advantages of ferrozine over other iron(II) selective reagents were higher molar absorption coefficient (2.8×10^4 L/mol/cm), water solubility, stability, and low viscosity. A potential problem with the classical ferrozine method is the incomplete reduction of organically complexed iron(III), (Luther et al., 1996) and also attributed a poor recovery of total iron to the precipitation of iron(III) and iron(II) humic complexes upon acidification of the sample by hydroxylamine hydrochloride.

Table 19.4 reports comparison of the analytical characteristics of flow-based methods with CL and spectrophotometric detections for the determination of vitamin E (α-tocopherol).

19.1.5 Miscellaneous Methods

Various analytical techniques such as spectrophotometry/colorimetry, FL and infrared spectrometry, voltammetry, thin-layer chromatography (TLC), gas chromatography (GC), and HPLC based on ultraviolet, diode array, or fluorometric detectors have been reported in the literature for analysis of vitamin E. Various critical and comprehensive reviews are available on vitamin E quantification in food and clinical samples (Ball, 1988, 1998; Lumley, 1993). The AOAC International *Official Methods of Analysis* (1995) provides several methods based on older, chemical approaches. The applications of these analytical techniques are briefly summarized below.

19.1.5.1 Spectrophotometry

Spectrophotometry is a simple analytical technique used to measure the amount of light that a sample absorbs. The instrument operates by passing a beam of light through a sample and measuring the intensity of light reaching a detector. Spectrophotometric analysis continues to be one of the most widely used analytical techniques available.

Emmerie and Engel (1938) devised the most widely used procedure for the determination of vitamin E in biological materials. The determination was based on the reduction of iron(III) to iron(II) ions in the presence of tocopherol and formed a colored complex with 2,2-bipyridine; the absorbance was measured at 520 or 534 nm and is an AOAC official method (1995). The existing spectrophotometric methods for vitamin E determination made use of the oxidizability of the 6-hydroxychroman ring of α-tocopherol to the

TABLE 19.4　Comparison of the Analytical Characteristics of Flow Injection Spectrophotometric and CL Methods Used for the Determination of Vitamin E (α-Tocopherol)

Flow Technique	Detection Technique	Sample/Matrix	Reaction	LOD	Linear Range	R^2	Calibration Equation	Samples/h	Reference
FIA	Spec.	Pharmaceuticals, infant milk formulations, and blood serum	Iron(III)–vitamin E-potassium ferricyanide, $\lambda_{max} = 735$ nm	0.04 μg/mL	0.1–40 μg/mL	0.9990	$A = 0.0281c + 0.0003$	12	Jadoon et al. (2010)
FIA	Spec.	Pharmaceuticals, infant milk formulations, and blood serum	Iron(III)–vitamin E-ferrozine–Triton X-100, $\lambda_{max} = 562$ nm	0.03 μg/mL	0.1–20 μg/mL	0.9993	$A = 0.0909c + 0.0015$	10	Saima et al. (2013)
FIA	Spec.	Pharmaceuticals, infant milk	Iron(III)–vitamin E-1,10-phenanthroline, $\lambda_{max} = 510$ nm	0.05 μg/mL	0.21–43 μg/mL	0.9990	$A = 0.0092c + 0.0009$	30	Rishi et al. (2011)
FIA	Spec.	Pharmaceuticals, infant milk powder	Mo(VI)–vitamin E-PO_4^{3-} –Triton X-100, $\lambda_{max} = 695$ nm	0.1 μg/mL	0.25–10 μg/mL	0.9989	$A = 0.0139c + 0.002$	50	Rishi et al. (2011)
FIA	Spec.	Pharmaceutical and facial oil	Iron(III)–vitamin E-1, 10-phenanthroline-SDS	2.0×10^{-6} M	$0.12–9.3 \times 10^{-4}$ M	0.993	$y = 0.05 + 996x$	60	Memon et al. (2004)
FIA	CL	Pharmaceutical preparations	Acidic $KMnO_4$–HCHO–vitamin E	5.0×10^{-9} mol/L	1.0×10^{-8}– 5.0×10^{-6} mol/L	0.9991	$I = 7.2181c + 1.170$	100	Waseem et al. (2009)
FIA	CL	Pharmaceutical formulations and blood serum	Luminol–V(V)–vitamin E CL	2.15 μg/L	5–4300 μg/L	0.9991	$I = 1.613c + 2.944$	20	Asgher et al. (2011a)
FIA	CL	Pharmaceutical formulations and blood serum	Lucigenin–Brij-35–vitamin E CL	4.31×10^{-4} mg/L	4.3×10^{-3}– 2.15 mg/L	0.9989	$I = 0.434c + 1.77$	120	Asgher et al. (2011b)

Note: FIA = flow injection analysis; Spec. = spectrophotometry; CL = chemiluminescence; LOD = limit of detection.

corresponding quinone, that is, α-tocopheryl quinone, by oxidizing agents, finally giving a colored product.

Tsen (1961) reported an improved spectrophotometric method for tocopherol determination using bathophenanthroline reagent based on the work of Emmerie and Engel (1938). The author used 4,7-diphenyl-1,10-phenanthroline in place of 2,2-bipyridine for determination of tocopherols with high sensitivity (2.5-fold) and maximal color intensity for tocopherol attained in 15 s compared with 4 min when 2,2′-bipyridine was used. Different concentrations of tocopherol (1–8 μg/mL) gave a linear calibration at 534 nm. It has been observed that excess iron(III) ions remained after the completion of reduction. The photochemical reduction of iron(III) ions took place readily. To prevent this photochemical reduction, Schuck and Floderer (1939) reported that the orthophosphoric acid was the most suitable reagent to add in the determination of tocopherols among the following reagents: 8-quinolinol, ethylenedinitrilo-tetraacetic acid, citric acid, metaphosphoric acid, sodium dihydrogen phosphate, disodium hydrogen phosphate, and acetic acid. The addition of orthophosphate to the assay reaction diminished the interference from reductones, creatinine, hydroxyacetone, reductic acid, glutathione, and cysteine in the plasma (Washko et al., 1992).

Sturm et al. (1966) separated individual tocopherols by TLC and determined their concentration by spectrophotometric detection. The method involved saponification of oil samples and the separation of tocopherols in the nonsaponifiable fraction by TLC. The α-, γ-, and δ-tocopherols were removed from the thin-layer plate and determined spectrophotometrically (Tsen, 1961) with modification of the method reported by Emmerie and Engel (1938).

Tutem et al. (1997) reported another spectrophotometric method for vitamin E (α-tocopherol) in pharmaceutical preparations using copper(II)-neocuproine reagent in neutral medium, that is, ammonium acetate buffer (pH 7, 0.1 mol/L) and obeyed Beer's law between 2.4×10^{-6} and 9×10^{-5} mol/L α-tocopherol with a correlation coefficient (R^2) of 0.996. The absorbance of the copper(I)-neocuproine complex as a result of vitamin E oxidation was stabilized after 30 min (monitored at 450 nm against the reagent blank) and was stable for at least a further 90 min.

Prieto et al. (1999) reported another spectrophotometric method for the quantitative determination of antioxidant capacity based on the reduction of Mo(VI) to Mo(V) by vitamin E in acidic conditions with incubation at 95°C for 90 min. The subsequent green phosphate/Mo(V) complex, after cooling at room temperature was monitored at 695 nm with a calibration range of $0.2–2 \times 10^{-4}$ M ($r^2 = 0.997$) and a detection limit of 0.135 μmol vitamin E. The method was applied for measurement of total antioxidant capacity of plant extracts and to determine vitamin E in a variety of grains and seeds, including corn and soybean. The recovery of vitamin E from seeds was determined by supplementing the samples with the different vitamin E isomers or α-tocopherol acetate as internal standard and applying both the proposed method and a standard HPLC assay (Huo et al., 1996) and yielded a recovery of 93%–97% tocopherols.

Amin (2001) reported a colorimetric method for vitamin E in pure form and multivitamin capsules based on the reduction of tetrazolium blue to formazan derivative by vitamin E in alkaline medium. The reaction mixture was incubated at $90 \pm 2°C$ for 10 min and the absorbance of the reaction product was monitored at 526 nm with relative standard deviation of 0.7%–1.5%, a limit of detection 0.012 mg/L and sample throughput of approximately 6/h. The reduction of tetrazolium blue to formazan derivative requires 3 h at room temperature and the color developed was stable for 3 h. EDTA was added to sample solution for masking any metal ions during analysis.

Ali and Iqbal (2008) reported a modified spectrophotometric method for determination of plasma levels of α-tocopherol in normal adults. The method was based on the work of Fabianek et al. (1968). Iron(III) was reduced to iron(II) in the presence of reducing agent (α-tocopherol) and then coupled with a chelating agent 4,7-diphenyl-1,10-phenanthroline to produce pink iron(II) colored complex. The absorbance of complex was monitored at 536 nm and was proportional to the concentration of α-tocopherol in the reaction over the range 0.5–4 µg/mL with a limit of detection 0.2 µg/mL. This modified assay involved extraction of α-tocopherol in n-hexane (less hazardous and improved precision of the assay) rather than xylene.

19.1.5.2 FL Spectroscopy

FL spectroscopy is a rapid, sensitive, and nondestructive analytical technique widely used in biological sciences. Razagul et al. (1992) devised a direct spectrofluorometric procedure for α-tocopherol in nutritional supplement products (single- and multivitamin tablets, capsules, and dietary formula preparations). The analytical procedure was carried out in a single reaction vessel and consisted of saponification of samples for 10 min in a water bath set at 85°C using absolute ethanol (5 mL) and 80% (w/v) potassium hydroxide solution (2 mL) as the saponification medium, with ascorbic acid (125 mg) as an antioxidant. The analyte was extracted in situ with n-hexane (20 mL). Treatment of the n-hexane layer with 60% (v/v) sulfuric acid (2 mL) was sufficient to remove interference by retinol, and the n-hexane layer was then used for direct FL measurement at 298/330 nm excitation and emission wavelengths, respectively, which showed no evidence of interference by any excipients or extraneous fluorescing lipid-soluble components.

Diaz et al. (2006) utilized fluorometric techniques and partial least squares (PLS-1) multivariate analysis for simultaneous determination of quaternary mixture of tocopherols (α-, β-, γ-, and δ-T) in vegetable oils dissolved in hexane:diethyl ether (70:30 v/v). In the proposed study, PLS-1 was applied to matrices made up of FL excitation and emission spectra (EEM) and with FL excitation, emission, and synchronous spectra (EESM) of tocopherols. When synthetic samples were analyzed, recoveries around 100% were obtained and detection limits were calculated using EEM and EESM. For the analysis of the oils, the samples, diluted in hexane, were cleaned in silica cartridges and tocopherols were eluted with hexane:diethyl ether (90:10 v/v). The results were satisfactory for α-, β-, and γ-tocopherol, but worse for δ-tocopherol.

Hossu et al. (2009) presented spectrofluorimetric methods for the determination of fat-soluble vitamin E in multivitamin pharmaceutical products. In method I, n-hexane was used as solvent for α-tocopherol assay, while in method II, ethanol was used as a carrier between the aqueous solution and the n-hexane fluorescent solution of α-tocopherol. At 290/306 nm excitation and emission wavelengths, method I was linear for α-tocopherol over the range 1–100 µg/mL ($R = 0.97687$), having a limit of detection 1 µg/mL and a limit of quantification 2 µg/mL, and method II was linear over the range 2–50 µg/mL ($R = 0.9709$) with a limit of detection 0.68 µg/mL and a limit of quantification 2.27 µg/mL.

Escuderos et al. (2009) described α-tocopherol determination in virgin olive oils without prior chromatographic separation by measuring the FL intensity at 370, 371, 378, 414, and 417 nm ($\lambda_{ex} = 350$ nm). The quantity of α-tocopherol was then calculated by using regression lines constructed by spiking olive oil samples with known amounts of α-tocopherol. The results were in good agreement with the official HPLC method.

19.1.5.3 Infrared Spectroscopy

Infrared spectroscopy (IR) is the measurement of the wavelength and intensity of the absorption of infrared light by a sample (Putzig et al., 1994). IR spectroscopy is a non-invasive and nondestructive technique (Karoui et al., 2004). The technique is rapid, relatively inexpensive and can be easily applied in fundamental research, in control laboratories, and in the factory to analyze food products. The introduction of the Fourier transform technique in IR (FTIR) has increased the use of IR in food analysis (McKelvy et al., 1998; Luykx and VanRuth, 2008).

Ahmed et al. (2005) reported FTIR-based methodology for quantitation of total tocopherols, tocotrienols, and plastochromanol-8 in vegetable oils and recorded the fingerprint infrared spectra of oils from several samples of canola, flax, soybeans, and sunflower seeds. The amounts of tocopherols, tocotrienols, and plastochromanol-8, collectively called chromanols, were determined using HPLC with fluorescent detection (at 290/330 nm, for excitation and emission, respectively). The sum of tocopherols, tocotrienols, and plastochromanol-8 amounts and the infrared spectra were subjected to PLS analysis and showed excellent correlation between the calculated and experimental values. The developed methodology may provide an alternative method for rapidly scanning vegetable oils for vitamin E type molecules.

Gotor et al. (2007) reported the use of near-infrared reflectance spectrometry (NIRS) to predict the contents of tocopherol and phytosterol in sunflower seeds. About 1000 samples of ground sunflower kernels were scanned by NIRS at 2 nm intervals from 400 to 2500 nm. For each sample, standard measurements of tocopherol and phytosterol contents were made. The total tocopherol and phytosterol contents were assessed by HPLC with a FL detector and GC, respectively. The calibration data set for tocopherol and phytosterol ranged from 175 to 1005 mg/kg oil (mean value around 510 ± 140 mg/kg oil) and from 180 to 470 mg/100 g oil (mean value 320 ± 50 mg 100/g oil), with R^2 values of 0.64 and 0.27, respectively. In this study, calibrations were obtained by a modified PLS method.

Silva et al. (2009) reported another FTIR methodology for the analysis of α-tocopherol in vegetable oils as an alternative to the HPLC methods. Thirteen vegetable oils (corn, peanut, soybean, sunflower, and mixtures) were analyzed by reversed-phase HPLC with FL detection (FD) (excitation and emission wavelengths 296/330 nm) in order to obtain standard values for α-tocopherol (linear range 1–90 mg/L) with repeatability 3.6% and accuracy within the range 70%–90%. FTIR spectra of the vegetable oils were acquired in the attenuated total reflection mode (45° and 60° crystals of ZnSe). To predict the α-tocopherol content in samples, calibration models were designed, and the PLS method was used to analyze data from the FTIR spectral region at 1472–1078 cm^{-1}. Results obtained showed that the calibration model implemented with a 45° crystal was more suitable for the proposed analysis. Five extra samples of vegetable oils were analyzed by HPLC/FD and FTIR. Using the calibration model implemented for FTIR (45° crystal), the α-tocopherol content in samples was determined. The results obtained by HPLC/FD and FTIR were compared. There were no significant differences among these methods. Results showed that FTIR can be used as an alternative method for rapid screening of α-tocopherol in vegetable oils without sample pretreatment.

19.1.5.4 Voltammetry

Voltammetry is one of the most universal electrochemical methods for determining trace amounts of substances. A limited number of voltammetric methods have been reported for the determination of vitamin E in food products and pharmaceuticals.

Smith et al. (1942) utilized a polarographic method to investigate the interference of sesame oil, fish oil, and cholesterol on the quantitative measurements of pure α-tocopherol in acetate buffer containing ethanol (75%). These experiments were repeated by Kolthoff and Lingane (1941) with the electropode. The curves obtained with α-tocopherol under their conditions were similar to described by Smith et al. (1942). A definite halfway potential at approximately +0.25 V at 31°C was shown compared with the reported value of +0.28 V at 25°C. Furthermore, the diffusion current was proportional to the concentration of tocopherol.

Breaver and Kaunitz (1944) reported that tocopherols in tissue extracts are associated with fats and cholesterol. Fats in concentrations of 2% were found to be insoluble in acetate and citrate buffers (pH 4.6) containing alcohol (75%) and no polarographic curves were obtained by emulsifying the fats with Arescap (0.1%, spreading agent). Finally, benzoate buffer 0.025 mol/L acetone (20 mL), and water (80 mL) gave a clear fat solution and the currents obtained with polarography were approximately 0.1% of those found with the acetate buffer containing alcohol (75%). The difference in the current at the halfway potential was about +0.35 V, with possibility to obtain readings with a 10^{-3} mol/L tocopherol in the presence of sesame and fish oils (2.5%–5%) or cholesterol (0.15%).

Taguchi et al. (2003) reported voltammetric determination of weak bases based on oxidation of α-tocopherol in an unbuffered solution using a plastic formed carbon electrode in methanol. A weak base solution in the presence of α-tocopherol gave rise to an oxidation prepeak resulting from the oxidation of α-tocopherol at a potential more negative than that of α-tocopherol. Methanol solution containing α-tocopherol (3 mmol/L) and $LiClO_4$ (50 mmol/L) gave prepeak current height linear to the weak base drug concentrations over the range 0.13–2.5 mmol/L ($r = 0.9962$) with a limit of detection ($S/N = 3$) of 0.05 mmol/L for each base compound and relative standard deviation of 1.19% ($n = 6$) for 0.498 mmol/L lidocaine.

Li et al. (2006) reported a procedure for α-tocopherol determination in vegetable oil by differential pulse voltammetry in the medium of 1,2-dichloroethane/ethanol on a platinum electrode modified with a polypyrrole coating. The use of the coated electrode allowed the attainment of better analytical parameters of determination. The calibration graph was linear in the range 5.0–300 μmol/L, with a limit of detection 1.5 μmol/L.

Mikheeva and Anisimova (2007) reported voltammetric procedures for fat-soluble vitamin E (α-tocopherol acetate) in multicomponent vitaminized mixtures by differential voltammetry in nonaqueous media using carbon electrodes of different types. The calibration graphs of electrooxidation current of α-tocopherol acetate in acetonitrile solutions of different supporting electrolytes were linear over the range 3–18 mg/L with limits of detection calculated from 3σ found to be 0.35 and 0.12 mg/L tocopherol, respectively. They used a glassy carbon indicator electrode, a PU-1 commercial polarograph and an STA voltammetric system (TU 4215-001-20694097-98). Voltammograms recorded under accumulating conditions of the STA analyzer improved the sensitivity for vitamin E, namely, 0.011 mg/L.

Ziyatdinova et al. (2007) proposed another voltammetric method for the determination of α-tocopherol in pharmaceuticals on a graphite electrode modified with carbon nanotubes in the presence of $HClO_4$ (0.1 mol/L) in acetonitrile. The use of a modified electrode was shown to substantially reduce the overvoltage and enhance the current of analyte oxidation. The lower limit of the analytical range for α-tocopherol was 5.43×10^{-5} mol/L. The analytical range varied between 6.51×10^{-5}–2.16×10^{-4} and 3.23×10^{-4}–2.01×10^{-3} mol/L.

Ziyatdinova et al. (2012a) determined α-tocopherol in drugs voltammetrically in the presence of surfactants with enhanced current of vitamin E oxidation. α-Tocopherol is electrochemically active on glassy carbon electrodes in the available range of anodic potentials in acetonitrile and its aqueous mixtures. The calibration curves for α-tocopherol oxidation in the presence of surfactants (0.1 mol/L $LiClO_4$ in acetonitrile/water 6:4) were linear over the range 2–140 μmol/L ($R = 0.9948$), 2–100 μmol/L ($R = 0.9938$), and 4.1–10 μmol/L ($R = 0.9982$) with limits of detection 1.02, 1.02, and 2.04 μmol/L using N-dodecylpyridinium bromide (1 μmol/L), Triton X-100 (1 μmol/L), and N-cetylpyridinium bromide (0.5 μmol/L) as surfactants, respectively. However, an increase in the water percentage up to 50% and beyond led to a substantial decrease in and then the total disappearance of the analytical signal.

Ziyatdinova et al. (2012b) also determined sterically hindered phenols (2,6-di-*tert*-butyl-4-methylphenol and its derivatives), which were irreversibly oxidized at +0.96–1.30 V on a glassy carbon electrode in $LiClO_4$ (0.1 mol/L) containing acetonitrile. α-Tocopherol gave an oxidation step at +0.4 V on voltammograms under the optimized conditions. The oxidation process led to formation of corresponding quinoid derivatives. Calibration graphs for all compounds under investigation were linear in the range studied, with limits of detection in the range 9.6–44.3 μmol/L. The approach has been applied to determination of butylated hydroxytoluene and α-tocopherol in vegetables, lubricating oils, pharmaceuticals, and cosmetics using preliminary single extraction of analytes with acetonitrile for 15 min at oil/extractant ratio of 1:2.5.

19.1.5.5 High-Performance Liquid Chromatography

HPLC has been used for measuring various compounds, for example, carbohydrates, vitamins, additives, mycotoxins, amino acids, proteins, triglycerides in fats and oils, lipids, chiral compounds, and pigments. Several sensitive and selective detectors such as ultraviolet–visible, FL, electrochemical, and diffractometric are available to utilize with HPLC depending on the compound to be analyzed. Various HPLC methods based on these detectors have been published for the measurement of vitamin E in biological and pharmaceutical samples and food products. Excellent literature reviews of HPLC based on various detectors in the analysis of vitamin E content in various matrices have been reported (Abidi, 2000; Aust et al., 2001; Ruperez et al., 2001; Lai and Franke, 2013). Table 19.5 reports several recent HPLC methods for the analysis of vitamin E and similar compounds in various matrices.

19.1.5.6 Gas–Liquid Chromatography

Gas–liquid chromatography (GLC) was a breakthrough for the separation of tocopherols. However, when applied to biological materials, other lipid constituents, particularly cholesterol, which is often present in large excess, interfered with the quantitation of tocopherols. In addition, β- and γ-tocopherols proved very difficult to separate. Both problems could be solved by combining GC with yet another chromatographic technique such as column chromatography and/or TLC. The resulting complex sample pretreatment schemes illustrate once again the need for a powerful chromatographic technique to quantitate vitamin E in biological materials. This requirement was fulfilled with the advent of high-resolution chromatographic techniques, such as capillary GC, and particularly HPLC. Excellent literature reviews of GLC based on various detectors in the analysis of vitamin E content of a wide variety of food, pharmaceutical, plants and biological samples have been reported (Nelis et al., 1985; Davidek and Velisek, 1986; Leenheer

TABLE 19.5 Summarized Characteristics of HPLC Methods Used for the Analysis of Vitamin E and Similar Compounds in Diverse Samples

Sample Matrix	Analyte	Column	Mobile Phase	Detection	Linear Range	LOD	R^2	Reference
Plasma	Vitamin D$_3$ Vitamin E acetate Vitamin K$_1$	Chromolith Performance, RP-18e column 100 × 4.6 mm	Gradient MeOH 95%, HCOOH 0.1% 1.0 mL/min	MS $m/z =$ 385.23, 473.47, 451.41	0.1–10 µg/mL 0.25–10 µg/mL 0.1–10 µg/mL	0.1 ng/mL 1.36 ng/mL 0.052 ng/mL	0.994 0.997 0.996	Abro et al. (2014)
Infant milk formulas	Vitamin A Vitamin E	Chromolith Performance, RP-18 e 100 × 4.6 mm	MeOH 1 mL/min	323 nm 292 nm	0.006–80 µg/mL 0.1–400 µg/mL	0.003 µg/mL 0.08 µg/mL	0.997 0.999	Yazdi and Yazdinezhad (2014)
Plant foods	α-Tocopherol β-Tocopherol γ-Tocopherol δ-Tocopherol	Dimethylpentaf-luorophenyl propyl 5 µm 15 cm × 0.46 cm	Isocratic MeOH:H$_2$0 (85:15) 1 mL/min	Fluorescence (APCI-MS) in negative ion mode	5–100 ng/mL 1–50 ng/mL 1–50 ng/mL 1–50 ng/mL	1.1 ng/mL 0.21 ng/mL 0.32 ng/mL 0.150 ng/mL	NR	Viñas et al. (2014)
Vegetable oils	Tocopherols (α, β + γ, and δ)	Alltima RP C-18 5 µm 250 × 4.6 mm	CH$_3$CN:MeOH (50:50) 1.0 mL/min	Fluorescence λ_{ex} = 290 nm λ_{em} = 325 nm	0.005–10 µg/g	9 ng/g	0.9956	Bele et al. (2013)
Foods, vitamin and mineral formulations, feed premixes, blood serum	Vitamin A Vitamin D$_3$ Vitamin E	Luna 5u C18(2) 150 × 3.0 mm	CH$_3$CN:H$_2$O (98:2) 0.420 mL/min	190–800 nm	0.1–160 µg/mL 0.1–160 µg/mL 1.25–2000 µg/mL	0.002 µg/mL 0.0025 µg/mL 0.04 µg/mL	≥0.9994	Chirkin et al. (2013)
Cereals	Tocopherols Tocotrienols	Cosmosil π-NAP 5 µm 250 mm × 46 mm	H$_2$O:MeOH: H$_3$CN (13:80:7) 1 mL/min	Fluorescence λ_{ex} = 295 nm λ_{em} = 330 nm	0.031–4.0 µg/mL	0.01–0.11 µg/mL	>0.9989	Shammugasamy et al. (2013)
Breast milk	Retinol α-Tocopherol	Kinetex C18 2.6 µm 100 × 4.6 mm	MeOH 100% 1.5 mL/min	325 nm 295 nm	0.0558–7.0 µmol/L 0.42–50.0 µmol/L	0.004 µmol/L 0.078 µmol/L	0.9999 0.9999	Plisek et al. (2013)

(Continued)

TABLE 19.5 (*Continued*) Summarized Characteristics of HPLC Methods Used for the Analysis of Vitamin E and Similar Compounds in Diverse Samples

Sample Matrix	Analyte	Column	Mobile Phase	Detection	Linear Range	LOD	R^2	Reference
Soybeans	Campesterol Stigmasterol β-Sitosterol α-Tocopherol δ-Tocopherol γ-Tocopherol Lutein	XTerra phenyl 3.5 μm 150 mm × 3.9 mm	Isocratic H3CN:MeOH:H2O (48:22.5:29.5)	ELSD (10 Hz) and Lutein at 450 nm	5–200 μg/mL 5–200 μg/mL 10–400 μg/mL 15–150 μg/mL 7.5–150 μg/mL 7.5–150 μg/mL 0.25–10 μg/mL	5 μg/mL 5 μg/mL 10 μg/mL 15 μg/mL 7.5 μg/mL 7.5 μg/mL 0.25 μg/mL	1.000 0.999 0.999 0.998 0.998 0.995 0.999	Slavin and Yu (2012)
Plasma and liver	α-, β-, γ-, δ-tocopherol and α-, β-, γ-, δ-tocotrienol	Phenomenex kinetex PFP 2.6 μm 150 × 4.6 mm	MeOH:H2O (85:15) 0.8 mL/min	Fluorescence λ_{ex} = 296 nm λ_{em} = 325 nm	27–156 pg	NR	0.9872–0.9999	Grebenstein and Frank (2012)
Wheat genotypes	Tocopherol and tocotrienol	Phenomenex Luna Silica 5 μm 250 × 4.6 mm	C6H14:C2H5CH3 COO:CH3 COOH (97.3:1.8:0.9) 1.6 mL/min	Fluorescence λ_{ex} = 290 nm λ_{em} = 330 nm	NR	NR	NR	Hussain et al. (2012)
Fruit, vegetable, human plasma, and adipose tissues	Retinol Tocopherol	YMC C30 250 × 4.6 mm	MeOH:(CH3)3 COCH3:H2O (96:2:2) 1 mL/min	325 nm 290 nm	0.25–10 nM	0.003 μg 50/μL 0.020 μg 50/μL	NR	Gleize et al. (2012)
Human breast milk	Retinol α-Tocopherol	Chromolith Performance, RP-18e, 100 mm × 4.6 mm monolithic	MeOH 100% 2.5 mL/min	325 nm 295 nm	NR	0.13 μmol/L 0.09 μmol/L	NR	Kasparova et al. (2012)

(*Continued*)

TABLE 19.5 (*Continued*) Summarized Characteristics of HPLC Methods Used for the Analysis of Vitamin E and Similar Compounds in Diverse Samples

Sample Matrix	Analyte	Column	Mobile Phase	Detection	Linear Range	LOD	R^2	Reference
Ready-to-eat green leafy vegetable products	Vitamin E Provitamin A	YMC C30 5 µm 250 × 4.6 mm	MeOH:H$_2$O: N(CH$_2$CH$_3$)$_3$ (90:10:0.1) and (CH$_3$)$_3$COCH$_3$: MeOH: H$_2$O: N(CH$_2$CH$_3$)$_3$ (90:6:4:0.1) 1 mL/min	295 nm 450 nm	2650–100,000 ng/mL 2650–250,000 ng/mL	170 ng/mL 70 ng/mL	0.999 0.999	Santos et al. (2012)
Vitamin-enriched soft drinks, oil samples, soybean oil	Vitamin A Vitamin E	Microsorb C18 5 µm 250 × 4.6 mm	SDS, 3% (w/v), C$_4$H$_9$OH,15.0% (v/v), PO$_4^{3-}$ buffer 0.02 mol/L, pH 7.0 2 mL/min	328 nm 280 nm	5.0–360 mg/L	0.234 mg/L 0.780 mg/L	0.9982 0.99997	Ortega et al. (2011)
Vegetable oils	α-Tocopherol γ-Tocopherol δ-Tocopherol	Lichrosorb RP-18 10 µm 150 × 4.0 mm	CTAB 0.01 M/ CH$_3$CH$_2$CH$_2$OH 65%	290 nm Fluorescence λ_{ex} = 296 nm λ_{em} = 398 nm	UV 0.20–36.5 mg/L 0.20–6.2 mg/L 0.20–18 mg/L FL 0.2–8.5 mg/L 0.30–2.5 mg/L 0.2–3.1 mg/L	UV 0.09 mg/L 0.05 mg/L 0.09 mg/L FL 0.08 mg/L 0.01 mg/L 0.07 mg/L	UV 0.9998 0.9989 0.9999 FL 0.9997 0.9996 0.9990	Andrés et al. (2011)
Vegetable oils	α-Tocopherol β-Tocopherol γ-Tocopherol δ-Tocopherol	HyPurity C18 5 µm 250 × 4.6 mm	(CH$_2$)$_4$O:MeOH (10:90) 1 mL/min	215 nm 240 nm 280 nm Fluorescence λ_{ex} = 296 nm λ_{em} = 330 nm	0.001–50 µg/g 0.001–50 µg/g 0.001–50 µg/g 0.001–50 µg/g	7 ng/g 6 ng/g 6 ng/g 6 ng/g	0.9970 0.9953 0.9971 0.9986	Chen et al. (2011)

Note: NR = not reported; LOD = limit of detection.

TABLE 19.6 Summarized Characteristics of GC Methods Used for the Analysis of α-Tocopherol and Similar Compounds in Diverse Samples

Sample Matrix	Analytes	Sample Treatment	Detection	Conditions	Linear Range and LOD	Reference
Corn oil, meat, and blood serum	Sterols and α-, δ-, and γ-tocopherols	Saponification (ascorbic acid with ethanol + KOH) ethanol precipitation, hexane/H_2O extraction–evaporation, treated with pyridine and Sylon BFT (BSTFE + TMCS, 99:1)	GC-FID	HP-5MS column, temperature from 180°C to 280°C	0–200 μg with r^2 values of 0.987 and 0.989 for α-, and δ-tocopherols (corn oil), respectively; 1 μg/mL	Du and Ahn (2002)
Standards	α-, β-, γ-, and δ-Tocopherols, tocopherol quinones, tocopherol hydroquinones	TMS derivatives	GC-MS	Rtx-5MS column, temperature 220–290°C	0–15 ng tocopherols; 40 pg/μL for all tocopherols	Melchert et al. (2002)
Human plasma	α-Tocopherol	Ethanol precipitation, hexane/dichloromethane (9:1) extraction—evaporation	GC-FID	HP-5 column, temperature from 150°C 320°C	1–30 μg/mL; 0.30 μg/mL	Demirkaya and Kadioglu (2007)
Oilseed rape	α-, β-, γ-, and δ-Tocopherols	Three modified sample preparation protocols used for extraction; (i) evaporation of solvent after extraction without silylation, (ii) direct supernatant collection after overnight extraction with drying and silylation, and (iii) trimethylsilylation with BSTFE	GC-FID	5Sil MS column, temperature from 180°C 280°C	100, 150, 200, 250, and 300 μg/mL with r values of 0.995, 0.995, 0.988, 0.994, and 0.995 for T-, α-, β, γ-, and δ-tocopherols, respectively; 0.35 μg/g sample	Hussain et al. (2013)
Human serum	Sterols and α-, β- γ-, and δ-tocopherols	Saponification with ethanol + KOH, precipitation, hexane extraction, treated with isooctane/pyridine (9:1) and acetonitrile/mixture of isooctane/TFECF (95:5)	GC-MS	DB-1HT column, temperature 180–330°C	0.75–30 μg/mL with r^2 values of 0.9796, 0.9986, 0.9531, and 0.9907 for α-, β- γ-, and δ-tocopherols, respectively; 0.15 μg/mL for tocopherols	Rimnacova et al. (2014)

Note: TMS = trimethylsilyl; BSTFE = *N,O*-bis(trimethylsilyl) trifluoroacetamide; Sylon BFT [BSTFA + TMCS (trimethylchlorosilane) = 99:1]; TFECF = trifluoroethyl chloroformate; GC = gas chromatography; MS = mass spectrometry; FID = flame ionization detector; LOD = limit of detection.

et al., 2000; Ruperez et al., 2001). Table 19.6 reports several recent GC methods used for the analysis of α-tocopherol and similar compounds in various matrices.

19.2 CONCLUSIONS

This chapter reviews various analytical methodologies used for monitoring vitamin E. Spectrophotometric and CL detectors are commonly combined with FIA, exhibiting the benefits of easy handling, cheap instrumentation, speed of analysis and accurate determination. In addition, CL methods are still developing, and further research is needed to focus on the development of more reliable mechanisms and stable reaction systems for fat-soluble vitamins.

REFERENCES

Abidi, S. L. 2000. Chromatographic analysis of tocol-derived lipid antioxidants. *J. Chromatogr.* A 881:197–216.

Abro, K., N. Memon, M. I. Bhanger, S. Abro, S. Perveen, and A. H. Laghar. 2014. Determination of vitamin E, D_3, and K_1 in plasma by liquid chromatography atmospheric pressure chemical ionization-mass spectrometry utilizing a monolithic column. *Anal. Lett.* 47:14–24.

Adcock, J. L., N. W. Barnett, C. J. Barrow, and P. S. Francis. 2014. Advances in the use of acidic potassium permanganate as a chemiluminescence reagent: A review. *Anal. Chim. Acta* 807:9–28.

Adcock, J. L., P. S. Francis, and N. W. Barnett. 2007. Acidic potassium permanganate as a chemiluminescence reagent—A review. *Anal. Chim. Acta* 601:36–67.

Ahmed, M. K., J. K. Daun, and R. Przybylski. 2005. FT-IR based methodology for quantification of tocopherols, tocotrienols and plastochromanol-8 in vegetable oils. *J. Food Compos. Anal.* 18:359–364.

Ali, N. M. and M. P. Iqbal. 2008. A micromethod for determination of plasma levels of α- tocopherol in Pakistani normal adults. *Pak. J. Pharm. Sci.* 21:361–365.

Amin, A. S. 2001. Colorimetric determination of tocopheryl acetate (vitamin E) in pure form and in multi-vitamin capsules. *Eur. J. Pharm. Biopharm.* 51:267–272.

Andrés, M. P. S., J. Otero, and S. Vera. 2011. High performance liquid chromatography method for the simultaneous determination of α-, β-, and γ-tocopherol in vegetable oils in presence of hexadecyltrimethylammonium bromide/*n*-propanol in mobile phase. *Food Chem.* 126:1470–1474.

AOAC International. 1995. *Official Methods of Analysis of AOAC* (16th ed.). Arlington, VA, USA, Association of Analytical Communities.

Asgher, M., A. Waseem, M. Yaqoob, and A. Nabi. 2011a. Flow injection chemiluminescence determination of retinol and α-tocopherol in blood serum and pharmaceuticals. *Anal. Lett.* 44:12–24.

Asgher, M., M. Yaqoob, A. Waseem, and A. Nabi. 2011b. Flow injection methods for the determination of retinol and a-tocopherol using lucigenin-enhanced chemiluminescence. *Luminescence* 26:416–423.

Aust, O., H. Sies, W. Stahl, and M. C. Polidori. 2001. Analysis of lipophilic antioxidants in human serum and tissues: Tocopherols and carotenoids. *J. Chromatogr.* A 936:83–93.

Azzi, A. 2007. Molecular mechanism of alpha-tocopherol action. *Free Radical Biol. Med.* 43:16–21.

Azzi, A. and A. Stocker. 2000. Vitamin E: Non-antioxidant roles. *Prog. Lipid Res.* 39:231–255.

Ball, G. F. M. 1988. *Chemical and Biological Nature of the Fat-Soluble Vitamins, in Fat Soluble Vitamin Assays in Food Analysis*, Chapters 2 and 8. New York: Elsevier.

Ball, G. F. M. 1998. Applications of HPLC to the determination of fat-soluble vitamins in foods and animal feeds. *J. Micronutr. Anal.* 4:255–283.

Ball, G. F. M. 2006. *Vitamins in Food: Analysis, Bioavailability, and Stability*. Boca Raton: Taylor & Francis.

Barcelo, D. 2008. Principles of flow injection analysis. In *Comprehensive Analytical Chemistry: Advances in Flow Injection Analysis and Related Techniques*, eds. S. D. Kolev and I. D. McKelvie, pp. 81–109. London: Elsevier.

Barnes, P. J. and P. W. Taylor. 1981. γ-Tocopherol in barley germ. *Phytochemistry* 20:1753–1755.

Barnett, N. W. and P. S. Francis. 2005. Chemiluminescence: Liquid-phase chemiluminescence. In *Encyclopedia of Analytical Science* (2nd ed.), eds. P. J. Worsfold, A. Townshend, and C. F. Poole. 511pp. Oxford: Elsevier.

Barnett, N. W., D. G. Rolfe, T. A. Bowser, and T. W. Paton. 1993. Determination of morphine in process streams using flow-injection analysis with chemiluminescence detection. *Anal. Chim. Acta* 282:551–557.

Barni, F., S. W. Lewis, G. M. Miskelly, and G. Lago. 2007. Forensic application of the luminol reaction as a presumptive test latent blood detection. *Talanta* 72:896–913.

Bauernfeind, J. C. 1980. Tocopherols in foods. In *Vitamin E, A Comprehensive Treatise*, ed. L. J. Machlin, pp. 99–167. New York: Marcel Dekker.

Bele, C., C. T. Matea, C. Raducu, V. Miresan, and O. Negrea. 2013. Tocopherol content in vegetable oils using a rapid HPLC fluorescence detection method. *Not. Bot. Horti Agrobot.* 41:93–96.

Bell, E. F. 1987. History of vitamin E in infant nutrition. *Am. J. Clin. Nutr.* 57:183–186.

Bezzi, S., S. Loupassaki, C. Petrakis, P. Kefalas, and A. Calokerinos. 2008. Evaluation of peroxide value of olive oil and antioxidant activity by luminol chemiluminescence. *Talanta* 77:642–646.

Blasco, A. J., A. G. Crevillen, M. C. Gonzalez, and A. Escarpa. 2007. Direct electrochemical sensing and detection of natural antioxidants and antioxidant capacity *in vitro* systems. *Electroanalysis* 19:2275–2286.

Booth, V. H. and M. P. Bradford. 1963. Tocopherol contents of vegetables and fruits. *Br. J. Nutr.* 17:575–581.

Booth, V. H. and H. A. Frohock. 1961. The α-tocopherol content of leaves as affected by growth rate. *J. Sci. Food Agric.* 12:251–256.

Boothby, L. A. and P. L. Doering. 2005. Vitamin C and E for Alzheimer's disease. *Ann. Pharmacother.* 39:2073–2080.

Bourassa, M. G. and J. C. Tardif. 2006. *Antioxidants and Cardiovascular Disease* (2nd ed.). New York: Springer Science + Business Media, Inc.

Boyle, E. A., B. Handy, and A. Geen. 1987. Cobalt determination in natural waters using cation-exchange liquid chromatography with luminol chemiluminescence detection. *Anal. Chem.* 59:1499–1503.

Bramley, P. M., I. Elmadfa, A. Kafatos et al. 2000. Review vitamin E. *J. Sci. Food Agric.* 80:913–938.

Breaver, J. J. and H. Kaunitz. 1944. The interference of sesame oil, fish oil and cholesterol with the polarographic determination of α-tocopherol. *J. Biol. Chem.* 152:363–365.

Campana, A. M. G. and W. R. G. Baeyens. 2001. *Chemiluminescence in Analytical Chemistry.* New York: Marcel Dekker.

Campana, A. M. G., F. J. Lara, L. G. Gracia, and J. F. H. Perez. 2009. Chemiluminescence detection coupled to capillary electrophoresis. *Trends Anal. Chem.* 28:973–986.

Chen, H., M. Angiuli, C. Ferrari, E. Tombari, G. Salvetti, and E. Bramanti. 2011. Tocopherol speciation as first screening for the assessment of extra virgin olive oil quality by reversed-phase high-performance liquid chromatography/fluorescence detector. *Food Chem.* 125:1423–1429.

Chirkin, V. A., S. I. Karpov, V. F. Selemenev, and N. I. Shumskiy. 2013. Determination of fat_soluble vitamins in foods, vitamin and mineral formulations, feed premixes, and blood serum by reversed-Phase HPLC. *J. Anal. Chem.* 68:748–753.

Christodouleas, D. C. Fotakis, K. Papadopoulos, D. Dimotikali, and A. C. Calokerinos. 2012. Luminescent methods in the analysis of untreated edible oils: A review. *Anal. Lett.* 45:625–641.

Combs, G. F. Jr. 2008. *The Vitamins: Fundamental Aspects in Nutrition and Health.* New York: Elsevier Academic Press.

Cort, W. M., W. Mergens, and A. Greene. 1978. Stability of alpha and gamma-tocopherol: Fe^{3+} and Cu^{2+} interactions. *J. Food Sci.* 43:797–798.

Cotton, F. A. and G. Wilkinson. 1988. *Advanced Inorganic Chemistry* (3rd ed.), 741pp. New York: John Wiley and Sons.

Davidek, J. and J. Velisek. 1986. Gas–liquid chromatography of vitamins in foods: The fat-soluble vitamins. *J. Micronutr. Anal.* 2:81–96.

Davies, B. K. J. 2007. Is vitamin E, an antioxidant, a regulator of signal transduction and gene expression, or a "junk" food? *Free Radical. Biol. Med.* 43:2–3.

Demirkaya, F. and Y. Kadioglu. 2007. Simple GC-FID method development and validation for determination of α-tocopherol (vitamin E) in human plasma. *J. Biochem. Biophys. Methods* 70:363–368.

Diaz, T. G., I. D. Meras, M. I. Rodriguez Caceres, and B. R. Murillo. 2006. Comparison of fluorimetric signal for simultaneous multivariate determination of tocopherols in vegetable oils. *Appl. Spectrosc.* 60:194–202.

Dowd, P. and Z. B. Zheng. 1995. On the mechanism of the anti-clotting action of vitamin E quinine. *Proc. Natl. Acad. Sci.* 92:8171–1875.

Driskell, W. J., W. J. Neese, C. C. Bryant, and M. M. Bashor. 1982. Measurement of vitamin A and vitamin E in human serum by high-performance liquid chromatography. *J. Chromatogr. B* 231:439–444.

Du, M. and D. U. Ahn. 2002. Simultaneous analysis of tocopherols, cholesterol, and phytosterols using gas chromatography. *J. Food. Sci.* 67:1696–1700.

Duval, C. and M. C., Poelman. 1995. Scavenger effect of vitamin E and derivatives on free radical generated by photoirradiated pheomelanin. *J. Pharm. Sci.* 84:107–110.

Emmerie, A. and C. Engel. 1938. Colorimetric determination of α-tocopherol. *Recl. Trav. Chim.* 57:1351–1355.

Epand, A. R. F. and R. M. Epand. 2008. Tocopherols and tocotrienols in membranes: A critical review. *Free Radical Biol. Med.* 44:739–764.

Escuderos, M. E., A. Sayago, M. T. Morales, and R. Aparicio. 2009. Evaluation of α-tocopherol in virgin olive oil by a luminiscent method. *Grasas Aceites* 60:336–342.

European Directive 95/2/EC. 1995. On food additives other than colours and sweeteners, February 20, 1995. http://eurlex.europa.eu/LexUriServ/LexUriServ.do?uri=CONSL EG:1995L0002:20060815:EN:PDF

Evans, H. M. and K. S. Bishop. 1922. On the existence of a hitherto unrecognized dietary factor essential for reproduction. *Science* 56:650–651.

Fabianek, J., J. DeFilippi, T. Rickards, and A. Herp. 1968. Micromethod for tocopherol determination in blood serum. *Clin. Chem.* 14:456–462.

Fan, A., Z. Cao, H. Li, M. Kai, and J. Lu. 2009. Chemiluminescence platforms in immunoassay and DNA analyses. *Anal. Sci.* 25:587–597.

Fletcher, P., K. N. Andrew, A. C. Calokerinos, S. Forbes, and P. J. Worsfold. 2001. Analytical applications of flow injection with chemiluminescence detection—A review. *Luminescence* 16:1–23.

Food Labels. 2012. *Live Well.* NHS Choices. Retrieved December 26, 2012. http://www.nhs.uk/Livewell/Goodfood/Pages/food-labelling.aspx

Food and Nutrition Board, Institute of Medicine. 2000. Dietary reference intakes for vitamin C, vitamin E, selenium, and carotenoids. *A Report of the Panel on Dietary Antioxidants and Related Compounds, Subcommittees on Upper Reference Levels of Nutrients and Interpretation and Uses of Dietary Reference Intakes, and the Standing Committee on the Scientific Evaluation of Dietary Reference Intakes.* Washington, DC: National Academy Press.

Garrett, R. H. and C. M. Grisham. 1999. *Biochemistry* (2nd ed.). Singapore: Thomson (Brooks/Cole).

Gleize, B., M. Steib, M. Andre, and E. Reboul. 2012. Simple and fast HPLC method for simultaneous determination of retinol, tocopherols, coenzyme Q10 and carotenoids in complex samples. *Food Chem.* 134:2560–2564.

Gorman, B. A., P. S. Francis, and N. W. Barnett. 2006. Tris(2,2′-bipyridyl)ruthenium(II), chemiluminescence. *Analyst* 131:616–639.

Gotor, A. A., E. Farkas, M. Berger et al. 2007. Determination of tocopherols and phytosterols in sunflower seeds by near-infrared spectroscopy. *Eur. J. Lipid Sci. Technol.* 109:525–530.

Gracia, L. G., A. M. G. Campana, J. J. S. Chinchilla, J. F. H. Perez, and A. G. Casado. 2005. Analysis of pesticides by chemiluminescence detection in the liquid phase. *Trends Anal. Chem.* 24:927–942.

Gracia, L. G., A. M. G. Campana, J. F. H. Pérez, and F. J. Lara. 2009. Chemiluminescence detection in liquid chromatography: Applications to clinical, pharmaceutical, environmental and food analysis—A review. *Anal. Chim. Acta* 640:7–28.

Grams, G. W., C. W. Blessin, and G. E. Inglett. 1970. Distribution of tocopherols within the corn kernel. *J. Am. Oil Chem. Soc.* 47:337–339.

Grebenstein, N. and J. Frank. 2012. Rapid baseline-separation of all eight tocopherols and tocotrienols by reversed-phase liquid-chromatography with a solid-core pentafluorophenyl column and their sensitive quantification in plasma and liver. *J. Chromatogr. A* 1243:39–46.

Hansen, E. H. 2004. The impact of flow injection on modern chemical analysis: Has it fulfilled our expectations? And where are we going? *Talanta* 64:1076–1083.

Hathcock, J. N., A. Azzi, J. Blumberg et al. 2005. Vitamins E and C are safe across a broad range on intakes. *Am. J. Clin. Nutr.* 81:736–745.

Hindson, B. J. and N. W. Barnett. 2001. Analytical applications of acidic potassium permanganate as a chemiluminescence reagent. *Anal. Chim. Acta* 445:1–19.

Hinze, W. L., T. E. Riehl, H. N. Singh, and Y. Baba. 1984. Micelle-enhanced chemiluminescence and application to the determination of biological reductants using Lucigenin. *Anal. Chem.* 5:2180–2191.

Hogarty, C. J., C. Ang, and R. R. Eitenmille. 1989. Tocopherol content of selected foods by HPLC/fluorescence quantitation. *J. Food Compos. Anal.* 2:200–209.

Holland, B., A. A. Welch, I. D. Unwin, D. H. Buss, A. A. Paul, and D. A. T. Southgate. 1991. *Mccance and Widdowson's The Composition of Food* (5th ed.). London: The Royal Society of Chemistry and MAFF.

Hossu, A. M., M. F. Maria, A. Stoica, M. Ilie, and M. Iordan. 2009. Determination of tocopherol in oil pharmaceuticals by using the spectrofluorimetry and method validation. *Annals Chem.* 20:53–56.

Huo, J., H. J. Nelis, P. Lavens, P. Sorgeloos, and A. P. De Leenheer. 1996. Determination of vitamin E in aquatic organisms by high-performance liquid chromatography with fluorescence detection. *Anal. Biochem.* 242:123–128.

Hussain, A., H. Larsson, M. E. Olsson, R. Kuktaite, H. Grausgruber, and E. Johansson. 2012. Is organically produced wheat a source of tocopherols and tocotrienols for health food? *Food Chem.* 132:1789–1795.

Hussain, N., Z. Jabeen, Y. L. Li et al. 2013. Detection of tocopherol in oilseed rape (*Brassica napus* L.) using gas chromatography with flame ionization detector. *J. Integr. Agric.* 12:803–814.

Iranifam, M. 2013. Revisiting flow-chemiluminescence techniques: Pharmaceutical analysis. *Luminescence* 28:798–820.

Jadoon, S., A. Waseem, M. Yaqoob, and A. Nabi. 2010. Flow injection spectrophotometric determination of vitamin E in pharmaceuticals, milk powder and blood serum using potassium ferricyanide-Fe(III) detection system. Chin. *Chem. Lett.* 21:712–715.

Kamidate, T., T. Kaneyasu, T. Segawa, and H. Watanabe. 1991. Effects of halide ions on micelle-enhanced chemiluminescence reaction of lucigenin with adrenaline. *Bull. Chem. Soc. Jpn.* 64:1991–1992.

Karlsson, J. 1997. Muscle metabolism and the antioxidant defense. *World Rev. Nutr. Diet.* 82:81–100.

Karoui, R., E. Dufour, L. Pillonel, D. Picque, T. Cattenoz, and J. O. Bosset. 2004. Determining the geographic origin of Emmental cheeses produced during winter and summer using a technique based on the concatenation of MIR and fluorescence spectroscopic data. *Eur. Food Res Technol.* 219:184–189.

Kasparova, M., J. Plíseka, D. Solichováb et al. 2012. Rapid sample preparation procedure for determination of retinol and α-tocopherol in human breast milk. *Talanta* 93:147–152.

Kolpf, L. L. and T. A. Nieman. 1985. Determination of conjugated glucuronic acid by combining enzymatic hydrolysis with lucigenin chemiluminescence. *Anal. Chem.* 57:46–51.

Kolthoff, I. M. and J. J. Lingane. 1941. *Polarography*, pp. 220–229. New York: Interscience.

Lai, J. F. and A. A. Franke. 2013. Analysis of circulating lipid-phase micronutrients in humans by HPLC: Review and overview of new developments. *J. Chromatogr. B* 931:23–41.

Lara, F. J., A. M. G. Campana, and J. J. Aaron. 2010. Analytical applications of photoinduced chemiluminescence in flow systems—A review. *Anal. Chim. Acta* 679:17–30.

Leenheer, A. P. D., W. E. Lambert, and J. F. V. Bocxlaer. 2000. *Modern Chromatographic Analysis of Vitamins* (3rd ed.), Revised and Expanded. New York: Marcel Dekker.

Lehman, J., H. L. Martin, E. L. Lashley, M. W., Marshall, and J. T. Judd. 1986. Vitamin E in foods from high and low linoleic acid diets. *J. Am. Diet. Assoc.* 86: 1208–1216.

Li, J. J., J. X. Du, and J. R. Lu. 2002. Flow injection electrogenerated chemiluminescence determination of vanadium and its application to environmental water sample. *Talanta* 57:53–57.

Li, S. G., W. T. Xue, and H. Zhang. 2006. Voltammetric behavior and determination of tocopherol in vegetable oil at a polypyrrol electrode. *Electroanalysis* 18:2337–1342.

Litwack, G. 2007. *Vitamin E: Vitamins and Hormones Advances in Research and Application*, Vol. 76. New York: Elsevier.

Liu, M., Z. Lin, and J. M. Lin. 2010. A review on applications of chemiluminescence detection in food analysis. *Anal. Chim. Acta* 670:1–10.

Lumley, I. D. 1993. *Vitamin Analysis in Foods, in the Technology of Vitamin in Foods*, Chapter 8, New York: Chapman and Hall.

Luther, III, G. W., P. A. Shellenbarger, and P. J. Brendel. 1996. Dissolved organic Fe(III) and Fe(II) complexes in salt marsh porewaters. *Geochim. Cosmochim. Acta* 60:951–960.

Luykx, D. M. A. M. and S. M. VanRuth. 2008. An overview of analytical methods for determining the geographical origin of food products. *Food Chem.* 107:897–911.

Mambro, V. M. D., A. E. C. S. Azzolini, Y. M. L. Valim, and M. J. V. Fonseca. 2003. Comparison of antioxidant activities of tocopherols alone and in pharmaceutical formulations. *Int. J. Pharm.* 262:93–99.

McKelvy, M. L., T. R. Britt, B. L. Davis, J. K. Gillie, F. B. Graves, and L. A. Lentz. 1998. Infrared spectroscopy. *Anal. Chem.* 70:119–177.

McLaughlin, P. J. and J. C. Weihrauch. 1979. Vitamin E content of foods. *J. Am. Diet. Assoc.* 75:647–665.

Melchert, H. U., D. Pollok, E. Pabel, K. Rubach, and H. J. Stan. 2002. Determination of tocopherols, tocopherolquinones and tocopherolhydroquinones by gas chromatography–mass spectrometry and preseparation with lipophilic gel chromatography. *J. Chromatogr. A* 976:215–220.

Memon, N., M. I. Bhanger, M. A. Memon, and M. H. Memon. 2004. Flow injection spectrophotometric determination of α-tocopherol in microemulsion reaction medium using Fe(III)-1,10 phenanthroline as an oxidant. *ACGC Chem. Res. Commun.* 17:52–59.

Mervartova, K., M. Polasek, and J. M. Calatayud. 2007. Recent applications of flow-injection and sequential-injection analysis techniques to chemiluminescence determination of pharmaceuticals. Review. *J. Pharm. Biomed. Anal.* 45:367–381.

Mestre, Y. F., L. L. Zamora, and J. M. Calatayud. 2001. Flow-chemiluminescence: A growing modality of pharmaceutical analysis. *Luminescence* 16:213–235.

Mikheeva, V. and L. S. Anisimova. 2007. Voltammetric determination of vitamin E (α-tocopherol acetate) in multicomponent vitaminized mixtures. *J. Anal. Chem.* 62:373–376.

Muller, D. P. 2010. Vitamin E and neurological function. *Rev. Mol. Nutr. Food Res.* 54:710–718.

Murphy, P. G., J. R. Bennett, D. S. Myers, M. J. Davies, and J. G. Jones. 1993. The effect of propofol anaesthesia on free radical-induced lipid peroxidation in rat liver microsomes. *Eur. J. Anaesthesiol.* 10:261–266.

Nakashima, K. 2003. Lohine derivatives as versatile analytical tools. *Biomed. Chromatogr.* 17:83–95.

Nelis, H. J., V. O. R. C. DeBevere, and A. P. De Leenheer. 1985. Vitamin E: Tocopherols and tocotrienols. In *Modern Chromatographic Anaylsis of the Vitamins*, eds. A. P. De Leenheer, W. E. Lambert, and M. G. M. DeRuyter, Chapter 3. New York: Marcel Dekker.

Ortega, A. C., D. C. da Silva, J. V. Visentainer, N. E. de Souza, V. de C. Almeida, and C. C. Oliveira. 2011. Determination of vitamins A and E exploiting cloud point extraction and micellar liquid chromatography. *Anal. Lett.* 44:778–786.

Ottaway, P. B. 2008. *Food Fortification & Supplementation*. New York: Taylor & Francis.

Percival, M. 1998. Antioxidants. *Clinical Nutition Insights*. NUT031 1/96 Rev. 10/98. Copyright © 1996 Advanced Nutrition Publications, Inc., Revised 1998. http://www.acudoc.com/Antioxidants.PDF.

Plisek, J., M. Kasparova, D. Solichova et al. 2013. Application of core-shell technology for determination of retinol and alpha-tocopherol in breast milk. *Talanta* 107:382–388.

Prieto, P., M. Pineda, and M. Aguilar. 1999. Spectrophotometric quantification of antioxidant capacity through the formation of a phosphomolybdenum complex: Specific application to the determination of vitamin E. *Anal. Biochem.* 269:337–341.

Putzig, C. L., M. A. Leugers, M. L. McKelvy et al. 1994. Infrared spectroscopy. *Anal. Chem.* 66:26–66.

Qin, W., Z. Zhang, and C. Zhang. 1997. Chemiluminescence flow system for vanadium(v) with immobilized reagents. *Analyst* 122:685–688.

Razagul, I. B., P. J. Barlow, and K. D. A. Taylor. 1992. A spectrofluorimetric procedure for the determination of α-tocopherol in nutritional supplement products. *Food Chem.* 44:221–226.

Rimnacova, L., P. Husek, and P. Simek. 2014. A new method for immediate derivatization of hydroxyl groups byfluoroalkyl chloroformates and its application for the determination of sterols and tocopherols in human serum and amniotic fluid by gas chromatography–mass spectrometry. *J. Chromatogr. A* 1339:154–167.

Rishi, L., S. Jadoon, A. Waseem, M. Yaqoob, and A. Nabi. 2011. Flow injection methods for the determination of α-tocopherol with spectrophotometric detection. *J. Chem. Soc. Pak.* 33:508–514.

Rose, A. L. and T. D. Waite. 2001. Chemiluminescence of luminol in the presence of iron(II) and oxygen: Oxidation mechanism and implication for its analytical use. *Anal. Chem.* 73:5909–5920.

Ruperez, F. J., D. Martin, E. Herrera, and C. Barbas. 2001. Chromatographic analysis of α-tocopherol and related compounds in various matrices. *J. Chromatogr. A* 935:45–69.

Ruzicka, J. and E. H. Hansen. 1975. Flow injection analysis, Part 1. A new concept of fast continuous flow analyses. *Anal. Chim. Acta* 78:145–157.

Ruzicka, J. and E. H. Hansen. 1978. Flow injection analysis: Part X. Theory, techniques and trends. *Anal. Chim. Acta* 99:37–76.

Ruzicka, J. and E. H. Hansen. 1988. *Flow Injection Analysis* (2nd ed.). New York: Wiley.

Saima, J., M. Arif, M. H. Qazi, and A. Mohammad. 2013. Spectrophotometric method for the determination of vitamin A and E using ferrozine–Fe(II) complex. *Asian J. Res. Chem.* 6:334–40.

Santos, J., J. A. Mendiola, M. B. P. P. Oliveira, E. Ibanez, and M. Herrero. 2012. Sequential determination of fat- and water-soluble vitamins in green leafy vegetables during storage. *J. Chromatogr. A* 1261:179–188.

Sattler, S. E., L. M. Saffrane, E. E. Farmer, M. Krischke, M. J. Mueller, and D. Dellapenna. 2006. Nonenzymatic lipid peroxidation reprograms gene expression and activates defense markers in arabidopsis tocopherol-defficient mutants. *Plant Cell* 18:3706–3720.

Schneider, C. 2005. Chemistry and biology of vitamin E. *Mol. Nutr. Food Res.* 49:7–30.

Schuck, E. and I. Floderer. 1939. Colorimetric determination of ferrous and ferric iron in the presence of aluminum, manganese, zinc, mercury, copper, phosphoric acid, or organic substances, with special emphasis on drug preparations. *Z. Anal. Chem.* 117:176.

Shammugasamy, B., Y. Ramakrishnan, H. M. Ghazali, and K. Muhammad. 2013. Combination of saponification and dispersive liquid-liquid micro-extraction for the determination of tocopherols and tocotrienols in cereals by reversed-phase high-performance liquid, chromatography. *J. Chromatogr. A* 1300:31–37.

Sheppard, A. J., J. A. T. Pennington, and J. L. Weihrauch. 1993. Analysis and distribution of vitamin E in vegetable oils and foods. In *Vitamin E in Health and Disease*, eds. L. Packer, and J. Fuchs, pp. 9–31. New York: Marcel Dekker.

Sies, H., W. Stahl, and A. R. Sundquist. 1992. Antioxidant functions of vitamins: Vitamins E, C, beta-carotene, and other carotenoids. *Ann. N. Y. Acad. Sci.* 669:7–20.

Silva, S. D., N. F. Rosa, A. E. Ferreira, L. V. Boas, and M. R. Bronze. 2009. Rapid determination of α-tocopherol in vegetable oils by FT-IR spectroscopy. *Food Anal. Methods* 2:120–127.

Slavin, M. and L. (L) Yu. 2012. A single extraction and HPLC procedure for simultaneous analysis of phytosterols, tocopherols and lutein in soybeans. *Food Chem.* 135:2789–2795.

Smith, L. I., L. J. Spillane, and I. M. Kolthoff. 1942. The chemistry of vitamin E. XXXV. The behavior of tocopherols at the dropping mercury electrode. *J. Am. Chem. Soc.* 64:447–451.

Sturm, P. A., R. M. Parkhurst, and W. A. Skinner. 1966. Qualitative determination of individual tocopherols by thin layer chromatography separation and spectrophotometry. *Anal. Chem.* 38:1244–1247.

Su, Y., H. Chen, Z. Wang, and Y. Lv. 2007. Recent advances in chemiluminescence. *Appl. Spectrosc. Rev.* 42:139–176.

Sure, B. 1942. Dietary requirements for reproduction: The existence of specific vitamin for Reproduction. *J. Biol. Chem.* 58:693–709.

Syvaoja, E. L., V. Piironen, P. V. Koivistoinen, and K. Salminen. 1985. Tocopherols and tocotrienols in Finnish foods: Human milk and infant formulas. *Int. J. Vitam. Nutr. Res.* 55:159–166.

Taguchi, J., S. Ohtsuki, and F. Kusu. 2003. Voltammetric determination of weak bases based on oxidation of α-tocopherol in an unbuffered solution. *J. Electroanal. Chem.* 557:91–97.

Traber, M. G. and H. Sies. 1996. Vitamin E in humans: Demand and delivery. *Annu. Rev. Nutr.* 16:321–347.

Trojanowicz, M. 2008. *Advances in Flow Analysis*. Weinheim, Germany: Wiley-VCH.

Tsen, C. C. 1961. An improved spectrophotometric method for the determination of tocopherols using 4,7-diphenyl-10-phenanthroline. *Anal. Chem.* 23:849–851.

Tutem, E., R. Apak, E. Gunayd, and K. Sozgen. 1997. Spectrophotometric determination of vitamin E (α-tocopherol) using copper(II)-neocuproine reagent. *Talanta* 44:249–255.

Villacorta, L., G. A. V. Souza, R. Ricciarelli, J. M. Zingg, and A. Azzi. 2003. α-Tocopherol induces expression of connective tissue growth factor and antagonizes tumor necrosis factor-α-mediated down regulation in human smooth muscle cells. *Circ. Res.* 92:104–110.

Viñas, P., M. B. Bravo, I. L. García, M. P. Belda, and M. H. Córdoba. 2014. Pressurized liquid extraction and dispersive liquid–liquid microextraction for determination of tocopherols and tocotrienols in plant foods by liquid chromatography with fluorescence and atmospheric pressure chemical ionization-mass spectrometry detection. *Talanta* 119:98–104.

Wang, X., M. L. Liu, X. L. Cheng, and J. M. Lin. 2009. Flow-based luminescence-sensing methods for environmental water analysis. *Trends Anal. Chem.* 28:75–87.

Waseem, A., L. Rishi, M. Yaqoob, and A. Nabi. 2009. Flow-injection determination of retinol and tocopherol in pharmaceuticals with acidic potassium permanganate chemiluminescence. *Anal. Sci.* 25:407–412.

Washko, P. W., R. W. Welch, K. Dhariwal, Y. H. Wangm, and M. Levine. 1992. Ascorbic acid and dehydroascorbic acid analysis in biological samples. *Anal. Biochem.* 204:1–14.

Wiberg, K. B. and K. A. Saegebarth. 1957. The Mechanisms of permanganate oxidation: IV. Hydroxylation of olefins and related reactions. *J. Am. Chem. Soc.* 79:2822–2824.

Worsfold, P. J., R. Clough, M. C. Lohan et al. 2013. Flow injection analysis as a tool for enhancing oceanographic nutrient measurements—A review. *Anal. Chim. Acta* 803:15–40.

Yaqoob, M., A. Waseem, L. Rishi, and A. Nabi. 2009. Flow-injection determination of vitamin A in pharmaceutical formulations using tris(2,2′-bipyridyl)Ru(II)–Ce(IV) chemiluminescence detection. *Luminescence* 24:276–281.

Yazdi, A. S. and S. R. Yazdinezhad. 2014. Simultaneous determination of vitamin A and E in infant milk formulas using semi-micro liquid-liquid extraction followed by HPLC-UV. *J. Liq. Chromatogr. Relat. Technol.* 37:391–403.

Yildiz, G. and A. T. Demiryurek. 1998. Ferrous iron-induced luminol chemiluminescence: A method for hydroxyl radical study. *J. Pharmacol. Toxicol. Methods* 39:179–184.

Zhang, G. F. and H. Y. Chen. 2000. Studies of micelle and trace non-polar organic solvent on a new chemiluminescence system and its application to flow injection analysis. *Anal. Chim. Acta* 409:75–81.

Zingg, J. M. and A. Azzi. 2004. Non-antioxidant activities of vitamin E. *Curr. Med. Chem.* 11:1113–1133.

Ziyatdinova, G. K., E. R. Giniyatova, and H. C. Budnikov. 2012a. Voltammetric determination of α-tocopherol in the presence of surfactants. *J. Anal. Chem.* 67:467–473.

Ziyatdinova, G. K., A. Khuzina, and H. Budnikov. 2012b. Determination of sterically hindered phenol and α-tocopherol by cyclic voltammetry. *Anal. Lett.* 45:1670–1685.

Ziyatdinova, G. K., N. A. Morozova, S. V. Simonova, H. C. Bodnikov, and T. I. Abdullin. 2007. *Abstract of papers, XVIII Mendeleeskii s'ezd po obshchei i prikladnoi khimii* (XVIII) (*Mendeleev Conference on General and Applied Chemistry*), Moscow, Vol. 4, p. 142.

CHAPTER 20

Determination of Phenolic Compounds (Gallic, Caffeic, Ferulic, and *p*-Coumaric Acids)

Semih Otles and Emine Nakilcioglu

CONTENTS

20.1 INTRODUCTION

Flow injection analysis (FIA) was developed by Ruzicka and Hansen in Denmark and Stewart and coworkers in the United States in the mid-1970s as a highly efficient technique for the automated analyses of samples (Skoog et al., 1998).

Flow injection analysis is an analytical technique based on injecting a known volume of sample into carrier or reagent streams that are transported in small-diameter conduits (typically 0.5–0.8 mm internal diameter) under laminar flow conditions (Ruzicka and Hansen, 1988; Taljaard and van Staden, 1998; McKelvie, 2008; van Staden and van Staden, 2012; Worsfold et al., 2013). In this moving stream, the sample does not decompose; it is physically and chemically converted into a detectable species that causes a detector response downstream of the injection point (Ruzicka and Hansen, 1988; Taljaard and van Staden, 1998; Saurina, 2008; Saurina, 2010; van Staden and van Staden, 2012). The results will be reproducible on condition that all critical parameters (reproducible injection, controlled reaction time, and controlled dispersion) are held within certain tolerance levels (Ruzicka and Hansen, 1988; Taljaard and van Staden, 1998; van Staden and van Staden, 2012). The basic FIA instrument is composed of a multichannel pump, an injection valve, a flow-through detector, and a signal output device (originally a recorder, lately a computer) (van Staden and van Staden, 2012). Before FIA, relevant variables such as flow rates, reactor dimensions, injection volume(s) and chemical (reaction) conditions are often optimized (Bosque-Sendra et al., 2003; Saurina, 2010). The FIA technique has among others (chromatography and electrophoresis, nonseparating techniques) a high potential for kinetic discriminations between analyte and interferences that are especially important (Horstkotte et al., 2010).

The advantages of FIA include a high throughput of sample, use of minimum sample and reagent, high-precision and high-accuracy measurement, and better sensitivity (Felix and Angnes, 2010; Worsfold et al., 2013). In the great majority of studies involving FIA, small volumes of sample were used. Furthermore, FIA creates sharp peaks with increase of fixed phase and decrease of the current, reflecting a transition of the analyte over the sensor surface (Felix and Angnes, 2010). FIA is also a well-rounded, flexible, economical technique, suitable for the rapid, routine analysis of large numbers of samples. The method is affordable and easy to use, It needs minimal operator intervention and requires a lower consumption of samples and reactants: this is an especially important consideration when working with toxic or expensive reactants (Sullivan et al., 1990; Ferreira et al., 1996; Pinho et al., 1998; Kazemzadeh and Ensafi, 2001; Andrade et al., 2003; Frenzel et al., 2004; Penteado et al., 2005; Ruiz-Capillas and Jimenez-Colmenero, 2008). It is installed online (Ruiz-Capillas and Jimenez-Colmenero, 2008). It can be congruent with numerous types of detectors (spectrophotometric [UV–Vis], chemiluminescence, bioluminescence, fluorescence, phosphorescence, atomic absorption spectrometry, inductively coupled plasma, electrochemical, Fourier transform infrared, Raman spectroscopy, mass spectrometry) (Karlberg and Pacey, 1989). The methodology meets the requirements of mechanization and automation in many laboratories that are faced with an increase in routine analyses (Sullivan et al., 1990; Ferreira et al., 1996; Pinho et al., 1998; Kazemzadeh and Ensafi, 2001; Andrade et al., 2003; Frenzel et al., 2004; Penteado et al., 2005; Ruiz-Capillas and Jimenez-Colmenero, 2008).

Although early studies of FIA focused on the determination of contaminants in water, nowadays the application of FIA extends to other fields (food, biomedicine, control of industrial processes, etc.) and more complex matrices (Ruiz-Capillas and Jimenez-Colmenero, 2008). FIA is also a current and preferred technique in the determination of phenolic compounds.

20.2 PHENOLIC COMPOUNDS

Bioactive compounds are nutritional constituents that occur naturally in small amounts in plants and food products (Kris-Etherton et al., 2002; Martins et al., 2011). The metabolism of plants is divided into primary and secondary. In primary metabolism the substances involved are common to living things and essential to cells maintenance (lipids, proteins, carbohydrates, and nucleic acids). In contrast, some substances that originate from several biosynthetic pathways and are restricted to certain groups of organisms are the results of secondary metabolism (Reis Giada, 2013). Secondary metabolites such as antibiotics, mycotoxins, alkaloids, food-grade pigments, plant growth factors, and phenolic compounds are identified in some of the most common bioactive compounds (Kris-Etherton et al., 2002; Hölker et al., 2004; Nigam, 2009; Martins et al., 2011). These compounds have many different endogenous and exogenous functions, serving, for example, as cell wall components, antioxidants, antimicrobial agents, antifeedants, insect attractants, and chemical signal producers. Most of the secondary metabolites have therapeutic and economic importance owing to their use as drugs, pharmaceutical and industrial precursors, flavorings, cosmetic ingredients, dyes, adhesives. Lately, some of these phytochemicals have been determined as having roles as health-protective dietary constituents (Hancianu and Aprotosoaie, 2012).

One of the biggest and widely distributed groups of secondary metabolites in nature is phenolic compounds or polyphenols (Scalbert and Williamson, 2000; Reis Giada, 2013). Phenolic compounds are a group usually found as esters or glycosides rather than as free compounds and are characteristic of many fruits, vegetables, and beverages (Vermerris and Nicholson, 2008). More than 8000 phenolic structures are currently known in the plant kingdom (Bravo, 1998). They consist of one or more hydroxyl groups attached directly to a benzene ring (Vermerris and Nicholson, 2008). Biogenetically, phenolic compounds are derived from two metabolic pathways: the shikimic acid pathway, in which mainly phenylpropanoids are formed; and the acetic acid pathway, where the main products are the simple phenol (Reis Giada, 2013). Most of the plant phenolic compounds are synthesized by the phenylpropanoid pathway (Hollman, 2001; Reis Giada, 2013). The flavonoids are the most plentiful group of phenolic compounds in nature and their formation occurs by a combination of both pathways. In addition, during the biosynthetic pathways to the synthesis of flavonoids, there are condensation and polymerization phases in which the condensed tannins or nonhydrolyzable tannins are formed that are not well elucidated. The derivatives of gallic acid or hexahydroxydiphenic acid are termed hydrolyzable tannins.

As phenolic compounds are found in a large number of heterogeneous structures that range from simple molecules to highly polymerized compounds, they can be classified in different ways. According to their chemical structure, phenolic compounds can be divided into two major classes (Reis Giada, 2013).

Flavonoids are one of the most common and widely distributed groups of plant phenolics. Their common structure is that of diphenylpropanes (C_6–C_3–C_6) and consists of two aromatic rings linked through three carbons that usually form part of an oxygen-containing heterocycle (Bravo, 1998). Variations in substitution patterns on ring C in the structure of these compounds generate the major flavonoid classes, such as flavonols, flavones, flavanones, flavanols, isoflavones, and anthocyanidins (Martins et al., 2011). On occasion, flavonoids occur in plants as aglycones, though they are most commonly found as glycoside derivatives.

Flavones (e.g., apigenin, luteolin, diosmetin), flavonols (e.g., quercetin, myricetin, kaempferol), and their glycosides are the most common compounds among the flavonoids. They exist widely in the plant kingdom, with the exception of algae and fungi.

Flavonols consist of O-glycosides, but flavone O- and C-glycosides are very common; they are defined by possession of a carbon–carbon linkage between the numeric carbon of a sugar molecule and the C–6 or C–8 carbon of the flavone nucleus. In contrast to O-glycosides, sugars in C-glycosides are not separated by acid hydrolysis.

Flavanones (e.g., naringenin, hesperidin) also can be formed as O- or C-glycosides. They are especially found in citrus foods and prunes. The variability of this group of flavonoids is notable, with about 380 flavonol glycosides and 200 different quercetin and kaempferol glycosides characterized to date.

Isoflavones (e.g., genistein, daidzein) occur especially in legumes and have ring B of the flavone molecule attached to the C–3 of the heterocycle

Although flavanols (e.g., catechin, epicatechin, gallocatechin) are also very common as free monomers, they are found as the monomeric constituents of the condensed tannins. In particular, they are present in varieties of tea (Bravo, 1998).

Anthocyanins, which are responsible for the colors of flowers and fruits of higher plants, are the most important group of water-soluble plant pigments. The anthocyanin is the form of the glycosides of anthocyanidin (e.g., pelargonidin, malvidin, cyanidin). In addition to glycosylation, common linkages between aromatic and aliphatic acids, as well as methyl ester derivatives, are also included. Anthocyanins and polymeric pigments derived from anthocyanins by condensation with other flavonoids are responsible for the color of red wine (Mazza, 1995; Bravo, 1998).

Phenolic acids also constitute an important class of phenolic compounds with bioactive functions like the flavonoids, usually found in plant and food products. According to their structure, phenolic acids can be divided in two subgroups: the hydroxybenzoic and the hydroxycinnamic acids. The most commonly found hydroxybenzoic acids include gallic, p-hydroxybenzoic, protocatechuic, vanillic and syringic acids; caffeic, ferulic, p-coumaric, and sinapic acids are classed in the hydroxycinnamic acids (Bravo, 1998; Martins et al., 2011).

Hydroxybenzoic acids are defined by the presence of a carboxyl group substituted on a phenol. Their structure is C_6–C_1 (Bravo, 1998). Hydroxybenzoic aldehydes, such as vanillin, have an aldehyde group instead of a carboxyl group.

The hydroxycinnamic acids, with a C_6–C_3 skeleton, are represented by six common cinnamic acids. All plants are thought to contain at least three of them. Hydroxycinnamic acids are commonly presented in plants as esters of quinic acid, shikimic acid, and tartaric acid. For instance, chlorogenic acid is an ester of caffeic acid and quinic acid (Vermerris and Nicholson, 2008).

As is well known, natural polyphenols contain a variety of compounds from simple molecules such as phenolic acids through to highly polymerized compounds such as tannins. They exist primarily in conjugated form, with one or more sugar residues linked to hydroxyl groups, but direct linkages of the sugar unit to an aromatic carbon atom occur. Monosaccharides, disaccharides, and even oligosaccharides are sugars that can be linked with the phenolic compounds. Glucose is the most common sugar residue, although galactose, rhamnose, xylose, and arabinose are also found, as well as glucuronic and galacturonic acids and many others. Phenolic compounds are widely linked with other compounds, such as carboxylic and organic acids, amines, and lipids (Bravo, 1998).

Phenolic compounds prevent oxidation due to their antioxidant properties. They are especially used for the prevention of lipid oxidation in food processing. Also, the browning of fruits after they have been cut is an example of the oxidation of phenolic compounds. Oxidation can end in the formation of metabolites that are toxic to animals and plants such as spoilage of foods in processing. However, toxic compounds from the oxidation of phenolics can inhibit pathogenic microorganisms. Erythorbic acid and sodium erythorbate, gallates, butylated hydroxyanisole, butylated hydroxytoluene, tertiary butylhydroquinone, and nordihydroguaiaretic acid are termed "artificial antioxidants" and are effective reducing agents that prevent oxidation of other molecules. The use of natural antioxidants such as phenolic compounds has been preferred due to the technological properties (Vermerris and Nicholson, 2008).

Phenolics are not only responsible for structural and protective functions in plants, they also contribute to flavor, color, astringency, and bitterness of fruits and vegetables. Their role in human health-related issues has been recognized. The demand for "functional" or "nutraceutical" foods increases continuously. Some applications related to natural phenolic compounds are based upon their antioxidant activity against reactive species involved in aging and in chronic, autoimmune, inflammatory, and degenerative diseases (German and Walzem, 2000; Erlund, 2004; Seifried et al., 2007; Soto et al., 2011). Although their antioxidant activity alone is not sufficient to explain all of their biological properties, their antioxidant properties may contribute to their potential cancer chemopreventive properties (Middleton et al., 2000; Havsteen, 2002; Bonfili et al., 2008; Halliwell, 2009; Soto et al., 2011). Phenolic compounds exhibit various activities, among others: reduction in the incidence of some degenerative diseases such as cancer and diabetes; reduction in risk factors for cardiovascular diseases; and antimutagenic, antiallergenic, anti-inflammatory, and antimicrobial effects. Because of these numerous beneficial characteristics for human health, interest in the phenolic compounds and phenolic compound–rich foods is increasing all the time (Kris-Etherton et al., 2002; Balasundram et al., 2006; Jiménez et al., 2008; Conforti et al., 2009; Kim et al., 2009; Parvathy et al., 2009; Martins et al., 2011).

20.2.1 Gallic Acid

Gallic acid (3,4,5-trihydroxybenzoic acid) can be formed by acid hydrolysis of hydrolyzable tannins. Differences among the ester derivatives are only in the number of carbon atoms in the aliphatic side-chain. These differences confer different physicochemical characteristics, especially in lipophilicity evaluated by the value of partition coefficient ($C \log P$). Chemical modifications in the gallic acid molecule can alter the pharmacokinetic and pharmacodynamic properties, such as changing the solubility and the degree of ionization. The names of the compounds are abbreviated according to the length of the side-chain (Locatelli et al., 2013).

Gallic acid is a yellowish white crystal (molecular mass 170.12 g mol^{-1}) with melting point 250°C and water solubility 1.1% at 20°C (Polewski et al., 2002; Verma et al., 2013).

Gallic acid is found abundantly in tea, grapes, berries, and other fruits as well as in wine as an endogenous plant polyphenol (Ma et al., 2003; Singh et al., 2004; Verma et al., 2013). Some hardwood plant species such as oak (*Quercus robur*) and chestnut (*Castanea sativa* L.) also contain gallic acid (Eyles et al., 2004; Murugananthan et al., 2005; Verma et al., 2013).

Gallic acid has several pharmacological and biochemical attributes such as antioxidant, anti-inflammatory, antimutagenic and anticancer properties (Kim et al., 2002; Verma et al., 2013).

20.2.2 Caffeic Acid

3,4-Dihydroxycinnamic acid, also called caffeic acid. is a common phenolic compound of the hydroxycinnamates group that is derived biosynthetically from phenylalanine in plants (Clifford, 1999; Medina et al., 2012).

Caffeic acid is a light yellow-white crystal (molecular mass 180.16 g mol^{-1}) with melting point range 211–213°C; its formula is $C_9H_8O_4$ (Sigma-Aldrich, 2014a).

Caffeic acid is found at extremely high levels in coffee, but it is also found in other plant foods at more modest levels and is ester-linked to quinic acid (soluble chlorogenic acid) (Clifford, 1999; Kroon and Williamson, 1999). It is found naturally in many agricultural products such as fruits, vegetables, wine, and olive oil (Clifford, 1999; Medina et al., 2012). Additionally, some of the by-products of plant food processing are rich sources of phenolic compounds (Moure et al., 2001; Medina et al., 2012). For example, caffeic acid is present in high proportions in artichoke blanching waters and olive milling wastes, which are a major potential source of phenolics considering that the annual production exceeds 7 million tonnes (Llorach et al., 2002; Visioli and Galli, 2003; Medina et al., 2012).

Caffeic acid and its derivatives have recently attracted widespread attention to their biological and pharmacological activities, including antioxidative activities (Taubert et al., 2003; Wu et al., 2007; Medina et al., 2012). The evaluation of antioxidant capacities requires information on thermodynamics (redox potentials) and kinetic rate constants with different types of radicals, the stability of the antioxidant-derived radicals, and stoichiometric properties. The molecular structure of caffeic acid, containing a catechol group with an α,β-unsaturated carboxylic acid chain is responsible for its efficient interaction with several types of oxidant radicals (Medina et al., 2012). The catechol group is the preferred binding site for trace metals, leading to significant chelating activity as well (Pietta, 2000; Medina et al., 2012). Thus, the catechol group in caffeic acid structure has potential antioxidant and chelating activity.

20.2.3 Ferulic Acid

4-Hydroxy-3-methoxycinnamic acid, known as ferulic acid (molecular mass 194.18 g mol^{-1}), is a solid with melting point range 168–172°C; its formula is $C_{10}H_{10}O_4$ (Sigma-Aldrich, 2014b).

Ferulic acid is a ubiquitous phenolic compound in plant tissues, and is therefore a bioactive ingredient of many foods. Among the rich sources of ferulic acid are many basic foods such as grain bran, whole grain foods, citrus fruits, bananas, beer, orange juice, eggplants, bamboo shoots, beetroot, cabbage, spinach, and broccoli (Clifford, 1999; Rechner et al., 2001; Sakakibara et al., 2003; Mattila et al., 2005; Mattila et al., 2006; Mattila and Hellstrom, 2007; Zhao and Moghadasian, 2008; Mancuso and Santangelo, 2014). In Western countries, extracts of spices and coffee are other major sources of ferulic acid, depending on the dietary intakes (Virgili et al., 2000; Zhao and Moghadasian, 2008). Ferulic acid, which is a caffeic acid derivative, is a component of Chinese medicinal herbs such as *Angelica sinensis*, *Cimicifuga racemosa*, and *Ligusticum chuangxiong* (Rechner et al., 2001; Ou and Kwok, 2004; Mancuso and Santangelo, 2014).

Interest in ferulic acid and other caffeic acid derivatives began the mid-1950s. These phenolic acids attracted attention because of their potential role in an adjuvant therapy for several free radical–induced diseases. As an antioxidant it offered a strong cytoprotective activity due to its ability both to scavenge free radicals and to activate the cell stress response. It inhibits the expression and/or activity of cytotoxic enzymes, including inducible nitric oxide synthase, caspases, and cyclooxygenase-2. On the basis of these activities, ferulic acid shows potential in the treatment of many disorders, including Alzheimer disease, cancer, cardiovascular diseases, diabetes mellitus, and skin diseases (Mancuso and Santangelo, 2014). Additionally, it is beneficial in prevention and/or treatment of disorders such as hypertension, atherosclerosis, and inflammatory diseases (Ohta et al., 1997; Fernandez et al., 1998; Dinis et al., 2002; Suzuki et al., 2002; Wang et al., 2004; Wang and Ou-Yang, 2005; Suzuki et al., 2007; Zhao and Moghadasian, 2008).

20.2.4 *p*-Coumaric Acid

p-Coumaric acid is a beige crystalline substance (molecular mass 164.16 g mol^{-1}) with melting point 214°C; its formula is $C_9H_8O_3$ (Sigma-Aldrich, 2014c).

p-Coumaric acid (3-[4-hydroxyphenyl]-2-propenoic acid, 4-hydroxycinnamic acid) is a common plant phenolic acid that is typically esterified to arabinoxylan residues of hemicellulose or to lignin in graminaceous plants. Maize, oats, and wheat contain it. It is also available in the form of an ester conjugate and as free acid in fruits and vegetables such as apples, grapefruits, oranges, tomatoes, potatoes, and spinach (Xing and White, 1997; Pan et al., 1998; Clifford, 1999; Konishi et al., 2003). It has been determined in wine (Luceri et al., 2007).

p-Coumaric acid is an intermediate product of the phenylpropanoid pathway of transformation from cinnamic acid in plants (Castelluccio et al., 1996; Zang et al., 2000).

It has been reported that *in vitro p*-coumaric acid can provide antioxidant protection to low-density lipoprotein as a consequence of the chain-breaking activity of *p*-coumaric acid (Castelluccio et al., 1996; Zang et al., 2000). Dietary supplementation with a crude extract of *p*-coumaric acid isolated from pulses showed that *p*-coumaric acid has some actions in reducing ester cholesterol and providing a protective mechanism against the development of atherosclerosis (Zang et al., 2000).

The compound has also been observed to have antioxidant and anti-inflammatory properties in rat colonic mucosa (Guglielmi et al., 2003; Luceri et al., 2007).

20.3 APPLICATION OF FIA IN PHENOLIC COMPOUNDS ANALYSIS

The use of glassy carbon electrodes (GCEs) modified with multiwalled carbon nanotube (CNT) films for the continuous monitoring of wine polyphenols in flow systems was evaluated by Sánchez Arribas et al. (2013). The performance of these modified electrodes was compared to that of bare GCEs by cyclic voltammetry experiments and by FIA with amperometric detection monitoring the response of gallic, caffeic, ferulic, and *p*-coumaric acids in 0.050 M acetate buffer pH 4.5 containing 100 mM NaCl. The GCE modified with multiwalled CNT dispersions in polyethyleneimine (PEI) exhibited lower overpotentials, higher sensitivity, and better signal stability under a dynamic regime than did bare GCEs. These properties allowed the estimation of the total polyphenol content in four kinds of red and five kinds of white wines with a remarkable long-term stability in the measurements despite the presence of potential fouling substances in the wine matrix. The versatility of the electrochemical methodology also allowed the selective estimation of the easily oxidizable polyphenol fraction as well as the total polyphenol content just by tuning the detection potential at +0.30 or +0.70 V, respectively. The significance of the electrochemical results was shown through correlation studies with the results obtained by conventional spectrophotometric assays for polyphenols with the Folin–Ciocalteu method at 280 nm and color intensity index. The excellent electroanalytical properties of these GCE/(CNT/PEI) suggested that useful methodologies for polyphenol mapping and the characterization of wine types could be developed in combination with high-performance liquid chromatography or capillary electrophoresis (Sánchez Arribas et al., 2013).

Flow injection methodology was used to estimate the total phenolic content of wine using acidic potassium permanganate chemiluminescence detection by Costin et al. (2003). Simple phenolic compounds such as quercetin, rutin, catechin, epicatechin, ferulic acid, caffeic acid, gallic acid, 4-hydroxycinnamic acid, and vanillin were examined analytically with chemiluminescence. Analysis of 12 different wines showed that detection limits were 2×10^{-8} M for caffeic acid, 3×10^{-8} M for ferulic acid, and 5×10^{-8} M for gallic acid. Comparison of the results of the chemiluminescence methodology and other total phenol/antioxidant assays showed that their correlation was good. Chemiluminescence detection in FIA was found to be selective with minimal interferences caused by the non-phenolic components in wine. The chemiluminescence method was a rapid and sensitive technique for evaluating the antioxidant or total phenolic content (Costin et al., 2003).

A variety of wine, tea, and fruit juice samples were examined by Nalewajko-Sieliwoniuk et al. (2010) for determination of the total phenolic content by a flow injection methodology with soluble manganese chemiluminescence detection. It was found when polyphenols were mixed with acidic soluble manganese in the presence of formaldehyde, chemiluminescence was significantly enhanced. Based on this evidence, a new flow injection chemiluminescence detection method was developed to estimate the total content of phenolic compounds (expressed as milligrams of gallic acid equivalent [GAE] per liter of drink) in a variety of wine, tea, and fruit juice samples. The developed method was sensitive to a detection limit of 0.02 ng mL^{-1} (gallic acid), offering a wide linear dynamic range (0.5–400 ng mL^{-1}) and high sample throughput (247 samples per hour). The relative standard deviations (RSD) of 15 measurements were 3.8% for 2 ng mL^{-1} and 0.45% for 10 ng mL^{-1} of GAE. In an analysis of 36 different samples, the results obtained by the proposed flow injection chemiluminescence detection method had a high correlation with those obtained by spectrophotometric methods commonly used for the evaluation of the total phenolic/antioxidant level. When these methods were compared with each other, the flow injection chemiluminescence detection method was found to be far simpler, more rapid, and selective, showing almost no interference from nonphenolic components of the samples examined (Nalewajko-Sieliwoniuk et al., 2010). Nalewajko-Sieliwoniuk et al. (2012) also carried out the determination of the total flavonoid compounds in inflorescence (expressed as mg L^{-1} of apigenin) and leaf (expressed as mg L^{-1} of linarin) extracts of *Cirsium oleraceum* and *Cirsium rivulare* species by a flow injection system with chemiluminescence detection. The strong enhancement in both plants by polyphenols of the chemiluminescence signal generated by the reaction of cerium(IV) with rhodamine 6G in a sulfuric acid medium was the basis of the method. For apigenin and linarin, the linear working ranges of 0.1–10 and 2.5–50 µmol L^{-1}, respectively, were obtained under optimized conditions. The developed method had detection limits of 38 nmol L^{-1} (apigenin) and 840 nmol L^{-1} (linarin), offering a high sample throughput (up to 300 samples per hour). In 10 measurements, the obtained RSD was 0.62% and 3.75%, for 5 µmol L^{-1} apigenin and linarin, respectively. The method was used to examine the flavonoid/antioxidant levels in aqueous and methanolic extracts from inflorescences and leaves of *Ci. oleraceum* and *Ci. rivulare*. For comparison, the antioxidant activities of *Ci. oleraceum* and *Ci. rivulare* extracts were also evaluated by a spectrophotometric 1,1-diphenyl-2-picryl-hydrazyl (DPPH) radical scavenging method (Nalewajko-Sieliwoniuk et al., 2012).

For the determination of gallic acid in olive fruits, an economical and environment friendly flow injection chemiluminescent method was developed by Wang et al. (2007). The inhibition of chemiluminescent emission of an alkaline luminol–KMnO$_4$ system by gallic acid was the basis of the method. The logarithm of the difference between the chemiluminescent intensity of the alkaline luminol–KMnO$_4$ system in the absence and in the presence of gallic acid was linear with the logarithm of the concentration of gallic acid in the range from 1.0×10^{-9} to 5.0×10^{-5} g mL^{-1}, with a detection limit of 2.2×10^{-10} g mL^{-1}. The RSD of 11 determinations of 1.0×10^{-6} g mL^{-1} gallic acid was 1.7% (Wang et al., 2007).

The method developed by Wang et al. (2008) was applied to the determination of ferulic acid in Taita Beauty Essence samples with satisfactory results. For the determination of ferulic acid a flow injection chemiluminescence method was developed based on the chemiluminescence reaction of ferulic acid with rhodamine 6G and ceric sulfate in sulfuric acid medium. When ferulic acid was injected into the acidic ceric sulfate solution in a flow cell, strong signal was observed. With this method the determination of ferulic acid was in the concentration range 8.0×10^{-6} to 1.0×10^{-4} mol L^{-1} and the detection limit for ferulic acid was 8.7×10^{-9} mol L^{-1}. The RSD was calculated as 2.4% for 10 replicate analyses of 1.0×10^{-5} mol L^{-1} ferulic acid (Wang et al., 2008).

For the direct determination of pyrogallol compounds in tea and coffee samples, Kanwal et al. (2009) developed a highly sensitive flow injection chemiluminescent method. The method included the enhanced effect of pyrogallol compounds on the chemiluminescence signals of a $KMnO_4$–H_2O_2 system in slightly alkaline medium. Three important pyrogallol compounds—pyrogallic acid, gallic acid, and tannic acid—were detected by this method, as the possible mechanism of the chemiluminescence reaction was also discussed. The properties of method were simple, convenient, rapid (60 samples per hour), and sensitive, having a linear range of 8×10^{-10} to 1×10^{-5} mol L^{-1}, for pyrogallic acid, with a detection limit of 6×10^{-11}; 4×10^{-8} to 5×10^{-3} mol L^{-1} for gallic acid with a detection limit of 9×10^{-10}; and 8×10^{-8} to 5×10^{-2} mol L^{-1} for tannic acid, with a detection limit of 2×10^{-9} mol L^{-1}. For 5×10^{-6} mol L^{-1} pyrogallic acid, gallic acid, and tannic acid, the RSD ($n = 15$) was 0.8%, 1.1%, and 1.3%, respectively (Kanwal et al., 2009).

A biamperometric method for the direct determination of pyrogallol compounds was based on the electrocatalytic oxidation of pyrogallol compounds at one pretreated platinum electrode and the reduction of platinum oxide at the other pretreated platinum electrode to form a biamperometric detection system with an applied potential difference of 10 mV. Three important compounds detected by the method were pyrogallol, gallic acid, and tannic acid. The linear relationships between currents and the concentrations of pyrogallol, gallic acid, and tannic acid were determined over the ranges 1.0×10^{-6}–1.0×10^{-4}, 1.0×10^{-6}–1.0×10^{-4} and 1.0×10^{-6}–2.0×10^{-4} mol L^{-1} with detection limits of 6.0×10^{-7}, 6.0×10^{-7}, and 8.0×10^{-7} mol L^{-1} ($S/N = 2$), respectively. For 30 successive determinations of 5.0×10^{-5} mol L^{-1} pyrogallol, gallic acid, and tannic acid, the RSDs observed were 1.9%, 2.5%, and 2.0%, respectively. Most ions and organic compounds tested were observed not to cause significant interference in the determinations. The method had been validated by the determination of pyrogallol compounds in tea and Chinese gall and was found to be simple, selective, and efficient (180 h^{-1}), performing well as a routine assay (Zhao et al., 2003).

Spectrographic graphite electrodes were modified by Haghighi et al. (2003) for the adsorption of laccase from *Trametes versicolor*. The laccase-modified graphite electrode was used as the working electrode in an amperometric flow-through cell for monitoring phenolic compounds in a single-line flow injection system. The experimental conditions were studied and optimized for bioelectrochemical determination of catechol. For the biosensor operation, the optimal conditions were 0.1 M citrate buffer solution (at pH 5.0), flow rate of 0.51 mL min^{-1}, and a working potential of −50 mV versus Ag/AgCl. The responses of the biosensor for various phenolic compounds were recorded and the sensor characteristics were calculated for these conditions and compared with those that are known for biosensors based on laccase from *Coriolus hirsutus*. For ferulic acid, the relative sensitivities were 386% for biosensors based on laccase from *T. versicolor* and 202% for biosensors based on laccase from *Co. hirsutus* (Haghighi et al., 2003).

Some of the most important chemicals naturally occurring in grapes and wines are phenolic compounds. The major quality properties of wines, such as flavor, color stability, and antioxidant capacity are strongly related to phenolic compounds. The study of Amárita et al. (2010) presented an alternative method for the rapid analysis of total phenolics in wines and juices based on an enzymatic FIA system. The design of the biosensor used the measurement of the oxygen consumed in the reaction catalyzed by the enzyme benzenediol:oxygen oxidoreductase (polyphenol oxidase, laccase) as a base. Laccase was immobilized in Eupergit C and was packed into a chromatographic minicartridge. The enzymatic reactor was then employed in a FIA system, comprising a peristaltic pump, flowing a carrier buffer (PBS, pH = 7.4), an injection valve, an oxygen microelectrode, and a signal transductor (oximeter). For the FIA system, the carrier flow rate and the possible interferences of different compounds naturally occurring in wine (ethanol, glucose, citric acid, malic acid, hydroxybenzoic acids) were optimized. Samples of red and white wines and fruit juices were analyzed by this system and the results were validated with standard methods of total phenolic content (Folin–Ciocalteu) and antioxidant capacity (Trolox equivalent antioxidant capacity [TEAC], DPPH). Gallic acid was used for calibrating the biosensor. The FIA system had a linear response from 50 to 500 ppm in 2 min and total phenolic compound concentrations could found in wines. There was a statistically significant correlation between the biosensor response and the standard methods for the analysis of total phenolics and antioxidant capacity ($R^2 > 0.84$). The FIA system was confirmed as a simple, inexpensive, and rapid technique for the analysis of total phenolics in wines and fruit juices and the device was designed for portability to facilitate its use in wineries (Amárita et al., 2010).

A new method coupling flow injection with capillary electrophoresis has been evaluated by Arce et al. (1998b) using diode array detection for measuring the concentration of *trans*-resveratrol, (–)-epicatechin, (+)-catechin, gentisic acid, salicylic acid, myricetin, quercetin, *p*-coumaric acid, caffeic acid, and gallic acid in wines. A flow injection system furnished with a C_{18} minicolumn was used to clean up the wines by solid phase extraction before capillary electrophoresis. The analytes were eluted from C_{18} with methanol and then given from the flow injection system to the autosampler of the capillary electrophoresis equipment by a programmable arm. The range of 3σ detection limit was from 0.05 mg L^{-1} (*trans*-resveratrol) to 0.36 mg L^{-1} ((–)epicatechin). The limit of detection values of *p*-coumaric acid, caffeic acid and gallic acid were 0.14, 0.34 and 0.30 mg L^{-1}, respectively. The method developed was shown to determine polyphenols at low levels, so that the detection limits for these polyphenols in wines were improved. The recoveries of added *trans*-resveratrol and other polyphenols from synthetic wines were from 92% to 110% (mean 99%). In 12 real wine samples, the recoveries had means of 99.92% for *p*-coumaric acid and 98.75% for caffeic acid. Comparing liquid/liquid extraction and liquid chromatography systems, the method was seen to be much faster and simpler (Arce et al., 1998b).

A novel flow injection procedure was developed for the determination of gallic acid in Jianmin Yanhou tablets based on the enhancement effect of gallic acid on the chemiluminescence of a luminol–$K_3Fe(CN)_6$ system. The known advantages of method such as higher sensitivity, higher selectivity, wider linear range, and simpler instrumentation were achieved. It is practicable for the determination of gallic acid in the range from 8.0×10^{-9} to 1.0×10^{-6} mol L^{-1} with a detection limit of 5.6×10^{-9} mol L^{-1}. In 11 samples the RSD was 2.2% for the determination of 1.0×10^{-7} mol L^{-1} gallic acid (Chenggen et al., 2002).

A capillary electrophoresis method was developed by Arce et al. (1998a) for the simultaneous determination of a number of major ingredients of green tea. The components found in green tea were caffeine, adenine, theophylline, epigallocatechin-3 gallate, epigallocatechin, epicatechin-3 gallate, (–)-epicatechin, (+)-catechin, gallic acid, quercetin and caffeic acid. Separation was performed using a fused capillary column with 0.15 M H_3BO_3 as buffer at a pH of 8.5, UV detection at 210 nm, and 20 kV of voltage. Samples were analyzed with a flow injection system which was coupled to capillary electrophoresis equipment via a programmable arm after extraction, filtration, and dilution treatments. These compounds were determined in less than 20 min. The standard addition method was also carried out. Limits of detection were 0.08 μg mL^{-1} for gallic acid and caffeic acid. The gallic acid content of eight green tea samples analyzed was reported as 32.37 μg mL^{-1} and caffeic acid was not found in them (Arce et al., 1998a).

A flow injection–electrochemiluminescent method was developed for the determination of gallic acid in a Chinese proprietary medicine—Jianming Yanhou Pian—by Xiang-Qin et al. (2004). The method was based on an inhibition effect on the $Ru(bpy)_3^{2+}$/tri-*n*-propylamine (TPrA) electrochemiluminescent system in pH 8.0 phosphate buffer solution. The determination limit and dynamic concentration range of the method were 9.0×10^{-9} and of 2×10^{-8}–2×10^{-5} mol L^{-1}, respectively. The RSD was 1.0% for 1.0×10^{-6} mol L^{-1} gallic acid ($n = 11$). The inhibition mechanism proposed for the quenching effect of the gallic acid on the $Ru(bpy)_3^{2+}$/TPrA electrochemiluminescent system was the interaction of electro-generated $Ru(bpy)_3^{2+*}$ and an *o*-benzoquinone derivative at the electrode surface. The electrochemiluminescent emission spectra and UV–Vis absorption spectra confirmed the mechanism (Xiang-Qin et al., 2004).

A FIA system for the determination of total phenolic compounds in tea was established by Leamsomrong et al. (2009). The detection method was based on the Folin–Ciocalteu reaction, which involves the reduction of a mixture of phosphotungstic and phosphomolybdic acids (Folin–Ciocalteu reagent) to tungsten and molybdenum oxides by phenolic compounds in basic medium and subsequent formation of a blue colored product. The standard or sample solutions were injected into a carrier stream (distilled water) to react with the Folin–Ciocalteu reagent and sodium carbonate. The blue colored product obtained was detected by spectrophotometry at 765 nm. Sample volume, flow rates of carrier and reagents, length of reaction coils, and concentration of reagents were optimized. In 20 replicates, the relative standard derivation of 5 ppm gallic acid was 0.72% and the detection limit was 0.0231 mg L^{-1}. The range of the linear calibration curve was from 0.5 to 100.0 mg L^{-1} and its regression equation was $y = 0.0123x + 0.021$, $R^2 = 0.9991$. The throughput of sampling was 32 samples per hour. The effects of potential interferences such as citric acid, fructose, and others were considered (Leamsomrong et al., 2009).

Two types of tea samples such as tea plants (four plants; two of *Camellia sinensis* var. *sinensis* and two of *Ca. sinensis* var. *assamica*) were divided into five parts: buds, leaves, branches, stems, roots and these tea leaf products (six samples; loose leaves, roll leaves, and bags) were then extracted by Suteerapataranon and Pudta (2008). A simple FIA–spectrophotometric system for the rapid determination of total polyphenols was developed for these samples. The method included the oxidation/reduction of polyphenolic compounds by the Folin–Ciocalteu reagent in alkaline medium, producing a blue molybdenum–tungsten complex. The linear range of the calibration graph was determined as 10–500 mg GAE L^{-1}. For five samples the RSD of the calibration graph was 2.93%. The precisions (%RSD, $n = 10$) of the determination were 2.74% for the 10 mg GAE L^{-1} and 3.02% for the 500 mg GAE L^{-1}. The detection limit was 7 mg L^{-1} and the sample throughput was 30 per hour. No significant difference was found comparing total phenolic contents determined by the developed system and the standard batch method (*t* test, at 95% confidence level). It was observed that the method overcame the drawbacks of the batch standard method by providing higher sample throughput and less reagent consumption with high precision over a wide concentration range, good sensitivity, and good accuracy (Suteerapataranon and Pudta, 2008).

The capabilities of different detection techniques such as UV, controlled-potential coulometry, and particle beam electron-impact mass spectrometry (PB-EI-MS) for the high-performance liquid chromatography (HPLC) analysis of 15 benzoic and cinnamic acid derivatives in alcoholic beverage samples were studied by Bocchi et al. (1996). A reversed-phase liquid chromatography (RP-LC) method was set up for the electro-chemical detector (ED), whereas normal-phase partition chromatography, on a cyano-bonded (CN) column, was used for UV and MS. Library-searchable electron impact (EI) mass spectra were found using the PB-MS technique with flow injection analysis. UV detection was provided at 280 nm, while measurements with the high-performance liquid chromatography system were carried out using a porous graphite electrode. The detector responses were compared in terms of linearity, precision, and limits of detection. To that end, the mass spectrometer was operated under selected-ion monitoring conditions. A linear dynamic range of at least 103 was determined for the HPLC method with electrochemical detection, with detection limits ranging from 1 to 5 pg injected; the RSD was 0.6%–3.0% at the 0.1 ng level for four samples. Using UV or PB-EI-MS detection, minimum amounts could be detected at 5–50 and 2–5 ng, respectively. The limit of detection obtained was at least 15 µg for most of the analytes by UV; the RSD of the peak areas detected in UV mode ranged from 1.2% to 3.1% at the 500 ng level ($n = 4$). Nonlinear behavior over the entire amount range studied (from 10 ng to 10 µg) was found using the LC-PB-MS technique, so that two different calibration graphs at low and high levels were calculated. Precision of the LC-PB-MS system was generally good (RSD between 0.5% and 1.8% at the 100 ng level, $n = 4$) other than for caffeic acid (RSD 7.5% at the 50 µg level, $n = 4$). In FIA mode, the linearity ranges of gallic acid, p-coumaric acid, ferulic acid, and caffeic acid were 5–250, 1–1000, 1–1000, and 5–1000 ng and their limits of detection were 0.3, 0.1, 0.1, and 0.7 ng, respectively, in the PB-EI-MS analysis. By combining selectivity and high sensitivity, electrochemical detection with FIA provided a method to simultaneously determine various phenolic acids in samples. LC-UV failed in the analysis of phenolic constituents in the examined sample because of the presence of numerous interferences and inadequate sensitivity (Bocchi et al., 1996).

A selective assay for the determination of one of the most important classes of phe-nolic compounds, namely trihydroxybenzoates (monomeric and polymeric compounds having at least one gallate moiety), based on their enhancing effect on the chemilumi-nogenic reaction between gold ions and luminol, was characterized for the first time by Giokas et al. (2013). In the presence of trihydroxybenzoate derivatives, the light emission formed when alkaline luminol was oxidized by gold ions was amplified several orders of magnitude compared to other common phenolic compounds that show minor reactivity or none at all (e.g., hydroxycinnamates, flavonols, benzenediols). Based on this, the experimental conditions were optimized to enable the determination of total trihydroxybenzoates in complex mixtures without resorting to separation techniques. According to this method, samples of teas of different composition (mainly red, black, and white teas and mixtures of green teas with the former) and herbal infusions (fruit teas, herbal teas, chamomile, and mixtures of them) as well as red and white wines were analyzed. Detection limits at the 10^{-7} mol L^{-1} level, intraday precision of 3.1%, interday precision less than 10%, and recoveries between 88.7% and 97.6% were determined. Also the total phenolic content ranges of teas, herbal infusions, and red wines and white wines were calculated as 40.3–70.4, 5.6–36.1, and 166.7–1412.2 mg g^{-1} respectively. The gallic acid content range of teas, herbal infusions, and red wines and white wines were 1.8–6.0, 1.3–2.3, and 6.5–94.0 mg g^{-1}, respectively. The strengths and weaknesses of the method were defined and discussed in relation to application in real samples (Giokas et al., 2013).

20.4 CONCLUSION

In phenolic compound analysis FIA is the preferred method rather than conventional chromatographic and spectrometric methods because it offers lower detection limit values, greater sensitivity, short time requirement, lower sample requirement, an easy technique, user-friendliness, and high linearity, recovery, and precision. This system represents a new contribution in the field of automation of analytical methods.

REFERENCES

Amárita, F., J. Ferrer, D. Pardal, and I. Tueros. 2010. Rapid analysis of total phenolics in wines and juices by an enzymatic FIA device. *J. Biotechnol. (Special Abstracts)* 150:1–576.

Andrade, R., C. O. Viana, S. G. Guadagnin, F. G. R. Reyes, and S. Rath. 2003. A flow-injection spectrophotometric method for nitrate and nitrite determination through nitric oxide generation. *Food Chem.* 80:597–602.

Arce, L., A. Ríos, and M. Valcárcel. 1998a. Determination of anti-carcinogenic polyphenols present in green tea using capillary electrophoresis coupled to a flow injection system. *J. Chromatogr. A* 827:113–120.

Arce, L., M. T. Tena, A. Rios, and M. Valcárcel. 1998b. Determination of trans-resveratrol and other polyphenols in wines by a continuous flow sample clean-up system followed by capillary electrophoresis separation. *Anal. Chim. Acta* 359:27–38.

Balasundram, N., K. Sundram, and S. Samman. 2006. Phenolic compounds in plants and agriindustrial by-products: Antioxidant activity, occurrence, and potential uses. *Food Chem.* 99:191–203.

Bocchi, C., M. Careri, F. Groppi, A. Mangia, E. Manini, and G. Moil. 1996. Comparative investigation of UV, electrochemical and particle beam mass spectrometric detection for the high-performance liquid chromatographic determination of benzoic and cinnamic acids and of their corresponding phenolic acids. *J. Chromatogr. A* 753:157–170.

Bonfili, L., V. Cecarini, and M. Amici. 2008. Natural polyphenols as proteasome modulators and their role as anti-cancer compounds. *REBS J.* 275:5512–5526.

Bosque-Sendra, J. M., L. Gámiz-Gracia, and A. M. García-Campana. 2003. An overview for qualimetrics strategies for optimization and calibration in pharmaceutical analysis using flow injection techniques. *Anal. Bioanal. Chem.* 377:863–874.

Bravo, L. 1998. Polyphenols: Chemistry, dietary sources, metabolism, and nutritional significance. *Nutr. Rev.* 56(11):317–333.

Castelluccio, C, G. P. Bolwell, C. Gerrish, and C. Rice-Evans. 1996. Differential distribution of ferulic acid to the major plasma constituents in relation to its potential as an antioxidant. *Biochem. J.* 316:691–694.

Chenggen, X., C. Hua, and L. Xiangqin. 2002. Flow injection analysis of gallic acid in jianmin yanhou tablets with enhanced chemiluminescent detection. *Chinese J. Anal. Chem.* 30:1316–1318.

Clifford, M. N. 1999. Chlorogenic acids and other cinnamates—Nature, occurrence and dietary burden. *J. Sci. Food Agr.* 79:362–372.

Conforti, F., F. Menichini, C. Formisano et al. 2009. Comparative chemical composition, free radical-scavenging and cytotoxic properties of essential oils of six *Stachys* species from different regions of the Mediterranean Area. *Food Chem.* 116:898–905.

Costin, J. W., N. W. Barnett, S. W. Lewis, and D. J. McGillivery. 2003. Monitoring the total phenolic/antioxidant levels in wine using flow injection analysis with acidic potassium permanganate chemiluminescence detection. *Anal. Chim. Acta* 499:47–56.

Dinis, T. C., C. L. Santosa, and L. M. Almeida. 2002. The apoprotein is the preferential target for peroxynitrite-induced LDL damage protection by dietary phenolic acids. *Free Radical Res.* 36:531–543.

Erlund, I. 2004. Review of the flavonoids quercetin, hesperetin, and naringenin: Dietary sources, bioactivities, bioavailability, and epidemiology. *Nutr. Res.* 24:851–874.

Eyles, A., N. W. Davies, T. Mitsunaga, R. Mihara, and C. Mohammed. 2004. Role of *Eucalyptus globulus* wound wood extractives: Evidence of superoxide dismutase-like activity. *Forest Pathol.* 34:225–232.

Felix, F. S. and L. Angnes. 2010. Fast and accurate analysis of drugs using amperometry associated with flow injection analysis. *J. Pharm. Sci.* 99(12):4784–4804.

Fernandez, M. A., M. T. Saenz, and M. D. Garcia. 1998. Anti-inflammatory activity in rats and mice of phenolic acids isolated from *Scrophularia frutescens*. *J. Pharm. Pharmacol.* 50:1183–1186.

Ferreira, I. M. P. L. V. O., J. L. F. C. Lima, M. C. B. S. M. Montenegro, R. P. Olmos, and A. Rios. 1996. Simultaneous assay of nitrite, nitrate and chloride in meat products by flow injection. *Analyst* 121:1393–1396.

Frenzel, W., J. Schulz-Brussel, and B. Zinvirt. 2004. Characterisation of a gas-diffusion membrane-based optical flow-through sensor exemplified by the determination of nitrite. *Talanta* 64:278–282.

German, J. B. and R. L. Walzem. 2000. The health benefits of wine. *Annu. Rev. Nutr.* 20:561–593.

Giokas, D. L., D. C. Christodouleas, I. Vlachou, A. G. Vlessidis, and A. C. Calokerinos. 2013. Development of a generic assay for the determination of total trihydroxy-benzoate derivatives based on gold-luminol chemiluminescence. *Anal. Chim. Acta* 764:70–77.

Guglielmi, F., C. Luceri, L. Giovannelli, P. Dolara, and M. Lodovici. 2003. Effect of 4-coumaric and 3,4-dihydroxybenzoic acid on oxidative DNA damage in rat colonic mucosa. *Br. J. Nutr.* 89:581–587.

Haghighi, B., L. Gorton, T. Ruzgas, and L. J. Jönsson. 2003. Characterization of graphite electrodes modified with laccase from *Trametes versicolor* and their use for bioelectrochemical monitoring of phenolic compounds in flow injection analysis. *Anal. Chim. Acta* 487:3–14.

Halliwell, B. 2009. The wanderings of a free radical. *Free Radical Biol. Med.* 46:531–542.

Hancianu, M. and A. C. Aprotosoaie. 2012. The effects of pesticides on plant secondary metabolites. In *Biotechnological Production of Plant Secondary Metabolites*, ed. I. E. Orhan, pp. 176–186. Bentham Science.

Havsteen, B. H. 2002. The biochemistry and medical significance of the flavonoids. *Pharmacol. Ther.* 96:67–202.

Hollman, P. C. H. 2001. Evidence for health benefits of plant phenols: Local or systemic effects? *J. Sci. Food Agr.* 81(9):842–852.

Hölker, U., M. Höfer, and J. Lenz. 2004. Biotechnological advances of laboratory-scale solid-state fermentation with fungi. *Appl. Microbiol. Biotechnol.* 64:175–186.

Horstkotte, B., C. M. Duarte, and V. Cerda. 2010. Response functions for SIMPLEX optimization of flow-injection analysis and related techniques. *Trends Anal. Chem.* 29(10):1224–1235.

Jiménez, J. P., J. Serrano, M. Tabernero et al. 2008. Effects of grape antioxidant dietary fiber in cardiovascular disease risk factors. *Nutrition* 24:646–653.

Kanwal, S., X. Fu, and X. Su. 2009. Highly sensitive flow-injection chemiluminescence determination of pyrogallol compounds. *Spectrochim. Acta Part A* 74:1046–1049.

Karlberg, B. and G. E. Pacey. 1989. *Flow Injection Analysis: A Practical Guide.* Amsterdam: Elsevier Science Technology Press.

Kazemzadeh, A. and A. A. Ensafi. 2001. Sequential flow injection spectrophotometric determination of nitrite and nitrate in various samples. *Anal. Chim. Acta* 442:319–326.

Kim, D. O., K. W. Lee, H. J. Lee, and C. Y. Lee. 2002. Vitamin C equivalent antioxidant capacity of phenolic phytochemicals. *J. Agr. Food Chem.* 50:3713–3717.

Kim, G. N., J. G. Shin, and H. D. Jang. 2009. Antioxidant and antidiabetic activity of Dangyuja (*Citrus grandis Osbeck*) extract treated with *Aspergillus saitoi. Food Chem.* 117:35–41.

Konishi, Y., S. Kobayashi, and M. Shimizu. 2003. Transepithelial transport of *p*-coumaric acid and gallic acid in coca-2 cell monolayers. *Biosci. Biotechnol. Biochem.* 67(11):2317–2324.

Kris-Etherton, P. M., K. D. Hecker, A. Bonanome et al. 2002. Bioactive compounds in foods: Their role in the prevention of cardiovascular disease and cancer. *Am. J. Med.* 113:71–88.

Kroon, P. A. and G. Williamson. 1999. Hydroxycinnamates in plants and food: Current and future perspectives. *J. Sci. Food Agr.* 79:355–361.

Leamsomrong, K., M. Suttajit, and P. Chantiratikul. 2009. Flow injection analysis system for the determination of total phenolic compounds by using Folin-Ciocalteu assay. *Asian J. Appl. Sci.* 2(2):184–190.

Llorach, R., J. C. Espín, F. A. Tomás-Barberán, and F. Ferreres. 2002. Artichoke (*Cynara scolymus* L.) by products as a potential source of health-promoting antioxidant phenolics. *J. Agr. Food Chem.* 50(12):3458–3464.

Locatelli, C., F. B. Filippin-Monteiro, and T. B. Creczynski-Pasa. 2013. Alkyl esters of gallic acid as anticancer agents: A review. *Eur. J. Med. Chem.* 60:233–239.

Luceri, C., L. Giannini, M. Lodovici. et al. 2007. *p*-Coumaric acid, a common dietary phenol, inhibits platelet activity *in vitro* and *in vivo. Br. J. Nutr.* 97:458–463.

Ma, J., X. D. Luo, and P. Protiva. 2003. Bioactive novel polyphenols from the fruit of *Manilkara zapota* (Sapodilla). *J. Nat. Prod.* 7:983–986.

Mankuso, C. and R. Santangelo. 2014. Ferulic acid: Pharmacological and toxicological aspects. *Food Chem. Toxicol.* 65:185–195.

Martins, S., S. I. Mussatto, G. Martínez-Avila, J. Montañez-Saenz, C. N. Aguilar, and J. A. Teixeira. 2011. Bioactive phenolic compounds: Production and extraction by solid-state fermentation. A review. *Biotechnol. Adv.* 29:365–373.

Mattila, P. and J. Hellstrom. 2007. Phenolic acids in potatoes, vegetables, and some of their products. *J. Food Comp. Anal.* 20:152–160.

Mattila, P., J. Hellstrom, and R. Torronen. 2006. Phenolic acids in berries, fruits, and beverages. *J. Agr. Food Chem.* 54:7193–7199.

Mattila, P., J. M. Pihlava, and J. Hellstrom. 2005. Contents of phenolic acids, alkyl- and alkenylresorcinols, and avenanthramides in commercial grain products. *J. Agr. Food Chem.* 53:8290–8295.

Mazza, G. 1995. Anthocyanins in grapes and grape products. *Crit. Rev. Food Sci. Nutr.* 35:341–371.

McKelvie, I. D. 2008. Principles of flow injection analysis. In *Advances in Flow Injection Analysis and Related Techniques*, eds. S. D. Kolev, and I. D. McKelvie, pp. 81–109. Amsterdam: Elsevier Science Technology Press.

Medina, I., I. Undeland, and K. Larsson. 2012. Activity of caffeic acid in different fish lipid matrices: A review. *Food Chem.* 131:730–740.

Middleton, E. J., C. Kandaswami, and T. C. Theoharides. 2000. The effects of plant flavonoids on mammalian cells: Implications for inflammation, heart disease and cancer. *Pharmacol. Rev.* 52:673–751.

Moure, A., J. M. Cruz, D. Franco, J. M. Domínguez, J. Sineiro, and H. Dominguez. 2001. Natural antioxidants from residual sources. *Food Chem.* 72:145–171.

Murugananthan, M., G. Bhaskar Raju, and S. Prabhakar. 2005. Removal of tannins and polyhydroxy phenols by electro-chemical techniques. *J. Chem. Technol. Biotechnol.* 80:1188–1197.

Nalewajko-Sieliwoniuk, E., J. Nazaruk, J. Kotowska, and A. Kojło. 2012. Determination of the flavonoids/antioxidant levels in *Cirsium oleraceum* and *Cirsium rivulare* extracts with cerium(IV)–rhodamine 6G chemiluminescence detection. *Talanta* 96:216–222.

Nalewajko-Sieliwoniuk, E., I. Tarasewicz, and A. Kojło. 2010. Flow injection chemiluminescence determination of the total phenolics levels in plant-derived beverages using soluble manganese(IV). *Anal. Chim. Acta* 668:19–25.

Nigam, P. S. 2009. Production of bioactive secondary metabolites. In *Biotechnology for Agro-Industrial Residues Utilization*, eds. P. S. Nigam, and A. Pandey, pp. 129–145. The Netherlands: Springer.

Ohta, T., N. Semboku, A. Kuchii, Y. Egashira, and H. Sanada. 1997. Antioxidant activity of corn bran cell-wall fragments in the LDL oxidation system. *J. Agr. Food Chem.* 45:1644–1648.

Ou, S. and K. C. Kwok. 2004. Ferulic acid: Pharmaceutical functions, preparation and applications in food. *J. Sci. Food Agr.* 84:1261–1269.

Pan, G. X., J. L. Bolton, and G. J. Learly. 1998. Determination of ferulic and p-coumaric acids in wheat straw and the amounts released by mild acid and alkaline proxide treatment. *J. Agr. Food Chem.* 46:5283–5288.

Parvathy, K. S., P. S. Negi, and P. Srinivas. 2009. Antioxidant, antimutagenic and antibacterial activities of curcumin-β-diglusoside. *Food Chem.* 115:265–271.

Penteado, J. C., L. Angnes, J. C. Masini, and P. C. C. Oliveira, P. C. C. 2005. FIA-spectrophotometric method for determination of nitrite in meat products: An experiment exploring color reduction of an azo-compound. *J. Chem. Edu.* 82:1074–1078.

Pietta, P. G. 2000. Flavonoids as antioxidants. *J. Nat. Prod.* 63:1035–1042.

Pinho, O., I. M. P. L. V. O. Ferreira, M. B. P. P. Oliveira, and M. A. Ferreira. 1998. FIA evaluation of nitrite and nitrate contents of liver pates. *Food Chem.* 62:359–362.

Polewski, K., S. Kniat, and D. Slawinska. 2002. Gallic acid, a natural antioxidant, in aqueous and micellar environment: Spectroscopic studies. *Curr. Top. Biophys.* 26:217–227.

Rechner, A. R., A. S. Pannala, and C. A. Rice-Evans. 2001. Caffeic acid derivatives in artichoke extract are metabolised to phenolic acids in vivo. *Free Radical Res.* 35:195–202.

Reis Giada, M. L. 2013. Food phenolic compounds: Main classes, sources and their antioxidant power. In *Oxidative Stress and Chronic Degenerative Diseases—A Role for Antioxidants*, ed. J. A. Morales-González, pp. 87–112. Rijeka, Croatia: InTech.

Ruiz-Capillas, C. and F. Jimenez-Colmenero. 2008. Determination of preservatives in meat products by flow injection analysis (FIA). *Food Addit. Contam.* 25(10):1167–1178.

Ruzicka, J. and E. H. Hansen. 1988. *Flow Injection Analysis.* New York: Wiley.

Sakakibara, H., Y. Honda, S. Nakagawa, H. Ashida, and K. Kanazawa. 2003. Simultaneous determination of all polyphenols in vegetables, fruits, and teas. *J. Agr. Food Chem.* 51:571–581.

Sánchez Arribas, A., M. Martínez-Fernández, M. Moreno, E. Bermejo, A. Zapardiel, and M. Chicharro. 2013. Analysis of total polyphenols in wines by FIA with highly stable amperometric detection using carbon nanotube-modified electrodes. *Food Chem.* 136:1183–1192.

Saurina, J. 2008. Multicomponent flow injection analysis. In *Advances in Flow Analysis,* ed. M. Trojanowicz, pp. 227–263. Weinheim: Wiley.

Saurina, J. 2010. Flow-injection analysis for multi-component determinations of drugs based on chemometric approaches. *Trend Anal. Chem.* 9:1027–1037.

Scalbert, A. and G. Williamson. 2000. Dietary intake and bioavailability of polyphenols. *J. Nutr.* 130:2073–2085.

Seifried, H. E., D. E. Anderson, E. I. Fisher, and J. A. Milner. 2007. A review of the interaction among dietary antioxidants and reactive oxygen species. *J. Nutr. Biochem.* 18:567–579.

Sigma-Aldrich. 2014a. Material safety data of caffeic acid. http://www.sigmaaldrich.com/catalog/search?interface=All&term=c0625&N=0&mode=match%20partialmax&focus=product&lang=en&region=TR (accessed May 1, 2014).

Sigma-Aldrich. 2014b. Material safety data of *trans*-ferulic acid. http://www.sigmaaldrich.com/catalog/search?interface=All&term=w518301&N=0&mode=match%20partialmax&focus=product&lang=en&region=TR (accessed May 1, 2014).

Sigma-Aldrich. 2014c. Material safety data of *p*-coumaric acid. http://www.sigmaaldrich.com/catalog/search?interface=All&term=c9008&N=0&mode=match%20partialmax&focus=product&lang=en&region=TR (accessed May 1, 2014).

Singh, J., G. K. Rai, A. K. Upadhyay, R. Kumar, and K. P. Singh. 2004. Antioxidant phytochemicals in tomato (*Lycopersicon esculentum*). *Indian J. Agr. Sci.* 74:3–5.

Skoog, D. A., F. J. Holler, and T. A. Nieman. 1998. *Principle of Instrumental Analysis.* Florida: Saunder College Press.

Soto, M. L., A. Moure, H. Dominguez, and C. J. Parajo. 2011. Recovery, concentration and purification of phenolic compounds by adsorption: A review. *J. Food Eng.* 105:1–27.

Sullivan, J. J., T. A. Hollingworth, M. M. Wekell et al. 1990. Determination of total sulfite in shrimp, potatoes, dried pineapple, and white wine by flow-injection analysis: Collaborative study. *J. AOAC* 73:35–42.

Suteerapataranon, S. and D. Pudta. 2008. Flow injection analysis-spectrophotometry for rapid determination of total polyphenols in tea extracts. *J. Flow Injection Anal.* 25(1):61–64.

Suzuki, A., D. Kagawa, A. Fujii et al. 2002. Short- and long-term effects of ferulic acid on blood pressure in spontaneously hypertensive rats. *J. Am. Soc. Hypertens.* 15:351–357.

Suzuki, A., M. Yamamoto, and H. Jokura et al. 2007. Ferulic acid restores endothelium-dependent vasodilation in aortas of spontaneously hypertensive rats. *J. Am. Soc. Hypertens.* 20:508–513.

Taljaard, R. E. and J. F. van Staden. 1998. Application of sequential-injection analysis as process analyzers. *Lab. Robotics Automat.* 10:325–337.

Taubert, D., T. Breitenbach, A. Lazar et al. 2003. Reaction rate constants of superoxide scavenging by plant antioxidants. *Free Radical Biol. Med.* 35:1599–1607.

Van Staden, J. F. and R. S. van Staden. 2012. Flow-injection analysis systems with different detection devices and other related techniques for the *in vitro* and *in vivo* determination of dopamine as neurotransmitter. A review. *Talanta* 102:34–43.

Verma, S., A. Singh, and A. Mishra. 2013. Gallic acid: Molecular rival of cancer. *Environ. Toxicol. Pharmacol.* 35:473–485.

Vermerris, W. and R. Nicholson. 2008. *Phenolic Compound Biochemistry*. New York, USA: Springer.

Virgili, F., G. Pagana, L. Bourne et al. 2000. Ferulic acid excretion as a marker of consumption of a French maritime pine (*Pinus maritima*) bark extract. *Free Radical Biol. Med.* 28:1249–1256.

Visioli, F. and C. Galli. 2003. Olives and their production waste products as sources of bioactive compounds. *Curr. Top. Nutraceut. Res.* 1(1):85–88.

Wang, B., J. Ou-Yang, Y. Liu et al. 2004. Sodium ferulate inhibits atherosclerogenesis in hyperlipidemia rabbits. *J. Cardiovasc. Pharm.* 43:549–554.

Wang, B. H. and J. P. Ou-Yang. 2005. Pharmacological actions of sodium ferulate in cardiovascular system. *Cardiovasc. Drug Rev.* 23:161–172.

Wang, J. P., N. B. Li, and H. Q. Luo. 2008. Chemiluminescence determination of ferulic acid by flow-injection analysis using cerium(IV) sensitized by rhodamine 6G. *Spectrochim. Acta Part A* 71:204–208.

Wang, X., J. Wang, and N. Yang. 2007. Flow injection chemiluminescent detection of gallic acid in olive fruits. *Food Chem.* 105:340–345.

Worsfold, P. J., R. Clough, M. C. Lohan et al. 2013. Flow injection analysis as a tool for enhancing oceanographic nutrient measurements—A review. *Anal. Chim. Acta* 803:15–40.

Wu, W. M., L. Lu, Y. Long et al. 2007. Free radical scavenging and antioxidative activities of caffeic acid phenethyl ester (CAPE) and its related compounds in solution and membranes: A structure-activity insight. *Food Chem.* 105:107–115.

Xiang-Qin, L., L. Feng, P. Yong-Qiang, and C. Hua. 2004. Flow injection analysis of gallic acid with inhibited electrochemiluminescence detection. *Anal. Bioanal. Chem.* 378:2028–2033.

Xing, Y. and P. J. White. 1997. Identification and function of antioxidants from oat groats and hulls. *J. Am. Oil Chem. Soc.* 74:303–307.

Zang, L. Y., G. Cosma, H. Gardner, X. Shi, V. Castranova, and V. Vallyathan. 2000. Effect of antioxidant protection by *p*-coumaric acid on low-density lipoprotein cholesterol oxidation. *Am. J. Physiol.-Endocrinol. Metab.* 279:954–960.

Zhao, C., J. Song, and J. Zhang. 2003. Flow-injection biamperometry of pyrogallol compounds. *Talanta* 59:19–26.

Zhao, Z. and M. H. Moghadasian. 2008. Chemistry, natural sources, dietary intake and pharmacokinetic properties of ferulic acid: A review. *Food Chem.* 109:691–702.

CHAPTER 21

Determination of Flavonoids

David González-Gómez, Juan Carlos Bravo, Alejandrina Gallego,
Rosa Mª Garcinuño, Pilar Fernández, and Jesús Senén Durand

CONTENTS

21.1 INTRODUCTION

21.1.1 General Overview of Polyphenols

Polyphenols are one of the commonest and most important groups of substances present in a great variety and number of plants, showing a wide range of structures and serving important functions in plants. The estimated number of plant polyphenols could include 8000 different structures [1] and their roles in plants involve protection against photosynthetic stress, free radical scavenging, and plant protection against herbivores. Besides these protective actions, they are also responsible for insect attraction due to their coloration and for or cell-to-cell signaling processes [2].

These compounds are present in all vegetative organs, flowers, and fruits and the health effects of polyphenols have been associated with consumption of diets rich in foods of plant origin. In recent years, the effects of dietary polyphenols on human health have been widely studied and found to be strongly related to the prevention of degenerative and cardiovascular diseases and cancers. The protective effect of polyphenols against diseases has generated great interest from the food and nutritional supplement industry regarding promotion and development of polyphenol-rich products [3]. The antioxidant properties of polyphenols have been reported in many studies, although the mechanisms of action of polyphenols are not completely elucidated [2]. Nevertheless, the antioxidant properties of polyphenols have been exploited in their use as natural preservatives in foods. Plant extracts as natural additives provide flavors, colors, antimicrobials, and antioxidants to foodstuffs. Oxidation affects the sensory attributes of food; taste, odor, and color change and the product life is shortened. Even though these natural antioxidants are less effective than other synthetic additives, the trend for use of preservatives is increase in demand for natural antioxidants over synthetic ones. Consumer prefers healthy or functional food and natural ingredients, even if they are more expensive.

21.1.2 Classification of Polyphenols

Polyphenols are plant secondary metabolites resulting from the phenylpropanoid pathway [4]. The key flavonoid precursors are phenylalanine, obtained via the shikimate and arogenate pathways, and malonyl-CoA, derived from citrate produced by the TCA (tricarboxylic acid) cycle. Their chemical structure is very diverse and complex, having molecular masses ranging from 500 to 3000–4000 Da and possessing 12–16 phenolic hydroxyl groups or five to seven aromatic rings per 1000 Da of relative molecular mass [5]. Polyphenols are found in plants associated with other complex biomolecules, such as sugars or organic acids. According to their structure, these compounds can be variously grouped from simple structures, such as phenolic acid derivatives, to complex polymeric forms with high molecular mass, but overall these compounds are characterized by featuring at least a phenolic ring [6].

Polyphenols can be classified according to their chemical structure into two important groups: flavonoids and nonflavonoid compounds (Figure 21.1).

21.1.2.1 Flavonoids

Flavonoids are polyphenols with a chemical structure of 15 carbon atoms disposed in the basic form C6–C3–C6 (two benzene rings with a chain of three carbon atoms of a γ-pyrone system). Thus, flavonoids are obtained by lengthening of the side-chain of cinnamic acids by the addition of one or more carbon units. In terms of the number of aromatic rings, unsaturation, and oxidation grade, the flavonoids group is subdivided in different categories: flavonols, flavan-3-ols, flavones, flavanones, anthocyanidins, and isoflavones. Flavonoids compounds can be isolated in their glycoside form, where the hydroxyl group of the basic structure of the flavonoid is joined with a hemiacetalic bond, generally through the C–1 carbon and with a bond of type β, to one or more sugars (Figure 21.2).

Flavonols: These compounds are widely distributed in plants and vegetal foods, the most widespread being quercetin, kaempferol, isorhamnetin, and myricin. Among them, quercetin is the most abundant dietary flavonol [7] since this compound is present in onion. Quercetin exhibits nonpolar properties, and thus in nature is mostly found conjugated with one or more sugar molecules.

Flavones: Although this flavonoid's structure is similar to that of flavonols (they differ only in the absence of hydroxylation at the 3-position on the C-ring), these compounds are less abundant in nature. In the diet the most relevant flavones are apigenin and luteolin, and they are found in significant amounts in celery, parsley, and artichoke [8].

Flavan-3-ols: Flavan-3-ols are the most abundant flavonoids in diet [9] and in addition are the most complex of the flavonoids found in nature [7]. Among them can be found a range of structures from simple monomeric structures such as (+)-catechin and its isomer (–)-catechin to more complex oligomeric structures and polymeric proanthocyanidins known as tannins. It has been reported [6] that these compounds can undergo esterification reactions with gallic acid to form catechin gallates and hydroxylation reactions to form gallocatechins. Flavan-3-ols are well distributed in the plant kingdom, and

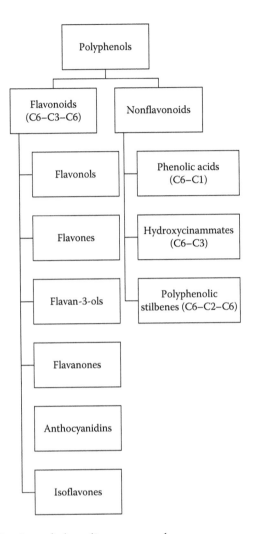

FIGURE 21.1 Classification of phenolic compounds.

important amounts of these compounds are found in a wide variety of fruit such as apricots, nectarines, cherries, grapes, berries, and green tea among others. Flavan-3-ols exhibit a wide range of biological activity, mostly due to their antioxidant characteristics. It has also been reported that they have a positive actions in cardiovascular diseases, confer neurological protection, and exhibit microbicidal activity [10,11].

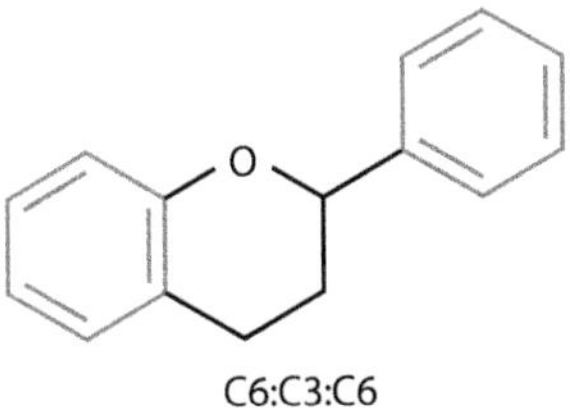

C6:C3:C6

FIGURE 21.2 Chemical structure of flavonoids.

Flavanones: These compounds are characterized for being highly reactive; they can undergo hydroxylation, glycoxylation, and O-methylation reactions. In the diet the input of flavanones comes from citrus fruit, mainly as hesperetin conjugates, which are also responsible for the bitter flavor [12].

Anthocyanidins: The water-soluble reddish, bluish, and purplish pigments of this group are found in a large number of fruits, vegetables, and flowers; they are found in fruit/plant/flower tissues since they are responsible for their coloration [6] and they are involved in protective processes against intense ultraviolet (UV) radiation by shading leaf mesophyll cells [7]. In nature, these pigments are normally present as sugar-conjugate derivate.

Isoflavones: These flavonoids are among the less-widely distributed flavonoids in nature, but they are the compounds most frequently tested for in humans [13]. The only important amount of these compounds were found in leguminous species, especially in soybeans, black beans, and green peas, the most common isoflavones being genistein, daidzein, and glycitein. In addition, these compounds undergo various reactions such as methylation, hydroxylation, or polymerization that lead from simple (isoflavanones, isoflavans, and isoflavanols) to more complex structures (rotenoids, pterocarpans, and coumestans) [12].

21.1.2.2 Nonflavonoids

The nonflavonoid polyphenols include the phenolic acids or hydroxybenzoates (basic chemical structure C6–C1), hydroxycinammates (basic chemical structure C6–C3), and polyphenolic stilbenes (basic chemical structure C6–C2–C6).

21.1.3 Antioxidant Properties of Flavonoids

Flavonoid compounds exhibit a great range of functions in plants and humans when they are consumed. In plants, due to their attractive colors, these compounds are used as visual signal for pollinating insects. In addition, they are also involved in plant defense; for instance, due to their astringency they discourage insects. Another important function in plants is their role as stress protectants by scavenging reactive oxygen species (ROS) produced by the photosynthetic electron transportation system and ROS generated by UV radiation [1]. Humans also use this antioxidant property after consumption of flavonoids from plants. Dietary flavonoids may have an additive effect to that of the endogenous scavenging compounds produced by the human immune system against ROS [14]. In fact, dietary flavonoids are able to interfere with different free radical-producing systems [15], countering the injury caused by free radicals in various ways. Due to the lower redox potential of flavonoids, mostly due to the high reactivity of the hydroxyl group [16], they are able to reduce highly oxidizing free radicals by transforming them into less reactive flavonoid radicals (direct radical scavenging). Other antioxidants properties of flavonoids are due to their capacity to scavenge superoxide and peroxynitrite. As result, dietary flavonoids can prevent low-density lipid (LDL) oxidation [17]. In addition, these compounds are able to scavenge nitric oxide [18] and they are therefore able to reduce the synthesis of the highly damaging peroxynitrites formed by nitric oxides and superoxide free radicals [19]. Another way flavonoids act as antioxidants is by inhibiting the xanthine oxidase activity [17], a source of oxygen free radicals. In addition, it has been also proved that the intake of flavonoids reduces the number of leukocytes during reperfusion and indirectly the formation of oxygen-derived free radicals [20,21]. Finally, the other possible mechanisms of these compounds acting as radical scavenging are through the interaction with diverse enzyme systems [15].

21.2 FLAVONOIDS AS FOOD ADDITIVES

Significant research efforts have been undertaken in recent decades into evidence that dietary flavonoids influence health [7,22–24]. The main dietary flavonoid intake source comes from the ingestion of foodstuffs with an important amount of these compounds, such as a great variety of vegetal food products [7]. On the other hand, plants spices and their extracts have long been used to improve the attributes of food without being considered as nutritionally significant ingredients. Polyphenolic compounds provide flavors and colors and preserve food products. More recently, these plant additives have become of increasing interest in the food industry because their antioxidant properties retard oxidative degradation of lipids and thereby conserve products and improve the quality of foodstuffs. Synthetic antioxidants such as butylated hydroxyanisole (BHA), butylated hydroxytoluene (BHT), and *tert*-butylhydroquinone (TBHQ) have been used widely as antioxidants in foods, but the demand now tends to be increasing for natural antioxidants. The antioxidant activities of phenolic compounds extracted from different botanical sources have been studied in order to establish the phenolic profiles of the different plants. Synthetic antioxidants have been substituted by natural additives, and the research on antioxidants has also focused on phenolic compounds, in particular the flavonoids and hydroxycinnamic acids [25]. The predominant flavonoids identified in plants were quercetin, luteolin, and apigenin [2,26].

The main flavonoids (Figure 21.3) identified in different plants used as culinary herbs are reviewed in Table 21.1.

Various government regulations have already included certain types of flavonoids extracted from purple corn as food additives [32], and the last update of the commission regulation of the European Union on Food and Additives (No. 1130/2011) includes rosemary extracts (E392) and anthocyanins (E163) as permitted food additives [33,34]. The European Food Safety Authority report [33] establishes that the rosemary extracts have relevant amounts of the flavonoid genkwanin, although the main applications of these extracts in foodstuff are due to the presence of phenolic acids, flavonoid diterpenoids, and triterpenoids such as carnosic acid and carnosol [35–37]. Rosemary extracts are increasingly used as an alternative to synthetic antioxidants and also provide flavor to foods. Usually both properties are valued, but the processing of rosemary extracts can be optimized to enhance the antioxidative function and to reduce that of flavoring.

Several authors reported the antioxidant effectiveness of rosemary extract in corn, soybean, peanut, and fish oil [38,39]. Similarly, the effectiveness of rosemary extracts combined with BHT, in different mixtures, has been studied in soybean oil [39–41]. The antioxidant and antimicrobial activities of rosemary extracts in chicken Frankfurters have been evaluated [37]. The effect of rosemary extract, added individually or in combination, on lipid oxidation and color stability of frozen beef burgers or fresh pork sausages has also been assessed [35,36].

The E163 additives (different types of plant/fruit extracts such as grape skin extracts, blackcurrant extracts, purple corn extract, and red cabbage extracts) are not strictly considered antioxidant additives; however, the antioxidant properties of anthocyanin flavonoids are already well described [42]. Anthocyanins constitute a natural pigment with functional properties [43] and they have been used in cheese manufacture [44].

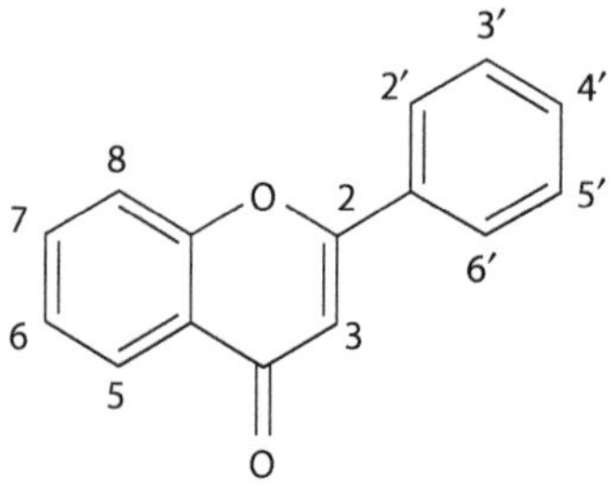

Apigenin (4′, 5, 7 -OH)
Cirsimaritin (4′, 5 -OH; 6, 7-OCH$_3$)
Cirsiliol (4′, 5′, 5 -OH; 6, 7-OCH$_3$)
Genkwanin (4′, 5-OH; 7-OCH$_3$)
Luteolin (3′, 4′, 5, 7 -OH)
Salvigenin (5 -OH; 4′, 6, 7-OCH$_3$)

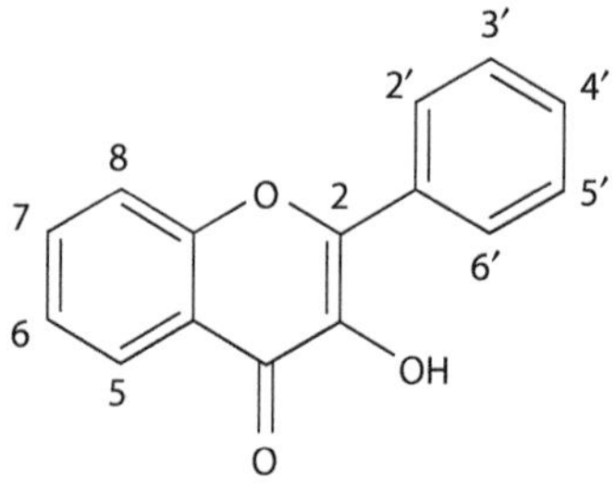

Isorhamnetin (4′, 5, 7 -OH; 3′-OCH$_3$)
Kaempferol (4′, 5, 7-OH)
Myricetin (3′, 4′, 5′, 5, 7-OH)
Quercetin (3′, 4′ 5, 7 -OH)

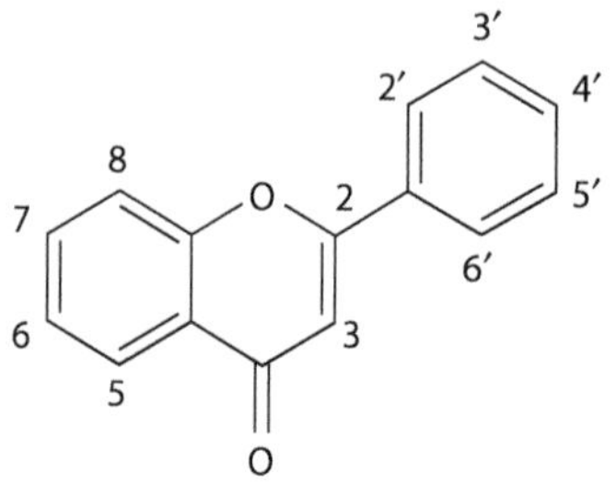

Hesperetin (3′, 5, 7 -OH; 4′-OCH$_3$)
Naringenin (4′, 5, 7 -OH)
Genkwanin (4′, 5-OH; 7-OCH$_3$)

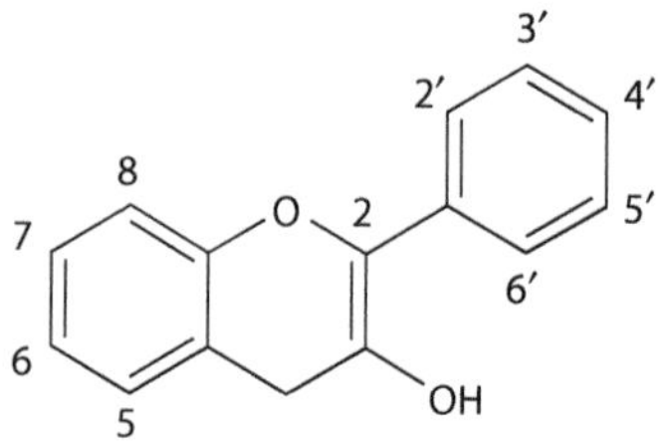

Catechin (4′, 5′, 5, 7 -OH)

FIGURE 21.3 Structures of the main flavonoids identified in culinary herbs.

21.3 ANALYSIS OF FLAVONOIDS FOOD ADDITIVES

Flavonoids are characterized by having an aromatic ring with different degrees of sub-stitution, including functional derivatives such as esters, methyl ethers, and glycosides. These structural characteristics determine the general physicochemical properties of these compounds.

Most flavonoid glycosides and aglycones exhibit highly polar characteristics, and therefore are alcohol-soluble or alcohol–water-soluble. This is an important consideration in extraction of flavonoids from plant material. Less polar flavonoids, such as isoflavones, flavanones, methylate flavones, and flavonols, could easily be extracted using nonpo-lar solvents such as chloroform, dichloromethane, or diethyl ether [45]. Based on their chemical structures, these compounds have two major UV bands: the first band is located between 320 and 380 nm, and the second between 240 and 270 nm. As was mentioned in Lee's work [46], the first absorption band is due to the flavonoid B-ring while the second band comes from absorption of the A-ring. This general UV spectral characteristic can be altered by the substituents of the aromatic ring, causing strong bathochromic shifts. Flavonoids spectra in ethanol, methanol, and acetonitrile are essentially the same. There are monographs reporting the UV spectral data of flavonoids, such as the work by Mabry et al. as cited by Marston [45].

TABLE 21.1 Flavonoids Identified in Culinary Herbs

Common Name	Botanical Name	Flavonoid	Reference
Basil, sweet	*Ocimum basilicum*	Apigenin, luteolin,	[2]
Cinnamon	*Cinnamomum*	Apigenin, catechin, kaempferol, quercetin	[26]
Coriander	*Coriandrum sativum*	Kaempferol, quercetin	[2,27]
Cumin	*Cuminum cyminum*	Catechin, hesperetin, kaempferol, quercetin	[26]
Dill	*Anethum graveolens*	Kaempferol, quercetin	[2,27]
Lemon balm	*Melissa officinalis*	Luteolin	[2]
Lovage	*Levisticum officinalis*	Kaempferol, quercetin	[2,27]
Marjoram	*Origanum marjorana*	Apigenin, naringenin	[2,28]
Mint	*Mentha* var.	Apigenin, hesperetin, luteolin	[2,27]
Oregano	*Origanum vulgare*	Apigenin, epicatechin, hesperetin, hesperidin, kaempferol, luteolin, quercetin	[2,26,27]
Parsley	*Petroselinum crispum*	Apigenin, kaempferol, luteolin, quercetin	[2,27,29]
Rosemary	*Rosmarinus officinalis*	Apigenin, epicatechin, hesperidin, hesperetin, kaempferol, luteolin, quercetin	[2,26,30]
Sage	*Salvia officinalis*	Apigenin, cirsiliol, cirsimaritin, genkwanin, hesperidin, luteolin, naringenin, quercetin, salvigenin	[2,26,28,31]
Tarragon	*Artemisia dracunculus*	Isorhamnetin, kaempferol, luteolin, quercetin	[27]
Thyme	*Thymus* var.	Apigenin, hesperetin, hesperidin, kaempferol, luteolin, naringenin, quercetin	[2,26–28]

21.3.1 Sample Treatment

The direct analysis of flavonoids is not possible due to the high complexity of the sample matrices. Sample pretreatment is a requirement in all flavonoid analysis to remove possible interferences, to increase the concentration of the analyte and hence the sensitivity of the analytical method, to adapt the analyte to the detection method, and to provide a more reproducible and robust analysis method. The selection of the proper sample treatment procedure depends on the type of flavonoids and the sample matrix (plant, food, or liquid sample such as biological fluids and drinks). Solid samples are normally lyophilized and homogenized before extraction; liquid samples are normally filtered and/or centrifuged before the isolation or extractive procedure.

Currently, direct solvent extraction (SE) constitutes the most common procedure. Flavonoid structure and polarity are the key factors for solvent selection; less polar flavonoids such as isoflavones, flavanones, and flavonols are extracted using nonpolar solvents, while more polar flavonoids (glycosides and polar aglycones) are extracted with polar solvents, mostly alcohol–water mixtures. Moreover, due to the great variety of flavonoids, sequential SE is a common practice in sample pretreatment. Thus, nonpolar solvents are used in a first stage to extract flavonoids aglycones and other nonpolar compounds, and secondly polar solvents, such as alcohol–water mixtures, are employed to complete the extraction. In the specific case of the flavan-3-ols, water can be directly used in SE. In the case of anthocyanidin pigments, the extraction is normally carried out using a cold alcohol–water mixture [47]. The presence of acid in the extraction mixture, such as acetic acid (7%) or trifluoroacetic acid (1%–3%) leads to the loss of the attached acyl groups and prevents the ionization of the flavonoids [48]. Soxhlet-SE is sometimes employed in solid samples [45], although this procedure is less frequently used because of the heat-sensitive compounds. On the other hand, liquid–liquid extraction (LLE) is normally used to treat liquid samples. In LLE the selection of the most suitable solvent is again chosen according to the target sample, and normally a small amount of acid is added to the extraction medium.

Besides direct SE, more complex extraction procedures are becoming of interest. Among these, solid-phase extraction (SPE) still constitutes a very reliable alternative for sample treatment. The most frequently used sorbents are alkyl-bonded silica or copolymer sorbents, such as C_{18}-bonded silica.

SE is normally followed by SPE to remove possible interferences and to increase the concentration or purify the analyte [49]. The presence of acid in the extraction solvent helps to prevent ionization of the flavonoids and hence to reduce their retention. In more complex studies, two-dimensional SPE (2D-SPE) can be applied for the extraction step [50]. Recently, molecular imprinted polymers (MIPs) have attracted attention for the SPE of flavonoids in food matrices [51], as in the determination of quercetin in red wines [52,53], myricetin in fruit samples [54], or a mixture of quercetin myricetin and amentoflavone from different plant extracts [55]; however the use of MIPs is limited because a specific MIP has to be designed for each target compound.

Other methods, such as matrix solid-phase dispersion (MSPD) have also been applied in sample pretreatment. Among other applications, MSPD was recently used for the analysis of seven flavonoids in citrus juice [56] or 18 flavonoids from *Ginkgo biloba* tablets combining extraction and cleanup in a single run [57].

21.3.2 Sample Separation and Detection

The flavonoid-extracted sample contains a number of different flavonoids together with other nonflavonoid species coextracted in the extraction process. Thus, the quantification of flavonoids requires a separation step. The great majority of analytical methods for flavonoid separation and analysis are liquid chromatography (LC)-based methods, followed by gas chromatography, capillary electrophoresis, and thin-layer chromatography [45,48]. The LC methods are normally performed in the reversed-phase mode using C_8 or C_{18} bonded silica columns, although other phases are also useful for the separation of flavonoids throughout LC systems. A gradient elution is the normal separation mode to achieve good resolutions, avoiding the use of phosphate buffer because of the fear of contamination of ion sources when MS (mass spectrometry) detection is used [58].

The most common detection systems used to quantify these compounds are UV absorbance, fluorescence, and MS detection. MS is the detection system that is gaining major attention in the most recent reports [59], especially MS/MS (tandem mass spectrometry) systems, since they are the most important technique for the identification of flavonoids and the structural characterization of unknown member of this class of compounds [60].

The presence of at least one aromatic ring in the flavonoid chemical structure ensures the absorption of UV radiation: as mentioned above, a first maximum occurs in the range of 240–285 nm, and a second one in the 300–550 nm range. Thus, UV detection is a satisfactory tool for use in screening and/or quantification studies. In contrast, the use of fluorescence detection is rarer since only few flavonoids exhibit native fluorescence, mostly isoflavones [61] and some flavones [62]. In other cases, derivatization processes, through the reaction between the flavonoid and metal cations, have been carried out for analysis of some flavonoids, such as quercetin, kaempferol, and morin, using florescence detection [63]. Lastly, other detection systems such as electrochemical detection (ED), are also practicable for flavonoid detections because most flavonoids have an electroactive characteristic due to the presence of phenolic groups. An example of the use of ED is the determination of isoflavones in soybean food or plant extracts [64,65].

As has been mentioned, nowadays MS/MS detection is widely used in the analysis of flavonoids, not only for its good sensitivity but also because of the identification and structural characterization capacity that these detectors have [66–69]. In almost all applications, atmospheric pressure ionization interfaces are used, such as APCI (atmospheric pressure chemical ionization) and electrospray ionization (ESI) [48].

21.4 DETERMINATION OF FLAVONOIDS USING FLOW INJECTION ANALYSIS

Flow injection analysis (FIA) has barely been used to analyze flavonoids compounds. Only a few studies were found in the literature [65,70–72].

Twelve flavonoids (fisetin, galangin, hesperetin, hesperidin, kaempferol, morin, myricetin, naringin, quercetin, quercitrin, rutin, and rhamnetin) were determined by flow injection/adsorptive stripping voltammetry using Nujol–graphite and diphenyl ether–graphite paste electrodes [70]. In this work, anodic stripping voltammetry (AdSV) was coupled with a FIA system, speeding up the analysis and reducing the analysis reagents. The ability of flavonoids to accumulate in carbon paste electrodes results in low determination limits (in the range of 10^{-8} M). This method was established to determine 12 flavonoids in a multivitamin complex.

FIA coupled with chemiluminescence (CL) has been used to determine rutin in pharmaceutical tablets [71]. In this research, potassium ferricyanide was used to catalyze the reaction of luminol with rutin to produce strong CL in alkaline solution. This reaction was easily adjusted to a FIA system, thus the luminol solution and potassium ferricyanide were first mixed, and then injected with the rutin sample. It was found that the ferricyanide catalyzed the CL reaction when the concentration was over 10 mM. The proposed method has a LOD of 30 ng mL^{-1}.

FIA-CL was also recently applied for the analysis of different flavonoids compounds in inflorescences and leaf extracts of *Cirsium oleraceum* and *Cirsium rivulare* species [72]. In this case, the CL signal was generated by the reaction of cerium(IV) with rhodamine 6G in a sulfuric acid medium. In the conditions optimized in this investigation, good linearity and sensitivity were achieved with detection limits of 38 nmol L^{-1} (apigenin) and 840 nmol L^{-1} (linarin). For the FIA analysis, cerium(IV) sulfate and rhodamine 6G were first mixed, and the sample was injected through a sodium hydroxide stream merging with the cerium/rhodamine mixture. The emitted light was collected with a photomultiplier tube (operated at 1100 V) placed in a light-tight box. The flow cell was a flat spiral PTFE coil of 1.0 mm i.d. (length of 25 cm in six windings) placed in front of a photomultiplier.

The electroactive property of flavonoids was used to optimize a FIA coupled to an electrochemical detection (FIA-EC) method to determine quercetin and rutin natural plant extracts and pharmaceutical preparations [65]. In this method, flavonoids were determined at normal (unheated) and hot platinum microelectrodes using cyclic voltammetry. The use of the FIA system causes an increase of the analytical signal by more than 6 times because of the increase of the temperature to about 76°C in a small zone close to the microelectrode. The limits of detection in this method are equal to 0.8 µmol L⁻¹ for quercetin and 2.2 µmol L⁻¹ for rutin.

FIA methods for determination of flavonoids are rapid, simple, and cheap, but the studies reviewed in this chapter suggest that the main interest of the research is actually the identification of phenolic profiles in plant sources; the assessment of the antioxidant effectiveness of flavonoids in foods; and the search for applications for the natural preservatives in the foodstuffs industries to satisfy consumer preferences and demands. These objectives require the use of the most sensitive and specific instrumentation with mass detection in combination with statistical tools.

REFERENCES

1. Pietta, P. 2000. Flavonoids as antioxidants. *J. Nat. Prod.* 63(7):1035–1042.
2. Andersen, O. M. and K. R. Markham. 2006. *Flavonoids: Chemistry, Biochemistry and Applications.* Boca Raton, FL: CRC Taylor & Francis.
3. Scalbert, A., I. J. Johnson, and M. Saltmarsh. 2005. Polyphenols: Antioxidants and beyond. *Am. J. Clin. Nutr.* 81(1):215S–217S.
4. Quideau, S., D. Deffieux, C. Douat-Casassus, and L. Pouysegu. 2011. Plant polyphenols: Chemical properties, biological activities, and synthesis. *Angew. Chem. Int. Ed.* 50(3):586–621.
5. Haslam, E. and Y. Cai. 1994. Plant polyphenols (vegetable tannins): Gallic acid metabolism. *Nat. Prod. Rep.* 11(11):41–66.
6. Ferrazzano, G. F., I. Amato, A. Ingenito, A. Zarrelli, G. Pinto, and A. Pollio. 2011. Plant polyphenols and their anti-cariogenic properties: A review. *Molecules* 16(2):1486–507.
7. Crozier, A., I. B. Jaganath, and M. N. Clifford. 2009. Dietary phenolics: Chemistry, bioavailability and effects on health. *Nat. Prod. Rep.* 26(8):1001–1043.
8. Crozier, A., D. Del Rio, and M. N. Clifford. 2010. Bioavailability of dietary flavonoids and phenolic compounds. *Mol. Aspects Med.* 31(6):446–467.
9. Monagas, M., M. Urpi-Sarda, F. Sanchez-Patan et al. 2010. Insights into the metabolism and microbial biotransformation of dietary flavan-3-ols and the bioactivity of their metabolites. *Food Funct.* 1(3):233–253.
10. Aron, P. M. and J. A. Kennedy. 2008. Flavan-3-ols: Nature, occurrence and biological activity. *Mol. Nutr. Food Res.* 52(1):79–104.
11. Geissman, T. A. and E. Hinreiner, 1952. Theories of the biogenesis of flavonoid compounds. *Bot. Rev.* 18(2):77–164.
12. Yokoyama, A., H. Sakakibara, A. Crozier et al. 2009. Quercetin metabolites and protection against peroxynitrite-induced oxidative hepatic injury in rats. *Free Radical Res.* 43(10):913–921.
13. Crozier, A. 2008. Plant secondary metabolites: Their role in the human diet. *Planta Med.* 74(3):315–315.
14. Halliwell, B. 1994. Free radicals, antioxidants, and human disease: Curiosity, cause, or consequence? *Lancet* 344(8924):721–724.
15. Nijveldt, R. J., E. van Nood, D. van Hoorn, P. G. Boelens, K. van Norren, and P. van Leeuwen. 2001. Flavonoids: A review of probable mechanisms of action and potential applications. *Am. J. Clin. Nutr.* 74(4):418–425.
16. Korkina, L. G. and I. B. Afanas'ev. 1997. Antioxidant and chelating properties of flavonoids. *Adv. Pharmacol.* 38:151–63.

17. Cos, P., M. Calomme, J. B. Sindambiwe et al. 2001. Cytotoxicity and lipid peroxidation-inhibiting activity of flavonoids. *Planta Med.* 67(6):515–9.
18. Corradini, E., P. Foglia, P. Giansanti, R. Gubbiotti, R. Samperi, and A. Lagana. 2011. Flavonoids: Chemical properties and analytical methodologies of identification and quantitation in foods and plants. *Nat. Prod. Res.* 25(5):469–495.
19. Shutenko, Z., Y. Henry, E. Pinard et al. 1999. Influence of the antioxidant quercetin in vivo on the level of nitric oxide determined by electron paramagnetic resonance in rat brain during global ischemia and reperfusion. *Biochem. Pharmacol.* 57(2):199–208.
20. Friesenecker, B., A. G. Tsai, C. Allegra, and M. Intaglietta. 1994. Oral administration of purified micronized flavonoid fraction suppresses leukocyte adhesion in ischemia-reperfusion injury: *In vivo* observations in the hamster skin fold. *Int. J. Microcirc. Clin. Exp.* 14(1–2):50–55.
21. Ferrandiz, M. L., B. Gil, M. J. Sanz et al. 1996. Effect of bakuchiol on leukocyte functions and some inflammatory responses in mice. *J. Pharmacol. Pharmacother.* 48(9):975–80.
22. Chen, A. Y. and Y. C. Chen. 2013. A review of the dietary flavonoid, kaempferol on human health and cancer chemoprevention. *Food Chem.* 138(4):2099–2107.
23. Pal, D. and P. Dubey, 2013. Flavonoids: A powerful and abundant source of antioxidants. *Int. J. Pharm. Pharm. Sci.* 5(3):95–98.
24. Del Rio, D., A. Rodriguez-Mateos, J. P. E. Spencer, M. Tognolini, G. Borges, and A. Crozier. 2013. Dietary (poly)phenolics in human health: Structures, bioavailability, and evidence of protective effects against chronic diseases. *Antioxid. Redox Signaling* 18(14):1818–1892.
25. Martínez-Valverde, I., M. J. Periago, G. Provan, and A. Chesson. 2002. Phenolic compounds, lycopene and antioxidant activity in commercial varieties of tomato (*Lycopersicum esculentum*). *J. Sci. Food Agric.* 82(3):323–330.
26. Vallverdú-Queralt, A., J. Regueiro, M. Martínez-Huélamo, J. F. Rinaldi Alvarenga, L. Neto Leal, and R. M. Lamuela-Raventos. 2014. A comprehensive study on the phenolic profile of widely used culinary herbs and spices: Rosemary, thyme, oregano, cinnamon, cumin and bay. *Food Chem.* 154(1):299–307.
27. Justesen, U. and P. Knuthsen. 2001. Composition of flavonoids in fresh herbs and calculation of flavonoid intake by use of herbs in traditional Danish dishes. *Food Chem.* 73(2):245–250.
28. Roby, M., M. Atef Sarhan, K. A. Selim, and K. I. Khalel. 2013. Evaluation of antioxidant activity, total phenols and phenolic compounds in thyme (*Thymus vulgaris L.*), sage (*Salvia officinalis L.*), and marjoram (*Origanum majorana L.*) extracts. *Ind. Crops Prod.* 43(1):827–831.
29. Justesen, U., P. Knuthsen, and T. Leth. 1998. Quantitative analysis of flavonols, flavones, and flavanones in fruits, vegetables and beverages by high-performance liquid chromatography with photo-diode array and mass spectrometric detection. *J. Chromatogr. A* 799(1–2):101–110.
30. Wojdyło, A., J. Oszmiański, and R. Czemerys. 2007. Antioxidant activity and phenolic compounds in 32 selected herbs. *Food Chem.* 105(3):940–949.
31. Farhat, M. B., A. Landoulsi, R. Chaouch-Hamada, J. A. Sotomayor, and M. J. Jordán. 2013. Characterization and quantification of phenolic compounds and antioxidant properties of Salvia species growing in different habitats. *Ind. Crops Prod.* 49(1):904–914.

32. Yokohira, M., K. Yamakawa, K. Saoo et al. 2008. Antioxidant effects of flavonoids used as food additives (purple corn color, enzymatically modified isoquercitrin, and isoquercitrin) on liver carcinogenesis in a rat medium-term bioassay. *J. Food Sci.* 73(7):C561–C568.

33. European Food Safety Authority. 2008. Use of rosemary extracts as a food additive—Scientific opinion of the panel on food additives, flavourings, processing aids and materials in contact with food. *Eur. Food Saf. Auth. J.* 6(6):1–29.

34. European Food Safety Authority Panel on Food Additives and Nutrient Sources added to food (ANS). 2013. Scientific opinion on the re-evaluation of anthocyanins (E 163) as a food additive. *Eur. Food Saf. Auth. J.* 11(4):3145–3196.

35. Georgantelis, D., I. Ambrosiadis, P. Katikou, G. Blekas, and G. A. Georgakis. 2007. Effect of rosemary extract, chitosan and α-tocopherol on microbiological parameters and lipid oxidation of fresh pork sausages stored at 4°C. *Meat Sci.* 76(1):172–181.

36. Georgantelis, D., G. Blekas, P. Katikou, I. Ambrosiadis, and D. J. Fletouris. 2007. Effect of rosemary extract, chitosan and α-tocopherol on lipid oxidation and colour stability during frozen storage of beef burgers. *Meat Sci.* 75(2):256–264.

37. Rižnar, K., Š. Čelan, Ž. Knez, M. Škerget, D. Bauman, and R. Glaser. 2006. Antioxidant and antimicrobial activity of rosemary extract in chicken frankfurters. *J. Food Sci.* 71(7):C425–C429.

38. Frankel, E. N., S. H. Huang, E. Prior, and R. Aeschbach. 1996. Evaluation of antioxidant activity of rosemary extracts, carnosol and carnosic acid in bulk vegetable oils and fish oil and their emulsions. *J. Sci. Food Agric.* 72(2):201–208.

39. Balasundram, N., K. Sundram, and S. Samman. 2006. Phenolic compounds in plants and agri-industrial by-products: Antioxidant activity, occurrence, and potential uses. *Food Chem.* 99(1):191–203.

40. Basaga, H., C. Tekkaya, and F. Acikel, 1997. Antioxidative and free radical scavenging properties of rosemary extract. *LWT—Food Sci. Technol.* 30(1):105–108.

41. Wanasundara, U. N. and F. Shahidi. 1998. Stabilization of marine oils with flavonoids. *J. Food Lipids* 5(3):183–196.

42. Azevedo, J., I. Fernandes, A. Faria et al. 2010. Antioxidant properties of anthocyanidins, anthocyanidin-3-glucosides and respective portisins. *Food Chem.* 119(2):518–523.

43. Silva, M. C., V. de Souza, M. Thomazini et al. 2014. Use of the jabuticaba (*Myrciaria cauliflora*) depulping residue to produce a natural pigment powder with functional properties. *LWT—Food Sci. Technol.* 55(1):203–209.

44. Prudencio, I. D., E. Schwinden Prudêncio, E. Fortes Gris, T. Tomazi, and M. T. Bordignon-Luiz. 2008. Petit suisse manufactured with cheese whey retentate and application of betalains and anthocyanins. *LWT—Food Sci. Technol.* 41(5):905–910.

45. Marston, A. and K. Hostettmann. 2006. Separation and quantification of flavonoids. In *Flavonoids. Chemistry, Biochemistry and Applications*, eds. O. M. Andersen, and K. R. Markham. Boca Raton, FL: CRC Press, Taylor & Francis Group, pp. 1–36.

46. Lee, H. S. 2000. HPLC analysis of phenolic compounds. In *Food Analysis by HPLC* (2nd ed.), ed. L. M. L. Nollet. New York: Marcel Dekker, Inc, pp. 775–824.

47. González-Gómez, D., M. Lozano, M. F. Fernández-León, M. J. Bernalte, M.C. Ayuso, and A. B. Rodríguez. 2010. Sweet cherry phytochemicals: Identification and characterization by HPLC-DAD/ESI-MS in six sweet-cherry cultivars grown in Valle del Jerte (Spain). *J. Food Compos. Anal.* 23(6):533–539.

48. de Rijke, E., P. Out, W. Niessen, F. Ariese, C. Gooijer, and U. A. T. Brinkman. 2006. Analytical separation and detection methods for flavonoids. *J. Chromatogr. A* 1112(1–2):31–63.
49. Vinha, A. F., F. Ferreres, B. M. Silva et al. 2005. Phenolic profiles of Portuguese olive fruits (*Olea europaea L.*): Influences of cultivar and geographical origin. *Food Chem.* 89(4):561–568.
50. Klejdus, B., D. Vitamvásová-Štěrbová, and V. Kubáň. 2001. Identification of isoflavone conjugates in red clover (*Trifolium pratense*) by liquid chromatography–mass spectrometry after two-dimensional solid-phase extraction. *Anal. Chim. Acta* 450(1–2):81–97.
51. He, C., Y. Long, J. Pan, K. Li, and F. Liu. 2007. Application of molecularly imprinted polymers to solid-phase extraction of analytes from real samples. *J. Biochem. Biophys. Methods* 70(2):133–150.
52. Molinelli, A., R. Weiss, and B. Mizaikoff. 2002. Advanced solid phase extraction using molecularly imprinted polymers for the determination of quercetin in red wine. *J. Agric. Food Chem.* 50 (7):1804–1808.
53. Song, X., J. Li, J. Wang, and L. Chen. 2009. Quercetin molecularly imprinted polymers: Preparation, recognition characteristics and properties as sorbent for solid-phase extraction. *Talanta* 80(2):694–702.
54. Zhong, S., Y. Kong, L. Zhou, C. Zhou, X. Zhang, and Y. Wang. 2014. Efficient conversion of myricetin from Ampelopsis grossedentata extracts and its purification by MIP-SPE. *J. Chromatogr. B* 945–946(1):39–45.
55. Bi, W., M. Tian, and K. Ho Row. 2013. Evaluation of molecularly imprinted anion-functionalized poly(ionic liquid)s by multi-phase dispersive extraction of flavonoids from plant. *J. Chromatogr. B* 913–914(1):61–68.
56. Barfi, B., A. Asghari, M. Rajabi, A. Barfi, and I. Saeidi. 2013. Simplified miniaturized ultrasound-assisted matrix solid phase dispersion extraction and high performance liquid chromatographic determination of seven flavonoids in citrus fruit juice and human fluid samples: Hesperetin and naringenin as biomarkers. *J. Chromatogr. A* 1311(1):30–40.
57. Liu, X. G., H. Yang, X. L. Cheng et al. 2014. Direct analysis of 18 flavonol glycosides, aglycones and terpene trilactones in Ginkgo biloba tablets by matrix solid phase dispersion coupled with ultra-high performance liquid chromatography tandem triple quadrupole mass spectrometry. *J. Pharm. Biomed. Anal.* 97:123–128.
58. Rauha, J., H. Vuorela, and R. Kostiainen. 2001. Effect of eluent on the ionization efficiency of flavonoids by ion spray, atmospheric pressure chemical ionization, and atmospheric pressure photoionization mass spectrometry. *J. Mass Spectrom.* 36(12):1269–1280.
59. Wu, H., J. Guo, S. Chen et al. 2013. Recent developments in qualitative and quantitative analysis of phytochemical constituents and their metabolites using liquid chromatography-mass spectrometry. *J. Pharm. Biomed. Anal.* 72:267–291.
60. Ma, Y. L., Q. M. Li, H. Van den Heuvel, and M. Claeys. 1997. Characterization of flavone and flavonol aglycones by collision-induced dissociation tandem mass spectrometry. *Rapid Commun. Mass Spectrom.* 11(12):1357–1364.
61. de Rijke, E., A. Zafra-Gómez, U. A. Freek Ariese, T. H. Brinkman, and C. Gooijer. 2001. Determination of isoflavone glucoside malonates in *Trifolium pratense L.* (red clover) extracts: Quantification and stability studies. *J. Chromatogr. A* 932(1–2):55–64.

62. Huck, C. and G. Bonn. 2001. Evaluation of detection methods for the reversed-phase HPLC determination of 3′,4′,5′-trimethoxyflavone in different phytopharmaceutical products and in human serum. *Phytochem. Anal.* 12(2):104–109.
63. Hollman, P. C. H., J. van Trijp, and M. Buysman. 1996. Fluorescence detection of flavonols in HPLC by postcolumn chelation with aluminum. *Anal. Chem.* 68(19):3511–3515.
64. Klejdus, B., J. Vacek, V. Adam et al. 2004. Determination of isoflavones in soybean food and human urine using liquid chromatography with electrochemical detection. *J. Chromatogr. B* 806(2):101–111.
65. Magnuszewska, J. and T. Krogulec. 2013. Application of hot platinum microelectrodes for determination of flavonoids in flow injection analysis and capillary electrophoresis. *Anal. Chim. Acta* 786:39–46.
66. Stobiecki, M. 2000. Application of mass spectrometry for identification and structural studies of flavonoid glycosides. *Phytochemistry* 54(3):237–256.
67. Careri, M., A. Mangia, and M. Musci. 1998. Overview of the applications of liquid chromatography–mass spectrometry interfacing systems in food analysis: Naturally occurring substances in food. *J. Chromatogr. A* 794(1–2):263–297.
68. González-Paramás, A. M., C. Santos-Buelga, M. Dueñas, and S. González-Manzano. 2011. Analysis of flavonoids in foods and biological samples. *Mini-Rev. Med. Chem.* 11(14):1239–1255.
69. March, R. and J. Brodbelt, 2008. Analysis of flavonoids: Tandem mass spectrometry, computational methods, and NMR. *J. Mass Spectrom.* 43(12):1581–1617.
70. Volikakis, G. J. and C. E. Efstathiou. 2000. Determination of rutin and other flavonoids by flow-injection/adsorptive stripping voltammetry using nujol-graphite and diphenylether-graphite paste electrodes. *Talanta* 51(4):775–785.
71. Du, J., Y. Li, and J. Lu. 2001. Flow injection chemiluminescence determination of rutin based on its enhancing effect on the luminol-ferricyanide/ferrocyanide system. *Anal. Lett.* 34(10):1741–1748.
72. Nalewajko-Sieliwoniuk, E., J. Nazaruk, J. Kotowska, and A. Kojło. 2012. Determination of the flavonoids/antioxidant levels in *Cirsium oleraceum* and *Cirsium rivulare* extracts with cerium(IV)–rhodamine 6G chemiluminescence detection. *Talanta* 96:216–222.

CHAPTER 22

Determination of Glutathione

L.K. Shpigun

CONTENTS

22.1 PHYSIOLOGICAL SIGNIFICANCE OF GLUTATHIONE

Glutathione (GSH) (L-γ-glutamyl-cysteinyl-glycine or (2S)-2-amino-4-{[(1R)-1-[(carboxy-methyl)-carbamoyl]-2-sulfanylethyl]-carbamoyl}butanoic acid) is a tripeptide composed of glycine, L-cysteine, and L-glutamate. This substance belongs to an important group of low-molecular-mass thiols that play unique roles in biological systems, food, pharmaceutical, and aroma industries. Structurally, its molecule contains an unusual peptide linkage between the carboxyl group of the glutamate side-chain and the amine group of L-cysteine.

GSH is known as the "mother of all antioxidants, maestro of detoxification, and chief-pilot of the immune system" in the human body [1]. From a survey of the literature, this substance fulfils significant physiological functions [2–5]. First of all, GSH is the major endogenous antioxidant that prevents damage to important cellular components caused by reactive oxygen species such as free radicals and peroxides [6,7]. Moreover, GSH takes part in the maintaining of other exogenous antioxidants such as vitamins C and E in their reduced (active) forms [8]. GSH is a pivotal component of the GSH–ascorbate cycle, a system that reduces poisonous hydrogen peroxide. In addition, it has a vital function in iron metabolism as well as in the regulation of the nitric oxide cycle, which is critical for life but can be problematic if unregulated. While the intracellular concentration of GSH ranges from 2 to 8 mM, the tripeptide occurs extracellularly in concentrations ranging from 5 to 15 µM [9]. The highest concentration of GSH is found in the liver (up to 5 mM) making it critically important in the detoxification and elimination of free radicals [10]. Depletion of GSH has been noted to accompany premature arteriosclerosis, leukemia, cervical cancer, diabetes, sepsis, liver disease, cataracts, and several other disorders [11]. In contrast, abnormally high levels of these agents have been recorded in patients suffering from AIDS (acquired immunodeficiency syndrome)-related dementia, Alzheimer disease, and Parkinson disease, and in the immediate aftermath of strokes [12,13]. Furthermore, there is increasing evidence that GSH might play a protective role in the treatment of patients with drug-resistant cancer; it could potentially increase side-effects of chemotherapy since GSH-based detoxification mechanisms are also present in normal host cells [14].

Several studies have been demonstrated that reduced GSH serves as a reducing agent in many biochemical reactions, being converted to oxidized glutathione or glutathione disulfide (GSSG) in which the cysteine residues of two GSH molecules are connected by a disulfide bridge.

GSH can be regenerated from GSSG by the enzyme glutathione reductase (GR) linked to the NADPH/NADP+ (nicotinamide adenine dinucleotide phosphate) system.

The GSH–GSSG system is being investigated extensively. The ratio of reduced GSH to GSSG within cells is often used as a measure of cellular toxicity. The normal bounds of glutathione in its oxidized and reduced forms in physiological fluids are succinctly summarized in a review [15]. In healthy cells and tissue, more than 90% of the total GSH pool is in the reduced form and less than 10% exists in the disulfide form. An increased GSSG-to-GSH ratio is considered indicative of oxidative stress, immune deficiency, and xenobiotic overload.

TABLE 22.1 Concentration Level of GSH in Various Foodstuffs

Product	Content of GSH	Reference
Grape juice	0–40 mg/L	[18]
	100–750 µM	[19]
Mangos	40–255 nmol/g	[20]
Kiwi	0.709 mol/g	[21]
Jujube	0.837 mol/g	
Physalis	1.417 mol/g	
White wine	2–5 mg/L	[18]
	1–27 mg/L	[22]
Tomatoes, cucumbers	3–300 mg/g	[23]
Tomato plants hydroponically grown	51–100 nmol/g	[24]
Vegetables (broccoli, potato, etc.)	0–60 nmol/g	[20]
	1–200 µM	[25]
Brussels sprouts	112 µmol/100 g	[26]
Wheat germ	180 mg/100 g	[27]
Maize plant	1–6 µg	[28]
Brown rice	520–550 nmol/g	[29]
Marine products	0.125–1.250 mg/L	[30]
Baker's yeast	2–5 mg/g	[31]
Brewer's yeast	5–100 mg/L	[32]
Meats	21–45 nmol/g	[20]

Although GSH can be synthesized by the cells, these days commercial supplements of GSH are widely popular. When the plasma GSH concentration is low, such as in patients with HIV infection, alcoholics, and patients with cirrhosis, increasing the availability of circulating GSH by oral administration might be of therapeutic benefit [16]. A number of synthetic antioxidants have been established on the basis of GSH in the form of biologically active food additives, radioprotectors, and drugs used for the treatment of atherosclerosis, coronary heart disease, and intoxication of the organism [17]. GSH is found naturally in many foodstuffs (Table 22.1).

22.2 GSH AS AN ANALYTICAL OBJECT

Due to increasing production and consumption of plant food, it is necessary to provide routine control of amounts of GSH and its metabolites in different objects. A review of the techniques used for identifying and determining GSH content in food samples is given in [33]. According to the literature, numerous methods are available for measuring GSH and GSSG, but few are suitable for direct analysis in routine use [28,34–37]. In almost all cases the determination of GSH and related thiols in food is quite complicated. The main difficulties in the assay of thiols lie in their unfavorable physicochemical properties. In particular, these compounds are highly polar and water soluble, which makes their extraction from complex matrices almost impossible. The absence of a chromophore group presents the next problem. Therefore, all methods, except those based on electrochemical and tandem mass spectrometry, include a step of derivatization of GSH. The derivatizing agents should react rapidly and specifically with thiols to form stable products. They typically include tagging labels introducing UV–Vis, fluorescent, or electrochemical properties to the analytes. Although GSH has three sites susceptible to derivatization—the carboxylic, amino, and thiol, functionalities—labeling of the thiol moiety is preferred for its specificity and for its function as protective group. In contrast, GSSG can undergo derivatization only on the amino and carboxylic groups. As a rule, the derivatization of GSH, GSSG, and their analogues with an appropriate derivatizing agent improves the sensitivity of their spectrophotometric or spectrofluorometric detection. It also suggests the possibility of their determination by means of indirect detection. The derivatization of thiols has previously been reviewed in detail [15,38–40].

Table 22.2 summarizes information about the most commonly used derivatizing agents described in the literature for GSH and related compounds. Among them, 5,5′-dithiobis(2-nitrobenzoic acid) (DTNB, Ellman's reagent) has been employed in several studies for the determination of both GSH and GSSG [57,58]. It should be noted that DTNB can serve as a thiol-masking agent only when used in a great molar excess [59]. Otherwise, other thiols react with the DTNB-derivatized GSH molecule to form disulfides, reactions utilized in the GR-coupled enzymatic recycling method. 4-DPS (4,4′-dithiodipyridine) was found to be a great advantage over DTNB due to its hydrophobic nature, its smaller size, and the possibility of its use at low pH, where thiol–disulfide exchange reactions normally do not occur [47]. FDNB (1-fluoro-2,4-dinitrobenzene), a derivatizing agent of the amino group, has the advantage of allowing the simultaneous evaluation of GSH and GSSG in a single run [15,41]. The alkyl esters of propiolic acid have been recognized as efficient and advantageous UV labels for these types of compounds [31,43–45]. The main advantages of these reagents include (i) the reaction kinetics are fast (reaction time less than 1 min), (ii) the reaction proceeds under mild

TABLE 22.2   Derivatizing Agents for GSH and Related Compounds

Derivatizing Agent	Detection Mode	Reference
FDNB; 1-fluoro-2,4-dinitrobenzene	UV, 355 nm	[41]
CDNB; 1-chloro-2,4-dinitrobenzene	UV	[42]
MP; methyl propiolate	UV, 285 nm	[31,43]
EP; ethyl propiolate	UV	[44,45]
DTNB; 5,5′-dithiobis(2-nitrobenzoic acid), Ellman's reagent	Vis, 402 nm	[36]
	MS	[46]
4-DPS; 4,4′-dithiodipyridine	UV	[47]
NEM; N-ethylmaleimide;	UV, 200 nm	[48]
OPA; o-phthalaldehyde	UV, 388 nm, FL	[49]
	FL	[18,50]
MBB; monobromobimane	FL	[51]
MMPB; 5-maleimidyl-2-(m-methylphenyl)-benzoxazole		[52]
NDA; naphthalene-2,3-dicarboxaldehyde		[53,54]
SBD-F; ammonium 7-fluorobenzo-2-oxa-1,3-diazole-4-sulfonate		[55]
ABD-F; 4-(aminosulfonyl)-7-fluoro-2,1,3-benzoxadiazole		[56]
Diazo-luminol	CL	[57]

Note: UV, ultraviolet absorbance detection; Vis, visible absorbance detection; FL, fluorescence detection; MS, mass spectrometric detection; CL, chemiluminescence detection.

conditions (room temperature, aqueous alkaline media), (iii) the reagents are commercially available and at relatively low cost [31].

Fluorophores that react with thiol groups are the most selective and also prevent artifacts arising from the autoxidation of GSH during sample pretreatment and analysis, but do not allow the detection of GSSG. o-Phthalaldehyde (OPA) is found to be an attractive agent for derivatization of thiols: it reacts with GSH with formation of a highly fluorescent tricyclic isoindole derivative without the need for a thiol coreagent [18,49,50]. Naphthalene-2,3-dicarboxaldehyde (NDA), a reagent similar to OPA, has been successfully used for the simultaneous assessment of ROS (reactive oxygen species) (such as superoxide, hydroxyl radical, or hydrogen peroxide) and the levels of GSH [54]. SBD-F (ammonium 7-fluorobenzo-2-oxa-1,3-diazole-4-sulfonate) is a low-reactivity reagent and requires drastic conditions to react quantitatively with thiols (pH 9.5 and 60°C for 1 h) [55]. In contrast, ABD-F (4-(aminosulfonyl)-7-fluoro-2,1,3-benzoxadiazole) offers fast and quantitative reaction under mild conditions, and is therefore suitable for online derivatization with thiols [56].

Generally, the following points must be checked in order to achieve accurate determination of GSH and GSSG: sample collection, reduction of disulfides, deproteinization, and derivatization of thiols. In particular, GSH in various food samples can be determined after extraction with an aqueous solution of perchloric acid, which is effective for removal of protein and stabilization of GSH in reduced form.

To date, the activities of all research groups that aim to improve methods for the determination of GSH and related compounds are greatly increased. To overcome the difficulties associated with sample pretreatment and conversion of the analytes to detectable derivatives, the application of automated methods, such as flow injection (FI) or sequential injection (SI) techniques, is especially promising and provides viable alternative for the determination of GSH and GSSG in a variety of sample matrixes [60]. The reported FI and SI methods for the determination of GSH are based on UV–Vis [61–64], calorimetric [65], chemiluminescence (CL) [66–69], and electrochemical [70–72] detection. In addition, flow cytometric methods with fluorometric detection that allow rapid measurement of cellular GSH content in clinical samples, namely in tumor tissue, would be useful for biochemical analysis [73,74].

22.3 DETERMINATION OF GSH BY FI- AND SI-SPECTROPHOTOMETRY

22.3.1 FI Methods Based on Color Redox Reactions

The occurrence of a thiol moiety in the GSH molecule endows it with the potential to act as a reductant by donating an electron to some endogenous acceptors, and to behave as a stabilizer of free radicals by donating a hydrogen atom to the latter species. That is why several variants of FI or SI procedures proposed for the determination of thiols are based on redox reactions. Among these reactions, oxidation of the analyte by Fe(III) or hexacyanoferrate(III) was found to be most suitable [75,76]. In particular, GSH could be determined using the complexes Fe(III)-1,10-phenanthroline (phen) or Fe(III)-2,2′-dipyridyl (dipy) as a chromogenic agent. Figure 22.1 shows a schematic diagram of the spectrophotometric flow system used. As can be seen, sample solution (200 µL) and reagent solution (200 µL) are injected simultaneously into separate carrier solution streams in a two-line manifold. The colored merged zone is then transported to

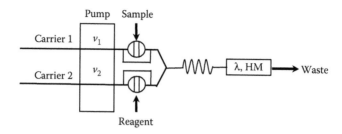

FIGURE 22.1   A Schematic diagram of the FI manifold used for indirect spectrophoto-metric determination of GSH.

a flow-through spectrophotometric cell. The recording of the absorbance signal is based on using the following indicator reaction between GSH and Tris-chelate of Fe(III), which accompanied with the formation of intensively colored cationic complexes of Fe(II)-phen ($\lambda_{max} = 510$ nm) or Fe(II)-dipy ($\lambda_{max} = 525$ nm):

$$2GSH + 2[L_3Fe^{III}]^{3+} \leftrightarrow 2[L_3Fe^{II}]^{2+} + GS - SG + 2H^+ \tag{22.1}$$

The reaction is comparatively slow. However, it was noticed that copper(II) greatly accelerates the reaction rate with GSH (Figure 22.2). The reduction of Cu^{2+} ions by GSH could give rise to the formation of a redox-active species capable of catalyzing the subsequent reduction of molecular oxygen into superoxide anion, and that of hydrogen peroxide into hydroxyl radical [77]:

$$2Cu^{2+} + 2GSH \rightarrow 2Cu^+ + GSSG$$

$$Cu^{2+} + O_2 \leftrightarrow Cu^{2+} + O_2^{\cdot -} \tag{22.2}$$

$$Cu^+ + H_2O_2 \rightarrow Cu^{2+} + HO^\cdot + HO^-$$

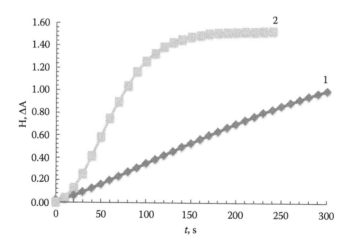

FIGURE 22.2   Kinetic curves demonstrating spectrophotometric behavior of GSH in the absence (1) and in the presence of Cu (2) in an acidic medium (pH 3.4).

Expressed in terms of measured absorbance units, the linear calibration function for the peak height can be summarized by the following equation:

$$H = 0.002 + 0.232 \, [GSH] \tag{22.3}$$

The most relevant parameters that affected on the residence time of the sample in the system were the flow rates and the reaction coil (RC) length. At the optimized conditions, the linear range for the determination of GSH was (0.0–1.0) mM with the detection limit (3σ) of 0.03 mM. The average sample throughput was 60–75 h⁻¹.

A FI spectrophotometric method has been proposed for the simultaneous determination of GSH and ascorbic acid (AA) in a single injection [61]. In contrast to GSH, AA rapidly reduces an Fe(III)-phen complex. Therefore, the use of the double flow cell that has two cell compartments in parallel makes it possible to detect simultaneously two analytes from a single shot, that is, AA can be detected at 510 nm in the absence of copper when an injected sample zone passes through the first compartment in the double flow cell, and then after the sample zone merges with a copper(II) solution, the total concentration of AA and GSH can be detected in the second compartment. Under the optimum conditions, AA and GSH can be determined simultaneously in the ranges 5×10^{-4}–0.5 mM for AA and 2×10^{-3}–0.1 mM for GSH with a sample throughput rate of 45 h⁻¹. The proposed FI method was successfully applied to the determination of these compounds in pharmaceutical preparations and a supplement.

A FI system for the indirect spectrophotometric detection of GSH through reaction with the intensively colored radical 2,2′-diphenyl-1-picryl-hydrazyl (DPPH•) has been described [78]. The reaction was carried out in a FI manifold with *in-valve* reactor (Figure 22.3). The test solution stream merged with the DPPH• solution stream (0.2 mM) and filled a coiled reactor that was placed instead of a sample loop in the injection valve. After 300 s, an aliquot of the reaction mixture (300 μL) was injected into the carrier stream (50% ethanol) which was pumped through a flow cell of the spectrophotometric detector. The detector registered the negative absorbance signal corresponding to the decrease of the concentration of DPPH• due to the following reaction schema [79]:

$$GSH + DPPH^{\bullet} \rightarrow GS^{\bullet} + DPPH-H$$

$$2RS^{\bullet} \rightarrow GS-SG \tag{22.4}$$

$$GS^{\bullet} + DPPH^{\bullet} \rightarrow DPPH-SR$$

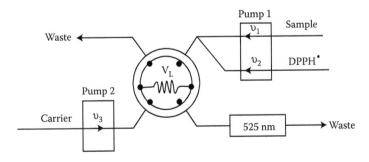

FIGURE 22.3   A schematic diagram of the FI manifold for the spectrophotometric determination of GSH by using reaction with free radical DPPH•.

Under the chosen conditions, Lambert–Beer's law was obeyed in the concentration range of 5×10^{-3}–0.05 mM GSH

$$H = -7.307[\text{GSH}] + 0.693 \,(r = 0.9999) \tag{22.5}$$

The average sample throughput was 60–75 h^{-1}.

22.3.2 SI Methods Based on the Formation of UV-Absorbing Thioacrylate Derivatives

Zacharis et al. [62] proposed a SI method based on the online formation of UV-absorbing thioacrylate derivatives ($\lambda_{max} = 285$ nm) upon reaction of GSH with ethyl propiolate (EP) in alkaline medium. The reaction mechanism is based on the nucleophilic attack of the thiolate ion on the α-carbon atom of the triple bond, resulting in a stable ethyl thioacrylate compound that absorbs in the region of 280–290 nm [15,44]. A schematic diagram of the SI manifold used is depicted in Figure 22.4. The various variables that typically affect such a protocol were studied. At the optimum conditions, the linear range for the determination of GSH was 0.15–70.0 mg/L with the detection limit (3σ) of 50 μg/L. As many as 100 sample injections were enabled per hour.

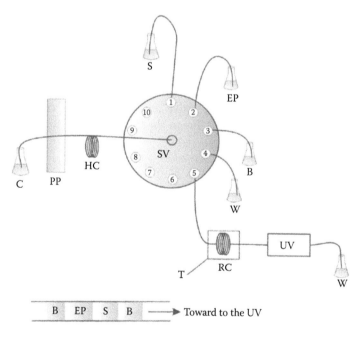

FIGURE 22.4 A typical SI manifold for the UV – detection of GSH-EP: C, carrier (water); PP, peristaltic pump; HC, holding coil (300 cm/0.5 mm i.d.); SV, selection valve; UV, detector (λmax = 285 nm); S, sample; B, Britton-Robinson buffer; RC, reaction coil; T, thermostat; W, waste. (Adapted from Zacharis, C. K. et al. 2009. *J. Pharm. Biomed. Anal.* 50:384–391.)

22.3.3 FI and SI Methods Using Enzymatic Recycling Reactions

Enzymatic assays are advantageous in contrast with other methods that involve a non-specific derivatization of the GSH sulfhydryl group. These assays basically fall into two categories: 5,5′-dithiobis(2-nitrobenzoic acid) (DTNB)–GR recycling assay [80] and the GSH-S-transferase 1-chloro-2,4-dinitrobenzene (CDNB) endpoint method [81]. DTNB (Ellman's reagent) is utilized for the derivatization of GSH and GSSG in the classical spectrophotometric GR-coupled enzymatic recycling or GSH-recycling assay developed by Tietze [82]. It is based on the progressive reduction of GSSG to GSH with GR in the presence of NADPH, while the rate of formation of the colored product 5-thionitrobenzoate is measured (Figure 22.5).

To date, two methods have been reported for GSH and/or GSSG determination by using FI- or SI-spectrophotometry based on enzymatic recycling reactions. A sensitive FI method for the determination of total ($\Sigma_{GSH+GSSG}$) and oxidized glutathione (GSSG) has been proposed by Redegeld et al. [63]. The method provided specific amplification of the response to GSH by combined use of GR and DTNB. GSSG was detected separately after alkylation of GSH with N-ethylmaleimide (NEM). The reduced GSH concentration was then obtained as the difference between the $\Sigma_{GSH+GSSG}$ and GSSG concentrations. A schematic diagram of the FI manifold is shown in Figure 22.6. A 0.4 M potassium phosphate buffer of pH 6.75 was used as carrier and as solvent for reagent solutions R1 (1 mM DTNB) and R2 (0.5 mM NADPH + 3 units GR/mL). Both carrier

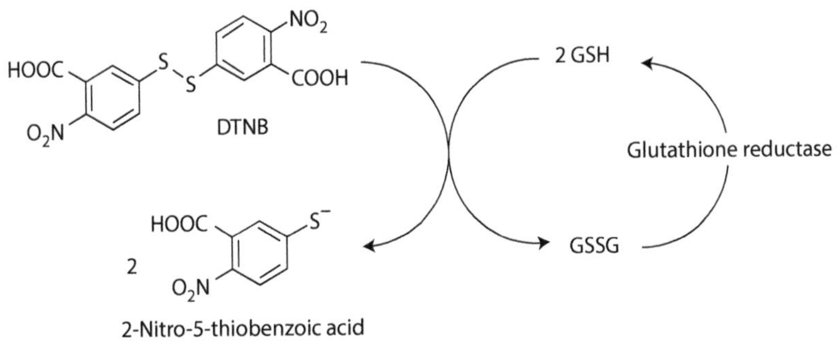

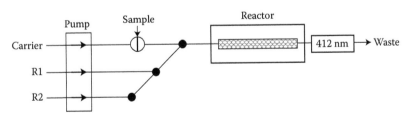

FIGURE 22.5 The DTNB-GSSG reductase recycling system. (Adapted from Araujo, A. R. T. S., M. L. M. F. S. Saraiva, and J. L. F. C. Lima. 2008. *Talanta* 74:1511–1519.)

FIGURE 22.6 A schematic diagram of the FI manifold used for the GSH determination based on an enzymatic recycling reaction. (Modified from Redegeld, F. A. M. et al. 1988. *Anal. Biochem.* 174(2):489–495.)

and reagent solutions were pumped with a flow rate of 0.2 mL/min. The enzymatic recycling reaction was performed in a single bead string reactor (2 m long, 1.5 mm i.d.) filled with 1.0-mm glass beads and fixed in a water bath thermostated at 25°C. The accumulation of the thiolate anions was detected spectrophotometrically at 412 nm. The influence of residence time, temperature, and enzyme concentration on the response characteristics has been studied and the optimum reaction conditions have been selected. Under standard conditions the mean residence time was 235 s, the amplification factor was 28, and the sample throughput was 30 h^{-1}. The peak height was linear up to 400 pM for GSH ($r = 0.9996$) and up to 200 pM for GSSG ($r = 0.9997$). The method has been evaluated by the quantification of GSH and GSSG in biological samples (isolated hepatocytes). A high correlation between the data obtained by this FI method and the original spectrophotometric batch assay has been found (slope = 1.039, INTERCEPT = 0.6, $n = 216$, $r = 0.977$).

Later, SI method has been developed for the enzymatic determination of ΣT_{GSH} and oxidized (GSSG) glutathione (Figure 22.7) [64]. The analysis was also based on the DTNB–GR recycling assay, followed by spectrophotometric detection of the 2-nitro-5-thiobenzoic acid (TNB) formed ($\lambda = 412$ nm). The implementation of this reaction in a SI system with an in-line dilution strategy permitted the necessary distinct application ranges for $\Sigma_{GSH+GSSG}$ and GSSG.

The recycling assay began with the aspiration of sample/standard to the holding coil (HC) followed by DTNB, enzyme, and NADPH solutions. After flow reversal, the four-stacked zones were sent through the RC to the thermostatically controlled bath where the flow was stopped. After a preset time period the pump was reactivated and the reaction zone was sent through the detector toward the flow cell, where the signal was acquired, after which the content of the flow cell was washed out. The HC and RC were respectively 1.5 and 3 m length, both of which had a figure-of-eight shaped configuration. The four solutions involved in this reaction were drawn into the HC in the following order: sample/standard, DTNB, enzyme, and NADPH. Although a complete overlap of the four zones had already been achieved on passage through the detector,

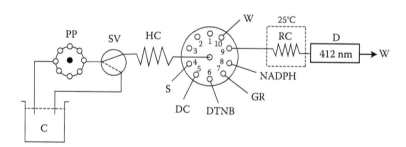

FIGURE 22.7 A schematic diagram of the SI manifold for enzymatic determination of GSH and GSSG: C, carrier (propulsion flow rate of 1.90 mL min^{-1}) – 100 mM phosphate buffer solution (pH 7.0); PP –peristaltic pump; SV – solenoid valve; HC – holding coil; DC – dilution coil; S – standard sample; RC – reaction coil (fixed in the same water baht); D – spectrophotometric detector; W – waste. Chemical conditions: DTNB = 3.0 mM DTNB solution; GR = 5.0 U/ml enzyme solution (immersed in a water bath thermostatically controlled at 25 °C); NADPH = 1.2 mM NADPH solution. (Adapted from Araujo, A. R. T. S., M. L. M. F. S. Saraiva, and J. L. F. C. Lima. 2008. *Talanta* 74:1511–1519.)

TABLE 22.3   Parameters of Equation $C_p = C_o + sC_c$ for Comparing the Results (μM) Obtained by the Proposed SI Method (C_p), the Comparison Batch Procedure (C_c), and the Results of the Student t-Test

Analyte	C_o	s	r	$t_{0.05}$ Calculated	$t_{0.05}$ Tabulated
Total GSH	8 ($\pm$105)	1.0 ($\pm$0.1)	0.9797	0.53	2.16
GSSG	0.1 ($\pm$0.2)	0.97 ($\pm$0.03)	0.9984	0.53	2.16

Source: Araujo, A. R. T. S., M. L. M. F. S. Saraiva, and J. L. F. C. Lima. 2008. *Talanta* 74:1511–1519.

Note: The values in parentheses are the limits at the 95% confidence.

sample and DTNB should be aspirated first as this minimizes the physical dispersion of enzyme and NADPH.

The influence on the sensitivity and performance of the SI system of parameters such as reagent concentrations, temperature, pH, flow rate of the carrier buffer solution, as well as RC length, etc., was studied. GR concentration was studied between 0.4 and 5.0 U/mL and an increment of signal was observed throughout the range.

Under the optimized conditions, the calibration curves

$$Abs(a.u.) = (0.123 \pm 0.002)[GSH](\mu M) + (0.097 \pm 0.003), \quad R^2 = 0.998$$

$$Abs(a.u.) = (0.245 \pm 0.003)[GSSG](\mu M) + (0.098 \pm 0.002), \quad R^2 = 0.9994$$

obtained showed linear responses in concentration ranges 0.00–3.00 μM and 0.00–1.50 μM for GSH and GSSG, respectively. Detection limits of 0.031 and 0.014 μM were achieved for GSH and GSSG, respectively. The developed method showed good precision, with a relative standard deviation (RSD) <5.0% ($n = 10$) for determination of both GSH forms. Statistical evaluation showed good compliance, for a 95% confidence level, between the results obtained with the SIA system and those furnished by the comparison batch procedure (Table 22.3) [64].

22.4   DETERMINATION OF GSH BY FI-CL

CL is an attractive means of detection for analysis of biological substances because of its high sensitivity (nanomolar amounts) and wide linear working ranges, with relatively simple instrumentation. The first papers in which CL was used in the measurement of organosulfur compounds were published in 1987–1989 [83]. It was demonstrated that the transient CL emission is suitable for detection of thiols in a wide dynamic range. Since then, bioluminescent FI systems for the determination of GSH have been developed including reactions of luminol–peroxynitrite [66], luminol–H_2O_2 [67,68], and luminol–$NaIO_4$ [69]. The FI manifolds used are shown in Figures 22.8 through 22.10. In all cases, CL detection was carried out using a luminometer built on-site, consisting of a flat-coil flow cell and a photomultiplier tube.

The possible enhanced CL mechanism of the reaction between GSH and hydrogen peroxide may be attributed to the following reactions in alkaline solution:

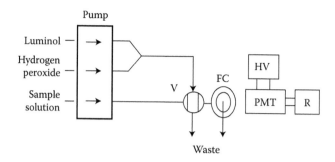

FIGURE 22.8   A schematic diagram of the «reversed» CL-FI system utilizing the luminol – hydrogen peroxide reaction: V – six-way injection valve; FC – flow cell; HV – high
voltage; PMT – photomultiplier tube; R – recorder. (Adapted from Du, J., Y. Li, and J. Lu.
2001. *Anal. Chim. Acta* 448(1–2, 3):79–83.)

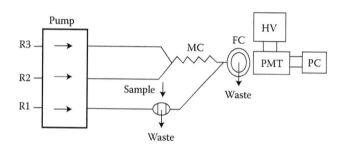

FIGURE 22.9   A schematic diagram of the FI system based on oxidation of luminol by
sodium periodate in alkaline solutions: R1, H_2O; R2, basic luminol; R3, sodium periodate; P, peristaltic pump; S, injection valve; MC, mixing coil; FC, flow cell; PMT, photomultiplier tube; PC, personal computer; HV, negative voltage. (Modified from Wang L., Y.
Li, D. Zhao, and C. Zhu. 2003. *Microchim. Acta* 141:41–45.)

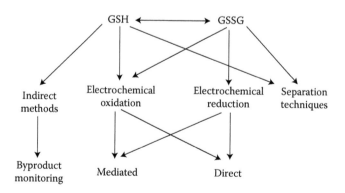

FIGURE 22.10   A map of the various routes that have been explored in the electroanalytical determination of thiols. (Adapted from Harfield, J. C., Ch. Batchelor-McAuley, and R.
G. Compton. 2012. *Analyst* 137:2285–2296.)

$$H_2O_2 + HO^- \leftrightarrow HO_2^- + H_2O$$

$$HO_2^- - e \rightarrow HO_2^{\cdot}$$

$$2HO_2^{\cdot} \rightarrow H_2O_2 + O_2^{\cdot}$$

$$2GSH + 2O_2^{\cdot} \rightarrow 2GS^{\cdot} + H_2O \qquad\qquad (22.6)$$

$$2GS^{\cdot} \rightarrow GSSG^*$$

$$GSSG^* + \text{aminophthalate} \rightarrow \text{aminophthalate}^* + GSSG$$

$$\text{aminophthalate}^* \rightarrow \text{aminophthalate} + \text{light (425 nm)}$$

The excited state of the 3-aminophthalate ion has been confirmed as an emitter.

Du et al. [67] developed a FI-CL method for the GSH determination based on this phenomenon. As shown in Figure 22.8, the luminol solution stream was merged with hydrogen peroxide solution stream prior to reaching a six-way injection valve. Then, 40 μL of the mixture of luminol and hydrogen peroxide solutions was injected into the stream of test solution, producing CL. CL intensity was directly proportional to the concentration of GSH in the range from 2.0×10^{-9} to 2.0×10^{-7} M with the detection limit (3σ) of 9×10^{-10} M.

The enhancing effect of GSH on the CL reaction of luminol with hydrogen peroxide in alkaline solution has also been utilized by Wang et al. [68] to develop a rapid FI method for the determination of this compound. The effect of the flow rate was found to be a critical parameter: the CL intensity was increased with the increasing flow rate in the range of 40–75 rotations/min. Lower flow rates result in lower CL emission; at high flow rates, more light is emitted, resulting in greater peak heights, probably due to better dispersion and mixing. A flow rate of 65 rotations/min was chosen as the optimum parameter because higher rates led to both high pressures and excessive consumption of reagents. Under optimal conditions, the maximum CL intensity (I_{CL}) was directly proportional to the concentration of GSH in the range from 3.0×10^{-7} to 2.0×10^{-5} M. The regression equation was $I_{CL} = 94.75 + 66.62 \times 10^6 [\text{GSH}]$ (M) with a correlation efficient of 0.9979 ($n = 9$). The detection limit was 6.8×10^{-8} M. The RSD was 3.4% for 5.0×10^{-6} M of GSH ($n = 11$). The influence of foreign substances on the determination of 5.0 μM GSH was studied (Table 22.4). The tolerance limit was estimated with a 5% relative error. When added at levels higher than those given, all interfering substances caused a positive error. The results of the determination of GSH in three artificial samples, in which a series of foreign substances had been added based on the tolerance level of foreign substances are displayed in Table 22.5. It can be seen that GSH in samples can be determined with satisfactory results.

A new sensitive FI-CL method for the determination of GSH at trace levels has been proposed by Ensafi et al. [69]. The method is based on the effect of GSH on the CL signal of the oxidation of luminol by sodium periodate in alkaline solutions. The schematic diagram of the flow system is shown in Figure 22.9. The flow lines were connected with a carrier H_2O (R1), 2.0×10^{-4} M luminol solution in Na_2CO_3 buffer (pH 10.5) (R2), and 1.5×10^{-3} M sodium periodate solution (R3). Luminol solution was first mixed with $NaIO_4$ solution through 20-cm silicon tubing (1.0 mm i.d.) to give a stable baseline. Then 250 μL of GSH solution was injected into the carrier stream. The mixture was

TABLE 22.4   Effect of Foreign Substances on the Determination of 5.0 μM GSH under Optimum Conditions

Interference Added	Maximum Tolerance Concentration[a] (mM)
K^+, Na^+, NO_3^-, SO_4^{2-}, Cl^-	10
CO_3^{2-}, HCO_3^-, PO_4^{3-}, Ba^{2+}, Ca^{2+}	1.0
Cd^{2+}, Mn^{2+}, Al^{3+}, Pb^{2+}, GSSG	0.2
Glucose, cysteine	0.04
Fe^{3+}, Co^{2+}, Cu^{2+}, Cr^{3+}, Fe^{2+}	1.0 μM
Human serum albumin	20 μg/mL
Bovine serum albumin	10 μg/mL

Source: Wang L. et al. 2003. *Microchim. Acta.* 141:41–45.
[a] Maximum concentration causing a relative error of less than 5%.

TABLE 22.5   Determination of GSH in Synthetic Samples ($n = 5$)

GSH in Samples (μM)	Main Additives	Found Value (μM)	Recovery Range (%)	RSD (%)
1.0	Ba^{2+}, Ca^{2+} (0.8 mM), K^+ (5.0 mM)	0.98	96.3–101.8	1.9
2.0	NO_3^-, Cl^- (5.0 mM) CO_3^{2-} (0.6 mM)	2.06	97.1–106.4	2.4
5.0	Glucose (1 μM) Cysteine (10 μM)	5.10	96.7–105.3	2.3

Source: Wang L. et al. 2003. *Microchim. Acta* 141:41–45.

passed through the CL cell, while the CL signal was recorded with the photo multiplier tube (PMT). At the optimized conditions, the linear range for the determination of GSH was 1.0×10^{-8} to 1.0×10^{-5} M with the detection limit (3σ) of 8×10^{-9} M. The RSD for 10 repeated measurements of 1.0×10^{-6} M of GSH was 4%. The results of the method were compared with the Ellman's reference method and no significant difference was found.

22.5 FI METHODS FOR THE ELECTROCHEMICAL DETERMINATION OF GSH

GSH can be oxidized to give GSSG disulfide by the loss of 2 protons and 2 electrons, with both species present in equilibrium in the body as in the following equation:

$$2GSH \underset{k2}{\overset{k1}{\rightleftharpoons}} GSSG + 2e^- + 2H^+ \tag{22.7}$$

Therefore, from theoretical point of view, the oxidation of GSH or reduction of GSSG electrochemically can be used for the determination of these compounds in untreated samples [36]. However, in practice several problems have been recognized for the direct determination of GSH by electroanalytical techniques [84]. First, GSH exhibits irreversible oxidations requiring high overpotential at unmodified electrodes. Furthermore, the voltammetric behavior of electroactive species such as ascorbate, urate, and cysteine may lead to interference. Their redox signals may obscure the GSH/GSSG oxidation or reduction waves due to the electrochemical potentials being nearly coincident. In addition, the complex composition of the majority of samples presents significant problems in terms of electrode fouling. A problem with direct amperometric detection of thiols at noble metal electrodes such as platinum and gold is passivation of the electrode surface. As a result, the direct electrochemistry at solid electrodes often needs large anodic potentials to see appreciable signals (formal potential of GSH 0.23 V versus NHE [normal hydrogen electrode]) which then causes responses from other oxidizable species present in solution. That is why the development of different approaches to solving this problem has received considerable interest in recent years. A map of the various routes that have been explored in the electroanalytical determination of thiols is detailed in Figure 22.10 [36].

Numerous studies have been reported of FIA based on immobilized enzyme columns and an electrode or thermistor. In particular, Saton et al. [70] proposed a FI method for the amperometric determination of GSH. The system comprises the immobilized enzyme column and a flow-through membrane-covered platinum/silver/silver chloride electrode pair for detection of hydrogen peroxide. The calibration graph for GSH was linear from 0.05 to 1.0 mM. The procedure is simple and selective. The assay took 3 min; the sample volume was 200 μL. The RSD for 0.5 mM GSH was 2% ($n = 10$).

The most common way to increase the amplitude of the electrochemical signal for GSH or to enhance the selectivity for its oxidation is modification of the electrode. To date, a variety of modified electrodes have been studied for the electroanalytical chemistry of GSH and GSSG with carbon as electrode material. The most popular carbon-based electrode is a glassy carbon electrode (GCE). In particular, the electrochemical behavior of thiols at unmodified and modified GCE with incorporated multiwalled carbon nanotubes and ruthenium complexes was investigated [71]. The electrocatalytic effect of the multiwall carbon nanotubes–nafion modified GCE containing Ru[(try)(bpy)Cl]PF_6 was exploited for the FI determination of GSH in the presence of coexisting AA. Under the optimized conditions the calibration graph was linear in the concentration range 0.01–1.0 mM.

FIA coupled with electrochemical detection has been applied to study and characterize the behavior of various forms of GSH (reduced GSH [GSH], oxidized glutathione [GSSG], and S-nitrosoglutathione [GSNO]) [72]. The optimized conditions were as follows: mobile phase consisted of acetate buffer (pH 3) with a flow rate of 1 mL/min. Based on results obtained, 850 and 1100 mV were chosen as the optimal potentials for detection of GSH and GSSG or GSNO, respectively (Figure 22.11). It should be noted that GSSG gave the lowest current responses in comparison with other compounds of interest. The concentration dependence of GSSG was linear ($y = 5.486x + 0.491$; $R^2 = 0.9939$) within the concentration range from 0.1 to 2.5 μg/mL, with a detection limit of 50 ng/mL. In case of GSNO, a strictly linear concentration dependence ($y = 0.005x + 0.210$; $R^2 = 0.9957$) was obtained within the range from 1 to 500 ng/mL, whereas the detection limit was 300 pg/mL. GSH concentration dependence was also strictly linear ($y = 11.315x - 0.5146$; $R^2 = 0.9969$) within the range from 0.5 to 4.0 ng/mL. The detection limit of GSH was 100 pg/mL.

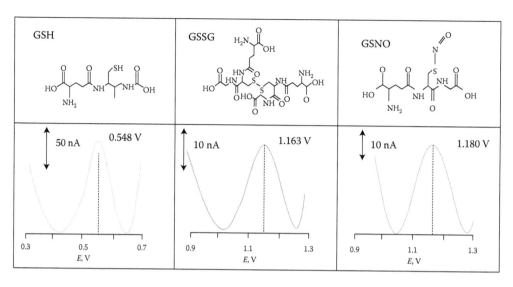

FIGURE 22.11   Square wave voltammograms (SWV) of GSH, GSSG and GSNO (100 µM) measured on carbon paste electrode using AUTOLAB Analyzer. The supporting electrolyte: Britton-Robinson buffer solution, pH 5. SWV parameters were as follows: the initial potential of -0.2 V, the end potential +1.5 V, frequency 200 Hz and step potential 5 mV. (Adapted from Zitka, O. et al. 2007. *Sensors* 7:1256–1270.)

Thus, FIA coupled with electrochemical detectors could be a suitable technique for detection of various forms of GSH. It seems to be very promising tool for such purposes because of very good sensitivity and relatively low cost.

22.6   MICROFLUIDIC SYSTEMS FOR THE DETERMINATION OF GSH

A method for the determination of GSH was developed by using a microfluidic chip coupled with electrochemical detection [85]. In this method, the cell injection, loading, and cytolysis, as well as the transportation and detection of intracellular GSH, were integrated in a microfluidic chip with a double-T injector and an end-channel amperometric detector. A single cell was loaded in the double-T-injector on the microfluidic chip. The GSH from the single cell was electrochemically detected at an Au/Hg electrode.

The FI system was developed for the determination of GSH, using a microfluidic platform with an optical sensor [86]. A schematic diagram of the microfluidic FI setup is shown in Figure 22.12. GSH was detected spectrophotometrically based on reaction with Elman's reagent or DTNB. The effects of various parameters were investigated. The optimum conditions were as follows: a DTNB concentration of 3.0 mM, a phosphate buffer concentration of 100 mM, a buffer pH of 7.0, a sample volume of 5.0 µL, and a flow rate of each stream of 200 µL/min.

The system was successfully employed for the determination of GSH in dietary supplement samples. A linear calibration graph in the range 5.0–60.0 mg/L GSH was obtained with a detection limit of 0.01 mg/L. The system provided a sample throughput of 48 h^{-1}, with microliter consumption of the reagent. The developed system represents a greener analytical system because it was fabricated using a simple and nonhazardous

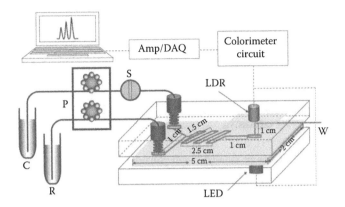

FIGURE 22.12 A schematic diagram of the microfluidic FI setup: C, PBS carrier stream; R, DTNB solution; S, sample (5 μL injection volume); P, peristaltic pump; LED, light-emitting diode; LDR, light-dependent resistor; Amp/ DAQ, amplifier and data acquisition unit; W, waste. (Adapted from Supharoek, S., Youngvises, N., and Jakmunee, J. 2012. *Analyt Sci* 28:651–656.)

technique, and it provides fast analysis with low consumption of the chemicals and low waste production.

REFERENCES

1. Dolphin, D., R. Poulson, and O. Avramovic., eds. 1989. *Glutathione—Chemical, Biochemical and Medical Aspects. Part A.* New York, NY: John Wiley & Sons.
2. Gilbert, H. E. 1989. *Glutathione Centennial Molecular Perspectives and Clinical Implications*, ed. A. Meister, pp. 73–87. San Diego: Academic Press.
3. Wu, G., Y. Z. Fang, S. Yang et al. 2004. Glutathione metabolism and its implications for health. *J. Nutr.* 134(3):489–492.
4. Townsend, D. M., K. D. Tew, and H. Tapiero. 2003. The importance of glutathione in human disease. *Biomed. Pharmacother.* 57:145–155.
5. Sies, H. 1999. Glutathione and its cellular functions. *Free Radicals Biol. Med.* 27:916–921.
6. Baskin, S. I. and H. Salem. 1997. *Oxidants, Antioxidants and Free Radicals*, pp. 173–174. Washington: Taylor & Francis.
7. Halliwell, B. 1994. Free radicals, antioxidants & human disease: Curiosity, cause or consequence? *Lancet* 344(8924):721–724.
8. Noctor, G. and C. H. Foyer. 1998. Ascorbate and glutathione: Keeping active oxygen under control. *Annu. Rev. Plant Physiol. Plant Mol. Biol.* 49:249–279.
9. Moran, L. K., J. M. C. Gutteridge, and G. J. Quinlan. 2001. Thiols in cellular redox signaling and control. *J Curr. Med. Chem.* 8:763–772.
10. Soboll, S., S. Gundel, J. Harris et al. 1995. The content of glutathione and glutathione S-transferases and the glutathione peroxidase activity in rat liver nuclei determined by a non-aqueous technique of liver cell fractionation. *Biochem. J.* 311:889–894.

11. Meister, A. 1991. Glutathione deficiency produced by inhibition of its synthesis, and its reversal: Applications in research and therapy. *Pharmacol. Ther.* 51:155–194.

12. Lang, C. A., B. Mills, W. Mastropaolo et al. 2000. Blood glutathione decreases in chronic diseases. *J. Lab. Clin. Med.* 135:402–405.

13. Kleinman, W. A. and J. P. Richie, Jr. 2000. Status of glutathione and other thiols and disulfides in human plasma. *Biochem. Pharm.* 60:19–29.

14. Locigno, R. and V. Castronovo. 2001. Reduced glutathione system: Role in cancer development, prevention and treatment (Review). *Int. J. Oncol.* 19:221–236.

15. Monostori, P., G. Wittmann, E. Karg, and S. Turi. 2009. Determination of glutathione and glutathione disulfide in biological samples: An in-depth review. *J. Chromatogr., B: Anal. Technol. Biomed. Life Sci.* 877:3331–3346.

16. Witschi, B., S. Reddy, B. Stoferet et al. 1992. The systemic availability of oral glutathione. *Eur. J. Clin. Pharmacol.* 43(6):667–669.

17. Tew, K. D. 1994. Glutathione-associated enzymes in anticancer drug resistance. *Cancer Res.* 64:4313–4320.

18. Park, S. K., R. B. Boulton, and A. C. Noble. 2000. Automated HPLC analysis of glutathione and thiol-containing compounds in grape juice and wine using precolumn derivatization with fluorescence detection. *Food Chem.* 68(4):475–480.

19. Cubukcu, M., F. N. Ertas, and U. Anik. 2012. Metal/metal oxide micro/nanostructured modified GCPE for GSH detection. *Curr. Anal. Chem.* 8(3):351–357.

20. Wei, Q.-K. and G.-Ch. Yen, 1995. Determination of glutathione contents in foods. *Zhongguo Nongye Huaxue Huizhi.* 33(3):355–362.

21. Piao, Yu., L. Shen, L. Han, and Sh. Cui. 2012. Reduced glutathione (GSH) and total thiol detection in some fruits by derivatizing methods. *Shipin Gongye Keji* 33(20):60–64.

22. Fracassetti, D., N. Lawrence, A. G. J. Tredoux et al. 2011. Quantification of glutathione, catechin and caffeic acid in grape juice and wine by a novel ultraperformance liquid chromatography method. *Food Chem.* 128:1136–1142.

23. Huang, Y., J. Duan, M. Yang et al. 2003. Determination of glutathione in tomatoes and cucumbers by capillary electrophoresis. *Sepu* 21(5):510–512.

24. Mendoza, J., T. Garrido, R. Riveros et al. 2008. Simultaneous determination of glutathione and glutathione disulfide in an acid extract of plant shoot and root by capillary electrophoresis. *J. Chil. Chem Soc.* 53(3):1626–1630.

25. Bener, M., M. Ozyurek, K. Guclu et al. 2013. Novel optical fiber reflectometric CUPRAC sensor for total antioxidant capacity measurement of food extracts and biological samples. *J. Agric. Food Chem.* 61(35):8381–8388.

26. Mills, B. J., C. T. Stinson, M. C. Liu, and C. A. Lang. 1997. Glutathione and cyst(e)ine profiles of vegetables using high performance liquid chromatography with dual electrochemical detection. *J. Food Compos. Anal.* 10(2):90–101.

27. Liu, Q., L. Chen, Y. Hu et al. 2012. Study on extraction and determination method of glutathione from wheat germ. *Zhongguo Liangyou Xuebao* 27(10): 104–108, 112.

28. Supalkova, V., D. Huska, V. Diopan et al. 2007. Electroanalysis of plant thiols. *Sensors* 7:932–959.

29. Wei, Q.-K., Sh.-F. Tsai, and G.-Ch. Yen. 1996. Determination of glutathione content in rice by HPLC. *Shipin Kexue* 23(4):538–543.

30. Pan, Yu., Sh. Chen, Q. Song et al. 2012. Determination of L-glutathione in marine products by the method of indirect fluorescence. *Shipin Gongye Keji* 33(20):70–72.

31. Tsardaka, E.-H., C. K. Zacharis, P. D. Tzanavaras et al. 2013. Determination of glutathione in baker's yeast by capillary electrophoresis using methyl propiolate as derivatizing reagent. *J. Chromatogr. A* 1300:204–208.
32. Zhu, L., Sh. Wei, F. Ge et al. 2011. Determination of glutathione in brewers yeast by high performance capillary electrophoresis. *Shipin Gongye Keji* 32(8):394–396.
33. Omca, D. and E. Nuran. 2012. Glutathione. In *Handbook of Analysis of Active Compounds in Functional Foods*, eds L.M.L. Nollet, and F. Toldra, pp. 69–85. Boca Raton, FL: CRC Press Taylor & Francis group.
34. Dalle-Donne, I. and R. Rossi. 2009. Analysis of thiols. *J. Chromatogr. Ser. B* 877(28):3271–3273.
35. Kandar, R., P. Zakova, H. Lotkova et al. 2007. Determination of reduced and oxidized glutathione in biological samples using liquid chromatography with fluorimetric detection. *J. Pharm. Biomed. Anal.* 43:1382–1387.
36. Harfield, J. C., Ch. Batchelor-McAuley, and R. G. Compton. 2012. Electrochemical determination of glutathione: A review. *Analyst* 137:2285–2296.
37. Liu, J., Ya. Wang, G. Liu, and B. Zhang. 2004. Comparison of three methods for determination of glutathione. *Beijing Huagong Daxue Xuebao, Ziran Kexueban* 31(3):35–38.
38. Lunn, G. and L. C. Hellwig, eds. 1998. *Handbook of Derivatization Reactions for HPLC*. New York, NY: John Wiley & Sons Inc.
39. Shimada, K. and K. Mitamura. 1994. Derivatization of thiol-containing compounds. *J. Chromatogr. Ser. B* 659:227–241.
40. Rosenfeld, J. M. 2003. Derivatization in the current practice of analytical chemistry. *Trends Anal. Chem.* 22:785–798.
41. Giustarini, D., I. Dalle-Donne, R. Colombo et al. 2003. An improved HPLC measurement for GSH and GSSG in human blood. *Free Radicals Biol. Med.* 35(11):1365–1372.
42. Hermsen, W. L. J. M., P. J. Mcmahon, and J. W. Anderson. 1997. Determination of glutathione in plant extracts as a 1-chloro-2,4-dinitrobenzene conjugate in the presence of glutathione S-transerase. *Plant Physiol. Biochem.* 35 (6):495–496.
43. Karakosta, T. D., P. D. Tzanavaras, and D. G. Themelis. 2013. Determination of glutathione and cysteine in yeasts by hydrophilic interaction liquid chromatography followed by on-line postcolumn derivatization. *J. Sep. Sci.* 36:1877–1882.
44. Zacharis, C. K., P. D. Tzanavaras, and D. G. Themelis. 2009. Ethyl-propiolate as a novel and promising analytical reagent for the derivatization of thiols: Study of the reaction under flow conditions. *J. Pharm. Biomed. Anal.* 50:384–391.
45. Zacharis, C. K., P. D. Tzanavaras, and A. Zotou. 2011. Ethyl propiolate as a post-column derivatization reagent for thiols: Development of a green liquid chromatographic method for the determination of glutathione in vegetables. *Anal. Chim. Acta* 690(1):122–128.
46. Guan, X., B. Hoffman, C. Dwivedi et al. 2003. A simultaneous liquid chromatography/mass spectrometric assay of glutathione, cysteine, homocysteine and their disulfides in biological samples. *J. Pharm. Biomed. Anal.* 31:251–261.
47. Hansen, R. E., H. Østergaard, P. Nørgaard, and J. R. Winther. 2007. Quantification of protein thiols and dithiols in the picomolar range using sodium borohydride and 4,4′-dithiodipyridine. *Anal. Biochem.* 363:77–82.
48. Yang, Q., C. Krautmacher, D. Schilling et al. 2002. Simultaneous analysis of oxidized and reduced glutathione in cell extracts by capillary zone electrophoresis. *Biomed. Chromatogr.* 16:224–228.

49. Tsikas, D., J. Sandmann, D. Holzberg et al. 1999. Determination of S-nitroso-glutathione in human and rat plasma by high-performance liquid chromatography with fluorescence and ultraviolet absorbance detection after pre-column derivatization with o-phthalaldehyde. *Anal. Biochem.* 273:32–40.

50. Ates, B., B. C. Ercal, K. Manda et al. 2009. Determination of glutathione disulfide levels in biological samples using thiol-disulfide exchanging agent, dithiothreitol. *Biomed. Chromatogr.* 23(2):119–123.

51. Sakhi, A. K., R. Blomhoff, Th. E. Gundersen. 2007. Simultaneous and trace determination of reduced and oxidized glutathione in minute plasma samples using dual mode fluorescence detection and column switching high performance liquid chromatography. *J Chromatogr A* 1142(2):178–184.

52. Liang, Sh.-C., H. Wang, Zh.-M. Zhang et al. 2002. Direct spectrofluorimetric determination of glutathione in biological samples using 5-maleimidyl-2-(m-methylphenyl)benzoxazole. *Anal Chim Acta* 451(2):211–219.

53. Qin, J., N. Ye, L. Yu et al. 2005. Simultaneous and ultra-rapid determination of reactive oxygen species and reduced glutathione in apoptotic leukemia cells by microchip electrophoresis. *Electrophoresis* 26(6):1155–1162.

54. Gao, N., L. Li, Zh. Shi et al. 2007. High-throughput determination of glutathione and reactive oxygen species in single cells based on fluorescence images in a microchannel. *Electrophoresis* 28(21):3966–3975.

55. Nolin, T. D., M. E. McMenamin, and J. Himmelfarb. 2007. Simultaneous determination of total homocysteine, cysteine, cysteinylglycine, and glutathione in human plasma by high-performance liquid chromatography: Application to studies of oxidative stress. *J. Chromatogr. B: Anal. Technol. Biomed. and Life Sci.* 852(1–2): 554–561.

56. Pastore, A., G. Federici, E. Bertini, and F. Piemonte. 2003. Analysis of glutathione: Implication in redox and detoxification. *Clin. Chim. Acta* 333(1):19–39.

57. Zhao, Sh., X. Li, and Yu.-M. Liu. 2009. Integrated microfluidic system with chemiluminescence detection for single cell analysis after intracellular labeling. *Anal. Chem.* 81(10):3873–3878.

58. Shaik, I. H. and R. Mehvar. 2006. Rapid determination of reduced and oxidized glutathione levels using a new thiol-masking reagent and the enzymatic recycling method: Application to the rat liver and bile samples. *Anal. Bioanal. Chem.* 385:105–112.

59. Camera, E. and Picardo, M. 2002. Analytical methods to investigate glutathione and related compounds in biological and pathological processes. *J. Chromatogr. B: Anal. Technol. Biomed. Life Sci.* 781(1–2):181–206.

60. Tzanavaras, P. D. and D. G. Themelis. 2010. Automated determination of pharmaceutically and biologically active thiols by sequential injection analysis: A review. *The Open Chem. Biomed. Methods J* 3:37–45.

61. Teshima, N., T. Nobuta, and T. Sakai. 2008. Simultaneous spectrophotometric determination of ascorbic acid and glutathione by kinetic-based flow injection analysis. *Bunseki Kagaku* 57(5):327–333.

62. Zacharis, C. K. and Tzanavaras, P. D. 2010. High throughput automated determination of glutathione based on the formation of a UV absorbing thioacrylate derivative. *Comb. Chem. High Throughput Screening* 13:461–468.

63. Redegeld, F. A. M., M. A. J. van Opstal, E. Houdkamp, and W. P. van Bennekom. 1988. Determination of gutathione in bological material by flow-injection analysis using an ezymatic recycling reaction. *Anal. Biochem.* 174(2):489–495.

64. Araujo, A. R. T. S., M. L. M. F. S. Saraiva, and J. L. F. C. Lima. 2008. Enzymatic determination of total and oxidized glutathione in human whole blood with a sequential injection analysis system. *Talanta* 74:1511–1519.

65. Saton, I., Sh. Arakawa, and A. Okamoto. 1991. Calorimetric flow-injection determination of glutathione with enzyme-thermistor detector. *Sens. Actuators* 5(1–4):245–247.

66. Wheatley, R. A., M. Sarıahmetoglu, and I. Çakıcı. 2000. Enhancement of luminol chemiluminescence by cysteine and glutathione. *Analyst* 125:1902–1904.

67. Du, J., Y. Li, and J. Lu. 2001. Investigation on the chemiluminescence reaction of luminal–H_2O_2–S^{2-}/R–SH system. *Anal. Chim. Acta* 448(1–2, 3):79–83.

68. Wang L., Y. Li, D. Zhao, and C. Zhu. 2003. A novel enhancing flow-injection chemiluminescence method for the determination of glutathione using the reaction of luminol with hydrogen peroxide. *Microchim. Acta* 141:41–45.

69. Ensafi, A. A., T. Khayamian, and F. H. Hasanpour. 2008. Determination of glutathione in hemolysed erythrocyte by flow injection analysis with chemiluminescence detection. *J. Pharm. Biomed. Anal.* 48:140–144.

70. Saton, I., Sh. Arakawa, and A. Okamoto. 1988. Flow-injection determination of glutathione with amperometric monitoring of the enzymatic reaction. *Anal. Chim. Acta* 214(1–2):415–419.

71. Shpigun, L. K., M. A. Suranova, N. A. Ryabova et al. 2010. Voltammetric determination of biologically important thiols in pharmaceutical preparations. *11th Eurasia Conf. on Chem. Sci.*, Jordan.

72. Zitka, O., D. Huska, S. Krizkova et al. 2007. An investigation of glutathione-platinum(II) interactions by means of the flow injection analysis using gassy carbon electrode. *Sensors* 7:1256–1270.

73. Treumer, J. and G. Valet. 1986. Flow-cytometric determination of glutathione alterations in vital cells by o-phtalldialdehyde (OPT) staining. *Exp. Cell Res.* 163(2):518–524.

74. Chow, S. and D. Hedley. 1995. Flow cytometric determination of glutathione in clinical samples. *Cytometry* 21:68–71.

75. Tzanavaras, P. D., D. E. Themelis, A. Economou, and G. Georgios Theodoridis. 2003. Flow and sequential injection manifolds for the spectrophotometric determination of captopril based on its oxidation by Fe(III). *Microchim. Acta* 142:55–62.

76. Suarez, W. T., O. D. Pssoa-Neto, B. C. Janegitz et al. 2011. Flow injection spectrophotometric determination of N-acetylcysteine and captopril employing Prussian blue generation reaction. *Anal. Lett.* 44:2394–2405.

77. Speiskya, H., C. M. Gómeza, C. Carrasco-Pozoa et al. 2008. Cu(I)–Glutathione complex: A potential source of superoxide radicals generation. *Biol. Med. Chem.* 16(13):6568–6574.

78. Shpigun, L. K., Ya. V. Shushenachev, and P. M. Kamilova. 2008. Automatic methods for testing antiradical capacity of biocompounds. In *Chemistry: The Global Science. 2nd EuChem Chemistry Congress.* Torino, Italy.

79. Yordanov, N. D. and A. G. Christova. 1997. Quantitative spectrophotometric and EPR-determination of 1,1-diphenil-2-picril-hydrazil (DPPH). *J. Anal. Chem.* 358:610–613.

80. Rachman, I., A. Kode, and S. K. Biswas. 2006. Assay for quantitate determination of glutathione and glutathione disulfide levels using enzymatic recycling method. *Nat. Protoc.* 1(6):3159–3165.

81. Brigelius, R., Ch. Muckel, T. P. M. Akerboom, and H. Sies. 1983. Identification and quantitation of glutathione in hepatic protein mixed disulfides and its relationship to glutathione disulfide. *Biochem. Pharm.* 32(17):2529–2534.
82. Tietze, F. 1969. Enzymic method for quantitative determination of nanogram amounts of total and oxidized glutathione: Applications to mammalian blood and other tissues. *Anal. Biochem.* 27(3):502–522.
83. Lancaster, J. S. and Worsfold, P. J. 1989. Flow injection determination of organosulfur compounds with chemiluminescence detection. *Anal. Proc.* 26(1):19–20.
84. White, P. C., N. S. Lawrence, J. Davis, and R. G. Compton. 2002. Electrochemical determination of thiols: A perspective. *Electroanalysis* 14(2):89–98.
85. Wang, W. and W. Jin. 2007. Determination of glutathione in a single human hepatocarcinoma cell using a microfluidic device coupled with electrochemical detection. *Sepu* 25(6):799–803.
86. Supharoek, S., N. Youngvises, and J. Jakmunee. 2012. A simple microfluidic integrated with an optical sensor for micro flow injection colorimetric determination of glutathione. *Anal. Sci.* 28:651–656.

SECTION V

Sweeteners

CHAPTER 23

Determination of
Aspartame and Sorbitol

Paula C.A.G. Pinto, Célia M.G. Amorim, Alberto N. Araújo,
M. Conceição B.S.M. Montenegro, and M. Lúcia M.F.S. Saraiva

CONTENTS

23.1 INTRODUCTION

The intake of sugar in the diet has become a worldwide concern since the consumption is high and is usually associated with weight gain and adverse effects on glycemia control and diabetes. In this context, some health organizations, especially in more developed countries have focused on recommending reducing the intake of sugar and replacing it with nonnutritive sweetener (NNS) products.

Artificial sweeteners have been classified as nutritive and nonnutritive depending on whether they are source of calories or not. The nutritive sweeteners can be derived from naturally occurring substances [1], including plants or sugar itself (e.g., sorbitol and others), in this case being approximately equivalent to sucrose in sweetness [2]. The NNSs, known as artificial sweeteners, are synthetic sugar substitutes, from different chemical classes, that interact with taste receptors and typically exceed the sweetness of sucrose by a factor of 30–13,000 times [3]. NNSs such as aspartame, cyclamate, saccharine, and acesulfame K are nowadays used as food additives in an increasingly wide range of low- or reduced-calorie food products, including baked goods, soft drinks, powdered drink mixes, candy, puddings, canned foods, jams and jellies, and dairy products, as well as in other foods and beverages. The intense sweetness of NNSs allows their utilization in small portions to yield sugarlike sweetness in food items, thus enabling the production of the so-called "sugar free" or "noncaloric" products. Their use and concentration depends on the food, the substitution level, and the initial fat content of the food.

In the United States, food additives are regulated by the Food and Drug Administration (FDA), which limits the conditions of use of substances that are reasonably expected to become food components, including artificial sweeteners. The European Union through EU Directive 94/35/CE amended by EU Directive 2003/115/CE and EU Directive 2006/52/CE

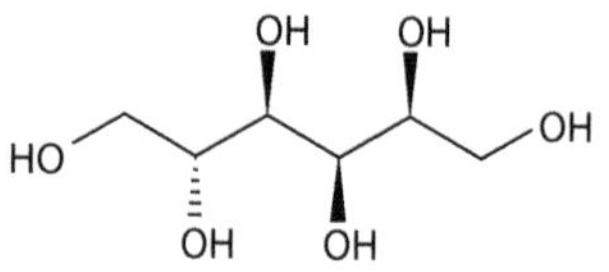

FIGURE 23.1 Chemical structure of D-sorbitol.

establishes the replacement of sugar by sweeteners for low- or reduced-calorie food products and defines the maximum dose per kilogram of distinct sweetener that is permitted in different foods.

Consumption of NNS-containing foods has increased among people of all ages, with reports of intake of about 28% of the total population [4]. This trend is highly prevalent among children, especially when it comes to beverage intake. Analysis of National Health and Nutrition Examination Survey data shows that the use of NNS-containing beverages increased from 6.1% to 12.5% among children and from 18.7% to 24.1% among adults in an 8-year period [5]. This high consumption brings leads to the dispersion of these compounds into the environment, so that they have recently been recognized as a class of emerging environmental contaminants [6]. Recent reports revealed the environmental presence of these compounds, especially sucralose, in concentrations of around micrograms per liter. There is also evidence of their environmental persistence after wastewater treatment, suggesting that their presence can constitute a marker for the study of the impact on source waters and drinking waters [6].

Although there is no literature data on risk assessment of sorbitol (Figure 23.1) and its potential toxicity in humans, this is not the case for aspartame.

Aspartame is a dipeptide artificial sweetener (Figure 23.2), 200 times sweeter than sucrose, composed of the amino acids phenylalanine and aspartic acid plus a small quantity of methanol [7]. This sweetener was first proposed for approval for dry foods in 1974. As a consequence of a series of investigations and hearings into the operations of G. D. Searle and studies supporting the safety of aspartame, a Public Board of Inquiry that was convened in 1980 concluded that aspartame should not be approved pending further studies. In 1981 the FDA approved the use of aspartame for dry products on the basis of evidence from additional studies. In 1983 it was further approved for use in carbonated drinks and subsequently, in 1996, for use in all foods and drinks being currently authorized worldwide [8]. Since its approval, aspartame has been used in more than 6000 products by hundreds of millions of people in countries all around the world [9] under the brand names Equal and NutraSweet.

Aspartame can be found in a wide variety of prepared foods including soft drinks, chewing gum, confections, gelatins, dessert mixes, puddings and fillings, multivitamins,

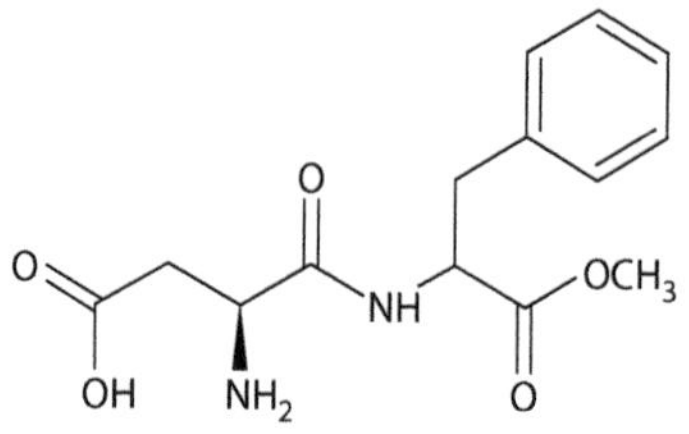

FIGURE 23.2 Chemical structure of aspartame.

breakfast cereals, tabletop sweeteners, and pharmaceuticals. Aspartame has been by far the most controversial artificial sweetener because of its potential toxicity. Internationally recognized authorities around the world including the FDA [8] and the Joint FAO (Food and Agriculture Organization)/WHO (World Health Organization) Expert Committee on Food Additives (JECFA) [10], defined health-based guidance values that regulate the use of aspartame in more than 100 countries. In Europe, the Scientific Committee for Food (SCF from 1974 until April 2003) [11–14], the European Food Safety Authority (EFSA) through the Panel on Food Additives, Flavourings, Processing Aids and Materials in contact with Food (from 2003 to 2008) [15–17] and currently the EFSA Panel on Food Additives and Nutrient Sources Added to Food (ANS, from 2008 to present) [18–22] are the scientific bodies responsible for advising on the safety of sweeteners. These advisory bodies and agencies, working independently, have set the acceptable daily intake (ADI) for aspartame. This value is greatly dependent on the amount of aspartame present in the food product, so it is necessary to have concrete data regarding this additive. Accordingly, different analytical methods have been proposed in the literature to assess the content of aspartame in different foods. In this chapter reference is made to the developed analytical methods with particular emphasis on those based on flow analysis techniques.

23.2 FLOW METHODOLOGIES FOR THE DETERMINATION OF ASPARTAME

Due to the increasing awareness, referred to above, of the eventual long-term effects associated with the continuous ingestion of aspartame [23] as a sweetener in dietetic products, its determination became mandatory in order to guarantee the safety of its ingestion. Several methods have been developed and proposed to perform the quantification of aspartame in food items and beverages, mainly spectrophotometric [24,25] and electrochemical [26–29] ones. However, the insufficient sensitivity or selectivity of the majority of these methodologies led to increased investment in the development of separative techniques for the quantification of aspartame generally simultaneously with other food additives. Approaches based on high-performance liquid chromatography (HPLC) [30,31], liquid chromatography (LC) [32], and thin-layer chromatography (TLC) [33], among others were successfully implemented. Even though these methodologies are generally highly sensitive and selective, the high cost of the equipment required hinders their application in laboratory routine. Similarly, the analysis time can also be an important factor when the number of samples to be analyzed is significant.

In this context, flow techniques exhibit suitable characteristics for the determination of aspartame in food and beverages. In fact, these systems enable the incorporation of sample treatment devices, allowing the implementation of online sample treatments with reduction of human intervention and associated errors. On the other hand, the possibility of coupling the flow system with a variety of detection systems opens immense possibilities regarding the analytical methodology to be implemented. Moreover, the consumption of reagents can be dramatically decreased due to the downsizing of the reaction environment through the use of reduced-diameter tubing. The analysis time can also be reduced as a consequence of making nonequilibrium measurements.

In this chapter, the existing methodologies for the determination of aspartame in food items and beverages will be described and discussed in terms of the analytical strategies implemented, their analytical performances, and the sample pretreatment approaches.

Table 23.1 summarizes the analytical features of nonenzymatic flow methods for the determination of aspartame. The first attempts to determine aspartame in flow systems resorted to simple spectrophotometric methodologies with well-known capabilities [34,35]. Generally, these methodologies are simple and easy to implement and also present advantages in terms of sampling frequency when compared to the most commonly used methods for the same purpose, namely, chromatographic ones.

TABLE 23.1 Analytical Features of Nonenzymatic Flow Methods for Aspartame Determination in Dietetic Foods and Beverages

Methodology	Concentration Range	Detection Limit	Samples	Sampling Rate	References
FIA—spectrophotometric; stopped flow	$0.06–0.6$ mmol L^{-1}	20 µmol L^{-1}	Colored beverages	$40\ h^{-1}$	[34]
FIA—spectrophotometric; ninhydrin	Up to 2.4 mmol^{-1}	38 µmol L^{-1}	Table sweetener Pudding Gelatin Refreshments	$36\ h^{-1}$	[35]
FIA—Sephadex CM-25 cationic exchanger	$5.0–600.0$ µg mL^{-1}	0.75 µg mL^{-1}	Chewing gum Custard Sugarless sweets Strawberry dietetic jam Iron supplement	$24\ h^{-1}$	[36]
FIA—separation with Sephadex DEAE A-25	$10–100$ µg mL^{-1}	5.65 µg mL^{-1}	Tabletop sweeteners	$24\ h^{-1}$	[37]
FIA—monolithic column (C_{18})	$0.94–45.0$ µg mL^{-1}	0.32 µg mL^{-1}	Cola drink	$9\ h^{-1}$	[38]
FIA—monolithic column (quaternary amine anion exchanger)	$9.5–130.0$ µg mL^{-1}	2.9 µg mL^{-1}	Cola drink Fruit juice Black tea	$7\ h^{-1}$	[39]
FIA—monolithic column (C_{18})	$10–200$ µg mL^{-1}	1.4 µg mL^{-1}	Sweets Soft drink Juice drink Strawberry dietetic jam	$10\ h^{-1}$	[40]
FIA—$Cu_3(PO_4)_2$ immobilized in polyester resin	$20–80$ µg mL^{-1}	2 µg mL^{-1}	Tabletop sweeteners	$70\ h^{-1}$	[41]
FIA—$Zn_3(PO_4)_2$ immobilized in polyester resin	$10–80$ µg mL^{-1}	4 µg mL^{-1}	Tabletop sweeteners	$70\ h^{-1}$	[42]
SIA—capillary electrophoresis	$20–1000$ µmol L^{-1}	6.5 µmol L^{-1}	Confectionary Soft drink	$19\ h^{-1}$	[43]
SIA—thin-layer paramagnetic stationary phase ($Fe_3O_4–SiO_2$)	$1–60$ mg L^{-1}	3 mg L^{-1}	Diet drinks	$7\ h^{-1}$	[44]

In this context, a stopped-flow approach was tested for the implementation of a micellar catalyzed reaction with 1-fluoro-2,4-dinitrobenzene for the determination of several compounds including pharmaceuticals and aspartame [34]. This reaction has been used in the determination of amino acids and primary and secondary amines [45,46], among others, with applicability in chromatographic methods [47]. Major advantages of the automation of this methodology in a 3-channel flow system (Figure 23.3) include the possibility of safe handling of the reagent, the simplification of the analytical procedure, the increase of sample throughput, and the enhancement of the precision of the determinations.

The possibility of analyzing colored samples, such as beverages, after a simple dilution in buffer solution is also one of the most important features of this methodology. The methodology and its sensitivity can be adjusted to particular situations by changing the reaction pH and the concentration of the surfactant, which makes it very attractive for the analysis of distinct samples with variable concentration of analyte.

Ninhydrin is also a well-known reagent for the detection of primary and secondary amines, through the formation of a purple compound, applicable to the determination of aspartame [48,49]. Resorting to flow injection analysis (FIA) it is possible to create the ideal conditions for the development of the reaction in terms of either time or temperature. Nóbrega and Fatibello-Filho developed a FIA manifold that incorporates a 200 cm reaction coil maintained in a water bath at 60°C, enabling the determination of aspartame up to 2.4 mmol L^{-1}, with a detection limit of 38 μmol L^{-1} and a sampling rate of 36 determinations per hour [35]. The reaction is performed in a methanol–isopropanol (1:1 v/v) solution that is used as solvent for the preparation of the reagent and as carrier in the flow system. The system was applied to the analysis of samples of gelatin, pudding, table sweetener, and refreshments that were dissolved and diluted in the same solvent prior to injection in the system by means of a homemade manual commutator. The results were validated through the comparison with a manual procedure and resort to recovery assays. The use of an organic solvent associated with the change of the detection wavelength from 412 to 603 nm prevents the interference of food dyes and other compounds such as cyclamate or citric acid.

Solid-phase chemistry associated with spectrophotometric detection in flow systems is also a good alternative for the determination of aspartame in food items. In this topic, one must consider not only the utilization of solid phase exchangers or adsorbents to build flow-through optosensors [36,37,40] but also the immobilization of reagents on a solid matrix in packed reactor format [41,42]. This approach has been successfully explored mainly for the analysis of analyte mixtures through the use of microcolumns

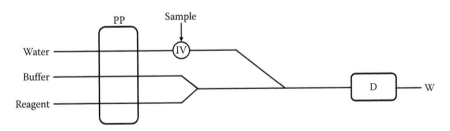

FIGURE 23.3 Schematic representation of the FIA system developed by Georgiou et al. [34] for the determination of aspartame in colored beverages. PP: peristaltic pump; IV: injection valve; D: detector; and W: waste.

with variable position in the flow system or through the insertion of the exchanger on the flow cell [50–52]. Several ion exchangers (Sephadex DEAE A-25, Sephadex QAE A-25, and Sephadex CM C-25) and hydrophilic and hydrophobic adsorbents (Sephadex G-25, Sephadex G-15, and C-18 silica adsorbents, respectively) were tested for solid phase retention of aspartame and other artificial sweeteners in the flow cell of a FIA system based on intrinsic absorbance measurements [36,37,40]. These are very interesting approaches since they do not require any reagent besides an appropriate buffer solution and can be implemented in single-channel flow systems equipped with a conventional spectrophotometer. As expected, these methodologies present slightly lower sampling frequencies due to the solid-phase retention step. For the single determination of aspartame in low-calorie and dietary products, the best results are obtained with the cationic exchanger Sephadex CM C-25, which provides good retention of the analyte and adequate elution by the carrier, avoiding the use of an extra eluent [36]. The resulting flow system is very simple and the analytical cycle comprises the insertion of sample on a buffer and its propulsion to the flow cell, where the analyte is retained with registration of an analytical signal proportional to the concentration of aspartame. The methodology enables the determination of aspartame in the range of 0.017–2.0 mmol L^{-1} with a detection limit of 2 µmol L^{-1}. Several samples were analyzed using this methodology, after dissolution and appropriate dilution in buffer, namely chewing gums, sugarless sweets, dietetic jams, and supplements. Accurate results were obtained either in recovery assays or in comparison with a HPLC method. No interference was caused by common components of dietary products in concentrations near those usually found in the analyzed products.

For the simultaneous determinations of aspartame and acesulfame-K the best results are obtained with ion exchanger Sephadex DEAE A-25 placed as before in the flow cell of a single channel FIA system incorporating a diode array spectrophotometer that allows the detection of both analytes [37]. In this case, the determination is based on the faster transient retention of acesulfame-K as the sample reaches the flow cell, which enables the separation and determination of the two analytes through measurement of the absorbance at the corresponding wavelengths and appropriate timing. As before, the methodology was validated by the analysis of several commercial tabletop sweeteners and recovery assays and comparison with a reference procedure. The reformulation of the methodology to adapt it to a multianalyte determination slightly compromises the detection limit of the determination of aspartame (20 µmol L^{-1} versus 2 µmol L^{-1}) without influence on the analysis of real samples. This strategy can be also applied to the simultaneous determination of aspartame and saccharin in a similar single-channel FIA system. For this, a C$_{18}$ silica gel minicolumn is used to perform the preconcentration of aspartame and Sephadex G-25 to retain saccharin on the flow cell [40]. Saccharin is eluted with methanol and detected after the transient retention of aspartame. This analytical strategy enables determination of aspartame between 34 µmol L^{-1} and 0.7 mmol L^{-1} with a detection limit of 5 µmol L^{-1} and a sampling rate of 10 h^{-1}. As before, the analysis of sweets and drinks after dissolution, filtration, and dilution indicates that the results are comparable to those provided by a reference assay.

Extensive research in the field of chromatographic determination of aspartame also provided the basis for the development of novel flow methodologies based on monolithic columns [38,39]. These separative columns are based on highly cross-linked porous monolithic polymers that provide excellent separation ability while offering exceptional chemical stability. Due to the presence of macropores, the monolithic column exhibits a porosity about 15% higher than conventional particulate HPLC columns. Thus, column back-pressure is significantly lower, enabling the implementation of chromatography separations without the need for high-pressure pumps [53]. These particular features permit the combination of monolithic columns with low-pressure flow systems, opening interesting perspectives regarding the analysis of multicomponent samples [54]. A 5 mm silica C_{18} monolithic column can be used to perform the separation of multianalyte food samples, including aspartame, resorting to a FIA approach. For this, the column must be inserted in the flow line, before a diode array spectrophotometer that enables the detection of the distinct analytes based on their spectroscopic properties. The procedure was implemented by using two rotary, one to perform the injection of sample in a carrier stream of acetonitrile/water and another to change the carrier solution, acting like a selection valve. The change of the carrier to a solution of methanol/water is a strategy to speed up the less polar analytes. In the optimized conditions, the method allows the separation of seven analytes without prior derivatization, with a peak of aspartame at 205 nm, 36 s after sample injection. The method responds to aspartame concentrations between 3 mmol L^{-1} and 0.15 mol L^{-1} in the presence of different ratios of the other additives and with a sampling frequency of 9 h^{-1}. The performance of the method with aspartame was evaluated through the analysis of samples of cola drinks after extraction of the analyte with an extractor cartridge and comparison of the results with a HPLC procedure. Recoveries of aspartame in the range 100.8%–102.1% were obtained. Even though the resolution of the FIA methodology is lower than that of the HPLC reference assay, it assures the determination of the analytes in the levels in which they are usually present in the analyzed samples. Similarly, a commercial anionic monolithic disk was inserted in a single-channel FIA system for the simultaneous determination of aspartame, acesulfame-K, and saccharin based on the transient retention of the three analytes [39]. The analytical cycle is based on the injection of the sample in a buffer carrier stream toward the monolithic disk where the analytes are separated. As before, aspartame is weakly retained and a peak is obtained 38 s after injection at 205 nm. Compared with the previously described method, this one is far more sensitive regarding aspartame determination, with the analytical signal varying linearly with the concentration between 0.03 µmol L^{-1} and 0.44 mmol L^{-1} and a detection limit of 10 nmol L^{-1}. However, the sampling frequency is lower. As in the previous example, due to the weak retention of aspartame, the determination of this analyte is affected by nonretained substances present in the sample. Thus, for this determination, samples must be previously purified by means of solid-phase extraction. The results obtained from the analysis of beverages were not significantly different from those provided by a HPLC reference procedure.

Still on the chromatographic perspective but in a distinct format, the determination of aspartame can be done using a thin-layer paramagnetic stationary phase retained on the inner wall of a minicolumn, which allows the implementation of a chromatographic modality in flow mode that once again does not require high pressures [44]. This strategy was successfully implemented in a sequential injection analysis (SIA) system consisting of a selection valve, a syringe pump, and a diode array spectrophotometer. In this format, the stationary phase (Fe_3O_4–SiO_2) is trapped in a Perspex minicolumn placed before the detector and connected to a power supply by means of a copper coil wrapped around it (Figure 23.4). This configuration enables the adjustment of the external magnetic field through the variation of the current applied to the coil. After online column preparation, the sample and a complexing agent are aspirated sequentially and sent to the detector, passing through the column over the course of 9 min.

The complexing agent is composed of $CuCl_2$ and 1,10-phenantroline and is used to confer paramagnetic properties on the analytes. The use of a methanol–phosphate buffer (25:75) pH 7 mobile phase enables the separation and detection of aspartame and its

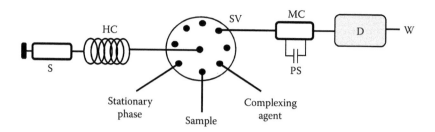

FIGURE 23.4   Schematic representation of the SIA system developed by Barrado et al. [44] for the determination of aspartame. S: syringe; HC: holding coil; SV: selection valve: MC: minicolumn; PS: power source; D: detector; and W: waste.

hydrolysis products L-aspartic acid and L-phenylalanine. This methodology presents characteristics that permit the determination of aspartame in the range 0.017–0.34 mmol L^{-1} with a detection limit of 0.01 mmol L^{-1}. The analysis of soft drink samples (diluted in buffer) and comparison with an HPLC procedure validated the methodology. The versatility of the SIA system is of the highest importance for the feasibility of this methodology as it allows not only the fully automated separation and determination of the analytes but also the renewal of the stationary phase when necessary. These features make this methodology a valid alternative for the determination of aspartame and it opens an interesting perspective for the separation and determination of other analytes such as amino acids and metalloproteins.

As stated before, the immobilization of reactants on polyester resins is also an interesting possibility for the determination of aspartame [41,42]. This scheme is based on the methyl ethyl ketone–catalyzed attachment of copper [41] or zinc [42] phosphate to a polyester matrix. The immobilized reagents can be packed in columns and incorporated in simple FIA systems equipped with a spectrophotometer. The passage of aspartame through the column promotes the release of the metal ions as metal–aspartame complexes that can be quantified by means of alizarin red S, a spectrophotometric reagent with the ability to form colored complexes with both copper and zinc. The methodologies based on this strategy enable the determination of aspartame up to 0.3 mmol L^{-1}, with detection limits of 7 and 14 µmol L^{-1} in the case of copper and zinc, respectively. The methodologies were validated by the analysis of several commercial tabletop sweeteners and comparison of the results with those obtained with the ninhydrin assay. For these methodologies, samples were prepared by simple dissolution and dissolution in adequate buffer solution. The most significant feature of these methods is their high determination frequency of about 70 determinations per hour.

Still in the field of separation techniques, capillary electrophoresis with contactless conductivity detection was very recently applied to the determination of several artificial sweeteners including aspartame in a SIA format [43]. The designed SIA manifold, incorporating a syringe pump and micro-graduated needle valve, allows hydrodynamic pumping with benefits in terms of separation efficiency and analysis time without the need to eliminate or reverse electroosmotic flow. Narrow capillaries must be used to avoid band broadening. This methodology exhibits relatively high sensitivity, being capable of determinations of aspartame in a large linear range from 20 to 1000 µmol L^{-1} with a detection limit of 6.3 µmol L^{-1}, for a separation time of 190 s. Still, the residence time and consequently the analytical performance of the methodology can be controlled by changing the pumping so that the conditions can be adapted to the specifics of the samples to be analyzed. Two samples of a confectionary and a soft drink were dissolved, filtered, and diluted in buffer prior to analysis by the procedure described.

Aspartame can also be quantified with the use of biosensors incorporated in flow systems [55–58], usually with satisfactory sensitivity and selectivity (Table 23.2).

Generally, batch methods using enzymes in solution are the basis of the majority of these flow enzyme-based assays for the determination of aspartame. The use of enzymes in analytical methodologies is recognized as a good way to enhance the selectivity and specificity of the determinations and has been widely applied to the determination of single species in complex samples by means of nonspecific detection techniques [59]. Moreover, the utilization of enzymes in analytical methodologies to replace hazardous chemical reagents generates cleaner processes that are in harmony with the current concerns of Green Chemistry due to the inherent biodegradability of enzymes. The major improvement in the flow approaches for the enzyme-based determination of aspartame is the use of enzymes immobilized on solid supports. Despite the advantages of using enzymes in solution, enzyme immobilization is currently recognized as a good alternative for routine implementation as it guarantees good catalytic activity and enzyme reutilization, leading to a significant reduction of analysis costs [60]. Curiously, the first flow system developed for the determination of aspartame in dietary food products incorporated an enzyme electrode and two enzyme columns operating in FIA mode [55]. Peptidase and aspartate aminotransferase were immobilized on aminopropyl glass beads after activation with glutaraldehyde and packed in Tygon tubing. After cleavage of the dipeptide bond of aspartame and subsequent transamination, the glutamate produced can be monitored using glutamate oxidase immobilized on a nylon membrane in combination with a hydrogen peroxide electrode. The developed flow system incorporates a peristaltic pump that is responsible for the circulation of both buffer and sample in the system. The sample aliquot is inserted in the buffer stream through an injection valve and is subsequently propelled to the enzyme columns and then to the amperometric electrode. The response of the biosensor is linear up to 1 mmol L^{-1} aspartame with a detection limit of 20 µmol L^{-1}. The system enables the analysis of 8 samples per hour with stability for more than 30 h of continuous use at room temperature. The immobilized enzymes are stable for at least

TABLE 23.2 Enzyme-Based Flow Methodologies for the Determination of Aspartame in Food Items

Enzymes	Detection System	Concentration Range	Detection Limit	Sample Throughput (per hour)	References
Peptidase Aspartate deaminase L-glutamate oxidase	Amperometry	0–1 mmol L^{-1}	20 µmol L^{-1}	8	[55]
Pronase L-amino-acid oxidase	Amperometry	0–1 mmol L^{-1}	25 µmol L^{-1}	15	[56]
α-Chymotrypsin Alcohol oxidase	Amperometry	0–1 mmol L^{-1}	1 µmol L^{-1}	–	[57]
α-Chymotrypsin Alcohol oxidase	Spectrophotometry	0–1.2 mmol L^{-1}	7.3 µmol L^{-1}	15.2	[58]

1 month if stored at 4°C. The system is adequate for the determination of aspartame in dietary food products after elimination of the proteins with trichloroacetic acid 10% and centrifugation. The analysis of dietary products, including beverages and desserts, demonstrated the method's adequacy as the obtained results are in good agreement with those reported by manufacturers. Besides the benefits of the flow approach, this methodology offers enhanced sensitivity compared with previously developed enzyme [29] and microbial [61] methodologies.

Similarly, an enzyme electrode incorporating L-amino-acid oxidase, immobilized on a nylon membrane, in combination with an amperometric electrode [56], was applied in a FIA system including also the enzyme pronase immobilized on aminopropyl glass beads after activation with glutaraldehyde. In this format, the phenylalanine formed through the passage of aspartame to the pronase reactor is converted by L-amino-acid oxidase, forming H_2O_2 that is determined amperometrically. The FIA system is similar to the one described before and only an additional channel was added to the peristaltic pump for the insertion of a proper buffer for the activity of L-amino-acid oxidase, in the last steps of the analytical cycle. In terms of analytical performance, this methodology is very similar to that described above [55]. The major improvement of this FIA biosensor is the use of pronase, which can be up to 10 times less expensive than peptidase. With this methodology, aspartame can be determined in soft drinks after dilution with appropriate buffer solution. As before, in solid samples aspartame must be extracted with trichloroacetic acid 10% followed by dilution with buffer solution to avoid the interference of proteins. In both cases the determined values are in good agreement with those reported by manufacturers.

The same detection system can be used for the determination of aspartame resorting to a bienzymic approach comprising equivalent activities of alcohol oxidase and α-chymotrypsin [57]. In this scheme, both enzymes are mixed with glutaraldehyde and this mixture is applied to the surface of a polycarbonate membrane. The resulting membrane can be then mounted on a platinum electrode that can in turn be inserted into a wall-jet flow cell to perform FIA measurements. This methodology is based on the cleavage of aspartame by α-chymotrypsin with formation of methanol that is then detected with the help of alcohol oxidase by the formation of H_2O_2 determined amperometrically. The low detection limit of the biosensor (1 μmol L^{-1}) is the distinguishing feature of this method. The major drawback of the method is the influence of both electrochemical and enzyme interferents. Electrochemical interferences can be easily eliminated through the use of a cellulose acetate membrane on the electrode. Regarding enzyme interferences, special attention must be devoted to molecules having alcoholic groups due to the broad specificity of alcohol oxidase. Several approaches were tested to eliminate this kind of interference, but the use of an extra alcohol electrode coupled with the bienzymic electrode proved to be the most effective solution. In this case, two signals are obtained, one related to the interfering compounds (alcohol oxidase) and other related to aspartame plus interferents (α-chymotrypsin and alcohol oxidase). To obtain the signal of aspartame the two signals must be subtracted. Liquid samples were analyzed after dilution 1:1 in buffer solution and solid samples were dissolved in the same buffer. The reliability of the methodology was assessed through recovery studies with acceptable results.

A similar strategy was implemented in SIA format with spectrophotometric detection [58]. In this case, 2,2′-azinobis(3-ethylbenzothiazoline-6-sulfonic acid) (ABTS) was used as electron donor of horseradish peroxidase for the detection of the H_2O_2 formed by the enzymatic conversion of aspartame by α-chymotrypsin and alcohol oxidase. Both enzymes were immobilized individually in aminopropyl glass beads and packed in Perspex columns that were linked sequentially to one of the lateral inlets of the selection valve.

A solenoid valve was used to insert the chromogenic reagent (ABTS and peroxidase) at a confluence point. The simple analytical cycle involved the aspiration of sample to the holding coil followed by its propulsion to the enzyme reactors, where aspartame was transformed in H_2O_2 through a sequence of two reactions catalyzed by α-chymotrypsin and alcohol oxidase. The activation of the selection valve caused the confluence of the product of the enzymatic reactions with the chromogenic reagent, with formation of a green compound that was detected spectrophotometrically at 420 nm. The analytical features of merit of this methodology are similar to those of the FIA biosensors described above (Table 23.2). This SIA methodology enables the analysis of aspartame in commercial sweetener tablets after simple dissolution in water and posterior filtration and dilution in buffer solution. The accuracy of the results obtained was evaluated through comparison with the values obtained using a HPLC methodology and through recovery assays. The SIA system proved to be a good automatic alternative for the routine determination of aspartame. The reduction of consumption of reagents (related to the flow reversal process) associated with the robustness and simplicity of the manifold make this strategy very suitable for implementation at industrial level. Moreover, the versatility of SIA enables the easy incorporation into the system of devices such as dialysis units or dilution chambers to implement, say, analyte extraction or sample dilution.

23.3 FLOW-BASED METHODOLOGIES FOR THE DETERMINATION OF SORBITOL

Although sorbitol is a naturally occurring alditol, its use as sweetener in sugar-free foods and drinks has resulted in increased ingestion of this compound by humans of all ages. Sorbitol is readily biodegradable, but it is known that excessive ingestion of it can cause gastric disturbances. In this context, it seems important to at least monitor the levels present in foods and beverages to guide consumers towards a healthy consumption of this additive. For this, food manufacturers and quality institutions need proper analytical tools to perform a fast and accurate evaluation of sorbitol levels in their final products.

As in the case of aspartame, in determination of sorbitol chromatographic methodologies have been preferred, taking into account that in normal-phase partition methods the elution of alditols overlaps that of monosaccharides. In this context, flow methodologies, mainly FIA, emerged as alternative tools for automation of the determination of sorbitol while creating online strategies for the elimination of interferents in different kinds of detection systems.

Despite the fact that these flow approaches have never been applied to the determination of sorbitol in foods and beverages, several examples of flow methods with eventual applicability in this field will be explored and discussed in this chapter. The majority of the flow methodologies developed for sorbitol determination are based on the enzymatic conversion of the analyte to products that can be detected either spectrophotometrically [62,63] or amperometrically [64–66].

The immobilization of pyranose oxidase and horseradish peroxidase in activated resins associated with a spectrophotometric reaction based on the formation of Bindshedler's green ($\lambda = 727$ nm) permits the determination of sorbitol in a FIA format with a detection limit of 1.2 μmol L^{-1} [62]. Prior to enzyme conversion, samples suffer an online treatment based on passage through a cleanup column filled with a strongly basic anion exchange resin, positioned after the injection valve. This step was of the utmost importance during the analysis of serum samples and can also be exploited for the analysis of foods and drinks that include substances that can be oxidized by pyranose oxidase, such as xylose or glucose. A similar FIA methodology based on sorbitol oxidase and horseradish peroxidase for the determination of sorbitol in biological fluids was patented in 1997 [63]. As in the above-mentioned methodology, in this case the incorporation of a cleanup column before the enzyme reactors obviates the need for further sample pretreatments.

The integration of sorbitol dehydrogenase and diaphorase in an amperometric biosensor can also constitute an interesting tool for the FIA determination of sorbitol [64]. The determination is NAD$^+$ dependent and is mediated by ferricyanide, which generates an oxidized mediator on the electrode surface and at the same time enables the conversion of the formed NADH (nicotinamide adenine dinucleotide hydride) to its oxidized form in the presence of diaphorase. The FIA system comprises a network of three injection valves for the insertion of sample, ferricyanide, and NAD$^+$ in such a manner that the streams of NAD$^+$ and ferricyanide merge in a T-shaped connector just before reaching the biosensor. This method determines sorbitol in a linear range up to 1 mmol L^{-1} with a detection limit of 20 μmol L^{-1} and a throughput rate of 10 h^{-1}. Although the method was validated by monitoring sorbitol consumption during L-sorbose production, it has characteristics that allow its application to the analysis of sorbitol in food and drinks after dissolution and dilution in the assay buffer. Due to the high selectivity of sorbitol dehydrogenase it is unlikely that the analysis of this kind of sample would suffer from significant interferences.

A simpler version of this method was also developed for the same purpose, requiring only the integration of sorbitol dehydrogenase into an electrode whose surface was modified with hyaluronic acid, carbon nanotubes, and toluidine blue [65]. Although the sensitivity of the determination is not significantly altered by the modifications, the determination rate increases enormously to 65 h^{-1}, so that this strategy becomes more attractive than the previous one.

Still in the field of amperometric detection, sorbitol can also be determined by a nonenzymatic methodology employing a nickel oxide-modified electrode that functions as a sensor for alditols [44].

23.4　CONCLUSIONS AND FUTURE TRENDS

The determination of the concentration of sweeteners in food items and beverages is of great importance for the safe and sustainable consumption of dietetic products. In particular, the assessment of the levels of aspartame is almost mandatory because of toxicity and accumulation issues. The analysis of series of samples using automated high-throughput methods would be very useful for the elaboration of databases of aspartame content that could constitute important tools for health professionals and consumers. The implementation of automated methods in the food industries could permit easy postproduction evaluation of sweetener contents to provide information that can contribute to nutritional tables.

REFERENCES

1. Sardesai, V. M. and T. H. Waldshan. 1991. Natural and synthetic intense sweeteners. *J. Nutr. Biochem.* 2:236–244.
2. Dills, W. L. 1989. Sugar alcohols as bulk sweeteners. *Annu. Rev. Nutr.* 9:161–186.
3. Whitehouse, C. R., J. Boullata, and L. A. McCauley. 2008. The potential toxicity of artificial sweeteners. *AAOHN J.* 56:251–259.

4. Shankar, P., S. Ahuja, and K. Sriram. 2013. Non-nutritive sweeteners: Review and update. *Nutrition* 29:1293–1299.
5. Sylvetsky, A. C., J. A. Welsh, R. J. Brown, and M. B. Vos. 2012. Low-calorie sweetener consumption is increasing in the United States. *Am. J. Clin. Nutr.* 96:640–646.
6. Lange, F. T., M. Scheurer, and H.-J. Brauch. 2012. Artificial sweeteners—A recently recognized class of emerging environmental contaminants: A review. *Anal. Bioanal. Chem.* 403:2503–2518.
7. Rencuzogullari, E., B. A. Tuylu, M. Topaktas et al. 2004. Genotoxicity of aspartame. *Drug Chem. Toxicol.* 27:257–268.
8. FDA. 2013. Food additives permitted for direct addition to food for human consumption. *Code of Federal Regulations Title 21—Food and Drugs.* Subpart I-Multipurpose Additives. Sector 172.804—Aspartame. Available in: http://www.accessdata.fda.gov/scripts/cdrh/cfdocs/cfCFR/CFRSearch.cfm?fr=172.804.
9. Butchko, H. H. and W. W. Stargel. 2001. Aspartame: Scientific evaluation in the postmarketing period. *Regul. Toxicol. Pharmacol.* 34:221–233.
10. JECFA. 1980. Aspartame. In *Evaluation of Certain Food Additives.* Geneva: Joint FAO/WHO Expert Committee on Food Additives. Technical Report Series 653.
11. SCF. 1985. Sweeteners, In *Report of the Scientific Commission for Food, Commission of the European Communities,* Luxembourg. Office for the Official Publication of the European Communities.
12. SCF. 1989. Colouring matters. Sweeteners. Quinine. Emulsifiers, stabilizers, thickeners and gelling agents. In *Reports of the Scientific Committee for Food, Commission of the European Communities,* Luxembourg. Office for the Official Publication of the European Communities.
13. SCF. 1997. *Minutes of the 107th Meeting of the Scientific Committee for Food,* Brussels. http://ec.europa.eu/food/fs/sc/oldcomm7/out13_en.html.
14. SCF. 2002. Opinion of the Scientific Committee on Food: Update on the safety of aspartame. *In European Commission—Health & Consumer Protection Directorate-General,* Brussels.
15. EFSA. 2005. Opinion of the Scientific Committee on a request from EFSA related to a harmonised approach for risk assessment of substances which are both genotoxic and carcinogenic. *EFSA J.* 282:1–31.
16. EFSA. 2006. Opinion of the scientific panel AFC related to new long-term carcinogenicity study on aspartame. *EFSA J.* 356:1–44.
17. EFSA. 2009. Use of the benchmark dose approach in risk assessment. *EFSA J.* 1150:1–72.
18. EFSA. 2011. Statement on two recent scientific articles on the safety of artificial sweeteners. *EFSA J.* 9:1996–2000.
19. EFSA. 2011. Statement on the scientific evaluation of two studies related to the safety of artificial sweeteners. *EFSA J.* 9:2089–2093.
20. EFSA. 2011. Overview of the procedures currently used at EFSA for the assessment of dietary exposure to different chemical substances. *EFSA J.* 9:2490–2522.
21. EFSA. 2012. Guidance for submission for food additive evaluations. *EFSA J.* 10:2760–2819.
22. EFSA. 2013. Scientific opinion on the re-evaluation of aspartame (E 951) as a food additive. *EFSA J.* 11:3496–3758.
23. Marinovich, M., C. L. Galli, C. Bosetti, S. Gallus, and C. La Vecchia. 2013. Aspartame, low-calorie sweeteners and disease: Regulatory safety and epidemiological issues. *Food Chem. Toxicol.* 60:109–115.

24. Turak, F. and A. U. Ozgur. 2013. Validated spectrophotometric methods for simultaneous determination of food colorants and sweeteners. *J. Chem.* 2013:1–9.
25. Lau, O. W., S. F. Luk, and W. M. Chan. 1988. Spectrophotometric determination of aspartame in soft drinks with ninhydrin as reagent. *Analyst* 113:765–768.
26. Medeiros, R.A., A. E. de Carvalho, R. C. Rocha-Filho, and O. Fatibello-Filho. 2008. Simultaneous square-wave voltammetric determination of aspartame and cyclamate using a boron-doped diamond electrode. *Talanta* 76:685–689.
27. Herzog, G., V. Kam, A. Berduque, and D. W. M. Arrigan. 2008. Detection of food additives by voltammetry at the liquid–liquid interface. *J. Agric. Food Chem.* 56:4304–4310.
28. Nikolelis, D. P. and U. J. Krull. 1990. Dynamic-response characteristics of the potentiometric carbon-dioxide sensor for the determination of aspartame. *Analyst* 115:883–888.
29. Fatibellofilho, O., A. A. Suleiman, G. G. Guilbault, and G. J. Lubrano. 1988. Bienzymatic electrode for the determination of aspartame in dietary products. *Anal. Chem.* 60:2397–2399.
30. Yang, D-J. and B. Chen. 2009. Simultaneous determination of nonnutritive sweeteners in foods by HPLC/ESI-MS. *J. Agric. Food Chem.* 57:3022–3027.
31. Gibbs, B. F., I. Alli, and C. N. Mulligan. 1996. Simple and rapid high-performance liquid chromatographic method for the determination of aspartame and its metabolites in foods. *J. Chromatogr. A* 725:372–377.
32. Grembecka, M. and P. Szefer. 2012. Simultaneous determination of caffeine and aspartame in diet supplements and non-alcoholic beverages using liquid-chromatography coupled to corona CAD and UV-DAD detectors. *Food Anal. Methods* 5:1010–1017.
33. Sherma, J., S. Chapin, and J. M. Follweiler. 1985. Quantitative TLC determination of aspartame in beverages. *Am. Lab.* 17:131–133.
34. Georgiou, C. A., M. A. Koupparis, and T. P. Hadjiioannou. 1991. Flow-injection stopped-flow kinetic spectrophotometric determination of drugs, based on micellar-catalyzed reaction with 1-fluoro-2,4-dinitrobenzene. *Talanta* 38:689–696.
35. Nobrega, J. D., O. Fatibello, and I. D. Vieira. 1994. Flow-injection spectrophotometric determination of aspartame in dietary products. *Analyst* 119:2101–2104.
36. Capitan-Vallvey, F., M. C. Valencia, and E. A. Nicolas. 2004. Flow-through spectrophotometric sensor for the determination of aspartame in low-calorie and dietary products. *Anal. Sci.* 20:1437–1442.
37. Garcia-Jimenez, J. F., M. C. Valencia, and L. F. Capitan-Vallvey. 2006. Improved multianalyte determination of the intense sweeteners aspartame and acesulfame-K with a solid sensing zone implemented in an FIA scheme. *Anal. Lett.* 39:1333–1347.
38. Garcia-Jimenez, J. F., M. C. Valencia, and L. F. Capitan-Vallvey. 2007. Simultaneous determination of antioxidants, preservatives and sweetener additives in food and cosmetics by flow injection analysis coupled to a monolithic column. *Anal. Chim. Acta* 594:226–233.
39. Garcia Jimenez, J. F., M. C. Valencia, and L. F. Capitan-Vallvey. 2009. Intense sweetener mixture resolution by flow injection method with on-line monolithic element. *J. Liq. Chromatogr. Relat. Technol.* 32:1152–1168.
40. Capitan-Vallvey, L. F., M. C. Valencia, E. A. Nicolas, and J. F. Garcia-Jimenez. 2006. Resolution of an intense sweetener mixture by use of a flow injection sensor with on-line solid-phase extraction. *Anal. Bioanal. Chem.* 385:385–391.

41. Fatibello-Filho, O., L. H. Marcolino, and A. V. Pereira. 1999. Solid-phase reactor with copper(II) phosphate for flow-injection spectrophotometric determination of aspartame in tabletop sweeteners. *Anal. Chim. Acta* 384:167–174.

42. Pereira, A. V., L. H. Marcolino, and O. Fatibello. 2000. Flow injection spectrophotometric determination of aspartame in sweeteners using a solid-phase reactor containing zinc phosphate immobilized. *Quim. Nova* 23:167–172.

43. Stojkovic, M., M. Thanh Duc, and P. C. Hauser. 2013. Determination of artificial sweeteners by capillary electrophoresis with contactless conductivity detection optimized by hydrodynamic pumping. *Anal. Chim. Acta* 787:254–259.

44. Barrado, E., J. A. Rodriguez, and Y. Castrillejo. 2006. Renewable stationary phase liquid magnetochromatography: Determining aspartame and its hydrolysis products in diet soft drinks. *Anal. Bioanal. Chem.* 385:1233–1240.

45. McIntire, F. C., L. M. Clements, and M. Sproull. 1953. 1-Fluoro-2,4-dinitrobenzene as a quantitative reagent for primary and secondary amines. *Anal. Chem.* 25: 1757–1758.

46. Athanasioumalaki, E., M. A. Koupparis, and T. P. Hadjiioannou. 1989. Kinetic determination of primary and secondary-amines using a fluoride-selective electrode and based on their reaction with 1-fluoro-2,4-dinitrobenzene. *Anal. Chem.* 61:1358–1363.

47. Souri, E., M. Eskandari, M. B. Tehrani, N. Adib, and R. Ahmadkhaniha. 2013. HPLC determination of pregabalin in bulk and pharmaceutical dosage forms after derivatization with 1-fluoro-2,4-dinitrobenzene. *Asian J. Chem.* 25:7332–7336.

48. Nagaraja, P., A. K. Shrestha, A. Shivakumar, and N. G. S. Al-Tayar. 2011. Molybdate assisted ninhydrin based sensitive analytical system for the estimation of drugs containing amine group. *J. Food Drug Anal.* 19:85–93.

49. Siddiqui, F. A., M. S. Arayne, N. Sultana et al. 2010. Spectrophotometric determination of gabapentin in pharmaceutical formulations using ninhydrin and pi-acceptors. *Eur. J. Med. Chem.* 45:2761–2767.

50. Barrales, P. O., A. D. Vidal, M. L. F. de Cordova, and A. M. Diaz. 2001. Simultaneous determination of thiamine and pyridoxine in pharmaceuticals by using a single flow-through biparameter sensor. *J. Pharm. Biomed. Anal.* 25:619–630.

51. Capitan-Vallvey, L. F., M. C. Valencia, and E. A. Nicolas. 2003. Simple resolution of butylated hydroxyanisole and *n*-propyl gallate in fatty foods and cosmetics samples by flow-injection solid-phase spectrophotometry. *J. Food Sci.* 68:1595–1599.

52. de Cordova, M. L. F., P. O. Barrales, G. R. Torne, and M. Diaz. 2003. A flow injection sensor for simultaneous determination of sulfamethoxazole and trimethoprim by using Sephadex SPC-25 for continuous on-line separation and solid phase UV transduction. *J. Pharm. Biomed. Anal.* 31:669–677.

53. Kika, F. S. 2009. Low pressure separations using automated flow and sequential injection analysis coupled to monolithic columns. *J. Chromatogr. Sci.* 47:648–655.

54. Fernandez, M., R. Forteza, and V. Cerda. 2012. Monolithic columns in flow analysis: A review of SIC and MSC techniques. *Instrum. Sci. Technol.* 40:90–99.

55. Male, K. B., J. H. T. Luong, and A. Mulchandani. 1991. Determination of aspartame in dietary food-products by a FIA biosensor. *Biosens. Bioelectron.* 6: 117–123.

56. Male, K. B., J. H. T. Luong, B. Gibbs, and Y. Konishi. 1993. An improved FIA biosensor for the determination of aspartame in dietary food-products. *Appl. Biochem. Biotech.* 38:189–201.

57. Compagnone, D., D. Osullivan, and G. G. Guilbault. 1997. Amperometric bienzymic sensor for aspartame. *Analyst* 122:487–490.

58. Peña, R. M., J. L. F. C. Lima, and M. L. M. F. S. Saraiva. 2004. Sequential injection analysis-based flow system for the enzymatic determination of aspartame. *Anal. Chim. Acta* 514:37–43.

59. Illanes, A. (ed.). 2008. *Enzyme Biocatalysis. Principles and Applications.* Netherlands: Springer Science + Business Media B.V.

60. Mateo, C., J. M. Palomo, G. Fernandez-Lorente, J. M. Guisan, and R. Fernandez-Lafuente. 2007. Improvement of enzyme activity, stability and selectivity via immobilization techniques. *Enzyme Microb. Technol.* 40:1451–1463.

61. Renneberg, R., K. Riedel, and F. Scheller. 1985. Microbial sensor for aspartame. *Appl. Microbiol. Biotechnol.* 21:180–181.

62. Tanabe, T., Y. Umegae, Y. Koyashiki et al. 1994. Fully automated flow-injection system for quantifying 1,5-anhydro-d-glucitol in serum. *Clin. Chem.* 40:2006–2012.

63. NIPPON KAYAKU KK (NIPK) IKEDA SHOKKEN KK (IKED-Non-standard). Determn. of D-sorbitol and reactors—Uses flow injection analysis where reactor to D-sorbitol oxidase is immobilised. Patent number: JP9023897-A.

64. Sefcovicova, J., A. Vikartovska, V. Patoprsty et al. 2009. Off-line FIA monitoring of D-sorbitol consumption during L-sorbose production using a sorbitol biosensor. *Anal. Chim. Acta* 644:68–71.

65. Sefcovicova, J., J. Filip, P. Tomcik et al. 2011. A biopolymer-based carbon nanotube interface integrated with a redox shuttle and a D-sorbitol dehydrogenase for robust monitoring of D-sorbitol. *Microchim. Acta* 175:21–30.

66. Cataldi, T. R. I. and D. Centonze. 1995. Nickel-oxide dispersed in a graphite/poly(vinyl chloride) composite matrix for an electrocatalytic amperometric sensor of alditols in flow-injection analysis. *Anal. Chim. Acta* 307:43–48.

CHAPTER 24

Determination of Acesulfame-K, Cyclamate, and Saccharin

M. Carmen Yebra-Biurrun

CONTENTS

24.1 INTRODUCTION

Acesulfame-K, cyclamate, and saccharin are noncaloric synthetic sweeteners [1–6] and, according to their molecular structure (Figure 24.1), are included in the group of sulfanil-amides/sulfamate artificial sweeteners [7–13].

Acesulfame-K was incidentally discovered in 1967 by Karl Clauss and Harald Jensen, scientists at the pharmaceutical company Hoechst (now Nutrinova), when they found a sweet-tasting compound that had not been synthesized before. This compound was known as 5,6-dimethyl-1,2,3-oxathiazin-4(3H)-one 2,2-dioxide. After further experimentation with this compound, 6-methyl-1,2,3-oxathiazine-4(3H)-one 2,2-dioxide was chosen as the most favorable for its sensory properties [14]. The World Health Organization (WHO) named the compound acesulfame potassium in 1978. Acesulfame potassium is a white crystalline powder with a molecular formula $C_4H_4KNO_4S$ and a molecular weight of 201.24, and it is also known as acesulfame-K or ace K, and marketed under the trade names "Sunett" and "Sweet One" [15]. Acesulfame-K has an excellent stability under high temperatures (it can and can be used in cooking and baking), and good solubility, which makes it suitable for numerous products [16]. It is approved for use in food and beverage products in more than 100 countries in more than 5000 products, including the United States, Switzerland, Norway, the United Kingdom, Canada, Australia, and the European Union (EU) where it is known under the E number (additive code) E950. This high-intensity sweetener is about 200 times sweeter than a 3% sucrose solution [17]. Toxicological studies on acesulfame-K developed by the Joint Expert

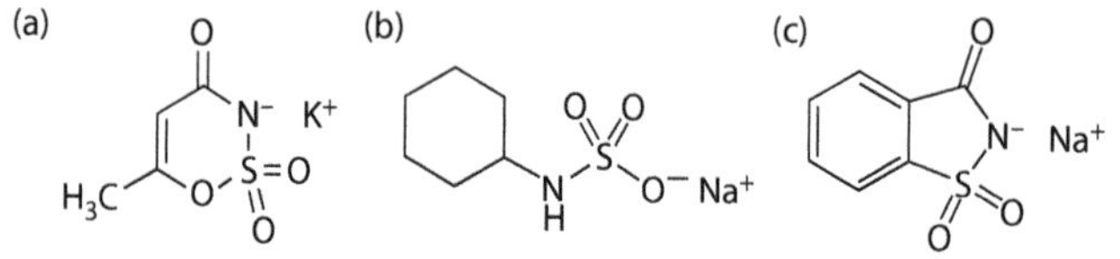

(a)   (b)   (c)

FIGURE 24.1 Chemical structures of (a) acesulfame-K, (b) sodium cyclamate, and (c) sodium saccharin.

Committee on Food Additives (JECFA), the Food and Agriculture Organization (FAO), and the WHO concluded that acesulfame-K is neither carcinogenic nor mutagenic, which demonstrates that the compound would be safe for use as an intense sweetener [18]. At the international level (JECFA), as well as for the U.S. Food and Drug Administration (FDA), the acceptable daily intake (ADI) for acesulfame-K has been set at 15 mg/kg body weight. At the European level, European Food Safety Authority (EFSA), the ADI has been set at 9 mg/kg body weight [19]. Although acesulfame-K can be used as an intense sweetener by itself, blends of acesulfame-K with many other sweet-tasting substances (aspartame, sodium cyclamate, or sucralose), show better sweetness properties than the single substances.

Cyclamate was accidentally discovered in 1937 by Michael Sveda, a graduate student at the University of Illinois while he was working with pharmaceutical compounds and detected that salts of cyclohexylsulfamic acid are sweet. Cyclamate in its acid form (cyclamic acid or cyclohexylsulfamic acid, molecular formula $C_6H_{13}NO_3S$) is a strong acid with pK_a of 1.71 and molecular weight of 179.20. Cyclamate is generally used in the form of a sodium salt because it is more soluble in water than the free acid. Both of these are colorless and odorless solids. The calcium salt is also used as a sweetener, but, for some applications, it is not suitable as it can cause gelation and precipitation. Sodium cyclamate exhibits good stability in the solid form and is also stable in soft drink formulations within the pH range 2–10. It has been reported that cyclamate is a potent carcinogen when it is converted into cyclohexylamine in the gastrointestinal tract [14]. Although animal studies fail to demonstrate that cyclamate is a carcinogen or a cocarcinogen, other issues must be resolved before cyclamate can be approved for commercial use as a food additive. Because the safety of cyclamate to human is not clear completely, the restricted content level is not the same in different countries. Thus, it is banned in the United States and it is approved for use in over 55 countries, including the EU (where it is known under the additive code E952), Australia, and Canada (for table-top and pharmaceutical use only) [18]. The ADI value for cyclamate has been set at 11 mg/kg body weight by JECFA and at 7 mg/kg body weight by EFSA [19]. Cyclamate is approximately 40 times as sweet as a 2% sucrose solution, but only 24 times as sweet as a 20% sucrose solution [17]. This trend may be at least partially due to the increasing levels of bitterness and aftertaste that characteristically appear at very high cyclamate concentrations. However, this off-taste is not a problem at concentrations normally used. According to its bitter and salty "off" taste attributes, as well as its low sweetness potency, cyclamate salts are most commonly employed in blends with other noncaloric sweeteners, principally saccharin, aspartame, and acesulfame-K [15].

Saccharin is commonly known as a widely used noncaloric synthetic sweetener. It is the oldest nonnutritive synthetic sweetener, having been on the market for more than 100 years. Saccharin was accidently discovered by Ira Remsen and Constantin Fahlberg, researchers at John Hopkins University in 1879. Saccharin is commercially available in three forms: acid saccharin, sodium saccharin, and calcium saccharin. Saccharin in its acid form (1,2-benzisothiazol-3(2*H*)-one 1,1-dioxide, benzoic sulfimide, *ortho*-sulfobenzamide, or 1,2-benzothiazol-3(2*H*)-one 1,1-dioxide, molecular formula $C_7H_5NO_3S$) is a strong acid with pK_a of 2.2 and molecular weight of 183.18 [14]. Sodium saccharin is the most commonly used form because of its high solubility and stability [14]. Saccharin and its salts in their solid form show good stability under conditions present in soft drinks. However, at low pH, they can slowly hydrolyze to 2-sulfobenzoic acid and 2-sulfamoylbenzoic acid. Saccharin is not metabolized in the gastrointestinal tract and therefore does not affect blood insulin levels, which makes saccharin a viable sugar substitute for patients with diabetes [15,16]. Saccharin has been the subject of extensive scientific studies and generated numerous debates. The totality of available research clearly shows that saccharin is safe for human consumption. However, there has been controversy over its safety based on the finding that consumption of large daily doses of sodium saccharin by male rats is associated with increased incidence in bladder tumors. However, in 2010, the EPA officially removed saccharin and its salts from their list of hazardous constituents and commercial chemical products and stated that saccharin is no longer considered a potential hazard to human health. Thus, saccharin continues to be used in food and drink formulations in more than 100 countries. In the EU, saccharin is also known by the additive code E954. The ADI for saccharin has been set at 5 mg/kg body weight by the JECFA and the EFSA [18,19]. Most publications characterize saccharin as 300 times sweeter than sugar; however, it was reported that at 1% sucrose concentration, saccharin is over 800 times sweeter than sugar; at 3% sucrose concentration, saccharin is about 500 times sweeter than sugar; and at 7% sucrose concentration, it is over 300 times sweeter than sucrose [17,20]. The use of a synergistic mixture with other sweeteners is strongly recommended in order to improve the saccharin taste. Saccharin presents excellent synergistic properties with cyclamate and aspartame, but it does not show synergism with acesulfame-K, probably owing to the similarities of their molecular structures.

24.2 ANALYTICAL TECHNIQUES FOR ACESULFAME-K, CYCLAMATE, AND SACCHARIN DETERMINATION

The analytical methods proposed for acesulfame-K, cyclamate, and saccharin determination in foods, drinks, dietary products, and pharmaceuticals can be grouped into methods for the determination of an individual artificial sweetener [21–27] and multianalyte approaches [28–38], sometimes also including other sweeteners and/or other food additives, such as colorants or preservatives [39–43]. High-performance liquid chromatography (HPLC) is the most frequently used technique for the determination of these sweeteners, and this is selected by international standard methods because of its multianalyte capability, compatibility with the physicochemical properties of sweeteners, high sensitivity, and robustness [44–47]. However, cyclamate requires chemical derivatization to make it detectable by the most commonly employed UV-absorption detector due to a lack of a chromophore, by conversion to dichlorohexylamine for UV detection or to a fluorescence derivative for fluorimetric detection. Another alternative for cyclamate detection is the postcolumn ion-pair extraction where the eluted sweetener is mixed with an appropriate dye (methyl violet or crystal violet), being detected by visible absorption. Furthermore, cyclamate can be detected directly by refractive index [4]. For this, few HPLC methods for the concurrent determination of these sweeteners exist and usually are based on detection by mass spectrometry [48–50]. However, acesulfame-K, cyclamate, and saccharin can be determined by all current analytical techniques. Thus, ion chromatography (IC) offers an attractive alternative to traditional HPLC methods for multisweetener analysis in food products. Instead of organic solvent-mediated separations, IC separations are performed using innocuous and inexpensive salt solutions as the eluents [51–53]. Capillary electrophoresis (CE) is another technique used for the simultaneous determination of multiple sweeteners [24,32,36,37,39]. The resolving power of IC and CE techniques is in many cases comparable with that of HPLC and, frequently, their running costs are lower. However, it seems that due to limited robustness, in the case of CE methods, and the modest choice of separation mechanisms, in the case of IC, these methods are less popular [44]. Other methods used for acesulfame-K, cyclamate, and saccharin determination are based on electrochemical [54–59], spectroscopic [12,24,60–63], gas chromatographic [64,65], or immunochemical analysis (enzyme-linked immunosorbent assay [ELISA]) [66] techniques.

24.3 FLOW ANALYSIS TECHNIQUES FOR ACESULFAME-K, CYCLAMATE, AND SACCHARIN DETERMINATION

Flow analysis techniques, as an automation tool for chemical analysis, have been used widely as a strategy to solve analytical problems, and were presented as an interesting alternative for use in sweetener determinations when only one analyte was determined in a large number of samples. Thus, in many instances, chromatographic methods can be replaced by fast flow analysis methodologies. In addition, the advantages provided by these procedures cannot be overlooked: high sample throughput, high reproducibility, simple automation of operation, low contamination risks, possible enhancement in selectivity by applying kinetic discrimination, and very limited laboratory bench space and equipment required [67–69]. Furthermore, flow analysis methodologies have a great potential for minimizing consumption of reagents and sample and generation of waste, as well as the ability to implement processes unreliable in batch to replace toxic chemicals, which contributes to the criteria of green analytical chemistry [70,71].

The ScienceFinder database was searched up to March 2014 for flow analysis methods for the determination of acesulfame-K, cyclamate, and saccharin. The keywords used as search criteria were "flow injection analysis," "sequential injection analysis," "lab-on-valve," "multicommutation," "multisyringe," "multipumping," as well as the individual sweetener names. As can be seen in Figure 24.2, only three flow analysis methodologies had been applied to the determination of these sweeteners, namely, flow injection analysis (FIA), sequential injection analysis (SIA), and multicommutation.

The distribution of flow analysis methods according to the type of determination (individual or multianalyte) is shown in Figure 24.3. This figure reveals that still there are no proposed methods for the individual determination of acesulfame-K by flow analysis.

The most important details of the published flow analysis procedures for acesulfame-K, cyclamate, and saccharin determination are shown in chronological order for each artificial sweetener in Table 24.1.

The following sections will describe the main features and characteristics of flow analysis methodologies proposed for the individual determination of cyclamate and saccharin, and for the multianalyte determination of acesulfame-K, cyclamate, and saccharin.

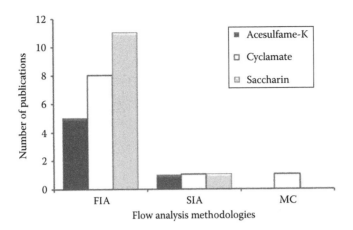

FIGURE 24.2   Distribution of flow analysis methodologies applied for the determination of acesulfame-K, cyclamate, and saccharin. FIA: flow injection analysis; SIA: sequential injection analysis; MC: multicommutation.

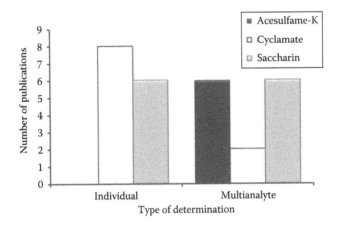

FIGURE 24.3   Distribution of flow analysis methodologies applied for the determination of acesulfame-K, cyclamate, and saccharin according to the type of determination.

24.3.1 Flow Injection Methodologies for Acesulfame-K

No methods could be found in the literature for the individual flow analysis determination of acesulfame-K. The first flow analysis method for acesulfame-K was proposed by Nikolelis et al. 2001 [72]. This method allowed the electrochemical flow injection monitoring and analysis of mixtures of acesulfame-K, cyclamate, and saccharin using stabilized systems of filter-supported bilayer lipid membranes (BLMs). Detection consisted of t time-dependent appearance of a transient ion current peak in which the time-dependence could be used to distinguish the presence of different artificial sweeteners, and the peak magnitude was related to the concentration of the artificial sweetener. The BLM-based system is able to monitor each artificial sweetener in mixtures. The apparatus for the formation of stabilized BLMs is shown in Figure 24.4. The method also offers response times of less than 1 min, which are the fastest times reported for any similar

TABLE 24.1   Acesulfame-K, Cyclamate, and Saccharin Determination Using Flow Analysis Methodologies

Sweetener	Flow Mode	Type of Determination	Detection Technique	Samples	SF (h^{-1})	LOD (µM)	RSD (%)	Reference
Acesulfame-K	FIA	Multianalyte (cyclamate, saccharin)	EB	Artificial sweetener tablets, diet soft drinks, wines, and yogurts	No data	0.5	<6	[72]
Acesulfame-K	FIA	Multianalyte (aspartame)	SP	Tabletop sweeteners	24	59.1	1.61	[73]
Acesulfame-K	FIA	Multianalyte (aspartame, saccharin, methylparaben, ethylparaben, propylparaben, butylparaben, propylgallate, butylhydroxyanisole)	SP	Soft drinks	9	0.5	1.5	[74]
Acesulfame-K	FIA	Multianalyte (aspartame, saccharin)	SP	Foods and soft drinks	No data	5	0.08	[75]
Acesulfame-K	FIA	Multianalyte (saccharin, caffeine, benzoic acid, and sorbic acid)	HPLC-UV	Soft drinks	4.3	0.5	0.2–3.8	[76]
Acesulfame-K	SIA	Multianalyte (aspartame, cyclamate, saccharin)	CE-CCD	Artificial sweetener tablets, soft drinks	No data	3.8	1.2–2.1	[77]
Cyclamate	FIA	Individual	CL	Not applied to real samples	100	0.4	1.0–2.8	[78]
Cyclamate	FIA	Individual	SP	Artificial sweetener tablets, soft drinks	24	30	>0.5	[79]
Cyclamate	FIA	Individual	BA	Diet products	90	2500	1.7	[84]

(Continued)

TABLE 24.1 (*Continued*) Acesulfame-K, Cyclamate, and Saccharin Determination Using Flow Analysis Methodologies

Sweetener	Flow Mode	Type of Determination	Detection Technique	Samples	SF (h^{-1})	LOD (μM)	RSD (%)	Reference
Cyclamate	FIA	Individual	BA	No data	No data	1300	2	[85]
Cyclamate	FIA	Individual	FAAS	Artificial sweetener tablets, soft drinks	35	1.2	3.1	[83]
Cyclamate	FIA	Individual	SP	Artificial sweetener tablets	No data	7.7	3.1	[80]
Cyclamate	FIA	Multianalyte (acesulfame-K, saccharin)	EB	Artificial sweetener tablets, diet soft drinks, wines, and yogurts	No data	1	< 6	[72]
Cyclamate	FIA	Individual	TB	Artificial sweetener tablets, soft drinks	45	298.2	5.9	[82]
Cyclamate	SIA	Multianalyte (aspartame, acesulfame-K, saccharin)	CE-CCD	Artificial sweetener tablets, soft drinks	No data	5	1.6–2.2	[77]
Cyclamate	MC	Individual	SP	Artificial sweetener tablets	60	30	1.7	[81]
Saccharin	FIA	Individual	P	Dietary products	60	No data	2.8	[86]
Saccharin	FIA	Individual	FAAS	Artificial sweeteners and pharmaceuticals	20	14.9	2.7	[88]
Saccharin	FIA	Individual	P	Dietary products	40	80	No data	[87]
Saccharin	FIA	Multianalyte (acesulfame-K, cyclamate)	EB	Artificial sweetener tablets, diet soft drinks, wines, and yogurts	No data	0.2	< 6	[72]
Saccharin	FIA	Individual	SP	Low-calorie and dietary products	20	1.1	0.78	[89]

(*Continued*)

TABLE 24.1 (*Continued*)   Acesulfame-K, Cyclamate, and Saccharin Determination Using Flow Analysis Methodologies

Sweetener	Flow Mode	Type of Determination	Detection Technique	Samples	SF (h^{-1})	LOD (μM)	RSD (%)	Reference
Saccharin	FIA	Multianalyte (aspartame)	SP	Low-calorie and dietary products	10	1.6	1	[92]
Saccharin	FIA	Multianalyte (aspartame, acesulfame-K)	SP	Foods and soft drinks	No data	4.9	0.09	[74]
Saccharin	FIA	Individual	TB	Liquid sweetener products	No data	3930	1.75	[90]
Saccharin	FIA	Multianalyte (aspartame, acesulfame-K, methylparaben, ethylparaben, propylparaben, butylparaben, propylgallate, butylhydroxyanisole)	SP	Soft drinks	9	4	0.93	[75]
Saccharin	FIA	Multianalyte (acesulfame-K, caffeine, benzoic acid, and sorbic acid)	HPLC-UV	Soft drinks	4.3	0.5	0.1–3.5	[76]
Saccharin	FIA	Individual	CL	No data	No data	1.08×10^{-5}	2.9	[91]
Saccharin	SIA	Multianalyte (aspartame, cyclamate, acesulfame-K)	CE-CCD	Artificial sweetener tablets, soft drinks	No data	4	2.1–2.9	[77]

Note: BA: biamperometric titration; CE-CCD: capillary electrophoresis with contactless conductivity detection; CL: chemiluminescence; EB: electrochemical biosensor; FAAS: flame atomic absorption spectrometry; FIA: flow injection analysis; HPLC-UV: high-performance liquid chromatography with UV detection; MC: multicommutation; P: potentiometry; SIA: sequential injection analysis; SP: spectrophotometry; TB: turbidimetry.

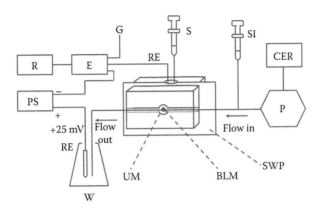

FIGURE 24.4 FIA manifold used for the analysis of mixtures of artificial sweeteners using filter-supported BLMs. BLM: bilayer lipid membrane; CER: carrier electrolyte reservoir; E: electrometer; G: ground; P: pump; PS: power supply; R: recorder; RE: reference electrode; S: syringe; SI: sample injector; SWP: Saran–Wrap partition; UM: ultrafiltration membrane; W: waste.

detectors. This method is faster and has a lower cost than that based on chromatographic techniques. Hence, this BLM-based biosensor can provide an attractive alternative to chromatographic devices or even for use as a detector in chromatographic columns. This work also demonstrates an example of a biosensor based on lipid films that can be used in nonaqueous media, such as wine samples. Interferences from constituents of artificial tablets and soft drinks were investigated, and no interferences were noticed from fructose, aspartame, lactose, dextrose, glucose, sodium citrate, bicarbonate and benzoate, caffeine, absolute ethanol up to 40% (w/w), and other wine constituents. The results obtained were in good agreement with the Official Methods of Analysis of the Association of Official Analytical Chemists.

García Jiménez et al. proposed the analysis of additive mixtures by FIA, including a monolithic minicolumn placed in Figure 24.5 or before the flow cell of a monochannel FIA setup with no prior derivatization reaction. Accordingly, they developed flow-through spectrophotometric sensors (UV absorption) based on the transient retention of analytes on several solid phases, namely, ion exchanger Sephadex DEAE A-25 [73], silica

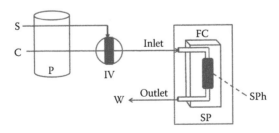

FIGURE 24.5 FIA manifold including a flow-through spectrophotometric sensor. C: carrier; FC: flow-through cell; IV: injection valve; P: pump; S: sample; SP: spectrophotometer; SPh: solid phase; W: waste.

C18 [74], and quaternary amine ion exchanger [75]. As carriers to achieve the separation in the FIA system were used orthophosphoric acid/sodium dihydrogen phosphate buffer 0.06 M; a mixture of ACN/water buffered with 10 mM pH 6.0 phosphate buffer and a methanol:water mixture; pH 9.0 Tris buffer 0.03 M, NaCl 0.4 M, and NaClO$_4$ 0.005 M. These investigations demonstrate that monolithic columns are powerful tools for achieving very good resolutions of complex mixtures and this technique shows the advantages of flow analysis methods in achieving good separations of several analytes in a very short time, working with inexpensive instruments and obtaining results comparable to those of more complex techniques such as HPLC.

Kritsunankul et al. [76] proposed flow injection online dialysis for sample pretreatment prior to the simultaneous determination of some food additives by HPLC and UV detection (FID-HPLC). For this, a liquid sample or mixed standard solution (900 μL) was injected into a donor stream (5%, w/v, sucrose) of a FID system and was pushed further through a dialysis cell, while an acceptor solution (0.025 mol/L phosphate buffer, pH 3.75) was held on the opposite side of the dialysis membrane. The dialysate was then flowed to an injection loop of the HPLC valve, where it was further injected into the HPLC system and analyzed under isocratic reversed-phase HPLC conditions and UV detection (230 nm) (Figure 24.6). The order of elution of five food additives was acesulfame-K, saccharin, caffeine, benzoic acid, and sorbic acid, with an analysis time of 14 min. This system has advantages of high degrees of automation for sample pretreatment, that is, online sample separation and dilution and low consumption of chemicals and materials.

Stojkovic et al. [77] used superimposed hydrodynamic pumping to determine artificial sweeteners by CE with contactless conductivity detection employing a sequential injection manifold based on a syringe pump (Figure 24.7). As a result, aspartame, cyclamate, saccharin, and acesulfame-K could be determined concurrently, without requiring any special sample pretreatment. The analyses were carried out in an aqueous running buffer consisting of 150 mM 2-(cyclohexylamino)ethanesulfonic acid and 400 mM tris(hydroxymethyl)aminomethane at pH 9.1 in order to render all analytes in the fully deprotonated anionic form. The use of surface modification to eliminate or reverse the electroosmotic flow was not necessary due to the superimposed bulk flow.

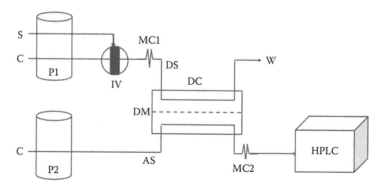

FIGURE 24.6   FIA manifold for online dialysis coupled with HPLC. AS: acceptor stream; C: carrier; DC: dialysis cell; DM: dialysis membrane; DS: donor stream; HPLC: high-performance liquid chromatography; IV: injection valve; MC1 and MC2: mixing coils; P1 and P2: pumps; S: sample; W: waste.

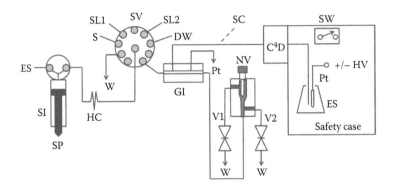

FIGURE 24.7 Sequential injection analysis–capillary electrophoresis manifold with contactless conductivity detection (SIA-CE-C⁴D). C⁴D: capacitively coupled contactless conductivity detector; DW: deionized water; ES: electrolyte solution; GI: grounded interface; HC: holding coil; HV: high-voltage power supply; ND: needle valve; S: sample; SC: separation capillary; SI: syringe; SL1: 1 M NaOH; SL2: 1 M HCl; SP: syringe pump; SS: safety switch; SV: selection valve; V1 and V2: solenoid valves; W: waste.

24.3.2 Flow Analysis Methodologies for Cyclamate

Most flow analysis methods for cyclamate have been proposed for its individual determination.

Psarellis et al. [78] investigated the sensitizing effect of sodium cyclamate on the chemiluminogenic oxidation of sulfite by cerium(IV) in sulfuric acid to determine this artificial sweetener in the range 1.00–50.0 µg/mL. Acetonitrile does not significantly alter the analytical characteristics of the method, which can therefore be proposed for the determination of the analyte after liquid chromatographic separation from other sweeteners and possible metabolites. Figure 24.8 shows the schematic diagram of the flow system used.

Cyclamate salts are not readily detected by spectroscopic techniques. So, a chemical derivatization is often performed in order to improve the characteristics of cyclamate for detection. Two spectrophotometric FIA methods were proposed. Gouveia et al. [79] proposed a procedure based on the reaction of cyclamate with an excess of nitrite solution. The unconsumed excess of nitrite is monitored spectrophotometrically using the Griess reaction; to eliminate interference caused by dye additives, an alumina column was incorporated into the flow system and used for both colorless and colored samples.

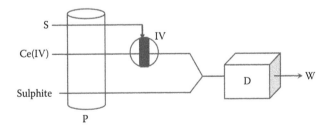

FIGURE 24.8 FIA manifold for chemiluminometric determination of sodium cyclamate. D: detector; IV: injection valve; P: pump; S: sample; W: waste.

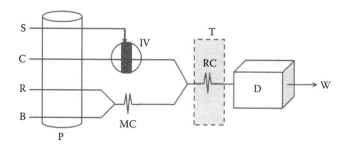

FIGURE 24.9 FIA manifold for spectrophotometric determination of cyclamate. B: buffer solution (0.05 M NaHCO$_3$ + 0.075 M NaOH); C: carrier; D: detector; IV: injection valve; MC: mixing coil; P: pump; R: reagent solution (2 × 10^{-3} M sodium 1,2-naphtho quinone-4-sulfonate + 0.1 M HCl); RC: reaction coil; S: sample; T: thermostatic bath; W: waste.

Cabero et al. [80] developed a method based on the conversion of cyclamate to cyclohexylamine and the subsequent reaction with 1,2-naphthoquinone-4-sulfonate, yielding a spectrophotometrically active derivative, which is detected at 480 nm; thus, other sweeteners, such as saccharin or aspartame, do not interfere in these determinations. The hydrolysis step is performed batchwise by treatment of cyclamate with hydrogen peroxide and hydrochloric acid, while the cyclohexylamine derivatization is carried out in the flow injection system (Figure 24.9). Rocha et al. [81] reported a flow system based on multicommutation for fast and clean determination of cyclamate. The procedure exploits the reaction of cyclamate with nitrite in an acidic medium and the spectrophotometric determination of the excess of nitrite by iodometry. The flow system was designed with a set of solenoid micropumps to minimize reagent consumption and waste generation (Figure 24.10).

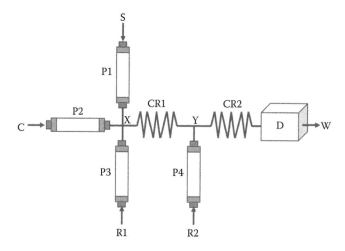

FIGURE 24.10 Multicommutated flow manifold for cyclamate determination. C: carrier; CR1 and CR2: coiled reactors; D: detector; P1, P2, P3, and P4: solenoid micro-pumps; R1: 0.20 mM NaNO$_2$; R2: 020 mM KI; S: sample; X and Y: confluence points; W: waste.

Lamas et al. [82] proposed a FIA turbidimetric method for cyclamate determination without sample pretreatment. The method is based on oxidation of the sulfamic group that is present in cyclamates to sulfate by using nitrite in acid medium. Afterwards, the sulfate ion is precipitated with barium in the presence of poly(vinyl alcohol) (PVA) in perchloric acid solution, at 30°C. The analytical signal was measured at 420 nm.

Yebra et al. [83] used a continuous-flow procedure for the indirect determination of sodium cyclamate by flame atomic absorption spectrometry (FAAS). This method is based on oxidation of the sulfamic group derived from cyclamate to sulfate in acidic conditions and in the presence of sodium nitrite. The procedure is adapted to a flow system with precipitate dissolution (Figure 24.11), where sulfate formed is continuously precipitated with lead ion. The lead sulfate formed is retained on a filter, washed with diluted ethanol, and dissolved in ammonium acetate (because of the formation of soluble lead acetate) for online FAAS determination of lead, the amount of which in the precipitate is proportional to that of cyclamate in the sample. In this work a home-made filtration device was used made of a Teflon tubing packed with a cotton pulp and the ends of the filter column were plugged with filter paper (chamber inner volume 141 μL). This precipitate collector was effective in retaining the precipitate and did not produce excessive back-pressure if the precipitate was dissolved following each precipitation cycle.

Biamperometric titration of cyclamate was exploited by Fatibello-Fiho et al. [84,85]. These investigators developed an inexpensive laboratory-built selectable voltage source specially constructed for use in biamperometric titrations. It was made of very simple electronic components and its performance was tested and compared with a potentiostat/galvanostat equipment. The titrant–reagent system was sodium nitrite in 1.0 M phosphoric acid and there was no interference from glucose, fructose, sucrose, lactose, sorbitol, saccharin, benzoic acid, salicylic acid, fumaric acid, Sunset Yellow, and Bordeaux-S.

Only two flow analysis methods have been published for multianalyte determination including cyclamate as analyte. Both methods determine cyclamate with other artificial sweeteners. One of them used stabilized systems of filter-supported BLMs in a FIA manifold [72], and the other is based on CE with contactless conductivity detection employing a sequential injection manifold based on a syringe pump [77].

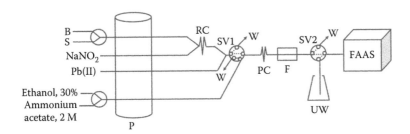

FIGURE 24.11   FIA manifold for the indirect atomic absorption determination of cyclamate. B: blank; F: filter; FAAS: flame atomic absorption spectrometer; P: pump; PC: precipitation coil; RC: reaction coil; S: sample; SV1 and SV2: switching valves; UW: ultrapure water; W: waste.

24.3.3 Flow Analysis Methodologies for Saccharin

A potentiometric determination of saccharin was proposed by Fatibello-Filho et al. [86]. In this method, saccharin was potentiometrically measured using a silver wire coated with a mercury film as the working electrode. With this, the main difficulty was the presence of a precipitate (mercurous saccharinate) that could adsorb on tube walls and the electrode surface. To avoid these undesirable effects, a relocatable filter unit was placed before the flow-through potentiometric cell and a surfactant was added to the carrier solution (Figure 24.12). The same investigation team reported the construction and analytical evaluation of a tubular ion-selective electrode coated with an ion pair formed between saccharinate anion and toluidine blue O cation incorporated on a poly(vinyl chloride) matrix [87]. This electrode was constructed and adapted in a FIA system. The optimum experimental conditions found were an analytical path of 120 cm, an injection sample volume of 500 μL, a pH of 2.5, a flow rate of 2.3 mL/min, and a tubular electrode length of 2.5 cm.

A FIA indirect method was described for saccharin determination by Yebra et al. [88]. This method was based on the precipitation of silver saccharinate in an acetic acid medium. The precipitate was dissolved in ammonia, and the dissolved silver was continuously measured by FAAS. The FIA precipitation–dissolution manifold used is similar to that used for cyclamate determination [83]; however, in this work, a Scientific System 05–105 column fitted with a removable screen-type stainless-steel filter (pore size 0.5 μm, chamber inner volume 580 μL, filtration area 3 cm²), originally designed as a cleaning device for liquid chromatography, was employed for filtration purposes.

Capitán-Vallvey et al. [89] developed a simple, rapid, and inexpensive monoparameter flow-through sensor for the determination of saccharin. The method is based on the transient adsorption of the sweetener on Sephadex G-25 solid phase packed to a height of 20 mm in the flow cell. The optimal transient retention of the synthetic sweetener, in terms of sensitivity and sampling frequency, was obtained when pH 2.75 citric acid–sodium citrate buffer 5×10^{-3} M was used as a carrier at a flow rate of 1.5 mL/min. Saccharin was determined by measuring its intrinsic absorbance at 217 nm at its residence time. The method does not need derivatization reactions and utilized a FIA monochannel manifold similar to those reported by the same investigation team as used to determine other artificial sweeteners [73–75].

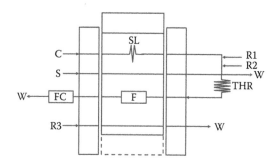

FIGURE 24.12 FIA manifold for potentiometric determination of saccharin with a relocatable filter. C: carrier solution (0.2 M sodium nitrate in 1 mM nitric acid); F: filter; FC: flow-through potentiometric cell; R1, R2, and R3: mercury(I) nitrate, 1 mM nitric acid, and 0.2 M sodium nitrate in 1 mM nitric acid, respectively; S: sample; THR: tubular helicoidal reactor; W: waste.

Bitencourt-Mendes et al. [90] described a method for saccharin determination in liquid sweetener products. The method is based on the precipitation reaction of Ag(I) ions with saccharin in aqueous medium (pH 3.0), using a FIA system with merging zones, the suspension was stabilized with 5 g/L Triton X-100. Based on interference studies performed with the substances commonly found in liquid sweeteners, such as sodium cyclamate, methylparaben, sodium aspartame, and benzoic and citric acids, at the analyte-to-interferent mole ratio of up to 1:10 no interference with the saccharin determination was observed. The presence of chloride ions interferes with the method, but a preceding liquid–liquid saccharin extraction with ethyl acetate was successfully employed to overcome this drawback.

Gao et al. [91] reported a FIA chemiluminescence determination of sodium saccharin based on the chemiluminescent reaction between luminol and hydrogen peroxide. In an acid medium, the sodium saccharin solution was reacted with potassium dichromate to reduce $Cr(VI)$ to $Cr(III)$, and the chemiluminescence intensity of the reaction was remarkably enhanced by the $Cr(III)$ formed. The content of $Cr(III)$ was then determined and converted into contents of sodium saccharin.

Saccharin is also determined with FIA methodologies, which allow multianalyte determination. These methodologies include methods cited previously for acesulfame-K and/or cyclamate: a method based on electrochemical detection of these three artificial sweeteners using stabilized systems of filter-supported BLMs [72]; online dialysis for sample pretreatment prior to the simultaneous determination of saccharin and caffeine, benzoic acid, and sorbic acid by HPLC-FID [76]; and molecular spectroscopic methods in the UV region based on the transient retention of analytes on solid phases, silica C_{18} [74], and quaternary amine ion exchanger [75].

Capitán-Vallvey et al. [92] proposed an integrated solid-phase spectrophotometry FIA method for the simultaneous determination of the mixture of saccharin and aspartame. The procedure is based on online preconcentration of aspartame on a C_{18} silica gel minicolumn and separation from saccharin, followed by measurement at 210 nm of the absorbance of saccharin, which is transiently retained on the adsorbent Sephadex G-25 placed in the flow-through cell of a monochannel FIA setup using pH 3.0 orthophosphoric acid–dihydrogen phosphate buffer, 3.75×10^{-3} M, as carrier. Subsequent desorption of aspartame with methanol enables its determination at 205 nm.

Saccharin was determined in multianalyte mode, as well as acesulfame-K and cyclamate, by CE with contactless conductivity detection employing a sequential injection manifold based on a syringe pump [77].

24.4 CONCLUSIONS

The determination of acesulfame-K, cyclamate, and saccharin individually or simultaneously with other artificial sweeteners and/or other food additives in foods, soft drinks, and tabletop sweeteners is very important for legal, health, and consumer safety aspects. Thus, reliable, simple, fast, sensitive, accurate, and robust analytical methods using low-cost equipment are essential to protect human health, meet the requirement to ensure product quality, and support the compliance and enforcement of laws and regulations pertaining to food safety. Flow analysis is shown to be a powerful analytical tool for the automated determination of acesulfame-K, cyclamate, and saccharin in food samples, and it is an interesting alternative for use in sweetener determinations when only one analyte is determined in a large number of samples. In the last few years, flow analysis methodologies have proved to be a reliable approach for multianalyte determinations, which is very important because in products of the food industry several types of additives are combined in order to obtain synergistic effects between them, such as with sweeteners and preservatives. There are still not many proposed methods, but it is hoped that in the future new flow analysis methodologies will be developed to determine these artificial sweeteners, especially methods exploiting the latest-generation flow analysis techniques.

REFERENCES

1. O'Brien Nabors, L. and R. C. Gelardi. 1991. Intense sweeteners. In *Handbook of Sweeteners*, eds. S. Marie, and J. R. Piggott, pp. 104–115. New York: Springer Science + Business Media.
2. Ager, D. J., D. P. Pantaleone, S. A. Henderson, A. R. Katritzky, I. Prakash, and D. E. Walters. 1998. Commercial, synthetic nonnutritive sweeteners. *Angew. Chem. Int. Ed.* 37:1802–1817.
3. Lipinski, G. W. and von R. 2003. Intense sweeteners: Status and new developments. *Int. Sugar J.* 105:308, 310–312.
4. Yebra-Biurrun, M. C. 2005. Sweeteners. In *Encyclopedia of Analytical Science* (2nd ed.), eds. P. Worsfold, A. Townshend, and C. Poole, pp. 562–572. Oxford: Elsevier.
5. Shankar, P., S. Ahuja, and K. Sriram. 2013. Non-nutritive sweeteners: Review and update. *Nutrition* 29:1293–1299.
6. Yebra-Biurrun, M. C. 2013. Sweeteners. In *Chemistry, Molecular Sciences and Chemical Engineering*. ed. J. Reedijk, pp. 562–572. Waltham: Elsevier.
7. Clauss, K. and H. Jensen. 1973. Oxathiazinone dioxides: A new group of sweetening agents. *Angew. Chem. Int. Ed.* 12:869–876.
8. Benson, G. A. and W. J. Spillane. 1976. Structure-activity studies on sulfamate sweeteners. *J. Med. Chem.* 19:869–872.
9. Spillane, W. J., C. A. Ryder, M. R. Walsh, P. J. Curran, D. G. Concagh, and S. N. Wall. 1996. Sulfamate sweeteners. *Food Chem.* 56:255–261.
10. Suami, T., L. Hough, T. Machinami, T. Saito, and K. Nakamura. 1998. Molecular mechanisms of sweet taste 8: Saccharin, acesulfame-K, cyclamate and their derivatives. *Food Chem.* 63:391–396.
11. Spillane, W. J., C. A. Ryder, P. J. Curran, S. N. Wall, L. M. Kelly, B. G. Feeney, and J. Newell. 2000. Development of structure-taste relationships for sweet and non-sweet heterosulfamates. *J. Chem. Soc. Perkin Trans.* 2(7):1369–1374.
12. Ni, Y., W. Xiao, and S. Kokot. 2009. A differential kinetic spectrophotometric method for determination of three sulphanilamide artificial sweeteners with the aid of chemometrics. *Food Chem.* 113:1339–1345.
13. Spillane, W. and J. B. Malaubier. 2014. Sulfamic acid and its N- and O-substituted derivatives. *Chem. Rev.* 114:2507–2586.
14. Salminen, S. and H. Hallikainen. 2001. Sweeteners. In *Food Additives* (2nd ed. revised and expanded). eds. A. L. Branen, P. M. Davidson, S. Salminen, and J. Thorngate, pp. 467–499. New York: Marcel Dekker.
15. Mitchell, H. 2006. *Sweeteners and Sugar Alternatives in Food Technology*. Oxford: Blackwell Publishing.
16. Wilson, R. 2007. *Sweeteners* (3rd ed.). Oxford: Blackwell Publishing.
17. O'Brien-Nabors, L. 2012. *Alternative Sweeteners* (4th ed.). Boca Raton: CRC Press.

18. Lipinski, R. 2002. Sweeteners. In *Food Chemical Safety, Vol. II: Additives.* ed. D. H. Watson, pp. 228–248. Boca Raton: CRC Press.
19. Tandel, K. R. 2011. Sugar substitutes: Health controversy over perceived benefits. *J. Pharmacol. Pharmacother.* 2:236–243.
20. DuBois, G. E., D. E. Walters, S. S. Schiffman et al. 1991. Concentration-response relationships of sweeteners. In *Sweeteners, Discovery, Molecular Design, and Chemoreception, ACS Symposium Series 450.* eds. D. E. Walters, F. T. Orthoefer, and G. E. DuBois, pp. 260–276. Washington DC: American Chemical Society.
21. Prodolliet, J. and M. Bruelhart. 1993. Determination of acesulfame-K in foods. *J. AOAC Int.* 76:268–274.
22. Shim, J. Y., I. K. Cho, H. K. Khurana, Q. X. Li, and S. Jun. 2008. Attenuated total reflectance-Fourier transform infrared spectroscopy coupled with multivariate analysis for measurement of acesulfame-K in diet foods. *J. Food Sci.* 73:C426–C431.
23. Santini, A. O., S. C. Lemos, H. R. Pezza, J. Carloni-Filho, and L. Pezza. 2008. Development of a potentiometric sensor for the determination of saccharin in instant tea powders, diet soft drinks and strawberry dietetic jam. *Microchem. J.* 90:124–128.
24. Horie, M., F. Ishikawa, M. Oishi, T. Shindo, A. Yasui, and K. Ito. 2007. Rapid determination of cyclamate in foods by solid-phase extraction and capillary electrophoresis. *J. Chromatogr. A* 1154:423–428.
25. Sheridan, R. and T. King. 2008. Determination of cyclamate in foods by ultra performance liquid chromatography/tandem mass spectrometry. *J. AOAC Int.* 91:1095–1102.
26. Scotter, M. J., L. Castle, D. P. T. Roberts, R. MacArthur, P. A. Brereton, S. K. Hasnip, and N. Katz. 2009. Development and single-laboratory validation of an HPLC method for the determination of cyclamate sweetener in foodstuffs. *Food Addit. Contam. A* 26:614–622.
27. Hashemi, M., A. Habibi, and N. Jahanshahi. 2010. Determination of cyclamate in artificial sweeteners and beverages using headspace single-drop microextraction and gas chromatography flame-ionization detection. *Food Chem.* 124:1258–1263.
28. Wasik, A., J. McCourt, and M. Buchgraber. 2007. Simultaneous determination of nine intense sweeteners in foodstuffs by high performance liquid chromatography and evaporative light scattering detection. Development and single-laboratory validation. *J. Chromatogr. A* 1157:187–196.
29. Cantarelli, M. A., R. G. Pellerano, E. J. Marchevsky, and J. M. Camina. 2009. Simultaneous determination of aspartame and acesulfame-K by molecular absorption spectrophotometry using multivariate calibration and validation by high performance liquid chromatography. *Food Chem.* 115:1128–1132.
30. Buchgraber, M. and A. Wasik. 2009. Determination of 9 intense sweeteners in foodstuffs by high-performance liquid chromatography and evaporative light-scattering detection: Interlaboratory study. *J. AOAC Int.* 92:208–222.
31. George, V., S. Arora, B. K. Wadhwa, and A. K. Singh. 2010. Analysis of multiple sweeteners and their degradation products in lassi by HPLC and HPTLC plates. *Int. J. Food Sci. Tech.* 47:408–413.
32. Bergamo, A. B., J. A. Fracassi da Silva, and D. Pereira de Jesus. 2011. Simultaneous determination of aspartame, cyclamate, saccharin and acesulfame-K in soft drinks and tabletop sweetener formulations by capillary electrophoresis with capacitively coupled contactless conductivity detection. *Food Chem.* 124:1714–1717.
33. Zygler, A., A. Wasik, and J. Namiesnik. 2010. Retention behavior of some high-intensity sweeteners on different SPE sorbents. *Talanta* 82:1742–1748.

34. Zygler, A., A. Wasik, A. Kot-Wasik, and J. Namiesnik. 2011. Determination of nine high-intensity sweeteners in various foods by high-performance liquid chromatography with mass spectrometric detection. *Anal. Bioanal. Chem.* 400:2159–2172.

35. Pierini, G. D., N. E. Llamas, W. D. Fragoso, S. G. Lemos, M. S. Di Nezio, and M. E. Centurion. 2013. Simultaneous determination of acesulfame-K and aspartame using linear sweep voltammetry and multivariate calibration. *Microchem. J.* 106:347–350.

36. Fernandes, V. N. O., L. B. Fernandes, J. P. Vasconcellos, A. V. Jager, F. G. Tonin, and M. A. Leal de Oliveira. 2013. Simultaneous analysis of aspartame, cyclamate, saccharin and acesulfame-K by CZE under UV detection. *Anal. Methods* 5:1524–1532.

37. Stojkovic, M., T. D. Mai, and P. C. Hauser. 2013. Determination of artificial sweeteners by capillary electrophoresis with contactless conductivity detection optimized by hydrodynamic pumping. *Anal. Chim. Acta* 787:254–259.

38. de Souza, V. R., P. A. Pereira, A. C. Pinheiro, M. H. A. Bolini, S. V. Borges, and F. Queiroz. 2013. Analysis of various sweeteners in low-sugar mixed fruit jam: Equivalent sweetness, time-intensity analysis and acceptance test. *Int. J. Food Sci. Tech.* 48:1541–1548.

39. Frazier, R. A., E. L. Inns, N. Dossi, J. M. Ames, and H. E. Nursten. 2000. Development of a capillary electrophoresis method for the simultaneous analysis of artificial sweeteners, preservatives and colors in soft drinks. *J. Chromatogr. A* 876:213–220.

40. Chen, Q. C. and J. Wang. 2001. Simultaneous determination of artificial sweeteners, preservatives, caffeine, theobromine and theophylline in food and pharmaceutical preparations by ion chromatography. *J. Chromatogr. A* 937:57–64.

41. Gao, H., M. Yang, M. Wang, Y. Zhao, Y. Cao, and X. Chu. 2013. Determination of 30 synthetic food additives in soft drinks by HPLC/electrospray ionization-tandem mass spectrometry. *J. AOAC Int.* 96:110–115.

42. Midey, A. J., A. Camacho, J. Sampathkumaran, C. A. Krueger, M. A. Osgood, and C. Wu. 2013. High-performance ion mobility spectrometry with direct electrospray ionization for the detection of additives and contaminants in food. *Anal. Chim. Acta* 804:197–206.

43. Grembecka, M., P. Baran, A. Blazewicz, Z. Fijalek, and P. Szefer. 2014. Simultaneous determination of aspartame, acesulfame-K, saccharin, citric acid and sodium benzoate in various food products using HPLC-CAD-UV/DAD. *Eur. Food Res. Technol.* 238:357–365.

44. Zygler, A., A. Wasik, and J. Namiesnik. 2009. Analytical methodologies for determination of artificial sweeteners in foodstuffs. *TrAC—Trend Anal. Chem.* 28:1082–1102.

45. Zhao, Y. G., X. H. Chen, S. S. Yao, S. D. Pan, X. P. Li, and M. C. Jin. 2012. Analysis of nine food additives in red wine by ion-suppression reversed-phase high-performance liquid chromatography using trifluoroacetic acid and ammonium acetate as ion-suppressors. *Anal. Sci.* 28:967–971.

46. Nollet, L. M. L. and F. Toldrá. 2012. *Handbook of Analysis of Active Compounds in Functional Foods*. Boca Raton: CRC Press.

47. European Commission Joint Research Centre. 2013. Analysis of needs in post-market monitoring of food additives and preparatory work for future projects in this field. http://www.efsa.europa.eu/en/supporting/doc/419e.pdf (accessed March 21, 2014).

48. Biemer, T. A. 1989. Analysis of saccharin, acesulfame-K and sodium cyclamate by high-performance ion chromatography. *J. Chromatogr. A* 463:463–468.

49. Lim, H. S., S. K. Park, I. S. Kwak, H. I. Kim, J. H. Sung, S. J. Jang, M. Y. Byun, and S. H. Kim. 2013. HPLC-MS/MS analysis of 9 artificial sweeteners in imported foods. *Food Sci. Biotechnol.* 22:233–240.

50. Da-jin, Y. and C. Bo. 2009. Simultaneous determination of nonnutritive sweeteners in foods by HPLC/ESI-MS. *J. Agric. Food Chem.* 57:3022–3027.

51. Chen, B. H. and S. C. Fu. 1995. Simultaneous determination of preservatives, sweeteners and antioxidants in foods by paired-ion liquid chromatography. *Chromatographia* 41:43–50.

52. Chen, P., Y. Zhang, W. Z. Wei, and S. Z. Yao. 2000. Determination of cyclamate radical by single-column ion-chromatography with a bulk acoustic wave detector. *J. Liq. Chromatogr. R. T.* 23:1961–1971.

53. Zhu, Y., Y. Guo, M. Ye, and F. S. James. 2005. Separation and simultaneous determination of four artificial sweeteners in food and beverages by ion chromatography. *J. Chromatogr. A* 1085:143–146.

54. Elmosallamy, M. A. F., M. M. Ghoneim, H. M. A. Killa, and A. L. Saber. 2005. Potentiometric membrane sensor for determination of saccharin. *Microchim. Acta* 151:109–113.

55. Medeiros, R. A., A. Evaristo de Carvalho, R. C. Rocha-Filho, and O. Fatibello-Filho. 2008. Voltammetric determination of sodium cyclamate in dietetic products using a boron-doped diamond electrode. *Quim. Nova* 31:1405–1409.

56. Medeiros, R. A., A. Evaristo de Carvalho, R. C. Rocha-Filho, and O. Fatibello-Filho. 2008. Simultaneous square-wave voltammetric determination of aspartame and cyclamate using a boron-doped diamond electrode. *Talanta* 76:685–689.

57. Herzog, G., V. Kam, A. Berduque, and D. W. M. Arrigan. 2008. Detection of food additives by voltammetry at the liquid-liquid interface. *J. Agric. Food Chem.* 56:4304–4310.

58. Alvarez-Romero, G. A., S. M. Lozada-Ascencio, J. A. Rodriguez-Avila, C. A. Galan-Vidal, and M. E. Paez-Hernandez. 2010. Potentiometric quantification of saccharin by a selective membrane formed by pyrrole electropolymerization. *Food Chem.* 120:1250–1254.

59. Pierini, G. D., N. E. Llamas, W. D. Fragoso, S. G. Lemos, M. S. Di Nezio, and M. E. Centurion. 2013. Simultaneous determination of acesulfame-K and aspartame using linear sweep voltammetry and multivariate calibration. *Microchem. J.* 106:347–350.

60. Armenta, S., S. Garrigues, and M. de la Guardia. 2004. FT-IR Determination of aspartame and acesulfame-K in tabletop sweeteners. *J. Agric. Food Chem.* 52:7798–7803.

61. Armenta, S., S. Garrigues, and M. de la Guardia. 2004. Sweeteners determination in table top formulations using FT-Raman spectrometry and chemometric analysis. *Anal. Chim. Acta* 521:149–155.

62. Weinert, P. L., H. R. Pezza, J. E. de Oliveira, and L. Pezza. 2004. Simplified spectrophotometric method for routine analysis of saccharin in commercial noncaloric sweeteners. *J. Agric. Food Chem.* 52:7788–7792.

63. Turak, F. and M. U. Ozgur. 2013. Validated spectrophotometric methods for simultaneous determination of food colorants and sweeteners. *J. Chem.* 2013:1–9. http://www.hindawi.com/journals/jchem/2013/127847/cta/

64. Nakaie, Y., T. Yogi, K. Kakehi, D. Inoue, H. Hirose, S. Hashimoto, and Y. Tonogai. 1999. Simultaneous and simple determination of saccharin and acesulfame K in foods by GC-NPD. *Shokuhin Eiseigaku Zasshi* 40:223–229.

65. Hashemi, M., A. Habibi, and N. Jahanshahi. 2011. Determination of cyclamate in artificial sweeteners and beverages using headspace single-drop micro-extraction and gas chromatography flame-ionization detection. *Food Chem.* 124:1258–1263.

66. Yu, W., X. Zhen-Lin, X. Yan-Yun et al. 2011. Development of polyclonal antibody-based indirect competitive enzyme-linked immunosorbent assay for sodium saccharin residue in food samples. *Food Chem.* 126:815–820.

67. Yebra-Biurrun, M. C. 2000. Flow injection determinations of artificial sweeteners: A review. *Food Addit. Contam.* 17:733–738.

68. Ruzicka, J. and E. H. Hansen. 1988. *Flow Injection Analysis* (2nd ed.). Chichester: Wiley.

69. Valcarcel, M. and M. D. Luque de Castro. 1987. *Flow-Injection Analysis: Principles and Applications.* Chichester: Ellis Horwood Ltd.

70. de la Guardia, M. and S. Garrigues. 2012. *Handbook of Green Analytical Chemistry.* Chichester: John Wiley & Sons.

71. Melchert, W. R., B. F. Reis, and F. R. P. Rocha. 2012. Green chemistry and the evolution of flow analysis: A review. *Anal. Chim. Acta* 714:8–19.

72. Nikolelis, D. P. and S. Pantoulias. 2001. Selective continuous monitoring and analysis of mixtures of acesulfame-k, cyclamate, and saccharin in artificial sweetener tablets, diet soft drinks, yogurts, and wines using filter-supported bilayer lipid membranes. *Anal. Chem.* 73:5945–5952.

73. Garcia-Jimenez, J. F., M. C. Valencia, and L. F. Capitan-Vallvey. 2006. Improved multianalyte determination of the intense sweeteners aspartame and acesulfame-K with a solid sensing zone implemented in an FIA scheme. *Anal. Lett.* 39:1333–1347.

74. Garcia-Jimenez, J. F., M. C. Valencia, and L. F. Capitan-Vallvey. 2007. Simultaneous determination of antioxidants, preservatives and sweetener additives in food and cosmetics by flow injection analysis coupled to a monolithic column. *Anal. Chim. Acta* 594:226–233.

75. Garcia-Jimenez, J. F., M. C. Valencia, and L. F. Capitan-Vallvey. 2009. Intense sweetener mixture resolution by flow injection method with on-line monolithic element. *J. Liq. Chromatogr. R. T.* 32:1152–1168.

76. Kritsunankul, O. and J. Jakmunee. 2011. Simultaneous determination of some food additives in soft drinks and other liquid foods by flow injection on-line dialysis coupled to high performance liquid chromatography. *Talanta* 84:1342–1349.

77. Stojkovic, M., T. D. Mai, and P. C. Hauser. 2013. Determination of artificial sweeteners by capillary electrophoresis with contactless conductivity detection optimized by hydrodynamic pumping. *Anal. Chim. Acta* 787:254–259.

78. Psarellis, I. M., E. G. Sarantonis, and A. C. Calokerinos. 1993. Flow-injection chemiluminometric determination of sodium cyclamate. *Anal. Chim. Acta* 272:265–270.

79. Gouveia, S. T., O. Fatibello-Filho, and J. de Araujo Nobrega. 1995. Flow-injection spectrophotometric determination of cyclamate in low-calorie soft drinks and sweetener. *Analyst* 120:2009–2012.

80. Cabero, C., J. Saurina, and S. Hernandez-Cassou. 1999. Flow-injection spectrophotometric determination of cyclamate in sweetener products with sodium 1,2-naphthoquinone-4-sulfonate. *Anal. Chim. Acta* 381:307–313.

81. Rocha, F. R. P., E. Ródenas-Torralba, A. Morales-Rubio, and M. de la Guardia. 2005. A clean method for flow injection spectrophotometric determination of cyclamate in table sweeteners. *Anal. Chim. Acta* 547:204–208.

82. Llamas, N. E., M. S. Di Nezio, M. E. Palomeque, and B. S. Band Fernandez. 2005. Automated turbidimetric determination of cyclamate in low calorie soft drinks and sweeteners without pre-treatment. *Anal. Chim. Acta* 539:301–304.

83. Yebra, M. C. and P. Bermejo. 1998. Indirect determination of cyclamate by an online continuous precipitation-dissolution flow system. *Talanta* 45:1115–1122.

84. Fatibello-Filho, O., M. D. Capelato, and S. A. Calafatti. 1995. Biamperometric titration and flow injection determination of cyclamate in low-calorie products. *Analyst* 120:2407–2412.

85. Fatibello-Fiho, O., I. C. Vieira, S. A. Calafatti, and N. S. M. Curi. 1996. Biamperometric titration of cyclamate of low-calorie products using a selectable voltage source. *An. Assoc. Bras. Quím.* 45:131–137.

86. Fatibello-Filho, O., J. A. Nobrega, and A. J. Guarita-Santos. 1994. Flow injection potentiometric determination of saccharin in dietary products with relocation of filtration unit. *Talanta* 41:731–734.

87. Fatibello-Filho, O. and C. Aniceto. 1999. Flow injection potentiometric determination of saccharin in dietary products using a tubular ion-selective electrode. *Lab. Robotics Automat.* 11:234–239.

88. Yebra, M. C., M. Gallego, and M. Valcarcel. 1995. Precipitation flow-injection method for the determination of saccharin in mixtures of sweeteners. *Anal. Chim. Acta* 308:275–280.

89. Capitan-Vallvey, L. F., M. Valencia, and E. A. Nicolas. 2004. Flow-through spectrophotometric sensor for the determination of saccharin in low-calorie products. *Food Addit. Contam.* 21:32–41.

90. Mendes, C. B., E. P. Laignier, M. R. P. L. Brigagao, P. O. Luccas, and C. R. T. Tarley. 2010. A simple turbidimetric flow injection system for saccharin determination in sweetener products. *Chem. Pap.* 64:285–293.

91. Gao, X. Y., J. M. Wei, S. Wang, and Z. R. Yin. 2011. Rapid determination of sodium saccharin by FI-chemiluminescence. *Huaxue Fence* 47:1446–1449.

92. Capitan-Vallvey, L. F., M. C. Valencia, E. A. Nicolas, and J. F. Garcia-Jimenez. 2006. Resolution of an intense sweetener mixture by use of a flow injection sensor with on-line solid-phase extraction. *Anal. Bioanal. Chem.* 385:385–391.

Colorants and Dyes

CHAPTER 25

Determination of Riboflavin

Leo M.L. Nollet

CONTENTS

25.1 RIBOFLAVIN: VITAMIN AND COLOR

The chemical names of riboflavin (RF) are 3,10-dihydro-7,8-dimethyl-10-[(2*S*,3*S*,4*R*)-2,3,4,5-tetrahydroxypentyl]benzo[g]pteridine-2,4-dione or 7,8-dimethyl-10-(1′-D-ribityl) isoalloxazine. RF or lactoflavin is known as vitamin B_2. The chemical formula is $C_{17}H_{20}N_4O_6$ and its CAS number is 83-88-5.

Riboflavin is the vitamin that imparts the orange color to solid B-vitamin preparations, the yellow color to vitamin supplement solutions, and the unusual fluorescent-yellow color to the urine of persons who are taking high-dose B-complex preparations.

Riboflavin is a water-soluble vitamin. It is synthesized by all plants and many microorganisms, but it is not produced by higher animals. Because it is a precursor of coenzymes that are necessary for the enzymatic oxidation of carbohydrates, RF is essential to basic metabolism.

Riboflavin can be also used intentionally as an orange-red food color additive, and as such is designated in Europe by the E number E101. We distinguish: E101—RF; E101 (i)—synthetic RF; E101 (ii) or E106—RF 5′-phosphate sodium; and E101 (iii)—RF from *Bacillus subtilis*.

Synthetic RF was evaluated by the Joint FAO/WHO Expert Committee on Food Additives (JECFA) Committee at its 13th meeting (Annex 1, reference 19), when an ADI of 0–0.5 mg/kg body weight (BW) was allocated on the basis of limited data. At the 25th meeting (Annex 1, reference 56), a group ADI of 0–0.5 mg/kg BW was allocated to RF and RF-5′-phosphate, expressed as RF, on the basis of a more extensive database including information on reproductive toxicity [1].

It is used in foods predominantly for fortification; however, quantities accounting for approximately 2% of the total sales are used for coloring, for example, for sweets.

25.2 ANALYSIS METHODS

No article on detection of RF as colorant has been found in the literature, but a number of papers on flow injection analysis (FIA) of RF as vitamin are discussed below.

An automatic method for the separation and determination of RF vitamin in food samples (chicken liver, tablet, and powder milk) is proposed by Zougagh and Rios [2]. The method is based on the online coupling of supercritical fluid extraction (SFE) with a continuous flow-CE system with guided optical fiber fluorometric detection (CF-CE-FD). The whole SFE-CF-CE-FD arrangement allowed the automatic treatment of food samples (cleanup of the sample followed by the extraction of the analytes), and the direct introduction of a small volume of the extracted material to the CE-FD system for the determination of RF vitamins. Fluorescence detection introduced an acceptable sensitivity and contributed to avoidance of interferences by nonfluorescent polar compounds coming from the matrix samples in the extracted material. Electrophoretic responses were linear within the 0.05–1 µg/g range, whereas the detection limits of RF vitamins were in the 0.036–0.042 µg/g range.

A unique stopped-FIA system for simultaneous synchronous spectrofluorometric online dissolution monitoring of multicomponent solid preparations that makes use of a fiberoptic sensor was presented by Li et al. [3]. A new means of *in vitro* therapeutic drug monitoring was developed by hyphenating the fiberoptic sensor technique and a chemometric method. An artificial neural network method was applied to construct the mathematical model for the simultaneous analysis of the mixture of vitamins B_1, B_2, and B_6 by synchronous spectrofluorometry. The selection of the wavelength interval, the pH of the carrier solution, and other experimental conditions were evaluated. The proposed method has been applied to the dissolution monitoring of vitamin B tablets with satisfactory results. The recovery was 97.8%–105%, and the relative standard deviation (RSD) was 1.1–7.5.

The fiberoptic spectrophotometer was composed of a bifurcated fiber optic (common leg bundle diameter 4.5 mm, individual leg bundle diameter 2.5 mm) and a fluorophotometer with a data recorder. Two arms of the bifurcated fiberoptic bundle were fixed between the excitation monochromator and the emission monochromator.

The FIA system was equipped with a peristaltic pump and a polyethylene flow cell (10 mm × 4 mm i.d.). They were connected with a polyethylene tube (1.0 mm i.d.). The volume from the end of the tube to the entrance to the flow cell was 0.80 mL. A metallized reflector (4 mm × 10 mm) was mounted in the flow cell and the common end of the fiberoptic bundle was placed perpendicular to the flow cell opposite the reflector. The gap between the fiberoptic bundle and the polyethylene was filled with cementing agent containing the epoxy resin for sealing the flow cell.

The FIA system was coupled with a paddle dissolution tester equipped with a dissolution vessel in a thermostated water bath. A stainless steel tube (0.5 mm i.d.) for sampling was connected to the FIA system by a polyethylene tube whose end was fixed between the surface of the dissolution fluid and the end of the paddle.

A multicommutated flow-through optosensor based on the direct fluorescence measurements of vitamins B_2 and B_6 using a nonpolar sorbent (C_{18} silica gel) as solid sensing zone (to accomplish the separation and subsequent preconcentration/detection of the target analytes) has been developed by Eulogio et al. [4]. The proposed flow system was controlled by home-made software written in Java and was designed using three-way solenoid valves for independent automated manipulation of sample and carrier solutions. The native fluorescence signal was simultaneously monitored at two pairs of excitation/emission wavelengths (450/519 and 294/395 nm for vitamins B_2 and B_6, respectively). The separation of the analytes was performed in the detection flow cell, using the differences in the sorption/elution process on the solid support between the two vitamins, due to their different polarities. Using an optimized sampling time, the analytical signal showed linearity in the range 0.01–0.4 and 0.15–3 µg/mL with detection limits of 0.003 and 0.045 µg/mL for B_2 and B_6, respectively, obtaining RSD (%) values better than 2% for both analytes. The proposed methodology was applied to different pharmaceutical preparations, obtaining remarkably good results with recoveries ranging from 96% to 107.5%.

The flow diagram of the system is shown in Figure 25.1. In the initial status, all valves are switched off and the carrier (a 10^{-2} M HCl solution [pH 2]), is flowing through the flow cell while all other solutions are recycling to their vessels. The sample containing RF and B_6 is introduced by simultaneously switching the valves V_1 and V_2 on for 40 s. B_6 develops its analytical signal when reaching the flow-through cell and is eluted by the carrier itself, while RF is strongly retained on the additional solid phase in the flow-through cell. After RF has developed its signal, by switching valves V_1 and V_4 the 20% MeOH eluting solution is introduced into the flowing system for 210 s. The portion of tubing placed between valves V_1 and V_3 was cleaned between samples in order to avoid any possible contamination. A scheme showing all the valve switching procedures is depicted in Figure 25.1.

A novel optical chemical sensor based on dynamic liquid drops combined with flow injection has been developed for continuous and sequential determinations of the mixture of vitamins B_1, B_2, and B_6 [5]. The dynamically growing and falling drops serve as windowless optical cells. The adsorption, desorption, and quantitative determinations of vitamin B_1, RF, and vitamin B_6 were carried out based on selective adsorption of Sephadex CMC-25 and different fluorescence characteristics of RF and vitamin B_6 in the basic solutions.

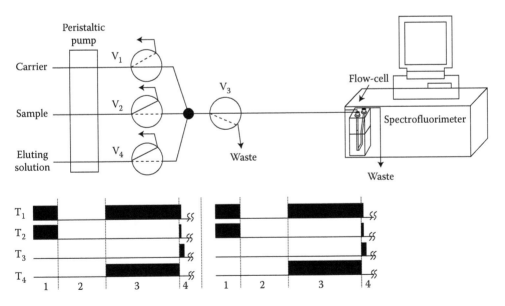

FIGURE 25.1 Multicommutated flow configuration; $V_{1,2,...}$: valves 1, 2,.... Valve scheme. T_1, T_2, T_3, and T_4 refer to the timing courses of solenoid valves V_1, V_2, V_3, and V_4, respectively. The shadow surface above the valve timing course line indicates that the corresponding valve was switched on. The steps were the following: (1) sample introduction; (2) B_6-elution; (3) B_2-elution; (4) cleaning step. (Adapted from Llorent-Martínez, E. J. et al. 2006. *Anal. Chim. Acta* 555:128–133.)

A sampler designed in the author's laboratory is fixed on the top of the chamber of the luminescence spectrometer. A stainless syringe needle (1.0 mm i.d., 1.5 mm o.d.) polished carefully before use is fixed on the elevator and inserted into an immobile septum injector. The sampler is composed of a plexiglass elevator and a stainless steel needle. The needle is inserted into the top hole of the plexiglass. The height between the needle bottom and the axis of the excited light beam can be adjusted by tightening screws. A buffer and a reagent solution are pumped by the syringe pump and form drops on the tip of the polytetrafluoroethylene tube. The volume of a drop is ~20 μL.

The light beam from a xenon discharge lamp illuminates the whole drop during the processing of the drop growing. The height of the drop head can be adjusted through the elevator. The fluorescence is detected by a photodiode of the luminescence spectrometer using the time drive function.

The spectrometer is interfaced with a microcomputer with appropriate software for spectral acquisition. Instrument excitation and emission slits were set at 15 and 10 nm, respectively. See Figure 25.2.

Under these conditions, linear calibration curves were obtained over the ranges 0.01–8.00, 0.01–10.00, and 0.01–3.00 μg/mL for vitamin B_1, RF, and vitamin B_6 with the limits of determination of 0.008, 0.005, and 0.006 μg/mL, respectively. This method has been applied with satisfactory results in the determination of synthetic mixture of vitamin B_1, RF, vitamin B_6, and vitamin B compound tablets. The technique described provides a simple, effective, and sensitive method for assay of the biological samples.

The sensitizing effect of RF and RF-5′-phosphate (FMN) on the photooxidation of dianisidine was studied [6]. The rate of the photochemical reaction was monitored spectrophotometrically at 460 nm. The method allows the determination of RF and FMN in the range 1×10^{-7}–5×10^{-6} mol/dm^3 with a RSD of about 0.68%. The method can be successfully adapted to flow injection. Manual and flow injection methods were satisfactorily applied by the authors to the determination of RF and FMN in fortified breads and pharmaceutical preparations.

A novel wall-jet cell with parallel dual cylinder (PDC) microelectrodes was constructed and used for FIA [7]. The detector takes advantage of "redox recycling" between

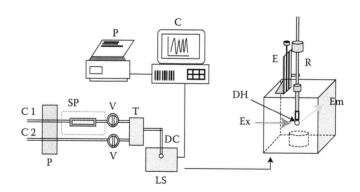

FIGURE 25.2   Schematic diagram of the drop-cell detection arrangement: C 1, channel 1; C2, channel 2; P, syringe pump; SP, Sephadex column; T, tee connector; DC, drop cell; LS, luminescence spectrometer; P, printer; C, computer; E, elevator; DH, drop head; R, stainless steel rod; Ex, excitation light; Em, emission light. (Adapted from Feng, F. et al. 2004. *Anal. Chim. Acta* 527:187–193.)

bipotentiostated microcylinder electrodes (−0.4 and 0.0 V vs. SCE [saturated calomel electrode], respectively) and shows high sensitivity and selectivity for RF which is electrochemically reversible. In 0.1 mol/L NaHSO$_3$ media and at a flow rate of 1.0 ml/L, 56.8% collection efficiency was found at a PDC device with cylinder diameter of 20 μm and an interelectrode gap of 2 μm. With this device, the FIA peak current at collector electrode had a linear relationship with RF concentration over a wide range (from 70 ng to 35 pg), and the detection limit was 35 pg (signal-to-noise ratio of 3). For 10 successive replicate injections of 35 and 3.5 ng of the analyte, RSDs were 1.7% and 2.8%, respectively. This method was applied to the direct determination of RF in multivitamin tablets.

A flow injection method for the determination of RF based on the inhibition of the intensity of chemiluminescence (CL) from the luminol–K$_3$Fe(CN)$_6$ system is described by Song and Wang [8]. When RF mixed with K$_3$Fe(CN)$_6$, the fast oxidation reaction between RF and K$_3$Fe(CN)$_6$ generated K$_4$Fe(CN)$_6$, which then inhibited the CL reaction of K$_3$Fe(CN)$_6$ and luminol in alkaline aqueous solution. The CL emission was correlated with the RF concentration in the range from 0.032 to 100 μg/mL, and the detection limit was 0.01 μg/mL (3σ). A complete analysis could be performed in 2 min with a RSD of less than 2.2%. The influence of foreign species was studied and the method has been applied successfully to the determination of RF in pharmaceutical samples; the recovery was from 98.0% to 102%.

A novel continuous-flow sensor based on CL detection was developed for the determination of RF at pg/mL levels by the immobilization of the reagents [9]. It was found that the CL intensity from the oxidation between luminol and periodate could be enhanced in the presence of RF. The increase of CL emission was correlated with the RF concentration in the range from 0.04 to 200 ng/mL, and the detection limit was 0.02 ng/mL (3σ). Considering the effective reaction ions, luminol, and IO$_4^-$ was immobilized on anion exchange resin. The system could produce an evident CL signal with water as eluent. The flow sensor could greatly improve the selectivity and sensitivity for determination of RF with a high signal-to-noise ratio. A complete analysis, including sampling and washing, could be performed in 0.5 min with a RSD of less than 3.0%. The flow sensor was applied successfully to the determination of RF in pharmaceutical preparations and human urine samples.

Zhang and Chen developed a sensitive CL method induced by light for the determination of RF with FIA [10]. RF shows a strong enhancement effect on the CL reaction of luminol oxidized by periodate after the photochemical reaction of RF in alkaline solution, and the enhancing effect of RF disappears under dark conditions. Oxygen can further improve the enhancing effect of the RF in this CL system, and those effects were also investigated. The range of the linear response for RF using this method was ca. 2.4×10^{-8}–1.0×10^{-6} mol/L, and the detection limit (S/N = 3) is 1.0×10^{-8} mol/L. The applicability of the method was demonstrated by the determination of RF in a pharmaceutical preparation.

A method based on the CL generated during the oxidation of luminol by N-bromosuccinimide (NBS) and N-chlorosuccinimide (NCS) in alkaline medium was developed [11]. It was found that RF could greatly enhance this CL intensity when present in the luminol solution. Based on this observation, a new flow injection CL method for the determination of RF was proposed in this paper. The detection limits were 7.5 and 3.5 ng/mL of RF for the NBS–luminol and NCS–luminol CL systems, respectively. The relative CL intensity was linear with RF concentration in the range 19–2000 mg/mL and 12–2000 mg/mL for the NBS–luminol CL system, and 12–200 and 200–2000 ng/mL for the NCS–luminol CL system. The results obtained for the assay of pharmaceutical preparations compared well with those obtained by the official method and demonstrated good accuracy and precision.

A weak CL emission was observed in the decomposition of peroxomonosulfate (HSO_5^-), which was accelerated in the presence of trace amounts of cobalt(II). The mechanism was due to the production of singlet oxygen ($^1O_2$) [12]. Interestedly, RF can enhance the CL, and the CL intensity was strongly dependent on RF concentration. Based on this phenomenon, a FIA CL method was established for the determination of RF. Additionally, a possible mechanism for the CL is proposed based on the kinetic curve of the CL reaction, CL spectra, UV–Vis spectra and fluorescent spectra. The CL intensity was correlated linearly with concentration of RF over the range 1.0×10^{-4}–1.0×10^{-8} g/mL; the detection limit was 9.0×10^{-9} g/mL (S/N = 3); the RSD was 1.4% for 9×10^{-7} g/mL RF ($n = 11$). This method was successfully applied to the determination of RF in real tablets and injections.

A flow-through optosensor for the sensitive determination of RF in some pharmaceutical preparations was developed [13]. The sensor was developed in conjunction with a flow injection analysis system and uses C_{18} silica gel as a substrate. The wavelengths of excitation and emission were 466 and 515 nm, respectively. The applicable concentration range was 2.2–91.8 ng/mL, with a detection limit (3σ) of 0.4 ng/mL. The RSD for 10 determinations of 15 ng/mL of RF was 1.1%.

Tymecki et al. developed two compact optoelectronic fluorometric devices operating according to the paired-emitter-detector-diode concept [14]. The fluorometric detector, fabricated with three light emitting diodes only, has been applied for the development of fluorometric optosensor by further integration with sensing solid phase. In these investigations RF and C_{18}-silica were chosen as respectively model analyte and model sensing layer useful for solid-phase spectrometry. Both developed analytical devices have been applied for nonstationary fluorometric measurements performed under conditions of FIA. This flow-through detector and sensor operating under given flow conditions offer RF determination in mg/L and µg/L ranges of concentration.

In the review of Lara et al. [15], the recent evolution and the state of the art of photochemical reactions coupled with CL processes are presented. Different CL systems have been considered together with suitable photochemical derivatization processes that can affect either the analyte of interest or even the chemiluminogenic reagent, producing some derivatives able to participate more efficiently in the CL reactions and enhancing the CL emission. The online integration of the photochemical reactions and the coupling of the resulting photoinduced CL (PICL) method with dynamic analytical systems, such as FIA, liquid or gas chromatography, and capillary electrophoresis, have been discussed. Important applications of PICL have been proposed in environmental, pharmaceutical, and food analysis.

REFERENCES

1. JECFA (Joint FAO/WHO Expert Committee on Food Additives), 1999. International Programme on Chemical Safety. World Health Organisation. Safety evaluation of certain food additives. WHO Food Additives Series, No 42. Prepared by the 51st meeting of the Joint FAO/WHO Expert Committee on Food Additives. Riboflavin derived by fermentation with genetically modified *Bacillus subtilis*. Geneva. Available online: http://www.inchem.org/documents/jecfa/jecmono/v042je05.htm
2. Zougagh, A. and A. Rios. 2008. Supercritical fluid extraction as an on-line clean-up technique for determination of riboflavin vitamins in food samples by capillary electrophoresis with fluorimetric detection. *Electrophoresis* 29(15):3213–3219.

3. Li, W., J. Chen, B. Xiang, and D. An. 2000. Simultaneous on-line dissolution monitoring of multicomponent solid preparations containing vitamins B1, B2 and B6 by a fiber-optic sensor system. *Anal. Chim. Acta* 408:39–47.

4. Llorent-Martínez, E. J., J. F. García-Reyes, P. Ortega-Barrales, and A. Molina-Díaz. 2006. A multicommuted fluorescence-based sensing system for simultaneous determination of Vitamins B2 and B6. *Anal. Chim. Acta* 555:128–133.

5. Feng, F., K. Wang, Z. Chen, Q. Chen, J. Lin, and S. Huang. 2004. Flow injection renewable drops spectrofluorimetry for sequential determinations of Vitamins B1, B2 and B6. *Anal. Chim. Acta* 527:187–193.

6. Perez-Ruiz, T., C. Martinez-Lozano, V. Tomas, and O. Val. 1994. Photochemical spectrophotometric determination of riboflavin and riboflavin 5′-phosphate by manual and flow injection methods *Analyst* 119:1199–1203.

7. Peng, W., H. Li, and E. Wang. 1994. Highly sensitive and selective determination of riboflavin by flow injection analysis using parallel dual-cylinder microelectrodes. *J. Electroanal. Chem.* 375(1–2):185–182.

8. Song, Z. and L. Wang. 2000. Determination of riboflavin using flow injection inhibitory chemiluminescence. *Anal. Lett.* 33(13):2767–2778.

9. Song, Z. and L. Wang. 2001. Reagentless chemiluminescence flow sensor for the determination of riboflavin in pharmaceutical preparations and human urine. *Analyst* 126(8):1393–1398.

10. Zhang, G.-F. and H.-Y. Chen. 2000. A sensitive photoinduced chemiluminescence method for the determination of riboflavin with flow injection analysis. *Anal. Lett.* 33(15):3285–3302

11. Safavi, A., M. A. Karimi, and M. R. Hormozi Nezhad. 2005. Flow injection analysis of riboflavin with chemiluminescence detection using a N-halo compounds–luminol system. *Luminescence* 20(3):170–175.

12. Wang, M., L. Zhao, M. Liu, and J. M. Lin. 2007. Determination of riboflavin by enhancing the chemiluminescence intensity of peroxomonosulfate–cobalt(II) system. *Spectrochim. Acta Part A: Mol. Biomol. Spectrosc.* 66(4–5):1222–1227.

13. Gong, Z. and Z. Zhang. 1997. An optosensor for riboflavin with C_{18} silica gel as a substrate. *Anal. Chim. Acta* 339(1–2):161–165.

14. Tymecki, L., M. Rejnis, M. Pokrzywnicka, K. Sttrzelak, and R. Koncki. 2012. Fluorimetric detector and sensor for flow analysis made of light emitting diodes. *Anal. Chim. Acta.* 721:92–96.

15. Lara, F. J., A. M. Garcia-Campaña, and J.-J. Aaron. 2010. Analytical applications of photoinduced chemiluminescence in flow systems—A review. *Anal. Chim. Acta* 679(1–2):17–30.

CHAPTER 26

Determination of Quinoline Yellow and Sunset Yellow

Juan Carlos Bravo, David González-Gómez, Alejandrina Gallego,
Rosa Mª Garcinuño, Pilar Fernández, and Jesús Senén Durand

CONTENTS

26.1 INTRODUCTION

The use of colorants as additives for different purposes and in different fields has been reported throughout the human history. The use of colorants in drugs, cosmetics, and foods by ancient Egyptians is well documented. Archaeologists have evidence that Egyptian women used green ore of copper as an eye shadow as early as 5000 BC and paintings in Egyptian tombs dating from 1500 BC depict the making of colored candy [1]. Colored spices have been used as culinary condiments contributing to the appearance, taste, and flavor of food for centuries around the world.

Color is one of the most important attributes of foodstuffs and is associated with freshness and quality. Color is also associated with expectation of flavor and taste, changing consumer perceptions. The basic use of colorants is to improve natural colors destroyed during storage or processing, to reinforce the colors already present in the food in order to obtain the best food appearance, and to ensure color uniformity.

Until the middle of the nineteenth century all colorants used were obtained from natural sources: animals, vegetables, and minerals. Then William Henry Perkin accidentally discovered the first synthetic organic dye in 1856, that was called "mauve," a purplish color. This substance dyed silk, which was stable when washed or exposed to light. Since then, synthetic colorants representing every color of the rainbow have been developed.

A color additive can be defined as a material that is a dye, pigment, or other substances made by a process of synthesis or similar artifice, or extracted, isolated, or otherwise derived, with or without intermediate or final change of identity, from a vegetable, animal, mineral, or other source and that, when added or applied to a food, drug, or cosmetic or to the human body or any part thereof, is capable (alone or through reaction with another substance) of imparting a color thereto [2].

In the United States, the Food and Drug Administration classifies color additives as straight colors, lakes, and mixtures. Straight colors are color additives that have not been mixed or chemically reacted with any other substance. Lake colorants are formed by chemically reacting straight colors with precipitants and substrata. They are made with aluminum cation as the precipitant and aluminum hydroxide as the substratum. Mixtures are color additives formed by mixing one color additive with one or more other color additives or noncolored diluents, without a chemical reaction [3].

The demand for cheaper and more stable colorants and colorants of higher tinctorial power has led to a more extended use of synthetic colorants as food additives, although natural colorants present the advantages of being perceived as safer and healthier than synthetic ones. Replacement of synthetic colorants with natural colorants is actually a marketing trend. Although the natural colorants have important limitations of use and applications as replacements for the synthetic colorants, food scientists are working to replacing artificial colorants with natural alternatives [4].

26.2 SAFETY AND EFFECTS OF COLOR ADDITIVES

Product quality is usefully described as a collection of attributes, including color; improper application of color additives could conceal the fact that a food product is adulterated or damaged. On the other hand, the extended use of color additives has been recognized as a potential risk to human health. Ingestion of a number of food color additives, especially when they are consumed excessively, has been suggested to provoke asthma, urticaria, and other diseases in humans and may even be carcinogenic [5,6]. Also, recent studies have considered a possible relationship between hyperactivity in children and the consumption of food colors.

In 2007, seven synthetic additives [Quinoline Yellow, Ponceau 4R, Allura Red, Azorubine (carmoisine), Tartrazine, Sunset Yellow, and Sodium Benzoate] were evaluated in a study conducted by the University of Southampton [7]. The study was intended to investigate the effects of these color additives in children from 3 to 9 years old. The study provided supporting evidence of a possible link between the test mixtures and hyperactivity in children, but the role of food color additives in affecting children's behavior is controversial.

Subsequently, the European Food Safety Authority (EFSA) has reevaluated the safety of the six color additives used in the Southampton study and concluded that the available scientific evidence does not substantiate a link between the color additives and behavioral effects [8,9]. However, after assessing all the available evidences, the Panel on Food Additives and Nutrient Sources added to Food (ANS) modified the acceptable daily intake

(ADI), defined as the amount of a substance that can be consumed every day throughout the lifetime of an individual without any appreciable health effects [10], for three of the colors in question, Quinoline Yellow (E104), Sunset Yellow FCF (E110), and Ponceau 4R (E124). As a result, the panel concluded that exposure to these additives through the diet could exceed the new ADI for both adults and children. In 2011, following completion of its refined exposure assessment of Sunset Yellow, EFSA found that high-level exposure estimates are below the temporary ADI of this substance except for U.K. preschool children, who might slightly exceed the ADI under the most conservative scenarios [11].

Consequently, the Joint FAO/WHO Expert Committee on Food Additives (JECFA) established a temporary ADI of 0–5 mg/kg body weight for Quinoline Yellow, incorporating an additional twofold safety factor, and withdrew the previous ADI of 0–10 mg/kg body weight [12]. In the same way, the report of JECFA established an ADI of 0–4 mg/kg body weight for Sunset Yellow and withdrew the previous ADI of 0–2.5 mg/kg body weight.

26.3 REGULATION OF FOOD COLOR ADDITIVES

The use of synthetic colorants is strictly controlled by different national legislations. Food colorants are tested for safety by various bodies around the world, which list the permitted substances that can be used for this purpose, while acceptable daily intakes (ADI) are evaluated by the World Health Organization (WHO) and the Food and Agricultural Organization of the United Nations (FAO).

In the United States, the Food and Drug Administration (FDA) is the agency that approves the use of colorants in food. All color additives required to be listed by the FDA fall into two categories: those that are subject to the FDA's certification process and those that are exempt from the certification process. Color additives subject to batch certification are synthetic organic dyes, lakes, or pigments. Color additives exempt from certification generally include those derived from plant or mineral sources.

In the European Union (EU), the safety of food colorants is evaluated by the EFSA. The results affect all member countries of the EU, while non-EU member states are regulated by their national authorities.

Figure 26.1 shows a map of the colorant status (for Quinoline Yellow and Sunset Yellow) by country, made with the data of the Feingold Association of United States report [13].

Currently, Quinoline Yellow is authorized for use in the EU with a maximum allowed use level of 50–500 mg/kg food for various foodstuffs. Quinoline Yellow is also allowed in beverages at levels up to 200 mg/L. Sunset Yellow is authorized for use in the EU with a maximal allowed use level of 50–500 mg/kg food for various foodstuffs. Sunset Yellow is also allowed in alcoholic beverages at levels up to 200 mg/L and nonalcoholic beverages up to 50 mg/L [14].

26.4 CHARACTERIZATION OF QUINOLINE YELLOW (E104) AND SUNSET YELLOW (E110)

26.4.1 Quinoline Yellow (E104)

Quinoline Yellow is a quinophthalone dye consisting essentially of a mixture of monosulfonates, disulfonates (principal component), and trisulfonates, and subsidiary coloring

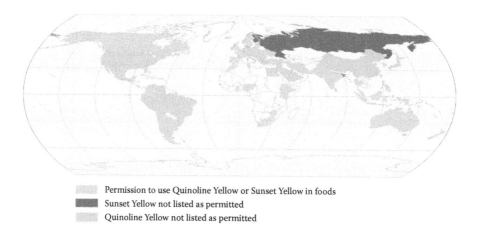

Permission to use Quinoline Yellow or Sunset Yellow in foods
Sunset Yellow not listed as permitted
Quinoline Yellow not listed as permitted

FIGURE 26.1   Permission status for Quinoline Yellow and Sunset Yellow in foods by country. (From Feingold Association of United States. 2014. *List of Colorants* 2014. http://www.feingold.org/Research/PDFstudies/List-of-Colorants.pdf (accessed July 1, 2014).)

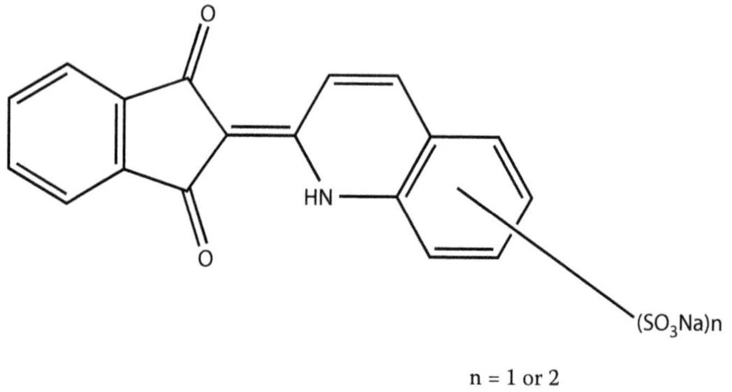

$n = 1$ or 2

FIGURE 26.2   Structural formula of Quinoline Yellow (E104).

matters together with sodium chloride and/or sodium sulfate as the principal uncolored components (Figure 26.2). In both the JECFA and the European Commission specifications, the purity is given as not less than 70% total coloring matters, calculated as the sodium salt.

Quinoline Yellow has a CAS number of 8004-92-0, a molecular formula of $C_{18}H_9NNa_2O_8S_2$ (principal component) and a molecular weight of 477.38 g/mol (principal component). Quinoline Yellow is soluble in water, slightly soluble in ethanol and practically insoluble in vegetable oils [15]. Quinoline Yellow is produced in two different forms, a fine yellow powder and a granular product (mainly for liquid application).

26.4.2 Sunset Yellow (E110)

Sunset Yellow (E110) is an azo dye consisting essentially of disodium 2-hydroxy-1-(4-sulphonatophenylazo)naphthalene-6-sulfonate and subsidiary coloring matters

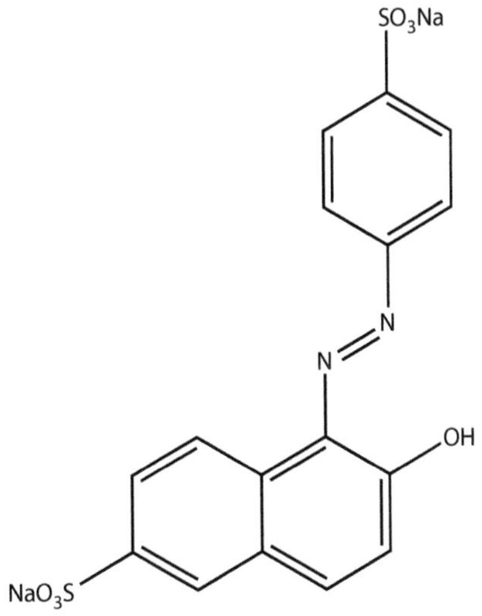

FIGURE 26.3 Structural formula of Sunset Yellow (E110) (principal component).

together with sodium chloride and/or sodium sulfate as the principal uncolored components (Figure 26.3). The purity is given as not less than 85% total coloring matters, calculated as the disodium salt.

Sunset Yellow (E110) has a CAS number of 2783-94-0, a molecular formula of $C_{16}H_{10}N_2Na_2O_7S_2$ (principal component) and a molecular weight of 452.37 g/mol (principal component). Sunset Yellow is soluble in water and slightly soluble in ethanol [15]. Sunset Yellow is produced in two different forms, a fine yellow powder and a granular product (mainly for liquid application).

26.5 ANALYSIS OF QUINOLINE YELLOW (E104) AND SUNSET YELLOW (E110)

Colorants are one of the most important additives that are found in foodstuffs. As mentioned previously, synthetic dyes are being replaced by natural colorants, though artificial dyes offer comparatively low cost and high stability. However, their negative effects in humans is a critical consideration for the necessary control of their content in food.

Tables 26.1 and 26.2 show maximum permitted levels and use levels of Quinoline Yellow and Sunset Yellow, respectively, obtained from different surveys for several food categories.

To ensure conformity with the regulation of food dyes and the need for quality control requirements, new analytical methods are necessary to determine nature and concentration of these colorants in food products.

There are many published methods for the determination of water-soluble dyes in foodstuffs. The general scheme for identifying dyes present in foods usually involves (1) preliminary treatment of the food, (2) extraction and purification of the dye from the prepared solution or extract of the food, (3) separation of mixed coloring if more than one is present, and (4) identification of the separated dyes [16].

TABLE 26.1 Maximum Permitted Levels of Use of Quinoline Yellow in Beverages and Foodstuffs According to Council Directive 94/36/EC and Use Levels

Maximum Permitted Levels	Use Level	Food Category
100 mg/L	1–60 mg/L	Nonalcoholic flavored drinks
200 mg/L	1–40 mg/L	Spirituous beverages
100 mg/L	12–100 mg/kg	Jam, jellies, and marmalades
200 mg/L	88–146 mg/kg	Fine bakery wares
150 mg/L	2–80 mg/kg	Desserts

Source: European Food Safety Authority. 2009. Scientific opinion on the re-evaluation of Quinoline Yellow (E 104) as a food additive. *EFSA J.* 7(11):1329, 40p.

TABLE 26.2 Maximum Permitted Levels of Use of Sunset Yellow in Beverages and Foodstuffs According to Council Directive 94/36/EC and Use Levels

Maximum Permitted Levels	Use Level	Food Category
50 mg/L	1–48 mg/L	Nonalcoholic flavored drinks
200 mg/L	17 mg/L	Spirituous beverages
100 mg/L	Not detected	Jam, jellies, and marmalades
50 mg/L	6–50 mg/ kg	Fine bakery wares
50 mg/L	1–10 mg/kg	Desserts

Source: European Food Safety Authority. 2009. Scientific opinion on the re-evaluation of Sunset Yellow FCF (E 110) as a food additive. *EFSA J.* 7(11):1330, 44pp.

26.5.1 Sample Treatment

Sample preparation is a very important procedure in analysis of food colorants to allow isolation of target dyes from the rest of the food matrix constituents. Since foodstuff formulations exhibit great variability in their composition, this step cannot normally be avoided. The first step of analysis is extraction of the colorant from the food sample, and it depends on the kind of food matrix and the other substances that the matrix contains. Therefore, a pretreatment stage is necessary prior to the instrumental analysis.

Different sample preparation methods have been reported, depending on the kind of sample and the determination technique. The simplest sample preparation processes are mainly used for beverages (alcoholic or not), jellies, jams, and sweets; these are dissolved in water and, if necessary, heat is applied. For more complex matrices, extraction with solvents and additional cleanup steps such as liquid–liquid extraction and/or solid-phase extraction (SPE) may be needed. Nowadays, this last technique, SPE, is the most common since it is simple, reproducible, and rapid. When liquid turbid samples are examined, filtration [17,18] and centrifugation [19] are common techniques, using ultrasonication to degas carbonated beverages [20].

A spectrophotometric method for the analysis of Sunset Yellow, Tartrazine, and Ponceau 4R in commercial foods was carried out using a simple liquid extraction with water [21]. Dimethyl sulfoxide (DMSO) as the extraction solvent has also been used for the determination of Sunset Yellow, Tartrazine, Amaranth, Ponceau 4R, and Sudan

(I–IV) by high-performance liquid chromatography with diode array detection and electrospray mass spectrometry (HPLC-DAD/ESI-MS) [22].

Polyamide powder has been proposed in batch experiments for the extraction and HPLC-DAD determination of Sunset Yellow and 13 other synthetic food colorants in fish roe and caviar [23]. This extraction method obtained better recovery rates than other solid materials (Al_2O_3, XAD-2, anion exchangers) also tested in this work. The sample preparation proposed by Kirschbaum consisted of the extraction of colorants from roe with 1 M aqueous ammonia, defatting with *n*-hexane, adjustment of the extract to pH 2, adsorption of colorants on polyamide, and desorption with a mixture of aqueous ammonia (25% v/v) and methanol (10:90% v/v) [23].

A SPE procedure with hydrophilic-lipophilic-balanced (HLB) sorbent has been used for extraction of Sunset Yellow and nine other dyes in shrimp flakes [24]. In this procedure, the elution step carried out with 5 mL methanol containing 0.1% v/v ammonia. A similar extraction method was used in the determination of 40 colorants, including Quinoline Yellow and Sunset Yellow, in soft drink samples [25]. In 2014, Hamedpour and Amjadi developed a simple, rapid and sensitive method using *in situ* surfactant-based SPE (ISS-SPE) combined with UV–Vis spectrophotometry for the preconcentration and determination of trace amounts of Quinoline Yellow in food and water samples [26]. A Box–Behnken design was employed to optimize the extraction efficiency.

Molecularly imprinted SPE (MISPE) is a type of SPE that is more time-consuming than simple SPE since the molecularly imprinted polymer (MIP) stationary phase has to be specifically synthesized for the target dye recognition. Recently a titania-based MIP was synthesized through a sol-gel process using Sunset Yellow as template, without use of a functional monomer [27]. This MIP was used as an SPE material for the isolation and enrichment of sulfonic acid dyes in beverages. The better cleanup ability demonstrated the capability of MIP for the isolation and enrichment of sulfonic acid dyes in complicated food samples.

26.5.2 Sample Separation and Detection

The synthetic colorants Quinoline Yellow and Sunset Yellow are determined individually or simultaneously with other food additives by several methods, mainly spectrophotometry and chromatography. At present, the most used technique for synthetic colorants determination is liquid chromatography. The most popular reported HPLC modes applied for synthetic colorants analysis are reversed-phase chromatography (RPC), ion pair chromatography (IPC), and ion chromatography (IC). In recent years, electrochemical techniques have attracted wide attention due to their characteristics such as low operational cost, rapid and sensitive detection, selectivity, and reproducibility.

26.5.2.1 Determination of Quinoline Yellow (E104)

Five colorants (Quinoline Yellow, Tartrazine, Yellow Orange, Azo-rubine, and Ponceau) were determined in soft drinks by direct injection RP-HPLC with visible detection at a single wavelength [28]. The method proposed is simple, rapid, and inexpensive and the only sample pretreatment was homogenization and degassing of the beverage solution.

A Quinoline Yellow determination procedure was developed using a direct potentiometric method [29]. A Quinoline Yellow–coated graphite electrode and a polymeric membrane electrode, based on the use of the ion-pair complex of tetraphenylphosphonium–Quinoline Yellow embedded in a poly(vinyl chloride) matrix as the electroactive substance, were successfully used in this analysis.

Tartrazine and Quinoline Yellow as Cr(III) complexes in a Tween-80 micellar medium were simultaneously determined using an H-point standard addition method [30]. The method proposed is based on the complexation of the food dyes with potassium chromate or Cr(III) as complexing agent at pH 6.0 and solubilizing the complexes in Tween-80 micellar media. Tartrazine and Quinoline Yellow can be determined simultaneously in the concentration ranges 2.67–12.0 and 0.954–2.39 μg/mL, respectively. This method has been applied for the determination of Tartrazine and Quinoline Yellow in mango flavor with satisfactory results.

26.5.2.2 Determination of Sunset Yellow (E110)

Several methods have been developed for determination of Sunset Yellow mixed with other colorants. A simple and sensitive kinetic spectrophotometric method has been developed for the simultaneous determination of Amaranth, Ponceau 4R, Sunset Yellow, Tartrazine, and Brilliant Blue in mixtures with the aid of chemometrics [31]. This work involved two coupled reactions, the reduction of iron(III) by the analytes to iron(II) in sodium acetate/hydrochloric acid solution (pH 1.71,) and the chromogenic reaction between iron(II) and hexacyanoferrate(III) ions to yield a Prussian blue peak at 760 nm. The spectral data were recorded over the 500–1000 nm wavelength range every 2 s for 600 s. The kinetic data were collected at 760 nm and 600 s, and linear calibration models were satisfactorily constructed for each of the dyes, with detection limits in the range 0.04–0.50 mg/L.

Llamas et al. developed other spectrophotometric methods for the determination of Amaranth, Sunset Yellow, and Tartrazine in beverages [32]. The spectra of the samples (simply filtered) were recorded between 359 and 600 nm, and mixtures of pure dyes, in concentrations between 0.01 and 1.8 mg/L for Amaranth, 0.08 and 4.4 mg/L for Sunset Yellow, and 0.04 and 1.8 mg/L for Tartrazine, were disposed in a column-wise augmented data matrix. This kind of data structure, analyzed by multivariate curve resolution–alternating least squares (MCR-ALS), makes it possible to exploit the so-called "second-order advantage." The MCR-ALS algorithm was applied to the experimental data under the nonnegativity and equality constraints. As a result, the concentration of each dye in the sample and their corresponding pure spectra were obtained.

Chromatographic methods were also developed for Sunset Yellow in a mixture of colorants. Eight synthetic food colorants (Sunset Yellow, Amaranth, Brilliant Blue, Indigo Carmine, New Red, Ponceau 4R, Tartrazine, and Allura Red) were determined by high-performance ion chromatography on an anion exchange analytical column with very low hydrophobicity and visible absorption detection in drinks and instant drink powders [33]. Gradient elution with hydrochloric acid–acetonitrile effected both the chromatographic separation of these colorants and the online cleanup of the analytical column, which was very advantageous for routine analysis. Alves et al. developed a relatively fast method applied to the determination of Sunset Yellow, Tartrazine, Amaranth, Brilliant Blue, and Red-40 in three different kinds of foodstuffs: solid juice powders, solid jelly powders, and soft drinks [18]. High-performance liquid chromatography with UV-DAD detection was employed. The developed chromatographic method employed an ODS Zorbax column (250 mm; 4.6 mm; 5 μm). In 2011, a rapid and highly sensitive method for the determination of 10 water-soluble dyes, including Sunset Yellow, in food was proposed [24]. The method combined HLB SPE column pretreatment with ultraperformance liquid chromatography–diode array detection–tandem mass spectrometry (UPLC-DAD-MS/MS). The developed method displayed a high analysis speed (less than 3 min), low limits of detection (LOD < 0.04 μg/mL, $S/N = 3$) and excellent recoveries (ranging from 92.1%

to 105.4%) for 10 dyes. This method was successfully applied to determining the concentration of the 10 dyes in shrimp flake and drinks. A green chromatographic method for determination of Tartrazine, Brilliant Blue, and Sunset Yellow in food samples was developed [34]. The method is based on the modification of a C_{18} column with a 0.25% (v/v) Triton X-100 aqueous solution at pH 7 and the use of the same surfactant solution as mobile phase without the presence of any organic solvent modifier. After the separation process on the chromatographic column, the colorants are detected at 430, 630, and 480 nm, respectively. The chromatographic procedure yielded precise results and is able to run one sample in only 8 min, consuming 15.0 mg of Triton X-100 and 38.8 mg of phosphate. Detection limits obtained for Brilliant Blue FCF, Sunset Yellow, and Tartrazine were 0.080, 0.143, and 0.125 mg/L, respectively.

Xu et al. synthesized a MIP by electropolymerization of pyrrole on a glassy carbon electrode in the presence of Sunset Yellow as the template, for the detection of Sunset Yellow in wine samples with an electrochemical sensor [35]. Comparing to the polypyrrole nonimprinted modified (NIP) electrode, the polypyrrole MIP electrode improved the electrochemical performance of the sensor significantly. Recently, a novel water-compatible molecularly imprinted ionic liquid polymer–ionic liquid functionalized graphene composite film coated glassy carbon electrode (MIP-rGO-IL/GCE) was produced and applied for the determination of Sunset Yellow in soft drinks [36]. The water-compatible MIP was prepared by free radical polymerization in a methanol–water system using Sunset Yellow as template and ionic liquid 1-(α-methyl acrylate)-3-allylimidazolium bromide (1-MA-3AI-Br) as functional monomer, which can interact with Sunset Yellow through π–π, hydrogen bonding, and electrostatic interactions. The resulting MIP-rGO-IL/GCE shows good performance when it is used for the differential pulse voltammetric determination of Sunset Yellow. Under the optimized conditions (i.e., pH 7.5, 0.1 M phosphate buffer, preconcentration under open-circuit for 570 s), the peak current is linear to Sunset Yellow concentration in the ranges 0.010–1.4 and 1.4–16 µM with sensitivities of 5.0 and 1.4 µA/µM mm², respectively; the detection limit is 4 nM ($S/N = 3$).

In 2013, a voltammetric sensor for Sunset Yellow was developed by polymerizing L-cysteine on the surface of a glassy carbon electrode [37]. The modified electrode was characterized by cyclic voltammetry and electrochemical impedance spectroscopy. The electrochemical behavior and kinetic parameters of Sunset Yellow in phosphate buffer solution (pH 6.6) were investigated by cyclic voltammetry and chronocoulometry using the proposed sensor, and results showed that the electrochemical response of Sunset Yellow was significantly improved. The peak current (differential pulse voltammetry) of Sunset Yellow linearly increased with the concentrations in the range of $8.0 \times 10^{-9} - 7.0 \times 10^{-7}$ mol/L, and the minimum detectable concentration of Sunset Yellow was estimated to be 4.0×10^{-9} mol/L. An expanded graphite paste electrode (EGPE) modified with attapulgite (ATP) was developed for the determination of Sunset Yellow in soft drinks using square-wave stripping voltammetry (SWSV) method [37]. The experimental results suggest that the ATP/EGPE exhibited higher electroanalytic activity toward the Sunset Yellow than did the bare EGPE. The fabricated electrode exhibited some advantages such as convenient preparation, good stability and high sensitivity toward the Sunset Yellow determination. Under optimal experimental conditions, the ATP/EGPE revealed broad linear ranges of $2.5 \times 10^{-9}–1.5 \times 10^{-6}$ M with a detection limit of 1.0 nM.

A novel methodology for the quantification of Sunset Yellow has been developed using image analysis (RGB histograms) and partial least squares (PLSs) regression [38]. The developed method presented many advantages compared with alternative methodologies such as HPLC and UV–Vis spectrophotometry. It was faster, did not require sample pretreatment steps or any kind of solvents and reagents, and used low cost equipment—a commercial flatbed scanner. This method was able to quantify Sunset Yellow in isotonic drinks and orange sodas, in the range 7.8–39.7 mg/L, with relative prediction errors lower than 10%.

26.5.2.3 Simultaneous Determination of Quinoline Yellow (E104) and Sunset Yellow (E110)

For the simultaneous determination of Quinoline Yellow and Sunset Yellow in foods, a derivative spectrophotometry and PLS method has been developed [39]. The colorants were isolated in Sephadex DEAE A-25 gel at pH 5.0, the gel–colorants system was packed in a 1 mm silica cell, and spectra were recorded between 400 and 600 nm against a blank.

Minioti et al. developed a reversed-phase high-performance liquid chromatographic method for the determination of 13 synthetic food colorants: Tartrazine, Quinoline Yellow, Sunset Yellow, Carmoisine, Amaranth, Ponceau 4R, Erythrosine, Red 2G, Allura Red, Patent Blue, Indigo Carmine, Brilliant Blue, and Green S [17]. A C_{18} stationary phase was used and the mobile phase contained an acetonitrile–methanol (20:80 v/v) mixture and a 1% (m/v) ammonium acetate buffer solution at pH 7.5. Successful separation was obtained for all the compounds using an optimized gradient elution within 29 min. A diode-array detector was used to monitor the colorants between 350 and 800 nm. The method was applied to the determination of colorants in various water-soluble foods, such as fruit flavored drinks, alcoholic drinks, jams, sugar confectionery, and sweets, with simple pretreatment (dilution or water extraction). Zatar developed a reversed-phase high-performance liquid chromatography method for the simultaneous determination of Quinoline Yellow, Sunset Yellow, Tartrazine, Carmoisine, Ponceau 4R, Erythrosine, and Brilliant Blue in various food products, such as carbonated fruit-flavored drinks, concentrated fruit-flavored drinks, jams, and sweets [40]. The method was based on using a C_{18} column for separation and a mobile phase containing N-acetyl-N,N,N-trimethylammonium bromide cationic surfactant as an ion pair to achieve complete separation of the food colorants. IPC was developed by Gianotti et al. to determine Tartrazine, Quinoline Yellow, Sunset Yellow, Carmoisine, Amaranth, New Coccine, Patent Blue Violet, and Brilliant Blue in soft drinks [41]. IPC consists in adding a hydrophobic ionic substance to a mobile phase, such as quaternary ammonium cation (e.g., ammonium acetate buffer); neutral ion pairs are formed with the target analytes, which then are separated in a reversed-phase system.

26.6 DETERMINATION OF QUINOLINE YELLOW (E104) AND SUNSET YELLOW (E110) USING FLOW INJECTION ANALYSIS

Very few works dealing with the determination of these synthetic colorants in foods have been published. In 2012, Medeiros et al. developed a single-line flow injection system for the simultaneous determination of two pairs of synthetic colorants Tartrazine–Sunset Yellow and Brilliant Blue–Sunset Yellow in several food products (juice drinks, gelatines, and nutrient-enhanced sports drinks), without a previous separation step, by coupling FIA with multiple pulse amperometry (MPA) with a boron-doped electrode [42]. The influence of flow conditions on the MPA response was evaluated by varying the flow rate from 0.45 to 3.5 mL/min, obtaining a maximum electrical current at

2.7 mL/min, and a high sampling rate of almost 80 determinations per hour. The effect of the injected sample volume was studied in the range 50–500 μL, obtaining a maximum signal with an injected sample volume of 350 μL. Detection limits obtained were 0.80 and 2.5 μmol/L for Sunset Yellow and Tartrazine and 0.85 and 3.5 μmol/L for sunset Yellow and Brilliant Blue.

Valencia et al. developed an integrated solid-phase spectrophotometry–FIA method for the determination of Sunset Yellow in drinks in the presence of its unsulfonated derivative Sudan I [43]. The procedure is based on the retention and the preconcentration of the low level Sudan I in the upper zone of a C_{18} silica gel packed cell, while Sunset Yellow is not retained and its optical signal at 487 nm is read directly over the packed cell. The applicable concentration range and detection limit for Sunset Yellow were 0.5–20.0 mg/L and 0.2 mg/L, respectively.

REFERENCES

1. Marmion, D. M. 1991. *Handbook of U.S. Colorants: Foods, Drugs, Cosmetics, and Medical Devices* (3rd ed.). United States: John Wiley & Son.
2. U.S. Food and Drug Administration. 2004. Color additives. Definitions. edited by Code of Federal Regulation 21 CFR 70.3(f). http://www.accessdata.fda.gov/scripts/cdrh/cfdocs/cfcfr/cfrsearch.cfm?fr=70.3 (accessed July 1, 2014).
3. Barrows, J. N., A. L. Lipman, and C. J. Bailey. 2003. Color Additives: FDA's Regulatory Process and Historical Perspectives. *Food Safety Magazine* (October-November). http://www.foodsafetymagazine.com/magazine-archive1/octobernovember-2003/color-additives-fdas-regulatory-process-and-historical-perspectives/
4. Wrolstad, R. E. and C. A. Culver. 2012. Alternatives to those artificial FD&C food colorants. *Annu. Rev. Food Sci. Technol.* 3(3):59–77.
5. Khanavi, M., M. Hajimahmoodi, A. M. Ranjbar, M. R. Oveisi, M. R. S. Ardekani, and G. Mogaddam. 2012. Development of a green chromatographic method for simultaneous determination of food colorants. *Food Anal. Method* 5(3):408–415.
6. Shahabadi, N. and M. Maghsudi. 2013. Gel electrophoresis and DNA interaction studies of the food colorant quinoline yellow. *Dyes Pigments* 96(2):377–382.
7. McCann, D., A. Barrett, A. Cooper et al. 2007. Food additives and hyperactive behaviour in 3-year-old and 8/9-year-old children in the community: A randomised, double-blinded, placebo-controlled trial. *Lancet* 370(9598):1560–1567.
8. European Food Safety Authority. 2009. Scientific opinion on the re-evaluation of Quinoline Yellow (E 104) as a food additive. *EFSA J.* 7(11):1329, 40p.
9. European Food Safety Authority. 2009. Scientific opinion on the re-evaluation of Sunset Yellow FCF (E 110) as a food additive. *EFSA J.* 7(11):1330, 44p.
10. JECFA. 1996. Summary of Evaluations Performed by the joint FAO/WHO Expert Committee on Food Additives. 1956–1996. Geneva: FAO/IPCS/WHO.
11. European Food Safety Authority. 2014. FAQ on colours in food and feed. http://www.efsa.europa.eu/en/faqs/faqfoodcolours.htm (accessed July 1, 2014).
12. WHO. 2011. Evaluation of certain food additives and contaminants Seventy-fourth report of the Joint FAO/WHO Expert Committee on Food Additives. http://whqlibdoc.who.int/trs/WHO_TRS_966_eng.pdf (accessed July 1, 2014).
13. Feingold Association of United States. 2014. *List of Colorants* 2014. http://www.feingold.org/Research/PDFstudies/List-of-Colorants.pdf (accessed July 1, 2014).

14. European Commision. 1994. European Parliament and Council Directive 94/36/EC on colours for use in foodstuffs. *Official Journal of the European Communities* L237:13–29.

15. Merck. 2006. *The Merck Index: An Encyclopedia of Chemicals, Drugs, and Biologicals* (14th ed.). New Jersey: Merck Inc.

16. Kirk, R. S. and R. Sawyer. 1991. *Pearson's Composition and Analysis of Foods* (9 ed.). United Kingdom: Longman Publishing Group.

17. Minioti, K. S., C. F. Sakellariou, and N. S. Thomaidis. 2007. Determination of 13 synthetic food colorants in water-soluble foods by reversed-phase high-performance liquid chromatography coupled with diode-array detector. *Anal. Chim. Acta* 583(1):103–110.

18. Alves, S. P., D. M. Brum, E. C. Branco de Andrade, and A. D. Pereira Netto. 2008. Determination of synthetic dyes in selected foodstuffs by high performance liquid chromatography with UV-DAD detection. *Food Chem.* 107(1):489–496.

19. Fuh, M. R. and K. J. Chia. 2002. Determination of sulphonated azo dyes in food by ion-pair liquid chromatography with photodiode array and electrospray mass spectrometry detection. *Talanta* 56(4):663–671.

20. Yoshioka, N. and K. Ichihashi. 2008. Determination of 40 synthetic food colors in drinks and candies by high-performance liquid chromatography using a short column with photodiode array detection. *Talanta* 74(5):1408–1413.

21. Berzas Nevado, J. J., J. Rodriguez Flores, C. Guiberteau Cabanillas, M. J. Villaseñor Llerena, and A. Contento Salcedo. 1998. Resolution of ternary mixtures of Tartrazine, Sunset Yellow and Ponceau 4R by derivative spectrophotometric ratio spectrum-zero crossing method in commercial foods. *Talanta* 46(5):933–942.

22. Ma, M., X. B. Luo, B. Chen, S. P. Sub, and S. Z. Yao. 2006. Simultaneous determination of water-soluble and fat-soluble synthetic colorants in foodstuff by high-performance liquid chromatography-diode array detection-electrospray mass spectrometry. *J. Chromatogr. A* 1103(1):170–176.

23. Kirschbaum, J., C. Krause, and H. Bruckner. 2006. Liquid chromatographic quantification of synthetic colorants in fish roe and caviar. *Eur. Food Res. Technol.* 222(5–6):572–579.

24. Ji, C., F. Feng, Z. X. Chen, and X. G. Chu. 2011. Highly sensitive determination of 10 dyes in food with complex matrices using spe followed by uplc-dad-tandem mass spectrometry. *J. Liq. Chromatogr. Rela. Technol.* 34(2):93–105.

25. Feng, F., Y. S. Zhao, W. Yong, L. Sun, G. B. Jiang, and X. G. Chu. 2011. Highly sensitive and accurate screening of 40 dyes in soft drinks by liquid chromatography-electrospray tandem mass spectrometry. *J. Chromatogr. B. Anal. Technol. Biomed. Life Sci.* 879(20):1813–1818.

26. Hamedpour, V. and M. Amjadi. 2014. Application of Box-Behnken design in the optimization of *in situ* surfactant-based solid phase extraction method for spectrophotometric determination of Quinoline Yellow in food and water samples. *Food Anal. Method* 7(5):1123–1129.

27. Li, M., R. Li, J. Tan, and Z. T. Jiang. 2013. Titania-based molecularly imprinted polymer for sulfonic acid dyes prepared by sol-gel method. *Talanta* 107:203–210.

28. Garcia-Falcon, M. S., and J. Simal-Gandara. 2005. Determination of food dyes in soft drinks containing natural pigments by liquid chromatography with minimal clean-up. *Food Control* 16(3):293–297.

29. Rouhani, S. and T. Haji-ghasemi. 2009. Novel PVC-based coated graphite electrode for selective determination of Quinoline Yellow. *J. Iran. Chem. Soc.* 6(4):679–685.

30. Amandeep, K. and G. Usha. 2012. Simultaneous determination of binary mixtures of Tartrazine and Quinoline Yellow food colorants in various food samples and cosmetic products in micellar media by H-Point Standard Addition Method (HPSAM). *Int. J. Res. Chem. Environ.* 2(1):293–300.

31. Ni, Y. N., Y. Wang, and S. Kokot. 2009. Simultaneous kinetic spectrophotometric analysis of five synthetic food colorants with the aid of chemometrics. *Talanta* 78(2):432–441.

32. Llamas, N. E., M. Garrido, M. S. Di Nezio, and B. S. F. Band. 2009. Second order advantage in the determination of Amaranth, Sunset Yellow FCF and Tartrazine by UV-vis and multivariate curve resolution-alternating least squares. *Anal. Chim. Acta* 655(1–2):38–42.

33. Chen, Q. C., S. F. Mou, X. P. Hou, J. M. Riviello, and Z. M. Ni. 1998. Determination of eight synthetic food colorants in drinks by high-performance ion chromatography. *J. Chromatogr. A* 827(1):73–81.

34. Vidotti, E. C., W. F. Costa, and C. C. Oliveira. 2006. Development of a green chromatographic method for determination of colorants in food samples. *Talanta* 68(3):516–521.

35. Xu, J., Y. Zhang, H. Zhou, M. Wang, P. Xu, and J. Zhang. 2012. An amperometric sensor for sunset yellow FCF detection based on molecularly imprinted polypyrrole. *Engineering* 5:159–162.

36. Zhao, L. J., F. Q. Zhao, and B. Z. Zeng. 2014. Preparation and application of Sunset Yellow imprinted ionic liquid polymer—Ionic liquid functionalized graphene composite film coated glassy carbon electrodes. *Electrochim. Acta* 115:247–254.

37. Zhang, K., P. L. Luo, J. J. Wu, W. J. Wang, and B. X. Ye. 2013. Highly sensitive determination of Sunset Yellow in drink using a poly (L-cysteine) modified glassy carbon electrode. *Anal. Method* 5(19):5044–5050.

38. Botelho, B. G., L. P. de Assis, and M. M. Sena. 2014. Development and analytical validation of a simple multivariate calibration method using digital scanner images for Sunset Yellow determination in soft beverages. *Food Chem.* 159:175–180.

39. Capitan Vallvey, L. F., M. D. Fernandez, I. de Orbe, J. L. Vilchez, and R. Avidad. 1997. Simultaneous determination of the colorants sunset yellow FCF and quinoline yellow by solid-phase spectrophotometry using partial least squares multivariate calibration. *Analyst* 122(4):351–354.

40. Zatar, N. A. 2007. Simultaneous determination of seven synthetic water-soluble food colorants by ion-pair reversed-phase high-performance liquid chromatography. *J. Food Technol.* 5(3):220–224.

41. Gianotti, V., S. Angioi, F. Gosetti, E. Marengo, and M. C. Gennaro. 2005. Chemometrically assisted development of IP-RP-HPLC and spectrophotometric methods for the identification and determination of synthetic dyes in commercial soft drinks. *J. Liquid Chromatogr. Rela. Technol.* 28(6):923–937.

42. Medeiros, R. A., B. C. Lourencao, R. C. Rocha-Filho, and O. Fatibello-Filho. 2012. Flow injection simultaneous determination of synthetic colorants in food using multiple pulse amperometric detection with a boron-doped diamond electrode. *Talanta* 99(0):883–889.

43. Valencia, M. C., F. Uroz, Y. Tafersiti, and L. F. Capitan-Vallvey. 2000. A flow-through sensor for the determination of the dye Sunset yellow and its subsidiary Sudan I in foods. *Quim. Anal.* 19(3):129–134.

SECTION VII

Flavor Enhancers

CHAPTER 27

Determination of Monosodium Glutamate

Carolina C. Acebal and Adriana G. Lista

CONTENTS

27.1 INTRODUCTION

The organoleptic properties of food are one of the characteristics that define food quality and govern food selection. Between the food additives that are added to increase its taste and palatability, free L-glutamic acid, as its sodium salt monosodium glutamate (MSG), imparts an intrinsic taste of its own, termed *umami*, the fifth taste. *Umami* means delicious in Japanese, and it is attributed to the sensory properties of MSG and some nucleotides, such as sodium 5′-inosinate (IMP) and sodium 5′-guanylate (GMP). MSG enhances the flavor of certain foods such as meat, poultry, seafood, and vegetables.

In 1866, MSG was isolated for the first time [1], and in 1908, it was found to be the component responsible for the enhancing effect in the taste of seaweed *Laminaria japonica* extracts, traditionally used in Japanese cuisine [2]. As a consequence of this, the Japanese started to add MSG to their meals, extending its use as a food additive worldwide.

In 1968, MSG was felt to be the cause of a series of reactions to Chinese food known as the Chinese restaurant syndrome, which is characterized by headache, a burning sensation along the back of the neck, chest tightness, nausea, and sweating [3]. Moreover, it has been reported that the consumption of food containing large amounts of MSG may cause allergic effects such as asthma [4,5], and an excess of MSG has been associated with diseases of the central nervous system [6,7]. Although many people identify themselves as sensitive to MSG, it has not been proven that MSG is the substance that causes this sensitivity and it continues to be used as an additive in some food products and meals.

Despite this, Food and Drug Administration (FDA) considers the addition of MSG to food to be "generally recognized as safe" (GRAS) [8]. In some countries, such as Argentina, the legislation establishes the maximum allowable concentrations of this additive in certain food [9], particularly to prevent fraud due to the addition of MSG in quantities that can, for example, hide a lower content of main ingredients, such as meat.

Therefore, because of the wide use of MSG in food technology, researchers have turned their attention for developing analytical methods to determine this analyte in food matrices.

In this chapter, we summarize the main flow strategies that have been applied to determine MSG in food samples. We briefly describe different techniques that are employed for determining this analyte, and then we give an overview of the main flow methodologies that are used. In order to give readers a better understanding, we classify the flow methods in enzymatic and chemical-based methods, namely those methods in which enzymatic reactions take place and those that are nonenzymatic. We then give an exhaustive description of flow enzymatic-based methods classified according to the enzyme that is employed.

Finally, we describe the flow chemical-based methods that quantify MSG by a chemical approach.

27.2 MSG DETERMINATION

Since the beginning of the 1990s, the determination of MSG has been a relevant topic among researchers and has received a great deal of attention because of the controversial concept that dietary ingestion of MSG might be responsible for a variety of health problems. During 1990s and at the beginning of 2000, different methods were developed for the quantification of glutamate in food samples. Figure 27.1 shows the tendency to use flow methodologies as determined from database sources.

Monosodium glutamate determination in food has been performed by chromatographic techniques [10]. However, these methods involve time-consuming procedures and require extensive and laborious sample treatment. Usually, derivatization pre- or postcolumn is needed to increase the sensitivity of the method [11]. In most, postcolumn derivatization strategy with o-phthalaldehyde-2-mercaptoethanol is used for this analyte in the isocratic elution mode due to a faster reaction [12]. Actually, chromatography and capillary electrophoresis techniques have mainly been employed to determine glutamate in biological samples in which low limits of detection are needed.

Alternative methods have been proposed based on spectrophotometric [13] and fluorometric [14] techniques. But undoubtedly, the most widespread methods for MSG quantification are electrochemical methods using enzymatic biosensors. The latter combine the advantages of the sensitivity of these techniques and the specificity of enzymatic reactions.

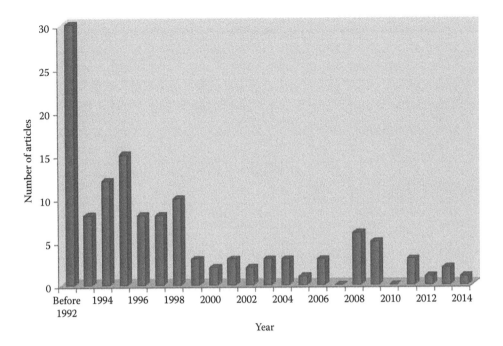

FIGURE 27.1 Tendency to use flow methodologies for the determination of monosodium glutamate (MSG) in food samples in accordance to the number of scientific articles published from 1992 to date.

27.3 FLOW METHODS APPLIED TO MSG DETERMINATION IN FOOD SAMPLES

Several continuous-flow procedures have been developed, taking into account the characteristic advantages of flow methodologies.

Flow methods are not the exception to the general trend of using enzymes for the determination of MSG. Therefore, most of the continuous methods are based on enzymatic reactions whose reaction products are spectrometrically, fluorometrically, and electrochemically detected. Figure 27.2 illustrates a general flow configuration that employs an enzyme-packed reactor. Generally, the immobilized enzyme reactor is placed in the carrier stream before the injection of the sample. The analyte passes through it, and the corresponding enzymatic reaction takes place.

Table 27.1 mentions diverse enzymes that are employed in flow methods for food analysis, in terms of food sample, sample preparation, flow methodology, detection technique, and analytical parameters such as linear range, limit of detection (LOD), and relative standard deviation (RSD).

On the other hand, there are few nonenzymatic flow approaches for determining MSG in food samples, probably because many compounds that interfere with these analyses are found in these complex matrices. Among nonenzymatic flow approaches, multicomponent analysis using chemometric techniques is used to determine MSG and other flavor enhancers [15].

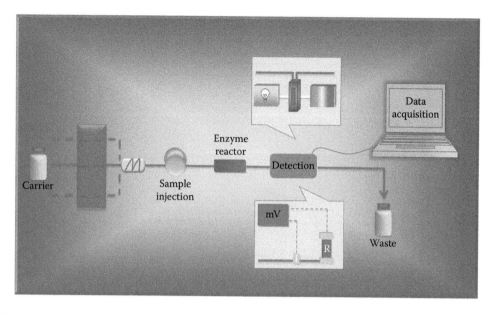

FIGURE 27.2   General flow configuration using enzyme packed reactor with an optical/electrochemical detection. Dashed lines represent channels that could be added to the system. I = indicator electrode, R = reference electrode.

27.3.1 Flow Enzymatic-Based Methods

The outstanding proposed flow enzymatic methods found in the literature are described below.

27.3.1.1 L-Glutamate Oxidase

In 1998, Mizutani et al. developed a flow injection analysis (FIA) system with a GlOD enzyme/polyion complex bilayer membrane–based electrode as a detector to determine L-glutamic acid [16]. The authors remark that the bilayer membrane is better than the others, because the analyte permeates freely to the layer and undergoes the enzymatic reaction, whereas the permeation of the interferents is strongly restricted. To form the bilayer membrane for L-glutamic acid determination, the authors used a gold electrode with a suitable treatment to modify it. The most important steps are:

1. The electrode modified with a 3-mercaptopropionic acid (MPA) is dipped into a solution containing poly-L-lysine and poly(4-styrenesulfonate) to form a polyion complex on the electrode surface; it is then necessary to dry it in air for 4 h and place it under vacuum.
2. An enzyme solution (potassium phosphate buffer at pH = 7) containing GlOD 2% w/v is placed on the polyion complex layer. The electrode is left to dry in the air for 4 h.

The designed single-line manifold system includes a double-plunger pump, a potentiostat, and a thermostat, with a flow cell and injection valve. The flow-through cell consists of the enzyme electrode, a reference electrode, and a stainless-steel auxiliary

TABLE 27.1 Different Enzymatic Flow Methods for Determining Monosodium Glutamate in Food Samples

Food	Enzymes	Sample Pretreatment	Flow Methodology	Detection	Sample Throughput (h^{-1})	Linear Range ($mmol\ L^{-1}$)	LOD ($mmol\ L^{-1}$)	RSD (%)	Ref.
Soy sauces	GlOD	No	FIA	Amperometric	180	3×10^{-3}–0.50	3×10^{-3}	N/A	16
Seasoning samples (powder and infusion)	GlOD	Boiled and filtered	FIA	Amperometric	N/A	0.05–1.0	N/A	N/A	17
Soy sauce	GlOD	No	FIA	Amperometric	N/A	0.5–8.0	N/A	10.2	18
Seasonings	GlOD	Dissolved in water, heated at 70°C for 20 min and filtrated	FIA	Amperometric	60	5×10^{-6}–1×10^{-3}	10^{-6}	0.7	19
Food formulas of commercial soups	GlAD	No	FIA	Potentiometric	N/A	2.5–75	N/A	N/A	20
Meat softener, meat soup, spices	GlAD	No	FIA	Potentiometric	40	10–100	N/A	N/A	21
Beef and chicken bouillon cubes, soy sauces	GlDH	No	FIA	Amperometric	N/A	0.01–1.0	N/A	N/A	22

(Continued)

TABLE 27.1 (*Continued*) Different Enzymatic Flow Methods for Determining Monosodium Glutamate in Food Samples

Food	Enzymes	Sample Pretreatment	Flow Methodology	Detection	Sample Throughput (h^{-1})	Linear Range ($mmol\ L^{-1}$)	LOD ($mmol\ L^{-1}$)	RSD (%)	Ref.
Beef and chicken bouillon cubes; soy sauces; apple, orange and tomato juice	GlOD	No	FIA	Amperometric	N/A	0.01–1.0	N/A	N/A	22
Fish, oyster, and soy sauces	GlDH, D-PhgAT	No	FIA	Spectrophotometric/fluorimetric	N/A	2.5×10^{-3}–0.050	4×10^{-4}	8.9	23
Food	GlDH, D-PhgAT	No	Microfluidic FIA/SIA	Fluorometric	N/A	3.1×10^{-3}–0.05	3×10^{-3}	6.6	24
Cream of asparagus soup, broccoli sauce	GlAD	pH adjustment to 4.0, ultrasonic extraction, filtered	Gas diffusion FIA	Spectrophotometric	30	2.0–20	N/A	2.8	27
Cream of asparagus soup, broccoli sauce	GlDH	pH adjustment to 9.0, ultrasonic extraction, filtered	FIA	Spectrophotometric	30	0.05–0.06	N/A	2.4	27
Cream of asparagus soup, broccoli sauce,	GlDH	pH adjustment to 9.0, ultrasonic extraction, filtered	Stop-flow FIA	Spectrophotometric	30	0.2–1.0	N/A	1.9	27

Note: FIA, flow injection analysis; LOD, limit of detection; RSD, relative standard deviation.

electrode. The injection volume is 5 μL. The detector provides rapid and highly selective measurements of enzyme substrate. The obtained LOD is 3×10^{-3} mmol L^{-1} and the linear relationship is up to 0.5 mmol L^{-1}. The proposed method is applied to soy sauces, and results are compared with those obtained by using F-kit method.

In other work, Matsumoto et al. designed a parallel FIA configuration for the simultaneous determination of IMP and MSG in food seasoning, taking into account the *umami* taste [17]. They used two immobilized enzyme reactors in the manifold and amperometric detection. The FIA system consists of two micro-tube pumps to introduce the carrier solutions, one for each analyte, and a fixed volume of samples mixed with coenzyme (NAD^+) is injected. A third pump is used for loading the samples. As the concentration of MSG in the samples is more than five times higher than IMP concentration, an online dilution is proposed. The flow cell is connected with a multichannel potentiostat and a recorder. To determine IMP, inosine monophosphate dehydrogenase (IMPDH) from *Bacillus cereus* and NADH oxidase (NOD) from *Bacillus licheniformis* are used. MSG is quantified by using GlOD from *Streptomyces* sp. immobilized onto controlled pore glass activated with 5% glutaraldehyde solution in 0.2 M carbonate buffer (pH = 10) for 2 h at 20°C. In both cases, the H_2O_2 produced during the enzymatic reaction is amperometrically monitored with a platinum electrode. To optimize the determination FIA conditions, the authors set a concentration ratio of IMP and MSG at a typical level of ordinary seasoning samples. The linear relationship between sensor responses and the concentration of each analyte are: 0.1–1.0 mmol L^{-1} for IMP and 1.0–10.0 mmol L^{-1} for MSG with $R^2 > 0.997$. The authors apply the method to real seasoning samples and the obtained results are compared to those obtained by a liquid chromatography method for IMP and by a glutamate-kit method for MSG. The correlation between the results is satisfactory. The authors propose the use of an electropolymerized film of 1,2-diaminobenzene (1,2-DAB) to solve the problems related to interferents and electrode fouling.

In 2003, Surareungchai et al. developed a FIA system with an amperometric GlOD electrode based on three layers of polymer films for the high selective determination of MSG in soy sauces [18]. The single-channel flow system includes an electrochemical homemade flow-through cell that contains the enzyme electrode, an Ag/AgCl reference electrode, and a platinum wire counterelectrode. The multilayer enzyme electrode is constructed of an inner membrane of electropolymerized 1,3-DAB, a middle layer containing the enzyme entrapped in photopolymerized poly(vinylferrocene)–poly(ethylene glycol) hydrogel polymer, and an outer dialysis membrane. In this way, a highly sensitive, stable, and low-potential operation for an amperometric method is obtained. Furthermore, electropolymerization of DAB allows the membrane to eliminate some interferences (such as ascorbic acid, cysteine, and methionine) by size exclusion. After optimization of FIA variables, the stability and reproducibility of the FIA-glutamate sensor is examined. The sensor appears to be stable, with the standard deviation in a range of 0.5%–2.0%, and the sensor response is 60% of the initial value after 16 days. The interferences of common compounds present in soy sauce are tested, and the obtained error values are acceptable. The authors test the applicability of the sensor by analyzing 20 commercial soy sauce samples. The results are compared with the commercial test analysis using a colorimetric method. A paired *t* test is applied to compare both results, and no significant differences between the methods are obtained for the mean L-glutamate content.

In 1990, Yao et al. proposed two different alternatives to determine MSG in seasonings using immobilized GlOD: a packed-bed reactor and a lab-made flow-through enzyme electrode used as a detector [19]. Both configurations are tested in a FIA system with a simple channel manifold to compare the performance of the enzyme with regard to sensitivity, sample throughput and long-term stability. In both systems, the product (H_2O_2) is detected amperometrically. In the case of the enzyme-reactor system, a flow-through platinum electrode is employed as a detector. The enzyme-reactor system presents higher sample throughput than the flow-through enzyme electrode (120 h^{-1} vs. 60 h^{-1}). Both systems have detection limits on the same order (10^{-3} mmol L^{-1}) and are stable with time, even after repetitive use for 6 months.

27.3.1.2 L-Glutamate Decarboxylase

In 2001, a Brazilian scientific group presented a work in which the enzymatic determination of MSG is done using GlAD obtained from green pepper (*Capsicum annuum*) [20]. The authors affirm that high amounts of enzyme can be extracted from green pepper at low cost. The enzyme is immobilized in aminopropyl-activated glass beads and is placed into a continuous-flow system coupled to a potentiometric detector. The optimization of the reactor length, the volume of the sample, and the carrier flow rate are performed with a factorial design. The linear range for the determination of MSG is 2.5–75 mmol L^{-1}, and the sample throughput is 60 h^{-1}. The authors also evaluate the stability of the reactor, proving that it remains unaltered over 72 days (638 determinations). This method is applied to chicken soup, meat softener, ready-made seasonings, and vegetable broth, and the results are compared with the conventional procedure (spectrophotometric determination). It is important to highlight that the sample does not need a previous treatment.

The same FIA system is employed by Arruda and coworkers to determine MSG in food samples [21]. As a difference, the GlAD enzyme is obtained from pumpkin (*Cucurbita maxima*), and the enzymatic reactor is prepared in a completely different way. As the preparation of enzymatic reactors based on enzyme extracts is known to be a laborious procedure, the authors propose to prepare the reactor with a naturally immobilized enzyme. For this purpose, a polyethylene column is filled with 200 mg of square pieces of the outer layer of *Cu. maxima*. A piece of the fruit is first cut and rinsed with distilled water. Then the outer layer is cut in square pieces with 2-mm sides and 1-mm thick and washed with phosphate buffer solution. This column is incorporated into the FIA system. The obtained linear range with the optimized FIA system is 10–100 mmol L^{-1}, and the analytical frequency is 40 h^{-1}. The method is applied to the analysis of real samples (meat softener, meat soup, and spices), and the results are compared with those obtained applying the conventional spectrophotometric method to the same samples. The implemented procedure has a comparable precision to the conventional method (deviation values lower than 5%). This enzymatic reactor can be used for 21 days (200 determinations) without significant changes.

27.3.1.3 L-Glutamate Dehydrogenase

Janarthanan and Mottola have demonstrated the advantages of using rotating bioreactors for online or in-line determinations in food analysis by means of the determination of MSG [22]. They use two enzymatic approaches, one of them using two enzymes: GlDH in the main enzymatic reaction and diaphorase in the indicator reaction, which involves NADH and hexacyanoferrate(III). The hexacyanoferrate(II) produced by the indicator reaction is amperometrically monitored with a platinum-ring electrode. The other approach utilizes GlOD, and the amperometrically monitored species is H_2O_2. For testing both approaches, the FIA system is constructed with three channels: a NAD$^+$ and Fe(CN)$_6^{3-}$ channel (only used with the double enzymatic system), a sample channel, and a third channel to introduce the buffer solution. A programmable pump and a valve unit to stop and restart the flow are used. In the stop-flow mode, the unit pumps a programmable volume of sample and reagent into the cell, and the mixture of sample and reagent remains stationary for a programmed time interval during which the measurement is performed. An electrochemical cell/bioreactor unit is placed before the potentiostat/amperometric detector. The interference of ascorbate is eliminated by placing a packed reactor containing ascorbate oxidase online.

Once FIA and chemical variables are optimized, the determination of MSG is possible in a wide concentration range between 0.010 and 1.0 mmol L^{-1}. Different sensitivity is observed between 0.01 and 0.10 mmol L^{-1} and 0.10 and 1.0 mmol L^{-1}, so the authors propose two different calibration curves. The method is applied to different food samples such as beef and chicken bouillon cubes and soy sauces.

In the second part of the work, the authors used only GlOD due to the simplicity of detecting hydrogen peroxide electrochemically. To determine MSG, the same FIA system is used, and after its optimization, the method is applied to the previous food samples and, in addition, to apple, orange, and tomato juice.

A different strategy using the same enzyme is proposed by Khampha et al. [23]. The proposed FIA method is based on an enzymatic recycling substrate with specific UV and fluorescence detection of MSG in fish, oyster, and soy sauce samples. The content of MSG in the sample is amplified by cycling between GlDH and D-phenylglycine aminotransferase (D-PhgAT). In this work, a coimmobilized GlDH/D-PhgAT reactor is placed in a three-line FIA manifold: (a) Tris–HCl buffer pH 8.0 solution, (b) d-4-hydroxyphenylglycine solution, and (c) NAD^+ solution. They merge in a mixing coil placed before the injection valve. Then, the sample is injected into this stream and passed through the enzymatic reactor where the reaction takes place. GlDH converts MSG into 2-oxoglutarate with concomitant reduction of NAD^+ to NADH. D-PhgAT transfers an amino group from 4-hidroxyphenylglycine to 2-oxoglutarate, thus recycling MSG. The NADH, amplified in the reactor according to the reaction cycle, is monitored both by fluorescence and UV detection and correlated to the concentration of MSG in the samples. The enzymes are covalently immobilized on aminopropyl controlled-pore glass, which is activated by glutaraldehyde. The authors explain that any cycling assay involves amplification of a reaction product signal allowing the detection of analytes with higher sensitivity. Therefore, the coimmobilized enzyme ratio is optimized, and the stability of the enzyme reactor and its cycling efficiency are evaluated. With the optimized FIA system, the calibration curves for MSG using fluorescence and UV detection are recorded between 2.5 and 100 µmol L^{-1}, and linearity is observed up to 50 µmol L^{-1}. The LOD using fluorescence is 0.4 µmol L^{-1} and for UV detection is 0.7 µmol L^{-1}. RSD values are calculated within day and between day for both detections, and the respective values are 4.3%–7.3% and 8.9% for fluorescence signals and 4.0%–10.9% and 13.0% for UV detection, respectively. Some other amino acids and ascorbic acid are tested as possible interferences, and the results show a good selectivity for the new method (recoveries for MSG between 95% and 103%). Samples only required a dilution before injection in the FIA system, and the obtained results are compared with those obtained by applying a commercial spectrophotometric method. A good correlation is obtained ($R^2 = 0.998$) with a relative error less than 4.0%.

Microfluidic systems based on the previous recycling reaction are proposed by Emnéus et al. for specific analysis of MSG with fluorescence detection [24]. GlDH and D-PhgAT are covalently immobilized on the surface of silicon microchips containing 32 porous flow channels for one of the systems, and on polystyrene Poros™ beads for the other. The first microfluidic chip is fabricated according to the literature [25,26]. The microfluidic channels of the microchip are open, and the chip is thus first covered with the Perspex lid (a sheet of latex film), placed in a Plexiglas holder, and then incorporated into the FIA system. The second microchip contains a single channel fabricated in 110-silicon anodically etched in KOH. Before the outlet, a grid of standing parallel walls is microfabricated in the channel. The grid is used as a frit to retain enzyme activated beads. The open microfludic channel is sealed by anodic bonding with a glass lid to form a robust and nonleaky microfluidic platform. Subsequently, capillaries are glued into the inlet and outlet for fluidic interfacing. The above-described single-channel microchip is integrated to the system by connecting the inlet capillary to an injection valve and the outlet capillary to the fluorescence detector.

Microchips with immobilized enzymes beads are incorporated in the flow system. The introduction of the beads' suspension into the microfluidic channels is performed manually using a syringe. The microchip with immobilized enzymes and open porous wall channels is first placed in a specially designed flow cell unit made of Plexiglas with the inlet and outlet tubing glued at the ends. The beads are then moved to the microfabricated grid frit region and retained within the channel. When a decrease in the assay signal is observed, the beads are removed from the microchip by introducing a buffer in a reverse direction. Thus, the microchip could be used repeatedly by reloading it with a fresh group of beads. MSG standard solutions at concentrations between 0 and 100 μmol L^{-1} are introduced into the FIA system via the injection port. In the case of the microchip with immobilized enzymes, a carrier flow rate of 20 μL min^{-1} was used. When the MSG standard reaches the microchip surface, the flow is stopped for 5 min, allowing the incubation of the analyte with enzymes. In the following step, the flow rate is increased to 40 μL min^{-1}, and an NADH peak is recorded by the fluorescence detector at 460 nm. For the micro-bead reactor, a flow rate of 20 μL min^{-1} is used throughout all experiments.

With silicon microchips, the MSG calibration curve following this protocol is linear between 3.1–50 μmol L^{-1} with a minimum detection limit of 3 μmol L^{-1}, calculated using threefold signal-to-noise ratio.

The microchips are coupled to an SIA system for fully automated determination of MSG. The system is constructed with a syringe pump, a 10-multiposition valve, an enzymatic microchip, and a fluorescence detector. The inlet capillary of the single-channel microchip is connected to a 10-position valve, but in this case, only four ports are used (only beads, sample, waste, and microchip position). The whole system is controlled by a computer using software developed in-house. The fluorescence of the NADH peak is recorded by a chart recorder.

The assay cycle is performed by aspirating the reaction mixture containing the substrates, aspirating MSG sample, and then dispensing the solutions to the microchip. The flow is stopped when the sample reaches the micro-beads reactor in order to demonstrate the recycling efficiency, and then registration of the corresponding fluorescence signal is performed.

The linear range for this device is 50–200 μmol L^{-1}, with an LOD of 3 μmol L^{-1} ($S/N = 3$) and an RSD value of 6.6%. Common interferences in food samples (glutamine, aspartate, alanine, phenylalanine, histidine, and tryptophan) are tested, and the recovery values of MSG are between 91% and 108%. Finally, the authors affirm that both proposed methods are suitable for determining MSG in food samples.

To compare the performance of GAD and GlDH enzymes immobilized in reactors, Shi and Stein proposed three different FI systems for determining MSG in cream of asparagus soup and broccoli sauce samples [27].

The GAD reactor system consists of a gas-diffusion FIA system in which GAD is immobilized in a reactor and placed in a pyridoxal 5'-phosphate/phosphate buffer pH 4.0 carrier stream. The sample is injected into the carrier stream and passed through the GAD reactor, where MSG is decarboxylated and CO_2 is released. This stream merges with a phosphoric acid solution and is diffused through the gas-permeable Teflon membrane. The CO_2 reacts with an acid–base indicator solution that acts as acceptor and is detected spectrophotometrically at 430 nm. The optimization of the sample flow rate is crucial, since it affects the contact time with the GAD enzyme and the diffusion time of the CO_2 through the membrane. Sample throughput is 30 h^{-1}.

The GlDH reactor is employed in two different FIA configurations: normal and stop-flow systems. In both, the enzymatic product, NADH, is detected spectrophotometrically at 340 nm. In the normal FIA approach, the sample is injected in a phosphate buffer pH 9.0/NAD^+ carrier solution and passed through the GlDH reactor. In the stop-flow approach, two carrier solutions are employed: phosphate buffer pH 9.0/NAD^+ and pure phosphate buffer. Here, the carrier containing NAD^+ is introduced into the system for 30 s, the sample solution is injected and the flow is stopped during 5 s. The flow is then restored, and the pure buffer is used as a carrier. In this way, the consumption of NAD^+, an expensive coenzyme, is reduced significantly, consequently decreasing the cost of the analysis. On the other hand, the linear range and the sensitivity of the normal FIA system are notably higher.

27.3.2 Flow Chemical-Based Methods

27.3.2.1 Potentiometric Determination Based on a Solid-State Electrode

Isa and Ghani proposed a reliable and low-cost FIA system to determine MSG by a potentiometric analysis using a solid-state electrode constructed with a 2B graphite pencil rod with chitosan as a coating material [28]. The FIA system consists of two channels, both for carrier solution (6×10^{-6} mol L^{-1} HCl, pH 5.2), at a flow rate of 1.5 mL min^{-1}. A low-pressure four-way rotary injection valve is placed in one of the carrier streams, and 75 µL of sample volume is injected. The solutions are propelled to a Perspex flow-through cell of the wall jet design with a built-in Ag/AgCl reference electrode. One of the carrier streams passes through the solid-state electrode and the reference electrode to perform the baseline, while the one with the sample passes by the chitosan solid-state electrode to measure the MSG signal.

As the reaction depends on pH, the potentiometric measurements are obtained using a pH/mV meter connected to single chart recorder and potentiometric peaks are performed.

All FIA variables are optimized, taking into account the magnitude of the signal as peak height. The recovery values of MSG in the presence of other amino acids and food additives and preservatives are investigated, obtaining recovery percentages between 88.9 and 99.2. The analysis of seasonings, chicken soups, and mushroom soup samples are proposed. The samples are used in their powdered form, and the fat content has to be decomposed by dissolving the sample and heating it at 70°C for 10 min. The solution is then filtered and diluted with pure water prior to the injection into the FIA system.

The chitosan electrode for MSG determination presents a linearity range of 0.01–1.00 mmol L^{-1}, with a detection limit of 0.008 mmol L^{-1}. Results of the proposed method are validated against an HPLC method.

27.3.2.2 Flow-Batch Analysis Approach

Determination of MSG in dehydrated broth was performed by Acebal et al. using flow-batch analysis (FBA) methodology with nephelometric detection [29]. The method is based on the decrease of the scattering signal due to the inhibitory effect of the MSG over the crystallization of L-lysine in a nonaqueous medium (isopropanol/acetone mixture). The FBA approach is a hybrid system that combines characteristics of flow (multicommutation system) and batch. The fluids are driven sequentially or simultaneously to a mixing chamber, where the reaction and even detection occur. Mainly, the FBA system design consists of two peristaltic pumps, five three-way solenoid valves, and a mixing chamber, as is shown in Figure 27.3. In this system, one peristaltic pump is used to propel solutions to the mixing chamber, and the other is used as an auxiliary pump to empty the mixing chamber by fixing the maximum speed of the pump with the larger available tube diameter (3.18 mm id). The solenoid valves are employed to introduce the adequate aliquot of each solution [L-lysine standard solution, MSG working solution or sample, 50:50 (v/v) isopropanol/acetone mixture, isopropanol, and water] in the appropriate order into a Teflon® lab-constructed mixing chamber, where the solutions are mixed. When the solenoid valves are switched off, the solutions return to their respective flasks. In this way, only the quantity of reagents and samples that are needed for reaction are employed, minimizing the amount of reagents used and consequently the amount of waste.

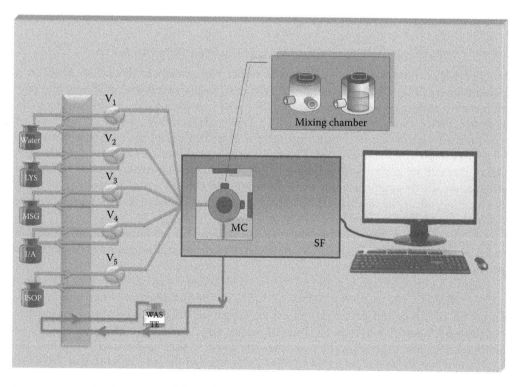

FIGURE 27.3   FBA system with nephelometric detection. V_1 to V_5: solenoid valves, Lys: L-lysine, I/A: isopropanol-acetone mixture, ISOP: isopropanol, MC: mixing chamber, SF: spectrofluorometer.

The mixing chamber has two quartz windows that are located at 90° from each other and is placed in a luminescence spectrometer, instead of the cell holder, to perform the nephelometric measurements at 420 nm.

The flow-batch manifold is controlled by an electronic actuator connected to a Pentium® four microcomputer through a Labview® 5.1 visual programming language computer program.

The selection of the FBA methodology, among others, lies in the importance of strictly controlling the variables that have a significant effect on the formation of the precipitate in order to obtain a better reproducibility in the measurements. Moreover, another advantage of the system is that the calibration curve is easily performed online.

The analytical curve is constructed over the range of 2.8×10^{-3} to 1.1×10^{-2} g L^{-1}. The detection limit (LOD) is 9.7×10^{-5} g L^{-1} calculated from the calibration curve, and the RSD value is 3.8% obtained from nine replicates of real samples of 1.1×10^{-2} g L^{-1} MSG concentration. The sample throughput was 9 h^{-1}.

The FBA system is satisfactory applied to quantify the amount of MSG in meat and vegetable dehydrated broths.

27.4 CONCLUDING REMARKS

It is clear that to determine MSG in food samples the flow enzymatic methods are preferred due to their selectivity. The enzymatic reactions can take place in a simple packed reactor that can be easily incorporated in a flow system.

Usually, the reviewed methods apply simple FIA configurations, in which the most laborious step is the enzyme immobilization. Different enzymes are evaluated in terms of their specificity and diverse strategies are proposed for the determination of MSG. However, several nonenzymatic FIA methods can be applied to determine MSG in these kinds of samples with satisfactory analytical parameters.

ACKNOWLEDGMENTS

The authors gratefully acknowledge the National University of the South (UNS) and National Council of Scientific and Technical Research (INQUISUR-CONICET).

REFERENCES

1. Ault, A. 2004. The monosodium glutamate story: The commercial production of MSG and other amino acids. *J. Chem. Educ.* 81:347–355.
2. Ikeda, K. 2002. New seasonings. *Chem. Senses* 27:847–849.
3. Williams, A. and K. Woessner. 2009. Monosodium glutamate 'allergy': Menace or myth? *Clin. Exp. Allergy* 39:640–646.
4. Bellisle, F. 1998. Effects of monosodium glutamate on human food palatability. *Ann. N. Y. Acad. Sci.* 855:438–441.
5. Allen, D. H., J. Delohery, and G. Baker. 1987. Monosodium L-glutamate-induced asthma. *J. Allergy Clin. Immunol.* 80:530–537.

6. Tapiero, H., G. Mathé, P. Couvreur, and K. Tew. 2002. Free amino acids in human health and pathologies. II. Glutamine and glutamate. *Biomed. Pharmacother.* 56:446–457.

7. Conn, P. J. and J. P. Pin. 1997. Pharmacology and functions of metabotropic glutamate receptors. *Annu. Rev. Pharmacol. Toxicol.* 37:205–237.

8. Database of Select Committee on GRAS Substances (SCOGS) Reviews. 1980. Monosodium L-glutamate. http://www.fda.gov/default.htm

9. Código Alimentario Argentino. 2014. ANMAT. http://www.anmat.gov.ar/alimentos/normativas_alimentos_caa.asp

10. Anderson, L. W., D. W. Zaharevitz, and J. M. Strong. 1987. Glutamine and glutamate: Automated quantification and isotopic enrichments by gas chromatography/ mass spectrometry. *Anal. Biochem.* 163:358–368.

11. Krishna Veni, N., D. Karthika, M. Surya Devi, M. Rubini, M. Vishalini, and Y. Pradeepa. 2010. Analysis of monosodium l-glutamate in food products by high-performance thin layer chromatography. *J. Young Pharm.* 2:297–300.

12. Lindroth, P. and K. Mopper. 1979. High performance liquid chromatographic determination of subpicomole amounts of amino acids by precolumn fluorescence derivatisation with o-phthaldialdehyde. *Anal. Chem.* 51:1667–1674.

13. Khampha, W., V. Meevootisom, and S. Wiyakrutta. 2004. Spectrophotometric enzymatic cycling method using L-glutamate dehydrogenase and D-phenylglycine aminotransferase for determination of L-glutamate in foods. *Anal. Chim. Acta* 520:133–139.

14. Chapman J. and M. Zhou. 1999. Microplate-based fluorometric methods for the enzymatic determination of l-glutamate: Application in measuring l-glutamate in food samples. *Anal. Chim. Acta* 402:47–57.

15. Acebal, C., A. Lista, and B. Fernández Band. 2008. Simultaneous determination of flavor enhancers in stock cube samples by using spectrophotometric data and multivariate calibration. *Food Chem.* 106:811–815.

16. Mizutani, F., Y. Sato, Y. Hirata et al. 1998. High-throughput flow-injection analysis of glucose and glutamate in food and biological samples by using enzyme/polyion complexbilayer membrane-based electrodes as the detectors. *Biosens. Bioelectron.* 13:809–815.

17. Matsumoto, K., W. Asada, and R. Murai. 1998. Simultaneous biosensing of inosine monophosphate and glutamate by use of immobilized enzyme reactors. *Anal. Chim. Acta* 358:127–136.

18. Na Nakorn, P., M. Suphantharika, W. Udomsopagit et al. 2003. Poly(vinylferrocene)–poly(ethylene glycol) glutamate oxidase electrode for determination of L-glutamate in commercial soy sauces. *World J. Microbiol. Biotechnol.* 19:479–485.

19. Yao, T., N. Kobayashi, and T. Wasa. 1990. Flow-injection analysis for L-glutamate using immobilized L-glutamate oxidase: Comparison of an enzyme reactor and enzyme electrode. *Anal. Chim. Acta* 231:121–124.

20. Oliveira, M., M. Pimentel, M. Montenegro et al. 2001. L-Glutamate determination in food samples by flow-injection analysis. *Anal. Chim. Acta* 448:207–213.

21. Arruda, J., N. Filho, M. Montenegro et al. 2003. Simple and inexpensive flow L-glutamate determination using pumpkin tissue. *J. Agric. Food Chem* 51:6945–6948.

22. Janarthanan, C. and H. Mottola. 1998. Enzymatic determinations with rotating bioreactors: Determination of glutamate in food products. *Anal. Chim. Acta* 369:147–155.

23. Khampha W., J. Yakovleva, D. Isarangkul et al. 2004. Specific detection of L-glutamate in food using flow-injection analysis and enzymatic recycling of substrate. *Anal. Chim. Acta* 518:127–135.
24. Laiwattanapaisal, W., J. Yakovleva, M. Bengtsson et al. 2009. On-chip microfluidic systems for determination of L-glutamate based on enzymatic recycling of substrate. *Biomicrofluidics* no. 3. http://dx.doi.org/10.1063/1.3098319
25. Laurell, T., J. Drott, L. Rosengren et al. 1996. Enhanced enzyme activity in silicon integrated enzyme reactors utilizing porous silicon as the coupling matrix. *Sens. Actuators B* 31:161–166.
26. Laurell, T., J. Drott, and L. Rosengren 1995. Silicon wafer integrated enzyme reactors. *Biosens. Bioelectron.* 10:289–299.
27. Shi, R. and K. Stein. 1996. Flow injection methods for determination of L-glutamate using glutamate decarboxylase and glutamate dehydrogenase reactors with spectrophotometric detection. *Analyst* 121:1305–1309.
28. Isa, I. and S. Ab Ghani. 2009. A non-plasticized chitosan based solid state electrode for flow injection analysis of glutamate in food samples. *Food Chem.* 112:756–759.
29. Acebal, C., M. Insausti, M. Pistonesi et al. 2010. A new automated approach to determine monosodium glutamate in dehydrated broths by using the flow-batch methodology. *Talanta* 81:116–119.

CHAPTER 28

Determination of Inosine Monophosphate and Guanosine Monophosphate

Kiyoshi Matsumoto

CONTENTS

28.1 INTRODUCTION

Flavor enhancers, "*umami* taste" compounds, play important roles in the taste, palatability, and acceptability of food, and they are widely used as additives to enhance the flavor of a variety of foods.

Umami (the Japanese term for deliciousness) is a characteristic taste imparted by monosodium glutamate (MSG) in seaweeds, inosine 5′-monophosphate (IMP) in meat and fish, and guanosine 5′-monophosphate (GMP) in mushrooms. A unique characteristic of *umami* is its synergism. Purine nucleotides, such as IMP and GMP, strongly enhance the intensity of *umami* taste. In human taste tests, IMP, which does not elicit any *umami* taste by itself, can increase one's sensitivity to *umami* taste in MSG. Inversely, the detection threshold of IMP is dramatically lowered in the presence of MSG. This suggests that IMP by itself has no *umami* taste and simply acts as an enhancer of a modulator, whereas MSG acts as an agonist for *umami* receptors in humans [1,2]. GMP also does not have the specific *umami* taste by itself but strongly enhances many other flavors.

Inosine monophosphate and GMP are synthesized in the human body and play diverse roles in cellular metabolism [3]. Dietary nucleotides are energetically useful to fulfill the liver's need for nucleotides [4], although people with high levels of uric acid (UA) in their blood and urine must avoid foods with these compounds, because degradation of purine nucleotides leads to the formation of UA [3]. In addition, these flavor enhancers are added by food manufacturers to improve the taste of food. Therefore, monitoring these compounds in foods or food seasonings is very important in food-quality or food-processing control.

Inosine monophosphate and GMP have been determined by liquid chromatographic (LC) methods equipped with various detectors, including mass spectrometry (MS), or by capillary electrophoresis methods. These methods are useful for determining the presence of various nucleotides, but they are time consuming. Flow injection analysis (FIA) methods are useful for determination of flavor enhancers and have high sample throughput. However, few articles have reported determination of flavor enhancers by FIA methods. One approach is the use of biospecific sensors or analyzers based on enzyme minireactors. IMP biosensing has been performed by sequential application of 5′-nucleotidase (or alkaline phosphatase), nucleoside phosphorylase, and xanthine oxidase, and oxygen consumption or UA produced by the final enzymatic reaction (xanthine oxidase) was monitored [5,6]. An FIA system using immobilized IMP dehydrogenase and NADH oxidase has also been reported [7]. Another technique to detect flavor enhancers is a spectrophotometric method using a diode-array detector and chemometric analysis. The biosensing techniques and spectrometric methods are introduced first in this chapter, and then ordinary LC methods and capillary electrophoresis methods are described.

28.2 PREPARATION OF SAMPLES

Sample preparation of meat and fish or other perishable materials involves, in general, a rapid cessation of enzymatic reactions and exclusion of protein to provide a clean supernatant. The fundamental method includes: (1) deproteinization of materials by precipitation with trichloroacetic acid or perchloric acid, (2) centrifugation, and (3) neutralization of the supernatant with sodium hydroxide. However, these procedures are tedious and are not always required for all food samples, so simpler methods are ordinarily adopted. The simpler method is based on the homogenization of the food material with an appropriate amount of distilled water. The homogenate is then filtered through a membrane filter (0.2-μm pore size). The filtrate is diluted to a certain volume with distilled water and used immediately as the sample. In some cases, the filtrate is subjected to a 30-kDa cutoff membrane pass before injection.

28.3 SIMULTANEOUS BIOSENSING OF IMP AND MSG BY USE OF IMMOBILIZED ENZYME REACTORS

In general, biosensing of IMP is performed by sequential applications of 5′-nucleotidase (or alkaline phosphatase), nucleoside phosphorylase, and xanthine oxidase, and the oxygen consumption or UA production by the final enzymatic reaction (xanthine oxidase) is monitored. The series of enzymatic reactions, however, is complicated, and the optimal conditions for each of these three enzymes are different. Matsumoto et al. [7] reported a simpler enzymatic system that used IMP dehydrogenase (IMPDH) from *Bacillus cereus* and NADH oxidase (NOD) from *Bacillus licheniformis*.

28.3.1 Principle

The method for quantitation of IMP is based on the following reactions:

$$IMP + NAD^+ \xrightarrow[K^+]{IMPDH} Xanthylic\ acid + NADH + H^+ \tag{28.1}$$

$$NADH + H^+ + O_2 \xrightarrow[FAD]{NOD} NAD^+ + H_2O_2 \tag{28.2}$$

where FAD is the flavin adenine dinucleotide.

Monosodium glutamate was also quantitated simultaneously in this system. The enzyme used was glutamate oxidase (GluOD) from *Streptomyces* sp., and the reaction scheme is shown below.

$$L\text{-glutamate} + O_2 + H_2O \xrightarrow{GluOD} \alpha\text{-ketoglutarate} + NH_3 + H_2O_2 \tag{28.3}$$

Hydrogen peroxide production was monitored amperometrically with a platinum electrode [7,8].

28.3.2 Flow System

The development of an FIA system for simultaneous quantitation of two flavor enhancers using biospecific sensors is very important from the standpoint of food quality control.

A schematic representation of the FIA system for quantitation of a single component (IMP or MSG) is shown in Figure 28.1a, and for simultaneous quantitation of IMP and MSG in Figure 28.1b. Each working solution (carrier I or II) in a reservoir was propelled by a microtube pump (P1 or P2) through an air damper, a sample-injection valve (six-way switching valve), and an immobilized enzyme reactor and it was then transported to an electrochemical flow-through cell and finally to a waste tank. The sample flow system consisted of another multichannel microtube pump (P3). The sample and coenzyme (NAD⁺) were mixed before introduction into the sample loop. Because the concentration of MSG is generally more than five times higher than that of IMP in an ordinary seasoning sample, it was desirable to dilute the sample six times before introduction into the sample loop of the MSG measuring valve (V1). As shown in Figure 28.1b, one channel

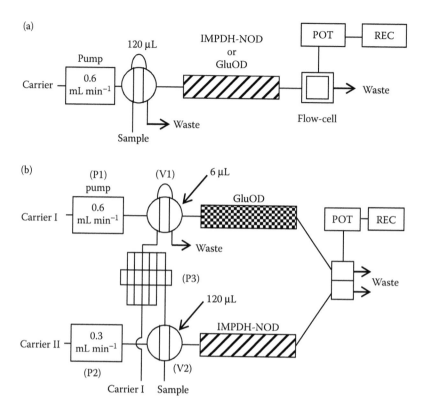

FIGURE 28.1 Schematic representation of the FIA system for quantitation of a single component (IMP or MSG [a]) and for the simultaneous quantitation of IMP and MSG (b). IMP, inosine monophosphate; MSG, monosodium glutamate; POT, multichannel potentiostat; REC, recorder; GluOD, immobilized glutamate oxidase reactor; IMPDH/NOD, immobilized inosine monophosphate dehydrogenase and NADH oxidase reactor; Carrier I, 50 mM phosphate buffer (pH 8.9); Carrier II, 0.3 M K_2HPO_4–KH_2PO_4 buffer (pH 8.0, containing 50 μM flavin adenine dinucleotide and 0.5% bovine serum albumin); V1, valve 1; V2, valve 2; P1, microtube pump 1; P2, microtube pump 2; P3, multichannel microtube pump 3. (Reprinted from *Anal. Chim. Acta*, 358, Matsumoto, K., W. Asada, and R. Murai, Simultaneous biosensing of inosine monophosphate and glutamate by use of immobilized enzyme reactors, 127–136, Copyright 1998, with permission from Elsevier.)

of pump 3 (P3) is allotted to draw up the sample and the other five channels of pump 3 (P3) to carrier I to dilute the sample. As a result, a sample for the quantitation of MSG was diluted six times. The sample loop for IMP quantitation was 120 μL and that for MSG was 6 μL. The flow-through cell was composed of two Pt electrodes (working and counter) and a reference electrode (Ag/AgCl). The design of the flow-through cell is described in Ref. [8]. The potential of the working electrode (a platinum plate) was set at +0.65 V versus Ag/AgCl [8]. To prevent interference with the sample, electropolymerization of 1,2-diaminobenzene on a platinum electrode was performed using this detection system. Enzymes were immobilized on controlled-pore glass beads and packed in minireactors.

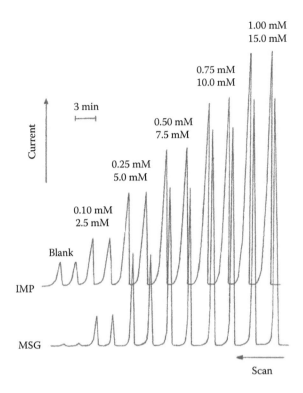

FIGURE 28.2   Typical FIA responses for IMP (upper concentrations) and MSG by the system shown in Figure 28.1. IMP, inosine monophosphate; MSG, monosodium gluta-mate. (Reprinted from *Anal. Chim. Acta*, 358, Matsumoto, K., W. Asada, and R. Murai, Simultaneous biosensing of inosine monophosphate and glutamate by use of immobilized enzyme reactors, 127–136, Copyright 1998, with permission from Elsevier.)

28.3.3 Simultaneous Quantitation of IMP and MSG

Figure 28.2 shows typical FIA responses for a standard mixture of IMP and MSG under the optimized conditions shown in Figure 28.1b. The ratio of the concentration of IMP to MSG was set at a moderate level typical of ordinary seasoning samples. The current from each channel increased rapidly just after injection of the sample and returned to baseline within about 3 min. The blank signal was subtracted from each sample signal. Linear relationships between sensor responses and the concentration of each compound were observed between 0.1 and 1.0 mM for IMP and between 1.0 and 10.0 mM for MSG with a correlation coefficient > 0.997 ($n = 5$). The IMPDH/NOD reactor was able to respond to 200 assays under the described conditions.

28.4   FLOW SYSTEM TO DETERMINE FISH FRESHNESS

Apart from the importance of IMP as a flavor enhancer, IMP has been quantified to determine fish freshness by many authors [9]. Freshness has been determined on the basis of indicators, such as adenosine triphosphate (ATP)-related compounds, that normally do not exist in the tissues of living fish. ATP-related compounds are degraded to UA after death by the following process:

$$ATP \rightarrow ADP \rightarrow AMP \rightarrow IMP \rightarrow HxR \rightarrow Hx \rightarrow UA$$

where ADP is adenosine diphosphate, AMP is adenosine monophosphate, HxR is inosine, and Hx is hypoxanthine.

In this degradation process, ATP, ADP, and AMP are rapidly degraded, resulting in accumulations of inosine and hypoxanthine. The percentages of inosine and hypoxanthine in the total nucleic acid–related compounds are highly correlated to the stage of freshness in fish [10]. Thus, to indicate fish freshness, a K value is defined as Equation 28.4.

$$K = \frac{HxR + Hx}{ATP + ADP + AMP + IMP + HxR + HX} \times 100 \tag{28.4}$$

However, ATP, ADP, and AMP generally disappear around 24 h after death, and fish are usually obtained from the market at least 24 h after death. Therefore, another freshness indicator, K_i, is defined as Equation 28.5.

$$K_i = \frac{HxR + Hx}{IMP + HxR + Hx} \times 100 \tag{28.5}$$

Okuma et al. [11] produced a simple and long-lived enzyme sensor for the determination of fish freshness. The measurement system consisted of two immobilized enzyme reactors with two oxygen electrodes positioned close to the reactors.

IMP, HxR, and Hx determinations are based on the following enzyme reactions:

$$IMP \xrightarrow{\text{NT}} HxR \xrightarrow{\text{NP}} Hx \xrightarrow{\text{XOD}} UA$$

where NT is 5′-nucleotidase, NP is nucleotide phosphorylase, and XOD is xanthine oxidase.

The concentrations of IMP, HxR, and Hx were determined by the current decrease corresponding to oxygen consumption in the final step. Therefore, the total concentrations of HxR and Hx were determined by the catalytic action of NP and XOD simultaneously immobilized on chitosan porous beads. Similarly, total concentrations of IMP, HxR, and Hx were determined by the catalytic action of NT, NP, and XOD simultaneously immobilized on chitosan beads. One assay can be completed in 5 min. The enzyme reactors showed good operational stability during 1 month for at least 200 assays [11].

28.5 SIMULTANEOUS DETERMINATION OF FLAVOR ENHANCERS BY USING SPECTROPHOTOMETRIC DATA AND CHEMOMETRICS

A suitable chemometric analysis of spectrophotometric data was used for simultaneous determinations of MSG, GMP, and IMP. The overlapping of absorption spectra from each flavor enhancer was resolved by multicomponent analysis using chemometric techniques.

Acebal et al. [12] reported a spectrophotometric method for simultaneous determination of MSG, GMP, and IMP in stock cubes using a diode-array detector and a partial least-square (PLS) analysis. Subsequently, the same authors reported simultaneous determination of three flavor enhancers by a multiple linear regression (MLR) analysis [13]. Multiple linear regression yields models, which are simpler and easier to interpret than principal component regression (PCR) and PLS, because these calibration techniques perform regression on latent variables without a physical meaning [14]. On the other hand, MLR is more dependent on spectral selection variables. To overcome this problem, Araújo et al. [14] proposed a novel variable-selection strategy for MLR calibration: the "successive projections algorithm" (SPA). Acebal et al. [13] showed the predictive ability of MLR-SPA models to simultaneously determine concentrations of three flavor enhancers in a flow system. The method is outlined in the following sections.

28.5.1 Procedure

The continuous-flow manifold is depicted in Figure 28.3. A selection valve (SV) was used to introduce standard mixtures or sample solutions into the system. These solutions merged with the buffer solution stream in the reaction coil (RC; PTFE tubing; 600 mm, 0.5 mm id) and reached the flow cell (18 µL; 1 cm optical path). At that moment, the flow was stopped and the absorption spectrum between 190 and 320 nm was recorded by a diode-array spectrophotometer (D). The sample solution was filtered online by a column (length 4.0 cm; 0.7 cm id) (AC) packed with acetate (material for cigarette filters) placed before the SV. Reagents and flow conditions are shown in the legend of Figure 28.3. Nine standard solutions were prepared for a calibration set. The concentration range for each analyte (IMP: 5.028 – 34.190; GMP: 5.028 – 34.190; MSG: 447.6 – 1398.6 µg mL^{-1}) was selected considering the component ratio present in this kind of food. Nine synthetic mixtures were also prepared for external validation [13].

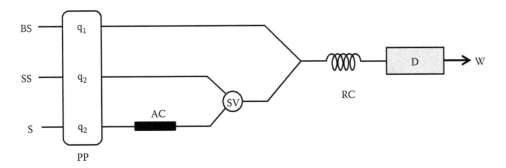

FIGURE 28.3   Flow system for the simultaneous determination of MSG, GMP, and IMP. MSG, monosodium glutamate; GMP, guanosine monophosphate, IMP, inosine monophosphate; AC, acetate column; BS, buffer solution; D, detector; PP, peristaltic pump; q_1, buffer solution flow rate (2.06 mL min^{-1}); q_2, standard and sample solution flow rate (1.08 mL min^{-1}); RC, reactor coil; S, sample; SS, standard solution; SV, selection valve; W, waste. (Reprinted from *Talanta*, 82, Acebal, C. C. et al., Successive projections algorithm applied to spectral data for the simultaneous determination of flavor enhancers, 222–226, Copyright 2010, with permission from Elsevier.)

28.5.2 Multiple Linear Regression-Successive Projections Algorithm, Application to Absorption Data

Prior to the application of MLR-SPA, the data were mean centered to remove constant background effects [13]. MLR-SPA calculations were performed using a routine developed by Araújo et al. [14] in MATLAB® 7.0 (MathWorks, South Natic, MA, USA) software. The MLR-SPA method is complicated, and a detailed explanation of the projection operations is given in the literature [14,15].

Figure 28.4a and b shows the pure spectra of MSG, GMP, and IMP, and the spectrum of a mixture of MSG, GMP, and IMP, respectively. When the MLR-SPA method was applied to the calibration set of nine standard solutions and the external validation set of nine synthetic mixtures, only nine, six, and four variables were selected by SPA for MSG, GMP, and IMP, respectively, as shown in Table 28.1. In the spectra shown in

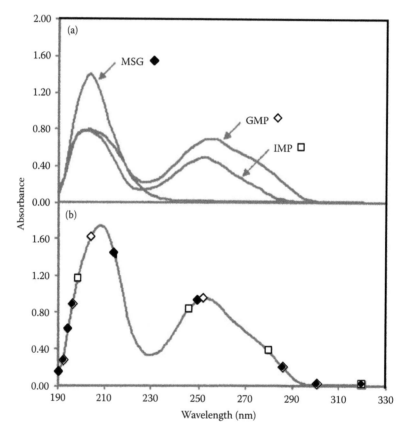

FIGURE 28.4 (a) Pure spectra of MSG (923.1 mg mL⁻¹), GMP (19.6 mg mL⁻¹), and IMP (19.6 mg mL⁻¹). (b) Spectrum of a mixture containing MSG (923.1 mg mL⁻¹), GMP (19.6 mg mL⁻¹), and IMP (19.6 mg mL⁻¹). MSG, monosodium glutamate; GMP, guanosine monophosphate; IMP, inosine monophosphate. In this spectrum, the selected variables by multiple linear regression-successive projections algorithm (MLR-SPA) are indicated. (Reprinted from *Talanta*, 82, Acebal, C. C. et al., Successive projections algorithm applied to spectral data for the simultaneous determination of flavor enhancers, 222–226, Copyright 2010, with permission from Elsevier.)

TABLE 28.1   Statistical Parameters Obtained for Calibration and Validation, and the Figures of Merit in Determination of Flavor Enhancer MSG, GMP, and IMP by MLR-SPA

Parameter	Flavor Enhancer		
	MSG	GMP	IMP
Selected variables (nm)	188, 190, 192, 194, 212, 248, 284, 296, 318	190, 194, 202, 250, 284, 296	196, 244, 278, 318
Calibration			
Concentration range ($\mu g\ mL^{-1}$)	447.6–1398.6	5.028–34.190	5.028–34.190
RMSE ($\mu g\ mL^{-1}$)	3.589	0.196	0.466
REP (%)	0.389	0.997	2.377
Validation			
Concentration range ($\mu g\ mL^{-1}$)	503.5–1202.8	10.056–30.168	10.056–30.168
RMSE ($\mu g\ mL^{-1}$)	18.83	0.187	0.364
REP (%)	2.29	0.908	1.763
Figures of merit			
Limit of detection ($\mu g\ mL^{-1}$)	168	2.0	1.1
Sensibility ($mL\ \mu g^{-1}$)	1.79×10^{-5}	1.5×10^{-3}	2.6×10^{-3}

Source: Reprinted from *Food Chem.*, 106, Acebal, C. C., A. G. Lista, and B. S. Fernández Band, Simultaneous determination of flavor enhancers in stock cube samples by using spectrophotometric data and multivariate calibration, 811–815, Copyright 2008, with permission from Elsevier.

Note: RMSE, root-mean-square error; REP, relative root-mean-square error.

Figure 28.4b, the variables selected by MLR-SPA are indicated. Table 28.1 also contains the root-mean square error (RMSE) and relative root-mean-square error (REP) for the calibration models and validation set, respectively.

RMSE was calculated as:

$$RMSE = \left(\frac{\sum_{i=1}^{l}(c_{nom} - c_{pred})^2}{l} \right)^{1/2}$$

where c_{nom} and c_{pred} represent nominal and predicted concentrations, respectively, and l is the total number of validation samples.

REP was calculated as:

$$REP = \frac{100}{c_{mean}} \left(\frac{\sum_{i=1}^{l}(c_{nom} - c_{pred})^2}{l} \right)^{1/2}$$

where c_{nom} and c_{pred} have the same meaning as that in the RMSE equation, l is the total number of calibration samples, and c_{mean} is the mean concentration.

The numbers in Table 28.1 indicate that the proposed method accurately quantifies artificial samples, as suggested by the low RMSE and REP values.

The proposed method [13] was applied to simultaneous determination of MSG, GMP, and IMP in dehydrated meat broths, and a validation test with spiked samples was also performed with satisfactory results.

28.6 HIGH- (OR ULTRA-HIGH-) PERFORMANCE LIQUID CHROMATOGRAPHIC DETERMINATION OF 5′-MONONUCLEOTIDES

Although the FIA technique does not include a separation procedure, the simultaneous determination of 5′-mononucleotides essentially requires a separation step before detection of these compounds. Thus, high-performance LC (HPLC) techniques were employed for the quantitation of 5′-mononucleotides. The 5′-mononucleotides were quantified not to determine flavor enrichment but as components of nutritional or clinical importance, especially in infant formulas [16]. Three main modes of LC are applied for nucleotide analysis: ion-exchange chromatography, reversed-phase LC (RP-LC), and ion-pair RP-LC [16].

Hartwick and Brown [17] successfully separated nucleotide mono-, di-, and triphosphates of adenosine, guanosine, inosine, xanthosine, cytidine, uridine, and thymidine by using a strong anion-exchange column and an acidic phosphate buffer gradient.

Separation of nucleotides by RP-LC with conventional C_{18} columns is somewhat limited because of inherently poor interaction of the highly polar analytes with the nonpolar C_{18} phase under the required conditions of low organic modifier content, resulting in poor retention and resolution [16]. In contrast, Gill and Indyk [18] developed a method for simultaneous analysis of nucleotide monophosphates and corresponding nucleosides by using a polymer-grafted silica Gemini C_{18} column and gradient elution with a phosphate buffer-methanol mobile phase.

The currently used techniques focus on (1) an ion-pair RP-LC and a photodiode-array (PDA) detection system and (2) an ion-pair RP-LC or polymer-grafted column with a polar-embedded C_{18} phase and an MS detection system [16,18–23].

28.6.1 Ion-Pair Reversed-Phase Liquid Chromatography with Photodiode-Array Detection

Ion-pair RP-LC has become the prevalent technique for analysis of nucleotides in recent years. Ferreira et al. [19] achieved the separation using a C_{18} column and a gradient elution with a mixture of two solvents: A and B. Solvent A consisted of water/glacial acetic acid/5 mM tetrabutylammonium hydrogen sulfate (TBAHS) in a 97.5:1.5:1.0 (v/v/v) ratio, and solvent B consisted of methanol/glacial acetic acid/5 mM TBAHS in a 97.5:1.5:1.0 (v/v/v) ratio. Gradient elution was carried out as follows: 0–5 min with 100% A, 5–20 min linear gradient to 10% B, 20–23 min 10% B, 23–29 min linear gradient to 40% B, 29–30 min 40% B, 30–34 min linear gradient back to 100% A (initial conditions), and a 2-min re-equilibration wash with 100% A. The flow rate was 1 mL min⁻¹. The effluent was monitored using a diode-array detector set at 260 nm.

The ionic nature of the phosphate ester facilitates strong interaction with a cationic ion-pair reagent (in this case: TBAHS) at the appropriate pH, thereby enhancing nucleotide retention and resolution [16]. Figure 28.5 shows a typical chromatogram for separation of five nucleotides using ion-pair RP-LC.

Perrin et al. [20] reported simultaneous quantitation of 5′-mononucleotides and nucleosides in infant formulas using two 25-cm reversed-phase columns (Nucleosil

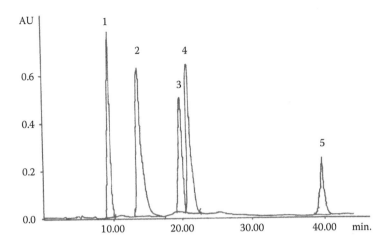

FIGURE 28.5   Typical chromatogram for separation of five nucleotides by ion-pair reversed-phase HPLC method. (1) cytidine 5′-monophosphate (CMP); (2) uridine 5′-monophosphate (UMP); (3) guanosine 5′-monophosphate (GMP); (4) inosine 5′-mono-phosphate (IMP); (5) adenosine 5′-monophosphate (AMP). The concentration of each nucleotide injected onto the column was 15 mg L^{-1}. (Reprinted from *Food Chem.*, 74, Ferreira, I. M. P. L. V. O. et al., The determination and distribution of nucleotides in dairy products using HPLC and diode-array detection, 239–244, Copyright 2001, with permission from Elsevier.)

120-C$_{18}$) coupled in series, followed by diode-array detection. The separation was per-formed by isocratic elution with a mobile phase consisting of 1 mM tetrabutylammonium hydrogen phosphate – 30 mM sodium phosphate buffer (pH 4.3) and 7% methanol. The detection wavelength was 257 nm for 5′-uridine monophosphate (5′-UMP), 5′-GMP, and 5′-AMP, adenosine, guanosine, and uridine; and 278 nm for 5′-cytidine monophos-phate (5′-CMP) and cytidine; and the spectrum acquisition range was between 200 and 300 nm. A column oven temperature of 40°C was necessary for complete separation. For analysis of hypoallergenic infant formulas, which contain hydrolyzed protein, some chromatographic interference substances were removed by using a strong anion-exchange solid-phase extraction column.

28.6.2 Ion-Pair Reversed-Phase Liquid Chromatography or Polymer-Grafted Column LC with Mass Spectrometry Detection

Simultaneous determination of 23 types of purine and pyrimidine nucleosides and nucle-otides in dietary foods and beverages was reported by using ion-pair LC-electrospray ionization-MS (ESI-MS) [21]. Dihexylammonium acetate (DHAA) was used as an ion-pairing agent, and an ultraperformance LC (UPLC^{TM}) system with a reversed-phase column and a gradient program was employed for separation of nucleosides and nucleo-tides. An ACQUITY UPLC^{TM} column HSS T3 (Waters, Milford, MA, USA) was used as the analytical column, and 1.25 mM DHAA in 10 mM HCOOH-HCOONH$_4$ (pH 5.0) and acetonitrile with a nonlinear three-step gradient program (concave) was used as the mobile phase for the separation of polar compounds. Positive-ion ESI-MS was applied for detection of nucleosides, and negative-ion ESI-MS was used for nucleotides. Single-ion monitoring (SIM) chromatograms of standard nucleotides are shown in Figure 28.6. The peaks corresponding to nucleosides and nucleotides are indicated with arrows. As shown in Figure 28.6, complete separation, including isotope compounds, was confirmed.

Conventional C_{18} columns are not suitable for separation of nucleotides, because the highly polar analytes have inherently poor interaction with the nonpolar C_{18} phase under highly aqueous conditions. The nucleotides with charged phosphate groups are relatively well retained on the hybrid and polymer-grafted columns and polar-embedded C_{18} phases [16]. Most previous studies used a phosphate buffer as the mobile phase and employed an ion-pair reagent to improve chromatographic resolution, although they are not compatible with MS [22]. Ren et al. [23] reported simultaneous determination of 5′-monophosphate nucleotides in infant formulas by HPLC-MS equipped with an ESI source. The complete chromatographic separation of five nucleotides was achieved through a Symmetry C_{18} column (150 mm length, 2.1 mm id, 3.5 μm particle size, Waters), after a binary gradient elution with water containing 0.1% formic acid and acetonitrile as the mobile phase. Multireaction monitoring was applied for tandem mass spectrometry analysis (MS/MS). This method avoids the disadvantages of employing an ion-pair reagent and phosphate buffer.

28.7 CAPILLARY ELECTROPHORETIC METHOD FOR NUCLEOTIDE ANALYSIS

Nucleotides are easily separated by capillary electrophoresis (CE) due to their negative charge. CE methods are broadly classified into two main categories: capillary zone electrophoresis (CZE) and micellar electrokinetic capillary chromatography [24]. Capillary electrophoresis analyses are generally faster than comparable HPLC analyses, the solvents used are inexpensive buffer salts, and smaller quantities of both buffer and sample are required. The separation of nucleotides by CZE is introduced in the following sections.

28.7.1 Capillary Zone Electrophoresis with UV Detection

Buffer pH plays a significant role in CE, because charge differences between analytes drive electrophoresis-based separations. There are two optimal pH regions with mostly different nucleotide mobilities that are moderately acidic at ~4 and alkaline at ~9. The use of an acidic region (pH < 5) offers excellent selectivity, because only a few compounds are negatively charged. However, nucleotides (mainly monophosphates) have only limited mobilities, which significantly prolong the separation. Friedecký et al. [25] used a buffer of pH 4.4 containing a cationic surfactant, cetyltrimethylammonium bromide (CTAB). The addition of CTAB reversed the direction of the electromotive force and significantly reduced analysis time. The concentration of CTAB, however, affected resolution of fast-migrating analytes, mainly nucleotide triphosphates. Using citrate-γ-aminobutyric acid (GABA) as a background electrolyte (BGE) results in a double-buffering system with the best UV transparency (UV 250 nm), yielding good separation properties. The final BGE consisted of 40 mM citric acid with 0.8 mM CTAB titrated by GABA to pH 4.4. The capillary used was an uncoated silica capillary (id/od 75/375 μm; effective/total length 90/97 cm). A complete resolution of 21 nucleotides and deoxynucleotides was obtained within 15 min.

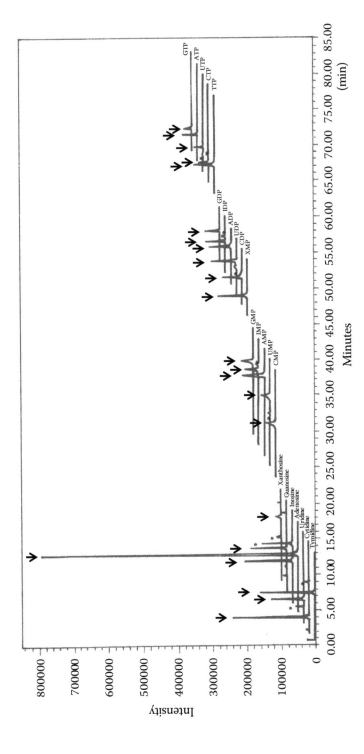

FIGURE 28.6 Single-ion monitoring (SIM) chromatograms of standard nucleosides and nucleotides. Twenty-five nanograms of each nucleoside and nucleotide were applied. The peaks corresponding to nucleoside and nucleotide are indicated with arrows. The isotope peaks are shown with asterisks. (Reprinted from *J. Chromatogr. B*, 878, Yamaoka, N. et al., Simultaneous determination of nucleosides and nucleotides in dietary foods and beverages using ion-pairing liquid chromatography-electrospray ionization-mass spectrometry, 2054–2060, Copyright 2010, with permission from Elsevier.)

28.7.2 Capillary Zone Electrophoresis with Electron-Spray Ionization–Mass Spectrometry Detection

Capillary electrophoresis-ESI-MS (CE-ESI-MS) was proposed for identification and simultaneous quantification of several 5′-monophosphate ribonucleotides in infant formula samples [26]. The BGE was a 30 mM ammonium formate-ammonia medium (pH 9.6). Electrophoretic separation was achieved with a voltage of 30 kV (positive CE mode), with an initial ramp of 7 s. The temperature of the capillary was kept constant at 25°C. The mass spectrometer was operated in the negative-ion mode. The optimized fragmenter voltage was 100 V for all analytes. For quantification, SIM mode was selected. The m/z values corresponding to the [M-H]$^-$ ions were 346 for AMP, 322 for CMP, 362 for GMP, 323 for UMP, and 347 for IMP. The results indicated that CE coupled to MS offered an alternative to other chromatographic methods for analysis of nucleotides and related compounds in infant formula samples.

REFERENCES

1. Yamaguchi, S. and K. Ninomiya. 2000. The use and utility of glutamates as flavoring agents in food. Umami and food palatability. *J. Nutr.* 130:921S–926S.
2. Yamaguchi, S. 1991. Basic properties of umami and effects on humans. *Phys. Behav.* 49:833–841.
3. Devlin, T. M. (ed.). 1997. Purine and pyrimidine nucleotide metabolism. In *Textbook of Biochemistry with Clinical Correlations* (4th ed.), Chapter 12, 489pp. New York: Wiley-Liss.
4. López Navarro, A. T., J. D. Bueno, A. Gil, and A. Sanchez Pozo. 1996. Morphological changes in hepatocytes of rats deprived of dietary nucleotides. *Brit. J. Nutr.* 76:579–589.
5. Karube, I., H. Matsuoka, S. Suzuki, E. Watanabe, and J. Toyama. 1984. Determination of fish freshness with an enzyme sensor system. *J. Agric. Food Chem.* 32:314–319.
6. Yao, T., Y. Matsumoto, and T. Wasa. 1988. Bio-FIA system for estimation of fish freshness. *Bunseki Kagaku* 37:236–241.
7. Matsumoto, K., W. Asada, and R. Murai. 1998. Simultaneous biosensing of inosine monophosphate and glutamate by use of immobilized enzyme reactors. *Anal. Chim. Acta* 358:127–136.
8. Matsumoto, K., H. Kamikado, H. Matsubara, and Y. Osajima. 1988. Simultaneous determination of glucose, fructose, and sucrose in mixtures by amperometric flow injection analysis with immobilized enzyme reactors. *Anal. Chem.* 60:147–151.
9. Okuma, H. and E. Watanabe. 2002. Flow system for fish freshness determination based on double mini-enzyme reactor electrodes. *Biosens. Bioelectro.* 17:367–372.
10. Saito, T., K. Arai, and M. Matsuyoshi. 1959. A new method for estimating the freshness of fish. *Bull. Jpn. Soc. Sci. Fish.* 24:749–750.
11. Okuma, H., H. Takahashi, S. Yazawa, and S. Sekimukai. 1992. Development of a system with double enzyme reactors for the determination of fish freshness. *Anal. Chim. Acta* 260:93–98.
12. Acebal, C. C., A. G. Lista, and B. S. Fernández Band. 2008. Simultaneous determination of flavor enhancers in stock cube samples by using spectrophotometric data and multivariate calibration. *Food Chem.* 106:811–815.

13. Acebal, C. C., M. Grünhut, A. G. Lista, and B. S. Fernández Band. 2010. Successive projections algorithm applied to spectral data for the simultaneous determination of flavor enhancers. *Talanta* 82:222–226.

14. Araújo, M. C. U., T. C. B. Saldanha, R. K. H. Galvão, T. Yoneyama, H. C. Chame, and V. Visani. 2001. The successive projections algorithm for variable selection in spectroscopic multicomponent analysis. *Chemom. Intel. Lab. Syst.* 57:65–73.

15. Galvão, R. K. H., M. F. Pimentel, M. C. U. Araújo, T. Yoneyama, and V. Visani. 2001. Aspects of the successive projections algorithm for variable selection in multivariate calibration applied to plasma emission spectrometry. *Anal. Chim. Acta* 443:107–115.

16. Gill, B. D. and H. E. Indyk. 2007. Determination of nucleotides and nucleosides in milks and pediatric formulas: A review. *J. AOAC Int.* 90:1354–1364.

17. Hartwick, R. A. and P. R. Brown. 1975. The performance of microparticle chemically-bonded anion-exchange resins in the analysis of nucleotides. *J. Chromatogr.* 112:651–662.

18. Gill, B. D. and H. E. Indyk. 2007. Development and application of a liquid chromatographic method for analysis of nucleotides and nucleosides in milk and infant formulas. *Int. Dia. J.* 17:596–605.

19. Ferreira, I. M. P. L. V. O., E. Mendes, A. M. P. Gomes, M. A. Faria, and M. A. Ferreira. 2001. The determination and distribution of nucleotides in dairy products using HPLC and diode array detection. *Food Chem.* 74:239–244.

20. Perrin, C., L. Meyer, C. Nujahid, and C. J. Blake, 2001. The analysis of 5′-mononucleotides in infant formulae by HPLC. *Food Chem.* 74:245–253.

21. Yamaoka, N., Y. Kudo, K. Inazawa et al. 2010. Simultaneous determination of nucleosides and nucleotides in dietary foods and beverages using ion-pairing liquid chromatography-electrospray ionization-mass spectrometry. *J. Chromatogr. B* 878:2054–2060.

22. Seifar, R. M., C. Ras, J. C. van Dam, W. M. van Gulik, J. J. Heijnen, and W. A. van Winden. 2009. Simultaneous quantification of free nucleotides in complex biological samples using ion pair reversed phase liquid chromatography isotope dilution tandem mass spectrometry. *Anal. Biochem.* 388:213–219.

23. Ren, Y., J. Zhang, X. Song, X. Chen, and D. Li. 2011. Simultaneous determination of 5′-monophosphate nucleotides in infant formulas by HPLC-MS. *J. Chromatogr. Sci.* 49:332–337.

24. Geldart, S. E. and P. R. Brown. 1998. Analysis of nucleotides by capillary electrophoresis. *J. Chromatogr. A* 828:317–336.

25. Friedecký, D., J. Tomková, V. Maíer, A. Janošťáková, M. Procházka, and T. Adam. 2007. Capillary electrophoretic method for nucleotide analysis in cells: Application on inherited metabolic disorders. *Electrophoresis* 28:373–380.

26. Gonzalo, E. R., J. Domínguez-Álvarez, M. Mateos-Vivas, D. García-Gómez, and R. Carabias-Martínez. 2014. A validated method for the determination of nucleotides in infant formulas by capillary electrophoresis coupled to mass spectrometry. *Electrophoresis* 35:1677–1684.

Antioxidant Capacity

CHAPTER 29

2,2-Diphenyl-1-Picrylhydrazyl Methods

*Raquel Maria Ferreira Sousa, Gracy Kelly Faria Oliveira,
Alberto de Oliveira, Sérgio Antônio Lemos de Moraes,
and Rodrigo Alejandro Abarza Munoz*

CONTENTS

29.1 INTRODUCTION

Many studies have indicated that the consumption of foods from natural product sources (cereal, vegetables, fruits) can reduce the risk of chronic diseases in humans caused by oxidative stress, such as coronary heart disease (Bhupathiraju and Tucker, 2011; Dalen and Devries, 2014), diabetes (Salas-Salvadó et al., 2011; Singh et al., 2013; Esposito and Giugliano, 2014), cancer (Steinmetz and Potter, 1996; Research and Fund, 2007; Boivin et al., 2009), and neurodegenerative diseases (Alcalay et al., 2012; Otaegui-Arrazola et al., 2014). Natural products contain substances produced by a plant's secondary metabolism and have antioxidant capabilities and the ability to overcome oxidative stress by neutralizing the overproduction of oxidant species (Dai and Mumper, 2010; Gülçin, 2012). Antioxidant activity can occur via several mechanisms, including free-radical scavenging; chelating pro-oxidative metals; quenching singlet oxygen and photosensitizers; and inactivating lipoxygenase, the enzyme responsible for lipid peroxidation (Choe and Min, 2009). The deterioration of foods is caused by lipid peroxidation, and some antioxidants, synthetic or naturally occurring, can be used as additives for food preservation (Niki, 1987; Niki et al., 2005). Phenolic compounds are the most common antioxidants that readily scavenge free radicals by donating hydrogen atoms. The antioxidant capability of phenol compounds involves donation of a phenolic hydrogen and the stabilization of the resulting antioxidant radical by electron delocalization by resonance throughout the phenolic ring structure and/or intramolecular hydrogen bonding or by further oxidation (Frankel and Meyer, 2000; Choe and Min, 2009).

Synthetic phenolic antioxidants used in food products to prevent or delay lipid oxidation during processing and storage of fats, oils, and lipid-containing foods are propyl gallate (PG), tertiary butyl hydroquinone (TBHQ), butylated hydroxyanisole (BHA), and butylated hydroxytoluene (BHT) (Saad et al., 2007).

Phenolic compounds present in natural products include simple phenols, phenolic acids, tocopherols, stilbenes, coumarins, flavonoids (flavonol, flavones, flavanols, anthocyanidins, proanthocyanidins), tannins, lignans, and lignins (Naczk and Shahidi, 2006; Folmer et al., 2014). Phenolic compounds are secondary plant metabolites that are synthesized in response to stress conditions, such as UV irradiation, nutrient deficiencies, infection, and wounding (Naczk and Shahidi, 2006; Ramakrishna and Ravishankar, 2011).

Evaluation of the antioxidant ability may use the terms "antioxidant activity" and "antioxidant capacity," which are often used interchangeably but have different meanings. The "activity" is a measure between the individual antioxidant and the free radical, and it refers to the rate constant of a reaction between a specific antioxidant and an oxidant. The "capacity" is a measure between an antioxidant solution and free radicals, and the antioxidant amount (as a mole, a concentration) may contain a mixture of antioxidant compounds, so in this case, the antioxidant capacity of each individual component is not measured (MacDonald-Wicks et al., 2006).

The determination of antioxidant capacity is explained and summarized in many reviews (Antolovich et al., 2002; Huang et al., 2005; MacDonald-Wicks et al., 2006; Magalhães et al., 2008; Liu, 2010; Gülçin, 2012; Alam et al., 2013). Several antioxidant capacity methods are based on free-radical scavenging ability of compounds such as 2,2-diphenyl-1-picrylhydrazyl (DPPH), 2,2'-azinobis-(3-ethylbenzothiazoline-6-sulfonate) (ABTS$^{\bullet+}$), and peroxyl superoxide (O2$^{\bullet-}$). Some works have highlighted (Tirzitis and Bartosz, 2010; Shalaby and Shanab, 2013) the difference between "antiradical" and "antioxidant" activity or capacity. According to Tirzitis and Bartosz (2010), antiradical activity characterizes the ability of compounds to react with free radicals (in a single free-radical reaction), but antioxidant activity represents the ability to inhibit the process of oxidation (which usually, at least in the case of lipids, involves a set of different reactions). However, according to Liu (2010), the ability of an antioxidant to scavenge radicals is a characteristic of the antioxidant capacity, because the interactions between antioxidants and radicals (stable or generated in situ) give direct evidence of the ability of antioxidants to trap radicals.

The DPPH scavenging radical method is one of the most frequently employed methods for analyzing antioxidant capacity in foods. The DPPH radical is one of the few stable organic nitrogen radicals, and it is commercially available. This method is based on a measure of the consumption of the DPPH radical by an antioxidant compound (generally phenolic compounds). The most commonly used method for evaluating the ability to consume the DPPH radical is the spectrophotometric measurement of the decrease in the absorbance of the DPPH radical after the reaction. However, other detection techniques have been applied to assess the consumption of the DPPH radical, such as electron spin resonance (EPR) and electrochemical techniques. In this review, many aspects of these methodologies are discussed, including reaction mechanism, stability of the DPPH radical, expression of the results, reaction conditions, detection techniques, and analyses in flow.

29.1.1 Characteristics and Stability of DPPH

DPPH is characterized as a stable free radical by a conjugative electron delocalization of the spare electron over the molecule as a whole (Valgimigli et al., 1995; Tsimogiannis and Oreopoulou, 2004; Kedare and Singh, 2011), so that the molecules do not dimerize (Kedare and Singh, 2011), like most other free radicals. This conjugative electron delocalization results in a special spectral absorbance, resulting in a violet color. The resonance structures of DPPH (Figure 29.1a) show the stabilization of the DPPH radical (Valgimigli et al., 1995; Tsimogiannis and Oreopoulou, 2004; Kedare and Singh, 2011).

DPPH is only soluble in organic solvents and is very stable in solution. The absorbance of DPPH radical in methanol for 90 min at 25°C decreases by only 0.01 units, but if the same experiment is performed with a DPPH solution just taken out of the refrigerator, after 90 min at 25°C, the absorbance decreases by 0.03 units. Therefore, it is better to keep the DPPH solution for a short period of time at room temperature than in a refrigerator (Deng et al., 2011). The solution can be stored under a nitrogen atmosphere and protected from light to obtain slow deterioration. Under these conditions, the loss of free-radical activity did not exceed 2%–4% per week. According to Blois (1958), DPPH solution is moderately stable with respect to changes in pH, except for highly alkaline solutions. However, for improved results, it is convenient to maintain pH from 5.0 to 6.5 with acetate buffers.

Although it is known that DPPH is more soluble in organic solvents, sometimes the modification of the solvent is necessary to solubilize the sample. This can often result in precipitation or degradation of DPPH, and then a study of the stability of DPPH in the solvent used for sample dissolution is necessary. Food samples can be extracted using different solvents. The type of solvent influences the properties of the antioxidants; thus, polar solvents (such as methanol and ethanol) extract hydrophilic antioxidants, whereas less polar solvents (such as chloroform or ethyl acetate) extract lipophilic antioxidants.

With respect to the solvent used to dissolve the DPPH, Tsimogiannis and Oreopoulou (2004) noted that reactivity of DPPH is much more enhanced in the presence of methanol than in ethyl acetate. This fact was stated based on the evaluation of the antioxidant capacity of the flavonoid, since the rate of the reaction and the quantity of the DPPH consumed were decreased when ethyl acetate was used instead of methanol.

The solvent's ability to donate or accept protons in a hydrogen bond influences the kinetics of the reaction with DPPH. As a rule, increasing the hydrogen bond acceptability of the solvent leads to a decrease in the reaction rate constant, a phenomenon that is independent of the abstracting radical (Valgimigli et al., 1995; Tsimogiannis and Oreopoulou, 2004). When DPPH is in methanol solution, the hydrogen bond involving methanol (Figure 29.1b) and DPPH increases the electron localization in N1, increasing its reactivity (Valgimigli et al., 1995; Tsimogiannis and Oreopoulou, 2004).

As previously mentioned, DPPH is soluble in organic solvents; however, depending on the type of compound to be analyzed (extracts, cosmetics, foods), the solvent polarity may need to be raised by adding water for analyte dissolution, but this can cause precipitation of the DPPH radical. Therefore, Staško et al. (2007) studied limits for water as a component of the mixed water–ethanol solvent in the analysis of antioxidant capacity by the DPPH method. They concluded that 50% water is a suitable choice for lipophilic and hydrophilic antioxidants, and the reaction rate increases considerably with increasing water ratios. However, in solutions that are more than 60% water, the antioxidant

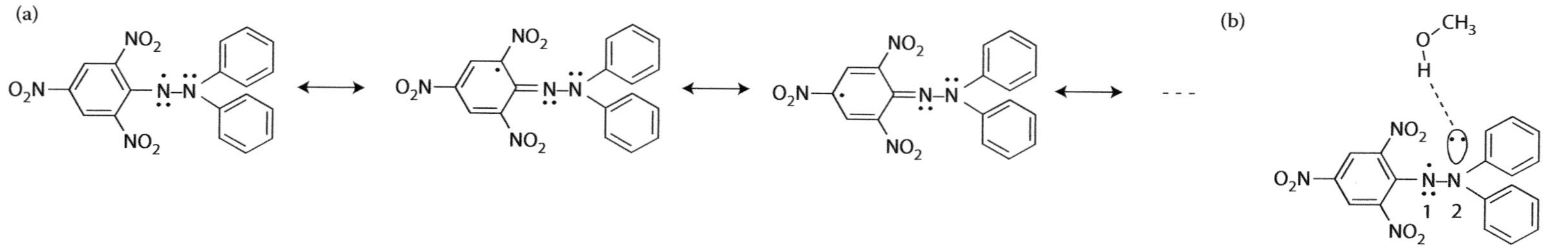

FIGURE 29.1 (a) Resonance structures of DPPH. (b) Hydrogen bond between methanol and DPPH. (Adapted from Valgimigli, L. et al. 1995. *Journal of the American Chemical Society* 117(40):9966–9971.)

capacity decreases, because part of the DPPH coagulates and is less accessible in the reaction with antioxidants.

The development of the DPPH method in aqueous solution using surfactant aggregates or micelles has also been studied (Noipa et al., 2011). When surfactants are added to DPPH radical aqueous solution at concentrations higher than the critical micelle concentration, they spontaneously form micelles with the hydrophobic tails pointed toward the center and the hydrophilic head groups pointed toward the surface of the micelle. Laguerre et al. (2014) utilized sodium dodecyl sulfate (SDS) as a surfactant. In this study, the concentration of SDS was varied (10, 20, and 60 mmol L^{-1}). After 25 min, a decrease of just ~5% of the DPPH radical initial concentration was observed at all SDS concentrations.

In buffered solution, the DPPH radical can show variations in its stability. Al-Dabbas et al. (2007) found that in methanol solution containing acetate buffer (pH 5.0), the absorbance of the DPPH radical was not reduced in a wide range of concentrations examined (0.01–0.2 mmol L^{-1}), while in phosphate buffer (pH 7.0), a reduction of the DPPH radical absorbance was observed at concentrations above 0.05 mmol L^{-1}. In other studies, Ozcelik et al. (2003) evaluated the variation in stability of the DPPH radical after 120 min. The absorbance of DPPH radical in potassium biphthalate buffer (pH 4.0) decreased by ~25% in methanol solution and by ~45% in acetone solution. DPPH radical in sodium bicarbonate buffer (pH 7) was stable in an acetone system (less than 10% reduction), but an ~30% decrease occurred in the absorbance in a methanol system. DPPH radical in potassium carbonate–potassium borate–potassium hydroxide buffer (pH 10) was stable in a methanol system (less than 10% reduction), but a decrease of ~70% occurred in the absorbance in an acetone system. Thus, the stability of DPPH in pH buffer solution mainly depends on the types of buffer and solvent used.

29.1.2 Reaction Mechanisms

The mechanisms that occur during the DPPH method are hydrogen atom transfer (HAT) and sequential proton loss electron transfer (SPLET). In HAT mechanisms, the ability of an antioxidant to quench free radicals by hydrogen donation to a DPPH radical is measured (Prior et al., 2005). The SPLET mechanism is a measurement of the ability of the phenolic anion (ArO$^-$) to transfer one electron to the DPPH radical (Foti et al., 2004; Litwinienko and Ingold, 2007; Liu, 2010). The mechanism is dependent on the solvent and pH. In ionizing solvents, such as methanol and ethanol, the rate-determining step for this reaction consists of a fast electron-transfer process from the phenoxide anions to the DPPH radical by SPLET mechanisms, because the hydrogen atom abstraction from neutral ArOH by DPPH (HAT mechanism) is very slow due to strong hydrogen bond–accepting solvents. In nonionizing solvents or in the presence of acids, the mechanisms will be by a slower HAT (Litwinienko and Ingold, 2004). Figure 29.2 shows the comparison between HAT and SPLET.

29.2 SPECTROPHOTOMETRIC METHODS

The spectrophotometric DPPH assay was the first method developed to determine antioxidant activity using the stable free-radical DPPH, which is based on the measurement of the scavenging of the DPPH radical by antioxidants. This method can be static (e.g., occurring in a cuvette of a spectrophotometer) or in flow. The unpaired electron of the

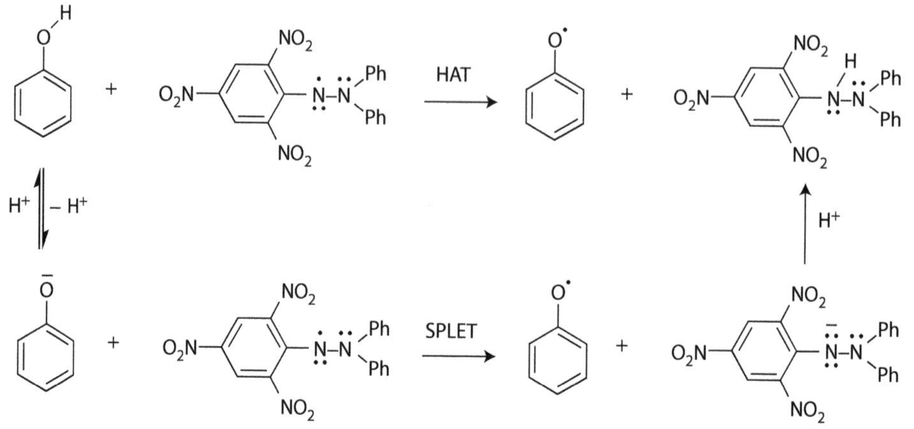

FIGURE 29.2   HAT and SPLET mechanisms between phenolic compounds and DPPH. (Adapted from Liu, Z. Q. 2010. *Chemical Reviews* 110(10):5675–5691.)

nitrogen atom in DPPH is reduced, either by receiving a hydrogen atom from antioxidants (HAT) or by an electron transfer of the phenoxide anions (SPLET), to form the corresponding hydrazine. Because of the high stability of the unpaired electron in resonance structures, the DPPH radical shows an absorbance maximum at between 515 and 528 nm, depending on the solvent (Brand-Williams et al., 1995; Molyneux, 2004); its solution has a violet color. Even in organic solvents, DPPH solubility is not high (~5×10^{-4} mol L^{-1} in alcoholic solutions; according to Blois [1958]), but due to high absorption of DPPH, even at low concentration, the Lambert–Beer law is obeyed in a useful range of concentrations.

The corresponding hydrazine (DPPH-H) that is formed after reaction with antioxidants (resulting in paired electrons) presents a yellow-colored solution, showing a minimum absorbance in the region where DPPH shows maximum absorbance (515–528 nm) (Figure 29.3). Thus, monitoring of the decrease in DPPH absorbance is measured in this spectral region because it is free of possible interference by hydrazine.

The spectrophotometric scavenging DPPH method was discovered 50 years ago by Marsden Blois (1958), when he worked at Stanford University and presented the method in a paper published in the journal *Nature* (Molyneux, 2004). In this study, he used the thiol-containing amino acid cysteine as an antioxidant and suggested maintaining a pH range between 5.0 and 6.5 with acetate buffer (Pyrzynska and Pekal, 2013). However, according to Pyrzynska (2013), the pH condition was abandoned later, due to a great uncertainty in the meaning of pH values in nonaqueous solutions, such as methanol or ethanol, which were mainly used as reaction media.

The results of the DPPH obtained by the spectrophotometric method can be expressed in many ways. The majority of studies express the results in terms of percentage inhibition ("reduction" or "quenching") of DPPH (I%), sometimes named antiradical activity (AA%). To express this result, the initial absorbance (A_0) of the DPPH solution free of antioxidant (considered 100% and sometimes named "control") and the absorbance after the reaction between the sample and DPPH (A_s) is measured, and calculated as I or AA (%) = $100 \times [(A_0 - A_s)/A_0]$. This equation can also be expressed as I or AA (%) = $100 \times [1 - (A_s/A_0)]$. Both ways of expressing the mathematical equation result in the same value. The percentage decrease in absorbance promoted by the sample concentration that was analyzed gives the result (Burda and Oleszek, 2001; Göktürk Baydar et al.,

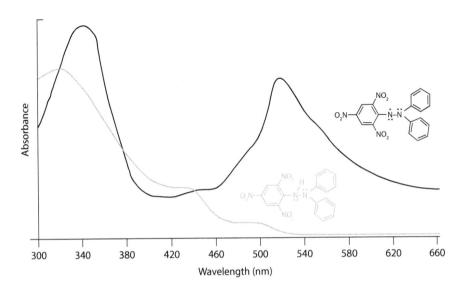

FIGURE 29.3   Absorption spectra of the DPPH radical and DPPH-H.

2007; Yamaguti-Sasaki et al., 2007; Paulino Zunini et al., 2010; Garcia et al., 2012; Alighourchi et al., 2013).

The other way to express the results of antioxidant capacity is in terms of IC_{50} ("inhibition concentration," also called EC_{50}, "efficient concentration" [Molyneux, 2004]), which is the antioxidant concentration necessary to inhibit 50% of the DPPH (Pyrzynska and Pekal, 2013). The lower the IC_{50} value, the higher the antioxidant capacity.

According to Molyneux (2004), this parameter was introduced by Brand-Williams (1995) (analogous with the "biological" parameter LD50, "lethal dose") and it was subsequently utilized for a number of studies. The IC_{50} value can be obtained by a graph of antioxidant concentration versus absorbance variation of DPPH (%) (Mensor et al., 2001; Moreira et al., 2005; Albayrak et al., 2008; Materska, 2012) or versus residual concentration of DPPH (%) (Brand-Williams et al., 1995; Sánchez-Moreno et al., 1998; Huang et al., 2005). The residual concentration of DPPH ($DPPH_r$) is calculated by $DPPH_r$ (%) $= 100 \times (DPPH_t/DPPH_0)$, where $DPPH_t$ and $DPPH_0$ are the DPPH concentrations after and before the reaction with an antioxidant, respectively (Molyneux, 2004; Huang et al., 2005; Pyrzynska and Pekal, 2013). The concentrations $DPPH_t$ and $DPPH_0$ are obtained from a calibration curve of the DPPH absorbance versus its concentration. Figure 29.4 shows how to calculate the IC_{50} in both cases.

The result of IC_{50} can be expressed in units of mass or molar concentration using the plot shown in Figure 29.4. Generally, for a sample in which the molar mass is not identified (e.g., extract of food, plant, or cosmetics), the antioxidant activity results are expressed by mass concentration. According to Reynertson et al. (2005), considering the IC_{50} in mass concentration, samples with values lower than 50 µg mL^{-1} are considered very active, 50–100 µg mL^{-1} are moderately active, 100–200 µg mL^{-1} are slightly active, and higher than 200 µg mL^{-1} are considered inactive. As well as IC_{50}, the results can also be expressed as IC_{100} and IC_{30}, which correspond to 100% and 30% inhibition of DPPH (Shimizu et al., 2010; Pattanaik et al., 2011).

The IC_{50} results can be expressed as the antioxidant to DPPH ratio. This ratio can be expressed in mass, for example, grams of antioxidant per kilogram of DPPH

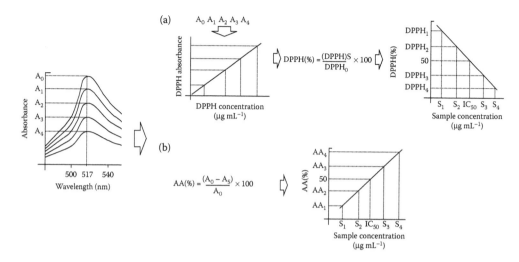

FIGURE 29.4 IC_{50} calculated by (a) residual concentration of DPPH and (b) absorbance of the reaction.

(Tsimogiannis and Oreopoulou, 2004), grams of antioxidant per gram of DPPH (Perez-Jimenez and Saura-Calixto, 2005; Noufou et al., 2012), and moles of antioxidant per mole of DPPH (when one molecule is analyzed) (Brand-Williams et al., 1995; Tsimogiannis and Oreopoulou, 2004; Lo Scalzo, 2008).

Another way to represent the results is in "equivalent antioxidant capacity" (EAC) of standard antioxidants, such as ascorbic acid, gallic acid, and Trolox. Usually, the acronym referring to the standard antioxidants is put before "EAC" (TEAC is acronym of Trolox, for example). A calibration curve of the standard antioxidant concentration versus $DPPH_r$ (%) or AA% is necessary to obtain the EAC. The result can be expressed as equivalents of the standard antioxidant (e.g., in grams) per gram of the sample. The results of the sample can be obtained in two ways:

- The $DPPH_r$ (%) or AA% for a certain concentration of the sample is calculated using the calibration curve of the standard antioxidant (mentioned above) and the concentration equivalents are obtained. Dicko et al. (2011) presented the results in milligrams of vitamin C equivalent antioxidant capacity (VCEAC) per 100 g of sample. Mungmai et al. (2014) expressed their results in milligrams of gallic acid equivalent antioxidant capacity (GEAC) per 100 g of sample. Wong et al. (2006) expressed the EAC in micromoles of Trolox equivalent (TEAC) per gram of sample.
- The IC_{50}(standard) and IC_{50}(sample) values are calculated and EAC is obtained following this equation: EAC = IC_{50}(standard)/IC_{50}(sample). Lim and Tee (2007) and Van De Velde et al. (2013) studied the antioxidant properties of several tropical fruits and two different strawberry species, respectively, and the results were also expressed in ascorbic acid equivalent antioxidant capacity (AEAC).

The other parameter to express antioxidant capacity is "antiradical efficiency" (AE), which is calculated by AE = $(1/IC_{50}) \times TIC_{50}$ (Sánchez-Moreno et al., 1998; Perez-Jimenez and Saura-Calixto, 2005). The TIC_{50} is the time needed to reach the steady state with IC_{50}. The larger the AE, the better the antioxidant capacity of the sample. In these results, the kinetics of the reaction is very important. According to Sánchez-Moreno and Saura-Calixto (1998), the kinetics of the antioxidant reaction with DPPH can be classified as follows: < 5 min (rapid), 5–30 min (intermediate), and > 30 min (slow) (Jiménez-Escrig et al., 2000), and this influences the AE obtained. De Souza and Giovani (2004) proposed that antioxidant activities can be classified by AE values of the compounds as follows: $AE \le 1 \times 104$ (low); $1 \times 104 < AE \le 5 \times 104$ (medium); $5 \times 104 < AE \le 10 \times 104$ (high); and $AE > 10 \times 104$ (very high).

Scherer and Godoy (2009) developed the antioxidant activity index (AAI) to evaluate antioxidant capacity. The AAI is calculated as follows: $AAI = DPPH_t/IC_{50}$. The $DPPH_t$ is the final concentration of DPPH in the control (without sample), and IC_{50} is in $\mu g\ mL^{-1}$. When $AAI < 0.5$, the antioxidant capacity is weak, $0.5 < AAI < 1.0$ is moderate, $1.0 < AAI < 2.0$ strong, and $AAI > 2.0$ very strong. Table 29.1 shows some results of antioxidant capacity in foods represented by several types of measures.

The DPPH spectrophotometric method is a very simple (only DPPH, solvent, and sample are necessary), rapid, and sensitive way to evaluate the antioxidant activity of several compounds. However, there are multiple ways of expressing the obtained results by this method, and consequently there are many conditions under which the method can be performed, as described in Table 29.2.

TABLE 29.1   Results of Antioxidant Activity in Foods by Several Types of Measures

Foods	Result Expressed	Value	Unit	Reference
Tropical fruits (guava, banana, dragon fruit, kedondong, langsat, mangosteen, papaya, star fruit, water apple, orange)	AEAC[a]	218, 27.8, 13.5, 37.6, 14.6, 32.3, 106, 98, 31, 69	mg AA per 100 g	Lim et al. (2007)
Strawberry (camarosa, selva)	AEAC	440.1 ± 8.1; 395.7 ± 7.2	mg AA per 100 g	Van De Velde (2013)
Pomegranate juice	AA[b]	74 ± 7	%	Alighourchi et al. (2013)
Propolis	AA[b]	94.96	%	Paulino Zunini et al. (2010)
Raw rice, boiled rice, wheat flour, French bread, oat bran, wheat bran	AE[c]	0.0016, 0.0027, 0.0052, 0.0014, 0.0012, 0.0009	–	Perez-Jimenez and Saura-Calixto (2005)
Apple, Araticum, Baru, Cagaita, Cajuzinho, Ingá, Jenipapo, Jurubeba, Lobeira, Mangaba, Tucum	AE[c]	0.4, 6.0, 0.2, 2.9, 0.9, 0.3, 0.5, 1.1, 16.3, 0.4, 4.3	–	Siqueira et al. (2013)
Wine: Tannat, Merlot, Shyraz, Syrah, Cabernnet Sauvignon, Peti Sirah, Ruby Cabernet	IC_{50}[d]	3.4, 5.9, 4.6, 6.3, 9.1, 8.4, 9.9	$\mu g\ mL^{-1}$	Lucena et al. (2010)
Papaya cultivars: Sunrise Solo, Red Lady, Tainung	IC_{50}[d]	52.1, 63.4, 71.8	$mL\ g^{-1}$ DPPH	Özkan et al. (2011)

[a] Ascorbic acid equivalent antioxidant capacity.
[b] Antiradical activity.
[c] Antiradical efficient.
[d] Inhibition concentration.

TABLE 29.2 Summary of Conditions Reported in Publications Using the DPPH Method

Condition	Reference
DPPH solution cuvette (μM)	
17	Göktürk Baydar et al. (2007); Paulino Zunini et al. (2010)
23–25	Mimica-Dukic et al. (2004); Kilic et al. (2014)
50–51	Karioti (2004); Meda et al. (2005)
67	Akowuah et al. (2005)
75–76	Umamaheswari et al. (2007); Huang et al. (2012); Van De Velde (2013)
94–98	Govindarajan et al. (2003); Zhao and Shah (2014)
100	Chen et al. (2005); Cui et al. (2005); Takebayashi et al. (2006); Elzaawely et al. (2007); Sharififar et al. (2007)
195	Xu et al. (2005)
250	Alma et al. (2003); Chung et al. (2005)
Reaction time (min)	
5	Govindarajan et al. (2003); Göktürk Baydar et al. (2007)
10	Cui et al. (2005)
15	Meda et al. (2005)
20	Chung et al. (2005)
30	Alma et al. (2003); Chen et al. (2005); Elzaawely et al. (2007); Sharififar et al. (2007); Umamaheswari et al. (2007); Huang et al. (2012); Kilic et al. (2014); Zhao and Shah (2014)
50	Paulino Zunini et al. (2010)
60	Karioti (2004); Mimica-Dukic et al. (2004); Akowuah et al. (2005)
90	Xu et al. (2005)
120	Takebayashi et al. (2006); Van De Velde (2013)
Solvent	
Methanol	Alma et al. (2003); Govindarajan et al. (2003); Mimica-Dukic et al. (2004); Akowuah et al. (2005); Meda et al. (2005); Sharififar et al. (2007); Göktürk Baydar et al. (2007); Van De Velde (2013); Zhao and Shah (2014)
Ethanol	Karioti (2004); Cui et al. (2005); Xu et al. (2005); Umamaheswari et al. (2007); Paulino Zunini et al. (2010); Huang et al. (2012); Kilic et al. (2014);
Methanol:acetate buffer (60:40%)	Chen et al. (2005); Elzaawely et al. (2007);
Ethanol:citrate buffer (60:40%)	Takebayashi et al. (2006)
Ethanol:tris–HCl buffer (60:40%)	Chung et al. (2005)

According to the results shown in Table 29.2, it can be observed that reaction time, solvent, and DPPH concentration vary according to the proposed method. Therefore, the comparison of antioxidant capacity obtained by the DPPH method for the same compounds is limited. Table 29.3 lists different IC_{50} (μM) values for some standard antioxidants, and the differences are probably due to the varied conditions used in each work.

TABLE 29.3 IC_{50} and Reaction Conditions Reported for Some Standard Antioxidants

	IC_{50} (μM)	Solvent	DPPH Solution (Cuvette, μM)	Reaction Time (min)	Reference
Gallic acid	5.7	Micelle system[a]	50	5	Noipa et al. (2011)
	4.2	Methanol	50	30	Noipa at al. (2011)
Ascorbic acid	13.9	Micelle system[a]	50	5	Noipa et al. (2011)
	11.6	Methanol	50	30	Noipa et al. (2011)
	11.8	Methanol	50	30	Sharma and Bhat (2009)
	11.5	Methanol:acetate buffer (60:40)	50	30	Sharma and Bhat (2009)
	55.9	Ethanol	100	30	Kano et al. (2005)
	629	Ethanol	85.7	10	Ricci et al. (2005)
Caffeic acid	10.9	Micelle system[a]	50	50	Noipa et al. (2011)
	9.7	Methanol	50	30	Noipa et al. (2011)
BHT	60	Methanol	50	30	Sharma and Bhat (2009)
	9.7	Methanol:acetate buffer (60:40)	50	30	Sharma and Bhat (2009)
	393	Ethanol	85.7	10	Ricci et al. (2005)
	24.4	Ethanol	23	60	Mimica-Dukic et al. (2004)

[a] CTAB: surfactant cetyltrimethylammonium bromide.

The data in Table 29.3 indicate the DPPH inhibition is influenced by the solvent. In the studies for BHT by Sharma and Bhat (2009), in which the reaction is characteristically slow, the IC_{50} obtained in 60:40 methanol:acetate buffer is smaller than in pure methanol; therefore, the buffer solution increased the percentage of DPPH consumption. On the other hand, for ascorbic acid, with which the reaction is instantaneous, the buffer solution did not influence the DPPH sequestration.

The solvent of the reaction not only influences the stability of DPPH (as already discussed) but also affects the reaction kinetics, since DPPH can be more or less available to receive an electron. In the work of Noipa et al. (2011), the use of a cationic surfactant (cetyltrimethylammonium bromide, CTAB) to solubilize the antioxidant standards (gallic acid, ascorbic acid, and caffeic acid) was studied, as shown in Table 29.3. The reaction between DPPH and gallic acid in CTAB surfactant showed that the rate constant was ~30 times higher than in methanol (k = 137,470 and 4520 M^{-1} s^{-1}, respectively), and the reaction times were 5 and 30 min in surfactant and methanol, respectively. According to the authors, this observation may be due to the positive charge of the surface of the CTAB micelle. The DPPH radical can interact with positively charged head groups of CTAB via electrostatic interactions. These attractive interactions allow the DPPH radical to remain inside the hydrophobic core of the CTAB micelle. Thus, the DPPH radical concentration inside the CTAB micelle is higher than in the bulk solution. Consequently, rates of scavenging of DPPH by the antioxidant were relatively high in the micelle system.

In all the above-mentioned studies, the analyses were performed statically in a spectrophotometer cuvette. However, over the years, with the discovery and refinement of analytical techniques, the development of automation systems and flow methods for the analysis of antioxidant capacity has been possible. Table 29.4 shows examples of the application of flow injection analysis (FIA), sequential injection analysis (SIA), multisyringe

TABLE 29.4 Flow Methods with Spectrophotometric Detection for Antioxidant Analysis in Food Samples

Flow Method	Food Sample	Results	Reference
Flow injection analysis (FIA)	Wines (Chardonay, Tannat, Merlot, Cabernet Sauvignon)	TEAC 1.23, 12.76, 11.34, 12.90[a]	Bukman et al. (2013)
Sequential injection analysis (SIA)	Mushroom (*Porphyrellus porphyrosporus, Leccinum caria, Matricaria discoidea*)	AA% 18.8, 28.9, 79.7	Polášek et al. (2004)
Multisyringe flow injection analysis (MSFIA)	Orange juice, isotonic drink, green tea, dark beer, lager beer, white wine, red wine	VCEAC 9.9, 10.1, 111, 36.1, 10.2, 24.8, 318[b]	Magalhães et al. (2006)
	Fruit wines (*Syzygium cumini* and *Cleistocaly nervosum*)	Reversed phase[c]	Nuengchamnong and Ingkaninan (2009)
Online HPLC-DPPH	Apples		Bandonienė and Murkovic (2002)
	Spices		Damasius et al. (2014)
	Tea (green, oolong, and white)		Zhang et al. (2013)
	Espresso coffees		Mnatsakanyan et al. (2010)
	Vegetable oil	Normal phase[c]	Zhang et al. (2009)
	Tea (*Orthosiphon grandiflorus*)	Reversed phase, coupled to ESI-MS/MS[c]	Nuengchamnong et al. (2011)
	Peanut shell	Reversed phase coupled to TOF/MS[c]	Qiu et al. (2012)
Online CE-DPPH	Honey	Capillary electrophoresis	Maruška et al. (2010)

[a] Trolox equivalent antioxidant capacity (mM).
[b] Vitamin C equivalent antioxidant capacity (mg of ascorbic acid per 100 mL).
[c] Antioxidant capacity was not identified; CE: capillary electrophoresis.

flow injection analysis (MSFIA), and high-performance liquid chromatography (HPLC) in food samples.

FIA is recognized as an excellent automated method due to ease of operation, potential of miniaturization, low consumption of sample and reagents, versatility; it involves a rapid injection of sample into a continuous stream (Magalhães et al., 2009). The use of the FIA system associated with the DPPH method is based on the injection of a sample plug into a flowing DPPH carrier stream. As the injected zone moves downstream, the sample solution disperses into the reagent, causing the reaction to occur. The (colorimetric) spectrophotometric detector is placed downstream and records the variation in absorbance (Mrazek et al., 2012). The change in absorbance allows the calculation of AA% (IC_{50} and equivalent antioxidant capacity) (Mrazek et al., 2012; Shpigun et al., 2012; Bukman et al., 2013), as illustrated in Figure 29.5.

SIA is a flow injection technique that replaces the typical injection valve by a multi-position selection valve in connection with a fully automated syringe and piston pump.

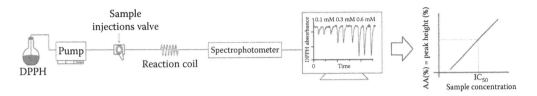

FIGURE 29.5   FIA system associated with the DPPH method and spectrophotometer detector.

The SIA technique consumes even smaller volumes, and reagents and samples can be drawn sequentially and stacked into the mixing coil before mixing while being pushed in the reverse direction into the detector; the system can also be programmed to stop for a desired period of time (Hartwell, 2012). This technique was adapted for the DPPH method (Polášek et al., 2004; Kolečkář et al., 2007; Rehakova et al., 2008).

Multisyringe FIA is an automatic-flow method that combines the multichannel operation of FIA and the flexibility of multicommutation flow systems. In this system, the antioxidant capacity can be determined by the DPPH method for several samples using the "stopped-flow" condition, which enables the monitoring of the absorbance decrease with time and evaluation of the reaction kinetics (Magalhães et al., 2006; Magalhães et al., 2009; Kedare and Singh, 2011). The absorbance in the presence and absence of an antioxidant is measured, and the AA% or IC_{50} can be calculated as in the static spectrophotometric method.

The DPPH assay was combined with HPLC for the rapid screening of antioxidants in complex mixtures. This approach was demonstrated in three different methods:

- Online postcolumn HPLC-DPPH analysis or HPLC-FIA: This method was developed for the detection of radical-scavenging components (screening antioxidants) in complex mixtures. A scheme of the instrumental setup is given in Figure 29.6. After HPLC sample separation and UV detection at the first detector (e.g., a diode-array detector), the separated compounds, eluted from the

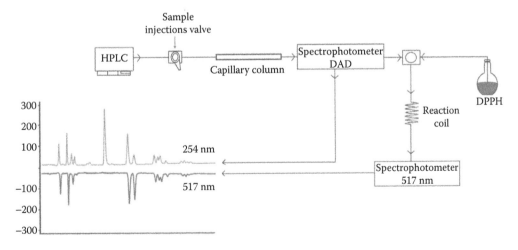

FIGURE 29.6   Instrumental setup for the online postcolumn HPLC-DPPH analysis.

HPLC column in the mobile phase, react with DPPH (postcolumn reaction). In sequence, the DPPH radical is detected at a second detector (515–528 nm). The antioxidant compounds in the sample consume DPPH radicals, resulting in the reduction of the net concentration of the DPPH radical, which lowers the absorbance, and then a negative peak is obtained on a constant background signal, while compounds without antioxidant activity do not change the constant background signal (Dapkevicius et al., 2001; Bandoniené and Murkovic, 2002; Bandoniené et al., 2002; Kosar et al., 2003; Koşar et al., 2004; Bandoniene et al., 2005; Xie et al., 2005; Pérez-Bonilla et al., 2006; Bartasiute et al., 2007; Wu et al., 2007; McDermott et al., 2010; Mnatsakanyan et al., 2010; Zhang et al., 2011; Shi et al., 2012; Inoue et al., 2012; Dai et al., 2013; Kraujalis et al., 2013; Ou et al., 2013; Šliumpaité et al., 2013; Zhang et al., 2013; Damasius et al., 2014; Yan et al., 2014). This method was first developed by Koleva et al. (2000), and from this pioneering work, many studies have exploited online postcolumn HPLC-DPPH. McDermott et al. (2010) published a detailed review about some aspects of this method (DPPH concentration, buffer utilized, type of DPPH sparging gas in degassing, length and thickness of the reaction coil, and temperature) to obtain higher resolution and sensitivity. In online postcolumn HPLC-DPPH, some works have used acronyms such as RP-HPLC-DPPH (Koleva et al., 2000; Wu et al., 2007) and NP-HPLC-DPPH (Zhang et al., 2009), where NP and RP are the column types: normal and reversed phase, respectively. Maruška et al. (2010) replaced the HPLC with capillary electrophoretic separation (CE), maintaining the spectrophotometric detection. Some works also coupled the online HPLC-DPPH system with a mass spectrophotometer (electrospray ionization or ESI, and time-of-flight mass or TOF) (Nuengchamnong et al., 2005; Tang et al., 2008; Shi et al., 2008; Nuengchamnong et al., 2009; Nuengchamnong and Ingkaninan, 2009; Nuengchamnong et al., 2011; Li et al., 2012; Qiu et al., 2012; Yao et al., 2012) in such a way that the radical-scavenging components in complex mixtures are accurately identified.

- DPPH-spiked HPLC analysis: This nonflow method is utilized to study oxidation products of the reaction between antioxidants and DPPH; thus, it is recommended for studies of isolated compounds (Yan et al., 2014).
- HPLC-based DPPH activity profiling: The sample is first analyzed by HPLC to obtain a screening of the compounds. Next, the sample reacts with DPPH and is injected into HPLC. The peak areas of antioxidant compounds are reduced or disappear in the HPLC chromatogram after their reaction with DPPH, and for those without antioxidant activities, the peak areas remain almost constant (Xie et al., 2005; Zhang et al., 2011; Shi et al., 2012). In some works, different amounts of the sample are reacted with DPPH to calculate the IC_{50}, and then the wavelength selected in the HPLC detector is that corresponding to the DPPH absorption (Chandrasekar et al., 2006; Helmja et al., 2009).

29.3 ELECTRON SPIN RESONANCE METHOD

The antioxidant capacity obtained by the DPPH method can be determined by other spectroscopic techniques such as ESR, also known as electron paramagnetic resonance (EPR). This technique directly detects paramagnetic compounds (i.e., containing unpaired electrons).

This concept is based on the interaction between electromagnetic radiation and magnetic moments of the electron. Each electron possesses an intrinsic magnetic dipole moment that arises from its spin. In most systems, electrons occur in pairs, such that the net moment is zero. Hence, only compound that contain one or more unpaired electrons possess the net spin moment necessary for interaction with an electromagnetic field (Carini et al., 2006). Electromagnetic wave radiation of the appropriate frequency under a given external magnetic field causes the excitation of unpaired electrons from the lower energy level to a higher level by the interaction of the magnetic moment of electron spin with the magnetic component of the electromagnetic wave (magnetic resonance). The ESR spectrometer detects the absorption of energy of the electromagnetic wave at the resonance frequency (Takeshita and Ozawa, 2004).

In the DPPH method with the ESR technique, the DPPH radical concentration in the sample is directly measured. The decrease in the ESR spectral amplitude becomes observable, indicating a decrease in the DPPH radical concentration. Figure 29.7 shows the ESR spectra of a DPPH solution before and after reaction with an antioxidant, allowing observation of the decrease in the amplitude.

The double integral of the ESR spectrum is proportional to the DPPH concentration; thus, the result can be expressed by relative concentration (r_{con}) using $r_{con} = [DIDPPH/DI_{ref}]t$, where DIDPPH represents the double integral of the reference signal (DPPH plus solvent) and DI_{ref} represents the double integral of the test signal (DPPH plus solvent and sample) (Calliste et al., 2001; Brezová et al., 2009). The DPPH antioxidant capacity can also be calculated by $I (\%) = [(II_0 - II)/II_0] \times 100$, where II_0 is the integral intensity of the DPPH signal for the sample control and II is the integral intensity measured after addition of the sample (Oszmianski et al., 2007; Polak et al., 2013). Table 29.5 shows some studies that have measured the antioxidant capacity of foods by the DPPH method using the ESR technique.

ESR is a more sensitive and effective method than absorption spectroscopy, because the former is not influenced by absorption of the antioxidant molecule or its decomposition products, since transitions of unpaired electrons in a magnetic field are measured instead of color intensity (Sanna et al., 2012; Polak et al., 2013). The only drawback of ESR is related to the aggregation phenomenon that occurs in the presence of solutions with high water contents (Sanna et al., 2012). Table 29.6 shows two studies that compared absorption with ESR spectroscopy. The results are not in complete agreement, and

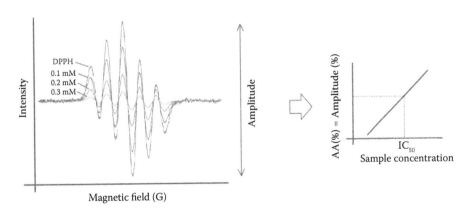

FIGURE 29.7   Antioxidant capacity expressed by IC_{50} with spectra ESR of DPPH.

TABLE 29.5   Antioxidant Capacity in Food Samples Obtained by the DPPH Method Using the ESR Technique

Food Sample	Method Parameters		Reference
	Solvent, DPPH, concentration, reaction time	Center field, microwave frequency, modulation frequency, modulation amplitude	
Apple	Methanol, 3.33 mM, 3 min	3368 G, 9.45 GHz, 100 kHz, 1 G	Oszmianski et al. (2007)
Nut	Methanol, 0.1 mM, 10 min	3495 G, 9.795 GHz, 86 kHz, 1.86 G	John and Shahidi (2010)
Propolis	Ethanol:water (1:1), 0.25 mM, 90 s	3500 G, 9.852 GHz, 100 kHz, 2 G	Benhanifia et al. (2013)
Colombian fruits	Methanol, 0.26 mM, 30 min	3473 G, 9.44 GHz, 100 kHz, 2 G	Santacruz et al. (2012)
Corozo fruit	Methanol, 0.14 mM, 30 min	3473 G, 9.44 GHZ, 100 kHz, 0.49 G	Osorio et al. (2011)

both authors reported probable interferences from the sample matrix, since the samples also absorb in the visible region.

FIA in association with the ESR technique was also applied for the determination of antioxidant capacity. Ukeda et al. (2002) developed an FIA system (Figure 29.8) for the analysis of tea, coffee, and wine samples. A double-line system was constructed in which a carrier stream containing the DPPH radical (solution B) was fed into the flat cell after confluence with a sample stream (solution A). Since the DPPH solution was continuously fed into the cell, a constant ESR signal corresponding to the baseline was observed. When the sample was injected into the carrier stream, the signal was suppressed, and a negative peak was therefore observed, proportional to the concentration of the radical scavenger. The results were expressed in Trolox equivalents.

TABLE 29.6   Antioxidant Capacity of Food Samples Obtained by the DPPH Method Using Absorption and ESR Spectroscopy Techniques

Food Sample		Spectroscopy Techniques		Antioxidant Capacity Unit	Reference
		Absorption	ESR		
Myrtle fruits	Barbara	41.08	34.24	IC_{50}	Sanna et al.
	Giovanna	61.05	32.15	(μg mL^{-1})	(2012)
	Maria Antonieta	60.73	42.46		
	Grazia	44.62	30.36		
	Angela	46.10	28.55		
Apple	Champion cloudy	446	677	TEAC	Oszmianski
	Champion clear	412	467	(μM Trolox	et al. (2007)
	Idared cloudy	225	359	per 100 mL)	
	Idared clear	202	196		

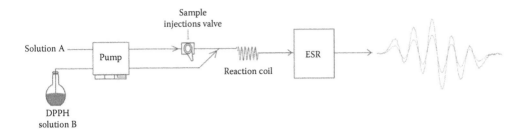

FIGURE 29.8   FIA system associated with the DPPH method and ESR detector.

29.4 ELECTROCHEMICAL METHODS

Electrochemical techniques have also been used for the determination of antioxidant capacity, since the DPPH radical can be electrochemically reduced and the current response generated from this process can be correlated with the DPPH concentration.

Cyclic voltammetry is the most commonly used technique to acquire qualitative information about electrochemical processes. The efficiency of this technique results in its rapid provision of information on the thermodynamics of redox processes, the kinetics of heterogeneous electron-transfer reactions, and the chemical reactions coupled with adsorptive processes (Pacheco et al., 2013).

Ahmed et al. (2012) investigated the electrochemical characteristics of DPPH. The cyclic voltammetry of DPPH (0.5 mM) was evaluated by using 0.1 mol L^{-1} tetrabutyl ammonium perchlorate (TBPA) in dimethyl sulfoxide (DMSO) at apparent pH 9.5 in the potential range between −0.2 and 0.45 V and at a scan rate of 50 mV s^{-1}. A reduction peak at 0.043 V was observed due to the one-electron transfer, which produces the monoanion species (DPPH⁻). Then the neutral radical is regenerated in the reverse scan (oxidation peak at 0.114 V). The potential difference (ΔE_p) was 71 mV and the peak current ratio (I_{pa}/I_{pc}) was in the range 1:02 to 1:05, indicating the system is electrochemically reversible and the electron transfer is controlled by diffusion. Figure 29.9 presents a typical cyclic voltammogram of the DPPH.

Other studies (Zhuang et al., 1999; Milardovic et al., 2006; Rodriguez Cid De León et al., 2011; Ahmed et al., 2012) have also reported the electrochemical behavior of this radical by cyclic voltammetry, showing that the redox process of DPPH is a monoelectron transfer and is fast, reversible, and controlled by diffusion.

Some studies have shown that factors such as variation in pH, solvent, electrolyte, reference electrode, and working electrode can change the values of oxidation and reduction potentials (Solon and Bard, 1964; Zhuang et al., 1999; Amatatongchai et al., 2012). Zhuang et al. (1999) used cyclic voltammetry to study the electrochemical behavior of DPPH adsorbed onto a graphite electrode immersed in aqueous electrolyte. A reversible reaction was observed with a reduction peak at 0.340 V. The effect of pH on the electrochemical behavior of DPPH was studied, and it was verified that the solubility of the radical is greater in basic solutions, and a shift in the reduction potential to more negative potentials was observed. Amatatongchai et al. (2012) performed cyclic voltammetry studies and evaluated two different working electrodes, a bare glassy-carbon electrode (GCE) and a GCE modified with multiwalled carbon nanotubes (CNT-GCE). The CNT-GCE exhibited an electrochemically reversible redox process with reduction and oxidation peaks at 0.305 and 0.327 V, respectively, while the GCE showed peaks at

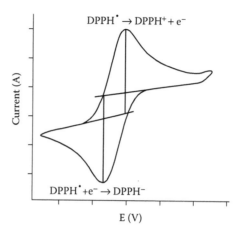

FIGURE 29.9   Typical cyclic voltammogram of DPPH.

0.362 and 0.310 V, respectively. The decrease in ΔE (from 52 to 22 mV) indicates a fast electron transfer provided by the carbon nanotubes. In this work, the effect of pH on the redox potential and peak current was also investigated in the pH range between 5 and 8 in 0.03 mol L^{-1} phosphate buffer, 0.03 mol L^{-1} KCl, and 40% (v/v) ethanol (Figure 29.10a). Figure 29.10 shows the cyclic voltammograms of these experiments and a plot of reduction potential as a function of pH. A linear correlation between pH and reduction potential was verified by the authors, as indicated in Figure 29.10b, with a slope of −56.3 mV, which indicates that the same number of protons and electrons are involved in this electrochemical process (one proton and one electron). They also verified a maximum peak current at pH 7.0, which was selected for the amperometric detection of DPPH under flow conditions.

Electrochemical techniques have been used to evaluate the antioxidant capacity in food samples using the scavenging DPPH method. Table 29.7 shows some examples.

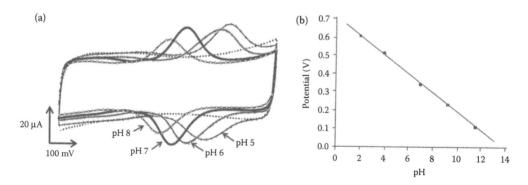

FIGURE 29.10   (a) Cyclic voltammetry of DPPH at various pH. (Adapted from Amatatongchai, M. et al. 2012. *Talanta* 97:267–272). (b) Plot of the peak potential versus pH. (Adapted from Zhuang, Q.-k., F. Scholz, and F. Pragst. 1999. *Electrochemistry Communications* 1(9):406–410.)

TABLE 29.7   Electrochemical Techniques Utilized for the Determination of Antioxidant
Capacity of Food Samples by the DPPH Method

Electrochemical Technique	Sample	Method Parameters (Electrolyte, Working Electrode, Applied Potential, DPPH Concentration, Expressing Result, Others)	Reference
Cyclic voltammetry	Fruit juices	Methanolic acetate buffer (80% methanol, pH 5–6.5); GCE; 259 mV; 1.27 mM; TEAC	Rodríguez Cid de León et al. (2011)
Differential pulse voltammetry	Foodstuff supplements	Methanol (0.03 M KCl); GCE; 180 mV; 0.1 mM; TEAC	Alvarez-Diduk et al. (2008)
Amperometric	Tea and fruits	Ethanolic phosphate buffer (EPBS, 40% ethanol, pH 7.4); pencil-rod working electrode; 100 mV; 0.2 mM; TEAC; microliter plate; automated system	Intarakamhang and Schulte (2012)
Amperometric	Tea, wine, and juice	Ethanolic phosphate buffer (EPBS, 40% ethanol, pH 7.4); GCE; 140 mV; 0.100 mM; TEAC	Milardovic et al. (2006)
Biamperometric	Juices	Ethanolic phosphate buffer (EPBS, 40% ethanol, pH 7.4); two identical Pt electrodes; 200 mV; 0.100 mM; TEAC	Pisoschi et al. (2009)
Biamperometric	Tea, wine, juice, and coffee	Ethanolic phosphate buffer (EPBS, 40% ethanol, pH 7.4); two identical GCEs; 200 mV; 0.100 mM; TEAC	Milardovic et al. (2005)

Note: GCE: glassy-carbon electrode.

The determination of the antioxidant capacity is based on the current generated by the electrochemical reduction of DPPH during analysis. Since the current is proportional to DPPH concentration, it is possible to evaluate the percentage of DPPH consumed by the antioxidant. Therefore, the analysis of antioxidant capacity is achieved by the decrease in the DPPH current measured at a constant potential selected by the cyclic voltammetry study (or by the hydrodynamic voltammogram). The current analysis can be performed by various electrochemical techniques, such as cyclic voltammetry, differential pulse voltammetry, and amperometry.

Ahmed et al. (2012) investigated the antioxidant capacity of flavonoid standards as a function of DPPH consumption using cyclic voltammetry. The antioxidants did not present any redox electrochemical process in the potential range at which the DPPH/DPPH$^-$ redox pair occurred. Therefore, the authors proposed the direct measurement of antioxidant capacity of flavonoids by detecting the consumption of DPPH using cyclic voltammetry. The IC_{30} and IC_{50} values were calculated by plotting the percentage of peak current reduction $\%I = 100 \times [(I_{po} - I_p)/I_{po}]$, where I_{po} and I_p are the peak currents of DPPH in the absence and presence of the flavonoid, respectively, versus the flavonoid concentration. Other studies have also used this approach to measure the antioxidant capacity (Zhuang et al., 1999; Rodriguez Cid De León et al., 2011; Tyurin et al., 2011).

Another electrochemical technique that is utilized in association with the DPPH method is differential pulse voltammetry (DPV), which presents a substantial increase in sensitivity in comparison with cyclic voltammetry once the residual current (capacitive current) is subtracted from the net current (Litescu and Radu, 2000; Alvarez-Diduk et al., 2008). Alvarez-Diduk et al. (2008) determined the antioxidant capacity of two dietary supplements using DPV to monitor the consumption of DPPH. The results were expressed in TEAC.

Another electrochemical technique used for the determination of antioxidant capacity is amperometry (Milardovic et al., 2006; Intarakamhang and Schulte, 2012). The current utilized in this technique is selected after a voltammetric analysis. Milardovic et al. (2006) determined the antioxidant activity of eight samples of different types of tea, wine, and other beverages. The amperometric method is based on the electrochemical reduction of DPPH at a glassy-carbon electrode. Initially, the voltammetric study of different standard antioxidants (caffeic acid and Trolox) was performed to obtain their respective reduction potentials. The potential selected for the analysis was 140 mV, to avoid possible interference from electrochemical processes of caffeic acid and Trolox. The concentration of DPPH significantly decreased after the antioxidant addition. The results were expressed as Trolox equivalents.

Biamperometry is another technique that can be utilized for the determination of antioxidant capacity based on the DPPH redox couple. According to Milardovic et al. (2005) and Pisoschi et al. (2009), this method is based on the measurement of the current flowing between two identical working electrodes, polarized via a small potential difference, immersed in the solution containing the reversible redox couple.

Comparing the spectrophotometric and electrochemical methods, the former is easily affected by the color and turbidity of a sample, requiring pretreatment of the sample, while the latter exhibits good chemical stability and colored or turbid samples do not affect the response of the detector. Some studies have compared spectrophotometric and electrochemical results, and good agreement was obtained, as shown in Table 29.8.

TABLE 29.8 Trolox Equivalent Antioxidant Capacity (TEAC) of the Food Samples by Spectrophotometric and Electrochemical Techniques Using the Scavenging DPPH Method

Food Samples		Technique		Antioxidant Capacity Unit	Reference
		Spectrophotometric	Electrochemical		
Juices	Jumex® orange	2.92×10^{-3}	1.97×10^{-3}	meq Tx per mL[a]	Rodríguez Cid de León et al. (2011)
	Boing® guava	1.52×10^{-3}	1.38×10^{-3}		
	Boing® mango	3.86×10^{-3}	2.67×10^{-3}		
	V8 Herdez®	2.07×10^{-3}	1.77×10^{-3}		
Juices	Orange juice	9.25	9.07	mM Trolox[b]	Pisoschi et al. (2009)
	Lemon juice	6.50	6.25		
	Fanta orange	0.71	0.70		
	Cappy grapefruit	0.064	0.0615		
	Frutti fresh Tutti	4.00	4.32		
	Fanta lemon	1.54	1.45		
	Prigat orange	2.40	2.46		
	Prigat peach	1.24	1.20		
Foodstuff supplements	Red yeast rice	1.91	1.96	meq Tx per g[c]	Alvarez-Diduk et al. (2008)
	Lycopene	0.63	0.38		

[a] Amperometric.
[b] Biamperometric.
[c] Differential pulse voltammetry.

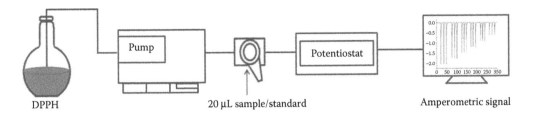

FIGURE 29.11 Instrumental setup for the determination of antioxidant capacity based on the DPPH detection in the FIA system. (Adapted from Amatatongchai, M. et al. 2012. *Talanta* 97:267–272.)

Flow analysis was also applied in the determination of antioxidant capacity using electrochemical detectors. Amatatongchai et al. (2012) developed an analytical method to determine the antioxidant capacity using an FIA with amperometric detection of the DPPH. Figure 29.11 illustrates the complete FIA system, which contains a reservoir for the DPPH solution, a pump, an injection valve, and the electrochemical detector (Amatatongchai et al., 2012).

After voltammetric experiments, Amatatongchai et al. (2012) selected 0.05 V (vs. Ag/AgCl) as the optimal potential for DPPH detection in the FIA system based on its electrochemical reduction. Since the solution of DPPH flowed continuously through the electrochemical cell at which its reduction occurred at the working electrode, negative transient signals were obtained after addition of antioxidants due to the consumption of DPPH. The antioxidant capacity was evaluated by Trolox equivalents. A calibration curve of the percentage of current reduction as a function of Trolox concentration was plotted in such a way that the antioxidant capacity of the samples was given in Trolox equivalents. The results showed a good correlation between the results obtained by both the proposed FIA amperometric and spectrophotometric methods.

REFERENCES

Ahmed, S., S. Tabassum, F. Shakeel, and A. Y. Khan. 2012. A facile electrochemical analysis to determine antioxidant activity of flavonoids against DPPH radical. *Journal of the Electrochemical Society* 159(5):F103–F109.

Akowuah, G. A., Z. Ismail, I. Norhayati, and A. Sadikun. 2005. The effects of different extraction solvents of varying polarities on polyphenols of *Orthosiphon stamineus* and evaluation of the free radical-scavenging activity. *Food Chemistry* 93(2):311–317.

Al-Dabbas, M. M., K. Al-Ismail, K. Kitahara, N. Chishaki, F. Hashinaga, T. Suganuma, and K. Tadera. 2007. The effects of different inorganic salts, buffer systems, and desalting of *Varthemia* crude water extract on DPPH radical scavenging activity. *Food Chemistry* 104(2):734–739.

Alam, M. N., N. J. Bristi, and M. Rafiquzzaman. 2013. Review on *in vivo* and *in vitro* methods evaluation of antioxidant activity. *Saudi Pharmaceutical Journal* 21(2):143–152.

Albayrak, S., A. Aksoy, and E. Hamzaoglu. 2008. Determination of antimicrobial and antioxidant activities of Turkish endemic *Salvia halophila* Hedge. *Turkish Journal of Biology* 32(4):265–270.

Alcalay, R. N., Y. Gu, H. Mejia-Santana, L. Cote, K. S. Marder, and N. Scarmeas. 2012. The association between Mediterranean diet adherence and Parkinson's disease. *Movement Disorders* 27(6):771–774.

Alighourchi, H. R., M. Barzegar, M. A. Sahari, and S. Abbasi. 2013. Effect of sonication on anthocyanins, total phenolic content, and antioxidant capacity of pomegranate juices. *International Food Research Journal* 20(4):1703–1709.

Alma, M. H., A. Mavi, A. Yildirim, M. Digrak, and T. Hirata. 2003. Screening chemical composition and *in vitro* antioxidant and antimicrobial activities of the essential oils from *Origanum syriacum* L. growing in Turkey. *Biological and Pharmaceutical Bulletin* 26(12):1725–1729.

Alvarez-Diduk, R., P. Ibarra-Escutia, J. L. Marty, A. Galano, A. Rojas-Hernández, and M. T. Ramírez-Silva. 2008. Electrochemical determination of the antioxidant capacity of organic compounds. *ECS Transactions* 15(1):471–478.

Amatatongchai, M., S. Laosing, O. Chailapakul, and D. Nacapricha. 2012. Simple flow injection for screening of total antioxidant capacity by amperometric detection of DPPH radical on carbon nanotube modified-glassy carbon electrode. *Talanta* 97:267–272.

Antolovich, M., P. D. Prenzler, E. Patsalides, S. McDonald, and K. Robards. 2002. Methods for testing antioxidant activity. *Analyst* 127(1):183–198.

Bandonienė, D. and M. Murkovic. 2002. On-line HPLC-DPPH screening method for evaluation of radical scavenging phenols extracted from apples (*Malus domestica* L.). *Journal of Agricultural and Food Chemistry* 50(9):2482–2487.

Bandonienė, D., M. Murkovic, W. Pfannhauser, P. Venskutonis, and D. Gruzdienė. 2002. Detection and activity evaluation of radical scavenging compounds by using DPPH free radical and on-line HPLC-DPPH methods. *European Food Research and Technology* 214(2):143–147.

Bandoniene, D., M. Murkovic, and P. R. Venskutonis. 2005. Determination of rosmarinic acid in sage and borage leaves by high-performance liquid chromatography with different detection methods. *Journal of Chromatographic Science* 43(7):372–376.

Bartasiute, A., B. H. C. Westerink, E. Verpoorte, and H. A. G. Niederländer. 2007. Improving the *in vivo* predictability of an on-line HPLC stable free radical decoloration assay for antioxidant activity in methanol–buffer medium. *Free Radical Biology and Medicine* 42(3):413–423.

Benhanifia, M., W. M. Mohamed, Y. Bellik, and H. Benbarek. 2013. Antimicrobial and antioxidant activities of different propolis samples from north-western Algeria. *International Journal of Food Science & Technology* 48(12):2521–2527.

Bhupathiraju, S. N. and K. L. Tucker. 2011. Coronary heart disease prevention: Nutrients, foods, and dietary patterns. *Clinica Chimica Acta* 412(17–18):1493–1514.

Blois, M. S. 1958. Antioxidant determinations by the use of a stable free radical. *Nature* 181(4617):1199–1200.

Boivin, D., S. Lamy, S. Lord-Dufour, J. Jackson, E. Beaulieu, M. Côté, A. Moghrabi, S. Barrette, D. Gingras, and R. Béliveau. 2009. Antiproliferative and antioxidant activities of common vegetables: A comparative study. *Food Chemistry* 112(2):374–380.

Brand-Williams, W., M. E. Cuvelier, and C. Berset. 1995. Use of a free radical method to evaluate antioxidant activity. *LWT—Food Science and Technology* 28(1):25–30.

Brezová, V., A. Šlebodová, and A. Staško. 2009. Coffee as a source of antioxidants: An EPR study. *Food Chemistry* 114(3):859–868.

Bukman, L., A. C. Martins, E. O. Barizão, J. V. Visentainer, and V. C. Almeida. 2013. DPPH assay adapted to the FIA system for the determination of the antioxidant capacity of wines: Optimization of the conditions using the response surface methodology. *Food Analytical Methods* 6(5):1424–1432.

Burda, S. and W. Oleszek. 2001. Antioxidant and antiradical activities of flavonoids. *Journal of Agricultural and Food Chemistry* 49(6):2774–2779.

Calliste, C.-A., P. Trouillas, D.-P. Allais, A. Simon, and J.-L. Duroux. 2001. Free radical scavenging activities measured by electron spin resonance spectroscopy and B16 cell antiproliferative behaviors of seven plants. *Journal of Agricultural and Food Chemistry* 49(7):3321–3327.

Carini, M., G. Aldini, M. Orioli, and R. M. Facino. 2006. Electron paramagnetic resonance (EPR) spectroscopy: A versatile and powerful tool in pharmaceutical and biomedical analysis. *Current Pharmaceutical Analysis* 2(2):141–159.

Chandrasekar, D., K. Madhusudhana, S. Ramakrishna, and P. V. Diwan. 2006. Determination of DPPH free radical scavenging activity by reversed-phase HPLC: A sensitive screening method for polyherbal formulations. *Journal of Pharmaceutical and Biomedical Analysis* 40(2):460–464.

Chen, Y. C., Y. Sugiyama, N. Abe, R. Kuruto-Niwa, R. Nozawa, and A. Hirota. 2005. DPPH radical-scavenging compounds from dou-chi, a soybean fermented food. *Bioscience, Biotechnology and Biochemistry* 69(5):999–1006.

Choe, E. and D. B. Min. 2009. Mechanisms of antioxidants in the oxidation of foods. *Comprehensive Reviews in Food Science and Food Safety* 8(4):345–358.

Chung, Y. C., S. J. Chen, C. K. Hsu, C. T. Chang, and S. T. Chou. 2005. Studies on the antioxidative activity of *Graptopetalum paraguayense* E. Walther. *Food Chemistry* 91(3):419–424.

Cui, Y., D. S. Kim, and K. C. Park. 2005. Antioxidant effect of *Inonotus obliquus*. *Journal of Ethnopharmacology* 96(1–2):79–85.

Dai, J. and R. J. Mumper. 2010. Plant phenolics: Extraction, analysis and their antioxidant and anticancer properties. *Molecules* 15(10):7313–7352.

Dai, X., Q. Huang, B. Zhou, Z. Gong, Z. Liu, and S. Shi. 2013. Preparative isolation and purification of seven main antioxidants from *Eucommia ulmoides* Oliv. (Du-zhong) leaves using HSCCC guided by DPPH-HPLC experiment. *Food Chemistry* 139(1–4): 563–570.

Dalen, J. E. and S. Devries. 2014. Diets to prevent coronary heart disease 1957–2013: What have we learned? *The American Journal of Medicine* 127(5):364–369.

Damasius, J., P. R. Venskutonis, V. Kaskoniene, and A. Maruska. 2014. Fast screening of the main phenolic acids with antioxidant properties in common spices using on-line HPLC/UV/DPPH radical scavenging assay. *Analytical Methods* 6(8):2774–2779.

Dapkevicius, A., T. A. van Beek, and H. A. G. Niederländer. 2001. Evaluation and comparison of two improved techniques for the on-line detection of antioxidants in liquid chromatography eluates. *Journal of Chromatography A* 912(1):73–82.

de Souza, R. F. V. and W. F. De Giovani. 2004. Antioxidant properties of complexes of flavonoids with metal ions. *Redox Report* 9(2):97–104.

Deng, J., W. Cheng, and G. Yang. 2011. A novel antioxidant activity index (AAU) for natural products using the DPPH assay. *Food Chemistry* 125(4):1430–1435.

Dicko, A., F. Muanda, D. Koné, R. Soulimani, and C. Younos. 2011. Phytochemical composition and antioxidant capacity of three Malian medicinal plant parts. *Evidence-Based Complementary and Alternative Medicine* 2011:1–8.

Elzaawely, A. A., T. D. Xuan, and S. Tawata. 2007. Essential oils, kava pyrones and phenolic compounds from leaves and rhizomes of *Alpinia zerumbet* (Pers.) B.L. Burtt. & R.M. Sm. and their antioxidant activity. *Food Chemistry* 103(2):486–494.

Esposito, K. and D. Giugliano. 2014. Mediterranean diet and type 2 diabetes. *Diabetes/ Metabolism Research and Reviews* 30(S1):34–40.

Folmer, F., U. Basavaraju, M. Jaspars, G. Hold, E. El-Omar, M. Dicato, and M. Diederich. 2014. Anticancer effects of bioactive berry compounds. *Phytochemistry Reviews* 13(1):295–322.

Foti, M. C., C. Daquino, and C. Geraci. 2004. Electron-transfer reaction of cinnamic acids and their methyl esters with the DPPH• radical in alcoholic solutions. *The Journal of Organic Chemistry* 69(7):2309–2314.

Frankel, E. N. and A. S. Meyer. 2000. The problems of using one-dimensional methods to evaluate multifunctional food and biological antioxidants. *Journal of the Science of Food and Agriculture* 80(13):1925–1941.

Garcia, E. J., T. L. C. Oldoni, S. M. de Alencar, A. Reis, A. D. Loguercio, and R. H. M. Grande. 2012. Antioxidant activity by DPPH assay of potential solutions to be applied on bleached teeth. *Brazilian Dental Journal* 23:22–27.

Göktürk, B., N., G. Özkan, and S. Yaşar. 2007. Evaluation of the antiradical and antioxidant potential of grape extracts. *Food Control* 18(9):1131–1136.

Govindarajan, R., S. Rastogi, M. Vijayakumar, A. Shirwaikar, A. K. S. Rawat, S. Mehrotra, and P. Pushpangadan. 2003. Studies on the antioxidant activities of *Desmodium gangeticum*. *Biological and Pharmaceutical Bulletin* 26(10):1424–1427.

Gülçin, I. 2012. Antioxidant activity of food constituents: An overview. *Archives of Toxicology* 86(3):345–391.

Hartwell, S. K. 2012. Flow injection/sequential injection analysis systems: Potential use as tools for rapid liver diseases biomarker study. *International Journal of Hepatology* 2012:1–8.

Helmja, K., M. Vaher, T. Püssa, and M. Kaljurand. 2009. Analysis of the stable free radical scavenging capability of artificial polyphenol mixtures and plant extracts by capillary electrophoresis and liquid chromatography–diode array detection–tandem mass spectrometry. *Journal of Chromatography A* 1216(12):2417–2423.

Huang, D., O. U. Boxin, and R. L. Prior. 2005. The chemistry behind antioxidant capacity assays. *Journal of Agricultural and Food Chemistry* 53(6):1841–1856.

Huang, W.-y., H.-c. Zhang, W.-x. Liu, and C.-yang Li. 2012. Survey of antioxidant capacity and phenolic composition of blueberry, blackberry, and strawberry in Nanjing. *Journal of Zhejiang University SCIENCE B* 13(2):94–102.

Inoue, K., E. Baba, T. Hino, and H. Oka. 2012. A strategy for high-speed countercurrent chromatography purification of specific antioxidants from natural products based on on-line HPLC method with radical scavenging assay. *Food Chemistry* 134(4):2276–2282.

Intarakamhang, S. and A. Schulte. 2012. Automated electrochemical free radical scavenger screening in dietary samples. *Analytical Chemistry* 84(15):6767–6774.

Jiménez-Escrig, A., I. Jiménez-Jiménez, C. Sánchez-Moreno, and F. Saura-Calixto. 2000. Evaluation of free radical scavenging of dietary carotenoids by the stable radical 2,2-diphenyl-1-picrylhydrazyl. *Journal of the Science of Food and Agriculture* 80(11):1686–1690.

John, J. A. and F. Shahidi. 2010. Phenolic compounds and antioxidant activity of Brazil nut (*Bertholletia excelsa*). *Journal of Functional Foods* 2(3):196–209.

Kano, M., T. Takayanagi, K. Harada, K. Makino, and F. Ishikawa. 2005. Antioxidative activity of anthocyanins from purple sweet potato, *Ipomoera batatas* cultivar Ayamurasaki. *Bioscience, Biotechnology, and Biochemistry* 69(5):979–988.

Karioti, A., D. Hadjipavlou-Litina, M. L. K. Mensah, T. C. Fleischer, and H. Skaltsa. 2004. Composition and antioxidant activity of the essential oils of *Xylopia aethiopica* (Dun) A. Rich. (Annonaceae) leaves, stem bark, root bark, and fresh and dried fruits, growing in Ghana. *Journal of Agricultural and Food Chemistry* 52(26):8094–8098.

Kedare, S. B. and R. P. Singh. 2011. Genesis and development of DPPH method of antioxidant assay. *Journal of Food Science and Technology* 48(4):412–422.

Kilic, I., Y. Yeşiloğlu, and Y. Bayrak. 2014. Spectroscopic studies on the antioxidant activity of ellagic acid. *Spectrochimica Acta Part A: Molecular and Biomolecular Spectroscopy* 130:447–452.

Kolečkář, V., D. Jun, L. Opletal, L. Jahodář, and Kamil Kuča. 2007. Assay of radical scavenging activity of antidotes against chemical warfare agents by DPPH test using sequential injection technique. *Journal of Applied Biomedicine* 5:81–84.

Koleva, I. I., H. A. G. Niederländer, and T. A. van Beek. 2000. An on-line HPLC method for detection of radical scavenging compounds in complex mixtures. *Analytical Chemistry* 72(10):2323–2328.

Koşar, M., D. Dorman, K. Başer, and R. Hiltunen. 2004. An improved HPLC post-column methodology for the identification of free radical scavenging phytochemicals in complex mixtures. *Chromatographia* 60(11–12):635–638.

Kosar, M., H. J. D. Dorman, O. Bachmayer, K. H. C. Baser, and R. Hiltunen. 2003. An improved on-line HPLC-DPPH method for the screening of free radical scavenging compounds in water extracts of Lamiaceae plants. *Chemistry of Natural Compounds* 39(2):161–166.

Kraujalis, P., P. R. Venskutonis, V. Kraujalienė, and A. Pukalskas. 2013. Antioxidant properties and preliminary evaluation of phytochemical composition of different anatomical parts of *Amaranth*. *Plant Foods for Human Nutrition* 68(3):322–328.

Laguerre, M., V. Hugouvieux, N. Cavusoglu, F. Aubert, A. Lafuma, H. Fulcrand, and C. Poncet-Legrand. 2014. Probing the micellar solubilisation and inter-micellar exchange of polyphenols using the DPPH free radical. *Food Chemistry* 149:114–120.

Li, Y.-J., J. Chen, Y. Li, and P. Li. 2012. Identification and quantification of free radical scavengers in the flower buds of *Lonicera* species by online HPLC-DPPH assay coupled with electrospray ionization quadrupole time-of-flight tandem mass spectrometry. *Biomedical Chromatography* 26(4):449–457.

Lim, Y. Y., T. T. Lim, and J. J. Tee. 2007. Antioxidant properties of several tropical fruits: A comparative study. *Food Chemistry* 103(3):1003–1008.

Litescu, S. C. and G. L. Radu. 2000. Estimation of the antioxidative properties of tocopherols an electrochemical approach. *European Food Research and Technology* 211(3):218–221.

Litwinienko, G. and K. U. Ingold. 2004. Abnormal solvent effects on hydrogen atom abstraction. 2. Resolution of the curcumin antioxidant controversy. The role of sequential proton loss electron transfer. *The Journal of Organic Chemistry* 69(18):5888–5896.

Litwinienko, G. and K. U. Ingold. 2007. Solvent effects on the rates and mechanisms of reaction of phenols with free radicals. *Accounts of Chemical Research* 40(3):222–230.

Liu, Z. Q. 2010. Chemical methods to evaluate antioxidant ability. *Chemical Reviews* 110(10):5675–5691.

Lo Scalzo, Roberto. 2008. Organic acids influence on DPPH scavenging by ascorbic acid. *Food Chemistry* 107(1):40–43.

Lucena, A. P. S., R. J. B. Nascimento, J. A. C. Maciel, J. X. Tavares, J. M. Barbosa-Filho, and E. J. Oliveira. 2010. Antioxidant activity and phenolics content of selected Brazilian wines. *Journal of Food Composition and Analysis* 23(1):30–36.

MacDonald-Wicks, L. K., L. G. Wood, and M. L. Garg. 2006. Methodology for the determination of biological antioxidant capacity in vitro: A review. *Journal of the Science of Food and Agriculture* 86(13):2046–2056.

Magalhães, L. M., M. Lúcio, M. A. Segundo, S. Reis, and J. L. F. C. Lima. 2009. Automatic flow injection based methodologies for determination of scavenging capacity against biologically relevant reactive species of oxygen and nitrogen. *Talanta* 78(4–5):1219–1226.

Magalhães, L. M., M. Santos, M. A. Segundo, S. Reis, and J. L. F. C. Lima. 2009. Flow injection based methods for fast screening of antioxidant capacity. *Talanta* 77(5):1559–1566.

Magalhães, L. M., M. A. Segundo, S. Reis, and J. L. F. C. Lima. 2006. Automatic method for determination of total antioxidant capacity using 2,2-diphenyl-1-picrylhydrazyl assay. *Analytica Chimica Acta* 558(1–2):310–318.

Magalhães, L. M., M. A. Segundo, S. Reis, and J. L. F. C. Lima. 2008. Methodological aspects about *in vitro* evaluation of antioxidant properties. *Analytica Chimica Acta* 613(1):1–19.

Maruška, A., M. Stankevičius, and Ž. Stanius. 2010. Coupling of capillary electrophoresis with reaction detection for the on-line evaluation of radical scavenging activity of analytes. *Procedia Chemistry* 2(1):54–58.

Materska, M. 2012. The scavenging effect and flavonoid glycosides content in fractions from fruits of hot pepper *Capsicum annuum* L. *Acta Scientiarum Polonorum, Technologia Alimentaria* 11(4):363–371.

McDermott, G. P., L. K. Noonan, M. Mnatsakanyan, R. A. Shalliker, X. A. Conlan, N. W. Barnett, and P. S. Francis. 2010. High-performance liquid chromatography with post-column 2,2′-diphenyl-1-picrylhydrazyl radical scavenging assay: Methodological considerations and application to complex samples. *Analytica Chimica Acta* 675(1):76–82.

Meda, A., C. E. Lamien, M. Romito, J. Millogo, and O. G. Nacoulma. 2005. Determination of the total phenolic, flavonoid and proline contents in Burkina Fasan honey, as well as their radical scavenging activity. *Food Chemistry* 91(3):571–577.

Mensor, L. L., F. S. Menezes, G. G. Leitão, A. S. Reis, T. C. dos Santos, C. S. Coube, and Suzana G. Leitão. 2001. Screening of Brazilian plant extracts for antioxidant activity by the use of DPPH free radical method. *Phytotherapy Research* 15(2):127–130.

Milardovic, S., D. Ivekovic, and B. S. Grabaric. 2006. A novel amperometric method for antioxidant activity determination using DPPH free radical. *Bioelectrochemistry* 68(2):175–180.

Milardovic, S., D. Iveković, V. Rumenjak, and B. S. Grabarić. 2005. Use of DPPH.|DPPH redox couple for biamperometric determination of antioxidant activity. *Electroanalysis* 17(20):1847–1853.

Mimica-Dukic, N., B. Bozin, M. Sokovic, and N. Simin. 2004. Antimicrobial and antioxidant activities of *Melissa officinalis* L. (Lamiaceae) essential oil. *Journal of Agricultural and Food Chemistry* 52(9):2485–2489.

Mnatsakanyan, M., T. A. Goodie, X. A. Conlan, P. S. Francis, G. P. McDermott, N. W. Barnett, D. Shock, F. Gritti, G. Guiochon, and R. A. Shalliker. 2010. High performance liquid chromatography with two simultaneous on-line antioxidant assays: Evaluation and comparison of espresso coffees. *Talanta* 81(3):837–842.

Molyneux, P. 2004. The use of the stable free radical diphenylpicrylhydrazyl (DPPH) for estimating antioxidant activity. *Songklanakarin Journal Science and Technology* 26(2):211–219.

Moreira, D. de L., S. G. Leitão, J. L. S. Gonçalves, M. D. Wigg, and G. G. Leitão. 2005. Antioxidant and antiviral properties of *Pseudopiptadenia contorta* (Leguminosae) and of quebracho (*Schinopsis* sp.) extracts. *Quimica Nova* 28:421–425.

Mrazek, N., K. Watla-iad, S. Deachathai, and S. Suteerapataranon. 2012. Rapid antioxidant capacity screening in herbal extracts using a simple flow injection-spectrophotometric system. *Food Chemistry* 132(1):544–548.

Mungmai, L., S. Jiranusornkul, Y. Peerapornpisal, B. Sirsithunyalug, and P. Leelapornpisid. 2014. Extraction, characterization and biological activities of extracts from freshwater macroalga [*Rhizoclonium hieroglyphicum* (C.Agardh) kützing] cultivated in Northern Thailand. *Chiang Mai Journal of Science* 41(1):14–26.

Naczk, M. and F. Shahidi. 2006. Phenolics in cereals, fruits and vegetables: Occurrence, extraction and analysis. *Journal of Pharmaceutical and Biomedical Analysis* 41(5):1523–1542.

Niki, E. 1987. Antioxidants in relation to lipid peroxidation. *Chemistry and Physics of Lipids* 44(2–4):227–253.

Niki, E., Y. Yoshida, Y. Saito, and N. Noguchi. 2005. Lipid peroxidation: Mechanisms, inhibition, and biological effects. *Biochemical and Biophysical Research Communications* 338(1):668–676.

Noipa, T., S. Srijaranai, T. Tuntulani, and W. Ngeontae. 2011. New approach for evaluation of the antioxidant capacity based on scavenging DPPH free radical in micelle systems. *Food Research International* 44(3):798–806.

Noufou, O., S. R. Wamtinga, T. André, B. Christine, L. Marius, H. A. Emmanuelle, K. Jean, D. Marie-Geneviève, and G. I. Pierre. 2012. Pharmacological properties and related constituents of stem bark of *Pterocarpus erinaceus* Poir. (Fabaceae). *Asian Pacific Journal of Tropical Medicine* 5(1):46–51.

Nuengchamnong, N., C. F. de Jong, B. Bruyneel, W. M. A. Niessen, H. Irth, and K. Ingkaninan. 2005. HPLC coupled on-line to ESI-MS and a DPPH-based assay for the rapid identification of anti-oxidants in *Butea superba*. *Phytochemical Analysis* 16(6):422–428.

Nuengchamnong, N., K. Krittasilp, and K. Ingkaninan. 2011. Characterisation of phenolic antioxidants in aqueous extract of *Orthosiphon grandiflorus* tea by LC-ESI-MS/MS coupled to DPPH assay. *Food Chemistry* 127(3):1287–1293.

Nuengchamnong, N. and K. Ingkaninan. 2009. On-line characterization of phenolic antioxidants in fruit wines from family Myrtaceae by liquid chromatography combined with electrospray ionization tandem mass spectrometry and radical scavenging detection. *LWT—Food Science and Technology* 42(1):297–302.

Nuengchamnong, N., K. Krittasilp, and K. Ingkaninan. 2009. Rapid screening and identification of antioxidants in aqueous extracts of *Houttuynia cordata* using LC–ESI–MS coupled with DPPH assay. *Food Chemistry* 117(4):750–756.

Osorio, C., J. G. Carriazo, and O. Almanza. 2011. Antioxidant activity of corozo (*Bactris guineensis*) fruit by electron paramagnetic resonance (EPR) spectroscopy. *European Food Research and Technology* 233(1):103–108.

Oszmianski, J., M. Wolniak, A. Wojdylo, and I. Wawer. 2007. Comparative study of polyphenolic content and antiradical activity of cloudy and clear apple juices. *Journal of the Science of Food and Agriculture* 87(4):573–579.

Otaegui-Arrazola, A., P. Amiano, A. Elbusto, E. Urdaneta, and P. Martínez-Lage. 2014. Diet, cognition, and Alzheimer's disease: Food for thought. *European Journal of Nutrition* 53(1):1–23.

Ou, Zong-Quan, D. M. S., T. Rades, L. Larsen, and A. McDowell. 2013. Application of an online post-column derivatization HPLC-DPPH assay to detect compounds responsible for antioxidant activity in *Sonchus oleraceus* L. leaf extracts. *Journal of Pharmacy and Pharmacology* 65(2):271–279.

Ozcelik, B., J. H. Lee, and D. B. Min. 2003. Effects of light, oxygen, and pH on the absorbance of 2,2-diphenyl-1-picrylhydrazyl. *Journal of Food Science* 68(2):487–490.

Özkan, A., H. Gübbük, E. Güneş, and A. Erdoğan. 2011. Antioxidant capacity of juice from different papaya (*Carica papaya* L.) cultivars grown under greenhouse conditions in Turkey. *Turkish Journal of Biology* 35(5):619–625.

Pacheco, W. F., F. S. Semaan, V. G. K. De Almeida, A. G. S. L. Ritta, and R. Q. Aucélio. 2013. Voltammetry: A brief review about concepts. *Revista Virtual de Química* 5(4):516–537.

Pattanaik, P., J. Ravi, B. Srinivas, C. Mohanty, O. P. Panda, and S. S. Nanda. 2011. Comparative study of antioxidant activity of alcoholic and ethyl acetate extract of the bark of *Pterospermum acerifolium*. *International Journal of Pharmacy and Technology* 3(3):3335–3344.

Paulino Z., M., C. Rojas, S. De Paula, I. Elingold, E. A. Migliaro, M. B. Casanova, F. I. Restuccia, S. A. Morales, and M. Dubin. 2010. Phenolic contents and antioxidant activity in central-southern Uruguayan propolis extracts. *Journal of the Chilean Chemical Society* 55:141–146.

Pérez-Bonilla, M., S. Salido, T. A. van Beek, P. J. Linares-Palomino, J. Altarejos, M. Nogueras, and A. Sánchez. 2006. Isolation and identification of radical scavengers in olive tree (*Olea europaea*) wood. *Journal of Chromatography A* 1112(1–2):311–318.

Perez-Jimenez, J. and F. Saura-Calixto. 2005. Literature data may underestimate the actual antioxidant capacity of cereals. *Journal of Agricultural and Food Chemistry* 53(12):5036–5040.

Pisoschi, A. M., M. C. Cheregi, and A. F. Danet. 2009. Total antioxidant capacity of some commercial fruit juices: Electrochemical and spectrophotometrical approaches. *Molecules* 14(1):480–493.

Polak, J., M. Bartoszek, and I. Stanimirova. 2013. A study of the antioxidant properties of beers using electron paramagnetic resonance. *Food Chemistry* 141(3):3042–3049.

Polášek, M., P. Skála, L. Opletal, and L. Jahodář. 2004. Rapid automated assay of antioxidation/radical-scavenging activity of natural substances by sequential injection technique (SIA) using spectrophotometric detection. *Analytical and Bioanalytical Chemistry* 379(5–6):754–758.

Prior, R. L., X. Wu, and K. Schaich. 2005. Standardized methods for the determination of antioxidant capacity and phenolics in foods and dietary supplements. *Journal of Agricultural and Food Chemistry* 53(10):4290–4302.

Pyrzynska, K. and A. Pekal. 2013. Application of free radical diphenylpicrylhydrazyl (DPPH) to estimate the antioxidant capacity of food samples. *Analytical Methods* 5(17):4288–4295.

Qiu, J., L. Chen, Q. Zhu, D. Wang, W. Wang, X. Sun, X. Liu, and F. Du. 2012. Screening natural antioxidants in peanut shell using DPPH– HPLC–DAD–TOF/MS methods. *Food Chemistry* 135(4):2366–2371.

Ramakrishna, A. and G. A. Ravishankar. 2011. Influence of abiotic stress signals on secondary metabolites in plants. *Plant Signaling and Behavior* 6(11):1720–1731.

Rehakova, Z., V. Koleckar, F. Cervenka, L. Jahodar, L. Saso, L. Opletal, D. Jun, and K. Kuca. 2008. DPPH radical scavenging activity of several naturally occurring coumarins and their synthesized analogs measured by the SIA method. *Toxicology Mechanisms and Methods* 18(5):413–418.

Research, American Institute for Cancer, and World Cancer Research Fund. 2007. Food, nutrition, physical activity and the prevention of cancer. A global perspective a project of World Cancer Research Fund International. Washington, D.C.: American Institute for Cancer Research.

Reynertson, K. A., M. J. Basile, and E. J. Kennelly. 2005. Antioxidant potential of seven myrtaceous fruits. *Ethnobotany Research & Applications* 3(x):25–35.

Ricci, D., D. Fraternale, L. Giamperi, A. Bucchini, F. Epifano, G. Burini, and M. Curini. 2005. Chemical composition, antimicrobial and antioxidant activity of the essential oil of *Teucrium marum* (Lamiaceae). *Journal of Ethnopharmacology* 98(1–2): 195–200.

Rodriguez Cid De León, G. I., M. Gómez Hernández, A. M. Domínguez y Ramírez, J. R. Medina López, G. Alarcón Ángeles, and A. Morales Pérez. 2011. Adaptation of DPPH method for antioxidant determination. *ECS Transactions* 36(1):401–411.

Saad, B., Y. Y. Sing, M. A. Nawi, N. Hashim, A. S. M. Ali, M. I. Saleh, S. F. Sulaiman, K. M. Talib, and K. Ahmad. 2007. Determination of synthetic phenolic antioxidants in food items using reversed-phase HPLC. *Food Chemistry* 105(1):389–394.

Salas-Salvadó, J., M. Á. Martinez-González, M. Bulló, and E. Ros. 2011. The role of diet in the prevention of type 2 diabetes. *Nutrition, Metabolism and Cardiovascular Diseases* 21, Supplement 2:B32-B48.

Sánchez-Moreno, C., J. A. Larrauri, and F. Saura-Calixto. 1998. A procedure to measure the antiradical efficiency of polyphenols. *Journal of the Science of Food and Agriculture* 76(2):270–276.

Sanna, D., G. Delogu, M. Mulas, M. Schirra, and A. Fadda. 2012. Determination of free radical scavenging activity of plant extracts through DPPH assay: An EPR and UV–vis study. *Food Analytical Methods* 5(4):759–766.

Santacruz, L., J. G. Carriazo, O. Almanza, and C. Osorio. 2012. Anthocyanin composition of wild Colombian fruits and antioxidant capacity measurement by electron paramagnetic resonance spectroscopy. *Journal of Agricultural and Food Chemistry* 60(6):1397–1404.

Scherer, R. and H. T. Godoy. 2009. Antioxidant activity index (AAI) by the 2,2-diphenyl-1-picrylhydrazyl method. *Food Chemistry* 112(3):654–658.

Shalaby, E. A. and S. M. M. Shanab. 2013. Comparison of DPPH and ABTS assays for determining antioxidant potential of water and methanol extracts of *Spirulina platensis*. *Indian Journal of Marine Sciences* 42(5):556–564.

Sharififar, F., M. H. Moshafi, S. H. Mansouri, M. Khodashenas, and M. Khoshnoodi. 2007. *In vitro* evaluation of antibacterial and antioxidant activities of the essential oil and methanol extract of endemic *Zataria multiflora* Boiss. *Food Control* 18(7):800–805.

Sharma, O. P. and T. K. Bhat. 2009. DPPH antioxidant assay revisited. *Food Chemistry* 113(4):1202–1205.

Shi, S., Y Ma, Y. Zhang, L. Liu, Q. Liu, M. Peng, and X. Xiong. 2012. Systematic separation and purification of 18 antioxidants from *Pueraria lobata* flower using HSCCC target-guided by DPPH–HPLC experiment. *Separation and Purification Technology* 89:225–233.

Shi, S., Y. Zhao, H. Zhou, Y. Zhang, X. Jiang, and K. Huang. 2008. Identification of antioxidants from *Taraxacum mongolicum* by high-performance liquid chromatography–diode array detection–radical-scavenging detection–electrospray ionization mass spectrometry and nuclear magnetic resonance experiments. *Journal of Chromatography* A 1209(1–2):145–152.

Shimizu, T., Y. Nakanishi, M. Nakahara, N. Wada, Y. Moro-oka, T. Hirano, T. Konishi, and S. Matsugo. 2010. Structure effect on antioxidant activity of catecholamines toward singlet oxygen and other reactive oxygen species *in vitro*. *Journal of Clinical Biochemistry and Nutrition* 47(3):181–190.

Shpigun, L. K., N. N. Zamyatina, Ya V. Shushenachev, and P. M. Kamilova. 2012. Flow-injection methods for the determination of antioxidant activity based on free-radical processes. *Journal of Analytical Chemistry* 67(10):801–808.

Singh, R., N. Kaur, L. Kishore, and G. K. Gupta. 2013. Management of diabetic complications: A chemical constituents based approach. *Journal of Ethnopharmacology* 150(1):51–70.

Siqueira, E. M. d. A., F. R. Rosa, A. M. Fustinoni, L. P. de Sant'Ana, and S. F. Arruda. 2013. Brazilian Savanna fruits contain higher bioactive compounds content and higher antioxidant activity relative to the conventional red delicious apple. *PLoS One* 8(8):e72826.

Šliumpaitė, I., P. R. Venskutonis, M. Murkovic, and A. Pukalskas. 2013. Antioxidant properties and polyphenolics composition of common hedge hyssop (*Gratiola officinalis* L.). *Journal of Functional Foods* 5(4):1927–1937.

Solon, Emanuel and Allen J. Bard. 1964. The electrochemistry of diphenylpicrylhydrazyl. *Journal of the American Chemical Society* 86(10):1926–1928.

Staško, A., V. Brezová, S. Biskupič, and V. Mišík. 2007. The potential pitfalls of using 1,1-diphenyl-2-picrylhydrazyl to characterize antioxidants in mixed water solvents. *Free Radical Research* 41(4):379–390.

Steinmetz, K. A. and J. D. Potter. 1996. Vegetables, fruit, and cancer prevention: A review. *Journal of the American Dietetic Association* 96(10):1027–1039.

Takebayashi, J., A. Tai, E. Gohda, and I. Yamamoto. 2006. Characterization of the radical-scavenging reaction of 2-O-substituted ascorbic acid derivatives, AA-2G, AA-2P, and AA-2S: A kinetic and stoichiometric study. *Biological and Pharmaceutical Bulletin* 29(4):766–771.

Takeshita, K. and T. Ozawa. 2004. Recent progress in *in vivo* ESR spectroscopy. *Journal of Radiation Research* 45(3):373–384.

Tang, D., H.-J. Li, J. Chen, C.-W. Guo, and P. Li. 2008. Rapid and simple method for screening of natural antioxidants from Chinese herb Flos Lonicerae Japonicae by DPPH-HPLC-DAD-TOF/MS. *Journal of Separation Science* 31(20):3519–3526.

Tirzitis, G. and G. Bartosz. 2010. Determination of antiradical and antioxidant activity: Basic principles and new insights. *Acta Biochimica Polonica* 57(2):139–142.

Tsimogiannis, D. I. and V. Oreopoulou. 2004. Free radical scavenging and antioxidant activity of 5,7,3',4'-hydroxy-substituted flavonoids. *Innovative Food Science & Emerging Technologies* 5(4):523–528.

Tyurin, V. Y., J. Zhang, A. Glukhova, and E. R. Milaeva. 2011. Electrochemical antioxidative activity assay of metalloporphyrins bearing 2,6-di-*tert*-butylphenol groups based on electrochemical DPPH-test. *Macroheterocycles* 4 (3):211–212.

Ukeda, H., Y. Adachi, and M. Sawamura. 2002. Flow injection analysis of DPPH radical based on electron spin resonance. *Talanta* 58(6):1279–1283.

Umamaheswari, M., K. Asokkumar, R. Rathidevi, A. T. Sivashanmugam, V. Subhadradevi, and T. K. Ravi. 2007. Antiulcer and *in vitro* antioxidant activities of *Jasminum grandiflorum* L. *Journal of Ethnopharmacology* 110(3):464–470.

Valgimigli, L., J. T. Banks, K. U. Ingold, and J. Lusztyk. 1995. Kinetic solvent effects on hydroxylic hydrogen atom abstractions are independent of the nature of the abstracting radical. Two extreme tests using vitamin E and phenol. *Journal of the American Chemical Society* 117(40):9966–9971.

Van De V., F., A. Tarola, D. Güemes, and M. Pirovani. 2013. Bioactive compounds and antioxidant capacity of camarosa and selva strawberries (*Fragaria* × *ananassa* Duch.). *Foods* 2(2):120–131.

Wong, S. P., L. P. Leong, and J. H. William Koh. 2006. Antioxidant activities of aqueous extracts of selected plants. *Food Chemistry* 99(4):775–783.

Wu, J.-H., C.-Y. Huang, Y.-T. Tung, and S.-T. Chang. 2007. Online RP-HPLC-DPPH screening method for detection of radical-scavenging phytochemicals from flowers of *Acacia confusa*. *Journal of Agricultural and Food Chemistry* 56(2):328–332.

Xie, C., H. Koshino, Y. Esumi, S. Takahashi, K. Yoshikawa, and N. Abe. 2005. Vialinin A, a novel 2,2-diphenyl-1-picrylhydrazyl (DPPH) radical scavenger from an edible mushroom in China. *Bioscience, Biotechnology, and Biochemistry* 69(12):2326–2332.

Xu, J., S. Chen, and Q. Hu. 2005. Antioxidant activity of brown pigment and extracts from black sesame seed (*Sesamum indicum* L.). *Food Chemistry* 91(1):79–83.

Yamaguti-Sasaki, E., L. A. Ito, V. C. D. Canteli, T. M. A. Ushirobira, T. Ueda-Nakamura, B. P. Dias Filho, C. V. Nakamura, and J. C. P. De Mello. 2007. Antioxidant capacity and *in vitro* prevention of dental plaque formation by extracts and condensed tannins of *Paullinia cupana*. *Molecules* 12(8):1950–1963.

Yan, R., Y. Cao, and B. Yang. 2014. HPLC-DPPH screening method for evaluation of antioxidant compounds extracted from Semen Oroxyli. *Molecules* 19(4):4409–4417.

Yao, H., Y. Chen, P. Shi, J. Hu, S. Li, L. Huang, J. Lin, and X. Lin. 2012. Screening and quantitative analysis of antioxidants in the fruits of *Livistona chinensis* R. Br using HPLC-DAD–ESI/MS coupled with pre-column DPPH assay. *Food Chemistry* 135(4):2802–2807.

Zhang, Q., E. J. C. van der Klift, H.-G. Janssen, and T. A. van Beek. 2009. An on-line normal-phase high performance liquid chromatography method for the rapid detection of radical scavengers in non-polar food matrixes. *Journal of Chromatography A* 1216(43):7268–7274.

Zhang, Y., Q. Li, H. Xing, X. Lu, L. Zhao, K. Qu, and K. Bi. 2013. Evaluation of antioxidant activity of ten compounds in different tea samples by means of an on-line HPLC–DPPH assay. *Food Research International* 53(2):847–856.

Zhang, Y., S. Shi, Y. Wang, and K. Huang. 2011. Target-guided isolation and purification of antioxidants from *Selaginella sinensis* by offline coupling of DPPH-HPLC and HSCCC experiments. *Journal of Chromatography B* 879(2):191–196.

Zhao, D. and N. P. Shah. 2014. Antiradical and tea polyphenol-stabilizing ability of functional fermented soymilk–tea beverage. *Food Chemistry* 158:262–269.

Zhuang, Q.-k., F. Scholz, and F. Pragst. 1999. The voltammetric behaviour of solid 2,2-diphenyl-1-picrylhydrazyl (DPPH) microparticles. *Electrochemistry Communications* 1(9):406–410.

CHAPTER 30

Ferric Reducing Antioxidant Power Method Adapted to FIA

Alessandro Campos Martins and Vitor de Cinque Almeida

CONTENTS

30.1 INTRODUCTION

Antioxidants are substances that protect living tissues from the damage induced by free radicals, which arise from aerobic breathing as a consequence of incomplete reduction of molecular oxygen and are directly related to several diseases, such as cancer, sclerosis, and aging skin (Krishnaiah et al., 2011). The human organism has the potential to inhibit the action of free radicals by enzyme complexes, but a daily ingestion of nonenzymatic antioxidant compounds that contribute to free radical scavenging is also necessary (Yilmaz and Toledo, 2004).

Vegetables, fruits, and beverages such as tea and wine are the main source of antioxidant compounds, and their ingestion is related to a healthy diet afar from free radicals. Hence, because of the importance of antioxidants, several chemical tests, such as Folin–Ciocalteu, 2,2-azinobis-(3-ethylbenzothiazoline-6-sulfonic acid) (ABTS), 2,2-diphenyl-1-picrylhydrazyl (DPPH), and ferric reducing antioxidant power (FRAP), have been proposed to evaluate the antioxidant capacity in foods and natural extracts. In a FRAP assay, the antioxidant property is evaluated from a redox reaction that occurs between the antioxidant (electron donor) and Fe^{3+} (electron acceptor), producing Fe^{2+} ions. This reaction is assessed spectrophotometrically by a change in color of the FRAP solution from translucent to dark blue, which absorbs in the wavelength of 593 nm (Benzie and Strain, 1996).

Flow injection analysis (FIA) systems are recognized as excellent solution managers (Krug et al., 1986). Among the FIA advantages, foremost are the possibilities for automation and miniaturization and their low cost, ease in operation, low sample and reagent consumption, versatility, and high analytical frequency. Also, the sample can be quantified before reaching the chemical equilibrium of the reaction (Ruzicka and Hansen, 1988).

Methodologies applied to determine antioxidant capacity such as ABTS, DPPH, Folin–Ciocalteu, and FRAP are routinely used; therefore, automation using flow injection (FI)-based methods can offer several advantages, since these systems provide a strict control of reaction conditions in both space and time, which is essential for the determination of species sensitive to environmental conditions (light, temperature, presence of O_2). In addition, sample throughput is enhanced when compared with conventional batch methods (Magalhães et al., 2009). In this respect, several FI arrangements have been adapted to antioxidant capacity assays, from a simple FIA system (Pellegrini et al., 2003; Mrazek et al., 2012) to more sophisticated ones, such as sequential injection analysis (SIA) (Lima et al., 2005) and multisyringe flow injection analysis (MSFIA) (Magalhães et al., 2006).

30.2 ANTIOXIDANTS

Antioxidants are natural or artificial substances that, in low concentrations, are able to inhibit or prevent biological substrate oxidation caused by free radicals (Halliwell et al., 1995). Free radicals, in turn, are species that can be produced from an endogenous source, for example, during aerobic respiration as a consequence of the incomplete reduction of oxygen molecules, or from an exogenous source, for example, toxic gas inhalation, smoking, and exposure to gamma or ultraviolet radiation (Halliwell et al., 1995). Most of these radicals are reactive oxygen species (ROS) such as superoxide ($O_2^{\bullet -}$), peroxide ($ROO^{\bullet}$), alkoxide ($RO^{\bullet}$), and nitric oxide ($NO^{\bullet}$). Among them, there are ROS that are not radicals but are also as harmful, such as singlet oxygen ($^1O_2$) and hydrogen peroxide (H_2O_2) (Halliwell and Gutteridge, 2007). The human organism is capable of inhibiting the action of free radicals by means of regulatory enzymes, such as dismutase superoxide, which converts a superoxide radical in hydrogen peroxide, and catalase, which converts the hydrogen peroxide into water and molecular oxygen (Hassain et al., 1995). However, this regulatory system is not completely effective, and natural antioxidant compounds from vegetable sources, such as vitamins, flavonoids, phenolates, and others, have to be ingested to help the organism scavenge the free radicals (Yilmaz and Toledo, 2004).

In complex samples, such as plant extracts, separating each component with antioxidant properties and studying it individually is costly and inefficient (Huang et al., 2005). Furthermore, the antioxidant capacity of all these compounds together is higher than the sum of each component separately, that is, a synergistic interaction among them improves the antioxidant capacity of the whole sample. Hence, chemical screening tests based upon total antioxidant capacity are more representative, besides being cost efficient, and are easy to perform using simple chromogenic tests.

Because of the importance of antioxidants, several chemical assays have been proposed to determine the antioxidant capacity of foods, beverages, and plant extracts. Since the antioxidant is not represented by a specific chemical group, these assays usually are based upon two properties, radical scavenging and electron transfer. Among those tests based upon radical scavenging, foremost are the DPPH and ABTS assays, in which antioxidant capacity is assessed by discoloration of a colorful solution of a stable radical. Among those based on electron transfer, foremost is FRAP, in which antioxidant capacity is also assessed by a change in color, but from the redox reaction of the pair Fe^{3+}/Fe^{2+} complexed to 2,4,6-tripyridyl-s-triazine (TPTZ). These tests have been applied separately or together to antioxidant screening of a wide range of samples such as natural foods, beverages, olive oil, and plant extracts (Pellegrini et al., 2003).

30.3 FRAP ASSAY PRINCIPLE

Classified as an electron transfer method, the FRAP assay is based upon the antioxidant capacity a sample has to reduce the oxidizing species (free radicals) responsible for the oxidative stress that propagates in biological substrates. This test simulates electron donation from an antioxidant sample to an oxidizing species, represented by Fe^{3+} in stoichiometric excess that becomes Fe^{2+} when reduced and assessed by color change if complexed with TPTZ at low pH (Figure 30.1). The antioxidant power values are obtained by comparing the absorbance change at 593 nm in test reaction mixtures with those containing ferrous ions or another antioxidant standard (ascorbic acid, gallic acid, 6-hydroxy-2,5,7,8-tetramethylchromane-2-carboxylic acid [TROLOX]) in known concentration from a calibration curve. Hence, these values are usually expressed as molar equivalent of iron or antioxidant standard by sample amount.

This assay was first applied for the direct measurement of antioxidant power in blood plasma by Benzie and Strain (1996) as an alternative to the other *in vivo* methods based upon inhibition of oxidative species purposefully generated in the reaction mixture. In inhibition assays, antioxidant action induces a lag phase; exhaustion of antioxidant power is denoted by a change in signal, such as rate of oxygen utilization, fluorescence, or chemiluminescence; however, these assays used to be technically demanding and less sensitive. The FRAP assay is simple, fast, inexpensive, and robust, and it has also been adapted and used for *in vitro* assays to determine the antioxidant power of vegetable and plant extracts (Benzie and Szeto, 1999).

30.4 METHODOLOGY

The FRAP method was followed according to Benzie and Szeto (1999) for *in vitro* assay with modification. The FRAP reagent is prepared by mixing acetate buffer (0.3 mol L^{-1}) at pH 6.0, TPTZ (10 mmol L^{-1}), and $FeCl_3$ (20 mmol L^{-1}) in a ratio of 10:1:1. A sample volume of 100 µL is mixed with 3 mL of the FRAP reagent in test tubes, which are kept in darkness at 37°C for 30 min. The absorbance measurements are performed at 593 nm. The calibration curve is set up with ferrous standard solution under the same

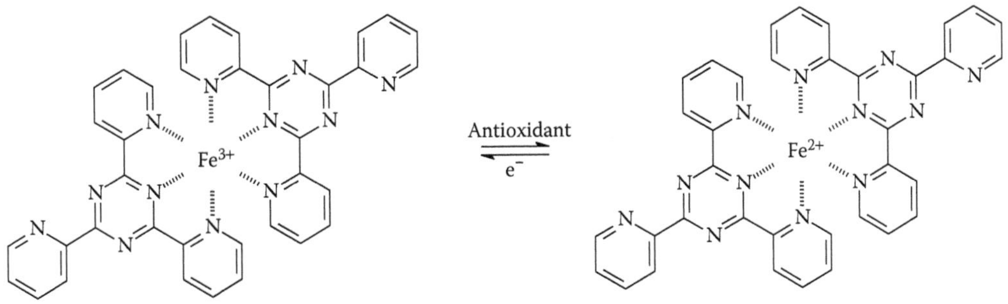

FIGURE 30.1   Ferric reducing reaction by antioxidant at presence of TPTZ. This reaction is screening at 593 nm.

conditions as the samples. After the dilution volume is corrected, the antioxidant capacity is expressed as molar amount of ferrous ion determined by calibration curve, by grams of dried sample or extract.

30.5 FIA SYSTEMS ADAPTED TO ANTIOXIDANT ASSAYS

The FIA system is defined as an automated method of an analytical procedure in which a liquid sample is injected into a continuous carrier flow, and this mixture is driven to the detection cell. During transport, the sample is subjected to several operations such as dilution, incubation, mixing, and dialysis (Ruzicka and Hansen, 1988). The monosegmented FIA is a kind of flow system in which the sample is transported between two air bubbles, but the flow is not continuous and the bubbles need to be purged before they enter the detection cell (Calatayud, 1996). The most sophisticated FIA manifolds are those based upon computer-controlled automated systems, such as SIA and MSFIA (Magalhães et al., 2009).

A common FIA manifold basically consists of an injector, peristaltic pump, analytical path (tubes, confluences), and detection cell. The injector has two positions, sampling and injection (Bergamin et al., 1978). The simpler FIA manifold is the single line, where the sample is injected directly into the reagent flow, homogenized appropriately in the coil, and driven to the detector (Figure 30.2a), as used by Mrazek et al. (2012) for a DPPH assay adapted to FIA. However, the manifold more frequently used is the double line, in which the sample is injected into a carrier flow, mixed with reagent flow through a confluence, and then homogenized in the coil and driven to the detector (Figure 30.2b), as used by Shpigun et al. (2006) for the determination of antioxidant activity of plant extracts based upon ferric reducing power. At the merging zone of the FIA manifold (Figure 30.2c), samples and reagents are injected simultaneously from distinct loopings, mixed at confluence, where the reaction starts, homogenized in the coil, and driven to the detector cell, as used by Martins et al. (2013) for FRAP assay adapted to FIA.

30.6 FIA DETECTION SYSTEM

A typical detector system in FIA usually records the signal against time in a continuous flow and must be sensible to variations of sample concentration. The most frequently used detector systems are potentiometric and spectrophotometric ones (Table 30.1) due to ease of assembly and their specificity, once the potential is proportional to concentration as well as to absorbance. Potentiometric detection was used by Shpigun et al. (2006) to assess antioxidant activity by iron reduction. The potential of the $[Fe(CN)_6]^{3-}$ was measured against time, when an antioxidant sample was injected, the potential decreased due to iron reduction to $[Fe(CN)_6]^{4-}$; a negative peak height (H) (measured in mV) has a logarithmic proportion to antioxidant concentration.

The height of transient signals recorded by an FIA system is proportional to the concentration of the sample. The sequence of a series of transient signals is called a "fiagram chart" (shown in Figure 30.3). These signals, however, must be well resolved, as thin peaks that characterize minimum dispersion can result from any factor affecting a sample dilution, such as a long analytical path, short looping, residency time, or injection of diluting samples. Even the laminar flow is a dispersion factor, because the sample must be spread in the coil to mix completely with the reagent, but in a turbulent flow regime, the sample does not react enough with the reagent (Calatayud, 1996).

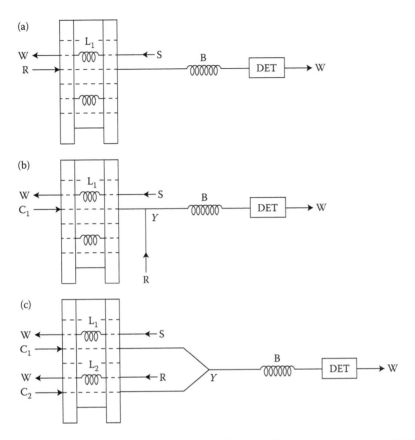

FIGURE 30.2   Single line FIA manifold (a) double line (b) and merging zone (c). The analytical path are composed by sample and reagent looping (L_1 and L_2), reagent flow (R), confluence (Y), coil (B), carrier flow (C_1 and C_2), waste (W) and detection cell.

To minimize the dispersion, these FIA parameters used to be widely analyzed univariately or multivariately in optimization studies and the height of the peak was commonly used as a response, usually the higher the better. Injections of a standard alone does not ensure the working range of the method; sensitivity measured from the calibration curve slope is more representative, since multiple injections of standards at different concentrations set a calibration curve. Figure 30.4 shows the deviation of the Beer–Lambert law occurring at a concentration higher than 0.5 mM of standard; an optimization study performed in this region and taking account of just the peak height would lead to biased results.

30.7 FRAP ASSAY ADAPTED TO MERGING ZONE FIA MANIFOLD

A simple manual acrylic injector, represented in Figure 30.5a, allows simultaneous injection from a merging zone FIA manifold. At the sampling position (Figure 30.3b), loops L_1 and L_2 are filled with the FRAP reagent and the antioxidant sample, respectively. At the insertion position (Figure 30.3c), both are carried by the carrier flow (C_1 and C_2) and mixed at the confluence (Y), where the reaction starts, homogenized in the coil (B), and driven to the detector cell, where the absorbance is recorded at 593 nm against time.

TABLE 30.1   Methods Based upon Ferric Reducing Adapted to Flow Injection Systems

Application	Oxidant	Detection	Manifold	Reference
Antioxidant activity of teas	$[Fe(TPTZ)_2]^{3+}$	Spectrophotometric	Merging zone	Martins et al. (2013)
Determination of ascorbic acid	$[Fe(TPTZ)_2]^{3+}$	Spectrophotometric	Triple line	Modun et al. (2012)
Determination of ascorbic acid	Fe(III)–DPPH	Spectrophotometric	Multiple line	Themelis et al. (2001)
Antioxidant activity of plant extract	$[Fe(CN)_6]^{3-}$	Potentiometric	Double line	Shpigun et al. (2006)
Determination of ascorbic acid	$Fe(III)–(C_5H_4N)_2$	Spectrophotometric	Double line	Kleszczewski and Kleszczewska (2002)
Determination of vitamin E	$[Fe(CN)_6]^{3-}$	Spectrophotometric	Single line	Jadoon et al. (2010)
Determination of total iron	$[Fe(TPTZ)_2]^{3+}$	Spectrophotometric	Triple line	Abdel-Azeem et al. (2013)
Determination of total iron	$Fe(II)–(C_{12}H_8N_2)$	Spectrophotometric	Triple line	Kozak et al. (2011)
Determination of amylase activity	$[Fe(CN)_6]^{3-}$	Potentiometric	Triple line	Ohura et al. (1998)
Determination of ascorbic acid	$Fe(II)–(C_{12}H_8N_2)$	Spectrophotometric	Single line	Zenki et al. (2004)
Determination of ascorbic acid	$[FeF_6]^{3-}$	FAAS	Multiple line	Jiang et al. (2001)
Determination of ascorbic acid	$Fe(II)–(C_{12}H_8N_2)$	Spectrophotometric	Single line	Pereira and Filho (1998)
Total curcuminoids	$[Fe(CN)_6]^{3-}$	Spectrophotometric	Triple line	Thongchai et al. (2009)

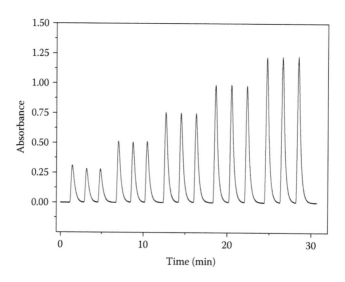

FIGURE 30.3 Transient signal recorder by triplicate injections of 5 (five) standards at crescent concentration.

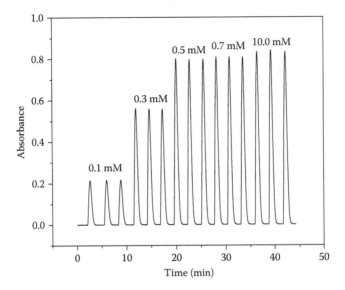

FIGURE 30.4   Calibration curve set up by five Fe^{2+} standards between 0.1–10 mmol L^{-1} injected in triplicate. Note the deviation of the Lambert-Beer Law at concentration higher than 0.5 mmol L^{-1}.

After detection, the resulting solutions are discarded at waste (W). In this adaptation, the FRAP reagent was prepared by mixing acetate buffer, $FeCl_3$, and TPTZ at the same concentration of the conventional method, but at proportions of 5:1:1.

Because of its high sensitivity, the FIA technique is more susceptible to interference, such as bubbles, sample turbidity, insoluble particles, and phase separation. Hence, the

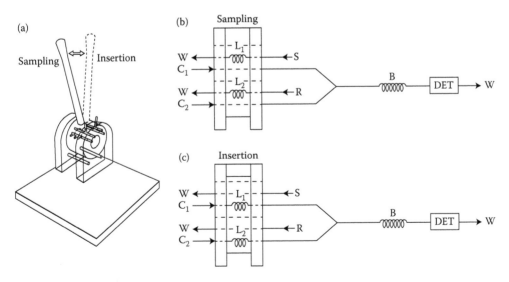

FIGURE 30.5   Commutator injector in acrylic (a) at sampling (b) and insertion (c) positions. L_1 and L_2 – looping; R – reactant; S – Sample; B – coil; W – waste; DET – detection system.

samples must be carefully prepared. Polar solvents such as aqueous or alcoholic extracts (Benzie and Strain, 1999) are recommended, once the FRAP reagent is an aqueous solution, that is, a polar medium. A homogeneous mix is also an important parameter in FIA, because a heterogeneous mix or residual nonsolubilized samples in the carrier flow causes light scattering in a spectrophotometric detection system.

Figure 30.6a shows a fiagram chart of 12 samples of tea aqueous extract injected in triplicate and five Fe^{2+} standards setting the calibration curve. The samples were diluted

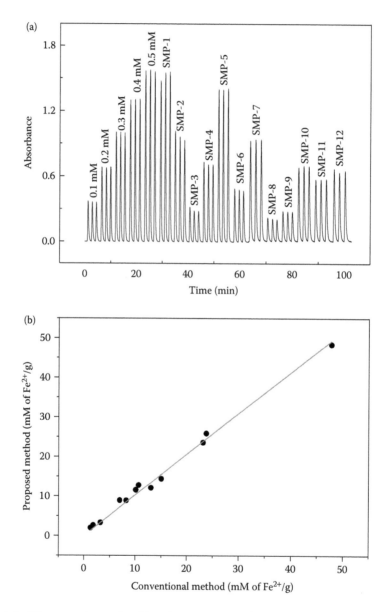

FIGURE 30.6   Fiagram chart of calibration curve standard and samples injected in triplicate (a) and the correlation between antioxidant capacity of the same sample get by FRAP essay conventional and proposed FIA method (b).

to fit in the work range, and the antioxidant power was determined as a concentration of Fe^{2+}, in mmol L^{-1}, by the amount of dried sample, in grams. In order to verify the applicability, Figure 30.6b shows a correlation between the results obtained by conventional and FIA method–adapted FRAP assay after correction by the conversion factor, as used by Shpigun et al. (2006). The conversion factor reflects any change of methodology conditions, such as dilution factor, reagent concentration, and equilibrium time. Otherwise, the FIA assays usually work under kinetic conditions, while the batch assays work under chemical equilibrium.

Since a few volumes of the FRAP reagent and sample are injected simultaneously, the assay is much more economical than for other FIA manifolds such as single or double lines, in which the FRAP reagent flow is constantly consumed even without a sample injection. At optimized conditions, this technique spares about 95% of the volume consumed of FRAP per analysis when compared with batch assays. Also, the analytical frequency of 30 sample/h is comparable to other antioxidant capacity assays adapted to the FIA system (Magalhães et al., 2009). In Figure 30.6a, the deviation among the heights of the peaks of triplicates is minimum for both standard and sample injection; low deviation implies good precision, and the correlation shown in Figure 30.6b implies that this technique reproduces the antioxidant capacity as well as the conventional assay; reproducibility implies that the method is accurate. Thus, besides accuracy and precision, this technique is also economic, fast, and easy to assemble in an FIA merging-zone manifold with spectrophotometric detection, making this technique much more suitable for routine antioxidant screening.

30.8 OTHER SIMILAR FIA ADAPTATIONS

Compared with other methodologies, there are only a few examples in the literature that describe the FRAP assay adapted to the FIA system. Drawbacks of the method, such as nonspecific reaction to some antioxidant compounds or irregular reaction times (Pulido et al., 2000; Huang et al., 2005; Prior et al., 2005), can be the main reasons for few studies of this assay. Nevertheless, other methods based on electron transfer, such as Folin–Ciocalteu, and those based on radical scavenging also have drawbacks and are widely studied and adapted to several flow injection systems (Magalhães et al., 2009).

FIA methods based on ferric reducing power similar to the FRAP principle reaction have been used for the determination of ascorbic acid, vitamin E, and total iron. Table 30.1 summarizes most of these studies. Spectrophotometric detection is also used in many of these studies because of the chromogenic effect from Fe oxidation, which changes with complex ligands, such as TPTZ, o-phenanthroline, cyanide, and DPPH. Different FIA manifolds have also been investigated and proposed, from single to multiple lines. Therefore, like other chromogenic antioxidant tests, the FRAP assay has the potential for several flow injection adaptations.

REFERENCES

Abdel-Azeem, S. M., N. R. Bader, H. M. Kuss, and M. F. El-Shahat. 2013. Determination of total iron in food samples after flow injection preconcentration on polyurethane foam functionalized with N,N-bis(salicylidene)-1,3-propanediamine. *Food Chem.* 138:1641–1647.

Benzie, I. F. F. and J. J. Strain. 1996. The ferric reducing ability of plasma (FRAP) as a measure of "antioxidant power": The FRAP assay. *Anal. Biochem.* 239:70–76.

Benzie, I. F. F. and Y. T. Szeto. 1999. Total antioxidant capacity of teas by the ferric reducing/antioxidant power assay. *J. Agric. Food Chem.* 47:633–636.

Benzie, I. F. and J. J. Strain. 1999. Ferric reducing/antioxidant power assay: Direct measure of total antioxidant activity of biological fluids and modified version for simultaneous measurement of total antioxidant power and ascorbic acid concentration. *Method Enzymol.* 299:15–27.

Bergamin, H., E. A. G. Zagatto, F. J. Krug, and B. F. Reis. 1978. Merging zones in flow injection analysis: Part 1. Double proportional injector and reagent consumption. *Anal. Chim. Acta* 101:17–23.

Calatayud, J. M. 1996. *Flow Injection Analysis of Pharmaceuticals: Automation in the Laboratory.* London: Taylor & Francis.

Halliwell, B., R. Aeschbach, J. Loliger, and O. I. Aruoma. 1995. The characterization of antioxidant. *Food Chem. Toxicol.* 33:601–617.

Halliwell, B. and J. M. C. Gutteridge, 2007. *Free Radicals in Biology and Medicine* (4th ed.), New York: Oxford University Press.

Hassain, S., W. Slikker, and S. F. Ali. 1995. Age-related changes in antioxidant enzymes, superoxide dismutase, catalase, glutathione peroxidase and glutathione in different regions of mouse brain. *Int. J. Dev. Neurosci.* 13:811–817.

Huang, D., B. Ou, and R. L. Prior. 2005. The chemistry behind antioxidant capacity assays. *J. Agric. Food Chem.* 53:1841–1856.

Jadoon, S., A. Waseem, M. Yaqoob, and A. Nabi. 2010. Flow injection spectrophotometric determination of vitamin E in pharmaceuticals, milk powder and blood serum using potassium ferricyanide–Fe(III) detection system. *Chinese Chem. Lett.* 21:712–715.

Jiang, Y. C., Z. Q. Zhang, and J. Zhang. 2001. Flow-injection, on-line concentrating and flame atomic absorption spectrometry for indirect determination of ascorbic acid based on the reduction of iron(III). *Anal. Chim. Acta* 435:351–355.

Kleszczewski, T. and E. Kleszczewska. 2002. Flow injection spectrophotometric determination of L-ascorbic acid in biological matter. *J. Pharm. Biomed. Anal.* 29:755–759.

Kozak, J., N. Jodlowska, M. Kozak, and P. Koscielniak. 2011. Simple flow injection method for simultaneous spectrophotometric determination of Fe(II) and Fe(III). *Anal. Chim. Acta* 702:213–217.

Krishnaiah, D., R. Sarbatly, and R. Nithyanandam. 2011. A review of the antioxidant potential of medicinal plant species. *Food Bioproducts Process.* 89:217–233.

Krug, F. J., H. F. Bergamin, and E. A. G. Zagatto. 1986. Comutation in flow analysis. *Anal. Chim. Acta* 179:103–118.

Lima, M. J. R., I. V. Tóth, and A. O. S. S. Rangel. 2005. A new approach for the sequential injection spectrophotometric determination of the total antioxidant activity. *Talanta* 68:207–213.

Magalhães, L. M., M. Santos, M. A. Segundo, S. Reis, and J. L. F. C. Lima. 2009. Flow injection based methods for fast screening of antioxidant capacity. *Talanta* 77:1559–1566.

Magalhães, L. M., M. A. Segundo, S. Reis, and J. L. F. C. Lima. 2006. Automatic method for determination of total antioxidant capacity using 2,2-diphenyl-1-picrylhydrazyl assay. *Anal. Chim. Acta* 558:310–318.

Martins, A. C., L. Bukman, A. M. M. Vargas et al. 2013. The antioxidant activity of teas measured by the FRAP method adapted to the FIA system: Optimizing the conditions using the response surface methodology. *Food Chem.* 138:574–580.

Modun, K. L., M. Biocic, and N. Radic. 2012. Indirect method for spectrophotometric determination of ascorbic acid in pharmaceutical preparations with 2,4,6-tripyridyl-*s*-triazine by flow-injection analysis. *Talanta* 96:174–179.

Mrazek, N., K. Watla-iad, S. Deachathai, and S. Suteerapataranon. 2012. Rapid antioxidant capacity screening in herbal extracts using a simple flow injection-spectrophotometric system. *Food Chem.* 132:544–548.

Ohura, H., T. Imato, Y. Asano, and S. Yamasaki. 1998. Potentiometric flow injection determination of amylase activity by using hexacyanoferrate(III)–hexacyanoferrate(II) potential buffer. *Talanta* 45:565–573.

Pellegrini, N., M. Serafini, B. Colombi et al. 2003. Total antioxidant capacity of plant, beverages and oils consumed in Italy assessed by three different *in vitro* assays. *Am. Soc. Nutr. Sci.* 133:2812–2819.

Pereira, A. V. and O. F. Filho. 1998. Spectrophotometric flow injection determination of L-ascorbic acid with a packed reactor containing ferric hydroxide. *Talanta* 47:11–18.

Prior, R. L., X. Wu, and K. Schaich. 2005. Standardized methods for the determination of antioxidant capacity and phenolics in foods and dietary supplements. *J. Agric. Food Chem.* 53:4290–4302.

Pulido, R., L. Bravo, and F. S. Calixto. 2000. Antioxidant activity of dietary polyphenols as determined by a modified ferric reducing/antioxidant power assay. *J. Agric. Food Chem.* 48:3396–3402.

Ruzicka, J. and E. H. Hansen, 1988. *Flow Injection Analysis* (2nd ed.). New York: John Wiley & Sons.

Shpigun, L. K., M. A. Arharova, K. Z. Brainina, and A. V. Ivanova. 2006. Flow injection potentiometric determination of total antioxidant activity of plant extracts. *Anal. Chim. Acta* 573:419–426.

Themelis, D. G., P. D. Tzanavaras, and F. S. Kika. 2001. On-line dilution flow injection manifold for the selective spectrophotometric determination of ascorbic acid based on the Fe(II)-2,2-dipyridyl-2-pyridylhydrazone complex formation. *Talanta* 55:127–134.

Thongchai, W., B. Liawruangrath, and S. Liawruangrath. 2009. Flow injection analysis of total curcuminoids in turmeric and total antioxidant capacity using 2,20-diphenyl-1-picrylhydrazyl assay. *Food Chem.* 112:494–499.

Yilmaz, Y. and R. T. Toledo. 2004. Health aspects of functional grape seed constituents. *Trends Food Sci. Technol.* 15:422–433.

Zenki, M., A. Tanishita, and T. Yokoyama. 2004. Repetitive determination of ascorbic acid using iron(III)-1.10-phenanthroline–peroxodisulfate system in a circulatory flow injection method. *Talanta* 64:1273–1277.

CHAPTER 31

2,2'-Azinobis-(3-Ethylbenzothiazoline-6-Sulfonic Acid)

L.K. Shpigun

CONTENTS

31.1 BASIC CONCEPTS

Intensive efforts are currently being made to discover natural or dietary antioxidants (AOXs) that significantly decrease the negative effects of reactive oxygen species on normal physiological functions in humans [1,2]. For example, it was stated that high levels of AOXs present in certain foods (fruits, vegetables, grains) and beverages (e.g., tea infusions) play a great role in protecting the gastrointestinal tract itself from oxidative damage and in delaying the development of stomach and rectal cancer [3]. In this connection, minor dietary constituents, especially plant-based foods, have come under intensive scrutiny. The literature on this subject is very extensive [4,5].

For estimating the *in vitro* antioxidative potential of biologically active substances, different experimental approaches have been reported [6,7]. Most of them are based on redox reactions with free radicals in aqueous or lipophilic media [8–11]. In particular, the reaction of AOXs with the resonance-stabilized radical monocation (or monoanion, if completely deprotonated sulfonate groups are considered) of 2,2′-azinobis(3-ethylben-zothiazoline-6-sulfonic acid), ABTS•+, is frequently employed [12–14], mainly due to its intensively blue-green color.

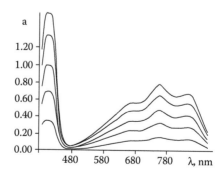

The ABTS•+ species have been characterized by different techniques such as cyclic voltammetry, UV-visible spectroscopy, electron spin resonance (ESR) spectroscopy, proton nuclear magnetic resonance (H-NMR) spectroscopy [13–16]. Figure 31.1 shows the UV–visible absorption spectrum of ABTS•+ (or monoanion, if completely deprot-onated sulfonate groups are considered) recorded in 0.1 M phosphate buffer solution (pH 7.4). As has been reported by some authors, its visible spectrum is characterized by rather high molar extinction coefficients at 414 nm ($\epsilon_{414\,nm} = 3.1 \times 10^4$ M^{-1} cm^{-1}), 645 nm ($\epsilon_{645\,nm} = 1.2 \times 10^4$ M^{-1} cm^{-1}), and 734 nm ($\epsilon_{734\,nm} = 1.34 \times 10^4$ M^{-1} cm^{-1}).

Because of their operational simplicity, ABTS•+-based assays are widely used in many research laboratories to study antioxidant capacity/activity. However, from an analytical point of view, many factors must be carefully controlled for reproducible results, as follows:

- The oxidizing agent used for the radical generation
- Age and storage conditions of ABTS•+ solution
- pH of ABTS•+ conversion
- Solvent phase and pH control agents
- Concentrations of ABTS•+ and AOX
- Reaction time between ABTS•+ and AOX
- Temperature of assay

FIGURE 31.1   Absorption spectrum of the ABTS radical cation. (Adapted from Re, R., et al. 1999. *Free Radic. Biol. Med.* 26:1231–1237.)

- Wavelength for monitoring ABTS$^{\bullet+}$
- AA/AC calculation method

It is well known that the ABTS$^{\bullet+}$ radical stability is dependent on the technique of its generation and on the experimental conditions under which it is maintained. In particular, its stability increases notably when temperature decreases. At room temperature (25°C), the radical ABTS$^{\bullet+}$ becomes relatively unstable and slowly disproportionate giving colorless 2,2'-azinobis(3-ethylbenzothiazoline-6-sulfonate) (ABTS) and the azo dication. Thus, the solution of the radical cation should be kept at 4°C (stable for several weeks) until use. In addition, its stability increases appreciably if the buffer solution is pretreated with Chelex-100.

31.1.1 Reactivity of the ABTS Radical Cation

Many AOXs can effectively scavenge the metastable ABTS$^{\bullet+}$ radical monocation in a concentration-dependent manner. During this interaction process, the green-blue radical is converted to its colorless form. The chemical schema of the reaction may be represented as follows:

$$n\text{ABTS}^{\bullet+} + \text{AOX} \rightarrow n\text{ABTS} + \text{P} \tag{31.1}$$

The process is usually monitored spectrophotometrically. Although other wavelengths such as 415 and 645 nm can be used in the ABTS-based assay, the near-infrared region at 734 nm is preferred due to less interference from plant pigments and sample turbidity [16]. Nevertheless, lower detection limits are achieved by monitoring the ABTS radical cation at 414 nm [13].

It might be noted that ABTS$^{\bullet+}$ reacts with AOXs in accordance with the electron transfer mechanism, unlike other free radicals, being an acceptor of hydrogen atom. Therefore, the relative strength of ABTS$^{\bullet+}$ can be evaluated using cyclic voltammetric measurements. Figure 31.2 shows typical cyclic voltammograms of the ABTS radical

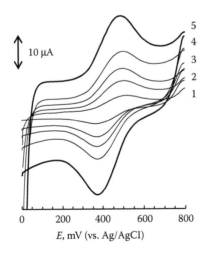

FIGURE 31.2 Cyclic voltammograms of ABTS$^{\bullet+}$ at a glassy-carbon electrode (PBS, pH 7.4). Scan rate (mV s^{-1}): 1–25; 2–50; 3–75; 4–100; 5–150.

recorded in 0.1 M phosphate buffer solution (pH 7.4) at a glassy-carbon electrode (GCE). As it can be seen, two well-defined peaks are observed in the potential range from 0.0 to 0.8 V (vs. Ag/AgCl). The cathodic peak is caused by the reduction of the radical to its original neutral form, ABTS. The anodic peak corresponds to the diffusion-controlled one-electron oxidation of ABTS back to ABTS$^{\bullet+}$. The formal redox potential ($E^{\circ\prime}$) for this redox pair was found to be +0.435 V at the scan rate υ of 100 mV s^{-1}.

Thermodynamically, any substance that has a redox potential lower than that of ABTS$^{\bullet+}$ may react with this radical. In other words, ABTS$^{\bullet+}$ is rather powerful one-electron oxidant for many common AOXs, such as L-ascorbic acid (Asc) and phenolic compounds (ArOH) (Figure 31.3). The reaction equations may be summarized as follows:

$$2\ ABTS^{\bullet+} + Asc \rightarrow 2\ ABTS + DHAsc \tag{31.2}$$

$$ABTS^{\bullet+} + ArOH \rightarrow ABTS + ArO^{\bullet} + H^+ \tag{31.3}$$

However, the rate of scavenging of ABTS$^{\bullet+}$ radicals by weak AOXs that have higher redox potentials or form stable intermediates was found to be very slow. This suggests that an ABTS$^{\bullet+}$-based assay may be selective for active AOXs in the presence of weak or inactive substances.

It should be kept in mind that antioxidant properties of various water-soluble and lipid-soluble AOXs are a complex function of many variables such as reaction kinetics, temperature, concentrations of an antioxidant, and pH [16–19]. Therefore, the influence of all these factors, especially the duration of interaction between AOXs and ABTS$^{\bullet+}$, must be taken into account when estimating their free radical–scavenging capacity. Moreover, it is important to consider that the initial rate of ABTS$^{\bullet+}$ scavenging can depend on the antioxidant concentration.

Some investigators have discussed structure–antioxidative ability relationships for flavonoids and phenolic acids as bioactive components of food [20–28]. For phenolic compounds, the scavenging capacity of ABTS$^{\bullet+}$ was found to be influenced by a number of structural features; the most significant being the position and degree of hydroxylation,

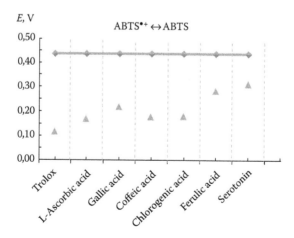

FIGURE 31.3 Experimentally recorded oxidation potentials for a variety of AOXs compared with the formal potential of the ABTS$^{\bullet+}$/ABTS redox couple. Background electrolyte: 0.1 M PBS, pH 7.4.

the number of double bonds in the heterocyclic ring, methoxy substitution, and steric hindrance. It was also shown that in the evaluation of the antioxidant power of bio-compounds, not only is the chemical structure important but also the type of reaction products that are formed, as they may show scavenging activity against $ABTS^{\bullet+}$.

Obviously, the concentrations of specific antioxidants cannot predict antioxidant status of complex samples that depends on a variety of AOXs, some of which might not be detected. Another shortcoming of quantitation of specific antioxidants results from possible synergistic or antagonistic effects of AOXs in the mixtures and the limited knowledge of the antioxidant substrates contained in some samples. For this reason, analytical chemists are interested in the determination of total antioxidant activity/capacity applied for use in the food industry [29].

31.1.2 Experimental Approaches to ABTS-Based Assay Development

The idea of employing the $ABTS^{\bullet+}$ radical cation for testing *in vitro* antioxidant abilities of natural and synthetic substances was first suggested by Miller et al. [12]. This provided a basis for the development of two types of assays that may be used for these purposes, namely, (1) decolorization assays involving the preliminary generation of radical cation $ABTS^{\bullet+}$, followed by measuring the ability of the samples to suppress the absorbance of this radical cation at a fixed time point; and (2) kinetic assays based on the radical cation formation in the presence of potential AOXs.

To date, there are few experimental approaches using the $ABTS^{\bullet+}$ radical cation for evaluating or predicting the antioxidative potency of individual biocompounds or complex matrices such as fruits, vegetables, and beverages. For decolorizing-type assays, the reaction between the preformed $ABTS^{\bullet+}$ and AOXs might proceed until the disappearance of these radicals. The extent of decolorization of the $ABTS^{\bullet+}$ radical cation is determined as a function of concentration and time. This actually makes the selection of the end-point time of an assay more arbitrary and hence more difficult.

In the case of inhibition assays, the order of addition of reagents and sample has been criticized as a major pitfall, because antioxidants (e.g., quercetin) can react with H_2O_2 and/or with oxidizing species that inhibit the $ABTS^{\bullet+}$ radical formation, leading to over-estimation of antioxidant capacity. Therefore, a post-addition assay or decolorization strategy seems to be more reliable and less susceptible to artifacts.

31.1.3 Antioxidant Activity/Capacity Values Calculations

It is notable that the terms "antioxidant activity" and "antioxidant capacity" have different meanings [11]: antioxidant activity deals with the kinetics of a reaction between an AOX and the radical; antioxidant capacity measures the thermodynamic conversion efficiency of an oxidant probe upon reaction with an AOX.

A variety of parameters can be used for expressing results of the evaluation of these characteristics. One of the commonly used parameters is the percent loss of color or the percentage of color remaining at a given time point. The $ABTS^{\bullet+}$-scavenging effect (P) can be calculated as

$$P(\%) = \left(A_{\text{blank}} - \frac{A_{\text{test}}}{A_{\text{blank}}} \right) \times 100 \qquad (31.4)$$

$$P(\%) = 100 - \left[\left(\frac{A_{test}}{A_{blank}} \right) \times 100 \right] \qquad (31.5)$$

where A_{test} is an absorbance value of the test solution obtained at a fixed time and A_{blank} is an absorbance value of the solution containing ABTS$^{\bullet+}$ itself in the absence of added AOXs.

The calculation of these parameters is derived from both antioxidant concentration and reaction time. Therefore, while measuring the abilities of testing compounds to scavenge ABTS$^{\bullet+}$ in comparison with the standard antioxidant, the variation of values with time should be taken into account.

The antioxidative activity can also be evaluated by calculating the area under the curve, derived from plotting the gradient of the percentage inhibition/concentration plots as a function of the reaction time. The length of the lag time to end-point change was measured; samples with higher antioxidant activity suppress the change far longer than those with less activity. The period of the lag phase before the reaction starts is proportional to the concentration of an antioxidant in the sample (Figure 31.4) [30]. This parameter can be determined by computer linear extrapolation of the steady-state rate to zero absorbance [13].

Some other features are used to measure antioxidant power, for example, an antiradical capacity (EC_{50} value), which represents the amount of AOX required to reduce the initial concentration (absorbance) of the radical by 50% [6]. This value can be determined graphically by plotting the absorbance against the AOX concentration or calculated by using the corresponding linear regression equation.

Time required to decrease the concentration of test free radical by 50% (T_{EC50}) can also be calculated [31]. Based on the T_{EC50} values, the kinetic behavior of AOXs can be classified as follows: < 5 min (rapid); 5–30 min (intermediate); > 30 min (slow). Antiradical efficiency (AE) [23] has also been proposed as a parameter to characterize AOXs where

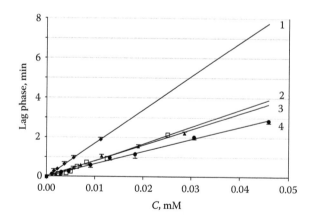

FIGURE 31.4 Lag phases of the four standard antioxidants: gallic acid (1), Trolox (2), ascorbic acid, (3) and uric acid (4), using different concentrations in the *TEAC* assay. (Adapted from Bondet, V., W. Brand-Williams, and C. Berset. 1997. *Food Sci. Technol.* 30:609–615.)

$$AE = \frac{1}{EC_{50}}(T_{EC50}) \qquad (31.6)$$

For a fixed set of conditions, AA could be defined, "independently" of the test method, as follows [6]:

$$AA = \frac{(t - t_{REF})}{[AOX]t_{REF}} \qquad (31.7)$$

where t is time for treated ABTS$^{\bullet+}$ to reach a set level of oxidation according to the test method; t_{REF} is time for untreated ABTS$^{\bullet+}$ or reference substrate to reach the same level of oxidation; and [AOX] is the concentration of AOX in suitable units. Consistent with this simple definition, AA would be zero if $t_{REF} = t$ and would become larger if t increased. Similar expressions could be written involving rates of oxidation.

A more meaningful parameter might be the relative antioxidant activity (R_{AA}). This can be expressed as

$$AOX_1 = \frac{AA_1}{AA_{REF}} \qquad (31.8)$$

where AA_1 and AA_{REF} are the activities of the test and reference antioxidants (AOX_1 and AOX_{REF}) at the same molar concentration, respectively.

Equation 31.8 can be rearranged to

$$AA_1 = R_{AA1} \times AA_{REF} \qquad (31.9)$$

This parameter gives the activity equivalence of a test substance relative to the reference substance, a common method of comparing activities.

Trolox (6-hydroxy-2,5,8-tetramethylchroman-2-carboxylic acid, Tr), a water-soluble analog of α-tocopherol, and L-ascorbic acid are usually used as standard reference compounds.

The Tr equivalent antioxidant capacity ($TEAC$) is the millimolar (mM) concentration of a Tr solution having the antioxidant capacity equivalent to a 1.0 mM solution of the sample (substance) under investigation (at the specific time point). In practice, $TEAC$ has been calculated in several ways.

As used by Miller and Rice-Evans [32], $TEAC$ reflects the relative ability of hydrogen- or electron-donating antioxidants to scavenge the ABTS radical cation compared with that of Tr. In the case of pure antioxidant compounds, the $TEAC$ values are determined according to the formula

$$TEAC = \frac{(\Delta A_{sample}\, c_{Tr})}{(\Delta A_{Tr}\, c_{sample})} \qquad (31.10)$$

where ΔA_{sample} and ΔA_{Tr} are changes of absorption after addition of sample and Tr solution, respectively; c_{Tr} and c_{sample} are the concentration of Tr standard solution and tested sample solution, respectively.

A *TEAC* value of an unknown sample can be calculated as follows:

$$TEAC = \frac{(\Delta A_{sample} \, c_{Tr})}{\Delta A_{Tr}} \qquad (31.11)$$

The L-ascorbic acid equivalent antioxidant capacity (*AEAC*) for food extracts (mg of Asc/100 g extract) can be calculated using the following equation:

$$AEAC = (\Delta A_{sample}/(\Delta A_{Asc}) \, c_{Asc}) \, V(100/W) \qquad (31.12)$$

where ΔA_{sample} and ΔA_{Asc} are the change of absorption after addition of plant extract and Asc, respectively; c_{Asc} is the concentration of the Asc standard solution (mg mL^{-1}); V is the volume of tested and standard solution (mL); and W is the mass of food extract (g). It should be noted, however, that the *TEAC* values are time dependent.

A number of experiments were performed to evaluate *in vitro* total antioxidant status of fruit juices, wines, olive oils, and various aqueous or organic food extracts [33–39]. For example, the ABTS radical-scavenging capacity of various solvent extracts of ginseng leaves with various concentrations was evaluated [35]. The ABTS radical-scavenging activity of ethanol extracts from ginseng leaves was found to be significantly higher than for their water extracts.

Scavenging of the ABTS radical cation was also applied to compare the total antioxidant activity (TAA) of several seasonings used in Asian cooking [36]. The *TEAC* activities of dark soy sauces were found to be exceptionally high.

The *TEAC* values of red wines were found to be 10 times higher than those of white wines, in accordance with their greater phenolic content, and there were no statistical differences between white and sherry wines (evaluated at 2 and 15 min) [38]. A comparison of the $TEAC_{2\,min}$ values for these wines with those obtained for other foods showed that one glass of red wine (125 mL) has the same antioxidant activity as 212 mL of grape juice [39], 190 mL of orange juice [40], 225 mL of black tea [40], 286 g of fresh spinach [41], or 926 g of tomatoes [41].

31.2 ANALYTICAL PROCEDURES FOR GENERATION OF THE ABTS RADICAL CATION

As it well known, the ABTS$^{\bullet+}$ generation process is based on the direct oxidation of a water-soluble colorless reagent—2,2′-azino-bis(3-ethylbenzothiazoline-6-sulfonate) diammonium salt (ABTS) [42]. The choice of the oxidation technique, which can be used for the transformation of this compound into the colored radical cation, is considered to be a critical point in the development of the antioxidant activity/capacity assays.

31.2.1 Chemical Generation Procedures

To date, a number of procedures have been proposed for the production of ABTS$^{\bullet+}$ from ABTS in solutions by using chemical reactions with a variety of oxidants. The original assay developed by Miller and Rice-Evans (sometimes referred to as *TEAC I*) utilized the metmyoglobin—hydrogen peroxide system to produce a highly reactive intermediate (HO$^{\bullet}$), to which ABTS donates an electron, generating ABTS$^{\bullet+}$ [12,22]. In this case, the sample to be tested is added before the formation of ABTS$^{•+}$. The test compounds/samples reduce the ABTS$^{•+}$ radicals formed, and the lag phase, which corresponds to the delay time in radical formation, is measured. However, faster reacting AOXs may contribute to the reduction of ferrylmyoglobin radical and lead to positive interference [43].

The enzymatic systems consisting of laccase or horseradish peroxidase (HRP), hydrogen peroxide, and ABTS have been investigated [13,34,44]. The overall process may be represented as follows:

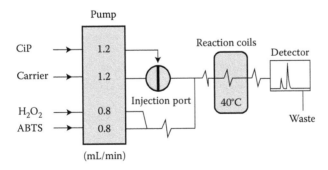

According to the manual procedure, the radical cations were generated by a reaction between 1.5 and 2.0 mM ABTS, 15–100 μM hydrogen peroxide, and 0.25 nM peroxidase in 50 mM glycine–HCl buffer solution (pH 4.5) [45].

To prepare enzyme-free ABTS$^{•+}$, 50 mL of 5 mM ABTS was added to 10 g of laccase immobilized on silica crystals and let stand at room temperature for approximately 7 d [46]. The oxidized ABTS solution was then separated from laccase by filtration and stored at 4°C.

Kadnikova and Kostic have described biocatalytical oxidation of ABTS by hydrogen peroxide using HRP encapsulated in sol–gel glass [47]. Campanella and coworkers have recently presented an electrochemical method for the determination of antioxidant capacity using a biosensor [48].

An automated flow injection (FI) method has been developed by performing the following catalytic reaction between ABTS, hydrogen peroxide, and *Coprinus cinereus* peroxidase (CiP) (pH 7.0) [49]:

$$2\ ABTS + H_2O_2 \rightarrow 2\ ABTS^{•+} + 2H_2O \tag{31.13}$$

The reaction was carried out by using an FI manifold (shown schematically in Figure 31.5). An aliquot of the CiP solution (40 μL) was injected into a carrier stream. The hydrogen peroxide and ABTS solutions were mixed in a 30-cm coil before they entered the carrier stream. The dehydrogenation of ABTS took place in a 200 cm coil at

FIGURE 31.5   FI system proposed for generating of ABTS$^{•+}$ by using an enzymatic reaction ABTS–H$_2$O$_2$. (Modified from Kadnikova, E. N. and N. M. Kostic. 2002. *J. Mol. Catal. B: Enzym.* 18:39–48.)

40°C ($t = 25$ s). Conditions for activation and stabilization of the enzyme were found, for example, ammonium sulfate acted as a peroxidase activator.

Production of ABTS$^{\bullet+}$ using hydrogen peroxide alone in acidic medium (the acetate buffer solution, pH 3.6) has been described [50]. In this case, the colored radical solution was stable for at least 6 months at 4°C.

The ABTS radical has been generated quantitatively by using a stoichiometric amount of potassium 12-tungstocobaltate(III) in 0.1 M citrate buffer solution (pH 5) [51]. The half-life of ABTS$^{\bullet+}$ was found to be 90 min under these conditions.

The ABTS radical cation has also been generated employing other oxidizing agents, such as bromine [52], periodate [53], or manganese dioxide [54,55]. Typically, the ABTS oxidation was performed by passing an aqueous solution of the substrate (pH 7.4) through manganese dioxide immobilized on a Whatman no. 5 filter paper [56]. Excess MnO_2 was removed from the filtrate by centrifugation and filtration. The resulting blue-green solution of the radical cation was kept at 4°C until use.

In the case of the reaction of ABTS with molecular bromine in water, the calculated rate constant for the stoichiometric process

$$2\ ABTS + Br_2 \rightarrow 2\ ABTS^{\bullet+} + 2\ Br^- \tag{31.14}$$

was 1.00×10^5 M^{-1} s^{-1} [52].

Several procedures have been proposed using the reaction between ABTS and the peroxyl radicals produced by the aerobic thermolysis of 2,2′-azobis(2-amidinopropane) (AAPH) at 45°C [57–59]. The oxidation by AAPH takes place with large activation energy and a low reaction order in ABTS. The mechanism of the process can be depicted in the following reactions [42]:

$$AAPH \rightarrow 2R^{\bullet} \tag{31.15}$$

$$R^{\bullet} + O_2 \rightarrow ROO^{\bullet} \tag{31.16}$$

$$ROO^{\bullet} + ABTS + H^+ \rightarrow ROOH + ABTS^{\bullet+} \tag{31.17}$$

Here, the reaction (31.15) is the rate-limiting step.

Regarding the use of the ABTS-based methodology for the evaluation of free-radical scavengers, the radical cations generated with the help of AAPH as an oxidant can be used only at low temperatures, under conditions in which further decomposition of the remaining AAPH is minimized.

The most common technique, proposed by Re et al. [14], involves the direct generation of ABTS$^{\bullet+}$ through the reaction between ABTS and peroxodisulfate anion as follows:

$$S_2O_8^{2-} + ABTS \rightarrow SO_4^{2-} + SO_4^{\bullet-} + ABTS^{\bullet+} \tag{31.18}$$

$$SO_4^{\bullet-} + ABTS \rightarrow SO_4^{2-} + ABTS^{\bullet+} \tag{31.19}$$

In practice, the radical cation can be prepared by mixing an aqueous solution of ABTS with potassium peroxodisulfate (PDS), followed by storage in the dark at room temperature for 12–16 h.

An extensive study of the stability of ABTS-derived radical cations has been performed employing PDS as an oxidant [42]. It was found that increasing the ratio of ABTS/PDS during the preparation of the radical cation increases the stability of the produced radical. The best results were obtained with the ABTS-derived radical generated in the reaction of PDS with an ABTS/PDS concentration ratio equal (or higher) to two. Under these conditions, about 60% of ABTS was oxidized into the radical cation form. However, it must be taken into account that a noticeable amount of unreacted ABTS remained, which in turn can influence the reaction rate of the radical with added AOXs.

The kinetics of ABTS$^{\bullet+}$ formation has been studied [47,60–62]. The results showed that the substrate oxidation rate depends on the structure and the concentration of the oxidant as well as on the temperature and pH of the reaction medium. In particular, Figure 31.6 demonstrates the rate of ABTS$^{\bullet+}$ formation in the presence of AAPH and PDS as oxidants [42]. Noticeable differences in the mechanisms of the investigated processes have been indicated. In particular, under fixed experimental conditions (45–50°C), initial rates of the radical cation formation were (0.96 ± 0.05) μM min^{-1} and (2.0 ± 0.3) μM min^{-1} for AAPH and PDS, respectively.

The ABTS$^{\bullet+}$-generation processes, which occur through interactions of ABTS with AAPH, PDS, and KIO$_4$, have been investigated [63]. The experiments were carried out by using an FI system (shown schematically in Figure 31.7). The change in the absorbance at 415 nm was measured to monitor the produced radical cation. According to the results obtained, the kinetics of ABTS oxidation is slow and the reactions take hours (Figure 31.8, curves 1–3). However, the periodate oxidation reaction was found to be considerably accelerated in the presence of manganese (II) [64].

Figure 31.8 (curve 3') demonstrates the effect of trace amounts of manganese (II) on the rate of ABTS$^{\bullet+}$ generation. The catalytic reaction can be expressed in terms of the following schema:

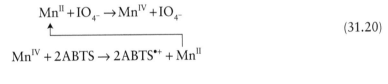

$$Mn^{II} + IO_{4^-} \rightarrow Mn^{IV} + IO_{4^-}$$

$$Mn^{IV} + 2ABTS \rightarrow 2ABTS^{\bullet+} + Mn^{II}$$

(31.20)

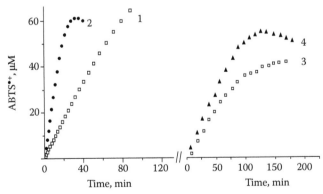

FIGURE 31.6 Increase in the ABTS$^{\bullet+}$ concentration when ABTS (150 μM) was incubated in the presence of PDS (300 μM) at 25°C (1) and at 50°C (2); incubated in the presence of AAPH (5 μM, 45°C) under air (3) and oxygen (4). (Modified from Henriquez, C., C. Aliaga, and E. Lissi. 2002. *Int. J. Chem. Kinet.* 34:659–665.)

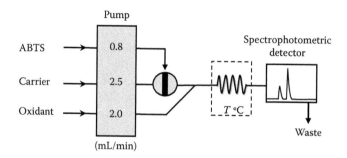

FIGURE 31.7   FI system used for the kinetic study of ABTS–oxidant interactions.

The results obtained showed that the presence of manganese ions increased the rate of ABTS$^{•+}$ generation in a concentration-dependent manner.

In order to choose optimal FI conditions for generating ABTS$^{•+}$, the effect of reaction temperature was examined over the range 20–60°C. It was found that absorbance intensity became larger as the temperature in RC_1 increased up to 40°C and then began to decrease. The relation of the peak height (H) to acidity in the range 4–9 looked like a curve with a maximum at pH 7.4. The optimum flow rates of 1.2 and 0.8 mL min^{-1} were obtained for the substrate and potassium periodate solution streams, respectively. A reactor of 120 cm was selected as appropriate. Under optimal experimental conditions, the dependence of H on the initial concentration of the substrate in the range 0.1–1.0 mM was described by the following linear equation [63]:

$$H(\text{units } A_{645\text{nm}}) = 1.086[\text{ABTS}] - 0.002 (r = 0.9998) \tag{31.21}$$

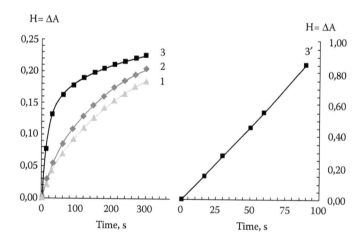

FIGURE 31.8   Comparison between kinetic curves obtained for the ABTS oxidation under the stopped-flow conditions in the FI system (Figure 31.7). Oxidant: 1—AAPH, 2—PDS, 3 and 3′—KIO$_4$ in the absence and in the presence of manganese (II) ions, respectively.

31.2.2 Electrochemical Generation Procedures

Published electrochemical studies have shown that ABTS could be electrochemically oxidized to both the radical cation (ABTS•+) and dication (ABTS^{2+}) [15,47,65–69]. As can be seen in Figure 31.9, the cyclic voltammogram exhibits two well-defined anodic peaks associated with two electrochemical processes. The first peak corresponds to the one-electron queasy-reversible oxidation of ABTS to ABTS•+. The second anodic peak corresponds to the oxidation of ABTS•+ to ABTS^{2+}, but these red-colored species are not stable. The overall electrode process may be represented by the schema shown in Figure 31.10.

The existence of a radical cation formed by cycling potential in the range 0.0–0.8 V was confirmed by registration of a characteristic absorption spectrum (Figure 31.1). No changes in the recorded spectra were observed when oxidation between 0.650 and 0.725 V, indicating that the radical cation was the only species stable in that potential range. At potentials greater than 725 mV, oxidation of the ABTS radical cation occurred, and the formation of a new species, marked with two absorption maxima at 264 and 295 nm, was observed.

An electrolytic system for ABTS•+ generation has been designed; it consists of a cathodic cell containing a saturated solution of $Zn(CH_3COO)_2$, a saline bridge containing a saturated solution of KCl, and an anodic cell with a solution of ABTS in a phosphate buffer medium (pH 6) [70].

ABTS•+ production has been described by using thin-layer spectroelectrochemistry [46]. Fifty microliters of ABTS in 0.1 M acetate buffer solution (pH 5) was oxidized in a quartz flat cell (0.1 cm width) containing an optical transparent thin-layer electrode (OTTLE system). The radical formation was measured with a potential scan from 0.65 to 0.70 V and returned to 0.65 V at the scan rate of 0.05 mV s^{-1}. The reactions were monitored spectrophotometrically every 30 s. Figure 31.11 shows the 3D plots obtained for spectral changes during ABTS electrolysis at different intervals. At the beginning, with ABTS as the sole species, two peaks were observed ($\lambda_1 = 214$ nm, $\lambda_2 = 340$ nm), but as the

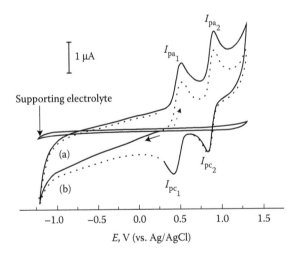

FIGURE 31.9   Typical cyclic voltammograms of 0.2 mM ABTS in a 0.1 M acetate buffer solution (pH 5) at a glassy-carbon electrode (0.0706 cm^2). (Modified from Thomas, J. H. et al. 2004. *Electroanalysis* 16:547–555.)

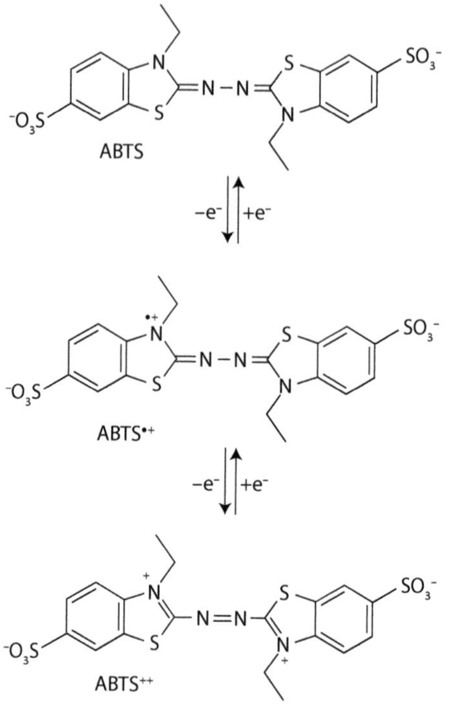

604   Flow Injection Analysis of Food Additives

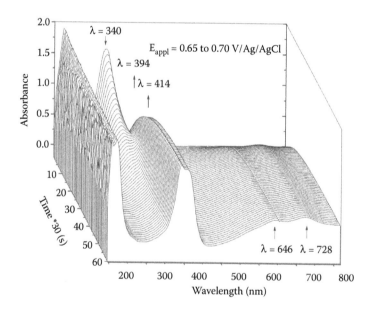

FIGURE 31.10 Chemical schema of the two oxidation steps of ABTS. (Modified from Branchi, B., C. Galli, and P. Gentili. 2005. *Org. Biomol. Chem.* 3:2604–2614.)

FIGURE 31.11 Spectral changes during the electrolysis of 50.0 μM ABTS in 0.1 M acetate buffer solution (pH 5). The spectra were acquired every 30 s during 1 h of electrolysis at the controlled potential in the range of 0.65–0.70 V. (Adapted from Solıs-Oba, M. et al. 2005. *J. Electroanal. Chem.* 579:59–66.)

electrolysis proceeded, four new peaks appeared (λ_3 = 394 nm, λ_4 = 414 nm, λ_5 = 646 nm, and λ_6 = 728 nm), whereas $A_{\lambda2}$ decreased. Also, interference appeared between ABTS and ABTS$^{\bullet+}$, with an absorbance near 414 nm. This interference was associated with the complex formation between ABTS and ABTS$^{\bullet+}$. Because of the presence of this complex, a correction of absorbance should be performed when spectrophotometric measurements are carried out in the 405–436 nm range.

Since the ABTS$^{\bullet+}$ radical cation can be reversibly reduced back to ABTS, the three-electrode flow-through cell was designed with anode and cathode compartments separated with a Nafion membrane (Figure 31.12) [71]. A tubular working electrode (WE) was designed as an assembly of 32 short bores concentrically drilled through the graphite cylinder serving as the anode. In such a way, a high surface area of the WE was obtained and, consequently, a small ohmic drop through the cell was achieved. The WE potential was 700 ± 5 mV to ensure that ABTS was oxidized only to ABTS$^{\bullet+}$. When a steady state was achieved, the radical concentration at the exit of the cell was determined spectrophotometrically by measuring the absorbance at 734 nm. The conversion efficiency decreases as the flow rate increases, but at flow rates less than 0.2 mL min^{-1}, a satisfactory high conversion (60%) can be achieved.

Although ABTS$^{\bullet+}$ can be generated either potentiostatically or galvanostatically, the operation of the cell in the galvanostatic mode has some advantages over the operation under constant-potential conditions. When a constant current (i, μA) is imposed on the cell and the exact value of the solution flow rate (v, mL min^{-1}) is known, the c(ABTS$^{\bullet+}$) concentration (mM) in the solution leaving the cell can be calculated using the formula derived from Faraday's law of electrolysis [71]:

$$c(ABTS^{\bullet+}) = \frac{60i}{96500v} \tag{31.22}$$

A good agreement was obtained between the conversion efficiency calculated from Equation 31.22 and the conversion efficiency determined experimentally. Thus, the concentration of the ABTS radical cation can be accurately predicted and, if needed, it can be changed simply by adjusting the current imposed on the cell to some other value. Of

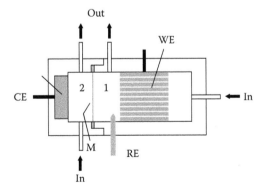

FIGURE 31.12 Schematic representation of the flow-through electrochemical cell used for online generation of the c(ABTS$^{\bullet+}$): 1, anode compartment; 2, cathode compartment; WE, working electrode; RE, reference electrode; CE, counter electrode; M, Nafion membrane. (Adapted from Ivekovič, D. et al. 2005. *Analyst* 130:708–714.)

course, to prevent the oxidation of the ABTS radical cation to the dication, care must be taken that the current imposed on the cell is never so large that the working electrode potential would exceed 750 mV.

Hence, it may be concluded that electrochemical techniques could be an attractive alternative to chemical procedures for generation of the the ABTS radical cation. Moreover, these techniques avoid the need to use other reagents and facilitate the control of the reaction, and thus provide great speed, reliability, and facility of application.

Overall, the ABTS•+-based assays offer many advantages that contribute to its widespread popularity in determining antioxidant status of a wide range of both pure bioactive compounds and foods. c(ABTS•+), being a singly positive charged radical, is soluble in both aqueous and organic solvents and is not affected by ionic strength, so it can been used in multiple media to determine both hydrophilic and lipophilic antioxidant activity/capacity. Moreover, the ABTS•+ scavenging can be evaluated over a wide pH range, which makes it useful to study the effect of pH on antioxidant mechanism. However, batch versions of the ABTS-based spectrophotometric assay are not appropriate for screening large series of samples.

Since a total antioxidant capacity (TAC) is thought to be one of the basic standards of biological foodstuff values, it is of great interest to develop automated variants of ABTS-based procedures. The methodology of stopped-flow analysis as well as FI analysis or its modifications introduced great possibilities for creating automated tools for these purposes [72,73].

31.3 STOPPED-FLOW MODIFICATION OF ABTS-BASED ASSAY

To study both fast- and slow-reacting antioxidants, a stopped-flow analyzer has been developed by Labrinea and Georgiou [17]. The principle of operation is based on the preliminary generation of ABTS•+, followed by measurement of the ability of the samples to supress the absorbance of this radical cation at a fixed time point. A schematic diagram of the laboratory-made analyzer is given in Figure 31.13. An antioxidant solution (5.0 or 10 µM) stream was mixed with an ABTS•+ stream while flowing to the spectrophotometric detector (18 µL flow cell, 1 cm path length). A combination of a 50 cm reaction coil and a total flow rate of 2.4 mL min⁻¹ was found to be sufficient for efficient mixing. After the flow was stopped, the ABTS•+ consumption was monitored using the absorption at 414 nm. This wavelength was selected among the available wavelengths for ABTS to achieve increased sensitivity. The MSDOS-based software was developed in C language and provides for fast, up to 60 kHz data acquisition by using direct memory access (DMA) data transfer from the AD converter.

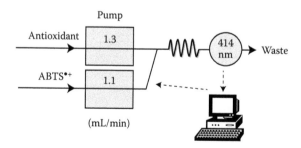

FIGURE 31.13 A laboratory-made stopped-flow analyzer for studying scavenging reactions. (Adapted from Lalbrinea, E. P., and C. A. Georgiou. 2004. *Anal. Chim. Acta* 526:63–68.)

TABLE 31.1   *TEAC* Values Obtained for Different Biocompounds (10.0 μM) by Using the Stopped-Flow Method

Antioxidant	pH 4.6		pH 7.4	
	10 s	360 s	10 s	360 s
L-Ascorbic acid	1.6 ± 0.1	1.6 ± 0.1	1.0 ± 0.1	1.0 ± 0.1
Caffeic acid	1.7 ± 0.1	1.7 ± 0.1	1.5 ± 0.1	1.9 ± 0.1
p-Coumaric acid	0.20 ± 0.02	1.8 ± 0.1	2.6 ± 0.1	3.3 ± 0.2
Ferulic acid	2.8 ± 0.2	3.1 ± 0.2	2.6 ± 0.2	3.4 ± 0.2
Gallic acid[a]	3.4 ± 0.3	4.9 ± 0.2	4.2 ± 0.3	5.8 ± 0.2
(+)-Catechin[a]	2.1 ± 0.1	4.3 ± 0.2	3.6 ± 0.1	5.3 ± 0.3
(−)-Epicatechin[a]	2.9 ± 0.1	5.7 ± 0.4	4.5 ± 0.2	6.4 ± 0.5
Quercetin[a]	2.7 ± 0.2	4.3 ± 0.2	4.5 ± 0.2	6.7 ± 0.4
Rutin	1.4 ± 0.1	1.6 ± 0.2	2.0 ± 0.2	3.7 ± 0.1
Glutathione	0.20 ± 0.03	2.1 ± 0.1	1.3 ± 0.1	2.1 ± 0.1
Albumin	0.20 ± 0.01	1.1 ± 0.1	1.0 ± 0.01	2.4 ± 0.1

Source: Data from Lalbrinea, E. P. and C. A. Georgiou. 2004. *Anal. Chim. Acta* 526:63–68.

[a] 5.0 μM.

ABTS•⁺ working solutions were prepared by mixing appropriate volumes of ABTS, H_2O_2, and HRP stock solutions in 0.02 M acetate buffer pH 4.6 and were stable for at least 2 d at 0–4°C. In order to obtain an ABTS•⁺ working solution of (0.9–1.1) absorbance units and to prevent possible reactions between AOXs and unreacted hydrogen peroxide, a 50-fold excess of ABTS was chosen.

The system was applied to monitor the radical-scavenging reaction for a number of AOXs at different pHs and reaction times. The *TEAC* values were found to be dependent on these parameters for almost all studied compounds (Table 31.1). Only Tr and Asc reacted rapidly (99% < 10 s) with ABTS•⁺, and their antioxidative capacity was not affected by the end-point time. Structurally similar compounds had the same pH-dependent behavior even if they differed significantly in *TEAC* values, for example, the *TEAC* values for (+)-catechin and (−)-epicatechin, caffeic acid, and ferulic acid. The same was observed for quercetin and its glycoside rutin; the higher activity of quercetin had been reported [74]. Independent of time and pH effects, quercetin, gallic acid, (+)-catechin, and (−)-epicatechin have shown higher activities.

The proposed stopped-flow method can be utilized for testing antioxidant compounds of unknown kinetics toward ABTS•⁺ at different pH values, and the results can be used to predict the total antioxidant capacity of structurally related compounds.

31.4 FLOW INJECTION METHODS FOR EVALUATION OF ABTS•⁺-SCAVENGING CAPACITY

The main problems in the development of FI methods for *in vitro* measurements of ABTS•⁺-scavenging capacity are associated with choosing the procedure of radical formation and the order of addition of reagents and sample. During the past years, researchers have focused their attention on automatic *TEAC* evaluation in a number of ways. At first, the radical cation was preformed prior to addition into an FI system.

31.4.1 FI Systems Based on Using Preformed ABTS Radical Cations

A *TEAC*-based FI method was first reported by Pellegrini and coworkers [40]. A single-channel manifold was used, in which a sample was injected directly into the ABTS$^{\bullet+}$ radical solution. After a sample containing AOXs was injected, a negative peak representing the decolorization of ABTS$^{\bullet+}$ was obtained, whose area was proportional to the concentration of ABTS$^{\bullet+}$ that was reduced. The *TEAC* value was calculated as the Tr concentration, providing a discoloration of ABTS$^{\bullet+}$ equal to that caused by the sample. The proposed method was shown to be very sensitive, with the detection limit of 4.14 μM of Tr. The proposed FI system demonstrated high repeatability and reproducibility and allowed the analysis of about 30 samples/h. It was applied to evaluate *in vitro* antioxidative ability of some pure compounds, beverages, and food extracts. *TEAC* values ranging from 0.09 mM for cola to 49.24 mM for espresso coffee were obtained. These results were correlated with those obtained by the original spectrophotometric ABTS-based assay.

An improved FI method was also developed by Bonpadre et al. [75], in order to evaluate the actual antioxidative features of complex samples. The method was demonstrated to be useful for screening total antioxidant activities of white wines rapidly, without dilution, with very limited handling of the sample, and with high repeatability. However, it was stated that the reaction time may be critical, depending on the nature of the substances employed.

An FI system in conjunction with a high-performance liquid chromatography separation module was constructed by Stewart et al. [76] and then modified by Takagaki et al. [77]. Each sample was injected and eluted with a 0.03% phosphoric acid stream. The eluting zone was then mixed with the ABTS radical solution, and the resulting mixture (pH 7.25) was continuously passed through a UV/Visible-spectrophotometric detector. The ABTS radical solution was preformed by dissolving ABTS (4 mM) in 50 mL of 3.5 mM $K_2S_2O_8$, which was kept for 30 min at room temperature, and then diluting it with 350 mL of 0.1 M phosphate buffer solution (pH 8.0). After standing overnight at room temperature, the resulting solution was maintained on ice in the dark. The applicability of the proposed technique was tested by measuring the antioxidative ability of common beverages. Tea phenolics from Kenyan green and black teas were identified using the relative radical-scavenging abilities of flavan-3-ols, caffeoylquinic acids, flavonols, and theaflavins in comparison with Tr. (−)-Epigallocatechin gallate (EGCg) was identified as the most potent antioxidant, with a *TEAC* value of 3.0, contributing approximately 30% of TAA of green tea. The ABTS radical-scavenging capacity of EGCg metabolites degraded by rat intestinal flora was also investigated [77].

The major drawback of the above-discussed FI methods is that the ABTS$^{\bullet+}$ solution must be prepared before the analysis, because the kinetics of ABTS oxidation are slow and the reaction takes many hours.

31.4.2 FI Systems Based on Using Online-Generated ABTS$^{\bullet+}$

Considerable progress for testing *in vitro* antioxidant capacity of food samples has been made by developing FI methods with *online* radical cation generation.

31.4.2.1 FI Systems Based on Enzymatic Production of Radicals

Ukeda [78] proposed a two-channel FI system with *online* biocatalytical oxidation of ABTS by hydrogen peroxide using a flow-through peroxidase reactor (2×35 mm^2) and spectrophotometric detection of the formed ABTS$^{\bullet+}$.

A fully automated FI method for antioxidant capacity assessment based on a low-cost laboratory-made analyzer was also reported [79]. Precision was better than 5% relative standard deviation (R.S.D.), and the linear range was 4–100 µM. The method was applied to pure compound, wine, and honey samples.

Later, Milardovic et al. [80,81] described FI systems employing continuous (bi)enzymatic production of ABTS$^{\bullet+}$ and its electrochemical detection (ECD) by using interdigitated electrodes (IDEs) as electrochemical sensors (Figure 31.14). One electron oxidation of ABTS into corresponding radical cation was achieved by interaction with hydrogen peroxide in the presence of peroxidase or glucose oxidase and peroxidase immobilized in polyacrylamide gel and packed into flow-through tubular reactors (4 mm in diameter and 4 cm in length). A sample (10 µL) was injected into the 50 mM phosphate buffer solution (pH 7.4) stream mixed with a stream of the reagent solution containing ABTS (2 mM) and H$_2$O$_2$ (120 µM). Electrochemical measurements were conducted at the potential imposed to IDEs of 100 mV (vs. Hg$_2$Cl$_2$| 3 M KCl). Figure 31.14c shows the signals obtained by injection of four common AOXs, such as Tr, Asc, glutathione, and uric acid. Biamperometric response of IDEs was linearly related to the Tr concentration up to 500 µM, the limit of detection was 6.5 µM. The time of analysis was about 85 s.

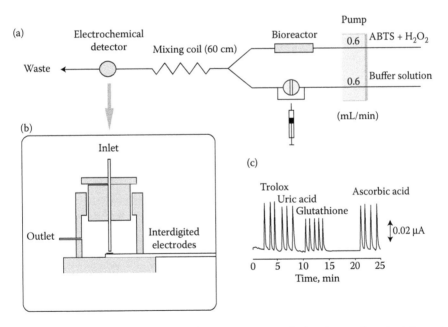

FIGURE 31.14 (a) Schematic diagram of an FI system with the electrochemical detection of online generation of ABTS$^{\bullet+}$. (b) Construction of a flow-through electrochemical detector. (c) FI signals obtained for water-soluble AOXs. The concentration of injected antioxidant solutions: 250 µM for Tr, Asc, and uric acid; 220 µM for reduced glutathione. (Modified from Milardovic, S., I. Kerekovic, and V. Rumenjak. 2007. *Food Chem.* 105:1688–1694.)

31.4.2.2 FI Systems Based on Chemical Production of ABTS•+

An FI method was developed for the spectrophotometric determination of free radical–scavenging capacity by using *online* generation of ABTS•+ [82]. A schematic diagram of the proposed system designed on the "reversed" FIA principles is shown in Figure 31.15. The radical generation based on the above-described procedure of the catalytic oxidation of ABTS diammonium salt by periodate ions took place in coil *RC1* (180 cm × 0.5 mm). An aliquot of the colored ABTS•+ solution (300 μL) was injected into the carrier stream, which then was mixed with a sample solution stream. The interaction of AOXs with ABTS•+ took place in coil *RC2* (120 cm × 0.5 mm). The ABTS radical cation absorbance at 645 nm was monitored by a detector placed after the reaction coil *RC2*. When the sample was mixed with an injected solution containing ABTS•+, the amount of the radicals scavenged by AOXs was determined by recording the absorbance decrease. The height *H* of the resulting "negative" peaks was proportional to the amount of the ABTS radical cation scavenged, and hence to the antioxidative ability of the sample. The proposed FI method offered a linear range between 1 and 20 μM of Tr and Asc according to the following regression equations:

$$H = (\Delta A_{645\,nm}) = -0.035[\text{Tr}] + 0.800 (r = 0.9996) \tag{31.23}$$

$$H = (\Delta A_{645\,nm}) = -0.023[\text{Asc}] + 0.788 (r = 0.9993) \tag{31.24}$$

The ABTS•+-scavenging capacity was evaluated as the percentage of free radicals scavenged as follows:

$$P(\%) = \left[\frac{(H_o - H_t)}{H_o} \right] \times 100 \tag{31.25}$$

where H_o and H_t are peak heights (in absorption units) measured without and with a sample at a fixed time of reaction *t*, respectively.

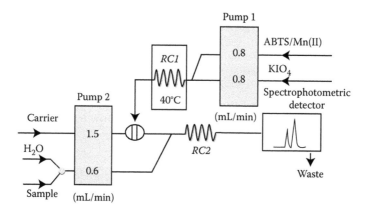

FIGURE 31.15   FI system based on catalytic chemical generation of ABTS•+ coupled with spectrophotometric monitoring of its interactions with AOXs.

31.4.2.3 FI Systems Based on Electrochemical Production of ABTS•+

To avoid the time-consuming step of the ABTS radical cation preparation and therefore to shorten the analysis time, an FI method was developed in which the radical cation was generated *online* by electrochemical oxidation of ABTS in the flow-through electrochemical cell forming a part of the system [71]. The configuration of the FI manifold is shown in Figure 31.16a. A 0.1 M phosphate buffer solution prepared in 35% (v/v) ethanol (or water, in the case of water-soluble AOXs) was used as a carrier stream. In the second stream, a solution of the ABTS radical cation was prepared by electrochemical oxidation of 1.0 mM ABTS in 0.1 M phosphate buffer solution (pH 7.40). The flow-through cell was operated under fixed-flow conditions, and the amount of ABTS•+ was controlled by the flow rate and the value of the current imposed on the cell. The desired ABTS radical cation/carrier ratio was set by changing the diameter of the pump tube pumping the reagent stream. The carrier and reagent streams were mixed by passing them through a reaction/mixing coil, and the ABTS radical cation absorbance at 734 nm was monitored by a detector placed after the coil. The absorbance intensity decreased with increasing antioxidant concentration in the reaction zone. Thus, negative peaks, whose heights correspond to the concentration of AOXs, were recorded.

To optimize the method, a set of experiments were performed in which the influence of the ABTS radical cation/carrier ratio, sample volume, flow rate, and mixing coil

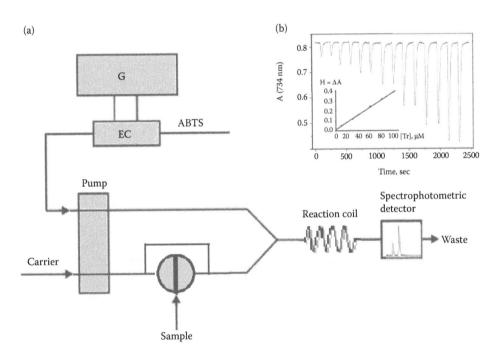

FIGURE 31.16   (a) FI system used for the evaluation of the antioxidant activity in which the ABTS radical cation was generated by electrochemical oxidation of ABTS in the flow-through electrolysis cell forming a part of the FIA system. EC, flow-through electrolysis cell; G, galvanostat. (b) FI peaks recorded for two injections of 10–100 μM Tr standard solutions. Insert: calibration graph for Tr in the range of 10–100 μM (Adapted from Iveković, D. et al. 2005. *Analyst* 130:708–714.)

TABLE 31.2   FI Methods Proposed for Evaluation of the ABTS Radical-Scavenging Capacity

Radical Generation Procedure	pH	Type of Detection	Sampling Rate (h^{-1})	R.S.D. (%)	Reference
Chemical oxidation by PDS	Unbuffered	Visible spectrophotometry (734 nm)	30	<1.7	[40]
	7.4		22	<2.7	[74]
	7.25		N/A	N/A	[76,77]
Enzymatic oxidation by H_2O_2	N/A	Visible spectrophotometry (417 nm)	60	2.0	[78]
	4.6	Visible spectrophotometry (414 nm)	120	<5.0	[79]
	7.4	Biamperometry	42	N/A	[80,81]
Oxidation by KIO_4 (Mn(II) as a catalyst)	7.4	Visible spectrophotometry (645 nm)	90	3–5	[82]
Electrochemical oxidation		Visible spectrophotometry (734 nm)	32	<2.0	[71]

volume on the FI response obtained upon the injection of a 20 μM Trolox solution was investigated. The injection volume of 100 μL was chosen as an optimum and employed in all further experiments. The peak width decreased as the flow rate increased, although at flow rates higher than 1.0 mL min^{-1}, the decrease was less pronounced than in the low–flow rates range. In the flow rate ranges 0.4–1.0 mL min^{-1} and 1.5–2.4 mL min^{-1}, the peak height increased with the increase of the flow rate, and it was almost independent of the flow rate in the range from 1.0 to 1.5 mL min^{-1}. Under the optimized conditions, a linear calibration graph for Tr was obtained over the range 10–100 μM, with a limit of detection of 1.6 μM (Figure 31.16b). Good reproducibility (R.S.D. 1.95%) and sample throughput (32 h^{-1}) were achieved.

TABLE 31.3   Comparison of Selected *TEAC* Values Obtained by the Different FI Methods and Classic Batch ABTS-based Assay

Compound	FI Method		TEAC (Batch Version)
	[71]	[80]	
Trolox	N/A	1.00	1.00
Ascorbic acid	1.00	0.99	0.96
Caffeic acid	1.04	N/A	0.94
Ferulic acid	1.83	N/A	1.89
Gallic acid	4.26	2.33	3.17
Quercetin	4.33	N/A	3.45
N-Acetyl-L-cysteine	1.16	1.25	1.48
Glutathione	1.14	0.82	0.92
Pyrogallol	2.87	N/A	2.25
Pyrocatechol	1.22	N/A	0.92
Ellagic acid	4.69	N/A	4.40

31.4.2.4 Analytical Features

The FI methods developed for automation of ABTS$^{•+}$-based assay are listed in Table 31.2. All the methods have been successfully employed for the rapid evaluation of the anti-oxidative abilities of pure compounds, beverages, and food extracts. The *TEAC* values were calculated according to formula (31.9) or (31.10) in case of pure compounds and unknown samples, respectively. As can be seen from the data presented in Table 31.3, the Tr equivalents obtained by the FI methods are in quite good agreement with each other and with the *TEAC* values determined by the classic spectrophotometric ABTS assay.

31.5 SEQUENTIAL INJECTION SYSTEMS BASED ON ABTS$^{•+}$ DECOLORIZATION ASSAY

Several works related to ABTS-based assays have been published using the methodology of sequential injection (SI) analysis with spectrophotometric or ECD (Table 31.4) [83–88]. A significant contribution to the development of this area was made by Portuguese researchers [73]. In particular, an SI system was designed to study the reduction of the preformed ABTS monocation radical (generated by oxidation of ABTS with PDS) in the presence of hydrogen donator antioxidant substances, resulting in a decrease in the initial absorbance of the solution at 734 nm [83]. The schematic diagram of the proposed system, incorporating a mixing chamber in the side port of the selection valve, is shown in Figure 31.17. The ABTS$^{•+}$ solution (~40 μL) and the sample (~21–62 μL) were sequentially aspirated (steps 1 and 2) to the holding coil (HC). The flow was then reversed, and the stacked zones were propelled to the mixing chamber (step 3). At this point, the flow was stopped for a time period between 15 and 240 s to enhance mixing (step 4). Afterward, the first part of the resulting solution was dispensed to waste (steps 5 and 6) to reduce bubble formation and improve repeatability. Consequently (steps 7 and 8), the other part of the resulting solution was propelled through the reaction coil (RC) to the detector, and spectrophotometric detection at 734 nm was carried out. The incorporation of the mixing chamber into the manifold made it possible to use one single

TABLE 31.4   Analytical Features of the SI Methods Developed for the Evaluation of ABTS Radical Cation-Scavenging Capacity

Type of Detection	pH	Sample Zone Volume (μL)	Sample	Sampling Rate (h^{-1})	R.S.D. (%) (n = 10–15)	Reference
Visible-spectrophotometric detection (734 nm)	5.4 7.4	20.6–61.8	Beer, milk Tea, juices	9–20	<0.33	[83]
	7.5	150–300	Wine	42	<2.4	[84]
	7.4	123–369	Etodolac	21	<4.7	[85][a]
	4.6	100	Beverages	12–18	<3.1	[86]
Amperometric detection (−0.10 V vs. Ag/AgCl)	7.0	37.5	Ginger infusions	40	2.28–4.11	[87]
		67		9–11	2.0–7.8	[88]

[a] The sequential analysis by using the MSFIA system.

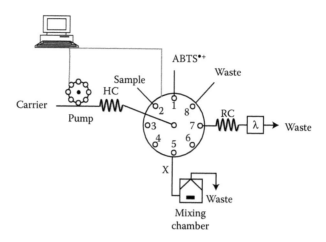

FIGURE 31.17   Sequential injection manifold for the determination of total antioxidants in food. HC, holding coil; RC, reaction coil. (Modified from Lima, M. J. R., I. V. Toth, and A. O. S. S. Rangel. 2005. *Talanta* 68:207–213.)

standard to perform the calibration procedure and simplified the manipulation without the necessity of increasing the analytical cycle.

The applicability of the developed method was tested by measurement of the antioxidant activity of pure compounds and by analyzing complex food and beverage samples. The antioxidant activity was calculated as L-ascorbic acid equivalence. The values obtained by the proposed SI method were not significantly different from the results obtained by the batch procedure (Table 31.5).

TABLE 31.5   Comparison between the *TEAC* Values Obtained by the Developed SIA Method and by Reference Procedure in Analysis of Food Samples of Different Origins

Carrier	Sample	SI Method	Batch Procedure
Water	Beer 1	0.54 ± 0.04	0.62 ± 0.09
	Beer 2	0.72 ± 0.05	0.77 ± 0.12
	Pasteurized milk 1	0.65 ± 0.02	0.85 ± 0.05
	Pumpkin	0.36 ± 0.04	0.42 ± 0.03
	Tea-based refreshing drink	0.85 ± 0.06	0.84 ± 0.08
Buffer solution, pH 7.4	Tea	0.37 ± 0.09	0.39 ± 0.02
	Yogurt, fortified	10.63 ± 1.9	10.68 ± 3.1
	Tomato juice	2.19 ± 0.5	2.81 ± 0.2
	Pasteurized milk 2	13.18 ± 0.02	12.3 ± 0.2
	Pumpkin	1.18 ± 0.04	1.33 ± 0.1
	Green tea	14.46 ± 1.1	12.23 ± 0.6
	Mandarin	5.00 ± 0.03	4.33 ± 0.3
	Apple juice	1.15 ± 0.03	1.6 ± 0.2

Source: Data from Lima, M. J. R., I. V. Toth, and A. O. S. S. Rangel. 2005. *Talanta* 68:207–213.

An automatic SI system that performs two analytical procedures, allowing the evaluation of the relative antioxidant capacity of wine samples, has been developed (Figure 31.18) [84]. One procedure was based on the decolorization of the ABTS radical cation, using a spectrophotometric detector. A second procedure consisted of the evaluation of the hydrogen peroxide–scavenging activity by measuring the oxidation of homovanylic acid (HVA) to its fluorescent dimer, using a fluorescent detector. The ABTS•+-scavenging activity of each sample was determined by measuring the decrease in the intensity of the blank signal corresponding to preformed ABTS•+ (Table 31.6). The analytical cycle proceeded with the intercalation of the radical solution zone (25 μL) between two equal sample segments (150 μL). The aspirated sequence was sent to the spectrophotometric detector. Before the beginning of the cycle, sample fragments were diluted. The sample dilution procedure was performed both in batch and directly in the sequential injection analysis (SIA) system by the use of two dilution coils placed in two of the inlets of the selection valve. The results of the analyses of the wine samples were expressed in *TEAC* values. The evaluation of the antioxidant power of 20 white and red wine samples, from different Portuguese wine-producing regions, was carried out

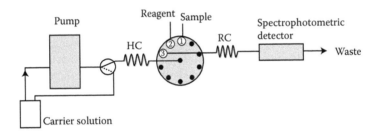

FIGURE 31.18   Schematic diagram of the SI system used for the determination and evaluation of antioxidative capacity of etodolac. HC, holding coil (4 m; 0.8 mm i.d.); RC, reaction coil (1.5 m; 0.8 mm i.d.) Reagent, ABTS monocation solution. (Adapted from Garcia, J. B., M. L. M. F. S. Saraiva, and J. L. F. C. Lima. 2006. *Anal. Chim. Acta* 573:371–375.)

TABLE 31.6   Percentage Inhibition of the ABTS Radical Cation (y) as a Function of the AOX Concentration (x, mM)

Antioxidant	Calibration Equation	Limit of Detection (mM)	R.S.D. (%)	TEAC (mM)
Trolox	$y = (8287 \pm 81)x$ $+ (4.2 \pm 0.5)$	8.4×10^{-7}	1.0–2.4	1.0
Gallic acid	$y = (27888 \pm 400)x$ $+ (81.0 \pm 0.7)$	1.8×10^{-7}	0.8–4.2	4.2 ± 0.2
Caffeic acid	$y = (8974 \pm 273)x$ $-(5.0 \pm 1.5)$	4.2×10^{-7}	1.7–3.5	0.78 ± 0.01
Ascorbic acid	$y = (8626 \pm 193)x$ $-(4.6 \pm 1.3)$	1.7×10^{-6}	1.0–2.3	0.68 ± 0.01
Catechin	$y = (21932 \pm 234)x$ $+ (0.6 \pm 0.4)$	2.1×10^{-7}	0.4–1.4	1.35 ± 0.01
Taxifolin	$y = 63750 (\pm 1366)x$ $+ (2.8 \pm 1.2)$	9.8×10^{-8}	1.3–2.1	8.8 ± 0.1

Source: Data from Pinto, P. C. A. G. et al. 2005. *Anal. Chim. Acta* 531:25–32.

sequentially, in the automatic system. A strong relation of the results with the type of wine (white/red) was found: namely, the red wines presented much higher antioxidant capacities associated with the high content of polyphenolic substances present in this type of wine.

A robust, accurate, and sensitive automated technique has been proposed to evaluate the antioxidant activity by measuring the absorption changes of ABTS$^{•+}$ solution after its interaction with AOXs [85]. The analytical cycle included three steps (Figure 31.18): it began with the sequential aspiration to the holding coil of 0.329 mL of sample (step 1) and 0.205 mL of a 4.9×10^{-4} M ABTS$^{•+}$ solution (step 2) at a flow rate of 0.68 mL min^{-1} through ports 1 and 2 (Figure 31.18), respectively. At step 3, the reaction zone was sent to the detector through port 3, at a flow rate of 1.20 mL min^{-1} during 125 s. Determinations started with the measurement of a blank signal by aspirating a 5% (v/v) ethanol solution (blank solution). The obtained blank signals correspond to the maximum absorbance signal in the absence of AOXs.

An SI system for the screening of antioxidant power based on ABTS$^{•+}$ scavenging has been developed [86]. Its schematic diagram is shown in Figure 31.18. Detection of the remaining green-blue-colored radical after reaction with AOXs was achieved using an in-house ECD. The developed system was applied to monitor the TAC in commercial ginger drinks in terms of gallic acid equivalent, GAE (mg g^{-1} sample), which then represents the total amount of antioxidant capacity in each sample. The correlation between the SIA and classical methods is good (slope $= 0.910 \pm 0.087$, intercept $= 0.115 \pm 0.064$, $R^2 = 0.956$).

An automatic SI system with amperometric detection of the preformed ABTS radical cation has recently been designed for the evaluation of TAC (Figure 31.19) [87,88]. A flow-through electrochemical cell (ECD) constructed in-house from transparent acrylic resin with the volume of 350 μL was used. The working electrode was a GCE (3-mm-diameter disk), the reference electrode was a Ag/AgCl (sat. KCl), and another GCE served as a counterelectrode.

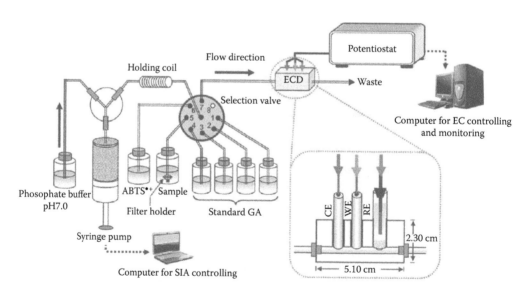

FIGURE 31.19 Schematic diagram of the SI system for the evaluation of TAC with an in-house flow-through ECD (1.9 cm width × 5.1 cm length × 2.3 cm height). CE, counterelectrode; WE, working electrode; RE, reference electrode. (Adapted from Chan-Eam, S. et al. 2011. *Talanta* 84:1350–1354.)

The SI procedure had four main steps for one cycle. First, phosphate buffer solution was aspirated into the system at a flow rate of 18 mL min^{-1}. Then, sample (or antioxidant standard solution) was aspirated in two segments (step 1 and 3), and the ABTS$^{\bullet+}$ solution was aspirated as the middle segment between the standard/sample segments (step 2). The phosphate buffer solution was aspirated to finish the segment sequence (step 4). Then, the syringe was moved forward and backward three times to induce zone mixing. Finally, the resulting solution was propelled to the detection cell through port 7 at the flow rate of 12 mL min^{-1}. The ABTS radical cation remaining after reaction with AOXs gave the cathodic current on the working GCE at the applied potential of -0.10 V vs. Ag/AgCl. The sample throughput rate of the SI system was 9 h^{-1} at a constant flow rate of 2.0 mL min^{-1}.

The standard antioxidant gallic acid (GA) showed a calibration curve of the linear equation in the concentration range of 0–70 ppm ($R^2 = 0.9969$):

$$H = (-6.208 \pm 0.202)[GA] + (609.85 \pm 8.30) \tag{31.26}$$

where H is peak current in nanoamperes.

Precision of peak current measurements from the reaction with 30 ppm GA gave an R.S.D. of 2.28% ($n = 10$). The detection limit was 5.6 ppm. The proposed method was applied to determine antioxidant capacity of instant ginger powder from three Thai brands. Antioxidant capacity tests of various commercial ginger samples were reported in gallic acid equivalent units. The results are presented in Table 31.2. The in-house written software and the fabricated electronic interface for the specified pump drive allowed the use of less expensive instruments and achieved the main advantage of SIA with significant reduction of reagent and sample consumption.

To summarize, natural products (vegetables, fruits, and cereals) are a very important part of our diet [89]. To find food additives with high antioxidant activity, most investigators are using a combination of antioxidant tests [90,91]. FI techniques, including their alternate versions, have proven to be suitable for routine use and screening purposes. By adapting ABTS-based assays to FI or SI systems and replacing the time-consuming step of chemical preparation of the ABTS radical cation by its online generation, a significant improvement in analysis time can be achieved.

REFERENCES

1. Zhou, B. and Zh.-L. Liu. 2005. Bioantioxidants: From chemistry to biology. *Pure Appl. Chem.* 77:1887–1903.
2. Sies, H. 1997. Oxidative stress: Oxidants and antioxidants. *Exp. Physiol.* 82:291–295.
3. Halliwell, B., K. Zhao, and M. Whiteman. 2000. The gastrointestinal tract: A major site of antioxidant action? *Free Radic. Res.* 33:819–830.
4. Rice-Evans, C. A., B. Halliwell, and G. G. Lunt. eds. 1995. In *Environment Drugs and Food Additives.* 276pp. London: Portland Press.
5. Miller, H. E., L. M. Rigelhof, A. Prakash et al. 2000. Antioxidant content of whole grain breakfast cereals, fruits and vegetables. *J. Am. College Nutr.* 19:312–319.
6. Antonovich, M., P. D. Prenzler, E. Patsalides et al. 2002. Methods for testing antioxidant activity. *Analyst* 127:183–198.
7. Prior, R. L., X. Wu, and K. Schaich. 2005. Standardized methods for the determination of antioxidant capacity and phenolics in foods and dietary supplements. *J. Agric. Food Chem.* 53:290–302.

8. Roginsky, V. and E. A. Lissi. 2005. Review of methods to determine chain-breaking antioxidant activity in food. *Food Chem.* 92:235–254.

9. Huang, D., B. Ou, and R. L. Prior. 2005. The chemistry behind antioxidant capacity assays. *J. Agric. Food Chem.* 53:1841–1856.

10. Magalhàes, L. M., M. A. Segundo, S. Ries et al. 2008. Methodological aspects about *in vitro* evaluation of antioxidant properties. *Anal. Chim. Acta* 613:1–19.

11. Apak, R., Sh. Gorinstein, V. Böhm et al. 2013. Methods of measurement and evaluation of natural antioxidant capacity/activity (IUPAC Technical Report). *Pure Appl. Chem.* 85:957–998.

12. Miller, N. J., C. Rice-Evans, M. J. Davies et al. 1993. A novel method for measuring antioxidant capacity and its application to monitoring the antioxidant status in premature neonates. *Clin. Sci.* 84:407–412.

13. Arnao, M. V., A. Cano, J. Hernandez-Ruiz et al. 1996. Inhibition by L-ascorbic acid and other antioxidants of 2,2′-azinobis(3-ethyl-benzthiazoline-6-sulfonatic acid) oxidation catalyzed by peroxidase: A new approach for the determining total antioxidant status of food. *Anal. Biochem.* 236:255–261.

14. Re, R., N. Pellegrini, A. Proteggente et al. 1999. Antioxidant activity applying an improved ABTS radical cation decolorization assay. *Free Radic. Biol. Med.* 26:1231–1237.

15. Scott, S. L., W.-J. Chen, A. Bakac et al. 1993. Spectroscopic parameters, electrode potentials, acid ionization constants, and electron exchange rates of the 2,2′-azinobis(3-ethylbenzothiazolineine-6-sulfonate) radicals and ions. *J. Phys. Chem.* 97:6710–6714.

16. Arnao, M. B. 2000. Some methodological problems in the determination of antioxidant activity using chromogen radicals: A practical case. *Trends Food Sci. Technol.* 11:419–421.

17. Lalbrinea, E. P. and C. A. Georgiou. 2004. Stopped-flow method for assessment of pH and timing effects on ABTS total antioxidant capacity assay. *Anal. Chim. Acta* 526:63–68.

18. Henriquez, C., C. Aliaga, and E. Lissi. 2004. Kinetic profiles in reaction of ABTS derived radicals with simple phenols and polyphenols. *J. Chil. Chem. Soc.* 49:65–68.

19. Walker, R. B. and J. D. Everette. 2009. Comparative reaction rates of various antioxidants with ABTS radical cation. *J. Agric. Food Chem.* 57:1156–1161.

20. Rice-Evans, C., N. J. Miller, and G. Paganga. 1996. Structure-antioxidant activity relationships of flavonoids and phenolic acids. *Free Radic. Biol. Med.* 20:933–956.

21. Benavente-Garcia, O., J. Castillo, J. Lorente et al. 2000. Antioxidant activity of phenolic extracted from *Olea europaea* L. leaves. *Food Chem.* 68:457–462.

22. Rice-Evans, C., N. J. Miller, P. G. Bolwell et al. 1995. The relative antioxidant activities of plant-derived polyphenolic flavonoids. *Free Radic. Res.* 22:375–383.

23. Sanchez-Moreno, C., J. A. Larrauri, and F. A. Saura-Calixto. 1998. A procedure to measure the antiradical efficiency of polyphenols. *J. Sci. Food Agric.* 76:270–276.

24. Salah, N., N. J. Miller, G. Paganga et al. 1995. Polyphenolic flavanols as scavengers of aqueous phase radicals and as chain-breaking antioxidants. *Arch. Biochem. Biophys.* 322:339–346.

25. Rice-Evans, C. A. and N. J. Miller. 1996. Antioxidant activities of flavonoids as bioactive components of food. *Biochem. Soc. Trans.* 24:790–795.

26. Miller, N. J., J. Sampson, L. P. Candeias et al. 1996. Antioxidant activities of carotenes and xanthophylls. *FEBS Lett.* 384:240–242.

27. Vinson, J. A., Y. Hao, X. Su et al. 1998. Phenol antioxidant quantity and quality in foods: Vegetables. *J. Agric. Food Chem.* 46:3630–3634.
28. Prior, R., G. Cao, A. Martin et al. 1998. Antioxidant capacity as influenced by total phenolic and anthocyanin content, maturity, and variety of *Vaccinium* species. *J. Agric. Food Chem.* 46:2686–2693.
29. Ghizelli, A., M. Serafini, F. Natella et al. 2000. Total antioxidant capacity as a tool to assess redox status: Critical view and experimental data. *Free Rad. Biol. Med.* 29:1106–1114.
30. Bondet, V., W. Brand-Williams, and C. Berset. 1997. Kinetics and mechanisms of antioxidant activity using the DPPH free radical method. *Food Sci. Technol.* 30:609–615.
31. Teixeira, D. M., C. Canelas, A. M. Do Canto et al. 2009. HPLC-DAD quantitation of phenolic compounds contributing to the antioxidant activity of *Maclure pomifera, Ficus carica and Fiqus elastica* extracts. *Anal. Lett.* 42:2986–3003.
32. Miller, N. J. and C. A. Rice-Evans. 1994. Total antioxidant status in plasma and body fluids. *Methods Enzymol.* 234:279–293.
33. Wang, H., G. Cao, and R. L. Prior. 1996. Total antioxidant capacity of fruits. *J. Agric. Food Chem.* 44:701–705.
34. Chen, I. C., H. C. Chang, H. W. Yang et al. 2004. Evaluation of total antioxidant activity of several vegetables and Chinese herbs: A fast approach with $ABTS/H_2O_2/$HRP system in microplates. *J. Food Drug Anal.* 12:29–33.
35. Kang, Ok-Ju. 2011. Antioxidant activities of various solvent extracts from ginseng (*Panax ginseng* C.A. Meyer) leaves. *J. Food Sci. Nutr.* 16:321–327.
36. Long, L. H., D. Kwee, and B. Halliwell, 2000. The antioxidant activities of seasonings used in Asian cooking. *Free Rad. Res.* 32:181–186.
37. Castillo, J., O. Benavente-Garcıa, J. Lorente et al. 2000. Antioxidant activity and radioprotective effects against chromosomal damage induced *in vivo* by X-rays of flavan-3-ols (procyanidins) from grape seeds (*Vitis vinifera*): Comparative study versus other phenolic and organic compounds. *J. Agric. Food Chem.* 48:1738–1745.
38. Villano, D., M. S. Fernandez-Pachon, A. A. M. Troncoso et al. 2004. The antioxidant activity of wines determined by the ABTS•+ method: Influence of sample dilution and time. *Talanta* 64:501–509.
39. Berg, R., G. R. M. M. Haenen, H. Berg et al. 1999. Applicability of an improved Trolox equivalent antioxidant capacity (TEAC) assay for evaluation of antioxidant capacity measurements of mixtures. *Food Chem.* 66:511–517.
40. Pellegrini, N., D. Del Rio, B. Colombi et al. 2003. Application of the 2,2'-azinobis-(3-ethylbenzothiazoline-6-sulfonic acid) radical cation assay to a flow injection system for the evaluation of antioxidant activity of some pure compounds and beverages. *J. Agric. Food Chem.* 51:260–264.
41. Buratti, S., N. Pellegrini, O. Brenna et al. 2001. Rapid electrochemical method for the evaluation of the antioxidant power of some lipophilic food extracts. *J. Agric. Food Chem.* 49:5136–5141.
42. Henriquez, C., C. Aliaga, and E. Lissi. 2002. Formation and decay of the ABTS derived radical cation: A comparison of different preparation procedures. *Int. J. Chem. Kinet.* 34:659–665.
43. Strube, M., G. R. Haenen, H. Van Den Berg et al. 1997. Pitfalls in a method for assessment of total antioxidant capacity. *Free Radic. Res.* 26(6):515–521.

44. Cano, A., J. Hernández-Ruiz, C. Garcıa-Cánovas et al. 1998. An end-point method for estimation of the total antioxidant activity in plant material. *Phytochem. Anal.* 9:196–202.
45. Fernandez-Pachon, M. S., D. Villano, A. M. Troncoso et al. 2006. Determination of the phenolic composition of sherry and table white wines by liquid chromatography and their relation with antioxidant activity. *Anal. Chim. Acta* 563:101–108.
46. Solıs-Oba, M., V. M. Ugalde-Saldırvar, I. Gonzarlez et al. 2005. An electrochemical–spectrophotometrical study of the oxidized forms of the mediator 2,2′-azino-bis-(3-ethylbenzothiazoline-6-sulfonic acid) produced by immobilized laccase. *J. Electroanal. Chem.* 579:59–66.
47. Kadnikova, E. N. and N. M. Kostic. 2002. Oxidation of ABTS by hydrogen peroxide catalyzed by horseradish peroxidase encapsulated into sol–gel glass: Effects of glass matrix on reactivity. *J. Mol. Catal. B: Enzym.* 18:39–48.
48. Campanella, L., E. Martini, and M. Tomasseti, 2005. Antioxidant capacity of the algae using a biosensor method. *Talanta* 66:902–911.
49. Holm, K. A. 1995. Automated determination of microbial peroxidase activity in fermentation samples using hydrogen peroxide as the substrate and 2,2′-azino-bis(3-ethylbenzothiazoline-6-sulfonate) as the electron donor in a flow injection system. *Analyst* 120:2101–2105.
50. Erel, O. 2004. A novel automated direct measurement method for total antioxidant capacity using a new generation, more stable ABTS radical cation. *Clin. Biochem.* 37:277–285.
51. Branchi, B., C. Galli, and P. Gentili. 2005. Kinetics of oxidation of benzyl alcohols by the dication and radical cation of ABTS. Comparison with laccase–ABTS oxidations: An apparent paradox. *Org. Biomol. Chem.* 3:2604–2614.
52. Maruthamuthu, P., L. Venkatasubramanian, and P. Dharmalingam. 1986. Reaction of halogens with 2,2′-azino-bis(3-ethylbenzothiazoline-6-sulfonate): Stopped flow kinetics of formation of radical cations and dications. *Proc. Indian Acad. Chem.* 97:213–218.
53. Mahuzier, G., B. S. Kirkacharian, and C. Harfouche-Obeika. 1975. Microdosage colorimétrique de l'acide periodique par l'acide 2,2′-azino-di(3-ethylbenzothiazole-6-sulfonique). *Anal. Chim. Acta* 76:79–83.
54. Lamien-Meda, A., C. E. Lamien, M. M. Y. Compaore et al. 2008. Polyphenol content and antioxidant activity of fourteen wild edible fruits from Burkina Faso. *Molecules* 13:581–594.
55. Aliaga, C. and E. Lissi. 2000. Reactions of the radical cation derived from 2,2′-azinobis(3-ethylbenzo thiazoline-6-sulfonic acid) (ABTS) with amino acids. Kinetics and mechanism. *Can. J. Chem.* 78:1052–1059.
56. Miller, N. J. and C. A. Rice-Evans. 1999. Factors influencing the antioxidant activity determined by the ABTY radical cation assay. *Free Rad. Res.* 26:195–199.
57. Campos, A. M. and E. Lissi. 1996. The total reactive antioxidant potential (TRAP) and total antioxidant reactivity (TAR) of *Ilex paraguayensis* extracts and red wine. *J. Braz. Chem. Soc.* 7:43–49.
58. Romay, C., C. Pascual, and E. Lissi. 1996. The reaction between ABTS radical cation and antioxidants and its use to evaluate the antioxidant status of serum samples. *J. Med. Biol. Res.* 29:175–183.
59. Bartosz, G., A. Janaszewska, D. Ertel et al. 1998. Simple determination of peroxyl radical-trapping capacity. *Biochem. Mol. Biol. Int.* 46:519–528.

60. Aliaga, C. and E. A. Lissi. 1998. Reaction of 2,2′-azinobis(3-ethylbenzothiazoline-6-sulfonic acid (ABTS) derived radicals with hydroperoxides. Kinetics and mechanism. *Int. J. Chem. Kinetics* 30:565–570.

61. Wolfenden, B. S. and R. L. Willson. 1982. Radical-cations as reference chromogens in kinetic studies of one-electron transfer reactions: Pulse radiolysis studies of 2,2′-azinobis-(3-ethylbenzthiazoline-6-sulphonate). *J. Chem. Soc. Perkin. Trans.* 2:805–812.

62. Venkatasubramanian, L. and P. Maruthamuthu. 1989. Kinetics and mechanism of formation and decay of 2,2′-azinobis(3-ethylbenzothiazole-6-sulfonate) radical cation in aqueous solution by inorganic peroxides. *Int. J. Chem. Kinetics* 21:399–421.

63. Shpigun, L. K., N. N. Zamyatina, P. M. Kamilova et al. 2010. Evaluation of antioxidative power of pharmaceutical substances by using oxidative systems forming free radicals. In *Book of Abstracts of 11th Eurasia Conference on Chemical Sciences.* 7-P-12. Amman: The University of Jordan.

64. Nakano, Sh., K. Tanaka, R. Oki et al. 1999. Flow-injection spectrophotometry of manganese by catalysis of the periodate oxidation of 2,2′-azinobis-(3-ethylbenzthiazoline-6-sulphonic acid). *Talanta* 49:1077–1082.

65. Bourbonnais, R., D. Leech, and M. G. Paice. 1998. Electrochemical analysis of the interactions of laccase mediators with lignin model compounds. *Biochim. Biophys. Acta* 1379:381–390.

66. Schröder, I., E. Steckhan, and A. Liese. 2003. *In situ* NAD$^+$ regeneration using 2,2′-azinobis(3-ethylbenzothiazoline-6-sulfonate) as an electron transfer mediator. *J. Electroanal. Chem.* 541:109–115.

67. Hsu, C. F., H. Peng, C. Basle et al. 2011. ABTS$^{•+}$ scavenging activity of polypyrrole, polyaniline and poly(3,4-ethylenedioxythiophene). *Polym. Int.* 60:69–77.

68. Fabbrini, M., C. Galli, and P. Gentili. 2002. Radical or electron-transfer mechanism of oxidation with some laccase/mediator systems. *J. Mol. Catal. B-Enzym.* 18:169–171.

69. Thomas, J. H., J. M. Drake, J. R. Paddock et al. 2004. Characterization of ABTS at a polymer-modified electrode. *Electroanalysis* 16:547–555.

70. Alonso, A. M., D. A. Guilln, G. Carmelo et al. 2003. Development of an electrochemical method for the determination of antioxidant activity. Application to grape-derived products. *Eur. Food Res. Technol.* 216:445–448.

71. Iveković, D., S. Milardović, M. Roboz et al. 2005. Evaluation of the antioxidant activity by flow injection analysis method with electrochemically generated ABTS radical cation. *Analyst* 130:708–714.

72. Ukeda, H. 2003. Flow injection analysis for estimating food function. *J. Flow Inj. Anal.* 20:21–25.

73. Magalhàes, L. M., M. Santos, M. A. Segundo et al. 2009. Flow injection based methods for fast screening of antioxidant capacity. *Talanta* 77:1559–1566.

74. Lopez, M., F. Martınez, C. Del Valle et al. 2003. Study of phenolic compounds as natural antioxidants by a fluorescence method. *Talanta* 60:609–616.

75. Bonpadre, S., L. Leone, A. Politi et al. 2004. Improved FIA-ABTS method for antioxidant capacity determination in different biological samples. *Free Rad. Res.* 38:831–838.

76. Stewart, A. J., W. Mullen, and A. Crozier. 2005. On-line high-performance liquid chromatography analysis of the antioxidant activity of phenolic compounds in green and black tea. *Mol. Nutr. Food Res.* 49:52–60.

77. Takagaki, A., Sh. Otani, and F. Nanjo. 2011. Antioxidative activity of microbial metabolites of (–)-epigallocatechin gallate produced in rat intestines. *Biosci. Biotechnol. Biochem.* 75:582–585.
78. Ukeda, H. 2004. Flow-injection analytical system for the evaluation of antioxidative activity. *Bunseki Kagaku* 53:221–231.
79. Labrinea, E. P. and C. A. Georgiou. 2005. Rapid, fully automated flow injection antioxidant capacity assay. *J. Agric. Food Chem.* 53:4341–4346.
80. Milardovic, S., I. Kerekovic, R. Derrico et al. 2007. A novel method for flow injection analysis of total antioxidant capacity using enzymatically produced ABTS•+ and biamperometric detector containing interdigitated electrode. *Talanta* 71:213–220.
81. Milardovic, S., I. Kerekovic, and V. Rumenjak. 2007. A flow injection biamperometric method for determination of total antioxidant capacity of alcoholic beverages using bienzymatically produced ABTS•+. *Food Chem.* 105:1688–1694.
82. Shpigun, L. K., N. N. Zamyatina, Ya. V. Shushenachev et al. 2012. Flow-injection methods for the determination of antioxidant activity based on free radical processes. *J. Anal. Chem.* 67:801–808. Pleiades Publishing, Ltd. Original Russian Text © published in *Zh Analitich Khim* 67:893–901.
83. Lima, M. J. R., I. V. Toth, and A. O. S. S. Rangel. 2005. A new approach for the sequential injection spectrophotometric determination of the total antioxidant activity. *Talanta* 68:207–213.
84. Pinto, P. C. A. G., L. M. F. S. Saraiva, R. Salette et al. 2005. Automatic sequential determination of the hydrogen peroxide scavenging activity and evaluation of the antioxidant potential by the 2,2-azinobis(3-ethylbenzothiazoline-6-sulfonic acid) radical cation assay in wines by sequential injection analysis. *Anal. Chim. Acta* 531:25–32.
85. Garcia, J. B., M. L. M. F. S. Saraiva, and J. L. F. C. Lima. 2006. Determination and antioxidant activity evaluation of etodolac, an anti-inflammatory drug, by sequential injection analysis. *Anal. Chim. Acta* 573:371–375.
86. Magalhàes, L. M., M. A. Segundo, S. Ries et al. 2007. Automatic flow system for sequential determination of ABTS•+ scavenging capacity and Folin-Ciocalteu index: A comparative study in food products. *Anal. Chim. Acta* 592:193–201.
87. Chan-Eam, S., S. Teerasong, K. Damwan et al. 2011. Sequential injection analysis with electrochemical detection as a tool for economic and rapid evaluation of total antioxidant capacity. *Talanta* 84:1350–1354.
88. Kongkedsuk, J., A. Hongwitayakorn, W. Bootnapang et al. 2013. Development of sequential injection analysis using peristaltic pump and electrochemical detection for antioxidant capacity test by ABTS assay. *Chiang Mai J. Sci.* 40:225–232.
89. Proteggente, A. R., A. S. Pannala, G. Paganga et al. 2002. The antioxidant activity of regularly consumed fruit and vegetables reflects their phenolic and vitamin C composition. *Free Radic. Res.* 36:217–233.
90. Vinson, J. A., X. Su, L. Zubik et al. 2001. Phenol antioxidant quantity and quality in foods, fruits. *J. Agric. Food Chem.* 49:5315–5321.
91. Bahorun, T., A. Luximon-Ramma, A. Crozier et al. 2004. Total phenol, flavonoid, proanthocyanidin and vitamin C levels and antioxidant activities of Mauritian vegetables. *J. Sci. Food Agric.* 84:1553–1561.

CHAPTER 32

Determination of Lipid Oxidation by Chemiluminescence Reagents

Andrei Florin Danet and Mihaela Badea-Doni

CONTENTS

Lipid oxidation by chemiluminescence (CL) reagents has been studied, especially for different types of edible oils (olive, seeds, etc.). It is well known that when left exposed to an oxygen-containing atmosphere, oils and fats undergo oxidative degradation. This process involves mainly the oxidation of unsaturated fatty acids or their derivatives present in oils and fats. The degree of unsaturation, presence of existing or added antioxidants, prooxidants, illumination, and thermal conditions of storage all affect the lipid oxidative stability. Evaluation of oxidative stability of lipids is an old and complex topic. Still, a generally applicable, fully satisfactory method is not yet available.

Lipid oxidation in foods is a complex chain of reactions that first consist of the introduction of a functional group containing two concatenated oxygen atoms (peroxides) into unsaturated fatty acids, in a free-radical chain reaction, that afterward gives rise to secondary oxidation products. Different pathways for lipid oxidation have been described: radical mechanism or autoxidation, singlet oxygen–mediated mechanism or photooxidation, and enzymatic oxidation.

Lipid autoxidation is generally believed to involve a free- radical chain mechanism: (1) initiation steps that lead to free radicals ($R^\bullet$), (2) propagation of the free radicals ($R^\bullet + O_2 \rightarrow ROO^\bullet$, $ROO^\bullet + RH \rightarrow ROOH + R^\bullet$), and (3) termination steps $R^\bullet + R^\bullet \rightarrow R—R$, $R^\bullet + ROO^\bullet \rightarrow ROOR$, $ROO^\bullet + ROO^\bullet \rightarrow O_2 + ROOR$ (or alcohol and carbonyl compound). The oxidation of lipids results in peroxides as primary oxidation products, which in turn degrade further to secondary oxidation products, including aldehydes, ketones, epoxides, hydroxy compounds, carboxylic acids, oligomers, and polymers.

Oxidation decreases the quality of foods by producing low-molecular-weight off-flavor compounds, as well as by destroying essential nutrients, and it produces toxic compounds and dimers or polymers of lipids and proteins, which in turn contribute to diseases and accelerate the aging process. Its measurement is a leading objective of food analysis. Evaluating lipid oxidation status is a challenging task due to a number of facts, such as the different compounds that are formed depending on time, extent of oxidation, and the mechanism involved.

Lipid oxidation can be evaluated in a variety of ways, such as the determination of peroxide value (PV), carbonyl compounds (malondialdehyde, hexanal, etc.) diene conjugation, and oxygen consumption. Various methods may be used to quantify the lipid oxidation such as iodometric titration (IT), UV–visible spectroscopy (ferrous oxidation method, iodide oxidation method), chromatography, CL, or fluorescence (FL) methods.

Chemiluminescence is defined as the production of light generated through a chemical reaction that is accompanied by an energy release of ≥ 45 kcal mol^{-1}. A CL reaction implies a molecular reaction of two (or more) ground-state molecules producing a final molecule(s) in an excited state. Potential energy from the reactants is translated to the product(s). Radical species are formed, which then interact and produce unstable intermediates that decompose with the formation of excited species. These are deactivated by either emission of light or transfer of energy to light-amplifier molecules (luminophore) with a high quantum yield. Light intensity that accompanies the oxidation of lipids is very low, so that light amplifiers should be used to increase it. Such light amplifiers are: luminol, lucigenin, or peroxalate.

Luminol (5-amino-2,3-dihydro-1,4-phthalazinedione) reacts with different reactive oxygen species (hydrogen peroxide, hydroperoxides, superoxide, etc.) in the presence of a base and a catalyst to produce an excited-state product (3-aminophthalate) that emits light at approximately 425 nm (Figure 32.1a). The CL intensity is proportional to the amount of reactive oxygen species in a sample.

Lucigenin (N,N'-dimethyl-9,9'-biacridinium dinitrate) is oxidized in alkaline medium containing ethanol or acetone and reactive oxygen species with emission of green light, which subsequently decays to greenish-blue and finally blue. The mechanism of the oxidation is presented in Figure 32.1b.

As can be seen in Figure 32.1b, lucigenin forms an unstable dioxetane, which decomposes to N-methylacridone in an electronically excited state. The excited acridone molecule emits light as it relaxes to a stable state.

In the peroxalate-enhanced luminescence, the initial excited-state product does not emit light, but it could transfer the energy to another molecule (fluorophore) that will emit light by losing the received energy. For example, bis(2,4,6-trichlorophenyl)oxalate (TCPO) can be oxidized by reactive oxygen species to produce at least one (and possible more) high-energy intermediate(s) capable of generating the excited state of a fluorophore, which by deactivating emits light (Figure 32.1c).

CL reagents can also be used for determining the total antioxidant capacity (TAC) of a sample. In this case, the CL produced is quenched by the antioxidants that scavenge active free radicals from the sample. The decrease of the intensity of emitted light depends on the antioxidant capacity of the sample.

The use of CL as a detection technique is characterized by several advantages, namely:

- CL is a "dark-field" detection technique, no excitation light source is required, and this results in lower background signals, very low detection limits, and wide dynamic ranges.
- CL requires simple instrumentation, since it makes use of no light source and, in most cases, no wavelength discriminator. Chemiluminescence detectors can be easily miniaturized and combined with flow-based approaches.

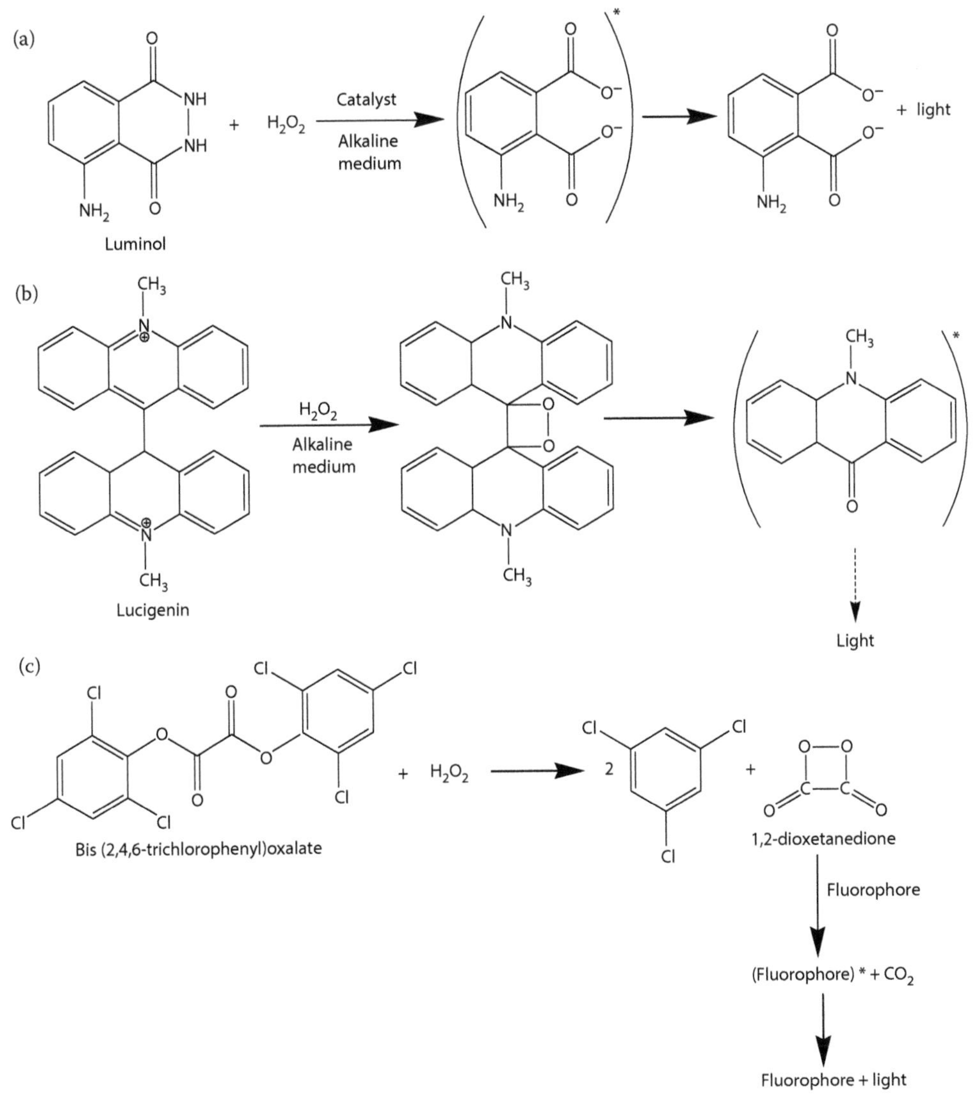

FIGURE 32.1   Chemiluminescence reactions of: a) luminol; b) lucigenin; c) bis(2,4,6- tri-chlorophenyl)oxalate.

- The CL methods require less time (taking only a few minutes), are highly sensitive (picomole levels have been assessed), have low sample requirements, are low cost, and are simpler when compared with other methods; in addition, the assay can be easily automated.

As for shortcomings, the method is not specific to lipids, since other oxidizing agents may give CL signals and the kinetic theory and mechanism for chemical processes involved in CL are not known in detail, which may complicate data interpretation. Versions of the method differ in the type of free radical produced and the manner of free-radical production, as well as in protocol details. While the majority of assays have been developed for testing biologically relevant samples, they can be easily applied to food testing too.

The CL methods can be applied with very good results for the analysis of lipid peroxides in food quality assessment and also to measure the ability of the natural or synthetic food antioxidants to quench free radicals in foods. For antioxidant capacity determination of foods, the measurements are based on competitive kinetics, in which the antioxidants compete with a CL reagent for the free radicals, resulting in a decrease in the light emission compared with the CL intensity obtained in the absence of the antioxidants.

Several relevant review papers can be found in the literature describing methods, including CL ones, for the determination of the lipid peroxides or antioxidant capacity of oils or other lipids (Navas and Jimenez 1996, 2007; Wheatley 2000; Roginsky and Lissi 2005; Christodouleas et al. 2012; Barriuso et al. 2013).

For quantification of the lipid peroxides, liquid chromatographic methods with CL detection are the most sensitive among the developed methods (Navas and Jimenez 1996; Rolewski et al. 2009). These methods have been applied only to aqueous extracts of oils.

Automation of CL assays by flow injection (FI) brings the inherent advantages of a strict control of mixing and timing, improving the assay reproducibility. When combined with an FI manifold system, CL detection acquires extra facilities, such as increased speed of analysis, high throughput, online sample-processing precision, and in-line multidetector possibilities.

The papers describing the determination of lipid oxidation and antioxidant capacity of different lipids by CL reagents are not too numerous, and only few of the developed methods were automated by using the FI methodology. The literature data for the determination of lipid oxidation by CL reagents were classified as follows:

- Methods for lipid hydroperoxide determination
- Methods for antioxidant capacity of lipids determination

32.1 METHODS FOR THE DETERMINATION OF LIPID HYDROPEROXIDES

The determination of lipid peroxides in lipids is an important issue, because lipid peroxides indicate that oils and fats can undergo oxidative deterioration. The methods for the determination of hydroperoxides present in lipids are based on the physical properties of the LOOH, for example, conjugated dienes, or on the chemical properties of the peroxide group, for example, PV. However, in some foods, the quantity of fats could be too low to apply the above-mentioned methods with accuracy. Chemiluminescence offers the sensitivity to overcome this problem.

The monohydroperoxides of triacylglycerols predominate in commercial oils (more than 90% of total peroxides), but the structure (and reactivity) of lipid hydroperoxides formed is related to the type of oil. For this reason, in some of the CL methods developed, the analytical signal depends not only on the concentration of peroxides present in the sample but also on the nature of the sample. This state was highlighted by Bunting and Gray (2003). The lipid hydroperoxides produced in oxidized oil during the Rancimat test have been monitored by the luminol CL reaction (Matthaus 1996). The obtained results were satisfactory, as the induction periods of the oils assessed by Rancimat and CL methods showed a significant linear correlation.

The CL methods for the determination of lipid hydroperoxides in fats are presented in Table 32.1. Some of them were automated by using FI methodology.

TABLE 32.1 Determination of Lipid Hydroperoxides in Fats by Chemiluminescence

Sample	Method Principle	Analytical Features	Comments	Reference
CL				
Walnut, safflower, sunflower, and rapeseed oils	Luminol/hemin-based CL reaction	–	Samples were dissolved in a mixture of acetone/ethanol (2:1) Results are influenced by the presence of antioxidants in oil by decreasing the CL signal	Matthaus (1996)
Linseed oil	Luminol/hemin-based CL reaction in Triton X-100–based emulsion	Linear range: 50 pM–4 μM LOOH	Triton X-100 was used to increase the emulsion stability and system efficiency Carotenes and phenols do not influence the LOOH determination	Rolewski et al. (2009)
Olive oil	Bis(2,4,6-trichlorophenyl) oxalate/Mn(II) as catalyst and 9,10-dimethylanthracene as fluorophore	Linear range: 0.6–100 meq. O_2 kg^{-1} RSD% = 1%–5% in the linear range	Oil dilution and reagent solutions were prepared in a mixture of ethyl acetate saturated with water and acetonitrile (9:1) Method allows the evaluation of total peroxides as an analytical index for the quality of olive oils	Stepanyan et al. (2005)
Olive oil	1. Luminol/Co(II) 2. Luminol/hemin–based CL reaction	Linear range: 2–30 meq. O_2 kg^{-1}oil	Three procedures were tested Samples were dissolved 10% (w/w) Tween solution or in a mixture of acetone–ethanol (2:1)	Bezzi et al. (2008)

(Continued)

TABLE 32.1 (*Continued*)   Determination of Lipid Hydroperoxides in Fats by Chemiluminescence

Sample	Method Principle	Analytical Features	Comments	Reference
Olive, corn sunflower, soybean, sesame oils, and a mixture of corn and sunflower oils	Luminol/Fe(III)	LOD = 0.23 mM di-*t*-butyl peroxide	Samples were dissolved in 1-propanol Di-*t*-butyl peroxide (di-*t*-BP) was used as peroxide standard Total hydroperoxide content expressed as di-*t*-BP ranged between 2.4–10.7 mM for olive oils, 20.0–27.2 mM for corn oils, 4.4–8.3 mM for sunflower oils, and was equal to 5.3 mM for soybean oil	Tsiaka et al. (2013)
FI–CL				
Low-density lipoprotein	Carrier: 0.1 M carbonate buffer, pH 10 CL reagents: luminol/microperoxidase in 0.1 M carbonate buffer, pH 10	LOD = 3 pmol linoleic acid hydroperoxide	Linoleic acid hydroperoxide for calibration was prepared by incubating soybean lipoxygenase with linoleic acid	Cominacini et al. (1993)
Corn, cottonseed, peanut, soybean, and wheat germ oils	Carrier: methanol/chloroform (9:1) mixture CL reagents: luminol/cytochrome *c* in borate buffer containing Triton X-100 (biphasic system)	0.1–4.0 mM *t*-BHPO 0.027–3.38 mM LOOH	*t*-BHPO was used as peroxide standard	Bunting and Gray (2003)
Wheat germ oil	Carrier: 100% methanol CL reagents: lucigenin in methanol/borate buffer (4:1) (monophasic system)	0.1–4.8 mM LOOH	Method is insensitive to α-tocopherol	Bunting and Gray (2003)

(*Continued*)

TABLE 32.1 (*Continued*)   Determination of Lipid Hydroperoxides in Fats by Chemiluminescence

Sample	Method Principle	Analytical Features	Comments	Reference
Soybean, rapeseed, corn, canola, and safflower oils	Carrier: 1-butanol/ methanol (2:1) FL reagents: diphenyl-1-pyrenylphosphine (DPPP) dissolved in 1-butanol/methanol (2:1) FL detection	LOD = 24 pmol (v_{inj} = 20 µL)	1-Myristoyl-2-(12-((7-nitro-2-1,3-benzoxadiazol-4-yl)amino) dodecanoyl)-*sn*-glycero-3-phosphocholine was used as standard	Sohn et al. (2005)

The first FI–CL system for lipid hydroperoxide determination, published in 1993, was based on luminol as the CL reagent and microperoxidase as the catalyst. The optimized system was used for measuring lipid hydroperoxides in native low-density lipoprotein (LDL) (Cominacini et al. 1993). The FI–CL system consisted of two pumps, an autosampler, and a chemiluminescent detector with a T-mixing coil. Samples were injected by using an autosampler and mixed with luminescent reagent (3 µM luminol and 1 µM microperoxidase in 0.1 M carbonate buffer, pH 10). The calibration curve was obtained using linoleic acid, and a detection limit of 3 pmol linoleic acid hydroperoxide was reached.

An FI–CL system for measuring lipid hydroperoxide (LOOH) in edible oils from plants is described by Bunting and Gray (2003). The FI–CL system was adapted from Auerbach and Gray (1999) by replacing one peristaltic pump with a high-performance liquid chromatography (HPLC) pump (Figure 32.2) to obtain better reproducibility.

Two working procedures were used, namely, an emulsion FI–CL procedure and a monophasic FI–CL procedure. For both procedures, the same FI–CL system was used. In the first case, an aliquot (50 µL) of sample was injected into a carrier phase of methanol/chloroform (9:1, v/v). The CL reagent consisted of cytochrome *c* (10 µg mL^{-1}) and luminol (1 µg mL^{-1}) in a borate buffer (25 mM, pH 9.3) containing Triton X-100 (0.5% w/v) delivered by HPLC pump. The CL measurements were done in the emulsion obtained by mixing the sample with reagents.

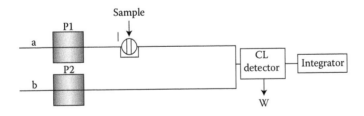

FIGURE 32.2   Flow diagram of the CL method in emulsion for measuring lipid hydroperoxides in edible oils. (a) chloroform/methanol (1:9, v/v), flow rate 1 mL min^{-1}; (b) 1 µg mL^{-1} luminol and 10 µg mL^{-1} cytochrome *c* in 25 mM borate buffer, pH 9.3, containing Triton X-100 (0.5% w/v), flow rate 1 mL min^{-1}; P1, HPLC pump; P2, peristaltic pump; I, injection valve, injection volume 50 µL; W – waste. (Adapted from Auerbach, R. H. and D. A. Gray. 1999. *J. Sci. Food Agric.* 79:385–389.)

In the case of the monophasic FI–CL procedure, the carrier consisted of methanol (100%) in which the sample (50 µL) was injected. The reagent flow was formed by lucigenin (200 µg mL^{-1}) in a 4/1 (v/v) methanol/borate buffer (25 mM, pH 10) solution. For both procedures, the analyzed oil samples were diluted with chloroform before analysis, and t-butyl hydroperoxide (t-BHPO) was utilized as a standard for LOOH. The flow rates of the carrier and CL reagent were set at 0.3 mL min^{-1}.

The results obtained by CL quantification of lipid hydroperoxides in different types of oils were compared with those obtained for PV with the IT method and the iron thiocyanate method. By using the emulsion FI–CL procedure, a nonlinear response between CL signal and the concentration of oils from analyzed samples was obtained. This was explained by differences in turbidity between samples (due to the different amount of oils contained in the analyzed sample). When samples that contained the same quantity of oil but increased concentration of LOOH were used, the relationship between LOOH concentration and CL intensity was linear (r = 0.9908).

In an analysis of different types of oil, the measured CL intensities associated with similar PV values were quite different; in some cases, the CL intensity associated with a sample was four times greater than the CL intensity associated with another one. The authors deduced that calibration of the emulsion FI–CL procedure would require an oil equivalent to calibrate the assay. Taking this condition into account, a good correlation between the results obtained for LOOH determination with the IT method and the emulsion FI–CL method was obtained. The assay was rapid (60 samples/h), sensitive (0.5 nmol LOOH), and reproducible (RSD% smaller than 10%). The conclusion was that quantitative data could be obtained using emulsion the FI–CL method, but only under certain conditions.

In order to refine the FI–CL system for quantification of LOOH in oil samples, another CL reagent, lucigenin, was studied, and the determinations were done in a monophasic system and not in an emulsion. Lucigenin was chosen because it is capable of reacting with hydrogen peroxide without generating free-radical intermediates. This may avoid the interference from chain-blocking antioxidants such as α-tocopherol (a natural component in oil). The effect of oil concentration in the injected sample on the CL signal was tested using the monophasic FI–CL system. A linear relationship was obtained between oil concentration (and LOOH concentration from oil) and CL intensity. However, the CL intensity relative to the actual PV was still oil dependent. The reaction kinetics may therefore differ with the lipid type. This hypothesis was confirmed by comparing the CL associated with oxidized linoleic acid, 1,3-dilinolein, and trilinolein. Experimental data have shown that CL intensity was dependent on the type of lipid. Despite the advantage of the monophasic system, the calibration still requires the use of an oil equivalent to that of the analyzed sample.

A calibration curve for the monophasic FI–CL method was drawn using wheat germ oil diluted in chloroform for LOOH concentrations between 0.1 and 4.8 µmol mL^{-1} for an injected sample of 50 µL. Analysis of the heated wheat germ oil samples by both the IT and monophasic FI–CL methods showed an overall increase in LOOH concentration with the heating time increase. These two methods did show slight differences in the trend of LOOH concentrations but were similar to those observed when using the emulsion system.

32.2 METHODS FOR THE DETERMINATION OF THE ANTIOXIDANT CAPACITY OF LIPIDS

By definition, the antioxidant capacity is the measure of the ability of a compound (or a mixture) to inhibit oxidative degradation, for example, lipid oxidation. The evaluation of the ability of antioxidants to counteract lipid oxidation was presented in a review paper (Laguerre et al. 2007). Phenolics are the main antioxidant components of food. In fats and plant oils, these are basically monophenolics, first of all tocopherols. In fruits, vegetables, tea, wine, and coffee, these are water-soluble polyphenols. The elevated reactivity of phenolics toward active free radicals is considered their principal mechanism of antioxidant activity.

Antioxidant activity is one of the most acclaimed properties of edible oils, and it has been extensively studied, especially for extra-virgin olive oil. Olive oil contains a high amount of natural antioxidants such as tocopherols, carotenoids, sterols, and phenolic compounds. Among phenolic compounds found in extra-virgin olive oils, o-hydroxy-phenolics are very potent antioxidants. The presence of antioxidants in olive oil is an important factor for their oxidative stability during storage and with respect to their thermal degradation. Polyphenol analysis is usually performed with HPLC methods. For olive oil quality control, it is more useful and practical to estimate TAC. This approach accounts for possible synergistic or antagonistic effects among antioxidant compounds and is more time and cost efficient. Total antioxidant capacity of oils is usually determined using CL by monitoring the scavenging of a free radical by antioxidants present in the analyzed sample. This results in inhibition of the chemical reaction or in changes of the reaction rate constants of the ongoing radical reaction.

The majority of CL methods in lipid analysis concern lipid extracts and not untreated lipids, because lipids are not miscible with water. However, any treatment of lipids prior to analysis, such as extraction, changes the chemical composition of the tested sample, which might lead to erroneous results. For this reason, the direct analysis of samples without any treatment would be preferable (Gokmen et al., 2009).

Some relevant CL methods for the evaluation of antioxidant capacity of lipids are presented in Table 32.2. Only a few of them had been automated by using FI methodology, although the automation of all of them could be achieved quite easily.

In the paper of Minioti and Georgiou (Minioti and Georgiou, 2008), an automated FI method for the determination of olive oil antioxidant capacity is presented. The method is based on the oxidation of luminol by hydrogen peroxide using horseradish peroxidase (HRP) as a catalyst and of p-iodophenol as a sensitizer of emission of light. Chemiluminescence is based on a reaction mechanism involving free radicals, and the antioxidant capacity is assessed through light emission inhibition due to free-radical consumption by the antioxidants from the analyzed sample. The method developed was applied for analysis of 50 olive oil samples from various regions of and varieties from Greece. The antioxidants from olive oils (0.70 g) were extracted twice with 0.7 mL from a mixture of methanol:water, 80:20 (v/v), and the two extracts were combined after 5 min centrifugation at 5000 rpm. For the FI method, the extracts were diluted 1:1 with solvent. The FI systems for olive oil TAC determination are presented in Figure 32.3.

Standards and samples are injected in the buffer channel that subsequently merges with a luminol–HRP–p-iodophenol reagent flux. The mixture then meets a stream of hydrogen peroxide just in front of the flow detector. By pumping the reagents in the FI system, and in the absence of an antioxidant in the injected sample, a high steady-state CL signal is registered. In the presence of an antioxidant in the injected sample, the CL signal is decreased, resulting in negative peaks, the height of these depending on the amount of antioxidant in the sample. The decrease of CL signal could be explained by the consumption by antioxidant molecules of the reactive oxygen species generated in the system.

From seven organic solvents tested for diluting olive oil extracts (methanol, ethanol, acetonitrile, dimethyl sulfoxide (DMSO), 1-propanol, 2-propanol, acetone) in different

TABLE 32.2   Determination of Antioxidant Capacity of Fats by Chemiluminescence

Samples	Method	Analytical Performance	Comments	Reference
CL				
Extra-virgin olive oils	Lucigenin/H_2O_2/HO^- in 2-propanol	LOD, 10^{-7}–10^{-5} for 18 pure antioxidants compounds. Straight calibration graph was obtained	A mechanism of CL was presented. The method was validated Hydrophilic and hydrophobic antioxidants could be determined together A comparison with DPPH method has been done	Christodouleas et al. (2009)
Vegetable oils (sunflower, corn, etc.)	Benzene solution of diphenyl methane oxidized by molecular oxygen in the presence of 2,2′-azobisisobutyronitrile and Eu(III)tris(thenoyltrifluoroacetonate)1,10-phenanthroline	Antioxidant analysis at nanomolar scale	A theoretical CL methodology demonstrated with some practical examples for studying the antioxidative properties of vegetable lipids is proposed	Fedorova et al. (2009)
Olive, sunflower, and corn oils	Lucigenin/H_2O_2/HO^-	–	Antioxidant capacity was determined for aqueous methanolic (60%) oil extracts	Papadopoulos et al. (2003)
Sunflower, corn, soy, sesame, pommace, oils, refined olive oils, and extra-virgin olive oils	Lucigenin/H_2O_2/HO^- in 2-propanol	0.04%–1.00% olive oil (w/v) 0.06%–2% seed oils (w/v) 0.2–4 µM gallic acid as a standard RSD% = 3.9% for the blank solution and 5.3% for the oils tested solutions Recovery 91.2%–109.3%	Method was fully validated for precision, trueness, robustness, additivity, and uncertainty Method is able to measure the antioxidant activity of untreated edible oils as well as their hydrophilic and lipophilic extracts	Christodouleas et al. (2011)

(*Continued*)

TABLE 32.2 (*Continued*) Determination of Antioxidant Capacity of Fats by
Chemiluminescence

Samples	Method	Analytical Performance	Comments	Reference
FI–CL				
Lipid extracts from oats	Carrier: methanol:chloroform; CL reagents: borate buffer/ luminol/cytochrome c	The method is two orders of magnitude more sensitive than the β-carotene method	Reverse FI peaks were registered IC_{50} (inhibition concentration) value was determined for analyzed samples	Auerbach and Gray (1999)
Extra-virgin olive oil	Carrier: buffer 1.0×10^{-2} M phosphate, pH 7.4; CL reagents: luminol/ H_2O_2/OH$^-$/HRP/ p-iodophenol in methanol:water 80:20 (v/v)	$LOD = 1.5 \times 10^{-7}$ M gallic acid. Linear range: 1.0×10^{-6}–1×10^{-4} M gallic acid RSD% = 2.8%– 5.6% (n = 4) Throughput: 180 samples per hour	Extracts of oil in methanol:water 80:20 (v/v) were analyzed Method was applied for the assessment of 50 extra-virgin olive oil	Minioti and Georgiou (2008)
Refined sunflower, soybean, corn, virgin olive, and sesame oils	Carrier: milli Q water; CL reagents: H_2O_2/Fe(II)/ luminol/OH$^-$	Linear calibration curve were obtained for 8–19 μg mL^{-1} gallic acid and 7–34 μg mL^{-1} butyl hydroxyanisole	IC_{50} value was determined for analyzed samples Samples were disolved in n-hexane Determinations were done in microemulsions of n-hexane in water	Pulgarin et al. (2010)

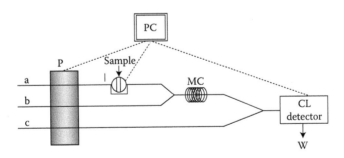

FIGURE 32.3 FI-CL system for olive oil total antioxidant capacity determination. (a) 1×10^{-2} M phosphate buffer, pH 7.4; (b) 2 mM luminol, 2 mM p-iodophenol and 0.64 IU mL^{-1} HRP in buffer; (c) 1 mM hydrogen peroxide in buffer; flow rate for each channel was 6.3 mL min^{-1}; P – peristaltic pump; I – injection valve, injection volume 17 μL; MC – mixing coil, 50 cm; W – waste. (Adapted from Minioti, K. S. and C. A. Georgiou. 2008. *Food Chem.* 109:455–461.)

solvent:water ratios, the best was a mixture of methanol:water, 80:20 (v/v). The other solvents generated either negative or positive blank peaks.

The method has been optimized on manifold, flow rates of the carrier and reagents, concentrations of reagents (hydrogen peroxide and p-iodophenol), and coil length. A calibration curve was traced using gallic acid. The sample antioxidant capacity is expressed in mmol L^{-1} of gallic acid equivalents per kilogram of oil. Limit of detection (LOD) was calculated at 1.5×10^{-7} M gallic acid, and the linear range was between 1.0×10^{-6} and 1×10^{-4} M. The sample throughput achieved with the fully automated method was 180/h. The RSD% was evaluated by multiple injections of gallic acid standards (n = 4) and were 2.8% (c = 1.0×10^{-6} M) and 1.1% (c = 1.0×10^{-4} M). Baseline noise measured for 2 h was lower than 5.3% RSD. The TAC of olive oils of different varieties and origins determined by means of the developed method was between 1.1 ± 0.1 and 100 ± 8 (n = 3) mM gallic acid kg^{-1} olive oil.

The proposed method, when tested against the 2,2'-azino-bis(3-ethylbenzothiazoline-6-sulphonic acid) (ABTS) and 2,2-diphenyl-1-picrylhydrazyl (DPPH) assays, did not show a clear correlation. The reasons could be as follows: (a) chemistry of the assay involved is different: reaction with free radicals generated from hydrogen peroxide in the case of described method and reaction with stable organic free radicals used in the other case; (b) the contact time between the reagents and antioxidants is also very different, about 1 s in the case of the described method and about 1 h in ABTS and DPPH methodology.

In the paper by Auerbach and Gray (1999), an FI–CL method is described for the determination of the antioxidant activity of lipid extracts from oats. The extraction was done by using methanol or propan-2-ol. The crude lipid extracts were fractionated by using a column of silicic acid. A first elution with petroleum ether was carried out to extract the hydrophobic material and a second one with methanol to extract polar lipid material.

The CL measurements were carried out by using an FI–CL assembly similar to that presented in Figure 32.2 and with the same reagents and working conditions as mentioned for the CL method in emulsion for measuring lipid hydroperoxides in edible oils (Bunting and Gray 2003). By pumping the reagents in the FI manifold, a high and steady-state CL signal was registered. By injecting a sample that contains an antioxidant, a reverse FI peak was obtained. Different dilutions using methanol were made of each oat extract (1:4000, 1:3000, 1:2000, 1:1500, 1:100, and 1:400), and t-BHPO was added to a concentration of 3 µmol mL^{-1}. The peak areas obtained at the injection of these dilutions were measured and compared with the area of an equivalent solution minus oat extract, which was set at 100%. For each analyzed sample, the IC_{50} (inhibition concentration) value was determined, which represents the antioxidant concentration that gives a peak area that is 50% of the value corresponding to the sample without antioxidant. IC_{50} values are expressed as the equivalent mass of oats required to inhibit the CL reaction by 50%.

The CL method has been proven to be a convenient, reliable, rapid, and sensitive method for measuring oat antioxidant capacity compared with the β-carotene bleaching method. The latter has been used to measure the antioxidant activity of oats and is considered to be convenient and rapid compared with conventional accelerated-storage trials. Repeated measurements of an oat extract by using the CL method gave consistent results with low variations. The method is two orders of magnitude more sensitive than the β-carotene method. It requires only 15 min of automated sampling per extract compared with 2 h for continuous spectrophotometric measurement of several samples.

This CL method provides absolute IC_{50} values that can be used to compare separate samples and experiments in contrast with the β-carotene method that gives results that fluctuate between experiments. The developed CL method offers the potential to screen oats rapidly for their antioxidant content and to monitor any industrial process designed to recover the active antioxidants from oats.

In a paper by Pulgarin et al. (2010), an FI–CL methodology was described for estimation of the radical-scavenging activity of edible oils as a measure of their antioxidant activity. The determinations were based on the inhibition of luminol CL induced by a Fenton's reagent in a microemulsion of n-hexane in water. The Fenton's reagent is used as the source of free radicals, mainly reactive oxygen species such as $HO^{\bullet}$ and $O_2^{\bullet-}$. These radicals react competitively either with the luminol to obtain CL emission or with antioxidants. The n-hexane is used to dissolve the oils or pure antioxidants. When it is injected in the FI–CL system, a microemulsion is formed, producing a significant increase in the light emission that is registered as an FI peak. The highest peak is obtained when pure n-hexane is injected. The increase in antioxidant concentration in the n-hexane phase entails a decrease of CL signal. The scheme of the FI assembly utilized for determinations is presented in Figure 32.4.

As can be seen from Figure 32.4, pump P1 propels the reagents through three channels. The first and the second channels were used to carry the solutions of hydrogen peroxide (0.3% (v/v) in water) and Fe(II) (0.1 mmol L^{-1}), respectively. Both solutions merged through a T-piece to generate reactive oxygen species. The third channel was used to propel the luminol solution (1 mmol L^{-1}) that merges through another T-piece with the carrier stream (milli Q water), which was propelled by the second pump. The injected samples and luminol met the reactive oxygen species in the flow cell placed in the front of the window of the photomultiplier tube.

The analyzed vegetable oils (refined sunflower, soybean, and corn, along with virgin olive oil and sesame oils) were dissolved in n-hexane to prepare solutions with concentrations between 10 and 225 µg mL^{-1}, which were subsequently injected (100 µL) into the FI system, and the obtained signals were recorded against time. Characteristic FI signals were obtained at the injection of different samples. The highest peak was obtained at the injection of a pure n-hexane sample (blank). At the injection of the n-hexane that contains dissolved vegetable oils or pure antioxidants, the CL emission decreased,

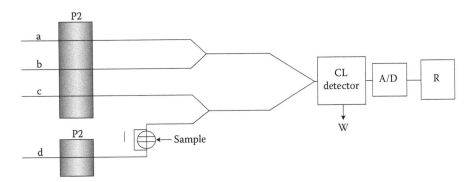

FIGURE 32.4   FI-CL system for the determination of the antioxidant capacity of vegetable oils in a two phase system (water-n-hexane). (a) 0.3 % (v/v) hydrogen peroxide; (b) 0.1 M Fe(II); (c) 1 mM luminol; (d) – water, flow rates of a, b and c are 0.2 mL min^{-1} each and flow rate of d is 3 mL min^{-1}; P1 and P2 – peristaltic pumps; I – injection valve, injection volume 100 µL; W – waste; R – data recording. (Adapted from Pulgarin, J. A. M., L. F. G. Bermejo, and A. C. Duran. 2010. *Eur. J. Lipid Sci. Technol.* 112:1294–1301.)

a fact evidenced by lower peaks. The decrease of CL could be explained by the consumption of reactive oxygen species by the antioxidants from samples. The inhibition of CL emission was measured in terms of the percentage of inhibition (%Inh) using the following equation:

$$\%\text{Inh} = \frac{I_{\text{max}} - I_{\text{oil solution}}}{I_{\text{max}}} \times 100$$

where I_{max} and $I_{\text{oil solution}}$ represent the maximum CL intensity for the blank and for the analyzed sample, respectively. The IC_{50} was calculated from the inhibition curves (obtained by plotting the %Inh against the concentrations of the oil solutions). The IC_{50} represents the concentration of the vegetable oil necessary to reduce the blank intensity to 50%.

By using solution in n-hexane of pure antioxidants, namely gallic acid (GA) and butyl hydroxyanisole (BHA), linear dependences of the %Inh with the antioxidant concentration were obtained that can be expressed by the following linear regression equations: %Inh = 3.39[GA] + 1.26; R^2 = 0.994 and %Inh = 1.96 [BHA] + 0.86; R^2 = 0.999.

Contrary to the results obtained for pure synthetic antioxidants, the inhibition curves for vegetable oils showed nonlinear dependence between the %Inh and oil concentration. By using the log transformation of the oil concentration values, a linear dependence with %Inh can be obtained, which leads to linear regression equations with very good determination coefficients. These equations were used to calculate IC_{50} values for different types of the studied oils.

The results obtained for the antioxidant capacity determination of the edible oils by CL were compared with those obtained by a standard DPPH test, but the results were not comparable. One of the possible explanations for these differences is the nature of the free radicals employed in the determinations. In the DPPH test, the 2,2-diphenyl-1-picrylhydrazyl radical, a very voluminous compound, is used, and its reaction with the antioxidants may be affected by steric impediments. In the case of the CL method, the small reactive oxygen species generated from the Fenton's reagent are the species that react with antioxidants, and they are also responsible for the luminol CL. Moreover, the DPPH test measures the antioxidant activity of the hydrophilic and hydrophobic compounds present in the oil, while the CL method accounts especially for the hydrophilic fraction.

The described FI–CL system provides a simple, sensitive, and accurate method for antioxidant capacity determination in lipidic media. Another advantage of this FI–CL method compared with the DPPH method is its rapidity of measurement; it requires less than 5 min for the complete evaluation of each sample, while the DPPH method needs about 15 min.

32.3 CONCLUSION

Generally, the application of CL to lipid oxidation analysis is limited, with the methods available applied especially to aqueous extracts. The use of the FI–CL methodology for lipid oxidation analysis is rather limited; only few papers have been published in this field. This is surprising if we consider the certain advantages of FI methods in the investigation of lipid oxidation. Moreover, the coupling of FI techniques with CL analysis presents obvious advantages, as was mentioned earlier.

It is expected that in the future, both methods for the determination of lipid oxidation, CL and FI–CL, will receive much more attention. We appreciate that the development of the FI–CL methods will be achieved by elaboration of new FI–CL methods, as well as by adaptation to FI of the existing batch CL methods.

REFERENCES

Auerbach, R. H. and D. A. Gray. 1999. Oat antioxidant extraction and measurement—Towards a commercial process. *J. Sci. Food Agric.* 79:385–389.

Barriuso, B., I. Astiasaran, and D. Ansorena. 2013. A review of analytical methods measuring lipid oxidation status in foods: A challenging task. *Eur. Food Res. Technol.* 236:1–15.

Bezzi, S., S. Loupassaki, C. Petrakis, P. Kefalas, and A. Calokerinos. 2008. Evaluation of peroxide value of olive oil and antioxidant activity by luminol chemiluminescence. *Talanta* 77:642–646.

Bunting, J. P. and D. A. Gray. 2003. Development of a flow injection chemiluminescent assay for the quantification of lipid hydroperoxides. *J. Am. Oil Chem. Soc.* 80:951–955.

Christodouleas, D., C. Fotakis, K. Papadopoulos, D. Dimotikali, and A. C. Calokerinos. 2012. Luminescent methods in the analysis of untreated edible oils: A review. *Anal. Lett.* 45:625–641.

Christodouleas, D., C. Fotakis, K. Papadopoulos, E. Yannakopoulou, and A. C. Calokerinos. 2009. Development and validation of a chemiluminogenic method for the evaluation of antioxidant activity of hydrophilic and hydrophobic antioxidants. *Anal. Chim. Acta* 652:295–302.

Christodouleas, D., K. Papadopoulos, and A. C. Calokerinos. 2011. Determination of total antioxidant activity of edible oils as well as their aqueous and organic extracts by chemiluminescence. *Food Anal. Methods* 4:475–484.

Cominacini, L., A. M. Pastorino, A. McCarthy, M. Campagnola, U. Garbin, A. Davoli, A. Desantis, and V. Locascio. 1993. Determination of lipid hydroperoxides in native low-density-lipoprotein by a chemiluminescent flow-injection assay. *Biochim. Biophys. Acta* 1165:279–287.

Fedorova, G. F., V. A. Menshov, A. V. Trofimov, and R. F. Vasil'ev. 2009. Facile chemiluminescence assay for antioxidative properties of vegetable lipids: Fundamentals and illustrative examples. *Analyst* 134:2128–2134.

Gokmen, V., A. Serpen, and V. Fogliano. 2009. Direct measurement of the total antioxidant capacity of foods: the 'QUENCHER' approach. *Trends Food Sci. Technol.* 20:278–288.

Laguerre, M., J. Lecomte, and P. Villeneuve. 2007. Evaluation of the ability of antioxidants to counteract lipid oxidation: Existing methods, new trends and challenges. *Progr. Lipid Res.* 46:244–282.

Matthaus, B. W. 1996. Determination of the oxidative stability of vegetable oils by Rancimat and conductivity and chemiluminescence measurements. *J. Am. Oil Chem. Soc.* 73:1039–1043.

Minioti, K. S. and C. A. Georgiou. 2008. High throughput flow injection bioluminometric method for olive oil antioxidant capacity. *Food Chem.* 109:455–461.

Navas, M. J. and A. M. Jimenez. 1996. Review of chemiluminescent methods in food analysis. *Food Chem.* 55:7–15.

Navas, M. J. and A. M. Jimenez. 2007. Chemiluminescent methods in olive oil analysis. *J. Am. Oil Chem. Soc.* 84:405–411.

Papadopoulos, K., T. Triantis, E. Yannakopoulou, A. Nikokavoura, and D. Dimotikali. 2003. Comparative studies on the antioxidant activity of aqueous extracts of olive oils and seed oils using chemiluminescence. *Anal. Chim. Acta* 494:41–47.

Pulgarin, J. A. M., L. F. G. Bermejo, and A. C. Duran. 2010. Evaluation of the antioxidant activity of vegetable oils based on luminol chemiluminescence in a microemulsion. *Eur. J. Lipid Sci. Technol.* 112:1294–1301.

Roginsky, V. and E. A. Lissi. 2005. Review of methods to determine chain-breaking antioxidant activity in food. *Food Chem.* 92:235–254.

Rolewski, P., A. Siger, M. Nogala-Kalucka, and K. Polewski. 2009. Chemiluminescent assay of lipid hydroperoxides quantification in emulsions of fatty acids and oils. *Food Res. Int.* 42:165–170.

Sohn, J. H., Y. S. Taki, H. Ushio, and T. Ohshima. 2005. Quantitative determination of total lipid hydroperoxides by a flow injection analysis system. *Lipids* 40:203–209.

Stepanyan, V., A. Arnous, C. Petrakis, P. Kefalas, and A. Calokerinos. 2005. Chemiluminescent evaluation of peroxide value in olive oil. *Talanta* 65:1056–1058.

Tsiaka, T., D. C. Christodouleas, and A. C. Calokerinos. 2013. Development of a chemiluminescent method for the evaluation of total hydroperoxide content of edible oils. *Food Res. Int.* 54:2069–2074.

Wheatley, R. A. 2000. Some recent trends in the analytical chemistry of lipid peroxidation. *Trends Anal. Chem.* 19:617–628.

CHAPTER 33

Determination of Antioxidant Activity by FIA and Microplates

Luís M. Magalhães, Inês Ramos, Ildikó V. Tóth,
Salette Reis, and Marcela A. Segundo

CONTENTS

Antioxidant activity includes several mechanisms of action, such as inhibition of oxidative enzymes responsible for the formation of reactive oxygen species (ROS) and reactive nitrogen species (RNS), induction of the expression of defense enzymes (superoxide dismutase, catalase), and metal-chelating action. Scavenging activity or reducing capacity acts as the second line of defense by removing or reducing reactive species before oxidative damage of biomolecules such as proteins, phospholipids, and DNA occurs (Valko et al., 2007; Lopez-Alarcon and Denicola, 2013). In fact, these two antioxidant mechanisms are one of the main active factors in the body's defense system (Niki, 2010). The available methodologies for assessing these two types of antioxidant action will be discussed in this chapter, highlighting their adaptation to microplate format and to automatic flow–based systems. A critical contribution of the different analytical approaches developed until the present moment for fast screening of antioxidant activity will also be discussed.

The determination of scavenging capacity of ROS/RNS (e.g., H_2O_2, O_2^-, $HO^{\bullet}$, $HOCl$, $NO^{\bullet}$, and $ONOO^-$) by a given compound or sample is not adequate for screening purposes, considering the wide variety of reactive species and the laborious analytical protocols (Huang et al., 2005; Magalhães et al., 2008). In this context, *in vitro* methods that use colored synthetic radicals, such as the $ABTS^{\bullet+}$ (2,2′-azinobis-(3-ethylbenzothiazoline-6-sulfonate) radical cation and the $DPPH^{\bullet}$ (2,2-diphenyl-1-picrylhydrazyl) radical, have been the first-line analytical procedures because of their reduced costs and easily implemented protocols. Moreover, the reducing capacity has been correlated with the antioxidant activity of food and biological samples (Niki, 2010; Gulcin, 2012). Several analytical methodologies have been widely used to measure the reducing properties, such as the Folin–Ciocalteu assay (Singleton et al., 1999), the cupric-reducing antioxidant capacity (CUPRAC) assay (Apak et al., 2004), the ferric-reducing antioxidant power (FRAP) assay (Benzie and Strain, 1996), and electrochemical assays (Chevion et al., 2000). Considering that all these methodologies are routinely applied in the agro-food industries and phytochemical research for screening purposes of antioxidant activity prior to characterization or isolation of bioactive compounds, their adaptation to a high-throughput microplate format or their automation exploiting flow injection–based methods is relevant to attain improved analytical assays suitable for routine analyses. Moreover, both analytical tools contributed to the implementation of greener analytical procedures. The flow-based methods developed for the measurement of the ability of compounds to scavenge ROS/RNS (Magalhães et al., 2009) will not be discussed in this chapter.

33.1 MICROPLATE ASSAYS FOR DETERMINATION OF ANTIOXIDANT ACTIVITY

The major achievements of implementation of antioxidant assays in microplate format instead of classical batchwise procedures are the reduction of reaction volumes (more than 100 times) and the increase of sample throughput. Hence, the reagent consumption, the volume of sample required for analysis, and the amount of waste produced are dramatically reduced (Sundberg, 2000). The large number of wells per microplate (from 24 to 384) combined with the short time required for the measurement of each well makes this analytical platform ideal for routine analysis. Moreover, the microplate readers allow different types of measurements such as absorbance, fluorescence, and chemiluminescence usually applied in almost all antioxidant assays (Figure 33.1). Temperature control and shaking before each measurement is also possible. All these features make this analytical tool suitable for high-throughput determination of antioxidant activity.

    The best example of the utility of a microplate platform is the analytical protocol adopted in the oxygen radical absorbance capacity (ORAC) assay (Ou et al., 2001). The principle of this assay is based on the ability of the sample to inhibit or retard the decay of the analytical signal (fluorescence or absorbance) of the probe along time under constant flux of peroxyl radicals generated from the thermal decomposition of 2,2′-azo-bis(2-amidinopropane) dihydrochloride (AAPH) in aqueous buffer (Huang et al., 2005). In the ORAC assay, the antioxidant capacity of the sample is determined, taking both the inhibition time and degree of inhibition into account as a single quantity. For this reason, kinetic measurements are performed, and the area under the curve (AUC) for both control (absence of antioxidant species) and for different antioxidant concentrations is determined (Figure 33.2). The increase of AUC (%) is further related to Trolox, and the results are expressed as Trolox equivalent antioxidant capacity (TEAC). Considering the issues of this widely used antioxidant assay and the features of microplate format, Huang et al. (2002) proposed a high-throughput instrument platform consisting of a robotic eight-channel liquid handling system and a microplate fluorescence reader in 96-well format. This took place because the ORAC assay had been implemented in the COBAS FARA II analyzer, an instrument discontinued by the manufacturer, and the manual sample preparation was time-consuming and labor intensive. In 2004, Davalos et al. (2004) extended the applicability of the ORAC assay to a manual liquid handling, using a conventional fluorescence microplate reader. Nevertheless, the work of Huang et al. (2002) represented a milestone for the other microplate–ORAC procedures, because further development of the analytical protocols basically modified the reaction conditions, for instance, (i) replacement of hydrophilic peroxyl radical generator (AAPH) by 2,2′-azobis-2,4-dimethyl

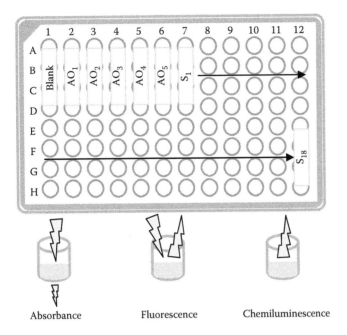

FIGURE 33.1 Schematic representation of the 96-well microplate distribution of experiments for antioxidant determination: AO_1 to AO_5, correspond to increasing concentrations of antioxidant standard solutions, each analyzed in quadruplicate; S_1 to S_{18}, correspond to different sample solutions, each analyzed in quadruplicate. The different modes of measurement mostly used in antioxidant assays, including the absorbance, fluorescence, and chemiluminescence, are represented at the bottom.

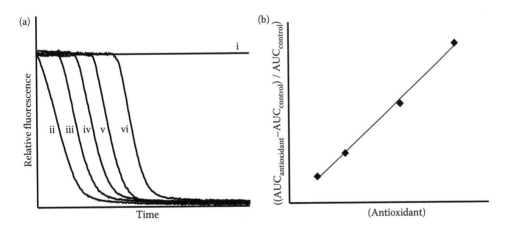

FIGURE 33.2 Illustration of kinetic matching approach. (a) Kinetic profile of fluorescent-probe oxidation mediated by peroxyl radicals: (i) analytical signal of probe along time; (ii) fluorescent-probe oxidation in the absence of antioxidant species; (iii) to (iv) fluorescent-probe oxidation in the presence of increasing concentrations of antioxidant species. (b) Relationship between the increase of the AUC and the concentration of the antioxidant species present in the reaction media.

valeronitrile (AMVN) lipophilic generator (Cho et al., 2007); (ii) replacement of oxidized fluorescent probe (fluorescein) by lipophilic fluorescent probes (Rodrigues et al., 2012), long-wavelength fluorimetric probe (Godoy-Navajas et al., 2011), and spectrophotometric probe (Ortiz et al., 2011); and (iii) the introduction of randomly methylated p-cyclodextrin (RMCD) as the water-solubility enhancer for lipophilic antioxidants (Folch-Cano et al., 2010; Ou et al., 2013). Recently, Ortiz et al. (2012) published the standardized conditions of the microplate–ORAC assay using pyrogallol red as the spectrophotometric probe. This alternative and complementary ORAC assay was based on the fact that kinetic profiles associated with ORAC–fluorescein are usually characterized by the presence of lag times, implying that this index is strongly influenced by stoichiometric factors rather than by the reactivity of antioxidants toward peroxyl radicals. For that reason, ORAC–pyrogallol values reflect the reactivity of the antioxidants toward peroxyl radicals, and among other probes (fluorescein and pyranine), pyrogallol showed a behavior compatible with near-gastric conditions (pH 2.0) (Atala et al., 2013).

The scavenging activity of colored synthetic radicals such as DPPH$^\bullet$ and ABTS$^{\bullet+}$ have also been adapted to a microplate format aiming to attain high-throughput results. Recently, a novel microplate method based on the DPPH$^\bullet$ dry reagent array was proposed to provide a simple analytical protocol with minimal solvent required (Musa et al., 2013). In this work, the radical was dried into a 96-well microplate and aimed to improve the sensitivity and determination throughput of antioxidant measurements. The feasibility of a microplate containing dried DPPH$^\bullet$ approach was tested for different standard antioxidants and food products such as banana, green tea, pink guava, and honeydew melon. The results obtained were comparable with those determined by the classical assay.

The DPPH$^\bullet$ assay is usually carried out in ethanolic or methanolic solution; nevertheless the increase of water to organic solvent ratio within 0%–50% (v/v) of water has shown that the reactivity of antioxidants toward radical species is increased (Stasko et al., 2007). Hence, some microplate procedures have been developed with ethanolic/water (50/50, v/v) solution (Magalhães et al., 2010) or with buffered medium (methanol: 10 mM Tris buffer pH 7.5, 1:1 v/v) (Abderrahim et al., 2013). Regarding the ABTS$^{\bullet+}$ assay, the analytical protocols developed are based on the addition of the preformed radical to microplate (Magalhães et al., 2010) or on the generation in-plate of radical species (Chen et al., 2004; Kambayashi et al., 2009). The former strategy is based on the measurement of absorbance decrease after a given time, while the second approach is based on the measurement of lag time in the absorbance increase due to delay in the formation of the colored ABTS$^{\bullet+}$ species by antioxidant compounds. In this context, a microplate procedure was developed to assess antioxidant activity of several vegetables and Chinese herbs using the ABTS/H_2O_2/horseradish peroxidase (HRP) system as generator of ABTS$^{\bullet+}$ species (Chen et al., 2004). A similar approach was further developed for human plasma, but in this case, the radical species were formed in the microplate well via the oxidation induced by myoglobin and H_2O_2 toward the ABTS reagent (Kambayashi et al., 2009). It was verified that the antioxidant pool of plasma increased for almost all volunteers after regular consumption of vegetable juice during 1 week.

Several microplate methods that measure the overall reducing/antioxidant capacity through a redox reaction between antioxidant compounds and an oxidizing agent such as tungstate–molybdate (FC assay), cupric ion (CUPRAC assay), and ferric ion (FRAP assay) have been developed. The FC assay relies on the transfer of electrons in alkaline medium from phenolic compounds to tungstate–molybdate acid complexes determined spectrophotometrically at 765 nm (Singleton et al., 1999). Ainsworth and Gillespie (2007)

proposed a microplate procedure to estimate the total phenolic content and other oxidation substrates in plant tissues; nevertheless, the reaction was performed in microcentri-
fuge tubes, and only the absorbance measurements were done in a 96-well plate reader,
enabling the analysis of 64 samples per day. After that, Magalhães et al. (2010) proposed
a microliter-scale procedure in which sodium hydroxide solution was used as alkaline
support instead of the carbonate buffer, since the reaction kinetics increased dramatically. Hence, it was possible to decrease the classical reaction time from 120 to 3 min,
providing a significant increase in the sample throughput. Similar to the original assay
(Singleton et al., 1999), this microplate method was not suitable for the determination of
the total phenolic content, especially in samples with low phenol levels, unless interfering
substances were considered or removed. Recently, the original FC assay was compared
with the microplate format and excellent correlations were obtained between the two
approaches (Attard, 2013).

The CUPRAC assay, which is based on reduction of the Cu(II)–neocuproine complex
to the highly colored Cu(I)–neocuproine complex measured spectrophotometrically at
450 nm (Apak et al., 2004), has been also adapted to microplate format. Ribeiro et al.
(2011) proposed a high-throughput CUPRAC assay to assess antioxidant activity in
human serum and urine samples. In this assay, uric acid was used as a calibration compound, and the reaction time was significantly reduced from 30 to 4 min, allowing a
determination throughput value of 288 h^{-1}. Recently, the influence of complexing agents,
reagent addition order, buffer types, and concentration on the determination of antioxidant data of biological samples has also been analyzed (Marques et al., 2014). Similar
study should be applied to food products to evaluate the influence of analytical conditions in the determination of antioxidant activity; this would minimize the variability
between laboratories.

The FRAP method (Benzie and Strain, 1996) is based on the measurement of the
ability of the substance to reduce Fe^{3+} to Fe^{2+}, which is determined spectrophotometrically through its colored complex with 2,4,6-tris(2-pyridyl)-s-triazine (TPTZ) at 595 nm.
This assay was also modified to be used in 96-well microplates for evaluation of the structure–antioxidant activity relationships of flavonoids (Firuzi et al., 2005). In this work,
a good correlation was observed between the FRAP assay and electrochemical results,
which confirms the reliability of the former method for the evaluation of the antioxidant
activity.

Considering that more than one method should be used to characterize the antioxidant activity of a given sample (Huang et al., 2005; Lopez-Alarcon and Denicola, 2013),
some works have reported different antioxidant assays in a microplate format aiming
to simplify the analytical protocols and to evaluate the potential correlations between
antioxidant data. Jimenez et al. (2008) adapted the ORAC, FRAP, and iron(II)-chelating
activity (ICA) assays to 96-well microplates to attain a comprehensive and high-throughput assessment of the antioxidant capacity of food extracts. These assays comprise three
different mechanisms of antioxidant activity, namely, peroxyl-radical scavenging activity, metal reduction, and metal chelation. Within the same area of challenge, different
assays such as Folin–Ciocalteu, CUPRAC, DPPH•, and ABTS•+ were performed in microplate format for the rapid assessment of end-point antioxidant capacity of red wines
(Magalhães et al., 2012). In this work, a kinetic matching approach was proposed to
circumvent the dependency of antioxidant data on the selected reaction time. Basically,
an antioxidant compound with a kinetic profile similar to the sample is selected as the
kinetic matching standard, allowing the assessment of antioxidant values independent of
the selected reaction time (Figure 33.3). Moreover, a way of converting the antioxidant

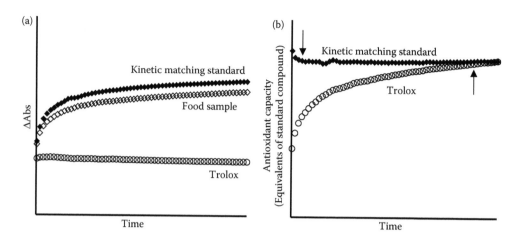

FIGURE 33.3 (a) Net absorbance values plotted against reaction time obtained for food samples, for kinetic matching standard, and for a conventional standard (Trolox). (b) Antioxidant capacity values, given as equivalents of a given standard compound at any time. The reaction time required to achieve antioxidant values that do not depend on analysis time is indicated in the graph by arrows.

results as equivalents of a given standard compound to a common standard (Trolox) was established by using the ratio between calibration sensitivities performed at end-point conditions. The results obtained showed that for the same type of food sample, the standard compound selected depends on the antioxidant assay applied, and there was no statistical difference between results attained by the kinetic matching approach (after <10 min of reaction) and that at end-point conditions (after 60–300 min) (Magalhães et al., 2012). Recently, this kinetic matching approach was applied to different categories of food products (espresso coffees, green teas, black teas, white and red wines) using a microplate ABTS$^{\bullet+}$ assay (Magalhães et al., 2014). The total antioxidant capacity can be determined in 5–15 min, corresponding to a sample throughput of 64–192 h^{-1}. Green teas and red wines contributed to higher antioxidant intake, with 765 ± 51 and 688 ± 71 mg of ascorbic acid per serving, respectively.

33.2 FLOW-BASED METHODS FOR DETERMINATION OF ANTIOXIDANT ACTIVITY

The automation of antioxidant assays by flow injection analysis (FIA) can offer several advantages, besides the enhancement of sample throughput, when compared with conventional batch methods. For antioxidant assessment, the features of FIA systems (Ruzicka and Hansen, 1975) provide strict control of reaction conditions in both space and time, which is essential when highly reactive species that are sensitive to environmental conditions (O$_2$, light, and temperature) are present in the reaction media (Magalhães et al., 2009). The antioxidant assays implemented in FIA systems have been adapted to other flow network strategies aiming to expand the benefits of automation, such as sequential injection analysis (SIA) (Ruzicka and Marshall, 1990), multisyringe flow injection analysis (MSFIA) (Albertus et al., 1999; Segundo and Magalhães, 2006), and a mesofluidic lab-on-valve (LOV) system (Wang and Hansen, 2003). These computer-controlled flow systems enable flexible access to reagent(s), sample, and carrier in any software-defined combination. Table 33.1 describes the main characteristics of FIA and other flow-based methods used to automate the widely used antioxidant assays such as ABTS$^{•+}$, DPPH$^{•}$, Folin–Ciocalteu, CUPRAC, FRAP, and amperometric measurements. Among all antioxidant assays that have been automated, more than half of the proposed applications are based on the utilization of colored, radical species (ABTS$^{•+}$ or DPPH$^{•}$), due to their simplicity, reduced costs, and broad application in the agrochemical area (Magalhães et al., 2009).

The automation of the ABTS$^{•+}$ assay is a good example of the evolution of flow injection–based methods applied to the automation of antioxidant assays (Figure 33.4). The first FIA–ABTS$^{•+}$ manifold was developed in 2003 by Pellegrini et al. (2003) and consisted of a single-channel flow system (Figure 33.4a). The injected sample containing antioxidant compounds reduced the radical species and provided a negative peak, whose area was proportional to the concentration of reduced ABTS$^{•+}$. The FIA system was applied for the evaluation of antioxidant capacity of several common beverages (beer, coffee, cola, fruit juices, and tea), and the results were not statistically different from the batch assay. However, this FIA–ABTS$^{•+}$ assay partially failed when more complex biological samples, such as human plasma, were analyzed. In this context, Bompadre et al. (2004) introduced temperature control of the reaction coil and proposed minor changes in the flow manifold, namely, sample volume and reaction coil configuration. Using this flow system, temperature was demonstrated as a critical point in the measurement of plasma antioxidant capacity, while its influence was less important for noncomplex biological samples (mouth rinse, white wines). The FIA–ABTS$^{•+}$ system with controlled temperature (35°C) was useful to screen rapidly and with high repeatability the antioxidant activity of both noncomplex biological mixtures and plasma samples.

The main limitation of single-channel systems is the depletion of reagent (ABTS$^{•+}$, in this case) in the central zone of the sample plug. Hence, a double-line FIA system was proposed for the ABTS$^{•+}$ assay (Labrinea and Georgiou, 2005), which allowed the addition of reagent to sample plug that was injected in a different channel (Figure 33.4b). Compared with the single-channel system, this flow manifold provided a higher sample throughput (four times higher, Table 33.1). As the flow injection signals are the output of two kinetic processes occurring simultaneously, that is, physical dispersion and chemical reaction, the concentration gradients formed along the injected sample bolus were exploited to obtain information on the reaction kinetics of ABTS$^{•+}$ scavenging. The antioxidant values of fast-reacting scavengers were independent from the time of measurement, while the antioxidant values of slow-reacting compounds increased using the peak tail readings because of their longer reaction time. The same authors had previously also concluded that the antioxidant values were dependent on reaction time as well as on pH value due to the differences in the rate of reaction for samples and for the reference compound (Trolox) (Labrinea and Georgiou, 2004). For that reason, the pH value of the FIA system was fixed at 4.6, similar to that of analyzed food products (Table 33.1).

The FIA–ABTS$^{•+}$ described above relies on the radical cation preformed offline by chemical or enzymatic oxidation of ABTS (Magalhães et al., 2008). In 2005, Ivekovic et al. (2005) proposed a novel FIA system that comprised a flow-through electrolysis cell to generate in-line the radical species by electrochemical oxidation of ABTS. This improvement represented a step further toward a fully automatic method, since the time-consuming procedure of ABTS oxidation by chemical or enzymatic reaction (16 and 3 h, respectively) was avoided. The amount of the ABTS$^{•+}$ generated in-line was established by the flow rate and the value of the current imposed on the cell. In-line enzymatic

TABLE 33.1   Determination of Antioxidant Activity by Flow-Based Methods

Antioxidant Assay	Flow Method	pH Value	Type of Sample	Detection System	Determination Rate (h^{-1})	Reference
ABTS[•+]	FIA	–	Beer, coffee, cola, juice, tea	Vis (734 nm)	30	Pellegrini et al. (2003)
	FIA	7.4	Mouthrinse, white wine, plasma	Vis (n.g.)	22	Bompadre et al. (2004)
	FIA	4.6	Honey, wine	Vis (414 nm)	120	Labrinea and Georgiou (2005)
	FIA	7.4	Coffee, red wine, tea	Vis (734 nm)	32	Ivekovic et al. (2005)
	SIA	7.4	Beer, juice, milk, tea	Vis (734 nm)	9–20	Lima et al. (2005)
	MSFIA	4.6	Beer, juice, tea, wine	Vis (734 nm)	12 or 18	Magalhães et al. (2007)
	LOV	4.6	Red wine	Vis (734 nm)	36	Ramos et al. (2014)
DPPH[•]	FIA	–	Coffee, red wine, tea	ESR	13	Ukeda et al. (2002)
	SIA	4.8	Herbal and mushroom extracts	Vis (525 nm)	45	Polasek et al. (2004)
	MSFIA	–	Beer, juice, tea, wine	Vis (517 nm)	13	Magalhães et al. (2006)
	FIA	–	Wine	Vis (515 nm)	41	Bukman et al. (2013)
Folin–Ciocalteu	MSFIA	>12	Beer, juices, tea, wine	Vis (750 nm)	12	Magalhães et al. (2006)
CUPRAC	FIA	7.0	Human serum, urine	Vis (450 nm)	15	Ribeiro et al. (2011)
FRAP	FIA	–	Tea	Vis (595 nm)	30	Martins et al. (2013)
Amperometric	FIA	7.5	Apple, pear, juices, wines	Amperometry	60	Blasco et al. (2005)

Note: FIA, flow injection analysis; Vis, visible; SIA, sequential injection analysis; MSFIA, multisyringe flow injection analysis; LOV, mesofluidic lab-on-valve system; CUPRAC, cupric reducing antioxidant capacity; FRAP, ferric reducing antioxidant power; ESR, electron spin resonance; n.g., not given.

strategies to generate the ABTS$^{\bullet+}$ have been also described in the literature (Milardovic et al., 2007a). In this case, the FIA system consisted of the continuous flow of ABTS/H$_2$O$_2$ solution through a tubular flow-through bioreactor containing immobilized HRP, which catalyzes the oxidation of ABTS. However, as the ABTS/H$_2$O$_2$ solution was unstable across the course of a day, another ABTS$^{\bullet+}$ generation system was proposed from a bienzymatic reaction of glucose oxidase and HRP enzymes, which were separately immobilized in tubular flow-through reactors (Milardovic et al., 2007b). The main drawback of these enzymatic strategies when compared with electrochemical generation of ABTS$^{\bullet+}$ is related to the interference of the unreacted H$_2$O$_2$.

Aiming to attain flexible automatic flow systems, Lima et al. (2005) proposed a novel SIA–ABTS$^{\bullet+}$ manifold that uses one single standard solution to perform the calibration procedure (Figure 33.4c). For this, a well-stirred mixing chamber, placed in a side port of the selection valve, was incorporated into the system, allowing a thorough mixing between the sample and ABTS$^{\bullet+}$ and diluting the antioxidant standard solution by aspiration of variable volumes of sample. The effect of reaction time and pH upon antioxidant activity of wine samples was also observed. Exploiting the high versatility of SIA systems, the ABTS$^{\bullet+}$ assay was automated along with other methodologies using the same manifold without the need for system reconfiguration (Pinto et al., 2005).

This analytical feature of flow-based methods meets the requirement that more than one method for antioxidant activity characterization should be applied due to the complexity of samples and the absence of a standard antioxidant assay (Huang et al., 2005). Following the same challenge, an automatic flow procedure based on MSFIA was proposed for the sequential determination of ABTS$^{\bullet+}$-scavenging capacity and Folin–Ciocalteu

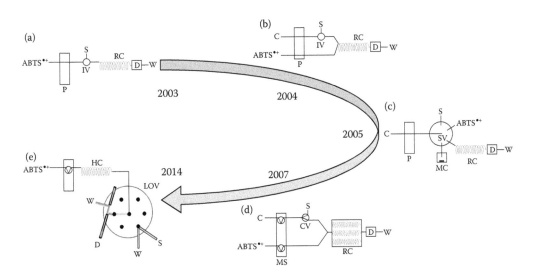

FIGURE 33.4 Schematic representation of flow-based systems developed from 2003 to 2014 for the determination of antioxidant activity by the ABTS$^{\bullet+}$ assay resorting to single (a) or double-line (b) flow injection analysis, sequential injection analysis (c), multisyringe flow injection analysis (d), and mesofluidic lab-on-valve (LOV) system (e) P, pump; IV, injection valve; S, sample; RC, reaction coil; D, detector; W, waste; C, carrier; SV, selection valve; MC, mixing chamber; MS, multisyringe; CV, commutation valve; HC, holding coil; LOV, lab-on-valve.

reducing capacity of a large number ($n = 72$) of beverages (Magalhães et al., 2007). The manifold configuration included several reaction coils for reaction development when the two methods were performed in tandem (Figure 33.4d). This manifold was particularly useful for the evaluation of the correlation between ABTS•+-scavenging capacity and -reducing capacity, showing that this correlation may vary according to the type of sample analyzed.

In flow injection–based procedures, the chemical equilibrium is seldom attained, and therefore the reaction time is usually lower than that applied in batch procedures, for which end-point conditions prevail, resulting in biased results in flow systems (Magalhães et al., 2009). In fact, the effect of the reaction time in the TEAC values is strongly evident in flow-based systems, because the reaction time selected is distant from end-point conditions, and the standard compound (Trolox) has a different kinetic profile from that obtained for antioxidant-rich samples (Labrinea and Georgiou, 2004). For screening purposes, this is not problematic, since the objective in this case is to obtain a "positive" or "negative" result in order to further investigate the compound/extract analyzed. However, when the total antioxidant capacity of the sample is the aim, the antioxidant data are underestimated, especially when slow-reacting antioxidants are present. Taking into account this issue, the ABTS•+ assay was recently automated on a miniaturized mesofluidic LOV system exploiting the kinetic matching approach to overcome the time-biased effect on antioxidant data (Ramos et al., 2014). The reduced dispersion of the sample in the LOV format due to the short distance between the injection point and the flow cell (Figure 33.4e) provided higher reagent/sample ratios suitable to fostering reaction completion. The sample volume was downscaled to 1 μL, and a stopped-flow approach was implemented for monitoring the reaction between pure compounds or red wines with ABTS•+ within the first minute. Since flow-based methods use Trolox, which has a kinetic profile different from samples, as a reference compound, the antioxidant data determined are dependent of selected reaction time. In this work, a standard compound (tannic acid) that has a similar kinetic profile to that obtained for samples was proposed. Hence, using a stopped-flow approach and tannic acid as the kinetic matching standard, the antioxidant capacity values at end-point conditions for red wines were attained in <1 min, providing a determination throughput of 36 h⁻¹ (Table 33.1).

The batchwise DPPH• assay is technically simple, but it is performed in organic or organic/aqueous media, which may represent an issue to flow-based systems. The first flow method for the determination of antioxidant activity by DPPH• assay was an FIA system coupled to electron spin resonance (ESR) spectrometry equipment (Ukeda et al., 2002). In this simple FIA system, the DPPH• methanolic solution was continuously fed into a flow-through flat cell, providing a constant ESR signal at a fixed magnetic field strength. When the sample was injected into the carrier stream, the signal was suppressed, and a negative peak appeared whose height was proportional to the antioxidant activity. The low determination rate (13 h⁻¹) was due to the low resistance to pressure of the flow cell, as flow rates higher than 0.32 mL min⁻¹ were not applicable.

In 2004, a spectrophotometric SIA–DPPH• methodology was proposed for routine screening of antioxidant capacity in a large series of lyophilized herbal and mushroom extracts (Polasek et al., 2004). A sandwich strategy, intercalation of a DPPH• plug between two plugs of the sample, was implemented to circumvent the mixing problems inherent to SIA manifolds. As described for ABTS•+ assays, lower antioxidant results were attained for samples when compared with the conventional batch assay due to the difference in the reaction time applied. Nevertheless, this SIA–DPPH• system was further applied for a screening study in a large number of plant extracts (Koleckar et al., 2008) and to several naturally occurring coumarins and their synthesized analogs (Rehakova et al., 2008).

An MSFIA–DPPH• system was further developed to monitor the absorbance decrease along time, aiming to evaluate the reaction kinetics upon the tested sample (Magalhães et al., 2006). For this, a stopped-flow approach was implemented, and for those samples that did not exhaust its scavenging capacity within the period of measurement, a mathematical model was successfully applied to estimate the total DPPH• consumed. The combination of automation and mathematical treatment of results enabled a significant reduction of analysis time (from 2 h to 3 min) and represented an alternative to the kinetic matching approach previously described. A similar MSFIA system was also developed to evaluate the DPPH• reaction conditions with respect to pH (unbuffered, 4.1, and 7.6) and solvent (methanol and ethanolic solution 50%, v/v) (Magalhães et al., 2007). Recently, Bukman et al. (2013) proposed an FIA–DPPH• system in which flow conditions were optimized with the response surface methodology using a central composite rotatable design. The parameters studied were the carrier flow rate, the length of the sampling, the reagent loops, and the reaction coil. The method proposed was applied to wine samples, and the determination rate was similar to that described for the SIA–DPPH• assay (Table 33.1).

Several flow-based methods that measure the overall reducing capacity have been developed during the last decade by automation of the Folin–Ciocalteu (FC) assay, CUPRAC assay, and FRAP assay, or through amperometric methods, which are based on the measurement of current intensity obtained at a fixed potential. The automation of the FC assay was implemented by an MSFIA system using spectrophotometric detection (Magalhães et al., 2006) (Table 33.1). The carbonate buffer solution used for pH adjustment in the batch procedure (Singleton et al., 1999) was replaced by sodium hydroxide solution, as the rate of reduction of tungstate–molybdate complexes was substantially increased in higher-alkalinity medium. This flow approach enabled the reduction of reaction time from 2 h (batch method) to 4 min. As described before, an MSFIA system was later implemented for the sequential determination of this parameter and ABTS•+ scavenging capacity (Magalhães et al., 2007).

Ribeiro et al. (2011) proposed an FIA system for the assessment of total antioxidant capacity in human urine and serum, exploiting the chemistry of the CUPRAC assay. The kinetic matching concept was applied, using a calibration compound (uric acid) with kinetic behavior similar to that shown by urine samples. Using this strategy, the reaction time was significantly reduced from 30 to 4 min, and the application to human serum reference and urine samples provided total antioxidant capacity (TAC) values that were in agreement with those of the end-point microplate procedure. A similar approach could be applied to food products by selecting an appropriate kinetic matching standard, as was done in microplate format for red wines (Magalhães et al., 2012). Recently, a FRAP assay was adapted to an FIA system for the determination of antioxidant activity of tea samples (Martins et al., 2013). As described before for FIA–DPPH (Bukman et al., 2013), the flow conditions were studied using the response surface methodology with a central composite rotatable design. Using the proposed method, about 95% less volume of FRAP reagent was consumed.

The amperometric detection has been widely implemented in diverse FIA systems to measure the total reducing capacity, which is correlated with the antioxidant activity (Magalhães et al., 2009). These flow systems do not require the use of reactive species, radicals or nonradicals, which is an important issue, as the antioxidant capacity assessed is strongly dependent on the oxidant species applied. Table 33.1 describes the features of the FIA system developed by Blasco et al. (2005) to determine the "electrochemical index," defined as the total polyphenolic content obtained by electrochemistry. For this, the amperometric current at different oxidation potentials (+0.8, +0.5, +0.3 V) was measured. This FIA electrochemical protocol was applied to several food products, and the results obtained were well correlated with those determined using the Folin–Ciocalteu assay. This approach was further applied to complex natural samples such as honey (Avila et al., 2006). Finally, a novel FIA system that generates the oxidant species in-line and determines the H_2O_2 by an amperometric biosensor was developed (Lates et al., 2011). In this automatic method, the oxidant species resulted from enzymatic reaction between xanthine and an immobilized xanthine oxidase, while an amperometric biosensor was based on Os-wired HRP.

The analytical features of flow-based methods were also exploited for chemiluminescence determination of antioxidant capacity (Fassoula et al., 2011; Pulgarin et al., 2012). In this case, the flow strategy has advantages when compared with microplate protocols due to the possibility of mixing the reagents and sample just before the measurement or even within the flow cell, improving the sensitivity and the repeatability of the measurements. This analytical strategy is particularly relevant when ROS/RNS are used as oxidant species associated with sensitive chemiluminescence detection (Magalhães et al., 2009).

33.3 CONTRIBUTIONS OF FLOW-BASED METHODS AND MICROPLATE FORMAT PROTOCOLS FOR ANTIOXIDANT ASSESSMENT

The automation of antioxidant assays presents several advantages, such as versatility, low cost, and high sample throughput suitable for routine determination in both pure compounds and complex matrices. Compared with microplate procedures, the reaction/determination takes place in a contained environment, minimizing operator exposure to organic solvents (e.g., DPPH• assay), and the reaction conditions (time, mixing, pH) are strictly controlled. These features improve the repeatability and reproducibility of the antioxidant data.

The antioxidant data obtained are dependent on the reaction time selected and this could be circumvented by both microplate and flow-based methods exploiting different strategies. The kinetic matching approach represents a valuable procedure to determine the total antioxidant capacity in short reaction times without the necessity of reaching end-point conditions (Magalhães et al., 2012). This approach was already applied to a microplate format and to flow-based systems (Magalhães et al., 2014; Ramos et al., 2014). Other strategies were implemented to obtain antioxidant data independent of the reaction time, such as the change of reaction media performed for the Folin–Ciocalteu assay (Magalhães et al., 2010) or the application of a mathematical model to the absorbance profile obtained in the first minutes of the reaction in order to predict the radical consumption at the end-point time (Magalhães et al., 2006). These strategies may contribute to the standardization of the reaction conditions of antioxidant assays, permitting the comparison of results within and between laboratories (Lopez-Alarcon and Denicola, 2013).

The microplate format has the limitation that the reproducibility of first measurements is dependent on the time taken between the additions of reagents to the microplate well and the start of the signal acquisition. Moreover, when the microplate is full, monitoring within the first minute of the reaction is not feasible. These issues are a strong limitation when kinetic or chemiluminescence measurements are required. The features of flow-based systems could overcome these limitations, due to reproducibility of time events and the proximity between injection/mixture point and detection system.

The simultaneous analysis of several samples can be performed in microplate format (for instance, in a 96-well microplate: 18 samples, 5 standard solutions, and a reagent blank can be analyzed in quadruplicate) (Figure 33.1). This point is relevant for kinetic measurements, as each well can be read within an interval lower than 1 min. This feature is explored in the microplate adaptation of ORAC assay, in which antioxidant activity is obtained by kinetic measurements and by the estimation of the AUC of probe oxidation (Ou et al., 2013). Flow-based methods could not compete with their determination rate; nevertheless, the strict control of reagent addition and further peroxyl radical generation in a controlled environment could be advantageous. On the other hand, the flexibility of flow techniques, especially those that are computer controlled, fostered the implementation of different antioxidant assays in the same manifold, enabling the analysis of the sample by different methods at the same time (Pinto et al., 2005; Magalhães et al., 2007). This methodology avoids the discrepancies that may arise due to sample modification over time and will provide a more reliable comparison between methods. The flow systems offer the possibility of generating the reactive species in-line (Ivekovic et al., 2005; Lates et al., 2011; Milardovic et al., 2007a,b), diluting the sample or standard solution in-line (Lima et al., 2005), and adjusting the pH in-line (Labrinea and Georgiou, 2004). In the case of the combination of FIA with amperometric detection, a rapid and reproducible way of measuring the reducing capacity of colored samples in the absence of a challenging oxidant species is provided (Blasco et al., 2005).

The analytical methods reported in the literature were mainly applied to pure compounds, to beverages, and to extracts of solid food products. In our opinion, the future challenge of both microplate- and flow-based methods is to determine the antioxidant activity of solid food samples simultaneous with the extraction procedure, with the aim of reducing both the laborious extraction protocols and the high variability between experiments. This is a critical point when digestive gastrointestinal fluids are used as extractor solvents to assess the fraction of antioxidant capacity that is really bioaccessible.

ACKNOWLEDGMENTS

I. V. Tóth and L. M. Magalhães thank FSE and Ministério da Ciência, Tecnologia e Ensino Superior (MCTES) for their financial support through the POPH-QREN program. This work received financial support from the European Union (FEDER funds through COMPETE) and National Funds (FCT, Fundação para a Ciência e Tecnologia) through project UID/Multi/04378/2013 and also from EU FEDER funds under the framework of QREN (Project NORTE-07-0124-FEDER-000067). I. I. Ramos and L. M. Magalhães thanks FCT for the financial support through the Grant SFRH/BD/97540/2013 and SFRH/BPD/101722/2014, respectively.

REFERENCES

Abderrahim, F., S. M. Arribas, M. C. Gonzalez, and L. Condezo-Hoyos. 2013. Rapid high-throughput assay to assess scavenging capacity index using DPPH. *Food Chem.* 141(2):788–794.

Ainsworth, E. A. and K. M. Gillespie. 2007. Estimation of total phenolic content and other oxidation substrates in plant tissues using Folin–Ciocalteu reagent. *Nat. Protoc.* 2(4):875–877.

Albertus, F., B. Horstkotte, A. Cladera, and V. Cerda. 1999. A robust multisyringe system for process flow analysis—Part I. On-line dilution and single point titration of protolytes. *Analyst* 124(9):1373–1381.

Apak, R., K. Guclu, M. Ozyurek, and S. E. Karademir. 2004. Novel total antioxidant capacity index for dietary polyphenols and vitamins C and E, using their cupric ion reducing capability in the presence of neocuproine: CUPRAC method. *J. Agric. Food Chem.* 52(26):7970–7981.

Atala, E., A. Aspee, H. Speisky, E. Lissi, and C. Lopez-Alarcon. 2013. Antioxidant capacity of phenolic compounds in acidic medium: A pyrogallol red-based ORAC (oxygen radical absorbance capacity) assay. *J. Food Compos. Anal.* 32(2):116–125.

Attard, E. 2013. A rapid microtitre plate Folin–Ciocalteu method for the assessment of polyphenols. *Central Eur. J. Biol.* 8(1):48–53.

Avila, M., A. G. Crevillen, M. C. Gonzalez et al. 2006. Electroanalytical approach to evaluate antioxidant capacity in honeys: Proposal of an antioxidant index. *Electroanalysis* 18(18):1821–1826.

Benzie, I. F. F. and J. J. Strain. 1996. The ferric reducing ability of plasma (FRAP) as a measure of "antioxidant power": The FRAP assay. *Anal. Biochem.* 239(1):70–76.

Blasco, A. J., M. C. Rogerio, M. C. Gonzalez, and A. Escarpa. 2005. "electrochemical index" as a screening method to determine "total polyphenolics" in foods: A proposal. *Anal. Chim. Acta* 539(1–2):237–244.

Bompadre, S., L. Leone, A. Politi, and M. Battino. 2004. Improved FIA-ABTS method for antioxidant capacity determination in different biological samples. *Free Radic. Res.* 38(8):831–838.

Bukman, L., A. C. Martins, E. O. Barizao, J. V. Visentainer, and V. D. Almeida. 2013. DPPH assay adapted to the FIA system for the determination of the antioxidant capacity of wines: Optimization of the conditions using the response surface methodology. *Food Anal. Methods* 6(5):1424–1432.

Chen, I. C., H. C. Chang, H. W. Yang, and G. L. Chen. 2004. Evaluation of total antioxidant activity of several popular vegetables and Chinese herbs: A fast approach with ABTS/H_2O_2/HRP system in microplates. *J. Food Drug Anal.* 12(1):29–33.

Chevion, S., M. A. Roberts, and M. Chevion. 2000. The use of cyclic voltammetry for the evaluation of antioxidant capacity. *Free Radic. Biol. Med.* 28(6):860–870.

Cho, Y. S., K. J. Yeum, C. Y. Chen et al. 2007. Phytonutrients affecting hydrophilic and lipophilic antioxidant activities in fruits, vegetables and legumes. *J. Sci. Food Agric.* 87(6):1096–1107.

Davalos, A., C. Gomez-Cordoves, and B. Bartolome. 2004. Extending applicability of the oxygen radical absorbance capacity (ORAC-fluorescein) assay. *J. Agric. Food Chem.* 52(1):48–54.

Fassoula, E., A. Economou, and A. Calokerinos. 2011. Development and validation of a sequential-injection method with chemiluminescence detection for the high throughput assay of the total antioxidant capacity of wines. *Talanta* 85(3):1412–1418.

Firuzi, O., A. Lacanna, R. Petrucci, G. Marrosu, and L. Saso. 2005. Evaluation of the antioxidant activity of flavonoids by "ferric reducing antioxidant power" assay and cyclic voltammetry. *Biochim. Biophys. Acta General Subjects* 1721(1–3):174–184.

Folch-Cano, C., C. Jullian, H. Speisky, and C. Olea-Azar. 2010. Antioxidant activity of inclusion complexes of tea catechins with beta-cyclodextrins by ORAC assays. *Food Res. Int.* 43(8):2039–2044.

Godoy-Navajas, J., M. P. A. Caballos, and A. Gomez-Hens. 2011. Long-wavelength fluorimetric determination of food antioxidant capacity using Nile Blue as reagent. *J. Agric. Food Chem.* 59(6):2235–2240.

Gulcin, I. 2012. Antioxidant activity of food constituents: An overview. *Archives Toxicol.* 86(3):345–391.

Huang, D. J., B. X. Ou, M. Hampsch-Woodill, J. A. Flanagan, and R. L. Prior. 2002. High-throughput assay of oxygen radical absorbance capacity (ORAC) using a multichannel liquid handling system coupled with a microplate flourescence reader in 96-well format. *J. Agric. Food Chem.* 50(16):4437–4444.

Huang, D. J., B. X. Ou, and R. L. Prior. 2005. The chemistry behind antioxidant capacity assays. *J. Agric. Food Chem.* 53(6):1841–1856.

Ivekovic, D., S. Milardovic, M. Roboz, and B. S. Grabaric. 2005. Evaluation of the antioxidant activity by flow injection analysis method with electrochemically generated ABTS radical cation. *Analyst* 130(5):708–714.

Jimenez-Alvarez, D., F. Giuffrida, F. Vanrobaeys et al. 2008. High-throughput methods to assess lipophilic and hydrophilic antioxidant capacity of food extracts *in vitro*. *J. Agric. Food Chem.* 56(10):3470–3477.

Kambayashi, Y., N. T. Binh, H. W. Asakura et al. 2009. Efficient assay for total antioxidant capacity in human plasma using a 96-well microplate. *J. Clin. Biochem. Nutr.* 44(1):46–51.

Koleckar, V., L. Opletal, E. Brojerova et al. 2008. Evaluation of natural antioxidants of Leuzea carthamoides as a result of a screening study of 88 plant extracts from the European Asteraceae and Cichoriaceae. *J. Enzyme Inhibition Med. Chem.* 23(2):218–224.

Labrinea, E. P. and C. A. Georgiou. 2004. Stopped-flow method for assessment of pH and timing effect on the ABTS total antioxidant capacity assay. *Anal. Chim. Acta* 526(1):63–68.

Labrinea, E. P. and C. A. Georgiou. 2005. Rapid, fully automated flow injection antioxidant capacity assay. *J. Agric. Food Chem.* 53(11):4341–4346.

Lates, V., J.-L. Marty, and I. C. Popescu. 2011. Determination of antioxidant capacity by using xanthine oxidase bioreactor coupled with flow-through H_2O_2 amperometric biosensor. *Electroanalysis* 23(3):728–736.

Lima, M. J. R., I. V. Toth, and A. O. S. S. Rangel. 2005. A new approach for the sequential injection spectrophotometric determination of the total antioxidant activity. *Talanta* 68(2):207–213.

Lopez-Alarcon, C. and A. Denicola. 2013. Evaluating the antioxidant capacity of natural products: A review on chemical and cellular-based assays. *Anal. Chim. Acta* 763:1–10.

Magalhães, L. M., L. Barreiros, M. A. Maia, S. Reis, and M. A. Segundo. 2012. Rapid assessment of endpoint antioxidant capacity of red wines through microchemical methods using a kinetic matching approach. *Talanta* 97:473–483.

Magalhães, L. M., L. Barreiros, S. Reis, and M. A. Segundo. 2014. Kinetic matching approach applied to ABTS assay for high-throughput determination of total antioxidant capacity of food products. *J. Food Compos. Anal.* 33(2):187–194.

Magalhães, L. M., M. Lucio, M. A. Segundo, S. Reis, and J. L. F. C. Lima. 2009. Automatic flow injection based methodologies for determination of scavenging capacity against biologically relevant reactive species of oxygen and nitrogen. *Talanta* 78(4–5):1219–1226.

Magalhães, L. M., M. Santos, M. A. Segundo, S. Reis, and J. L. F. C. Lima. 2009. Flow injection based methods for fast screening of antioxidant capacity. *Talanta* 77(5):1559–1566.

Magalhães, L. M., F. Santos, M. A. Segundo, S. Reis, and J. L. F. C. Lima. 2010. Rapid microplate high-throughput methodology for assessment of Folin–Ciocalteu reducing capacity. *Talanta* 83(2):441–447.

Magalhães, L. M., M. A. Segundo, S. Reis, and J. L. F. C. Lima. 2006. Automatic method for determination of total antioxidant capacity using 2,2-diphenyl-1-picrylhydrazyl assay. *Anal. Chim. Acta* 558(1–2):310–318.

Magalhães, L. M., M. A. Segundo, S. Reis, and J. L. F. C. Lima. 2008. Methodological aspects about *in vitro* evaluation of antioxidant properties. *Anal. Chim. Acta* 613(1):1–19.

Magalhães, L. M., M. A. Segundo, S. Reis, J. L. F. C. Lima, and A. O. S. S. Rangel. 2006. Automatic method for the determination of Folin–Ciocalteu reducing capacity in food products. *J. Agric. Food Chem.* 54(15):5241–5246.

Magalhães, L. M., M. A. Segundo, S. Reis, J. L. F. C. Lima, I. V. Toth, and A. O. S. S. Rangel. 2007. Automatic flow system for sequential determination of ABTS scavenging capacity and Folin–Ciocalteu index: A comparative study in food products. *Anal. Chim. Acta* 592(2):193–201.

Magalhães, L. M., M. A. Segundo, C. Siquet, S. Reis, and J. L. F. C. Lima. 2007. Multi-syringe flow injection system for the determination of the scavenging capacity of the diphenylpicrylhydrazyl radical in methanol and ethanolic media. *Microchim. Acta* 157(1–2):113–118.

Marques, S. S., L. M. Magalhães, I. V. Tóth, and M. A. Segundo. 2014. Insights on antioxidant assays for biological samples based on the reduction of copper complexes— The importance of analytical conditions. *Int. J. Mol. Sci.* 15:11387–11402.

Martins, A. C. et al. 2013. The antioxidant activity of teas measured by the FRAP method adapted to the FIA system: Optimising the conditions using the response surface methodology. *Food Chem.* 138(1):574–580.

Milardovic, S., I. Kerekovic, R. Derrico, and V. Rumenjak. 2007a. A novel method for flow injection analysis of total antioxidant capacity using enzymatically produced ABTS and biamperometric detector containing interdigitated electrode. *Talanta* 71(1):213–220.

Milardovic, S., I. Kerekovic, and V. Rumenjak. 2007b. A flow injection biamperometric method for determination of total antioxidant capacity of alcoholic beverages using bienzymatically produced ABTS($\bullet$+). *Food Chem.* 105(4):1688–1694.

Musa, K. H., A. Abdullah, B. Kuswandi, and M. A. Hidayat. 2013. A novel high throughput method based on the DPPH dry reagent array for determination of antioxidant activity. *Food Chem.* 141(4):4102–4106.

Niki, E. 2010. Assessment of antioxidant capacity *in vitro* and *in vivo*. *Free Radic. Biol. Med.* 49(4):503–515.

Ortiz, R., M. Antilen, H. Speisky, M. E. Aliaga, and C. Lopez-Alarcon. 2011. Analytical parameters of the microplate-based ORAC-pyrogallol red assay. *J. AOAC Int.* 94(5):1562–1566.

Ortiz, R., M. Antilen, H. Speisky, M. E. Aliaga, C. Lopez-Alarcon, and S. Baugh. 2012. Application of a microplate-based ORAC-pyrogallol red assay for the estimation of antioxidant capacity: First action 2012.03. *J. AOAC Int.* 95(6):1558–1561.

Ou, B. X., T. Chang, D. J. Huang, and R. L. Prior. 2013. Determination of total antioxidant capacity by oxygen radical absorbance capacity (ORAC) using fluorescein as the fluorescence probe: First action 2012.23. *J. AOAC Int.* 96(6):1372–1376.

Ou, B. X., M. Hampsch-Woodill, and R. L. Prior. 2001. Development and validation of an improved oxygen radical absorbance capacity assay using fluorescein as the fluorescent probe. *J. Agric. Food Chem.* 49(10):4619–4626.

Pellegrini, N., D. Del Rio, B. Colombi, M. Bianchi, and F. Brighenti. 2003. Application of the 2,2′-azinobis(3-ethylbenzothiazoline-6-sulfonic acid) radical cation assay to a flow injection system for the evaluation of antioxidant activity of some pure compounds and beverages. *J. Agric. Food Chem.* 51(1):260–264.

Pinto, P. C. A. G., M. L. M. F. S. Saraiva, S. Reis, and J. L. F. C. Lima. 2005. Automatic sequential determination of the hydrogen peroxide scavenging activity and evaluation of the antioxidant potential by the 2,2′-azinobis(3-ethylbenzothiazoline-6-sulfonic acid) radical cation assay in wines by sequential injection analysis. *Anal. Chim. Acta* 531(1):25–32.

Polasek, M., P. Skala, L. Opletal, and L. Jahodar. 2004. Rapid automated assay of anti-oxidation/radical-scavenging activity of natural substances by sequential injection technique (SIA) using spectrophotometric detection. *Anal. Bioanal. Chem.* 379(5–6):754–758.

Pulgarin, J. A. M., L. F. G. Bermejo, and A. C. Duran. 2012. Use of the attenuation of luminol-perborate chemiluminescence with flow injection analysis for the total antioxidant activity in tea infusions, wines, and grape seeds. *Food Anal. Methods* 5(3):366–372.

Ramos, I. I., M. A. Maia, S. Reis, L. M. Magalhães, and M. A. Segundo. 2014. Lab-on-valve combined with a kinetic-matching approach for fast evaluation of total antioxidant capacity in wines. *Anal. Methods* 6(11):3622–3628.

Rehakova, Z., V. Koleckar, F. Cervenka et al. 2008. DPPH radical scavenging activity of several naturally occurring coumarins and their synthesized analogs measured by the SIA method. *Toxicol. Mech. Methods* 18(5):413–418.

Ribeiro, J. P. N., L. M. Magalhães, S. Reis, J. L. F. C. Lima, and M. A. Segundo. 2011. High-throughput total cupric ion reducing antioxidant capacity of biological samples determined using flow injection analysis and microplate-based methods. *Anal. Sci.* 27(5):483–488.

Rodrigues, E., L. R. B. Mariutti, R. C. Chiste, and A. Z. Mercadante. 2012. Development of a novel micro-assay for evaluation of peroxyl radical scavenger capacity: Application to carotenoids and structure-activity relationship. *Food Chem.* 135(3):2103–2111.

Ruzicka, J. and E. H. Hansen. 1975. Flow injection analyses.1. New concept of fast continuous-flow analysis. *Anal. Chim. Acta* 78(1):145–157.

Ruzicka, J. and G. D. Marshall. 1990. Sequential injection–A new concept for chemical sensors, process analysis and laboratory assays. *Anal. Chim. Acta* 237(2): 329–343.

Segundo, M. A. and L. M. Magalhães. 2006. Multisyringe flow injection analysis: State-of-the-art and perspectives. *Anal. Sci.* 22(1):3–8.

Singleton, V. L., R. Orthofer, and R. M. Lamuela-Raventos. 1999. Analysis of total phenols and other oxidation substrates and antioxidants by means of Folin–Ciocalteu reagent. In *Oxidants and Antioxidants, Part A*. eds. L. Packer. San Diego: Elsevier Academic Press Inc.

Stasko, A., V. Brezova, S. Biskupic, and V. Misik. 2007. The potential pitfalls of using 1,1-diphenyl-2-picrylhydrazyl to characterize antioxidants in mixed water solvents. *Free Radic. Res.* 41(4):379–390.

Sundberg, S. A. 2000. High-throughput and ultra-high-throughput screening: Solution- and cell-based approaches. *Curr. Opin. Biotechnol.* 11(1):47–53.

Ukeda, H., Y. Adachi, and M. Sawamura. 2002. Flow injection analysis of DPPH radical based on electron spin resonance. *Talanta* 58(6):1279–1283.

Valko, M., D. Leibfritz, J. Moncol, M. T. D. Cronin, M. Mazur, and J. Telser. 2007. Free radicals and antioxidants in normal physiological functions and human disease. *Int. J. Biochem. Cell Biol.* 39(1):44–84.

Wang, J. H. and E. H. Hansen. 2003. Sequential injection lab-on-valve: The third generation of flow injection analysis. *TrAC—Trends Anal. Chem.* 22(4):225–231.

Antimicrobial Effects

CHAPTER 34

Determination of Volatile Nitrogenous Compounds
Ammonia, Total Volatile Basic Nitrogen, and Trimethylamine

Claudia Ruiz-Capillas, Ana M. Herrero,
and Francisco Jiménez-Colmenero

CONTENTS

34.1 INTRODUCTION

Muscle foods such as seafood and meat/meat products are highly perishable. Spoilage is caused by both intrinsic (species, composition, etc.) and extrinsic (processing, storage conditions, additives, etc.) factors. Once an animal dies or is slaughtered, various postmortem autolytic changes take place, initially due to breakdown of the cellular structure and biochemistry and later on due to the growth of microorganisms that are either associated naturally with the fish or that become part of the flora because of contamination during handling. These processes give rise to the formation of various compounds in fish muscle and in meat, such as nucleotides, nucleosides and their metabolites (adenosine monophosphate [AMP], adenosine diphosphate [ADP], adenosine triphosphate [ATP], inosine monophosphate [IMP], hypoxanthine [Hx], etc.), trimethylamine (TMA), dimethylamine (DMA), ammonia, free amino acids, volatile acids, indole, histamine, biogenic amines.

Trimethylamine is one of the more important volatile compounds in fish muscle. It is formed from the degradation of trimethylamine oxide (TMAO) during spoilage of refrigerated fish and fish products by bacteria reduction. In these conditions, TMAO is degraded mainly to TMA and to a small extent to DMA. However, when the fish is frozen, the TMAO is degraded to DMA and formaldehyde (FA) by endogenous enzymes. TMAO is a part of the body's normal buffer system and is used for osmoregulation in fish.

Other volatile compounds that are produced include ammonia and small amounts of monomethylamine and DMA; also, total volatile reducing substances appear in significant amounts during the different stages of spoilage of meat-based foods.

34.1.1 Additives and Formation of Volatile Nitrogenous Compounds

Various strategies have been used to control the formation of these compounds and so limit spoilage of fish and meat (and avoid economic loss): chilling, freezing, fermenting and canning, protective atmospheres, high-pressure irradiation, and others, including the use of additives. In this regard, various chemical additives have been used to preserve fish- and meat-based foods over long periods, among them mainly antimicrobial substances such as nitrites, sulfides and organic (acetic, sorbic, lactic, etc.) acids, sodium lactate (Ray, 2004; Chipley, 2005; Ruiz-Capillas and Jiménez-Colmenero 2008, 2009). These preservatives are substances used to prolong the shelf life of muscle food by reducing microbial proliferation during slaughtering, transportation, processing, and storage (Rahman, 1999, Davidson et al., 2005; Ruiz-Capillas and Jiménez-Colmenero, 2008, 2009; Dave and Ghaly, 2011). Such additives influence the formation of compounds such as volatile amines (TMA, DMA, etc.) and ammonia during spoilage of myosystems by inhibiting growth of the microorganisms that contribute to their formation.

34.1.2 Importance of Determining Volatile Nitrogenous Compounds

Because of their prominent role in myosystem spoilage and chemical characteristics, these compounds have long been used as indicators of quality and freshness in fish, and more recently in meat and meat products. One of the most widely used parameters in this regard is the presence and concentration of volatile amines such as TMA or TVB-N, ammonia, and other substances (Dyer, 1945; Moral, 1987; Connell, 1990; Gill, 1992; Civera et al., 1993; Botta, 1995; Huss, 1995; Oehlenschlälager, 1997; Ruiz-Capillas and Horner, 1999; Ruiz-Capillas et al., 2001a; Ruiz-Capillas and Moral, 2001a,b; Etienne, 2005). Levels of such compounds correlate well with sensory and microbial analyses. The amount of each component varies greatly depending on the freshness of the fish muscle. However, it is important to note that their degradation is quite species specific.

Trimethylamine has been used as an indicator of general fish spoilage, mainly in gadoids. It is useful as a rapid means of objective measurement in the mid- to late phases of spoilage but cannot be used as a freshness indicator, because the main source of TMA is bacterial growth (products), so levels are almost constant during the first days of iced storage. Trimethylamine is known for its characteristic fishy odor. Many researchers have conducted sensory tests, particularly on off-odors, along with TMA determination to demonstrate the correlation between the degree of spoilage and the TMA concentration in fish (Ruiz-Capillas and Moral, 2001a). Hebard et al. (1982) showed that the fishy odor always appeared at a TMA level of 4–6 mg N/100 mg. Trimethylamine has also been associated with free amino acids and biogenic amines (Ruiz-Capillas and Moral, 2001b,c, 2002).

It has been proposed that TMA levels between 5 and 10 mg/100 g tissue should be considered the maximum allowable levels in international trading. At a level of 10 mg N/100 g, there are definite off-odors. However, in some species (freshwater fish), the TMA value does not work as a spoilage index at all, because they do not contain enough of the precursor TMAO (Huss, 1995).

On the other hand, TMA, as would be expected of a bacterial product, is not useful in determining quality deterioration during frozen storage. That being so, DMA content is more suitable as a quality indicator to measure deterioration of fish in frozen storage, mainly for gadoid species. During freezing of fish, TMAO is converted (equimolecularly) by enzyme action to DMA and FA. Formaldehyde combines readily with proteins and other substances in fish, and so it is very difficult to measure the total amount of FA formed during frozen storage. Therefore, DMA is the best index of enzymatic deterioration during frozen storage, and TMA is the best index of prefreezing quality (Castell et al., 1974). For this reason, a combination of DMA and TMA testing for quality control of frozen fish has been suggested. Then again, DMA has also proven valuable in the early stages of spoilage in refrigerated fish, while TMA is more sensitive as an indicator of the later stages of spoilage, again mainly in gadoid fish (Hebard et al., 1982; Connell and Howgate, 1986).

TMA determination has also been used recently for quality control in meat and meat products. Although the source of these compounds in meat is not in principle clear, they could come from animal feeds (hence of dietary origin), from which the TMA enters the animal organism and is deposited in the muscle. Also, there have been studies on progress in the use of urea as a protein replacer for ruminants, since these have the ability to convert nonprotein nitrogen (NPN) into protein (Helmer and Bartley, 1971). Other authors have found that in general, the formation of volatile amines during chilled storage of chicken meat under packaging conditions seemed to be in good agreement with increases in microbiological counts and sensory taste scores (Balamatsia et al., 2007).

In some cases, *ammonia* content is also used as a parameter to control fish quality. Ammonia is also generated in fish muscle during processing and storage, mainly from free amino acids. Like other volatile compounds, levels of ammonia production are species dependent. This indicator is important mainly for quality control in elasmobranch fishes, as their flesh contains large amounts of urea. Shellfish may also develop more ammonia than most marine fish, and at an earlier stage. Thus, ammonia could potentially serve as an objective quality indicator for fish that degrade autolytically rather than primarily through bacterial spoilage (LeBlanc and Gill, 1984; Botta, 1995).

Total volatile basic nitrogen (TVB-N) is one of the most widely used measurements of seafood quality, together with TMA. Total volatile basic nitrogen represents the sum of ammonia, DMA, TMA, and other basic nitrogenous volatile compounds associated with seafood spoilage, and is an indicator of spoilage in most fish species, including red fish, flat fish, gadoids, and hake; therefore, legal requirements have been established for TVB-N. Moreover, as in the case of TMA, TVB-N is a good indicator of later spoilage of fish but cannot be used as a freshness indicator for the measurement of spoilage during the first 10 d of chilled storage when the level remains constant (Rehbein and Oehlenschläger, 1982; Vyncke et al., 1987; Ruiz-Capillas and Moral, 2001a,c). In freshly caught fish, TVB-N content is generally greater than 10 mg/100 g and not more than 15 mg/100 g, except for pelagic species. The TVB-N and TMA levels increase in parallel. In cod and other gadoid species, TMA accounts for most of the so-called TVB-N until spoilage.

34.1.3 Legislation of These Compounds

In view of the importance of these compounds for quality control in fish and fishery products, the EU and other institutions (governments, official laboratories, etc.) have set legal ceilings so as to limit their presence in various products.

A specific study of TVB-N and TMA is reported in Council Directive 91/493/EEC of July 22, 1991, laying down the health conditions for the production and the placing on the market of fishery products. Chapter V of Annex I of the directive deals, among other things, with chemical checks for health control and monitoring of production conditions. The levels of these parameters must be specified for each category of species, and the examinations must be carried out in accordance with reliable, scientifically recognized methods in accordance with the procedure recommended in this directive. The limit value indicated by the EU for TMA is 12 mg N/100 g. Then Commission Decision 95/149/EC (OJ L97, p84, 29/04/1995) of March 8, 1995, fixing the TVB-N limit values for certain categories of fishery products and specifying the analysis methods to be used, provides that unprocessed fishery products belonging to the different species categories shall be regarded as unfit for human consumption, where, organoleptic assessment having raised doubts as to their freshness, chemical checks reveal that the following TVB-N limits are exceeded:

1. 25 mg of N/100 g of flesh for *Sebastes* spp. (*Helicolenus dactylopterus* and *Sebastichthys capensis*)
2. 30 mg of N/100 g of flesh for the Pleuronectidae family (with the exception of halibut: *Hippoglossus* spp.)
3. 35 mg of N/100 g of flesh for *Salmo salar* and species belonging to the Merlucciidae and Gadidae families

There are also other more recent regulations dealing with these parameters. For instance, Regulation (EC) No. 853/2004 of the European Parliament and of the Council of April 29, 2004, and Commission Regulation (EC) No. 2074/2005 also refer to the TVB-N limit values fixed for the different species of fish and the reference method of analysis to be used in the event of dispute. Another example is the Food Hygiene (Scotland) Amendment Regulations 2012 that provide for the enforcement, in Scotland, of Regulation (EC) 1022/2008 amending Regulation (EC) 2074/2005 as regards the TVB-N limits. This legislation also recommends that member states and their official laboratories use these parameters, and specify the analysis methods to be used for these determinations.

Therefore, the development of different analytical methods for the control of these compounds is very important, given the paramount interest in practical methods for determining the degree of spoilage so that fish caught in offshore waters can be graded for processing and distribution.

34.2 DETERMINATION OF VOLATILE BASIC NITROGEN COMPOUNDS

There is a range of methods used to measure these volatile compounds based on steam distillation and titration, Conway microdiffusion (Conway, 1962), colorimetric assays such as the reaction of TMA with picric acid based on the Dyer (1945) (AOAC, 2002a), enzymatic methods, gas chromatographic methods, simultaneous measurement of TMA and DMA after a preliminary separation, high-performance liquid chromatographic methods, ion-selective electrode method for the determination of ammonia and volatile bases in fish (AOAC, 2002b), sensors, and spectroscopic techniques, among others (Keay and Hardy, 1972; Chang et al., 1976; Gill and Thompson, 1984; Wong and Gill, 1987; Lee et al., 1992; Veciana-Nogues et al., 1996; Lista et al., 2001; Ruiz-Capillas and Moral 2001a; Timm and Jørgensen, 2002; Adhoum et al., 2003; Mitsubayashi et al., 2004; Etienne, 2005; Armenta et al., 2006; Chan et al., 2006; Pena-Pereira et al., 2010). Before most of these methods are used, the sample generally needs to be deproteinated with acids (Ruiz-Capillas and Horner, 1999; Etienne, 2005). Some of these methods have drawbacks: they are time consuming and require toxic reagents; chromatographic methods require expensive specialized equipment and specialized staff; they normally increase costs, and cannot be performed on-site; these chromatographic methods pose inherent problems in the handling of low-molecular-weight amines due to their high water solubility and volatility. The current tendency of research in this field is to look for methods that ideally fulfill a series of requirements such as speed, nondestructiveness, portability, and applicability online. Hence, to establish a suitable measurement of food quality poses a major problem not only for the industry (producers and manufacturers) but also for distributors, retailers, consumers, and authorities. Flow injection analysis (FIA) methods have thus gained importance in the evaluation of food quality in that they could solve some of the problems posed by the methods traditionally used to evaluate muscle food quality, since FIA techniques are fast, simple, low cost, and reasonably reproducible and repeatable (Ruzicka and Hansen 1975; Wekell et al., 1987; Valcarcel and Luque de Castro, 1990; León et al., 1994; Cerdá et al. 1995; Zhi et al., 1995; Sadok et al., 1996; García-Garrido and Luque de Castro, 1997; Ruiz-Capillas and Horner, 1999; Baixas-Nogueras et al., 2001; Ruiz-Capillas et al., 2001a; Adhoum et al., 2003; Ruzicka, 2014).

34.3 DETERMINATION OF VOLATILE BASIC NITROGEN COMPOUNDS BY FIA

FIA has also been proposed for the determination of many different chemical compounds, including substances to be used for process control and metabolites produced during the storage of products, such as volatile compounds (TMA, ammonia, TVB-N, etc.) characteristic of fish spoilage, which in many cases are quality and freshness indicators (Cerdá et al., 1995; Zhi et al., 1995; Sadok et al., 1996; García-Garrido and Luque de Castro, 1997; Ruiz-Capillas and Horner, 1999; Pons-Sánchez-Cascado et al., 2000; Baixas-Nogueras et al., 2001; Ruiz-Capillas et al., 2001a; Adhoum et al., 2003).

Determination of spoilage volatile compounds by FIA requires two clearly defined steps: first, the extraction of volatile compounds, normally by acid, similar to the procedure followed in other methods used to determine these compounds; and second, quantification as by means such as spectrophotometry, fluorescence, or chemiluminescense.

34.3.1 Extraction of Volatile Nitrogen Compounds for FIA Determination

In general, systems for extraction of volatile amines, TMA, DMA, ammonia, and other compounds for FIA determination are similar to the systems used for extraction of NPN in the muscle of which they are a part.

Irrespective of the methodology followed for the determination of the various components of NPN in fish muscle or muscle foods in general, the preparation of extracts for such determination has traditionally involved homogenization of the muscle with an acid followed by centrifugation to precipitate the proteins and retain NPN in the supernatant. It is from this NPN that the various components (TMA, DMA, ammonia, TVB-N, etc.) are determined by means of specific methods.

Trichloroacetic acid (TCA) is generally used to determine TMA (Dyer, 1945; Conway, 1962; Ruiz-Capillas and Moral, 2001a; AOAC, 2002a) and perchloric acid (PCA) to determine TVB-N (Lundstrom and Raciot, 1983; Gill and Thompson, 1984; Vyncke et al., 1987; Antonacopoulos and Vyncke, 1989; Ruiz-Capillas and Horner, 1999; AOAC, 2002b).

In the case of FIA, both TCA and PCA have been assayed for the preparation of extracts to be injected into the FIA for subsequent determination of TMA and TVB-N (Sadok et al., 1996; Ruiz-Capillas and Horner, 1999; Baixas-Nogueras et al., 2001). Specific assays have been conducted to evaluate the use of alternative acids (TCA and PCA) to extract TMA and TVB-N for determination by FIA (Ruiz-Capillas and Horner, 1999). Given that the same FIA system, based on flow injection/gas diffusion (FIGD) would be used to determine both TMA and TVB-N, it was decided to conduct a study to validate the same extraction with a single acid, either PCA or TCA, to determine both TMA and TVB-N (Ruiz-Capillas and Horner, 1999). These studies were conducted on cod, monkfish, and skate. There was an excellent correlation between the data on identical samples extracted using the two different acids (TCA and PCA) (Figures 34.1 and 34.2). Using the same extract for the two analyses would produce a saving of sample quantity and time and reduce the probability of errors when making different extracts from the same sample.

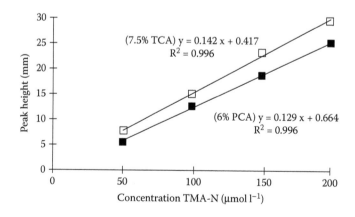

FIGURE 34.1 Regression curves for trimethylamine (TMA) using trichloroacetic acid (TCA) (7.5%) and perchloric acid (PCA) (6%). (Adapted from Ruiz-Capillas, C. and W. F. A. Horner. 1999. *J. Sci. Food Agric.* 79:1982–1986.)

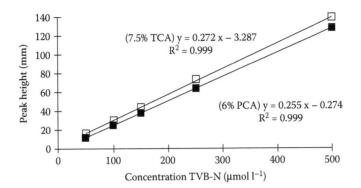

FIGURE 34.2 Regression curves for total volatile basic nitrogen (TVB-N) using tri-chloroacetic acid (TCA) (7.5%) and perchloric acid (PCA) (6%). (Adapted from Ruiz-Capillas, C. and W. F. A. Horner. 1999. *J. Sci. Food Agric.* 79:1982–1986.)

Determination of TMAO by FIGD requires prior reduction of TMAO to TMA in the fish extract. This is done with 15% TiCl at 100°C (Ruiz-Capillas et al., 2001a). In most cases, prior sample preparation is not required for the determination of these compounds in liquid solutions; the sample is frequently injected directly into the FIA system without any prior extraction treatment, for instance, in the case of ammonia analysis in water (Hunter and Uglow, 1993). In the case of liquid solutions presenting problems of dirt (wastewater treatment plant or river water samples) with suspended solids, prior filtration is important to prevent dirt from entering the FIA system. Contact with these particles can cause clogging and/or deterioration of the membrane, rendering it ineffective. In this connection, different procedures for sample preparation have been assayed, for example, pervaporation (PV). Almeida et al. (2011) developed the methodology with a new design for a membraneless gas diffusion (MGD) unit coupled to a multisyringe flow injection system, so that untreated environmental samples can be analyzed without prior filtration.

Direct injection into the FIA system has also been assayed in the case of TMA and TVB-N determination in liquid samples, for example, fish sauce; however, the results were disappointing, and it was apparent that prior extraction in acid would be required, as for the procedure used for muscle (Ruiz-Capillas et al., 2000). The outcome was similar when an attempt was made to analyze TMA and TVB-N by FIA directly in fish exudate (from applying pressure to the fish muscle) to obviate the need for prior extraction and be able to do the analysis directly in the FIGD system (QUALPOISS 2 project) so as to render FIA determination suitable for online or line-to-line use. The reason for this unsatisfactory result was that the exudate also contained other components, such as lipids and proteins, which significantly interfered with the analysis.

34.3.2 Analytical Detection of Volatile Basic Nitrogen Compounds by FIA

The most common FIA technique for the determination of TVB-N, TMA, and ammonia is based on the gas diffusion system. This technique is known as flow injection/gas diffusion (FIGD) or gas diffusion/flow injection analysis (GDFIA). The gas diffusion unit (Figure 34.3) generally consists of two liquid streams, a strong acidic donor solution (the acid medium enables the release of TMA, ammonia, DMA gas from the stream with the

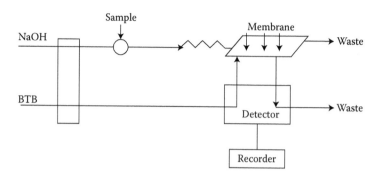

FIGURE 34.3   Diagram of flow injection/gas diffusion (FIGD) for the determination of total volatile basic nitrogen (TVB-N). (Adapted from Ruiz-Capillas, C. and W. F. A. Horner. 1999. *J. Sci. Food Agric.* 79:1982–1986.)

test solution), and an acceptor solution (to collect the released TMA, ammonia, DMA, etc.) containing an acid–base indicator separated by a gas-permeable membrane.

34.3.2.1 Detection of Volatile Basic Nitrogen Compounds and Ammonia

The system depicted in Figure 34.3 for TVB-N and ammonia determination (Ruiz-Capillas and Horner, 1999) is based on the magnitude of the color change in an acid–base indicator (bromothymol blue [BTB]) when ammonia and ammonia-like volatiles or amines are released from water samples or extracts from samples and passed through a semipermeable membrane separating the sample from the acceptor stream (BTB solution). This causes a pH change and subsequent color changes that are measured spectrophotometrically in a 635-nm laser diode system. The system uses NaOH and BTB as donor and acceptor streams, respectively. NaOH is used as a carrier and is responsible for the release of ammonia-like bases to the indicator by neutralizing the acid extract. BTB is a pH-dependent indicator, the degree of its color change is directly proportional to the concentration of volatile bases released from the extract as represented by the peak height.

This technique has been used for the measurement of total ammonia (NH_3^+, NH_4^+) in small volumes (100–450 µL) of seawater and hemolymph. Hunter and Uglow (1993) report the detection of up to 0.20 µmol ammonia/L and 30 determinations/h. The technique has also proven effective for ammonium/ammonia determination in complex environmental samples, such as surface and tap waters, marine waters, estuarine waters, river waters, wastewaters, and compost and fertilizer (Klimundová et al., 2003; Almeida et al., 2011). However, in the case of "dirty" samples or samples containing suspended solids, it is important to avoid their coming into contact with the gas-permeable membrane, since this may cause clogging and/or deterioration of the membrane, rendering it ineffective. To avoid this, prior sample preparation systems have been assayed, such as PV or modifications to the FIA system as such (Almeida et al., 2011). In this connection, these authors proposed a new design for an MGD unit coupled to a multisyringe flow injection system (multicommutation flow injection analysis—MCFIA), which allows untreated samples to be injected directly into the flow system. This new system achieved a detection limit of up to 2.20 mg/L NH_4^+ with a determination frequency of 11 samples/h. Simple frequency was about 4/h considering three replicate determinations for each sample. This new system is more robust, allows direct analysis of untreated samples, and reduces the cost per analysis.

The FIA system has also been used in combination with enzyme immobilization to determine ammonia in food samples. The coupling of an immobilized enzyme reactor (IMER) with flow injection provides increased rapidity, precision, and convenience compared with classical enzymatic procedures and reduces the cost per assay. A number of immobilized enzymes have been coupled with an FI methodology for the analysis of foods. Canale-Gutierrez et al. (1990) proposed determining ammonia in food samples by its reaction in an IMER containing glutamate dehydrogenase (GIDH) in a flow injection system, in which the decrease in the absorbance of ultraviolet radiation was measured by reduced nicotinamide adenine dinucleotide (NADH). This procedure combines the advantages of a flow injection method with the high specificity of an enzymatic method together with rapidity, precision, simplicity, cost savings, and a minimum of sample treatment.

Flow injection gas diffusion has also been coupled to ion chromatography for the analysis of ammonia in estuarine and freshwaters. Alkaline EDTA is added online to a flowing sample to achieve a sufficiently high pH (>12.0) to deprotonate >95% of the amines to their uncharged volatile forms. In this system, the amines also diffuse selectively across a gas-permeable microporous Teflon (polytetrafluoroethylene—PTFE) membrane into a recirculating flow of acidic "acceptor" in which they are reprotonated and preconcentrated. The acceptor solution is then injected onto an ion chromatograph (IC) where NH_4^+ and methylamine (MA) cations are separated within 15 min and detected by chemically suppressed conductimetry using cyclopropylamine as an internal standard for quantification (Gibb et al., 1995).

In TVB-N determination by FIA (Figure 34.3), known concentrations of ammonia are used as standards for calibration of the methodology, since the TVB-N index includes ammonia and other volatile amines present in muscle. This methodology (Figure 34.3) has been modified and adapted for the determination of TVB-N in various fish species (halibut, hake, monkfish, skate, cod, anchovy, etc.) (Wekell et al., 1987; Hollingworth et al., 1994; Ruiz-Capillas and Horner, 1999; Pons-Sánchez-Cascado et al., 2000; Ruiz-Capillas et al., 2000, 2001a,b; Baixas-Nogueras et al., 2001). Ruiz-Capillas and Horner (1999) proposed an adaptation of the method used to determine ammonia in aquatic liquid and hemolymph (neutral or slightly basic) (Hunter and Uglow, 1993). The adaptation was based on neutralization of the acid fish extracts to assure complete release of the ammonia and volatile amines from the sample. To ensure their release, the NaOH concentration was raised from 0.6 to 1 M. This was necessary, since the pH of the aquatic liquid and hemolymph samples was neutral or slightly basic and in any case higher than in the acid fish extracts or myosystems used to determine TVB-N, which is around 1.5. Lower NaOH concentrations produced a signal that was split into two peaks instead of one. The reason for this was that the NaOH concentration was not high enough to neutralize the acid in the sample extract. Concentrations in excess of 1 M produced no improvement in the peak definition (Ruiz-Capillas and Horner, 1999). A comparison of the effectiveness of this technique with the TVB-N determination procedure reported by Antonacopoulos and Vyncke (1989) proved satisfactory (Ruiz-Capillas and Horner, 1999), so one could be used in place of the other.

Over the years, there have been many advances in the determination of ammonia in liquids and not so many advances in the FIA determination of TVB-N in myosystems. In this connection, Dhaouadi et al. (2007) reported a potentiometric flow injection method using a gas diffusion cell for the determination of TVB-N in sardine, red mullet, mackerel, and hake. The method was based on the change of potential of a tungsten oxide electrode when volatile basic compounds, released from the fish extract sample, diffuse via a permeable membrane into a phosphate buffer acceptor stream and locally shift the pH.

Nowadays, both TVB-N and ammonia in muscle are generally determined conventionally by the Kjeldahl distillation method and the procedure reported by Antonacopoulos and Vyncke (1989), which is very time consuming, produces toxic vapors, and, moreover, gives yield values higher than expected as a result of the hydrolysis of the protein amino groups. Determination of TVB-N by FIA is therefore a major step forward in terms of time, money, and handling of toxic substances.

34.3.2.2 Detection of Trimethylamine, Dimethilamine, and Trimethylamine Oxide

When FIA was first used to determine TMA, the intention was to automate the AOAC (2002a) method (974.14) commonly used in fish (León et al., 1994), but nowadays the most commonly used procedure is very similar to the one used for TVB-N, based on FIGD systems with spectrophotometric determination (Sadok et al., 1996; García-Garrido and Luque de Castro, 1997; Ruiz-Capillas and Horner, 1999; Pons-Sánchez-Cascado et al., 2000; Baixas-Nogueras et al., 2001).

Trimethylamine determination is shown schematically in Figure 34.4. It differs that system used for TVB-N and ammonia determination in that there is a third channel running through the peristaltic pump that brings FA solution into a mixing coil, where it meets the injected sample (extraction), and from there it sequesters all the non-TMA volatile bases, ammonia, and nontertiary amines in the FA prior to alkalinization. TMA is thus left as the only base diffusing across the semipermeable membrane. A similar system has been used to determine TMAO in fish extracts (Ruiz-Capillas et al., 2001a). Bromocresol purple has been assayed as an indicator for FIA determination of DMA.

Because the presence of this FA is very important to avoid interferences, experiments have been conducted to assess the effect of its concentration and establish what levels are necessary for the determination of TMA in different types of fish (Ruiz-Capillas and Horner, 1999). Whereas Sadok et al. (1996) used 120 g/L of FA to determine TMA in rainbow trout, other authors (Ruiz-Capillas and Horner, 1999) proposed that levels should be adjusted in extracts from other types of fish such as elasmobranchs or samples with high levels of ammonia, or in samples with very high levels of TMA or samples that had been stored to their limit of fitness for consumption (Ruiz-Capillas and Horner, 1999). To establish the most suitable FA levels for a wide variety of situations, studies were conducted in fish extracts spiked with $100\,\mu mol/L$ and standard TMA hydrochloride and different concentrations of standard ammonium chloride solution ($50–200\,\mu mol/L$), using five different concentrations of FA (120, 200, 250, 300, and

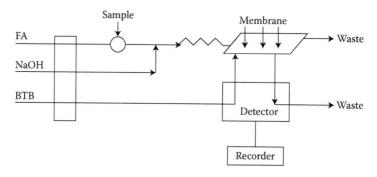

FIGURE 34.4 Diagram of the flow injection/gas diffusion (FIGD) for the determination of trimethylamine TMA. (Adapted from Ruiz-Capillas, C. and W. F. A. Horner. 1999. *J. Sci. Food Agric.* 79:1982–1986.)

360 g/L) to determine which FA concentration was sufficient to sequester the ammonia (Ruiz-Capillas and Horner, 1999). On the basis of this series of experiments, 200 g/L (instead of 120 g/L) was established as the most suitable and reliable FA concentration, as well as being environmentally and industrially friendly and retaining the particular advantages of the method (precision, rapidity, etc.).

Interference color assays of samples in FGID determination of TMA have also been conducted, in which dark-colored liquid fish sauce samples were analyzed. However, no interferences were detected in the color reaction for FIA determination of TMA due to sample color, although some differences in TMA levels were detected when compared with the conventional Dyer method (Ruiz-Capillas et al., 2000). FIGD determination of TMA in various fish species (cod, monkfish and skate, hake, anchovies, etc.) as affected by freshness stages and different stages of spoilage have also been compared with the routinely used conventional procedure based on a liquid–liquid extraction of TMA with toluene and its subsequent reaction with picric acid reagent to form a yellow complex (Dyer, 1945). These studies indicated a good correlation between determinations using official (AOAC, 2002a, based on Dyer's method) and FIA methods, with no apparent significant differences (Ruiz-Capillas and Horner, 1999).

The greatest advantage of FIGD over AOAC (2002a) for the determination of TMA in fish samples based on Dyer's method (Dyer, 1945) is its sensitivity, rapidity, and absence of hazardous compounds such as toluene. The conventional method has poor analytical sensitivity and involves several time-consuming steps as well as the use of large amounts of hazardous reagents. There has even been a report of an overestimation of the total TMA values with this conventional method (Ruiz-Capillas et al., 2000).

Other FIA systems, as well as FIGD, have been used to determine TMA. In this connection, Adhoum et al. (2003) developed a simple FIGD method for application in seafood with potentiometric detection using a tungsten oxide electrode, similar to the one reported for TVB-N. The method is based on the diffusion of TMA through a PTFE membrane from a sodium hydroxide donor stream to a phosphate buffer acceptor stream. The TMA in the acceptor stream passes through an electrochemical flow cell containing a tungsten oxide wire and a silver/silver chloride electrode, where TMA is sensitively detected. The results using this method were in close agreement with those obtained by the official method.

Trimethylamine determination has also been done by FIA coupled with biosensors or specific sensors based on selective identification of these compounds (Lüdi et al., 1990). Mitsubayashi et al. (2004) developed an FMO3 (flavin-containing monooxygenase type 3) immobilized biosensor in which the TMA is catalyzed to TMAO by the FMO3, with NADPH as the coenzyme.

34.4 CONCLUSION

FIA procedures for the determination of volatile nitrogenous compounds such as TMAO, TMA, and TVB-N in fish and meat products and in other foods and beverages generally offer considerable advantages over conventional methods, which in many cases are used as quality indicators for fish and fishery products, and more recently for quality control of meat and meat products. This system offers a number of advantages: analysis is much less time consuming than in conventional methods; several determinations can be performed on a single sample in a brief period of time; there is generally no need to handle the extract; and as a semi-online system, being less time consuming and nontoxic, it can be readily used in quality-control laboratories and industrial applications where reduced raw material usage and rapid quality assessments of end products are critical.

FIA determination of these compounds has been successfully correlated with conventional methods commonly used to determine the quality of muscle foods. FIA is therefore a suitable analytical method for quality control of muscle-based foods, which focuses on substances that in large amounts in fish tissues during spoilage and in some cases arise from the use of additives in these products. Rapid measurement of these substances is very important for research and for the industry.

ACKNOWLEDGMENTS

This research was supported by projects AGL 2010-1915, AGL 2011-29644-C02-01 of the Plan Nacional de Investigación Científica, Desarrollo e Innovación Tecnológica (I + D + I) Ministerio de Ciencia y Tecnología. The author wishes to thank the University of Hull for introduction to the field of FIA.

REFERENCES

Adhoum, N., L. Monser, S. Sadok, A. El-Abed, G. Greenway, and R. Uglow. 2003. Flow injection potentiometric detection of trimethylamine in seafood using tungsten oxide electrode. *Anal. Chim. Acta* 478:53–58.

Almeida, M. I. G. S., J. M. Estela, M. A. Segundo, and V. Cerdà. 2011. A membraneless gas diffusion unit—Multisyringe flow injection spectrophotometric method for ammonium determination in untreated environmental samples. *Talanta* 84:1244–1252.

Antonacopoulos, N. and W. Vyncke. 1989. Determination of volatile basic nitrogen in fish: A third collaborative study by the Western European Fish Technologists Association (WEFTA). *Z. Lebensm Unters Forsch.* 189:309–316.

AOAC. 2002a. Official Method 971.14. Trimethylamine nitrogen in seafood. In *AOAC Official Methods of Analysis of AOAC International*, Volume II, Chapter 35, 9.

AOAC. 2002b. Official Method 999-01. Volatile bases in fish, ammonia ion selective electrode method. In *AOAC Official Methods of Analysis of AOAC International*, Volume II, Chapter 35, 34–35.

Armenta, S., N. M. M. Coelho, R. Roda, S. Garrigues, and M. de la Guardia. 2006. Seafood freshness determination through vapour phase Fourier transform infrared spectroscopy. *Anal. Chim. Acta* 580:216–222.

Baixas-Nogueras, S., S. Bover-Cid, M. C. Vidal-Carou, M. T. Veciana-Nogués, and A. Mariné-Font. 2001. Trimethylamine and total volatile basic nitrogen determination by flow injection/gas diffusion in Mediterranean hake (*Merluccius merluccius*). *J. Agric. Food Chem.* 49:1681–1686.

Balamatsia, C. C., A. Patsias, I. N. Kontomina, and I. N. Savvaidis. 2007. Possible role of volatile amines as quality-indicating metabolites in modified atmosphere-packaged chicken fillets: Correlation with microbiological and sensory attributes. *Food Chem.* 104(4):1622–1628.

Botta, J. R. 1995. Chemical methods of evaluating freshness quality. In *Evaluation of Seafood Freshness Quality*. ed. J. R. Botta, pp. 9–33. New York, USA: VCH Publisher Inc.

Canale-Gutierrez, L., A. Maquieira and R. Puchades. 1990. Enzymic determination of ammonia in food by flow injection. *Analyst* 115(9):1243–1246.

Castell C. H., B. Smith, and W. J. Dyer 1974. Simultaneous measurements of trimethylamine and dimethylamine in fish and their use for estimating quality of frozen stored gadoid fillets. *J. Fish Res. Board Can.* 31:383–389.

Cerdà, A., M. T. Oms, R. Forteza, and V. Cerdà. 1995. Evaluation of flow injection methods for ammonium determination in wastewater samples. *Anal. Chim. Acta* 311:165–173.

Chan, S. T., M. W. Y. Yao, Y. C. Wong, T. Wong, and C. S. Mok. 2006. Evaluation of chemical indicators for monitoring freshness of food and determination of volatile amines in fish by headspace solid-phase microextraction and gas chromatography–mass spectrometry. *Eur. Food Res. Technol.* 224:67–74.

Chang G. W., W. L. Chang, and B. K. Lew. 1976. Trymethylamine specific electrode for fish quality control. *J. Food Sci.* 41:723–724.

Chipley, J. R. 2005. Sodium benzoate and benzoic acid. In *Antimicrobial in Food*, eds. P. M. Davidson, J. N. Sofos and A. L. Branen, pp. 11–48. New York: CRC Press.

Civera, T., R. M. Turi, C. Bisio, G. Parisi, and G. Fazio. 1993. Sensory and chemical assessment of marine teleosteans—Relationship between total volatile basic nitrogen, trimethylamine and sensory characteristics. *Sci. Aliment.* 13:109–117.

Commission Decision 95/149/EC of 8 March 1995 fixing the total volatile basic nitrogen (TVB-N) limit values for certain categories of fishery products and specifying the analysis methods to be used.

Commission Regulation (EC) No 853/2004 of the European Parliament and of the Council of 29 April 2004.

Commission Regulation (EC) No 2074/2005 of the European Parliament and of the Council 5 December 2005.

Commission Regulation (EC) No 1022/2008 of 17 October 2008 amending Regulation (EC) No 2074/2005 as regards the total volatile basic nitrogen (TVB-N) limits.

Connell, J. J. 1990. *Control of Fish Quality.* London: Fishing News (Boock) Ltd.

Connell, J. J. and P. F. Howgate. 1986. Fish and fish products. In *Quality Control in the Food Industry.* eds. S. M. Herschdoerfer, pp. 347–405. London: Academic Press.

Conway, E. J. 1962. Determination of volatile amines. In *Microdiffusion Analysis and Volumetric Error* (5th ed.), pp. 195–200. London, UK: Crosby Lockwood.

Council Directive 91/493/EEC of 22 July 1991 laying down the health conditions for the production and the placing on the market of fishery products. Official Journal L 268, 24/09/1991, pp. 0015–0034.

Dave, D. and A. E. Ghaly. 2011. Meat spoilage mechanisms and preservation techniques: A critical review. *Am. J. Agric. Biol. Sci.* 6(4):486–510.

Davidson, P. M., J. N. Sofos, and A. L. Branen. 2005. *Antimicrobials in Food* (3rd ed.), pp. 12–17, 29–68, 116–151, 460–469. Boca Raton, FL: CRC Press.

Dhaouadi, A., L. Monser, S. Sadok, and N. Adhoum. 2007. Validation of a flow-injection-gas diffusion method for total volatile basic nitrogen determination in seafood products. *Food Chem.* 103(3):1049–1053.

Dyer, W. J. 1945. Amines in fish muscle I. Colorimetric determination of trimethylamine as the picrate salt. *J. Fish Res. Board Can.* 6:351–358.

Etienne, M. 2005. Volatile amines as criteria for chemical quality assessment. SEAFOODplus (http://archimer.ifremer.fr/doc/2005/rapport-6486.pdf).

García-Garrido J. A. and M. D. Luque de Castro. 1997. Determination of trimethylamine in fish by pervaporation and photometric detection. *Analyst* 122(7):663–666.

Gibb, S. W., R. F. C. Mantouraa, and P. S. Liss. 1995. Analysis of ammonia and methylamines in natural waters by flow injection gas diffusion coupled to ion chromatography. *Anal. Chim. Acta* 316(3):291–304.

Gill, T. A. 1992. Biochemical and biochemical indices of seafood quality. In *Quality Assurance in the Fish Industry*, eds. H. H. Huss, M. Jakobsen, and J. Liston, pp. 377–388. Amsterdam, Holland: Elsevier Science Publishers B.V.

Gill, T. A. and J. W. Thompson. 1984. Rapid, automated analysis of amines in seafood by ion-moderated partition HPLC. *J. Food Sci.* 49:603–606.

Hebard, C. E., G. J. Flick, and R. E. Martin. 1982. Occurrence and significance of trimethylamine oxide and its derivatives in fish and shellfish. In *Chemistry and Biochemistry of Marine Food Products*. eds. R. E. Martin, G. J. Flick, C. E. Herbard, and D. R. Ward, pp. 149–304. Westport, Connecticut: AVI Publishing Company.

Helmer, L. G. and E. E. Bartley. 1971. Progress in the utilization of urea as a protein replacer for rumiants. A review. *J. Dairy Sci.* 54:25–51.

Hollingworth, T. A., J. M. Hungerford, J. D. Barnett, and M. M. Wekell. 1994. Total volatile acids: Temperature dependent decomposition indicator in halibut determined by flow injection analysis. *J. Food Protect.* 57(6):505–508.

Hunter, D. A. and R. F. Uglow. 1993. A technique for the measurement of total ammonia in small volumes of seawater and haemolymph. *Ophelia* 37:31–40.

Huss, H. H. 1995. Quality and quality changes in fresh fish. FAO. Fisheries Technical Paper 348. FAO, Rome, Italy.

Keay, J. N. and R. Hardy. 1972. The separation of aliphatic amines in dilute aqueous solutions by gas chromatography and the application of this technique to the quantitative analysis of tri- and dimethylamine in fish. *J. Sci Fd. Agric.* 23:9–19.

Klimundová, J., R. Forteza, and V. Cerdà. 2003. A multisyringe flow injection system coupled with a gas diffussion cell for ammonium determination. *Int. J. Environ. Anal. Chem.* 83:233.

LeBlanc, R. J. and T. A. Gill. 1984. Ammonia as an objective quality index in squid. *Can. Inst. Food Sci. Technol. J.* 17:195–201.

Lee, Y. C., R. P. Singh, and N. F. Haard. 1992. Changes in freshness of chilli pepper rockfish (*Sebastes goodei*) during storage as measured by chemical sensors and biosensors. *J. Food Biochem.* 16:119–129.

León, A., A. Chica, C. Chih-Ming, and F. Centrich. 1994. Determination of trimethylamine in fish by flow injection analysis. *Quím. Anal.* 13:78–81.

Lista, A. G., L. Arce, A. Ríos, and M. Valcárcel. 2001. Analysis of solid samples by capillary electrophoresis using a gas extraction sampling device in a flow system. *Anal. Chim. Acta* 438:315–322.

Lüdi, H., M. B. Garn, P. Bataillard, and H. M. Widmer. 1990. Flow injection analysis and biosensors: Applications for biotechnology and environmental control. *J. Biotechnol.* 14:71–79.

Lundstrom, R. C. and L. D. Raciot. 1983. Gas chromatographic determination of dimethylamine and trimethylamine in seafoods. *J. Assoc. Off. Anal. Chem.* 66:158–163.

Mitsubayashi, K., Y. Kubotera, K. Yano et al. 2004. Trimethylamine biosensor with flavin-containing monooxygenase type 3 (FMO3) for fish-freshness analysis. *Sensor. Actuat. B* 103:463–467.

Moral, A. 1987. Métodos fisico-químicos de control de calidad de pescados. *Alimentación Equipos y Tecnología* 5–6:115–122.

Oehlenschlälager, J. 1997. Suitability of ammonia-N, dimethylamine-N, trimethyl-amine-N, trimethylamine oxide-N and total volatile basic nitrogen as freshness indicators in seafood. In *Methods to Determine the Freshness of Fish in Research and Industry*. Nantes Conference, November 12–14.

Pena-Pereira, F., I. Lavilla, and C. Bendicho. 2010. Colorimetric assay for determination of trimethylamine-nitrogen (TMA-N) in fish by combining headspace-single-drop microextraction and microvolume UV–Vis spectrophotometry. *Food Chem.* 119:402–407.

Pons-Sánchez-Cascado, S., M. Izquierdo-Pulido, A. Mariné-Font, M.T. Veciana-Nogués, and M. C. Vidal-Carou. 2000. Reliability of trimethylamine and total volatile basic nitrogen determinations by flow injection-gas diffusion techniques in pelagic fish: *Engraulis encrasicholus*. *Quím. Anal.* 19:165–170.

Qualpoiss 2: http://www2.matis.is/media/utgafa/SKYRSLA28-01.pdf

Rahman, S. F. 1999. Post-harvest handling of foods of animal origin. In *Handbook of Food Preservation*. ed. S. F. Rahman, pp. 47–54. New York: Marcel Dekker.

Ray, B. (ed.). 2004. Control by antimicrobial preservatives. In *Fundamental Food Microbiology* (3rd ed.), pp. 439–506. Boca Raton, FL: CRC Press.

Rehbein, H. and J. Oehlenschläger. 1982. Zur Zusammensetzung der TVB-N-Fraktion (Flüchtige Basen) in sauren Extrakten und alkalischen Destillaten von Seefischfilet. *Arch. Lebensmittelhyg.* 33(2):44–48.

Ruiz-Capillas, C. and W. F. A. Horner. 1999. Determination of trimethylamine-nitrogen and total volatile basic-nitrogen in fresh fish by flow injection analysis. *J. Sci. Food Agric.* 79:1982–1986.

Ruiz-Capillas, C., C. M. Gillyon, and W. Horner. 2000. Determination of volatile basic nitrogen and trimethylamine nitrogen in fish sauce by flow injection analysis. *Eur. Food Res. Technol.* 210:434–436.

Ruiz-Capillas, C., C. M. Gillyon, and W. F. A. Horner. 2001a. Determination of different volatile base components as quality control indices in fish by official methods and flow injection analysis. *J. Food Biochem.* 25(6):541–553.

Ruiz-Capillas, C., W. F. A. Horner, and C. M. Gillyon. 2001b. Effect of packaging on the spoilage of king scallop (*Pecten maximus*) during chilled storage. *Eur. Food Res. Tech. (Z. Lebensm Unters Forsch)* 213(2):95–98.

Ruiz-Capillas, C. and F. Jiménez-Colmenero. 2008. Determination of preservatives in meat products by flow injection analysis (FIA). *Food Addit. Contam. Part. A* 25:1167–1178.

Ruiz-Capillas, C. and F. Jiménez-Colmenero. 2009. Application of flow injection analysis for determining sulphites in food and beverages: A review. *Food Chem.* 112:487–493.

Ruiz-Capillas, C. and A. Moral. 2001a. Correlation between biochemical and sensory quality indices in hake stored in ice. *Food Res. Int.* 34(5):441–447.

Ruiz-Capillas, C. and A. Moral. 2001b. Production of biogenic amines and their potential use as quality control indices for hake (*Merluccius merluccius* L.) stored in ice. *J. Food Sci.* 66(7):1030–1032.

Ruiz-Capillas, C. and A. Moral. 2001c. Formation of biogenic amines in bulk stored chilled hake (*Merluccius merluccius* L.) packed under atmospheres. *J. Food Protec.* 64(7):1045–1050.

Ruiz-Capillas, C. and A. Moral. 2002. Relation between the free amino acids, anserine and the total volatile basic nitrogen produced in muscle of hake (*Merluccius merluccius* L.) during iced storage. *J. Food Biochem.* 26(1):37–48.

Ruzicka, J. 2014. Tutorial on flow injection analysis. www.flowinjectiontutorial.com

Ruzicka, J. and E. H. Hansen. 1975. Flow injection analyses. 1. New concept of fast continuous-flow analysis. *Anal. Chim. Acta* 78:145–157.

Sadok, S., R. F. Uglow, and S. J. Haswell. 1996. Determination of trimethylamine in fish by flow injection analysis. *Anal. Chim. Acta* 321:69–74.

Timm, M. and B. M. Jørgensen. 2002. Simultaneous determination of ammonia, dimethylamine, trimethylamine and trimethylamine-n-oxide in fish extracts by capillary electrophoresis with indirect UV-detection. *Food Chem.* 76:509–518.

Valcarcel, M. and M. D. Luque de Castro. 1990. Flow injection analysis: A useful alternative for solving analytical problems. *Fresen. J. Anal. Chem.* 337(6):662–666.

Veciana-Nogues, M. T., M. S. Albala-Hurtado, M. Izquierdo-Pulido, and M. C. Vidal-Carou. 1996. Validation of a gas-chromatographic method for volatile amine determination in fish samples. *Food Chem.* 51:513–569.

Vyncke, W., J. Luten, K. BrUnner, and R. Moermans. 1987. Determination of total volatile bases in fish: A collaborative study by the West European Fish Technologistśassosiation (WEFTA). *Z. Lebensm. Unters. Forsch.* 184:110–114.

Wekell, M. M., T. A. Hollingworth, and J. J. Sullivan. 1987. Application of flow injection analysis (FIA) to the determination of seafood quality. In *Seaffod Quality Determination: Developments in Food Science.* eds. D. E. Kramer, J. Liston, Volum 15, pp. 17–25. New York: Elsevier Science Publishing Company Inc.

Wong, K. and T. A. Gill. 1987. Enzymatic determination of trimethylamine and its relationship to fish quality. *J. Food Sci.* 52:1–3.

Zhi, Z. L., A. Rios, and M. Valcarcel. 1995. Direct determination of trimethylamine in fish in the flow-reversal injection mode using a gas extraction sampling device. *Anal. Chem.* 67(5):871–877.

CHAPTER 35

Determination of Biogenic Amines

Claudia Ruiz-Capillas, Ana M. Herrero,
and Francisco Jiménez-Colmenero

CONTENTS

35.1 INTRODUCTION

Biogenic amines (BAs) are compounds that commonly occur in foods and beverages such as meat, fish, vegetables, and wine and are formed mainly by decarboxylation of free amino acids (FAAs) from the action of decarboxylase enzymes (Figure 35.1) (Ruiz-Capillas and Jiménez-Colmenero, 2004). The most important BAs in foods are histamine, tyramine, putrescine, cadaverine, and 2-phenylethylamine, which are produced respectively from histidine, tyrosine, ornithine/arginine, lysine, and phenylalanine. Agmatine, tryptamine, and serotonin are the main minor amines. Some of these (putrescine, spermidine, and spermine, and in plants, also cadaverine and agmatine) are also called polyamines; they are formed naturally by animal, plant, and microorganism metabolism and play important roles in their physiological functions (Halász et al., 1994; Bardócz, 1995; Ruiz-Capillas and Jiménez-Colmenero, 2004, 2009a; Kalač and Krausová, 2005; Kalač, 2014).

The formation of BAs is influenced by a number of factors such as FAA content and availability; microorganisms capable of producing decarboxylases; the nature of the medium; processing; and storage conditions (Figure 35.1). All these factors are interdependent and act together in various combinations. The combined action of these factors mainly determines final concentrations of BAs, because it either directly or indirectly affects the microbial growth and then determines the presence and the activity

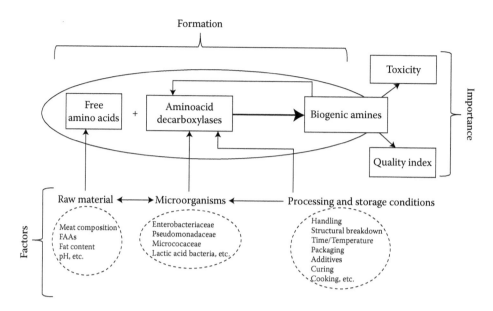

FIGURE 35.1 Formation of biogenic amines and factors that influence this formation. (Adapted from Ruiz-Capillas, C. and F. Jiménez-Colmenero. 2004. *Crit. Rev. Food Sci. Nutr.* 44:489–599.)

of substrate (FAA) and enzyme (Ruiz-Capillas and Jiménez-Colmenero, 2004, 2009a; Roig-Sagués et al., 2009).

A number of studies have associated the use of additives with the formation of BAs in foods and beverages. These are mainly preservatives such as sulfites, nitrates, nitrites, organic acids, and bacteriocins that have been used to prolong product shelf-life (Ruiz-Capillas and Jiménez-Colmenero, 2009b; Directive 2006/52/EC). Sulfur dioxide is one of the preservatives widely used in fish, meat, and beverages, chiefly by the wine industry (Loureiro and Querol, 1999; Bover-Cid et al., 2001; Ruiz-Capillas and Jiménez-Colmenero, 2009b, 2010; Ruiz-Capillas et al., 2011; Triki et al., 2013a; Prabhakar and Mallika, 2014) and has been associated with the reduction of BAs in such products. The use of SO$_2$ also produces a selection of specific types of flora, which in some cases may give rise to increased production of a specific BA (Bover-Cid et al., 2001; Ruiz-Capillas and Jiménez-Colmenero, 2010; Ruiz-Capillas et al., 2012; Triki et al., 2013a). Nitrates and nitrites are also widely used as additives in the preparation of foods, mainly meat products (Ruiz-Capillas et al., 2006, 2007; Delgado-Pando et al., 2011; Triki et al., 2013b), and have also been used in strategies to reduce the production of BAs (Ruiz-Capillas et al., 2006, 2011, 2012; Kurt and Zorba, 2010) and the formation of nitrosamine compounds with teratogenic, mutagenic, and carcinogenic effects highly dangerous for human health (Cassens, 1997; Ruiz-Capillas et al., 2011; Pegg, 2013; De Mey et al., 2014; Kalač, 2014).

35.1.1 Importance of Biogenic Amine Determination

BAs are important for toxicological reasons and for their role as possible quality indicators in food products (Vidal-Carou et al., 1990; Ruiz-Capillas and Moral, 2001; Ruiz-Capillas and Jiménez-Colmenero, 2004, Kalač, 2014). High levels of these amines may be toxic for certain consumers. Tyramine and histamine are the most potentially toxic BAs. The consumption of foods containing high concentrations of tyramine can produce high blood pressure, rapid heart rate, rapid breathing, glycemia, release of norepinephrine, and cerebral and cardiac hemorrhaging, known as the "cheese reaction" (Ruiz-Capillas and Jiménez-Colmenero, 2004; Karovičová and Kohajdová, 2005; Standarová et al., 2008; Pegg, 2013). The most common form of food poisoning from BAs is produced by histamine. This is known as histamine poisoning and "scombroid fish poisoning," owing to its association with the consumption of fish of that family, such as tuna, mackerel, and sardines (Ruiz-Capillas and Moral, 2001; Ruiz-Capillas and Jiménez-Colmenero, 2009a). The main symptoms of histamine poisoning are headache, low blood pressure, irregular heartbeat, anaphylaxis, itching, urticaria and reddening, diarrhea, nausea, and vomiting. Migraine has also recently been associated with high levels of BAs, particularly histamine (Vidal-Carou et al., 2010). Histamine is a subject of major concern in clinical and food chemistry. It is a powerful biological marker in the study of allergic responses and a variety of pathological conditions.

Although not in themselves highly toxic, putrescine and cadaverine enhance the toxicity of other amines, favoring intestinal absorption and hindering histamine and tyramine detoxification. Also, putrescine and cadaverine are implicated in the formation of nitrosamines, potentially carcinogenic compounds, as noted earlier. In normal circumstances, the human organism possesses systems for detoxification of these BAs, mainly in the intestine through the action of monoamine oxidase (MAO; CE 1.4.3.4), diamine oxidase (DAO; CE 1.4.3.6), and polyamine oxidase (PAO; CE 1.5.3.11). However, in certain cases, this mechanism can be disturbed by a variety of factors or circumstances, among them consumption of IMAO/IDAO (monoamine oxidase and diamine oxidase inhibitors), alcohol, immune deficiency, and gastrointestinal disorders (Bardócz, 1995). Individuals whose systems are thus disturbed are at high risk of poisoning in the presence of BAs (EFSA, 2011).

Biogenic amines have been used to establish quality indexes in various foods in order to signal the degree of freshness and/or deterioration of meat, fish, wines, and other foods and beverages (Mietz and Karmas, 1977; Hernández-Jover et al., 1996a; Ruiz-Capillas and Moral, 2001; Silva and Glória, 2002; Vinci and Antonelli, 2002; Ruiz-Capillas and Jiménez-Colmenero, 2004; Ruiz-Capillas et al., 2007, 2011). A food with toxic levels of amines such as histamine or tyramine often appears organoleptically "normal," like tuna or fermented chorizo, so that unacceptable and toxic levels of histamine are undetectable prior to consumption (Vidal-Carou et al., 1990; López-Sabater et al., 1996).

35.1.2 Legislation of Biogenic Amines in Food and Beverages

Biogenic amine content in foods is subject to European legislation under Directive 91/439/EEC, which allows up to a limit of 100 mg/kg of histamine in fishery products. The U.S. Food and Drug Administration (FDA, 2011) has set the limit of histamine tolerance in foods in general at 50 mg/kg. The European Food Safety Authority (EFSA, 2011) has also set a limit for meat products of 50 mg/kg for histamine and up to 600 mg/kg for tyramine.

Determination of BAs is a very important issue in view of both their utility as food quality indicators and their potential implications for consumer health. It is an issue that affects (a) the food industry (BAs help to evaluate food quality and the way that the industry deals with certain processing and/or storage conditions); (b) consumers (who are highly sensitive to all aspects affecting their health); (c) public authorities (for purposes of control and regulation); and (d) scientists (to help understand existing processes and develop responses to new ones). For all these reasons, it is essential to have accurate analytical methods.

35.2 ANALYTICAL METHODS FOR BIOGENIC AMINE DETERMINATION

Numerous analytical procedures have been developed for the determination of BAs in various foods. These range from more conventional methods such as the Association of Official Analytical Chemistry (1995a,b) colorimetric (No 957.07) and fluorimetric (No 977.13) (Taylor et al., 1978) procedures to other more novel ones such as immunoenzymatics and biosensors. There are various fast commercial kits based on the enzyme-linked immunosorbent assay (ELISA) enzyme immunoassay for detecting histamine in fish, but these are generally used to determine only histamine rather than a number of amines (EFSA, 2011). Methods have also been assayed based on capillary electrophoresis and various types of chromatography such as gas chromatography (CG) (Staruszkiewicz and Bond, 1981; Khuhawar et al., 1999; Lapa Guimaraes and Pickova, 2004; Zhang et al., 2004; Önal, 2007), high-performance liquid chromatography (HPLC), and high-performance thin layer chromatography (HPTLC). Of all these, HPLC is the most popular and frequently reported for the separation and quantification of BAs. This procedure offers high resolution, sensitivity, and versatility, and sample treatments are generally simple. Moreover, it offers the advantage that several BAs can be analyzed simultaneously (Hernández-Jover et al., 1996b; Ruiz-Capillas and Moral, 2001; Önal, 2007; Sánchez and Ruiz-Capillas, 2012; Triki et al., 2012). These methods generally require costly equipment that is too complex for use in routine analyses. Moreover, many such systems require prior derivatization, which is time-consuming and frequently poses interference problems. In this context, flow injection analysis (FIA) offers major advantages for BA determination, such as easy control of the chemical reaction, rapid analysis, low operating cost, environmentally friendly reagents, accuracy and precision, and suitability for use by untrained personnel, and it is therefore highly suitable for various kinds of analyses, including the more routine sort. This chapter provides an overview of available FIA methodologies for determining BAs in foods and beverages.

35.3 DETERMINATION OF BIOGENIC AMINES BY FLOW INJECTION ANALYSIS

Flow injection analysis has been widely used for BA determination owing to the advantages the procedure offers, and in fact many conventional procedures (AOAC, 1995a,b) for BA analysis have been automated using this methodology.

Generally speaking, there are two main stages in FIA methods for determining BAs: (a) extraction of BAs, frequently followed by a cleanup procedure to eliminate interferences; and (b) separation, identification, and quantification.

35.3.1 Extraction of Biogenic Amines

Analytical determination of BAs is not simple because of the complexity of the food matrices to be analyzed. Food samples frequently require extraction, separation, or concentration procedures prior to analysis. Whether such steps are required depends on the type of matrix. The various food matrices are not homogeneous but contain high protein levels and a wide variety of fat content (0.5%–30%). The complexity of these matrices is a critical consideration for adequate recovery of all BAs. Moreover, foods and beverages do not contain only a single BA; there are generally several in different ranges of concentration, and in some cases, their precursor amino acids are present and cause interference. The process becomes even more complex when several BAs have to be determined simultaneously. In this regard (as in the case of other methodologies), interfering components must be removed and the analyte isolated before FIA analysis proper can be performed (Ruiz-Capillas and Jiménez-Colmenero, 2009a).

Many different solvents, such as hydrochloric acid, trichloroacetic acid, and perchloric acid, as well as organic solvents such as methanol, dichloromethane, acetone, and acetonitrile have been used to extract BAs from food samples and for subsequent FIA determination. Which solvent is used depends on the subsequent FIA analysis system to be used, on whether separation columns are used, whether there is derivatization or combination with biosensors, and so on.

The simplest sample preparation processes for FIA determination are the ones for beverages. For fluorimetric determination of histamine in wine and cider using an anion-exchange column–FIA system, the samples only required appropriate dilution with hydrochloric acid and degasification in the case of cider, followed by filtration through a 0.22-μm nylon membrane microfilter to retain microparticulate matter (mainly tannic compounds) (del Campo et al., 2006).

In the case of food matrices, most extraction processes follow the AOAC methodology (1995a,b), which basically consists of homogenization or grinding of the sample, following which a few grams (5–10) of this sample are blended with TCA or methanol. Hungerford et al. (1990) blended fresh fish fillets or canned fish for 2 min with methanol (10 g/50 mL methanol); samples were then heated for 15 min at 60°C, cooled, diluted with methanol to 100 mL, and filtered. The extract was also filtered through 0.45-μm nylon syringe filters prior to injection in the FIA system.

These same authors (Hungerford and Arefyev, 1992) also used hydrochloric acid for the determination of BAs in thawed fish. The extract was prepared with a 5 g portion of thawed homogenate blended for 2 min with 45 mL of 0.02 M HCl solutions at pH 3 so as to achieve highly acidic pH of around 2–3. Extraction with acid assists precipitation of the proteins in the sample, so the BAs can be retrieved from the supernatant.

Other authors have used trichloroacetic acid (Katayama et al., 1993) for sample deproteination and determination of histamine by the chemiluminescence FLA method. Extraction from minced tuna fish (10 g) has been carried out first with water (15 mL) followed by extraction with 20 mL of 10% (w/v) trichloroacetic acid solution and filtration.

Deproteination has also been performed with perchloric acid (3%) (10 g/20 mL) in beefsteak, followed by 10 min centrifugation at 3000 g. The resulting supernatant was adjusted to pH 6.8, and the precipitate was removed by 10 min centrifugation at 3000 g. The supernatant was then made up to 25 mL with distilled water, followed by purification and concentration for FIA analysis (Yano et al., 1995).

In many cases, the extraction process is followed by purification of the sample or even by derivatization. Derivatization makes it possible to augment the sensitivity to these amines during subsequent detection, since most BAs exhibit neither satisfactory absorption nor significant fluorescence due to their low volatility and lack of chromophores. There are a number of different derivatizing agents for BAs, but in the case of

FIA analysis, the tendency has been to use mostly o-phthalaldehyde (OPA) (Hungerford et al., 1990, 2001; Hungerford and Arefyev, 1992; del Campo et al., 2006). Biogenic amine derivatization may be performed after column reaction chromatographic separation (Hungerford et al., 1990; del Campo et al., 2006). Some authors have even developed systems for derivatization of samples with OPA, obviating the need for a column (Hungerford and Arefyev, 1992). In most FIA systems that use OPA as a derivatizer, the reaction is carried out in the FIA system itself rather than prior to injection. This entails less handling and more control of the reactions, and the analysis time is shorter. Derivatization occurs at the same time, enhancing the reproducibility and sensitivity of the analysis. Prederivatization involves more sample preparation steps, which can cause problems in the analysis later on.

Then again, although methods using columns or derivatization are very sensitive and selective (Hungerford et al., 1990, 2001; del Campo et al., 2006), the step is time-consuming and slows down the analytical process. In this connection, an FIA–chemiluminescence method has been proposed that does not require derivatization (Katayama et al., 1993). The method is even more sensitive than chromatographic methods for determination of histamine. However, it requires prior sample preparation using a Sep-pack C18 ENV plus cartridge. In this case, the acidic extract is neutralized with sodium hydroxide, passed through the cartridge with 2% (v/v) acetonitrile, and injected into the FIA system to separate the histamine.

In the case of FIA determination in combined systems with electrodes, the extraction procedure is simpler (Takagi and Shikata, 2004). For instance, in the case of analysis of amines in muscle, the tissue was homogenized with 30 mL of deionized water (5 g/30 mL), microwaved for about 3 min, and then cooled and made up to 50 mL with deionized water. After 10 min centrifugation at 10,000 g, the supernatants were filtered through a membrane filter unit, and then the clear supernatant was directly injected for the FIA experiments. In other studies of this type, extraction has been done with phosphate buffer (pH 7.2) followed by centrifugation (Mei et al., 2007) or phosphate buffer mixed in a vortex with the tissue and later placed in an ultrasound bath and centrifuged to recover the supernatant (Frébort et al., 2000; Carelli et al., 2007).

35.3.2 Separation, Identification, and Quantification of Biogenic Amines by FIA

The first FIA method for the determination of histamine in fresh and canned fish, tuna, mackerel, and mahi-mahi was developed in 1990 by Hungerford et al. (1990). This system was based on the AOAC method for determining histamine in fish, by means of the reaction of histamine/OPA used to determine this and other amines with detection by fluorescence. The reaction of the OPA with histamine takes place in the manifold during transportation of the reagents.

The analytical technique proposed by Hungerford et al. (1990) is based on microfluidic manipulation of samples and reagents, whereby samples are injected into a carrier/reagent solution that transports the sample to a detector. The first stage of this FIA system combines an HPLC composed of an automatic injector, a pump, and an oven to act as a thermostat for the reaction of the amines with the OPA. One channel of the HPLC was used to propel the HCl stream, and the two reactor pumps were used to push the OPA and phosphoric acid reagent streams. The fluorescence detector was then connected. This was operated at excitation and emission wavelengths of 365 and 450 nm, respectively (Figure 35.2).

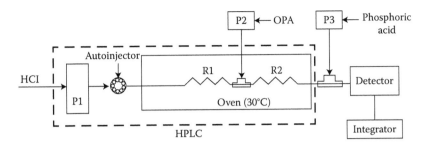

FIGURE 35.2 Flow injection analysis (FIA) system for the determination of histamine in fish and biological samples (P1, P2, P3: pumps; R1, R2: reactors). (Adapted from Hungerford, J. M. et al. 1990. *Anal. Chem.* 62:1971–1976.)

The FIA method described (Figure 35.2) can be divided into four stages:

1. First, the extract is injected (by an automatic injector) into a flowing stream of aqueous HCl (driven by pump P1), where it is diluted in the HCl reactor (R1). This step in the scheme replaces the manual dilution and pipetting step in batch procedures, with the advantage that this entails in terms of time and sample handling. The principal function of the HCl reactor was to level the pH of the injected sample zone to that of the HCl stream by dispersing the sample in this stream. Both online sample dilution and pH adjustment occurred concurrently and were achieved through control of the volume of sample injected and the HCl reactor dimensions. The authors (Hungerford et al., 1990) found that in this step, a high-dispersion HCl reactor was needed instead of a low-dispersion HCl reactor to eliminate the apparent matrix effects (the extracts of decomposed mackerel and mahi-mahi) if they were injected directly into the low dispersion reactor.
2. After this first step, the stream passes into a first mixing tee, where the acidified and diluted sample encounters alkaline OPA reagent (P2), and the alkaline part of the histamine–OPA reaction proceeds for a time determined by the OPA reactor volume (R2).
3. The alkaline part of the reaction is stopped downstream of the second mixing tee, when phosphoric acid reagent (driven by pump P3) reduces the pH to below 2. The acidic and final part of the reaction proceeds very rapidly in the transfer tubing (R3) between the second mixing tee and the detector.
4. Finally, fluorescence from the product is detected in the flow cell. A fluorescence detector with a smaller pressure drop is needed for this FIA system, which operates at low pressure using peristaltic pumps.

Many of the variables that affected the selectivity of the reaction and the peak heights of histamine, such as concentrations of OPA and NaOH, reagent flow rate, length of the OPA reactor, reaction time, and temperature, were analyzed to establish the optimum conditions for this system (NaOH [0.15 M], HCl [6.0 mM], OPA [24×10^{-4} M OPA, 0.3 M NaOH], phosphoric acid [0.37 M]).

In interference studies carried out on this system assaying the spike of free histidine and small histidyl peptides, Hungerford et al. (1990) suggested the use of an anion-exchange minicolumn (MC) placed in-line with the carrier stream, which, along with control of the OPA reaction temperature, was one of the critical points in the system. The minicolumn was placed between the injector and the first reactor (R1) (Figure 35.3) to eliminate major interferences in fish. It has been shown that the FIA-controlled OPA reaction is about 27 times more selective than the lengthy low-temperature procedure described by other authors (Hakanson and Ronnberg, 1974). Also, when the determinative step used in the official method (AOAC, 1995a) was followed without the ion-exchange cleanup step, selectivity for histamine was about 20 times less than in the FIA-controlled reaction. Other interference studies have also been conducted with solutions of indole, octopamine, putrescine, tryptamine, and tyramine, but they produced no detectable response, even at millimolar levels, and cadaverine and cysteine produced only weak responses. Therefore, to judge by its performance, this method (Hungerford et al., 1990) provides a set of excellent, simplified, rugged procedures suitable for the determination of histamine, with detection and quantitation limits near 0.8 and 2.4 mg/kg, and linearity to approximately 340 mg/kg. Each sample produces a response in < 1 min. The method is faster and more practical than other proposed screening methods, including alternative chemistries, HPLC methods, and immunoassays. With this method, a fluorescence signal can be obtained from the histamine/histidine reaction with the OPA in 32 s, and 89%–105% histamine recovery. The results of histamine levels in fish extracts from this system were compared with the data obtained using the AOAC fluorometric method (AOAC, 1995b).

The method of Hungerford et al. (1990) was recently proposed and improved by other authors (del Campo et al., 2006) for the determination of histamine in wine and cider. In this revised methodology (Figure 35.4), an FIA system was used with only one peristaltic pump, rather than the two used by Hungerford et al. (1990), and only three channels. An anion-exchange column (minicolumn filled with Dowex 1 × 8 resin in OH form) was also used to eliminate sample matrix interferences in the histamine/o-phthalaldehyde reaction. Factorial design was used to determine which operational parameters should be included in the optimization. The method developed shows good selectivity for histamine determination in alcoholic beverages. The detection and quantification limits were 30 and 101 μg/L, respectively. The recoveries achieved in wine and cider samples were close to 100%, making it possible to analyze 24 samples/h (del Campo et al., 2006) (Figure 35.4).

Some modifications have also been made to the original method by its authors in later studies (Hungerford et al., 2001), so the original high-pressure FIA method could be adapted to commercially available low-pressure FIA systems. These included

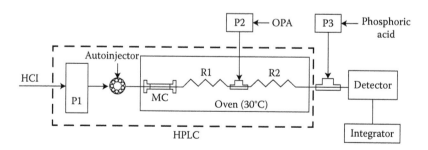

FIGURE 35.3 Flow injection analysis (FIA) system with the minicolumn (MC) for the determination of histamine in fish and biological samples (P1, P2, P3: pumps; R1, R2: reactors). (Adapted from Hungerford, J. M. et al. 1990. *Anal. Chem.* 62:1971–1976.)

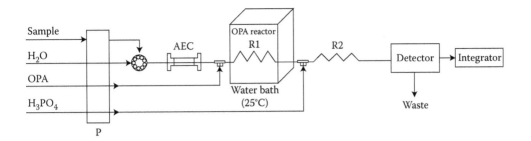

FIGURE 35.4 Flow injection analysis (FIA) system for the determination of histamine in wine and cider (P: pump; R1, R2: reactors; AEC: anion-exchange column). (Adapted from del Campo, G., B. Gallego, and I. Berregi. 2006. *Talanta* 68:1126–1134.)

incorporation of an OPA reactor, thermostatted at 40°C, with no need of a column, all adapted to a commercial Foss Tecator 5010 Analyser (Hungerford et al., 2001). This provided the means to develop a simpler and more precise FIA method with only peristaltic pumps and low-pressure injectors (Figure 35.5). This method is also selective for histamine and is consistent with the official AOAC fluorescence method (AOAC, 1995b). The correlation between the proposed FIA method and the AOAC fluorescence method is good, with $R^2 = 0.99$. The method can detect approximately 0.8 mg/kg histamine. This is >50 times lower than the threshold level of 50 mg/kg, providing a considerable safety margin.

Flow-injection analysis affords excellent control of convection, timing, and enhanced selectivity and has therefore also been used extensively in combination with immobilized enzymes and electrodes or reactors. Many studies of FIA and sensors or biosensors have also been reported for BAs in food, based on commercial or home-purified enzymes (Yang and Rechnitz, 1995; Chemnitius and Bilitewski, 1996; Male et al., 1996; Draisci et al., 1998; Esti et al., 1998; Tombelli and Mascini 1998; Carsol and Mascini, 1999; Wimmerová and Macholán 1999; Niculescu et al., 2000a,b; Compagnone et al., 2001; Lange and Wittmann, 2002; Serra et al., 2007; Alonso-Lomillo et al., 2010; Kivirand and Rinken, 2011; Bóka et al., 2012).

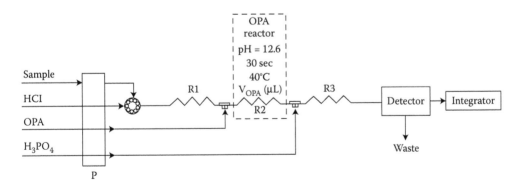

FIGURE 35.5 Flow injection analysis (FIA) system for the determination of histamine in fish samples (P: pump; R1, R2, R3: reactors; AEC: anion-exchange column). (Adapted from Hungerford, J. M., T. A. Hollingworth, and M. M. Wekell. 2001. *Anal. Chim. Acta* 438:123–129.)

These systems use different enzymes for development of the biosensor. One of the most widely used is DAO. Hungerford and Arefyev (1992) proposed an FIA-optimized method applied to detecting toxic enzyme inhibitors in fish and used in combination with rapid on-site sampling and extraction. This system proposed a flow injection procedure with DAO inhibition using a manifold based on double injection, the immobilized enzyme, and selective fluorimetric detection of endogenous histamine substrates (Hungerford and Arefyev, 1992). The flow-injection system described here can be used simultaneously to screen for the presence of DAO inhibition in fish and to determine histamine levels. Sample extracts are double injected; one of the two resulting sample zones is passed through a reactor containing DAO immobilized to Sepharose®.

Flow enzyme reactors for BAs (putrescine and histamine) have also been tested with spectrophotometric detection of enzymatically produced hydrogen peroxide by a peroxidase/guaiacol system. The system was based on the immobilized amine oxidase (EC 1.4.3.6) from grass pea (*Lathyrus sativus*) (Frébort et al., 2000).

Electrochemical biosensors for the determination of BA content have also been assembled using DAO and amperometric detectors. The enzyme was immobilized from DAO directly on the surface of a platinum electrode and included in an FIA (Male et al., 1996; Bouvrette et al., 1997; Tombelli and Mascini, 1998; Akbari-adergani et al., 2010; Kivirand and Rinken, 2011; Bóka et al., 2012). Esti et al. (1998) also assayed DAO and PAO covalently immobilized onto polymeric membranes for the determination of BA content in fruits and vegetables. In that study, in the presence of their substrates, both enzymes produced H_2O_2, which was detected on a platinum electrode polarized at +650 mV versus Ag/AgCl. A similar system using DAO has also been used to determine BAs in salted anchovy samples (Draisci et al., 1998). The assay is based on a platinum electrode that senses the hydrogen peroxidase produced in the reaction catalyzed by the DAO enzyme immobilized on the electrode surface. More recent studies have also reported the development and application of an amperometric biosensor for the determination of total BA content in food samples (cheese and anchovy) using a commercial DAO (from Porcine kidney E.C. 1.4.3.6) entrapped by glutaraldehyde on an electrosynthesized bilayer film (Carelli et al., 2007). The biosensor showed good sensitivity, stability, and complete suppression of electroactive interferences, even in flow injection systems. The values achieved using this optimized biosensor were not significantly different from those achieved using an ion chromatography system.

Niculescu et al. (2000a) also reported an amperometric biosensor for histamine detection based on the amine oxidase enzyme. The biosensor was coupled to an FIA line operated at +200 mV (vs. Ag/AgCl/0.1 M KCl).

Horseradish peroxidase has also been used coimmobilized and coupled to a glassy-carbon electrode with ferrocene monocarboxylic acid as mediator dissolved in the carrier stream. Low detection limits were achieved in all cases using these systems, with some differences, although the selectivity was quite similar in the different assemblies (Tombelli and Mascini, 1998). Horseradish peroxidase has also been used by other authors (Lerke et al., 1983; Castilho et al., 2005).

Putrescine oxidase and xanthine oxidase have also been used to develop an FIA biosensor system (two lines) combining the two enzymes immobilized on their respective electrodes for quality control of vacuum-packed beef by simultaneous measurement of putrescine and cadaverine and hypoxanthine and xanthine (Yano et al., 1995). Okuma et al. (2000) developed an enzyme reactor system with amperometric detection for the determination of putrescine, cadaverine, and spermidine in chicken. The system had a reactor in which putrescine oxidase was immobilized on chitosan porous beads, and an amperometric electrode was used for the detection of hydrogen peroxide by the enzyme reaction.

An FIA system has also been proposed for the determination of histamine in fish samples using a histamine dehydrogenase (HmDH)-based electrode for amperometric detection (Takagi and Shikata, 2004). The histamine dehydrogenase was immobilized on a glassy-carbon electrode, using $Os^{2+/3+}$ in the osmium-derivatized polymer as an electron transfer mediator. This electrode exhibits high selectivity to histamine and is not sensitive to other primary amines, including common BAs such as putrescine, cadaverine, and tyramine. An amperometric biosensor has recently been proposed in FIA using pea seedling amine oxidase (PSAO) for the determination of putrescine, cadaverine, and tyramine in chicken samples (Telsnig et al., 2012). In that study, PSAO was immobilized with Nafion®-containing film that displayed unspecific entrapment of the biocomponent.

Studies on the determination of histamine in fish have also been conducted using a flow-injection method for amines based on aryl oxalate-sulforhodamine 101 chemiluminescence (Katayama et al., 1993). As eluent, this system uses R1 acetonitrile–water containing 0.02 M hydrogen peroxide (9 + 1, v/v) and R2 (5.0×10^{-4} M TDPO-1.0×10^{-7} M sulforhodamine in acetonitrile), both applied at a flow rate of 1.5 mL/min. The sample (20 μL) was diluted with eluent R1 and injected into the FIA system. This method is particularly suitable for determining amines lacking strong absorbance (e.g., putrescine) and tertiary amines for which derivatization reactions are not applicable (e.g., triethylamine). The advantage of the chemiluminescence detection method proposed by these authors is its sensitivity, the simplicity of the apparatus, and the lack of a derivatization reaction (Katayama et al., 1993). A chemiluminescent plant tissue–based biosensor system employing sequential injection analysis (SIA) was recently used for the analysis of putrescine and cadaverine in fish. The pea-seedling tissue was used an active source of DAO, which is packed in a mini-PTFE column incorporated in the SIA system. In this system, amine analysis is based on enzymatic conversion in the column to produce hydrogen peroxide, which is detected through a chemiluminescence reaction involving luminol and Co^{2+} (Mei et al., 2007).

35.4 CONCLUSION

The determination of BAs in food and beverages is important in two ways: for the detection of potential sources of food poisoning and as a measure of quality. In both cases, it is important to analyze the BAs in both raw materials and finished products.

Flow injection analysis has been successfully used for the determination of BAs. This methodology offers a number of advantages, such as easy control of the chemical reaction, rapid reaction in the system, and all reagent additions performed automatically. With this system, conventional methods of BA analysis can be automated. Many FIA BA analysis systems offer high selectivity in the determination of BAs, in many cases obviating the need for extensive sample cleanup and allowing relatively simple extraction of the analyte or analytes. Many of the FIA methods of BA determination involve a reaction with OPA to enhance sensitivity to the amines in the process of identification and quantification.

Moreover, the FIA method has been extensively used in combination with immobilized enzymes and electrodes or reactors using various enzymes (amine oxidase, peroxidase, histaminase, etc.) for the determination of BAs in various elements by amperometry or chemiluminescence. This has been a major step forward in the biosensor-assisted FIA determination of BAs.

Flow injection analysis determination methods have been used in various foods such as seafood, meat, and fruit, and also in beverages (wine, cider, etc.). Also, many of these methods could be applied to other biological samples (brain tissue, plasma) in addition to food matrices. Flow injection analysis determination of BAs has been successfully correlated with traditional methods.

ACKNOWLEDGMENTS

This research was supported by projects AGL2010-19515 and AGL 2011-29644-C02-01 of the Plan Nacional de Investigación Científica, Desarrollo e Innovación Tecnológica (I + D + I) Ministerio de Ciencia y Tecnología. The author thanks Felicitas Pérez Calvo for the unconditional support.

REFERENCES

Akbari-adergani, B., P. Norouzi, M. R. Ganjali, and R. Dinarvand. 2010. Flow-injection electrochemical method for determination of histamine in tuna fish samples. *Food Res. Int.* 43:1116–1122.

Alonso-Lomillo, M. A., O. Dominguez-Renedo, P. Matos, and M. J. Arcos-Martinez. 2010. Disposable biosensors for determination of biogenic amines. *Anal. Chim. Acta* 665:26–31.

AOAC (Association of Official Analytical Chemistry). 1995a. Histamine in seafood: Chemical method. Sec. 35.5.31, Method 957.07. In *Official Methods of Analysis of AOAC International* (16th ed.), ed. P. A. Cunniff, pp. 15–16. Gaithersburg, MD: AOAC International.

AOAC (Association of Official Analytical Chemistry). 1995b. Histamine in seafood: Fluorometric method. Sec. 35.1.32, Method 977.13. In *Official Methods of Analysis of AOAC International* (16th ed.), ed. P. A. Cunniff, pp. 6–17. Gaithersburg, MD: AOAC International.

Bardócz, S. 1995. Polyamines in food and their consequences for food quality and human health. *Trends Food Sci. Tech.* 6:341–346.

Bóka, B., N. Adányi, J. Szamos, D. Virág, and A. Kiss. 2012. Putrescine biosensor based on putrescine oxidase from *Kocuria rosea*. *Enzyme Microb. Tech.* 51:258–262.

Bouvrette, P., K. B. Male, J. H. T. Luong, and B. F. Gibbs. 1997. Amperometric biosensor for diamine using diamine oxidase purified from porcine kidney. *Enzyme Microb. Tech.* 20:32–38.

Bover-Cid, S., M. J. Miguélez-Arrizado, and M. C. Vidal-Carou. 2001. Biogenic amine accumulation in ripened sausages affected by the addition of sodium sulphite. *Meat Sci.* 59:391–396.

Carelli, D., D. Centonze, C. Palermo, M. Quinto, and T. Rotunno. 2007. An interference free amperometric biosensor for the detection of biogenic amines in food products. *Biosensor. Bioelectron.* 23:640–647.

Carsol, M. A. and M. Mascini. 1999. Diamine oxidase and putrescine oxidase immobilized reactors in flow injection analysis: A comparison in substrate specificity. *Talanta* 50:141–148.

Cassens, R. G. 1997. Residual nitrite in cured meats. *Food Tech.* 51:53–55.

Castilho, T. J., P. T. Sotomayor, and L. T. Kubota. 2005. Amperometric biosensor based on horseradish peroxidase for biogenic amine determinations in biological samples. *J. Pharmaceut. Biomed. Anal.* 37:785–791.

Chemnitius, G. C. and U. Bilitewski. 1996. Development of screen-printed enzyme electrodes for the estimation of fish quality. *Sens. Actuators B* 32:107–113.

Compagnone, D., G. Isoldi, D. Moscone, and G. Palleschi. 2001. Amperometric detection of biogenic amines in cheese using immobilised diamine oxidase. *Anal. Lett.* 34:841–854.

De Mey, E., K. De Klerck, H. De Maere, L. Dewulf, G. Derdelinckx, M-Ch. Peeters, I. Fraeye et al. 2014. The occurrence of N-nitrosamines, residual nitrite and biogenic amines in commercial dry fermented sausages and evaluation of their occasional relation. *Meat Sci.* 96(2, Part A):821–828.

del Campo, G., B. Gallego, and I. Berregi. 2006. Fluorimetric determination of histamine in wine and cider by using an anion-exchange column-FIA system and factorial design study. *Talanta* 68:1126–1134.

Delgado-Pando, G., S. Cofrades, C. Ruiz-Capillas, M. T. Solas, M. Triki, and F. Jiménez-Colmenero. 2011. Low-fat frankfurters formulated with a healthier lipid combination as functional ingredient: Microstructure, lipid oxidation, nitrite content, microbiological changes and biogenic amine formation. *Meat Sci.* 89:65–71.

Directive 2006/52/EC. https://www.fsai.ie/uploadedFiles/Directive_2006_52_EC.pdf.

Directive 91/439/EEC. Directive of 22 July 1991 establishing standards to be applied to the production and commercialization of fishery products. *Off. J. Eur. Comm.* L268:15–34.

Draisci, R., G. Volpe, L. Lucentini, A. Cecilia, R. Federico, and G. Palleschi. 1998. Determination of biogenic amines with an electrochemical biosensor and its application to salted anchovies. *Food Chem.* 62:225–232.

EFSA (European Food Safety Authority). 2011. Scientific opinion on risk based control of biogenic amine formation in fermented foods. *EFSA J.* 9:2393–2486.

Esti, M., G. Volpe, L. Massignan, D. Compagnone, E. La Notte, and G. Palleschi. 1998. Determination of amines in fresh and modified atmosphere packaged fruits using electrochemical biosensors. *J. Agric. Food Chem.* 46(10):4233–4237.

FDA (Food and Drug Administration). 2011. *Fish and Fishery Products Hazards and Controls Guidance* (4th ed.) http://www.fda.gov/downloads/food/guidance complianceregulatoryinformation/ guidancedocuments/seafood/ucm251970.pdf.

Frébort, I., Skoupa, L., and Peč, P. 2000. Amine oxidase-based flow biosensor for the assessment of fish freshness. *Food Control* 11:13–18.

Hakanson, R. and A. L. Ronnberg. 1974. Improved fluorometric assay of histamine: Condensation with o-phthalaldehyde at –20°C. *Anal. Biochem.* 60:560–567.

Halász, A., A. Baráth, L. Simon-Sarkadi, and W. Holzapfel. 1994. Biogenic amines and their production by microorganisms in food. *Trends Food Sci. Tech.* 5:42–49.

Hernández-Jover, T., M. Izquierdo-Pulido, M. T. Veciana-Nogues, and M. C. Vidal-Carou. 1996a. Biogenic amine sources in cooked cured shoulder pork. *J. Agric. Food Chem.* 44:3097–3101.

Hernández-Jover, T., M. Izquierdo-Pulido, M. T. Veciana-Nogués, and M. C. Vidal-Carou. 1996b. Ion-pair high-performance liquid chromatographic determination of biogenic amines in meat products. *J. Agric. Food Chem.* 44:2710–2715.

Hungerford, J. M. and A. A. Arefyev. 1992. Flow-injection assay of enzyme inhibition in fish using immobilized diamine oxidase. *Anal. Chim. Acta* 261:351–359.

Hungerford, J. M., T. A. Hollingworth, and M. M. Wekell. 2001. Automated kinetics-enhanced flow injection method for histamine in regulatory laboratories: Rapid screening and suitability requirements. *Anal. Chim. Acta* 438:123–129.

Hungerford, J. M., K. D. Walker, M. M. Wekell, J. E. LaRose, and H. R. Throm. 1990. Selective determination of histamine by flow injection analysis. *Anal. Chem.* 62: 1971–1976.

Kalač, P. 2014. Health effects and occurrence of dietary polyamines: A review for the period 2005–mid 2013. *Food Chem.* 161(15):27–39.

Kalač, P. and P. Krausová. 2005. A review of dietary polyamines: Formation, implications for growth and health and occurrence in foods. *Food Chem.* 90:219–230.

Karovičová, J. and Z. Kohajdová. 2005. Biogenic amines in food. *Chem. Papers* 59:70–79.

Katayama, M., H. Takeuchi, and H. Taniguchi. 1993. Determination of amines by flow-injection analysis based on aryl oxalate-sulphorhodamine 101 chemiluminescence. *Anal. Chim. Acta* 281:111–118.

Khuhawar, M. Y., A. A. Memon, P. D. Jaipal, and M. I. Bhanger. 1999. Capillary gas chromatographic determination of putrescine and cadaverine in serum of cancer patients using trifluoroacetylacetone as derivatizing reagent. *J. Chromatogr. B* 723:17–24.

Kivirand, K. and T. Rinken. 2011. Biosensors for biogenic amines: The present state of art mini-review. *Anal. Lett.* 44:2821–2833.

Kurt, S. and O. Zorba. 2010. Biogenic amine formation in Turkish dry fermented sausage (sucuk) as affected by nisin and nitrite. *J. Sci. Food Agric.* 90:2669–2674.

Lange, J. and C. Wittmann. 2002. Enzyme sensor array for the determination of biogenic amines in food samples. *Anal. Bioanal. Chem.* 372:276–283.

Lapa Guimaraes, J. and J. Pickova. 2004. New solvent systems for thin-layer chromatographic determination of nine biogenic amines in fish and squid. *J. Chromatogr. A* 1045:223–232.

Lerke, P. A., P. N. Martina, and B. C. Henry. 1983. Screening test for histamine in fish. *J. Food Sci.* 48:155–157.

López-Sabater, E. I., J. J. Rodríguez-Jerez, M. Hernández-Herrero, A. X. Roig-Sagués, and M. T. Mora-Ventura. 1996. Sensory quality and histamine formation during controlled decomposition of tuna (*Thunnus thynnus*). *J. Food Protec.* 59:167–174.

Loureiro, V. and A. Querol. 1999. The prevalence and control of spoilage yeasts in foods and beverages. *Trends Food Sci. Tech.* 10–11:356–365.

Male, K. B., P. Bouvrette, J. H. T. Luong, and B. F. Gibbs. 1996. Amperometric biosensor for total histamine, putrescine and cadaverine using diamine oxidase. *J. Food Sci.* 61:1012–1016.

Mei, Y., L. Ran, X. Ying, Z. Yuan, and S. Xin. 2007. A sequential injection analysis/chemiluminescent plant tissue-based biosensor system for the determination of diamine. *Biosensor. Bioelectron.* 22:871–876.

Mietz, J. L. and E. Karmas. 1977. Chemical quality index of canned tuna as determined by high pressure liquid chromatography. *J. Food Sci.* 42:155–158.

Niculescu, M., I. Frébort, P. Peč, P. Galuszka, B. Mattiasson, and E. Csöregi. 2000a. Amine oxidase based amperometric biosensor for histamine detection. *Electroanalysis* 12:369–375.

Niculescu, M., C. Nistor, I. Frébort, P. Peč, B. Mattiasson, and E. Csöregi. 2000b. Redox hydrogel-based amperometric bienzyme electrodes for fish freshness monitoring. *Anal. Chem.* 72:1591–1597.

Okuma, H., W. Okazaki, R. Usami, and K. Horikoshi. 2000. Development of the enzyme reactor system with amperometric detection and application to estimation of the incipient stage of spoilage of chicken. *Anal. Chim. Acta* 411:37–43.

Önal, A. 2007. A review: Current analytical methods for the determination of biogenic amines in foods. *Food Chem.* 103:1475–1486.

Pegg, A. E. 2013. Toxicity of polyamines and their metabolic products. *Chem. Res. Toxicol.* 26:1782–1800.

Prabhakar, K. and E. N. Mallika. 2014. Permitted preservatives—Sulfur dioxide. *Encyclopedia of Food Microbiology* (2nd ed.), eds. C. A. Batt and Mary L. Tortorello, pp. 108–112. London, UK: Elsevier (Academic Press).

Roig-Sagués, A. X., C. Ruiz-Capillas, D. Espinosa, and M. Hernández. 2009. The decarboxylating bacteria present in foodstuffs and the effect of emerging technologies on their formation. In *Biological Aspects of Biogenic Amines, Polyamines and Conjugates*, Chapter 8, ed. G. Dandrifosse, pp. 201–230. Kerala, India: Transworld Research Network.

Ruiz-Capillas, C., P. Aller-Guiote, J. Carballo, and F. Jiménez-Colmenero. 2006. Biogenic amine formation and nitrite reactions in meat batter as affected by high-pressure processing and chilled storage. *J. Agric. Food Chem.* 54:9959–9965.

Ruiz-Capillas, C., A. M. Herrero, and F. Jiménez-Colmenero. 2011. Reduction of biogenic amines levels in meat and meat products. In *Natural Antimicrobials in Food Quality and Food Safety*, eds. M. Rai and M. L. Chikindas, pp. 154–166. UK: CAB International.

Ruiz-Capillas, C. and F. Jiménez-Colmenero. 2004. Biogenic amines in meat and meat products. *Crit. Rev. Food Sci. Nutr.* 44:489–599.

Ruiz-Capillas, C. and F. Jiménez-Colmenero. 2009a. Biogenic amines in seafood products. In *Handbook of Seafood and Seafood Products Analysis*, Chapter 46, eds. L. M. L. Nollet and F. Toldra, pp. 833–850. Boca Raton, USA: CRC Press, Taylor & Francis Group.

Ruiz-Capillas, C. and F. Jiménez-Colmenero. 2009b. Application of flow injection analysis for determining sulphites in food and beverages: A review. *Food Chem.* 112:487–493.

Ruiz-Capillas, C. and F. Jiménez-Colmenero. 2010. Effect of an argon-containing packaging atmosphere on the quality of fresh pork sausages during refrigerated storage. *Food Control* 21:1331–1337.

Ruiz-Capillas, C., F. Jiménez-Colmenero, A. V. Carrascosa, and R. Muñoz. 2007. Biogenic amines production in Spanish dry-cured "chorizo" sausage treated with high pressure and kept in chilled storage. *Meat Sci.* 77:365–371.

Ruiz-Capillas, C. and A. Moral. 2001. Production of biogenic amines and their potential use as quality control indices for hake (*Merluccius merluccius*, L.) stored in ice. *J. Food Sci.* 66:1030–1032.

Ruiz-Capillas, C., M. Triki, A. M. Herrero, and F. Jiménez-Colmenero. 2012. Biogenic amines in low-and reduced-fat dry fermented sausages formulated with konjac gel. *J. Agr. Food Chem.* 60:9242–9248.

Sánchez, J. A. and C. Ruiz-Capillas. 2012. Application of the simplex method for optimization of chromatographic analysis of biogenic amines in fish. *Eur. Food Res. Tech.* 234:285–294.

Serra, B., A. J. Reviejo, and J. M. Pingarrón. 2007. Application of electrochemical enzyme biosensors for food quality control. *Compr. Anal. Chem.* 49:255–298.

Silva, C. M. G. and M. B. A. Glória. 2002. Bioactive amines in chicken breast and thigh after slaughter and during storage at $4 \pm 1°C$ and in chicken-based meat products. *Food Chem.* 78:241–248.

Standarová, E., Borkovcová, I., and Vorlová, L. 2008. The occurrence of biogenic amines in dairy products on the Czech market. *Acta Scientiarum Polonorum, Med. Vet.* 7:35–42.

Staruszkiewicz, Jr. and J. F. Bond. 1981. Gas chromatographic determination of cadaverine, putrescine and histamine in food. *J. AOAC.* 64:584–591.

Takagi, K. and S. Shikata. 2004. Flow injection determination of histamine with a histamine dehydrogenase-based electrode. *Anal. Chim. Acta* 505:189–193.

Taylor, S. L., E. R. Lieber, and M. A. Leatherwood. 1978. A simplified method for histamine analysis of food. *J. Food Sci.* 43:247–250.

Telsnig, D., V. Kassarnig, C. Zapf, G. Leitinger, K. Kalcher, and A. Ortner, 2012. Characterization of an amperometric biosensor for the determination of biogenic amines in flow injection analysis. *Int. J. Electrochem.* 7:10476–10486.

Tombelli, S. and M. Mascini. 1998. Electrochemical biosensors for biogenic amines: A comparison between different approaches. *Anal. Chim. Acta* 358:277–284.

Triki, M., A. M. Herrero, F. Jiménez-Colmenero, and C. Ruiz-Capillas. 2013a. Storage stability of low-fat sodium reduced fresh merguez sausage prepared with olive oil in konjac gel matrix. *Meat Sci.* 94:438–446.

Triki, M., A. M. Herrero, L. Rodriguez-Salas, F. Jiménez-Colmenero, and C. Ruiz-Capillas. 2013b. Chilled storage characteristics of low-fat, n-3 PUFA-enriched dry fermented sausage reformulated with a healthy oil combination stabilized in a konjac matrix. *Food Control* 31:158–165.

Triki, M., F. Jiménez-Colmenero, A. M. Herrero, and C. Ruiz-Capillas. 2012. Optimisation of a chromatographic procedure for determining biogenic amine concentrations in meat and meat products employing a cation-exchange column with a post-column system. *Food Chem.* 130:1066–1073.

Vidal-Carou, M. C., M. L. Izquierdo, M. C. Matín-Morro, and A. Marine-Font. 1990. Histamine and tyramine in meat products: Relationship with meat spoilage. *Food Chem.* 37:239–249.

Vidal-Carou, M. C., F. Titus, and R. Guayta-Escolies. 2010. Evaluación del déficit de diaminooxidasa en pacientes con migraña. (Estudio MigraDAO). http://www.drhealthcare.com/estudio_migradao.pdf.

Vinci, G. and M. L. Antonelli. 2002. Biogenic amines: Quality index of freshness in red and white meat. *Food Control* 13:519–524.

Wimmerová, M. and L. Macholán. 1999. Sensitive amperometric biosensor for the determination of biogenic and synthetic amines using pea seedlings amine oxidase: A novel approach for enzyme immobilization. *Biosensor. Bioelectron.* 14:695–702.

Yang, X. and G. A. Rechnitz. 1995. Dual enzyme amperometric biosensor for putrescine with interference suppression. *Electroanalysis* 2:105–108.

Yano, Y., N. Miyaguchi, M. Watanabe, T. Nakamura, T. Youdou, J. Miyai, M. Numata, and Y. Asano. 1995. Monitoring of beef aging using a two-line flow injection analysis biosensor consisting of putrescine and xanthine electrodes. *Food Res. Int.* 28:611–617.

Zhang, L. Y., Y. M. Liu, Z. L. Wang, and J. K. Cheng. 2004. Capillary zone electrophoresis with pre-column NDA derivatization and amperometric detection for the analysis of four aliphatic diamines. *Anal. Chim. Acta* 508:141–145.

Acidity

CHAPTER 36

Determination of Acid Content

*Akira Kotani, Fumiyo Kusu, Hideki
Hakamata, and Kiyoko Takamura*

CONTENTS

36.1 INTRODUCTION

Various acids contained in food show significant effects on both taste and aroma. Acids also influence the ability of microorganisms to grow, so some, such as citric acid, lactic acid, or phosphoric acid, are used as common food additives for preservation or pH control. Therefore, the determination of acidity and acid content is important for the taste and quality assessment of food, regardless of whether acids exist in the original foods or in the additives. For example, fats and oils tend to decompose slowly upon storage in contact with the atmosphere and to release their fatty acid constituents, resulting in a lowering of their quality. Thus, a quality and freshness assessment of fats and oils can be made based on the determination of their free fatty acid content. In the case of beverages such as coffee and fruit juice, various organic acids are natural ingredients. pH control agents and acidulants are food additives used to achieve the stability and/ or flavor of foods. Even though the content of each acid is small, the acid amount has a significant effect on the taste and aroma of the beverages. Therefore, an acid assay is quite important for processing and quality control in the manufacturing and preservation of foods and beverages.

A commonly used method for determining acid content is alkali titration using an appropriate indicator [1–4]. However, in this method, there is the possibility of error due to an inability to detect a subtle color change of the indicator in the transition range, especially in cases of colored samples. In addition, the titration method is not sufficiently sensitive to detect low acid content, and it requires large sample amounts and a great deal of time. Potentiometric titration is commonly recommended as an alternative method.

We have developed a new voltammetric method using a quinone reagent [5]. Based on the voltammetry of quinone and acids, we assembled an electrochemical detector for measuring acid concentration at a given potential. The flow injection analysis with electrochemical detection (FIA-ECD) method we developed is preferable, as it is a simpler, more sensitive, and rapid method for acid determination. This chapter assesses this method for determining the acid content in foods and beverages.

36.2 QUINONE-MEDIATED AMPEROMETRIC SIGNALING FOR ACID DETERMINATION

Quinone itself gave a well-defined reduction peak on a potential sweep voltammogram using a glassy-carbon working electrode in an aqueous or an ethanol solution containing $LiClO_4$. Following the addition of acid into the solution, a new peak (termed a prepeak) was found at a more positive potential than that of the reduction peak of quinone. The prepeak is regarded as a quinone-mediated amperometric signal corresponding to the acid concentration. Electrochemical reduction of p-quinone is generally known to involve a two-electron transfer coupled with a two-proton transfer, represented as the following reaction (36.1), where Q and QH_2 denote p-quinone and its corresponding hydroquinone, respectively:

$$Q + 2HB + 2e^- \rightleftarrows QH_2 + 2B^-$$

$$(36.1)$$

The occurrence of the prepeak can be ascribed to the increased availability of protons from the added acid (HA), compared with the solvent molecules (HB). Instead of Equation 36.1, the electrochemical reduction of p-quinone in the presence of acid HA is shown by Equation 36.2. An increase in the proton supply from the added acid results in a lowering of energy for the reduction of p-quinone. Consequently, the reduction potential corresponding to Equation 36.2 shifts to a more positive direction than that of Equation 36.1.

$$Q + 2HA + 2e^- \rightleftarrows QH_2 + 2A^-$$

$$(36.2)$$

In general, since most acids are less active electrochemically, it is hard to measure the current signals due to the discharge of acids by conventional voltammetric techniques. However, the prepeak is found to appear on the voltammogram of quinone in the presence of acid in an unbuffered solution. The prepeak height is virtually independent of the pK_a but proportional to the total acid concentration up to twice the quinone concentration [6,7]. Based on this, the prepeak is regarded as a quinone-mediated amperometric signal corresponding to the total acid concentration. Essentially, the same prepeak is generally observed for various kinds of inorganic and organic acids, and its height (i_H) is found to be proportional to the total acid concentration ranging from 8×10^{-6} to

6×10^{-3} M, when the quinone concentration is at 3×10^{-3} M. Based on this, the total acid determination can be made by measuring the prepeak height.

The half-peak potential of the prepeak shifted to a negative value along with an increase in pK_a of the added acid. In the presence of mixed acids in the solution, stepwise prepeaks may appear when their pK_a values are different enough from each other. On the other hand, when the acid strengths and diffusion coefficients were nearly the same for all the acids, only one prepeak could be observed at an identical potential with a height proportional to the sum of the total acid concentration for all the acids. In such cases, the total acid determination can be made by measuring the prepeak height.

Choice of a quinone reagent should be made by considering its solubility, stability, and reduction potential in a test solution. The use of 1,4-benzoquinone or 1,2- benzoquinone is not favorable, because both of them are rather unstable in aqueous and alcoholic solutions. Instead, 2-methyl-1,4-naphthoquinone (so-called vitamin K_3, abbreviated as VK_3) and 3,5-di-t-butyl-1,2-benzoquinone (DBBQ) are preferred, since they facilitate acid determination due to their stability and solubility in solvents. Each of them gives a well-defined reduction peak on its voltammogram (see Figure 36.1, curve a) and is available commercially at a reasonable cost.

In Figure 36.1, voltammograms of VK_3 and DBBQ (lower and upper curves, respectively) obtained with and without palmitic acid (as curves b and a, respectively) in ethanol containing 0.1 M $LiClO_4$ (as a supporting electrolyte) are compared regarding their reduction potentials and peak shapes. In both cases, well-defined prepeaks caused by the addition of acid are clearly observed. In addition, the potential range (more negative than -0.4 V vs. SCE) for the discharge of oxygen dissolved in the solution is also shown at the bottom of Figure 36.1, because the dissolved oxygen shows a big reduction peak on

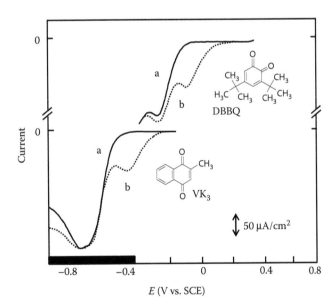

FIGURE 36.1 Voltammograms of 3,5-di-t-butyl-1,4-benzoquinone (DBBQ, upper) and 2-methyl-1,4-naphthoquinone (VK_3, lower) with (b) and without (a) palmitic acid obtained in ethanol. The black line at the bottom indicates the potential range of the discharge of oxygen dissolved in the ethanol solution.

the voltammogram. Therefore, the removal of oxygen from the solution is needed prior to the measurement of the signal current of VK_3 to eliminate interference arising from the overlap of the reduction peak of the dissolved oxygen. The removal of the dissolved oxygen from the electrolytic solution can be usually made by bubbling nitrogen gas into the solution in a voltammetric cell or by using a degasser in the flow line of the FIA system. To avoid overlap of the prepeak with the reduction peak of the dissolved oxygen, Figure 36.1 suggests that the use of quinones reduced at potentials more positive than −0.4 V vs. SCE is desirable.

Among the quinones examined, DBBQ was found to be the most suitable to fit this requirement. As seen in Figure 36.1, DBBQ exhibited a well-defined prepeak at −0.2 V vs. SCE, indicating that dissolved oxygen causes no interference in measuring the prepeak height. As a result, the use of DBBQ leads to a considerable reduction in operation time.

The stability of VK_3 and DBBQ in ethanol was demonstrated by time-course measurements of their voltammograms and absorption spectra. Both showed no appreciable change upon storage at room temperature for more than 1 month, indicating both VK_3 and DBBQ to be quite favorable as quinone reagents for acid determination.

36.3 FLOW INJECTION ANALYSIS SYSTEM AND ELECTROCHEMICAL DETECTION CELLS

36.3.1 System Components and Conditions

The system of flow injection analysis with electrochemical detection (FIA-ECD) for determining acid contents in foods and beverages consists of a degasser, a pump, a sample injector, an electrochemical detector, and a recorder. The deaerated carrier solution, in which was dissolved a quinone (VK_3 or DBBQ) and a supporting electrolyte in an unbuffered protoic solvent, was made to flow by the pump. A 5 μL aliquot of the sample solution was injected into the FIA-ECD system and allowed to merge with the carrier solution stream to mix together in the flow line. In the electrochemical detector, the quinone can be reduced at the working electrode, which is maintained at an applied potential. The electrochemical detector measures the electrical current generated by the quinone-mediated amperometric signaling corresponding to the acid concentration in the FIA eluent on the working electrode in the flow-type cell [5].

Examples of the components of the FIA-ECD system are as follows:

- Degasser: DG-980-50 (Jasco, Tokyo, Japan).
- Pump: 301M pump (Flom, Tokyo, Japan).
- Injector: 7725 sample injector fitted with 5 μL sample loop (Reodhyne, Cotati, California, USA).
- Electrochemical detector: LC-4C electrochemical detector (BAS, Tokyo, Japan).
- Electrochemical flow cell: the radial flow cell (BAS) constructed from a glassy-carbon working electrode (φ 6 mm), an Ag/AgCl reference electrode, and a stainless-steel counter electrode. The 25 μm thickness of the polytetrafluoroethylene (PTFE) gasket was set between the flow-cell block and the working electrode in the electrochemical cell; the cell volume was therefore 0.7 μL.

The FIA-ECD components and conditions of the carrier solution, flow rate, and applied potential should be optimized with respect to each sample and analyte. In

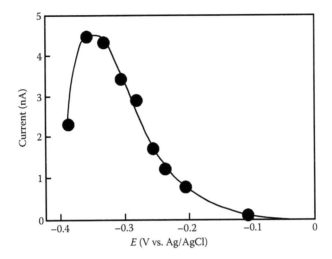

FIGURE 36.2 Potential dependence of the flow signal for 0.1 mM palmitic acid obtained. FIA-ECD conditions: carrier solution, ethanol solution containing 3 mM VK$_3$ and 38 mM LiClO$_4$; flow rate, 0.6 mL/min; injection volume, 5 μL. (Used and reproduced with permission from Takamura, K., T. Fuse, and F. Kusu. 1995. *Anal. Sci.* 11:979–982.)

particular, an applied potential was set based on the hydrodynamic voltammogram shown in Figure 36.2. Examples of the FIA-ECD conditions are listed in Table 36.1.

36.3.2 Assessment Strategy of FIA-ECD with Precision for Routine Checks

An assessment method is useful for establishing a stable FIA-ECD system with satisfactory precision to obtain highly reproducible and sensitive results. The measurement precision is described in terms of the standard deviation (SD) and/or relative standard deviation (RSD) measurements [8]. For example, when repetitive measurements ($n < 30$) are performed, measured data sets are obtained as $x_1, x_2, ..., x_n$. In this case, the equation of SD is

$$SD = \sqrt{[(x_1 - \bar{x})^2 + (x_2 - \bar{x})^2 + \cdots + (x_n - \bar{x})^2]/(n - 1)}$$

where

$$\bar{x} = (x_1 + x_2 + \cdots + x_n)/n$$

And the equation of RSD (%) is RSD = (SD/$\bar{x}$) × 100. In general, these are standard statistical applications, in which five or six runs of FIA-ECD experiments are carried out for obtaining an exact estimate of the measurement SD and RSD. On the other hand, ISO 11843-7, which provides an SD and RSD of measurement to examine detection limits from stochastic aspects of the signal and noise in a chromatogram based on the function of mutual information (FUMI) theory [9], has been proposed in recent years. When an FIA signal has a peak height, H, a measurement RSD can be estimated based on the noise

TABLE 36.1 Typical Examples of FIA–ECD Systems and Conditions for Determining Acid Content in Fat, Oil, Coffee, Orange Juice, Vinegar, Tomato Ketchup, and pH Control Agent

Apparatus	Fat, Oil, and Coffee	Orange Juice, Vinegar, and Tomato Ketchup	pH Control Agent
Pump	DMX-2200-T (SNK. Ind., Tokyo)	301 M (Flom, Tokyo)	DMX-2200-T (SNK. Ind, Tokyo)
Flow line	PTFE (0.5 mm, ID)	PEEK (0.13 mm, ID)	PTEF (0.5 mm, ID)
Sample injector	7125 (Rheodyne, Cotati, CA)	7725 (Rheodyne, Cotati, CA)	7125 (Rheodyne, Cotati, CA)
Electrochemical detector	HECS 311B Potentiostat (Huso, Kanagawa)	LC-4C Amperometric detector (BAS, Tokyo)	HECS 312 Potentiostat (Huso, Kanagawa)
Type of flow cell (cell volume)	Wall-jet type (2.7 µL)	Radial flow type (0.7 µL)	Cross flow type (0.7 µL)
Working electrode	Glassy carbon (Tokai Carbon)	Glassy carbon (BAS)	Glassy carbon (Jasco, Tokyo)
Reference electrode	Ag/AgCl	Ag/AgCl	Ag/AgCl
Counter electrode	Stainless steel	Stainless steel	Stainless steel

Conditions	Fat, Oil, and Coffee	Orange Juice, Vinegar, and Tomato Ketchup	pH Control Agent
Carrier solution	Ethanol solution containing 3 mM vitamin K_3 and 38 mM $LiClO_4$	Ethanol–water (1:1, v/v) mixture solution containing 3 mM 3,5-di-t-butyl-1,2-benzoquinone and 50 mM NaCl	Ethanol–water (1:1, v/v) mixture solution containing 3 mM vitamin K_3 and 38 mM $LiClO_4$
Flow rate	0.6 mL/min	0.1 mL/min	0.6 mL/min
Injection volume	5 µL	5 µL	5 µL
Applied potential	–0.33 V vs. Ag/AgCl	–0.15 V vs. Ag/AgCl	–0.24 V vs. Ag/AgCl

Abbreviations: FIA–ECD, flow injection analysis with electrochemical detection; PTFE, polytetra-fluoroethylene; PEEK, polyetheretherketone.

parameters of the SD ($\tilde{w}$) of white noise, and the SD ($\tilde{m}$) and the retention parameter (ρ) of the Markov process, respectively, as follows:

$$\text{RSD}^2 = \frac{\tilde{w}^2}{H^2} + \frac{\tilde{m}^2}{(1-\rho)^2 H^2}\left(1 - 2\rho\frac{1-\rho}{1-\rho} + \rho^2\frac{1-\rho^2}{1-\rho^2}\right) + I^2 \qquad (36.3)$$

where I is the RSD of the volume error of the sample injector. And then, $\tilde{w}$, $\tilde{m}$ and ρ can be determined by the least-squares fit of the model power spectrum of the white noise and Markov process to the real power spectrum of baseline noises in FIA-ECD. Detailed explanations of FUMI theory have already been given in Refs. [9,10]. The FUMI theory can be applied to estimate a measurement precision in flow analysis as high-performance liquid chromatography with electrochemical detection (HPLC-ECD) [10]. Using chromatographic baseline noise and peak of analyte, the predicted RSD can be calculated based on the FUMI theory. Meanwhile, using five chromatograms of the analyte, the experimentally observed RSD is statistically calculated. When the predicted RSD is within the 95% confidence interval of the statistically obtained RSD, the measurement precision by ISO 11843-7 is applicable to estimate the precision in HPLC-ECD without repeated measurements of real samples [10].

The measurement precision by the FUMI theory would be useful for circumventing the above-mentioned problem in the present FIA-ECD. In order to verify the precision of measurements, the FUMI theory can be applied to provide the present FIA-ECD. Theoretical RSDs predicted by the FUMI theory were compared with an experimentally observed RSD by repetitive measurements. Precision profiles of citric acid were examined as an example. The experimentally observed RSD at 0.05 mM was 2.1% ($n = 6$). Meanwhile, the theoretical predicted RSD of citric acid at 0.05 mM was 1.6%. This value was within the 95% confidence interval of the statistically obtained RSD at 0.05 mM citric acid ($n = 6$), ranging from 1.3% to 5.2%. Thus, it was found the measurement precision by ISO 11843-7 is applicable to estimate the precision in FIA-ECD without repeated measurements of real samples. The present assessment strategy using the prediction of measurement precision by ISO 11843-7 is useful for routine checks of FIA-ECD, saving not only considerable amounts of chemicals but also experimental time.

36.4 DETERMINATION OF ACID CONTENTS

The FIA-ECD method described above was successfully applied to the determination of free fatty acids in fat and oil, and various organic acids in coffee, fruit juice, vinegar, tomato ketchup, and the pH control agent. In some sections, the voltammograms of acids and acid mixtures are also shown, to give a better understanding of the origins of the FIA signal appearing for the analytical samples.

36.4.1 Free Fatty Acid Content in Fat and Oil

Since fats and oils tend to decompose slowly upon storage in contact with the atmosphere and to release their fatty acid constituents, quality and freshness assessments of fats and oils can be made based on the determination of free fatty acid content. A commonly used method for determining free fatty acid content is titration, in which a fat or oil sample in a mixture of ethanol and diethyl ether is neutralized with KOH or NaOH using a phenolphthalein indicator. However, in such a method, there is the possibility of error due to an inability to detect a subtle color change of the indicator in the transition range, especially in cases of colored samples. Actually, many fat and oil samples were yellow. For such cases, a potentiometric measurement is recommended for end-point matching. However, the adhesion of insoluble fatty acid salts (soaps) is often produced during titration on the indicator electrode surface, along with a consequent lowering of the response rate. In addition, titration methods require large sample sizes and long operation times [5,11–13].

Application of our voltammetric method to the determination of total free fatty acid content in edible oils seems favorable, because their constituent acids are mostly restricted to long-chain monocarboxylic acids ($C_{16} - C_{20}$) with a straight chain of carbon atoms, and their acid strengths and diffusion coefficients might be nearly the same [5]. The presence of long-chain fatty acids, such as linoleic acid and palmitic acid, in an ethanol solution containing VK_3 was found to give rise to a prepeak on the voltammogram of VK_3 at a potential more positive than that for the reduction peak of VK_3, as has been shown in Figure 36.1 (curve b in the lower figure). The peak potentials of the prepeaks for different free fatty acids in oils were essentially the same, and the current, per a unit acid concentration, was basically the same in all cases. Therefore, on the voltammogram of a mixture of long-chain fatty acids in an ethanol solution containing VK_3, only one prepeak with a height proportional to the total acid concentration was seen, from which the free acid content of fats and oils could be determined.

As a simpler and more rapid method, the FIA-ECD method is preferable for practical uses. A 5 μL aliquot of the standard solution containing palmitic acid (or linoleic acid) was injected into the flow injection system. From the voltammetric measurement of the prepeak heights at different potentials obtained for VK_3 with palmitic acid (Figure 36.2), the detection potential for monitoring free fatty acid content was decided at −0.33 V so as to obtain the maximum current of flow signals [13]. A flow signal appeared for each amount of palmitic acid injected (Figure 36.3). The response was linear in the concentration range between 5.0×10^{-6} and 3.0×10^{-4} M (25–1500 pmol/test) [13].

The present FIA method was applied to the determination of the free fatty acid content in different kinds of fat and oil samples (Table 36.2) [13]. To prepare the test solution, an appropriate amount of the sample (4–400 mg) was completely dissolved in a 5 mL ethanol solution containing 3 mM VK_3 and 38 mM $LiClO_4$. When the sample was not soluble in ethanol, it was first dissolved in ether and then transferred to ethanol. Recovery tests were made with the addition of standard linoleic acid in appropriate amounts, in each case being nearly the same as the acid amount in each sample. The recovery data of the added linoleic acid were in the range from 96% to 102%, indicating the reliability of the obtained data [13].

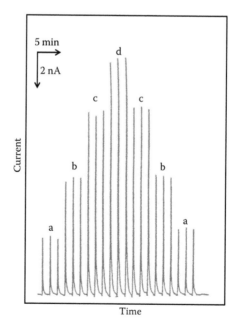

FIGURE 36.3 Flow signals obtained with (a) 0.05, (b) 0.1, (c) 0.15, and (d) 0.2 mM palmitic acid. FIA-ECD conditions: carrier solution, ethanol solution containing 3 mM VK_3 and 38 mM $LiClO_4$; flow rate, 0.6 mL/min; injection volume, 5 μL; applied potential −0.33 V vs. Ag/AgCl. (Used and reproduced with permission from Takamura, K., T. Fuse, and F. Kusu. 1995. *Anal. Sci.* 11:979–982.)

TABLE 36.2   Free Fatty Acid Content in Fat and Oil Obtained by the Present FIA–ECD and the Conventional Titration Methods

| | FIA–ECD | | Titration | | | |
| | | | Potentiometry | | Color Change[a] | |
Sample	Content (mol/g)	RSD[b] (%)	Content (mol/g)	RSD[b] (%)	Content (mol/g)	RSD[b] (%)
Camellia oil	4.0×10^{-5}	1.9	4.1×10^{-5}	2.0	4.1×10^{-5}	2.3
Cacao butter	4.2×10^{-5}	2.0	4.1×10^{-5}	2.7	4.2×10^{-5}	4.7
Glyceryl monostearate	1.5×10^{-5}	1.0	1.6×10^{-5}	1.5	1.8×10^{-5}	3.7
Mentha oil	3.6×10^{-6}	1.3	3.6×10^{-6}	2.5	3.9×10^{-6}	3.4
Corn oil	2.6×10^{-6}	0.54	2.3×10^{-6}	2.4	2.9×10^{-6}	4.8

Source: Used and reproduced with permission from Fuse, T., F. Kusu, and K. Takamura. 1995. *Bunseki Kagaku* 44:29–33.

Abbreviations: FIA–ECD, flow injection analysis with electrochemical detection; RSD, relative standard deviation.

[a] Using phenolphthalein indicator.

[b] $n = 5$.

The results obtained by the present FIA method were compared with those obtained using conventional titration (using KOH) methods (Table 36.2). In the titration methods, the end points were determined based on the color change of phenolphthalein and potentiometry. In Table 36.2, the values of the RSD of the data by the present FIA-ECD method are less than 2.0%, the best of the three methods. The RSD was greatest in conventional titration with phenolphthalein. Such a decreased reproducibility is considered to be due to an indicator error. However, no marked differences in RSD could be detected between the present and potentiometric titration methods. Potentiometric titration does not show any indicator error [13].

In addition, the FIA operating conditions made possible the processing of 30 samples/h [13]. The present FIA method is thus shown to be practical and useful for the determination of the acid content of fats and oils in quality-control processes in factories [14].

36.4.2 Acid Content in Coffee

The main acid components in coffee are chlorogenic, caffeic, and quinic acids, etc. Slight changes in their content readily lead to differences in aroma and the quality of the coffee beans, and content changes easily through processing, roasting, and extraction. The determination of coffee acid content is thus important for its quality control. Common methods of acid content assay, involving alkali titration and pH measurement, are not sufficiently sensitive to follow the slight changes of the acid content in coffee.

Flow signals obtained for chlorogenic acid are shown in Figure 36.4. The response was linear between 5.0×10^{-6} and 3.0×10^{-4} M [5,15]. The FIA-ECD method was examined in detail for determining the acid content in coffee. In the test solution preparation, 250 mL of hot water was poured onto 20 g coffee powder on a filter paper, and the filtrate was diluted with an ethanol solution containing 3 mM VK$_3$ and 38 mM LiClO$_4$ [5,15].

Since the sour taste of coffee depends on the roasting of coffee beans, changes in the acid content of coffee beans with roasting time were followed by the FIA-ECD method. During the course of roasting, the color of the coffee bean powder gradually changed

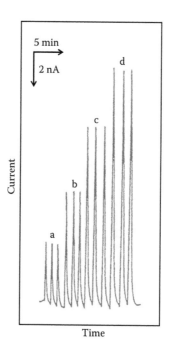

FIGURE 36.4   FIA signals for (a) 0.05, (b) 0.1, (c) 0.15, and (d) 0.2 mM chlorogenic acid. FIA-ECD conditions: carrier solution, ethanol solution containing 3 mM VK$_3$ and 38 mM LiClO$_4$; flow rate, 0.6 mL/min; injection volume, 5 μL; applied potential −0.33 V vs. Ag/AgCl. (Used and reproduced with permission from Fuse, T., F. Kusu, and K. Takamura. 1997. *J. Agric. Food Chem.* 45:2124–2127.)

from pale yellow to dark brown. In Figure 36.5a(i), the flow signal current obtained for Guatemalan coffee beans is plotted against roasting time. The current values essentially remain constant and low up to 11 min; afterward they increase markedly and attain a maximum at about 13 min, and then decrease gradually. Results of the sensory test for the same coffee obtained by seven participants clearly show that the acid taste intensity changed with the roasting time in the order of 13 > 15 > 16.5 > 0 min. Both results show excellent agreement, suggesting that the current signal of the FIA-ECD serves as an indication of sour taste. The changes in the titratable acidity obtained by the potentiometric titration of and pH changes in coffee extracts from the same coffee beans were also followed during roasting, and the results are shown in Figure 36.5a as (ii) and (iii), respectively. In the pH measurement, only a slight change is noted throughout the roasting period in contrast to Figure 36.5a(i), because the potential of the glass electrode cell responds only to the logarithm of the concentration of dissociated protons in the sample solution. In Figure 36.5a(ii), the plots of the titratable acidity against roasting time give a similar curve to that in Figure 36.5a(i) [5,15]. Similar results to those in Figure 36.5b were also obtained for coffee extracts from other coffee beans, such as Kilimanjaro. Titration required a large sample amount and a long time, that is, 50 mL coffee extract and about 10 min for each plot in Figures 36.5a(ii) and 36.5b(ii). In comparison with this, only 5 μL of 40-fold diluted coffee and about 1.5 min were required for each plot in Figures 36.5a(i) and 36.5b(i), indicating the FIA-ECD has a 400,000 times higher sensitivity than conventional methods [5,15].

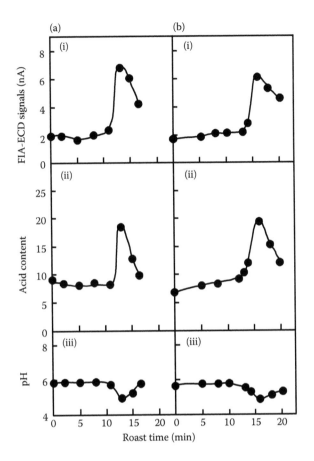

FIGURE 36.5   Effects of roasting time on acid content in coffee measured by (i) FIA-ECD, (ii) titration, and (iii) pH meter with coffee beans of (a) Guatemalan and (b) Kilimanjaro. (Used and reproduced with permission from Fuse, T., F. Kusu, and K. Takamura. 1997. *J. Agric. Food Chem.* 45:2124–2127.)

36.4.3 Acid Content in Fruit Juice

Various organic acids, such as tartaric, malic, succinic, and citric acids, are found in fruit juice. Acid content differs depending on the particular origin and growth conditions of the fruit. Moreover, acid content changes during the processing of the fruit juice. In addition to the naturally occurring acids, various kinds of acids are often added to fruit juice as preservatives and antioxidants. These acids are responsible for the taste and aroma of fruit juice; therefore, the determination of total acid content is essential for processing and quality control in fruit juice production. In this section, we discuss FIA-ECD as applied to the acid content determination for fruit juices such as orange juice.

As shown in Figure 36.6, the addition of various organic acids such as acetic, malic, tartaric, succinic, or citric acid to the individual DBBQ ethanol solutions gave rise to a single prepeak at 0–0.1 V (vs. SCE). In all cases, each acid concentration was the same, but the prepeak height differed according to the number of carboxyl groups of the acids added. The prepeak height was found to be linearly related to the equivalent concentration of each acid, ranging from 10^{-5} to 10^{-3} M [16]. In Figure 36.6, the potential difference

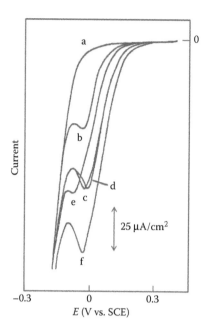

FIGURE 36.6 Voltammograms of 3.0 mM DBBQ with no acid (a) and in the presence of (b) 1.2 mM acetic acid, (c) 1.2 mM malic acid, (d) 1.26 mM tartaric acid, (e) 1.2 mM succinic acid, and (f) 1.2 mM citric acid, obtained in ethanol containing 0.1 M LiClO$_4$. (Used and reproduced with permission from Ohtsuki, S. et al. 2001. *Electroanalysis* 13:404–407.)

of the prepeak caused by the acids is rather slight, and accordingly, only one prepeak may appear at an identical potential on the voltammogram when a mixture of these acids is added. In such a case, it should thus be possible to determine the total equivalent concentration of acids by measuring the prepeak height. Voltammetry was then applied for acid content measurement in orange juice. As is evident in Figure 36.7, one prepeak was caused by the addition of 200 μL of orange juice to 20 mL of the DBBQ solution (curve b). From the standard addition method using a standard citric acid solution (cf., curves c–e), the acid content in orange juice can be calculated [17]. The color and turbidity of the juice cause no interference with the analytical results. Further, sugars contained in fruit juice, such as sucrose, dextrose, levulose, and saccharose, are found to have no effect on the total acid values determined by the voltammetric method [17].

The FIA-ECD method was utilized for the determination of acid in orange juice. A 5 μL aliquot of the standard solution containing citric acid was injected into the flow injection system. From the measurement of a hydrodynamic voltammogram of citric acid, the detection potential for monitoring the acid content was determined at −0.15 V vs. Ag/AgCl. A flow signal appeared for each amount of citric acid injected (Figure 36.8a). The response in the present FIA-ECD was found to be linearly related to a citric acid concentration ranging from 2.0×10^{-6} to 1.0×10^{-4} M.

The present FIA-ECD method was used for determining the acid content in four orange juices. The test solution was prepared by dissolving a 20 μL sample of juice in a 10 mL ethanol–water mixture solution (1:1, v/v) containing 3 mM DBBQ and 50 mM NaCl. The results for the acid content are shown as concentration of citric acid, since the value of the acid content in juices is usually referred to by the major acid component. As shown

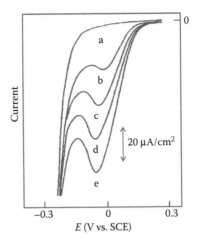

FIGURE 36.7 Voltammograms for the standard addition method for determining the acid content in orange juice. Initial volume of the DBBQ solution: 20 mL; concentration of standard citric acid: 0.100 M. Added amount of orange juice: 0 μL (curve a), 200 μL (curves b–e), final concentration of citric acid: 0 M (curves a and b), 1.98×10^{-4} M (curve c), 3.94×10^{-4} M (curve d), 5.91×10^{-4} M (curve e). (Used and reproduced with permission from Takamura, K., S. Ohtsuki, and F. Kusu. 2001. *Anal. Sci.* 17(Suppl.):i737–i739.)

in Table 36.3, the results by FIA-ECD were in good agreement with those acquired using titration ($r = 0.928$). The FIA-ECD processing was 30 samples/h. The present FIA-ECD method is superior in sensitivity and requires small sample volumes (20 μL), unlike titration, which requires relatively large sample volumes (10 mL). Although the precision of the present FIA-ECD method is less than that of titration, the RSD values of the data in

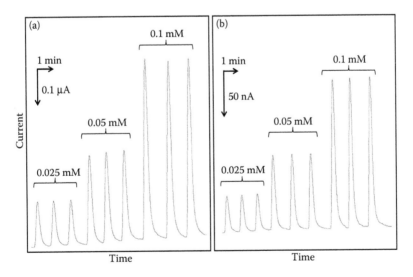

FIGURE 36.8 FIA signals for 0.025, 0.05, and 0.1 mM (a) citric acid and (b) acetic acid. FIA conditions: carrier solution, ethanol–water (1:1, v/v) mixture solution containing 3 mM DBBQ and 50 mM NaCl; flow rate, 0.1 mL/min; injection volume, 5 μL; applied potential −0.15 V vs. Ag/AgCl.

TABLE 36.3 Acid Content of Various Orange Juices, Vinegars, and Tomato Ketchups by FIA–ECD and Titration

Sample		FIA–ECD		Titration	
		Content (g/dL)	RSD[a] (%)	Content (g/dL)	RSD[a] (%)
Orange juice[b]	A	0.82	2.5	0.77	0.37
	B	0.64	2.1	0.68	0.14
	C	0.71	2.7	0.68	0.18
	D	0.68	1.6	0.68	0.15
Vinegar[c]	Rice	4.84	1.7	4.60	0.44
	Cereal	4.00	2.0	4.23	0.13
	Cider	5.22	2.2	5.06	0.08
	Wine	5.45	2.2	5.08	0.15
Tomato ketchup[b]	A	1.55	4.2	1.42	0.12
	B	1.64	3.6	1.68	0.26
	C	1.51	1.9	1.40	0.62
	D	1.32	4.4	1.23	0.42

Abbreviations: FIA–ECD, flow injection analysis with electrochemical detection; RSD, relative standard deviation.

[a] $n = 5$.
[b] Acid contents are shown as concentration of citric acid.
[c] Acid contents are shown as concentration of acetic acid.

the former are lower than 3%. The present FIA-ECD method proved to be suitable as an alternative to the conventional potentiometric titration method.

36.4.4 Acid Content in Vinegar

The detection potential for monitoring acetic acid by FIA-ECD was maintained at −0.15 V vs. Ag/AgCl. The flow rate of the carrier solution was 0.6 mL/min. A 5 µL aliquot sample was injected into the flow injection system. The flow signals obtained for acetic acid by the FIA-ECD system are shown in Figure 36.8b. The response in the present FIA-ECD was found to be linearly related to the acetic acid concentration ranging from 2.0×10^{-6} to 1.0×10^{-4} M. The present FIA-ECD method was applied to determine acid content in vinegars. To prepare the test solution for determining acid content in vinegars, vinegar samples were diluted with an ethanol–water mixture solution (1:1, v/v) containing 3 mM DBBQ and 50 mM NaCl. A 5 µL test solution was injected into the present FIA-ECD to observe a flow signal. The acids were monitored by measuring the current height of the flow signal in the FIA-ECD. The absolute standard curve method was used for determining acid content in vinegar.

The acid content of four vinegars was determined by the FIA-ECD and potentiometric titration methods. The content is referred to as the concentration of acetic acid (acetic acid g/100 mL vinegar). As shown in Table 36.3, the results by FIA-ECD were in good agreement with those obtained by titration ($r = 0.973$). The RSD of the results using the present FIA-ECD method was less than 3% ($n = 5$). The pretreatment of a 10,000-fold dilution was very simple, and there was no carryover between samples, even at a rate of 45 samples/h.

36.4.5 Acid Content in Tomato Ketchup

Tomato ketchup is a sauce commonly made from tomato, vinegar, sugar, seasonings, and spices, and contains various organic acids, all of which contribute to its taste and aroma. Thus, acid content determination is important for quality control in its manufacturing.

A tomato ketchup sample was well homogenized and was diluted 20-fold with deionized water, and the diluted solution was filtered using a membrane filter (pore size, 0.45 μm). The filtrated solution was diluted 100-fold with an ethanol–water mixture solution (1:1, v/v) containing 3 mM DBBQ and 50 mM NaCl. A 5 μL test solution of tomato ketchup was injected into the present FIA-ECD to observe a flow signal. The acids were monitored by measuring the current height of the flow signal in the FIA-ECD. The absolute standard curve method was used to determine acid content in tomato ketchup. The acid content (citric acid g/100 mL tomato ketchup) of four tomato ketchups was determined using FIA-ECD. As shown in Table 36.3, the results by FIA-ECD were in good agreement with those obtained using a potentiometric titration method ($r = 0.935$). The values of the RSD of the content by the present FIA-ECD method were less than 4.5% and were a little greater than those obtained by titration. However, the present FIA-ECD is a quick and a sensitive method, is easy to operate, and needs a simple pretreatment of the sample.

36.4.6 Acid Content in pH Control Agent

pH control agents are a kind of food additive used for the inhibition of the growth of bacteria in foods. The pH control agent, called DP-7L, is composed of 34% lactic acid, 8.5% phosphoric acid, 0.50% gluconic acid, and 50% of other materials, and is used in the food processing of noodles and fish sausage to reduce bacterial growth and prolong the shelf life of those foods. However, DP-7L in high concentrations makes those foods lose their flavor. Therefore, the control of the acid content of DP-7L is important in the processing of foods.

On the voltammograms of VK_3 with lactic acid, phosphoric acid, or gluconic acid, each prepeak appeared at about −0.1 V vs. SCE (Figure 36.9a). The presence of DP-7L was also found to give rise to a prepeak of the voltammogram of VK_3 (Figure 36.9b). The prepeak current is proportional to the concentration of DP-7L [18].

For acid determination by FIA-ECD, the test solution was prepared by diluting appropriate amounts of the sample solution with an ethanol–water mixture solution (1:1, v/v) containing 3.0 mM VK_3 and 38 mM $LiClO_4$. The detection potential was set at −0.24 V vs. SCE. Using FIA-ECD, the peak current height was proportional to the concentration of DP-7L from 12 to 100 ppm ($r = 0.997$) (Figure 36.10). The RSD of DP-7L at 75 ppm was determined to be 3.8% ($n = 5$).

When the acid determinations of DP-7L solutions of 12, 16, 27, 50, 75, and 100 ppm were carried out using the FIA-ECD and potentiometric titration methods, the results using FIA-ECD agreed with those obtained by titration ($r = 0.989$). The acidity of the DP-7L solution could be determined by the present method [18]. Therefore, FIA-ECD would be useful for monitoring the solution of pH control agents in the manufacturing process of boiled noodles.

36.5 CONCLUSIONS

This chapter dealt with the FIA-ECD method applied to total acid determination in foods, beverages, and pH control agents. Acid content in food is one of the most important

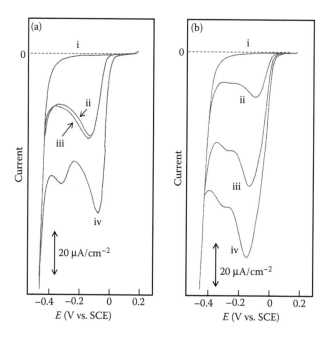

FIGURE 36.9 (a) Linear sweep voltammograms of 3.0 mM VK$_3$ with no acid (i) and in the presence of (ii) 0.7 mM gluconic, (iii) 0.75 mM lactic, and (iv) 0.7 mM phosphoric acid in ethanol–water (1:1, v/v) mixture solution containing 38 mM LiClO$_4$ and (b) linear sweep voltammograms of 3.0 mM VK$_3$ in the absence (i) and presence of (ii) 0.0.13, (iii) 0.049, and (iv) 0.082 g/dL pH control agent, called DP-7L, in ethanol–water (1:1, v/v) mixture solution containing 38 mM LiClO$_4$. A plastic-formed carbon, SCE, and platinum wire were used as working, reference, and counterelectrodes, respectively. Initial potential: 0.2 V vs. SCE; final potential: −0.5 V vs. SCE; scan rate: 20 mV/s. (Used and reproduced with permission from Takahashi, K. et al. 2004. *Bunseki Kagaku* 53:271–274.)

indications for the quality and taste assessment of food. However, acid content has not always been easy to check, especially in field tests, because a convenient probe device has yet to be developed for measurements based on an acid–base titration principle. Our method of acid determination was developed based on quinone-mediated amperometric signaling for the whole amount of acid components contained in food samples. This method made it possible to fabricate an ECD for detecting acids, and it was successfully used as a signal detector in flow injection analysis (FIA-ECD). Commonly used pH measurement is not adequate for this purpose, because the obtained pH value corresponds not to the total acid amounts but only to the activity value of the protons dissociated from the acids.

The acid contents obtained by the present FIA-ECD method show a good correlation with the titratable acidity obtained by conventional potentiometric titration. Further, the present FIA-ECD method is superior in sensitivity and rapidity, and accordingly, it requires smaller sample volumes and shorter analysis times, unlike a conventional potentiometric titration method, which requires relatively larger sample volumes and a longer time. The fact that the sample color causes no interference in the detection of acids permits the present FIA-ECD method to be widely applied to colored beverages such as

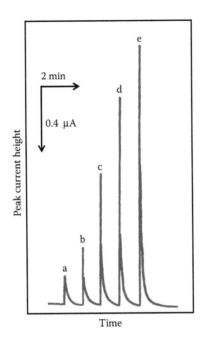

FIGURE 36.10 Dose-response signals of DP-7L. FIA conditions: carrier solution, ethanol–water (1:1, v/v) mixture solution containing 3 mM VK_3 and 38 mM $LiClO_4$; flow rate, 0.6 mL/min; injection volume, 5 µL; applied potential −0.24 V vs. Ag/AgCl. Concentrations of pH control agent, called DP-7L: (a) 16, (b) 27, (c) 50, (d) 75, and (e) 10 ppm. (Used and reproduced with permission from Takahashi, K. et al. 2004. *Bunseki Kagaku* 53:271–274.)

coffee and tomato ketchup. Because of its ease of use, the method is practical and useful for checking the maturity during the cultivation and fermentation of fruits.

In addition to the FIA-ECD, the ECD desribed in this chapter has also been noted to be practical and useful as a detector in HPLC. Using the HPLC-ECD method, each organic acid contained in food can be determined separately. In fact, the HPLC-ECD method was successfully applied to the determination of individual free fatty acid content in oils [19,20]. Furthermore, because the HPLC-ECD method is a simple and rapid procedure that does not require the derivatization of analyte acids, it was shown to be effective for monitoring individual acid content changes during the bacterial fermentation of food, such as wine brewing [21] and the production of yogurt [22].

In the end, our method for total acid determination was established based on the quinone-mediated amperometric signaling for the total amount of acid species. The method was thus shown to have potential in a broad range of applications for checking the quality of food materials.

REFERENCES

1. AOCS Official Method Cd 3d-63. 2009. *Official Methods and Recommended Practices of the AOCS*. Champaign, IL: American Oil Chemists' Society.

2. JOCS Official Method 2.3.1. 2013. *Standard Methods for the Analysis of Fats, Oils, and Related Materials*. Tokyo: Japan Oil Chemists' Society.

3. General Chapters, chemical tests, 401 Fats and fixed oils. 2013. *United States Pharmacopoeia* (37th ed.). Rockville, MD: United States Pharmacopeial Convention.

4. General tests, processes and apparatus, 1.13 Fats and fatty oil test. 2011. *Japanese Pharmacopoeia* (16th ed.). Tokyo: The Ministry of Health, Labour and Welfare.

5. Takamura, K., T. Fuse, K. Arai, and F. Kusu. 1999. A review of a new voltammetric method for determining acids. *J. Electroanal. Chem.* 468:53–63.

6. Takamura, K. and Y. Hayakawa. 1968. Polarographic determination of acids by means of the reduction wave of quinones in methyl cellosolve solution. *Anal. Chim. Acta* 43:273–279.

7. Takamura, K. and Y. Hayakawa. 1971. Effects of proton donors on the polarographic reduction of methyl-*p*-benzoquinone in aqueous and methyl cellosolve solutions. *J. Electroanal. Chem.* 31:225–232.

8. Miller, J. C. and J. N. Miller. 1988. *Statistics for Analytical Chemistry*. West Sussex, UK: Eills Horwood.

9. Hayashi, Y. and R. Matsuda. 1994. Deductive prediction of measurement precision from signal and noise in liquid chromatography. *Anal. Chem.* 66:2874–2881.

10. Kotani, A., S. Kojima, Y. Hayashi, R. Matsuda, and F. Kusu. 2008. Optimization of capillary liquid chromatography with electrochemical detection for determining femtogram levels of baicalin and baicalein on the basis of the FUMI theory. *J. Pharm. Biomed. Anal.* 48:780–787.

11. Kusu, F., T. Fuse, and K. Takamura. 1994. Voltammetric determination of acid values of fats and oils. *J. AOAC Int.* 77:1686–1689.

12. Fuse, T., F. Kusu, and K. Takamura. 1995. Voltammetric determination of free fatty acid content in fats and oils. *Bunseki Kagaku* 44:29–33.

13. Takamura, K., T. Fuse, and F. Kusu. 1995. Determination of free fatty acid content in fats and oils by flow injection analysis with electrochemical detection. *Anal. Sci.* 11:979–982.

14. Fuse, T., F. Kusu, and K. Takamura. 1997. Determination of acid values of fats and oils by flow injection analysis with electrochemical detection. *J. Pharm. Biomed. Anal.* 15:1515–1519.

15. Fuse, T., F. Kusu, and K. Takamura. 1997. Determination of acidity of coffee by flow injection analysis with electrochemical detection. *J. Agric. Food Chem.* 45:2124–2127.

16 Ohtsuki, S., N. Kunimatsu, K. Takamura, and F. Kusu. 2001. Determination of the total acid content in wine based on the voltammetric reduction of quinone. *Electroanalysis* 13:404–407.

17. Takamura, K., S. Ohtsuki, and F. Kusu. 2001. Development of a new amperometric sensor for probing the total acid of beverages. *Anal. Sci.* 17(Suppl.):i737–i739.

18. Takahashi, K., A. Kotani, S. Ohtsuki, and F. Kusu. 2004. Flow-injection analysis with electrochemical detection for determining the titratable acidity of a pH adjuster for foods. *Bunseki Kagaku* 53:271–274.

19. Fuse, T., F. Kusu, and K. Takamura. 1997. Determination of higher fatty acids by high-performance liquid chromatography with electrochemical detection. *J. Chromatogr. A* 764:177–182.

20. Kotani, A., F. Kusu, and K. Takamura. 2002. New electrochemical detection method in high-performance liquid chromatography for determining free fatty acids. *Anal. Chim. Acta* 465:199–206.

21. Kotani, A., Y. Miyaguchi, E. Tomita, K. Takamura, and F. Kusu. 2004. Determination of organic acids by high-performance liquid chromatography with electrochemical detection during wine brewing. *J. Agric. Food Chem.* 52:1440–1444.
22. Kotani, A., Y. Miyaguchi, and F. Kusu. 2001. Determination of organic acids by high-performance liquid chromatography with electrochemical detection. *Anal. Sci.* 17(Suppl.):a141–a143.

Index